CHEMICAL ENGINEERING

VOLUME 6

**Related Pergamon Titles in the CHEMICAL ENGINEERING Series by
J M COULSON & J F RICHARDSON**

*Chemical Engineering, Volume 1, Fourth edition
Fluid Flow, Heat Transfer and Mass Transfer
(with J R Backhurst and J H Harker)

*Chemical Engineering, Volume 2, Fourth edition
Unit Operations
(with J R Backhurst and J H Harker)

Chemical Engineering, Volume 3, Second edition
Chemical Reactor Design, Biochemical Reaction Engineering
including Computational Techniques and Control
(edited by J F Richardson and D G Peacock)

Chemical Engineering, Volume 4
Solutions to the Problems in Volume 1
(J R Backhurst and J H Harker)

Chemical Engineering, Volume 5
Solutions to the Problems in Volume 2
(J R Backhurst and J H Harker)

*In preparation

Related Pergamon Journals

CHEMICAL ENGINEERING SCIENCE

COMPUTERS & CHEMICAL ENGINEERING

INTERNATIONAL COMMUNICATIONS IN HEAT AND
MASS TRANSFER

INTERNATIONAL JOURNAL OF HEAT AND MASS TRANSFER

Full details of all Pergamon publications/free specimen copy of any Pergamon
journal available on request from your nearest Pergamon office.

CHEMICAL ENGINEERING

VOLUME 6

An Introduction to Chemical Engineering Design

BY

R. K. SINNOTT

Department of Chemical Engineering, University College of Swansea

PERGAMON PRESS
Member of Maxwell Macmillan Pergamon Publishing Corporation
OXFORD · NEW YORK · BEIJING · FRANKFURT
SÃO PAULO · SYDNEY · TOKYO · TORONTO

U.K.	Pergamon Press plc, Headington Hill Hall, Oxford OX3 0BW, England
U.S.A.	Pergamon Press Inc., Maxwell House, Fairview Park, Elmsford, NY 10523, U.S.A.
PEOPLE'S REPUBLIC OF CHINA	Pergamon Press, Room 4037, Qianmen Hotel, Beijing, People's Republic of China
FEDERAL REPUBLIC OF GERMANY	Pergamon Press GmbH, Hammerweg 6, D-6242 Kronberg, Federal Republic of Germany
BRAZIL	Pergamon Editora Ltda, Rua Eça de Queiros, 346, CEP 04011, Paraiso, São Paulo, Brazil
AUSTRALIA	Pergamon Press Australia Pty Ltd, P.O. Box 544, Potts Point, N.S.W. 2011, Australia
JAPAN	Pergamon Press, 5th Floor, Matsuoka Central Building, 1-7-1 Nishishinjuku, Shinjuku-ku, Tokyo 160, Japan
CANADA	Pergamon Press Canada Ltd., Suite No. 271, 253 College Street, Toronto, Ontario, Canada M5T 1R5

First edition 1983
Reprinted with corrections 1985
Reprinted 1986, 1989
Reprinted with corrections 1991

Library of Congress Cataloging in Publication Data

Coulson, J. M. (John Metcalfe)
Chemical engineering.
Vol. published in Oxford, New York.
Vols. 1-2 lack series statement.
Includes bibliographical references and indexes.
Contents: v. 1. Fluid flow, heat transfer, and mass transfer—v. 2. Unit operations— —v.6. Design by R. K. Sinnott.
I. Chemical engineering. I. Richardson, J. F. (John Francis)
II. Title.   III. Series.
TP145.C78   660.2   54–14486

British Library Cataloguing in Publication Data

Coulson, J. M.
Chemical engineering.
Vol. 6: Design
1. Chemical engineering
I. Title   II. Richardson, J. F.
III. Sinnott, R. K.
660.2   TP145.C78
ISBN 0–08–022969–7 (Hardcover)
ISBN 0–08–022970–0 (Flexicover)

Printed in Great Britain by BPCC Wheatons Ltd, Exeter

Author's Preface

THIS book has been written primarily for students on undergraduate courses in Chemical Engineering and has particular relevance to their design projects. It should also be of interest to graduates of other disciplines who are working in the chemical and process industries. In writing it, I have drawn on my experience of teaching design at the University College of Swansea, and on some years of experience in the process industries.

Books on design tend to fall into two categories. There are those written by academics, that are largely philosophical discussions of the nature and methodology of the design process, and which are usually of little practical use. And there are handbooks (cookbooks) covering design methods, information, and data, which are often derided by academics. As this book is intended to be used, the emphasis has been put on providing useful design methods and techniques. Clearly, it is not possible to cover in detail in one book the whole range of techniques and methods needed for the design of a chemical manufacturing process. Nor could this be within the range of experience and expertise of any one author. The approach taken has been to give sufficient detail for the preliminary design of processes and equipment, and to back this up with references both to authoritative texts and articles that cover the topics more thoroughly and to those that give detailed design methods.

The explanations that are given of the fundamental principles underlying the design methods are necessarily very brief. The scientific principles and unit operations of Chemical Engineering are covered in Volumes 1, 2 and 3 of this work, and in other textbooks cited in this volume, to which the reader is referred.

The chapters in this book can be grouped under three main topics. Chapters 1 to 9 and 14 cover process design, and include a brief explanation of the design method, including considerations of safety, costing, and materials selection. Chapters 10, 11 and 12 cover equipment selection and design. Chapter 13 covers the mechanical design of process plant.

Chapters 1 to 12 can be used as a text for courses on process and equipment design, omitting Chapter 2 and the first part of Chapter 3 for students who are familiar with material and energy balance calculations. Chapter 13 will give Chemical Engineering students some appreciation of the mechanical aspects of equipment design.

The art and practice of design cannot be learnt from books. The intuition and judgement necessary to apply theory to practice will come only from practical experience. I trust that this book will give its readers a modest start on that road.

In closing, I would like to express my appreciation to all those friends and colleagues who have influenced my own development as a professional engineer, and so contributed to this book.

R. K. SINNOTT

Preface

THE earlier volumes of this Series (Volumes 1, 2 and 3) dealt with the theoretical background to chemical engineering processes and operations and with the functioning of particular pieces of equipment. This volume completes the series and extends the treatment of the subject, by showing how a complete process is designed and how it must be fitted into the environment. It therefore includes material on flow-sheeting, piping, mechanical construction, safety and costing. It relies heavily on the earlier volumes for a discussion of the background theory, though in order to make the work complete in itself it includes illustrations of equipment items which have already featured in the previous works; furthermore, the treatment of distillation and heat exchanger design is expanded. Whilst the book is directed primarily to undergraduate students of chemical engineering, it should also be valuable to chemical engineers in industry (and particularly to those studying for the Design Project) and to chemists and mechanical engineers who have to tackle problems arising in the Process Industries.

The design engineer must use a wide range of information taken from a variety of sources and must take into account many conflicting requirements—technical, economic and environmental. Within the time span available to him, he must effect a satisfactory compromise and it should always be borne in mind that there is never a unique "best" solution to any design problem. Furthermore, what is a satisfactory design for one location may be totally unsuitable elsewhere. Although it is impossible to convey a complete philosophy through the medium of a single book, an attempt has been made to make the reader aware of many of the diverse factors which must be incorporated into any one design. This volume provides only an introduction but gives an indication of sources of more detailed information on individual branches of the subject.

J. F. RICHARDSON
J. M. COULSON

Acknowledgement

Material from British Standards is reproduced by permission of the British Standards Institution, 2 Park Street, London W1A 2BS from whom complete copies of the Standards can be obtained.

Contents

5. Piping and Instrumentation ... 148

6. Costing and Project Evaluation ... 181

9. Safety and Loss Prevention   274

12. Heat-transfer Equipment   511

13. Mechanical Design of Process Equipment

CHAPTER 1

Introduction to Design

1.1. Introduction

This chapter is an introduction to the nature and methodology of the design process, and its application to the design of chemical manufacturing processes.

1.2. Nature of design

This section is a general, somewhat philosophical, discussion of the design process; how a designer works. The subject of this book is chemical engineering design, but the methodology of design described in this section applies equally to other branches of engineering design.

Design is a creative activity, and as such can be one of the most rewarding and satisfying activities undertaken by an engineer. It is the synthesis, the putting together, of ideas to achieve a desired purpose. The design does not exist at the commencement of the project. The designer starts with a specific objective in mind, a need, and by developing and evaluating possible designs, arrives at what he considers the best way of achieving that objective; be it a better chair, a new bridge, or for the chemical engineer, a new chemical product or a stage in the design of a production process.

When considering possible ways of achieving the objective the designer will be constrained by many factors, which will narrow down the number of possible designs; but, there will rarely be just one possible solution to the problem, just one design. Several alternative ways of meeting the objective will normally be possible, even several best designs, depending on the nature of the constraints.

These constraints on the possible solutions to a problem in design arise in many ways. Some constraints will be fixed, invariable, such as those that arise from physical laws, government regulations, and standards. Others will be less rigid, and will be capable of relaxation by the designer as part of his general strategy in seeking the best design. The constraints that are outside the designer's influence can be termed the external constraints. These set the outer boundary of possible designs; as shown in Fig. 1.1. Within this boundary there will be a number of plausible designs bounded by the other constraints, the internal constraints, over which the designer has some control; such as, choice of process, choice of process conditions, materials, equipment.

Economic considerations are obviously a major constraint on any engineering design: plants must make a profit.

Time will also be a constraint. The time available for completion of a design will usually limit the number of alternative designs that can be considered.

The stages in the development of a design, from the initial identification of the objective

1

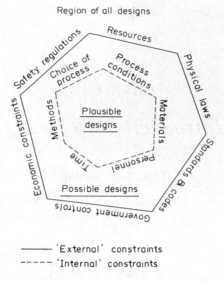

FIG. 1.1. Design constraints

to the final design, are shown diagrammatically in Fig. 1.2. Each stage is discussed in the following sections.

Figure 1.2 shows design as an iterative procedure; as the design develops the designer will be aware of more possibilities and more constraints, and will be constantly seeking new data and ideas, and evaluating possible design solutions.

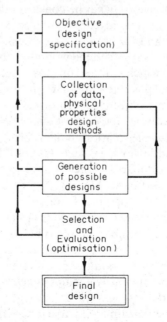

FIG. 1.2. The design process

1.2.1. The design objective (the need)

Chaddock (1975) defined design as, the conversion of an ill-defined requirement into a satisfied customer.

The designer is creating a design for an article, or a manufacturing process, to fulfil a particular need. In the design of a chemical process, the need is the public need for the product, the commercial opportunity, as foreseen by the sales and marketing organis-ation. Within this overall objective the designer will recognise sub-objectives; the requirements of the various units that make up the overall process.

Before starting work the designer should obtain as complete, and as unambiguous, a statement of the requirements as possible. If the requirement (need) arises from outside the design group, from a client or from another department, then he will have to elucidate the real requirements through discussion. It is important to distinguish between the real needs and the wants. The wants are those parts of the initial specification that may be thought desirable, but which can be relaxed if required as the design develops. For example, a particular product specification may be considered desirable by the sales department, but may be difficult and costly to obtain, and some relaxation of the specification may be possible, producing a saleable but cheaper product. Whenever he is in a position to do so, the designer should always question the design requirements (the project and equipment specifications) and keep them under review as the design progresses.

Where he writes specifications for others, such as for the mechanical design or purchase of a piece of equipment, he should be aware of the restrictions (constraints) he is placing on other designers. A tight, well-thought-out, comprehensive, specification of the requirements defines the external constraints within which the other designers must work.

1.2.2. Data collection

To proceed with a design, the designer must first assemble all the relevant facts and data required. For process design this will include information on possible processes, equipment performance, and physical property data. This stage can be one of the most time consuming, and frustrating, aspects of design. Sources of process information and physical properties are reviewed in Chapter 8.

Many design organisations will prepare a basic data manual, containing all the process "know-how" on which the design is to be based. Most organisations will have design manuals covering preferred methods and data for the more frequently used, routine, design procedures.

The national standards are also sources of design methods and data; they are also design constraints.

The constraints, particularly the external constraints, should be identified early in the design process.

1.2.3. Generation of possible design solutions

The creative part of the design process is the generation of possible solutions to the problem (ways of meeting the objective) for analysis, evaluation and selection. In this activity the designer will largely rely on previous experience, his own and that of others. It is doubtful if any design is entirely novel. The antecedence of most designs can usually be easily traced. The first motor cars were clearly horse-drawn carriages without the horse; and the development of the design of the modern car can be traced step by step from these early prototypes. In the chemical industry, modern distillation processes have developed from the ancient stills used for rectification of spirits; and the packed columns used for gas absorption have developed from primative, brushwood-packed towers. So, it is not often that a process designer is faced with the task of producing a design for a completely novel process or piece of equipment.

The experienced engineer will wisely prefer the tried and tested methods, rather than possibly more exciting but untried novel designs. The work required to develop new processes, and the cost, is usually underestimated. Progress is made more surely in small steps. However, whenever innovation is wanted, previous experience, through prejudice, can inhibit the generation and acceptance of new ideas; the "not invented here" syndrome.

The amount of work, and the way it is tackled, will depend on the degree of novelty in a design project.

Chemical engineering projects can be divided into three types, depending on the novelty involved:

1. Modifications, and additions, to existing plant; usually carried out by the plant design group.
2. New production capacity to meet growing sales demand, and the sale of established processes by contractors. Repetition of existing designs, with only minor design changes.
3. New processes, developed from laboratory research, through pilot plant, to a commercial process. Even here, most of the unit operations and process equipment will use established designs.

The first step in devising a new process design will be to sketch out a rough block diagram showing the main stages in the process; and to list the primary function (objective) and the major constraints for each stage. Experience should then indicate what types of unit operations and equipment should be considered. Jones (1970) discusses the methodology of design, and reviews some of the special techniques, such as brainstorming sessions and synectics, that have been developed to help generate ideas for solving intractable problems.

A good general reference on the art of problem solving is the classical work by Polya (1957).

The generation of ideas for possible solutions to a design problem cannot be separated from the selection stage of the design process; some ideas will be rejected as impractical as soon as they are conceived.

1.2.4. Selection

The designer starts with the set of all possible solutions bounded by the external constraints, and by a process of progressive evaluation and selection, narrows down the range of candidates to find the "best" design for the purpose.

The selection process can be considered to go through the following stages:

Possible designs (credible) – within the external constraints.
Plausible designs (feasible) – within the internal constraints.
Probable designs – likely candidates.
Best design (optimum) – judged the best solution to the problem.

The selection process will become more detailed and more refined as the design progresses from the area of possible to the area of probable solutions. In the early stages a coarse screening based on common sense, engineering judgement, and rough costings will usually suffice. For example, it would not take many minutes to narrow down the choice of raw materials for the manufacture of ammonia from the possible candidates of, say, wood, peat, coal, natural gas, and oil, to a choice of between gas and oil, but a more detailed study would be needed to choose between oil and gas. To select the best design from the probable designs, detailed design work and costing will usually be necessary. However, where the performance of candidate designs is likely to be close the cost of this further refinement, in time and money, may not be worth while, particularly as there will usually be some uncertainty in the accuracy of the estimates.

The mathematical techniques that have been developed to assist in the optimisation of designs, and plant performance, are discussed briefly in Section 1.10.

Rudd and Watson (1968) and Wells (1973) describe formal techniques for the preliminary screening of alternative designs.

1.3. The anatomy of a chemical manufacturing process

The basic components of a typical chemical process are shown in Fig. 1.3, in which each block represents a stage in the overall process for producing a product from the raw materials. Figure 1.3 represents a generalised process; not all the stages will be needed for any particular process, and the complexity of each stage will depend on the nature of the process. Chemical engineering design is concerned with the selection and arrangement of the stages, and the selection, specification and design of the equipment required to perform the stage functions.

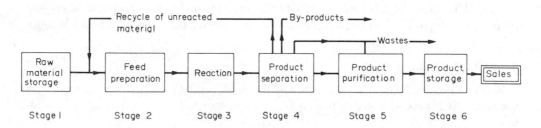

FIG. 1.3. Anatomy of a chemical process

Stage 1. Raw material storage

Unless the raw materials (also called essential materials, or feed stocks) are supplied as intermediate products (intermediates) from a neighbouring plant, some provision will have to be made to hold several days, or weeks, storage to smooth out fluctuations and interruptions in supply. Even when the materials come from an adjacent plant some provision is usually made to hold a few hours, or even days, supply to decouple the processes. The storage required will depend on the nature of the raw materials, the method of delivery, and what assurance can be placed on the continuity of supply. If materials are delivered by ship (tanker or bulk carrier) several weeks stocks may be necessary; whereas if they are received by road or rail, in smaller lots, less storage will be needed.

Stage 2. Feed preparation

Some purification, and preparation, of the raw materials will usually be necessary before they are sufficiently pure, or in the right form, to be fed to the reaction stage. For example, acetylene generated by the carbide process contains arsenical and sulphur compounds, and other impurities, which must be removed by scrubbing with concentrated sulphuric acid (or other processes) before it is sufficiently pure for reaction with hydrochloric acid to produce dichloroethane. Liquid feeds will need to be vaporised before being fed to gas-phase reactors, and solids may need crushing, grinding and screening.

Stage 3. Reactor

The reaction stage is the heart of a chemical manufacturing process. In the reactor the raw materials are brought together under conditions that promote the production of the desired product; invariably, by-products and unwanted compounds (impurities) will also be formed.

Stage 4. Product separation

In this first stage after the reactor the products and by-products are separated from any unreacted material. If in sufficient quantity, the unreacted material will be recycled to the reactor. They may be returned directly to the reactor, or to the feed purification and preparation stage. The by-products may also be separated from the products at this stage.

Stage 5. Purification

Before sale, the main product will usually need purification to meet the product specification. If produced in economic quantities, the by-products may also be purified for sale.

Stage 6. Product storage

Some inventory of finished product must be held to match production with sales. Provision for product packaging and transport will also be needed, depending on the nature of the product. Liquids will normally be dispatched in drums and in bulk tankers (road, rail and sea), solids in sacks, cartons or bales.

The stock held will depend on the nature of the product and the market.

Ancillary processes

In addition to the main process stages shown in Fig. 1.3, provision will have to be made for the supply of the services (utilities) needed; such as, process water, cooling water, compressed air, steam. Facilities will also be needed for maintenance, firefighting, offices and other accommodation, and laboratories; see Chapter 14.

1.4. The organisation of a chemical engineering project

The design work required in the engineering of a chemical manufacturing process can be divided into two broad phases.

Phase 1. Process design, which covers the steps from the initial selection of the process to be used, through to the issuing of the process flow-sheets; and includes the selection, specification and chemical engineering design of equipment. In a typical organisation, this phase is the responsibility of the Process Design Group, and the work will be mainly done by chemical engineers. The process design group may also be responsible for the preparation of the piping and instrumentation diagrams.

Phase 2. The detailed mechanical design of equipment; the structural, civil and electrical design; and the specification and design of the ancillary services. These activities will be the responsibility of specialist design groups, having expertise in the whole range of engineering disciplines.

Other specialist groups will be responsible for cost estimation, and the purchase and procurement of equipment and materials.

The sequence of steps in the design, construction and start-up of a typical chemical process plant is shown diagrammatically in Fig. 1.4 and the organisation of a typical project group in Fig. 1.5. Each step in the design process will not be as neatly separated from the others as is indicated in Fig. 1.4; nor will the sequence of events be as clearly defined. There will be a constant interchange of information between the various design sections as the design develops, but it is clear that some steps in a design must be largely completed before others can be started.

A project manager, often a chemical engineer by training, is usually responsible for the co-ordination of the project, as shown in Fig. 1.5.

As was stated in Section 1.2.1, the project design should start with a clear specification defining the product, capacity, raw materials, process and site location. If the project is based on an established process and product, a full specification can be drawn up at the start of the project. For a new product, the specification will be developed from an economic evaluation of possible processes, based on laboratory research, pilot plant tests and product market research.

The organisation of chemical process design is discussed in more detail by Rase and Barrow (1964) and Baasel (1974).

Some of the larger chemical manufacturing companies have their own project design organisations and carry out the whole project design and engineering, and possibly construction, within their own organisation. More usually the design and construction, and possibly assistance with start-up, is entrusted to one of the international contracting firms.

The operating company will often provide the "know-how" for the process, and will work closely with the contractor throughout all stages of the project.

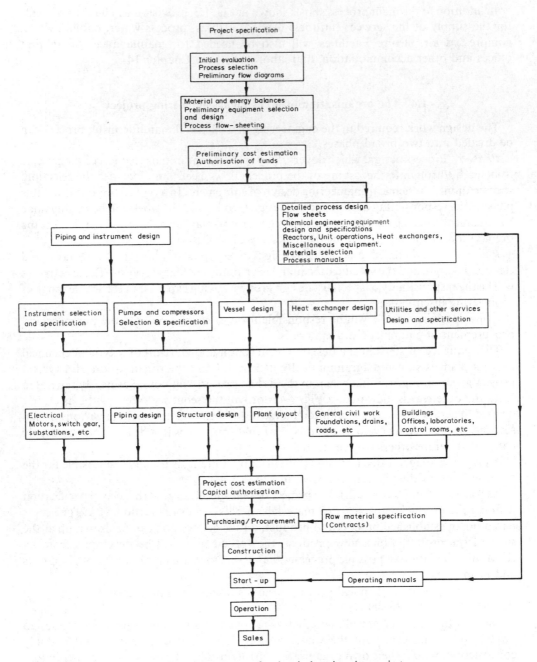

FIG. 1.4. The structure of a chemical engineering project

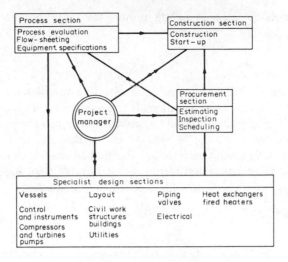

FIG. 1.5. Project organisation

1.5. Project documentation

As shown in Fig. 1.5 and described in Section 1.4, the design and engineering of a chemical process requires the co-operation of many specialist groups. Effective co-operation depends on effective communications, and all design organisations have formal procedures for handling project information and documentation. The project documentation will include:

1. General correspondence	within the design group and with:
	government departments
	equipment vendors
	site personnel
	the client
2. Calculation sheets	design calculations
	costing
	computer print-out
3. Drawings	flow-sheets
	piping and instrumentation diagrams
	layout diagrams
	plot/site plans
	equipment details
	piping diagrams
	architectural drawings
	design sketches
4. Specification sheets	for equipment, such as:
	heat exchangers
	pumps
5. Purchase orders	quotations
	invoices

All documents should be assigned a code number for easy cross referencing, filing and retrieval.

Calculation sheets

The design engineer should develop the habit of setting out calculations so that they can be easily understood and checked by others. It is good practice to include on calculation sheets the basis of the calculations, and any assumptions and approximations made, in sufficient detail for the methods, as well as the arithmetic, to be checked. Design calculations are normally set out on standard sheets. The heading at the top of each sheet should include: the project title and identification number and, most importantly, the signature (or initials) of the person who checked the calculation.

Drawings

All project drawings are normally drawn on specially printed sheets, with the company name; project title and number; drawing title and identification number; draughtsman's name and person checking the drawing; clearly set out in a box in the bottom right-hand corner. Provision should also be made for noting on the drawing all modifications to the initial issue.

Drawings should conform to accepted drawing conventions, preferably those laid down by the national standards, BS 308. The symbols used for flow-sheets and piping and instrument diagrams are discussed in Chapter 4. Drawings and sketches are now normally made on detail paper (semi-transparent) in pencil, so modifications can be easily made, and prints taken.

Specification sheets

Standard specification sheets are normally used to transmit the information required for the detailed design, or purchase, of equipment items; such as, heat exchangers, pumps, columns.

As well as ensuring that the information is clearly and unambiguously presented, standard specification sheets serve as check lists to ensure that all the information required is included.

Process manuals

Process manuals are often prepared by the process design group to describe the process and the basis of the process design. Together with the flow-sheets, they provide a complete technical description of the process.

Operating manuals

Operating manuals give the detailed, step by step, instructions for operation of the process and equipment. They would normally be prepared by the operating company personnel, but may also be issued by a contractor as part of the contract package for a less experienced client. The operating manuals would be used for operator instruction and training, and for the preparation of the formal plant operating instructions.

1.6. Codes and standards

The need for standardisation arose early in the evolution of the modern engineering industry; Whitworth introduced the first standard screw thread to give a measure of interchangeability between different manufacturers in 1841. Modern engineering standards cover a much wider function than the interchange of parts. In engineering practice they cover:

1. Materials, properties and compositions.
2. Testing procedures for performance, compositions, quality.
3. Preferred sizes; for example, tubes, plates, sections.
4. Design methods, inspection, fabrication.
5. Codes of practice, for plant operation and safety.

The terms STANDARD and CODE are used interchangeably, though CODE should really be reserved for a code of practice covering say, a recommended design or operating procedure; and STANDARD for preferred sizes, compositions, etc.

All of the developed countries, and many of the developing countries, have national standards organisations, responsible for the issue and maintenance of standards for the manufacturing industries, and for the protection of consumers. In the United Kingdom preparation and promulgation of national standards are the responsibility of the British Standards Institution (BSI). The Institution has a secretariat and a number of technical personnel, but the preparation of the standards is largely the responsibility of committees of persons from the appropriate industry, the professional engineering institutions and other interested organisations.

In the United States the government organisation responsible for coordinating information on standards is the National Bureau of Standards; standards are issued by Federal, State and various commercial organisations. The principal ones of interest to chemical engineers are those issued by the American National Standards Institute (ANSI), the American Petroleum Institute (API), the American Society for Testing Materials (ASTM), and the American Society of Mechanical Engineers (ASME) (pressure vessels). Burklin (1979) gives a comprehensive list of the American codes and standards.

The International Organisation for Standardisation (ISO) coordinates the publication of international standards.

In this book reference is made to the appropriate British Standard where relevant. All the published standards are listed, and their scope and application described, in the *British Standards Year Book*; which the designer should consult.

As well as the various national standards and codes, the larger design organisations will have their own (in-house) standards. Much of the detail in engineering design work is routine and repetitious, and it saves time and money, and ensures a conformity between projects, if standard designs are used whenever practicable.

Equipment manufacturers also work to standards to produce standardised designs and size ranges for commonly used items; such as electric motors, pumps, pipes and pipe fittings. They will conform to national standards, where they exist, or to those issued by trade associations. It is clearly more economic to produce a limited range of standard sizes than to have to treat each order as a special job.

For the designer, the use of a standardised component size allows for the easy integration of a piece of equipment into the rest of the plant. For example, if a standard range of centrifugal pumps is specified the pump dimensions will be known, and this facilitates the design of the foundations plates, pipe connections and the selection of the drive motors; standard electric motors would be used.

For an operating company, the standardisation of equipment designs and sizes increases interchangeability and reduces the stock of spares that have to be held in maintenance stores.

Though there are clearly considerable advantages to be gained from the use of standards in design, there are also some disadvantages. Standards impose constraints on the designer. The nearest standard size will normally be selected on completing a design calculation (rounding-up) but this will not necessarily be the optimum size; though as the standard size will be cheaper than a special size, it will usually be the best choice from the point of view of initial capital cost. Standard design methods must, of their nature, be historical, and do not necessarily incorporate the latest techniques.

The use of standards in design is illustrated in the discussion of the pressure vessel design standards (codes) in Chapter 13.

1.7. Factors of safety (design factors)

Design is an inexact art; errors and uncertainties will arise from uncertainties in the design data available and in the approximations necessary in design calculations. To ensure that the design specification is met, factors are included to give a margin of safety in the design; safety in the sense that the equipment will not fail to perform satisfactorily, and that it will operate safely: will not cause a hazard. "Design factor" is a better term to use, as it does not confuse safety and performance factors.

In mechanical and structural design, the magnitude of the design factors used to allow for uncertainties in material properties, design methods, fabrication and operating loads are well established. For example, a factor of around 4 on the tensile strength, or about 2·5 on the 0·1 per cent proof stress, is normally used in general structural design. The selection of design factors in mechanical engineering design is illustrated in the discussion of pressure vessel design in Chapter 13.

Design factors are also applied in process design to give some tolerance in the design. For example, the process stream average flows calculated from material balances are usually increased by a factor, typically 10 per cent, to give some flexibility in process operation. This factor will set the maximum flows for equipment, instrumentation and piping design. Where design factors are introduced to give some contingency in a process design, they should be agreed within the project organisation, and clearly stated in the project documents (drawings, calculation sheets and manuals). If this is not done, there is a danger that each of the specialist design groups will add its own "factor of safety"; resulting in gross, and unnecessary, over-design.

When selecting the design factor to use, a balance has to be made between the desire to make sure the design is adequate and the need to design to tight margins to remain competitive. The greater the uncertainty in the design methods and data, the bigger the design factor that must be used.

1.8. Systems of units

To be consistent with the other volumes in this series, SI units have been used in this book. However, in practice the design methods, data and standards which the designer will use are normally only available in the traditional scientific and engineering units. Chemical engineering has always used a diversity of units; embracing the scientific CGS and MKS systems, and both the American and British engineering systems. Those engineers in the older industries will also have had to deal with some bizarre traditional units; such as degrees Twaddle (density) and barrels for quantity. Desirable as it may be for industry world-wide to adopt one consistent set of units, such as SI, this is unlikely to come about for many years, and the designer must contend with whatever system, or combination of systems, his organisation uses. For those in the contracting industry this will also mean working with whatever system of units the client requires.

It is usually the best practice to work through design calculations in the units in which the result is to be presented; but, if working in SI units is preferred, data can be converted to SI units, the calculation made, and the result converted to whatever units are required. Conversion factors to the SI system from most of the scientific and engineering units used in chemical engineering design are given in Appendix E.

Some license has been taken in the use of the SI system in this volume. Temperatures are given in degrees Celsius (°C); degrees Kelvin are only used when absolute temperature is required in the calculation. Pressures are often given in bar (or atmospheres) rather than in the Pascals (N/m^2), as this gives a better feel for the magnitude of the pressures. In technical calculations the bar can be taken as equivalent to an atmosphere, whatever definition is used for atmosphere.

For stress, N/mm^2 have been used, as these units are now generally accepted by engineers, and the use of a small unit of area helps to indicate that stress is the intensity of force at a point (as is also pressure). For quantity, kmol are generally used in preference to mol, and for flow, kmol/h instead of mol/s, as this gives more sensibly sized figures, which are also closer to the more familiar lb/h.

For volume and volumetric flow, m^3 and m^3/h are used in preference to m^3/s, which gives ridiculously small values in engineering calculations. Litres per second are used for small flow-rates, as this is the preferred unit for pump specifications.

Where, for convenience, other than SI units have been used on figures or diagrams, the scales are also given in SI units, or the appropriate conversion factors are given in the text. The answers to some examples are given in British engineering units as well as SI, to help illustrate the significance of the values.

Some approximate conversion factors to SI units are given below. These are worth committing to memory, to give some feel for the units, for those more familiar with the traditional engineering systems. The exact conversion factors are given in brackets.

Energy 1 Btu $\simeq$ 1 kJ (1·05506)
Specific enthalpy 1 Btu/lb $\simeq$ 2 kJ/kg (2·326)
Heat capacity 1 Btu/lb°F = 1 cal/g°C $\simeq$ 4kJ/kg°C (4·1868)
Heat transfer coefficient 1 Btu/h ft^2 °F = 1 CHU/h ft^2 °C $\simeq$ 6 W/m^2 °C (5·678)
(CHU = Centigrade heat unit)
Viscosity 1 centipoise = 1 mN s/m^2

Pressure 1 atm $\simeq$ 1 bar $= 10^5\,N/m^2 = 10^5\,Pa$
    1 psi $\simeq 7\,kN/m^2 = 7\,kPa$ (6·894)
    1 ft water $\simeq 3\,kN/m^2 = 3\,kPa$ (2·989)
Mass 1 tonne $= 1000\,kg \simeq 1$ Imperial ton (0·98410)
Density 1 lb/ft$^3 \simeq 16\,kg/m^3$ (16·0190)
Note: 1 g/cm$^3 = 10^3\,kg/m^3$
Volume: 1 US gal $= 0·84$ Imperial gal $\simeq 3·7 \times 10^{-3}\,m^3$ ($3·7854 \times 10^{-3}$)
1 barrel (oil) $= 50$ US gal $\simeq 0·19\,m^3$ (0·1893)
Flow: 1 UK gpm $\simeq 16\,m^3/h$ (16·366)

When using American engineering literature and equipment catalogues, *beware of US gallons*; a pump capacity quoted in gpm (US) will have only 80 per cent of the required capacity if this is specified in gpm (Imp).

1.9. Degrees of freedom and design variables.
The mathematical representation of the design problem

In Section 1.2 it was shown that the designer in seeking a solution to a design problem works within the constraints inherent in the particular problem.

In this section the structure of design problems is examined by representing the general design problem in a mathematical form.

1.9.1. Information flow and design variables

A process unit in a chemical process plant performs some operation on the inlet material streams to produce the desired outlet streams. In the design of such a unit the design calculations model the operation of the unit. A process unit and the design equations representing the unit are shown diagrammatically in Fig. 1.6. In the "design unit" the flow of material is replaced by a flow of information into the unit and a flow of derived information from the unit.

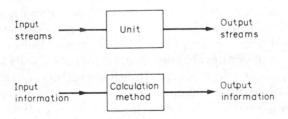

FIG. 1.6. The "design unit"

The information flows are the values of the variables which are involved in the design; such as, stream compositions, temperatures, pressure, stream flow-rates, and stream enthalpies. Composition, temperature and pressure are intensive variables: independent of the quantity of material (flow-rate). The constraints on the design will place restrictions on the possible values that these variables can take. The values of some of the variables will be fixed directly by process specifications. The values of other variables will be determined by "design relationships" arising from constraints. Some of the design relationships will be in the form of explicit mathematical equations (design equations); such as those arising from material and energy balances, thermodynamic relationships, and equipment performance parameters. Other relationships will be less precise; such as those arising from the use of standards and preferred sizes, and safety considerations.

The difference between the number of variables involved in a design and the number of design relationships has been called the number of "degrees of freedom"; similar to the use of the term in the phase rule. The number of variables in the system is analogous to the number of variables in a set of simultaneous equations, and the number of relationships analogous to the number of equations. The difference between the number of variables and equations is called the variance of the set of equations.

If N_v is the number of possible variables in a design problem and N_r the number of design relationships, then the "degrees of freedom" N_d is given by:

$$N_d = N_v - N_r \tag{1.1}$$

N_d represents the freedom that the designer has to manipulate the variables to find the best design.

If $N_v = N_r$, $N_d = 0$ and there is only one, unique, solution to the problem. The problem is not a true design problem, no optimisation is possible.

If $N_v < N_r$, $N_d < 0$, and the problem is over defined; only a trivial solution is possible.

If $N_v > N_r$, $N_d > 0$, and there is an infinite number of possible solutions. However, for a practical problem there will be only a limited number of feasible solutions. The value of N_d is the number of variables which the designer must assign values to solve the problem.

How the number of process variables, design relationships, and design variables defines a system can be best illustrated by considering the simplest system; a single-phase, process stream.

Process stream

Consider a single-phase stream, containing C components.

Variable	Number
Stream flow-rate	1
Composition (component concentrations)	C
Temperature	1
Pressure	1
Stream enthalpy	1
Total, $N_v =$	$C+4$

Relationships between variables	Number
Composition[1]	1
Enthalpy[2]	1
Total, $N_r =$	2

Degrees of freedom $N_d = N_v - N_r = (C+4) - 2 = \underline{\underline{C+2}}$

(1) The sum of the mass or mol, fractions, must equal one.
(2) The enthalpy is a function of stream composition, temperature and pressure.

Specifying $(C + 2)$ variables completely defines the stream.

Flash distillation

The idea of degrees of freedom in the design process can be further illustrated by considering a simple process unit, a flash distillation. (For a description of flash distillation see Volume 2, Chapter 11.)

The unit is shown in Fig. 1.7, where:

F = stream flow rate,
P = pressure,
T = temperature,
x_i = concentration, component i,
q = heat input.

Suffixes, 1 = inlet, 2 = outlet vapour, 3 = outlet liquid.

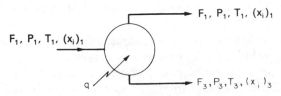

F_1, P_1, T_1, $(x_i)_1$

F_1, P_1, T_1, $(x_i)_1$

q

F_3, P_3, T_3, $(x_i)_3$

FIG. 1.7. Flash distillation

Variable	Number
Streams (free variables)[1]	$3(C + 2)^1$
Still	
pressure	1
temperature	1
heat input	1
$N_v =$	$3C + 9$

Relationship	Number
Material balances (each component)	C
Heat balance, overall	1
v–le relationships[2]	C
Equilibrium still restriction[3]	4
	$2C + 5$

Degrees of freedom $N_d = (3C + 9) - (2C + 5) = \underline{\underline{C + 4}}$

(1) The degrees of freedom for each stream. The total variables in each stream could have been used, and the stream relationships included in the count of relationships.

This shows how the degrees of freedom for a complex unit can be built up from the degrees of freedom of its components. For more complex examples see Kwauk (1956).

(2) Given the temperature and pressure, the concentration of any component in the vapour phase can be obtained from the concentration in the liquid phase, from the vapour–liquid equilibrium data for the system.

(3) The concept (definition) of an equilibrium separation implies that the outlet streams and the still are at the same temperature and pressure. This gives four equations:

$$P_2 = P_3 = P$$
$$T_2 = T_3 = T$$

Though the total degrees of freedom is seen to be $(C+4)$ some of the variables will normally be fixed by general process considerations, and will not be free for the designer to select as "design variables". The flash distillation unit will normally be one unit in a process system and the feed composition and feed conditions will be fixed by the upstream processes; the feed will arise as an outlet stream from some other unit. Defining the feed fixes $(C+2)$ variables, so the designer is left with:

$$(C+4)-(C+2)=2$$

as design variables.

Summary

The purpose of this discussion was to show that in a design there will be a certain number of variables that the designer must specify to define the problem, and which he can manipulate to seek the best design. In manual calculations the designer will rarely need to calculate the degrees of freedom in a formal way. He will usually have intuitive feel for the problem, and can change the calculation procedure, and select the design variables, as he works through the design. He will know by experience if the problem is correctly specified. A computer, however, has no intuition, and for computer-aided design calculations it is essential to ensure that the necessary number of variables is specified to define the problem correctly. For complex processes the number of variables and relating equations will be very large, and the calculation of the degrees of freedom very involved. Kwauk (1956) has shown how the degrees of freedom can be calculated for separation processes by building up the complex unit from simpler units. Smith (1963) uses Kwauk's method, and illustrates how the idea of "degrees of freedom" can be used in the design of separation processes.

1.9.2. Selection of design variables

In setting out to solve a design problem the designer has to decide which variables are to be chosen as "design variables"; the ones he will manipulate to produce the best design. The choice of design variables is important; careful selection can simplify the design calculations. This can be illustrated by considering the choice of design variables for a simple binary flash distillation.

For a flash distillation the total degrees of freedom was shown to be $(C+4)$, so for two components $N_d = 6$. If the feed stream flow, composition, temperature and pressure are fixed by upstream conditions, then the number of design variables will be:

$$N_d' = 6-(C+2) = 6-4 = 2$$

So the designer is free to select two variables from the remaining variables in order to proceed with the calculation of the outlet stream compositions and flows.

If he selects the still pressure (which for a binary system will determine the vapour–liquid equilibrium relationship) and one outlet stream flow-rate, then the outlet compositions can be calculated by simultaneous solution of the mass balance and equilibrium relationships (equations). A graphical method for the simultaneous solution is given in Volume 2, Chapter 11.

However, if he selects an outlet stream composition (say the liquid stream) instead of a flow-rate, then the simultaneous solution of the mass balance and v–l–e relationships would not be necessary. The stream compositions could be calculated by the following step-by-step (sequential) procedure:

1. Specifying P determines the v–l–e relationship (equilibrium) curve from experimental data).
2. Knowing the outlet liquid composition, the outlet vapour composition can be calculated from the v–l–e relationship.
3. Knowing the feed and outlet compositions, and the feed flow-rate, the outlet stream flows can be calculated from a material balance.
4. An enthalpy balance then gives the heat input required.

The need for simultaneous solution of the design equations implies that there is a recycle of information. Choice of an outlet stream composition as a design variable in effect reverses the flow of information through the problem and removes the recycle; this is shown diagrammatically in Fig. 1.8.

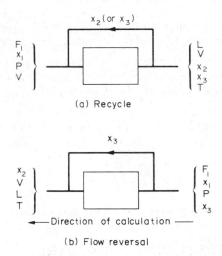

Fig. 1.8. Information flow, binary flash distillation calculation

1.9.3. Information flow and the structure of design problems

It was shown in Section 1.9.2, by studying a relatively simple problem, that the way in which the designer selects his design variables can determine whether the design calculations will prove to be easy or difficult. Selection of one particular set of variables can lead to a straightforward, step-by-step, procedure, whereas selection of another set can force the need for simultaneous solution of some of the relationships; which often requires an iterative procedure (cut-and-try method). How the choice of design variables, inputs to the calculation procedure, affects the ease of solution for the general design problem can be illustrated by studying the flow of information, using simple information flow diagrams. The method used will be that given by Lee *et al.* (1966) who used a form of directed graph; a biparte graph, see Berge (1962).

The general design problem can be represented in mathematical symbolism as a series of equations:

$$f_i(v_j) = 0$$

where $j = 1, 2, 3, \ldots, N_v$,

$i = 1, 2, 3, \ldots, N_r$

Consider the following set of such equations:

$$f_1(v_1, v_2) = 0$$
$$f_2(v_1, v_2, v_3, v_5) = 0$$
$$f_3(v_1, v_3, v_4) = 0$$
$$f_4(v_2, v_4, v_5, v_6) = 0$$
$$f_5(v_5, v_6, v_7) = 0$$

There are seven variables, $N_v = 7$, and five equations (relationships) $N_r = 5$, so the number of degrees of freedom is:

$$N_d = N_v - N_r = 7 - 5 = 2$$

The task is to select two variables from the total of seven in such a way as to give the simplest, most efficient, method of solution to the seven equations. There are twenty-one ways of selecting two items from seven.

In Lee's method the equations and variables are represented by nodes on the biparte graph (circles), connected by edges (lines), as shown in Fig. 1.9.

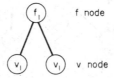

FIG. 1.9. Nodes and edges on a biparte graph

Figure 1.9 shows that equation f_1 contains (is connected to) variables v_1 and v_2. The complete graph for the set of equations is shown in Fig. 1.10.

The number of edges connected to a node defines the local degree of the node p. For example, the local degree of the f_1 node is 2, $p(f_1) = 2$, and at the v_5 node it is 3, $p(v_5) = 3$. Assigning directions to the edges of Fig. 1.10 (by putting arrows on the lines) identifies one

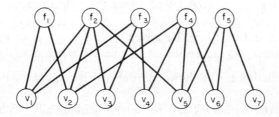

FIG. 1.10. Biparte graph for the complete set of equations

possible order of solution for the equations. If a variable v_j is defined as an output variable from an equation f_i, then the direction of information flow is from the node f_i to the node v_j and all other edges will be oriented into f_i. What this means, mathematically, is that assigning v_j as an output from f_i rearranges that equation so that:

$$f_i(v_1, v_2, \ldots, v_n) = v_j$$

v_j is calculated from equation f_i.

The variables selected as design variables (fixed by the designer) cannot therefore be assigned as output variables from an f node. They are inputs to the system and their edges must be oriented into the system of equations.

If, for instance, variables v_3 and v_4 are selected as design variables, then Fig. 1.11 shows one possible order of solution of the set of equations. Different types of arrows are used to distinguish between input and output variables, and the variables selected as design variables are enclosed in a double circle.

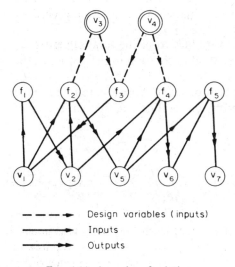

Fig. 1.11. An order of solution

Tracing the order of the solution of the equations as shown in Fig. 1.11 shows how the information flows through the system of equations:

1. Fixing v_3 and v_4 enables f_3 to be solved, giving v_1 as the output. v_1 is an input to f_1 and f_2.
2. With v_1 as an input, f_1 can be solved giving v_2; v_2 is an input to f_2 and f_4.
3. Knowing v_3, v_1 and v_2, f_2 can be solved to give v_5; v_5 is an input to f_4 and f_5.
4. Knowing v_4, v_2 and v_5, f_4 can be solved to give v_6; v_6 is an input to f_5.
5. Knowing v_6 and v_5, f_5 can be solved to give v_7; which completes the solution.

This order of calculation can be shown more clearly by redrawing Fig. 1.11 as shown in Fig. 1.12.

With this order, the equations can be solved sequentially, with no need for the simultaneous solution of any of the equations. The fortuitous selection of v_3 and v_4 as design variables has given an efficient order of solution of the equations.

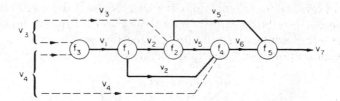

FIG. 1.12 Figure 1.11 redrawn to show order of solution

If for a set of equations an order of solution exists such that there is no need for the simultaneous solution of any of the equations, the system is said to be "acyclic", no recycle of information.

If another pair of variables had been selected, for instance v_5 and v_7, an acyclic order of solution for the set of equations would not necessarily have been obtained.

For many design calculations it will not be possible to select the design variables so as to eliminate the recycle of information and obviate the need for iterative solution of the design relationships.

For example, the set of equations given below will be cyclic for all choices of the two possible design variables.

$$f_1(x_1, x_2) = 0$$
$$f_2(x_1, x_3, x_4) = 0$$
$$f_3(x_2, x_3, x_4, x_5, x_6) = 0$$
$$f_4(x_4, x_5, x_6) = 0$$

$$N_d = 6 - 4 = 2$$

The biparte graph for this example, with x_3 and x_5 selected as the design variables (inputs), is shown in Fig. 1.13.

One strategy for the solution of this cyclic set of equations would be to guess (assign a value to) x_6. The equations could then be solved sequentially, as shown in Fig. 1.14, to produce a calculated value for x_6, which could be compared with the assumed value and the procedure repeated until a satisfactory convergence of the assumed and calculated value had been obtained. Assigning a value to x_6 is equivalent to "tearing" the recycle loop at x_6 (Fig. 1.15). Iterative methods for the solution of equations are discussed in Volume 3, Chapter 4; see also Henley and Rosen (1969).

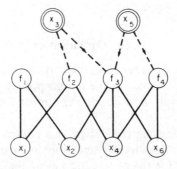

FIG. 1.13

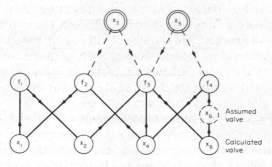

FIG. 1.14

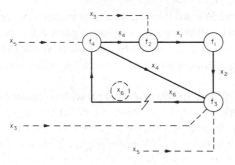

FIG. 1.15

When a design problem cannot be reduced to an acyclic form by judicious selection of the design variables, the design variables should be chosen so as to reduce the recycle of information to a minimum. Lee and Rudd (1966) and Rudd and Watson (1968) give an algorithm that can be used to help in the selection of the best design variables in manual calculations.

The recycle of information, often associated with the actual recycle of process material, will usually occur in any design problem involving large sets of equations; such as in the computer simulation of chemical processes. Efficient methods for the solution of sets of equations are required in computer-aided design procedures to reduce the computer time needed. Several workers have published algorithms for the efficient ordering of recycle loops for iterative solution procedures, and some references to this work are given in the chapter on flow-sheeting, Chapter 4.

1.10. Optimisation

Design is optimisation: the designer seeks the best, the optimum, solution to a problem.

Much of the selection and choice in the design process will depend on the intuitive judgement of the designer; who must decide when more formal optimisation techniques can be used to advantage.

The task of formally optimising the design of a complex processing plant involving several hundred variables, with complex interactions, is formidable, if not impossible. The task can be reduced by dividing the process into more manageable units, identifying the key variables and concentrating work where the effort involved will give the greatest benefit. Sub-division, and optimisation of the sub-units rather than the whole, will not necessarily give the optimum design for the whole process. The optimisation of one unit may be at the expense of another. For example, it will usually be satisfactory to optimise the reflux ratio for a fractionating column independently of the rest of the plant; but if the column is part of a separation stage following a reactor, in which the product is separated from the unreacted materials, then the design of the column will interact with, and may well determine, the optimisation of the reactor design.

In this book the discussion of optimisation methods will, of necessity, be limited to a brief review of the main techniques used in process and equipment design. The extensive literature on the subject should be consulted for full details of the methods available, and their application and limitations; see Beightler and Wilde (1967), Beveridge and Schechter (1970), Stoecker (1971), Rudd and Watson (1968).

The book by Rudd and Watson (1968), which deals with the development of optimum design strategies, is particularly recommended to students.

1.10.1. General procedure

When setting out to optimise any system, the first step is clearly to identify the objective: the criterion to be used to judge the system performance. In engineering design the objective will invariably be an economic one. For a chemical process, the overall objective for the operating company will be to maximise profits. This will give rise to sub-objectives, which the designer will work to achieve. The main sub-objective will usually be to minimise operating costs. Other sub-objectives may be to reduce investment, maximise yield, reduce labour requirements, reduce maintenance, operate safely.

When choosing his objectives the designer must keep in mind the overall objective. Minimising cost per unit of production will not necessarily maximise profits per unit time; market factors, such as quality and delivery, may determine the best overall strategy.

The second step is to determine the objective function: the system of equations, and other relationships, which relate the objective with the variables to be manipulated to optimise the function. If the objective is economic, it will be necessary to express the objective function in economic terms (costs).

Difficulties will arise in expressing functions that depend on value judgements; for example, the social benefits and the social costs that arise from pollution.

The third step is to find the values of the variables that give the optimum value of the objective function (maximum or minimum). The best techniques to be used for this step will depend on the complexity of the system and on the particular mathematical model used to represent the system.

A mathematical model represents the design as a set of equations (relationships) and, as was shown in Section 1.9.1, it will only be possible to optimise the design if the number of variables exceeds the number of relationships; there is some degree of freedom in the system.

1.10.2. Simple models

If the objective function can be expressed as a function of one variable (single degree of freedom) the function can be differentiated, or plotted, to find the maximum or minimum. This will be possible for only a few practical design problems. The technique is illustrated in the derivation of the formula for optimum pipe diameter in Chapter 5; also in what could be considered the classical example of optimisation in chemical engineering design, the determination of optimum reflux ratio; which is discussed in Volume 2, Chapter 11.

1.10.3. Multiple variable problems

The general optimisation problem can be represented mathematically as:

$$f = f(v_1, v_2, v_3, \ldots, v_n) \tag{1.2}$$

where f is the objective function and v_1, v_2, v_3, . . . , v_n are the variables.

In a design situation there will be constraints on the possible values of the objective function, arising from constraints on the variables; such as, minimum flow-rates, maximum allowable concentrations, and preferred sizes and standards.

Some may be equality constraints, expressed by equations of the form:

$$\Phi_m = \Phi_m(v_1, v_2, v_3, \ldots, v_n) = 0 \tag{1.3}$$

Others as inequality constraints:

$$\psi_p = \psi_p(v_1, v_2, v_3, \ldots, v_n) \leqslant P_p \tag{1.4}$$

The problem is to find values for the variables v_1 to v_n that optimise the objective function: that give the maximum or minimum value, within the constraints.

Analytical methods

If the objective function can be expressed as a mathematical function the classical methods of calculus can be used to find the maximum or minimum. Setting the partial derivatives to zero will produce a set of simultaneous equations that can be solved to find the optimum values. For the general, unconstrained, objective function, the derivatives will give the critical points; which may be maximum or minimum, or ridges or valleys. As with single variable functions, the nature of the first derivative can be found by taking the second derivative. For most practical design problems the range of values that the variables can take will be subject to constraints (equations 1.3 and 1.4), and the optimum of the constrained objective function will not necessarily occur where the partial derivatives of the objective function are zero. This situation is illustrated in Fig. 1.16 for a two-dimensional problem. For this problem, the optimum will lie on the boundary defined by the constraint $y = a$.

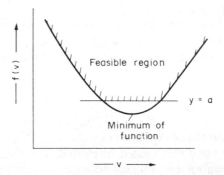

FIG. 1.16. Effect of constraints on optimum of a function

The method of Lagrange's undetermined multipliers is a useful analytical technique for dealing with problems that have equality constraints (fixed design values). Examples of the use of this technique for simple design problems are given by Stoecker (1971), Peters and Timmerhaus (1968) and Boas (1963a).

Search methods

The nature of the relationships and constraints in most design problems is such that the use of analytical methods is not feasible. In these circumstances search methods, that require only that the objective function can be computed from arbitrary values of the independent variables, are used. For single variable problems, where the objective function is unimodal, the simplest approach is to calculate the value of the objective function at uniformly spaced values of the variable until a maximum (or minimum) value is obtained. Though this method is not the most efficient, it will not require excessive computing time for simple problems. Several more efficient search techniques have been developed, such as the method of the golden section which is discussed in Volume 3, Chapter 4; see also Boas (1963b).

Efficient search methods will be needed for multi-dimensional problems, as the number of calculations required and the computer time necessary will be greatly increased, compared with single variable problems; see Himmelblau (1963), Stoecker (1971), Beveridge and Schechter (1970), and Baasel (1974).

Two variable problems can be plotted as shown in Fig. 1.17. The values of the objective function are shown as contour lines, as on a map, which are slices through the three-dimensional model of the function. Seeking the optimum of such a function can be likened to seeking the top of a hill (or bottom of a valley), and a useful technique for this type of problem is the *method of steepest ascent* (*or descent*) mentioned in Volume 3, Chapter 4.

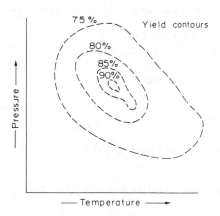

FIG. 1.17. Yield as a function of reactor temperature and pressure

1.10.4. *Linear programming*

Linear programming is an optimisation technique that can be used when the objective function and constraints can be expressed as a linear function of the variables; see Driebeek (1969), Williams (1967) and Dano (1965).

The technique is useful where the problem is to decide the optimum utilisation of resources. Many oil companies use linear programming to determine the optimum schedule of products to be produced from the crude oils available. Algorithms have been developed for the efficient solution of linear programming problems and the SIMPLEX algorithm, Dantzig (1963), is the most commonly used. The mathematical statement of the general problem in linear programming and the simplex algorithm is given in Volume 3, Chapter 4.

Examples of the application of linear programming in chemical process design and operation are given by Allen (1971), Rudd and Watson (1968) and Stoecker (1971).

1.10.5. Dynamic programming

Dynamic programming is a technique developed for the optimisation of large systems; see Nemhauser (1966), Bellman (1957) and Aris (1963).

The basic approach used is to divide the system into convenient sub-systems and optimise each sub-system separately, while taking into account the interactions between the sub-systems. The decisions made at each stage contribute to the overall systems objective function, and to optimise the overall objective function an appropriate combination of the individual stages has to be found. In a typical process plant system the possible number of combinations of the stage decisions will be very large. The dynamic programming approach uses Bellman's "Principle of Optimality",† which enables the optimum policy to be found systematically and efficiently by calculating only a fraction of the possible combinations of stage decisions. The method converts the problem from the need to deal with "N" optimisation decisions simultaneously to a sequential set of "N" problems. The application of dynamic programming to design problems is well illustrated in Rudd and Watson's book; see also Wells (1973).

1.10.6. Optimisation of batch and semicontinuous processes

In batch operation there will be periods when product is being produced, followed by non-productive periods when the product is discharged and the equipment prepared for the next batch. The rate of production will be determined by the total batch time, productive plus non-productive periods.

$$\text{Batches per year} = \frac{8760 \times \text{plant attainment}}{\text{batch cycle time}} \qquad (1.5)$$

where the "plant attainment" is the fraction of the total hours in a year (8760) that the plant is in operation.

Annual production = quantity produced per batch × batches per year.

$$\text{Cost per unit of production} = \frac{\text{annual cost of production}}{\text{annual production rate}} \qquad (1.6)$$

With many batch processes, the production rate will decrease during the production

---

† Bellman's (1957) principle of optimality: "An optimal policy has the property that, whatever the intial state and the initial decision are, the remaining decisions must constitute an optimal policy with regard to the state resulting from the first decision."

period; for example, batch reactors and plate and frame filter presses, and there will be an optimum batch size, or optimum cycle time, that will give the minimum cost per unit of production.

For some processes, though they would not be classified as batch processes, the period of continuous production will be limited by gradual changes in process conditions; such as, the deactivation of catalysts or the fouling of heat-exchange surfaces. Production will be lost during the periods when the plant is shut down for catalyst renewal or equipment clean-up, and, as with batch process, there will be an optimum cycle time to give the minimum production cost.

The optimum time between shut-downs can be found by determining the relationship between cycle time and cost per unit of production (the objective function) and using one of the optimisation techniques outlined in this section to find the minimum.

The factors that determine the optimum cycle time for batch reactors are discussed in Volume 3, Chapter 1.

With discontinuous processes, the period between shut-downs will usually be a function of equipment size. Increasing the size of critical equipment will extend the production period, but at the expense of increased capital cost. The designer must strike a balance between the savings gained by reducing the non-productive period and the increased investment required.

1.11 References

ALLEN, D. H. (1971) *Brit. Chem. Eng.* **16**, 685. Linear programming models.

ARIS, R. (1963) *Discrete Dynamic Programming* (Blaisdell).

BAASEL, W. D. (1965) *Chem. Eng., Albany* **72** (Oct. 25th) 147. Exploring response surfaces to establish optimum conditions.

BAASEL, W. D. (1974) *Preliminary Chemical Engineering Plant Design* (Elsevier).

BEIGHTLER, C. S. and WILDE, D. J. (1967) *Foundations of Optimisation* (Prentice-Hall).

BELLMAN, R. (1957) *Dynamic Programming* (Princeton University, New York).

BERGE, C. (1962) *Theory of Graphs and its Applications* (Wiley).

BEVERIDGE, G. S. G. and SCHECHTER, R. S. (1970) *Optimisation: Theory and Practice* (McGraw-Hill).

BOAS, A. H. (1963a) *Chem. Eng., Albany* **70** (Jan. 7th) 95. How to use Lagrange multipliers.

BOAS, A. H. (1963b) *Chem. Eng., Albany* **70** (Feb. 4th) 105. How search methods locate optimum in univariate problems.

BURKLIN, C. R. (1979) *The process Plant Designers Pocket Handbook of Codes and Standards* (Gulf).

CHADDOCK, D. H. (1975) Paper read to S. Wales Branch, Institution of Mechanical Engineers (Feb. 27th). Thought structure, or what makes a designer tick.

DANO, S. (1965) *Linear Programming in Industry* (Springer-Verlag).

DANTZIG, G. B. (1963) *Linear Programming and Extensions* (Princeton University Press).

DRIEBEEK, N. J. (1969) *Applied Linear Programming* (Addison-Wesley).

HENLEY, E. J. and ROSEN, E. M. (1969) *Material and Energy Balance Computations* (Wiley).

HIMMELBLAU, D. M. (1963) *Ind. Eng. Chem. Process Design & Development* **2**, 296. Process optimisation by search techniques.

JONES, C. J. (1970) *Design Methods: Seeds of Human Futures* (Wiley).

KWAUK, M. (1956) *AIChE Jl* **2**, 240. A system for counting variables in separation processes.

LEE, W. CHRISTENSEN J. H. and RUDD, D. F. (1966): *AIChE Jl* **12**, 1104. Design variable selection to simplify process calculations.

LEE, W. and RUDD, D. F. (1966) *AIChE Jl* **12**, 1185. On the ordering of recycle calculations.

MITTEN, L. G. and NEMHAUSER, G. L. (1963) *Chem. Eng. Prog.* **59**, (Jan.) 52. Multistage optimization.

NEMHAUSER, G. L. (1966) *Introduction to Dynamic Programming* (Wiley).

PETERS, M. S. and TIMMERHAUS, K. D. (1968) *Plant Design and Economics for Chemical Engineers*, 2nd ed. (McGraw-Hill).

POLYA, G. (1957) *How to Solve It*, 2nd ed. (Doubleday).

RASE, H. F. and BARROW, M. H. (1964) *Project Engineering* (Wiley).

RUDD, D. F. and WATSON, C. C. (1968) *Strategy of Process Design* (Wiley).
SMITH, B. D. (1963) *Design of Equilibrium Stage Processes* (McGraw-Hill).
STOECKER, W. F. (1971) *Design of Thermal Systems* (McGraw-Hill).
WELLS, G. L. (1973) *Process Engineering with Economic Objective* (Leonard Hill).
WILDE, D. J. (1964) *Optimum Seeking Methods* (Prentice-Hall).
WILLIAMS, N. (1967) *Linear and Non-linear Programming in Industry* (Pitman).

British Standards

BS 308:   ----- Engineering drawing practice.
           Part 1: 1972 General principles.
           Part 2: 1972 Dimensioning and tolerancing of size.
           Part 3: 1972 Geometrical tolerancing.

1.12 Nomenclature

		Dimensions in $\mathbf{MLT}\,\theta$
C	Number of components	—
F	Stream flow rate	$\mathbf{MT^{-1}}$
f	General function	—
f_i	General function (design relationship)	—
$f_1, f_2 \ldots$	General functions (design relationships)	—
N_d	Degrees of freedom in a design problem	—
N_d'	Degrees of freedom (variables free to be selected as design variables)	—
N_r	Number of design relationships	—
N_v	Number of variables	—
P	Pressure	$\mathbf{ML^{-1}T^{-2}}$
P_p	Inequality constraints	—
q	Heat input, flash distillation	$\mathbf{ML^2T^{-3}}$
T	Temperature	θ
v_j	Variables	—
$v_1, v_2 \ldots$	Variables	—
$x_1, x_2 \ldots$	Variables	—
ϕ	Equality constraint function	—
ψ	Inequality constraint function	—

Suffixes
1   Inlet, flash distillation
2   Vapour outlet, flash distillation
3   Liquid outlet, flash distillation

CHAPTER 2

Fundamentals of Material Balances

2.1. Introduction

Material balances are the basis of process design. A material balance taken over the complete process will determine the quantities of raw materials required and products produced. Balances over individual process units set the process stream flows and compositions.

A good understanding of material balance calculations is essential in process design.

In this chapter the fundamentals of the subject are covered, using simple examples to illustrate each topic. Practice is needed to develop expertise in handling what can often become very involved calculations. More examples and a more detailed discussion of the subject can be found in the numerous specialist books written on material and energy balance computations. Several suitable texts are listed under the heading of "Further Reading" at the end of this chapter.

The application of material balances to more complex problems is discussed in "Flowsheeting", Chapter 4.

Material balances are also useful tools for the study of plant operation and trouble shooting. They can be used to check performance against design; to extend the often limited data available from the plant instrumentation; to check instrument calibrations; and to locate sources of material loss.

2.2. The equivalence of mass and energy

Einstein showed that mass and energy are equivalent. Energy can be converted into mass, and mass into energy. They are related by Einstein's equation:

$$E = mc^2 \tag{2.1}$$

where E = energy, J,

m = mass, kg,

c = the speed of light *in vacuo*, 3×10^8 m/s.

The loss of mass associated with the production of energy is significant only in nuclear reactions. Energy and matter are always considered to be separately conserved in chemical reactions.

2.3. Conservation of mass

The general conservation equation for any process system can be written as:

Material out = Material in + Generation − Consumption − Accumulation

For a steady-state process the accumulation term will be zero. Except in nuclear processes, mass is neither generated nor consumed; but if a chemical reaction takes place a particular chemical species may be formed or consumed in the process. If there is no chemical reaction the steady-state balance reduces to

$$\text{Material out} = \text{Material in}$$

A balance equation can be written for each separately identifiable species present, elements, compounds or radicals; and for the total material.

Example 2.1

2000 kg of a 5 per cent slurry of calcium hydroxide in water is to be prepared by diluting a 20 per cent slurry. Calculate the quantities required. The percentages are by weight.

Solution

Let the unknown quantities of the 20% solution and water be X and Y respectively. Material balance on $Ca(OH)_2$

$$\begin{array}{ccc} In & & Out \\ X\dfrac{20}{100} & = & 2000 \times \dfrac{5}{100} \end{array} \qquad (a)$$

Balance on water

$$X\frac{(100-20)}{100} + Y = 2000\frac{(100-5)}{100} \qquad (b)$$

From equation (a) $X = 500$ kg.

Substituting into equation (b) gives $Y = \underline{\underline{1500 \text{ kg}}}$
Check material balance on total quantity:

$$X + Y = 2000$$
$$500 + 1500 = 2000, \text{ correct}$$

2.4. Units used to express compositions

When specifying a composition as a percentage it is important to state clearly the basis: weight, molar or volume.

The abbreviations w/w and v/v are used to designate weight basis and volume basis.

Example 2.2

Technical grade hydrochloric acid has a strength of 28 per cent w/w, express this as a mol fraction.

Solution

Basis of calculation 100 kg of 28 per cent w/w acid.

Molecular weights: water 18, HCl 36·5

Mass HCl	$= 100 \times 0.28 = 28$ kg	
Mass water	$= 100 \times 0.72 = 72$ kg	
kmol HCl	$= \dfrac{28}{36·5}$	$= 0.77$
kmol water	$= \dfrac{72}{18}$	$= 4.00$
Total mols		$= 4.77$
mol fraction HCl	$= \dfrac{0.77}{4.77}$	$= 0.16$
mol fraction water	$= \dfrac{4.00}{4.77}$	$= 0.84$
Check total		1.00

Within the accuracy needed for technical calculations, volume fractions can be taken as equivalent to mol fractions for gases, up to moderate pressures (say 25 bar).

Trace quantities are often expressed as parts per million (ppm). The basis, weight or volume, needs to be stated.

$$\text{ppm} = \frac{\text{quantity of component}}{\text{total quantity}} \times 10^6$$

Note. 1 ppm $= 10^{-4}$ per cent.

Minute quantities are sometimes quoted in ppb, parts per billion. Care is needed here, as the billion is usually an American billion (10^9), not the UK billion (10^{12}).

2.5. Stoichiometry

Stoichiometry (from the Greek *stoikeion*–element) is the practical application of the law of multiple proportions. The stoichiometric equation for a chemical reaction states unambiguously the number of molecules of the reactants and products that take part; from which the quantities can be calculated. The equation must balance.

With simple reactions it is usually possible to balance the stoichiometric equation by inspection, or by trial and error calculations. If difficulty is experienced in balancing complex equations, the problem can always be solved by writing a balance for each element present. The procedure is illustrated in Example 2.3.

Example 2.3

Write out and balance the overall equation for the manufacture of vinyl chloride from ethylene, chlorine and oxygen.

Solution

Method: write out the equation using letters for the unknown number of molecules of each reactant and product. Make a balance on each element. Solve the resulting set of equations.

$$A\,(C_2H_4) + B\,(Cl_2) + C\,(O_2) = D\,(C_2H_3Cl) + E\,(H_2O)$$

Balance on carbon

$$2A = 2D, \qquad A = D$$

on hydrogen

$$4A = 3D + 2E$$
substituting $D = A$ gives $E = A/2$

on chlorine

$$2B = D, \text{ hence } B = A/2$$

on oxygen

$$2C = E, \qquad C = E/2 = A/4$$

putting $A = 1$, the equation becomes

$$C_2H_4 + \tfrac{1}{2}Cl_2 + \tfrac{1}{4}O_2 = C_2H_3Cl + \tfrac{1}{2}H_2O$$

multiplying through by the largest denominator to remove the fractions

$$4C_2H_4 + 2Cl_2 + O_2 = 4C_2H_3Cl + 2H_2O$$

2.6. Choice of system boundary

The conservation law holds for the complete process and any sub-division of the process. The system boundary defines the part of the process being considered. The flows into and out of the system are those crossing the boundary and must balance with material generated or consumed within the boundary.

Any process can be divided up in an arbitrary way to facilitate the material balance calculations. The judicious choice of the system boundaries can often greatly simplify what would otherwise be difficult and tortuous calculations.

No hard and fast rules can be given on the selection of suitable boundaries for all types of material balance problems. Selection of the best sub-division for any particular process is a matter of judgement, and depends on insight into the structure of the problem, which can only be gained by practice. The following general rules will serve as a guide:

1. With complex processes, first take the boundary round the complete process and if possible calculate the flows in and out. Raw materials in, products and by-products out.
2. Select the boundaries to sub-divide the process into simple stages and make a balance over each stage separately.
3. Select the boundary round any stage so as to reduce the number of unknown streams to as few as possible.
4. As a first step, include any recycle streams within the system boundary (see Section 2.14).

Example 2.4

Selection of system boundaries and organisation of the solution.

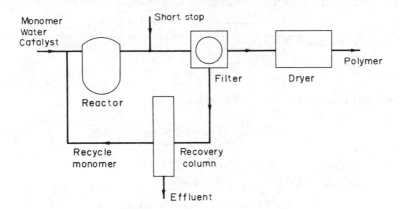

The diagram shows the main steps in a process for producing a polymer. From the following data, calculate the stream flows for a production rate of 10,000 kg/h.

Reactor, yield on polymer   100 per cent

        slurry polymerisation   20 per cent monomer/water

        conversion   90 per cent

        catalyst 1 kg/1000 kg monomer

        short stopping agent   0·5 kg/1000 kg unreacted monomer

Filter, wash water approx. 1 kg/1 kg polymer

Recovery column, yield 98 per cent (percentage recovered)

Dryer, feed ~ 5 per cent water, product specification 0·5 per cent H_2O

    Polymer losses in filter and dryer ~ 1 per cent

Solution

Only those flows necessary to illustrate the choice of system boundaries and method of calculation are given in the Solution.

Basis: 1 hour

Take the first system boundary round the filter and dryer.

```
                    ┌ ─ ─ ─ ─ ┐
                    │ Filter  │        Product
        Input ──────┤  and    ├──────▶ 10,000 kg polymer
                    │ Dryer   │        0·5 % water
                    └ ─ ─ ─ ─ ┘
                              ╲
                               ╲▶ Losses
```

With 1 per cent loss, polymer entering sub-system

$$= \frac{10,000}{0\cdot99} = \underline{\underline{10,101 \text{ kg}}}$$

Take the next boundary round the reactor system; the feeds to the reactor can then be calculated.

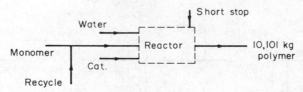

At 90 per cent conversion, monomer feed

$$= \frac{10,101}{0\cdot9} = \underline{\underline{11,223 \text{ kg}}}$$

Unreacted monomer $= 11,223 - 10,101 = \underline{\underline{1122 \text{ kg}}}$
Short-stop, at $0\cdot5$ kg/1000 kg unreacted monomer

$$= 1123 \times 0\cdot5 \times 10^{-3} = \underline{\underline{0\cdot6 \text{ kg}}}$$

Catalyst, at 1 kg/1000 kg monomer

$$= 11,223 \times 1 \times 10^{-3} = \underline{\underline{11 \text{ kg}}}$$

Let water feed to reactor be F_1, then for 20 per cent monomer

$$0\cdot2 = \frac{11,223}{F_1 + 11,223}$$

$$F_1 = \frac{11,223 \,(1 - 0\cdot2)}{0\cdot2} = \underline{\underline{44,892 \text{ kg}}}$$

Now consider filter-dryer sub-system again.
Water in polymer to dryer, at 5 per cent (neglecting polymer loss)

$$= 10,101 \times 0\cdot05 = \underline{\underline{505 \text{ kg}}}$$

Balance over reactor-filter-dryer sub-system gives flows to recovery column.

water, $44,892 + 10,101 - 505 \qquad = \underline{\underline{54,448 \text{ kg}}}$

monomer, unreacted monomer, $\quad = \underline{\underline{1123 \text{ kg}}}$

Now consider recovery system

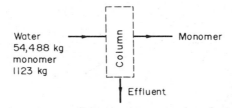

With 98 per cent recovery, recycle to reactor

$$= 0\cdot98 \times 1123 = \underline{\underline{1100 \text{ kg}}}$$

Composition of effluent 23 kg monomer, 54,488 kg water.
Consider reactor monomer feed

Balance round tee gives fresh monomer required

$$= 11,223 - 1100 = \underline{\underline{10,123 \text{ kg}}}$$

2.7. Choice of basis for calculations

The correct choice of the basis for a calculation will often determine whether the calculation proves to be simple or complex. As with the choice of system boundaries, no all-embracing rules or procedures can be given for the selection of the right basis for any problem. The selection depends on judgement gained by experience. Some guide rules that will help in the choice are:

1. Time: choose the time basis in which the results are to be presented; for example kg/h, tonne/y.
2. For batch processes use one batch.
3. Choose as the mass basis the stream flow for which most information is given.
4. It is often easier to work in mols, rather than weight, even when no reaction is involved.
5. For gases, if the compositions are given by volume, use a volume basis, remembering that volume fractions are equivalent to mol fractions up to moderate pressures.

2.8. Number of independent components

A balance equation can be written for each independent component. Not all the components in a material balance will be independent.

Physical systems, no reaction

If there is no chemical reaction the number of independent components is equal to the number of distinct chemical species present.

Consider the production of a nitration acid by mixing 70 per cent nitric and 98 per cent sulphuric acid. The number of distinct chemical species is 3; water, sulphuric acid, nitric acid.

Chemical systems, reaction

If the process involves chemical reaction the number of independent components will not necessarily be equal to the number of chemical species, as some may be related by the chemical equation. In this situation the number of independent components can be calculated by the following relationship:

Number of independent components = Number of chemical species −
                                                        Number of independent
                                                        chemical equations   (2.2)

Example 2.5

If nitration acid is made up using oleum in place of the 98 per cent sulphuric acid, there will be four distinct chemical species: sulphuric acid, sulphur trioxide, nitric acid, water. The sulphur trioxide will react with the water producing sulphuric acid so there are only three independent components

Reaction equation $SO_3 + H_2O \rightarrow H_2SO_4$

No. of chemical species	4
No. of reactions	1
No. of independent equations	3

2.9. Constraints on flows and compositions

It is obvious, but worth emphasising, that the sum of the individual component flows in any stream cannot exceed the total stream flow. Also, that the sum of the individual molar or weight fractions must equal 1. Hence, the composition of a stream is completely defined if all but one of the component concentrations are given.

The component flows in a stream (or the quantities in a batch) are completely defined by any of the following:

1. Specifying the flow (or quantity) of each component.
2. Specifying the total flow (or quantity) and the composition.
3. Specifying the flow (or quantity) of one component and the composition.

Example 2.6

The feed stream to a reactor contains: ethylene 16 per cent, oxygen 9 per cent, nitrogen 31 per cent, and hydrogen chloride. If the ethylene flow is 5000 kg/h, calculate the individual component flows and the total stream flow. All percentages are by weight.

Solution

$$\text{Percentage HCl} = 100 - (16 + 9 + 31) = \underline{\underline{44}}$$

$$\text{Percentage ethylene} = \frac{5000}{\text{total}} \times 100 \quad = \underline{\underline{16}}$$

$$\text{hence total flow} = 5000 \times \frac{100}{16} \quad = \underline{\underline{31{,}250 \text{ kg/h}}}$$

$$\text{so, oxygen flow} = \frac{9}{100} \times 31{,}250 \quad = \underline{\underline{2813 \text{ kg/h}}}$$

$$\text{nitrogen} = 31{,}250 \times \frac{31}{100} \quad = \underline{\underline{9687 \text{ kg/h}}}$$

$$\text{hydrogen chloride} = 31{,}250 \times \frac{44}{100} \quad = \underline{\underline{13{,}750 \text{ kg/h}}}$$

General rule: the ratio of the flow of any component to the flow of any other component is the same as the ratio of the compositions of the two components.

The flow of any component in Example 2.6 could have been calculated directly from the ratio of the percentage to that of ethylene, and the ethylene flow.

$$\text{Flow of hydrogen chloride} = \frac{44}{16} \times 5000 = \underline{\underline{13{,}750 \text{ kg/h}}}$$

2.10. General algebraic method

Simple material-balance problems involving only a few streams and with a few unknowns can usually be solved by simple direct methods. The relationship between the unknown quantities and the information given can usually be clearly seen. For more complex problems, and for problems with several processing steps, a more formal algebraic approach can be used. The procedure is involved, and often tedious if the calculations have to be done manually, but should result in a solution to even the most intractable problems, providing sufficient information is known.

Algebraic symbols are assigned to all the unknown flows and compositions. Balance equations are then written around each sub-system for the independent components (chemical species or elements).

Material-balance problems are particular examples of the general design problem discussed in Chapter 1. The unknowns are compositions or flows, and the relating equations arise from the conservation law and the stoichiometry of the reactions. For any problem to have a unique solution it must be possible to write the same number of independent equations as there are unknowns.

Consider the general material balance problem where there are N_s streams each containing N_c independent components. Then the number of variables, N_v, is given by:

$$N_v = N_c \times N_s \tag{2.3}$$

If N_e independent balance equations can be written, then the number of variables, N_d, that must be specified for a unique solution, is given by:

$$N_d = (N_s \times N_c) - N_e \qquad (2.4)$$

Consider a simple mixing problem

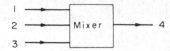

Let F_n be the total flow in stream n, and $x_{n,m}$ the concentration of component m in stream n. Then the general balance equation can be written

$$F_1 x_{1,m} + F_2 x_{2,m} + F_3 x_{3,m} = F_4 x_{4,m} \qquad (2.5)$$

A balance equation can also be written for the total of each stream:

$$F_1 + F_2 + F_3 = F_4 \qquad (2.6)$$

but this could be obtained by adding the individual component equations, and so is not an additional independent equation. There are m independent equations, the number of independent components.

Consider a separation unit, such as a distillation column, which divides a process stream into two product streams. Let the feed rate be 10,000 kg/h; composition benzene 60 per cent, toluene 30 per cent, xylene 10 per cent.

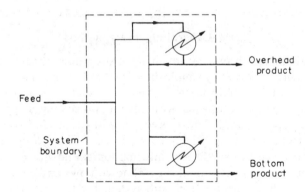

There are three streams, feed, overheads and bottoms, and three independent components in each stream.

Number of variables (component flow rates) = 9
Number of independent material balance
equations   = 3
Number of variables to be specified for
a unique solution   = 9 − 3 = 6

Three variables are specified; the feed flow and composition fixes the flow of each component in the feed.

Number of variables to be specifed by designer $= 6 - 3 = 3$. Any three component flows can be chosen.

Normally the top composition and flow or the bottom composition and flow would be chosen.

If the primary function of the column is to separate the benzene from the other components, the maximum toluene and xylene in the overheads would be specified; say, at 5 kg/h and 3 kg/h, and the loss of benzene in the bottoms also specified; say, at not greater than 5 kg/h. Three flows are specified, so the other flows can be calculated.

Benzene in overheads = benzene in feed − benzene in bottoms.

$$0.6 \times 10{,}000 - 5 = \underline{\underline{5995 \text{ kg/h}}}$$

Toluene in bottoms = toluene in feed − toluene in overheads

$$0.3 \times 10{,}000 - 5 = \underline{\underline{2995 \text{ kg/h}}}$$

Xylene in bottoms = xylene in feed − xylene in overheads

$$0.1 \times 10{,}000 - 3 = \underline{\underline{997 \text{ kg/h}}}$$

2.11. Tie components

In Section 2.9 it was shown that the flow of any component was in the same ratio to the flow of any other component, as the ratio of the concentrations of the two components. If one component passes unchanged through a process unit it can be used to tie the inlet and outlet compositions.

This technique is particularly useful in handling combustion calculations where the nitrogen in the combustion air passes through unreacted and is used as the tie component. This is illustrated in Example 2.8.

This principle can also be used to measure the flow of a process stream by introducing a measured flow of some easily analysed (compatible) material.

Example 2.7

Carbon dioxide is added at a rate of 10 kg/h to an air stream and the air is sampled at a sufficient distance downstream to ensure complete mixing. If the analysis shows 0.45 per cent v/v CO_2, calculate the air-flow rate.

Solution

Normal carbon dioxide content of air is 0.03 per cent

Basis: kmol/h, as percentages are by volume.

$$\text{kmol/h } CO_2 \text{ introduced} = 10/44 = 0.2273$$

Let X be the air flow.

Balance on CO_2, the tie component

$$CO_2 \text{ in} = 0.0003\,X + 0.2273$$
$$CO_2 \text{ out} = 0.0045\,X$$
$$X\,(0.0045 - 0.0003) = 0.2273$$
$$X = 0.2273/0.0042 = 54 \text{ kmol/h}$$
$$= 54 \times 29 = 1560 \text{ kg/h}$$

Example 2.8

In a test on a furnace fired with natural gas (composition 95 per cent methane, 5 per cent nitrogen) the following flue gas analysis was obtained: carbon dioxide 9.1 per cent, carbon monoxide 0.2 per cent, oxygen 4.6 per cent, nitrogen 86.1 per cent, all percentages by volume.

Calculate the percentage excess air flow (percentage above stoichiometric).

Solution

$$\text{Reaction: } CH_4 + 2O_2 \rightarrow CO_2 + 2H_2O$$

Note: the flue gas analysis is reported on the dry basis, any water formed having been condensed out.

Nitrogen is the tie component.

Basis: 100 mol, dry flue gas; as the analysis of the flue gas is known the mols of each element in the flue gas (flow out) can be easily calculated and related to the flow into the system.

Let the quantity of fuel (natural gas) per 100 mol dry flue gas be X.

Balance on carbon, mols in fuel = mols in flue gas

$$0.95\,X = 9.1 + 0.2, \text{ hence } X = \underline{9.79 \text{ mol}}$$

Balance on nitrogen (composition of air O_2 21 per cent, N_2 79 per cent).
Let Y be the flow of air per 100 mol dry flue gas.

$$N_2 \text{ in air} + N_2 \text{ in fuel} = N_2 \text{ in flue gas}$$
$$0.79\,Y + 0.05 \times 9.79 = 86.1, \text{ hence } Y = \underline{108.4 \text{ mol}}$$

Stoichiometric air; from the reaction equation 1 mol methane requires 2 mol oxygen,

$$\text{so, stoichiometric air} = 9.79 \times 0.95 \times 2 \times \frac{100}{21} = \underline{88.6 \text{ mol}}$$

$$\text{Percentage excess air} = \frac{(\text{air supplied} - \text{stoichiometric air})}{\text{stoichiometric air}} \times 100$$

$$= \frac{108.4 - 88.6}{88.6} = \underline{22 \text{ per cent}}$$

Note: the quantity of water in the flue gases can be calculated by carrying out a balance on oxygen or more directly, from the calculated quantity of fuel, water = 17.9 mol.

2.12. Excess reagent

In industrial reactions the components are seldom fed to the reactor in exact stoichiometric proportions. A reagent may be supplied in excess to promote the desired reaction; to maximise the use of an expensive reagent; or to ensure complete reaction of a reagent, as in combustion.

The percentage excess reagent is defined by the following equation:

$$\text{Per cent excess} = \frac{\text{quantity supplied} - \text{stoichiometric}}{\text{stoichiometric quantity}} \times 100 \qquad (2.7)$$

It is necessary to state clearly to which reagent the excess refers. This is often termed the limiting reagent.

Example 2.9

To ensure complete combustion, 20 per cent excess air is supplied to a furnace burning natural gas. The gas composition (by volume) is methane 95 per cent, ethane 5 per cent. Calculate the mols of air required per mol of fuel.

Solution

Basis: 100 mol gas, as the analysis is volume percentage.

$$\text{Reactions: } CH_4 + 2O_2 \rightarrow CO_2 + 2H_2O$$
$$C_2H_6 + 3\tfrac{1}{2}O_2 \rightarrow 2CO_2 + 3H_2O$$

Stoichiometric mols O_2 required $= 95 \times 2 + 5 \times 3\tfrac{1}{2} = \underline{207 \cdot 5}$

With 20 per cent excess, mols O_2 required $= 207 \cdot 5 \times \dfrac{120}{100} = \underline{\underline{249}}$

Mols air (21 per cent O_2) $= 249 \times \dfrac{100}{21} = \underline{\underline{1185 \cdot 7}}$

Air per mol fuel $= 1185 \cdot 7/100 = \underline{11 \cdot 86 \text{ mol}}$

2.13. Conversion and yield

It is important to distinguish between conversion and yield (see Volume 3, Chapter 1). Conversion is to do with reactants (reagents); yield with products.

Conversion

Conversion is a measure of the fraction of the reagent that reacts.

To optimise reactor design and to minimise by-product formation, the conversion of a particular reagent is often less than 100 per cent. If more than one reactant is used, the reagent on which the conversion is based must be specified.

Conversion is defined by the following expression:

$$\text{Conversion} = \frac{\text{amount of reagent consumed}}{\text{amount supplied}}$$

$$= \frac{(\text{amount in feed stream}) - (\text{amount in product stream})}{(\text{amount in feed stream})}$$

(2.8)

This definition gives the total conversion of the particular reagent to all products. Sometimes figures given for conversion refer to one specific product, usually the desired product. In this instance the product must be specified as well as the reagent. This is really a way of expressing yield.

Example 2.10

In the manufacture of vinyl chloride (VC) by the pyrolysis of dichloroethane (DCE), the reactor conversion is limited to 55 per cent to reduce carbon formation, which fouls the reactor tubes.

Calculate the quantity of DCE needed to produce 5000 kg/h VC.

Solution

Basis: 5000 kg/h VC (the required quantity).

$$\text{Reaction: } C_2H_4Cl_2 \rightarrow C_2H_3Cl + HCl$$
mol weights DCE 99, VC 62·5

$$\text{kmol/h VC produced} = \frac{5000}{62\cdot5} = \underline{80}$$

From the stoichiometric equation, 1 kmol DCE produces 1 kmol VC. Let X be DCE feed kmol/h:

$$\text{Per cent conversion} = 55 = \frac{80}{X} \times 100$$

$$X = \frac{80}{0\cdot55} = \underline{\underline{145\cdot5 \text{ kmol/h}}}$$

In this example the small loss of DCE to carbon and other products has been neglected. All the DCE reacted has been assumed to be converted to VC.

Yield

Yield is a measure of the performance of a reactor or plant. Several different definitions of yield are used, and it is important to state clearly the basis of any yield figures. This is often not done when yield figures are quoted in the literature, and the judgement has to be used to decide what was intended.

For a reactor the yield (i.e. *relative yield*, Volume 3, Chapter 1) is defined by:

$$\text{Yield} = \frac{\text{mols of product produced} \times \text{stoichiometric factor}}{\text{mols of reagent converted}}$$

(2.9)

Stoichiometric factor = Stoichiometric mols of reagent required per mol
of product produced

With industrial reactors it is necessary to distinguish between "Reaction yield" (chemical yield), which includes only chemical losses to side products; and the overall "Reactor yield" which will include physical losses.

If the conversion is near 100 per cent it may not be worth separating and recycling the unreacted material; the overall reactor yield would then include the loss of unreacted material. If the unreacted material is separated and recycled, the overall yield *taken over the reactor and separation step* would include any physical losses from the separation step.

Plant yield is a measure of the overall performance of the plant and includes all chemical and physical losses.

Plant yield (applied to the complete plant or any stage)

$$= \frac{\text{mols product produced} \times \text{stoichiometric factor}}{\text{mols reagent fed to the process}} \qquad (2.10)$$

Where more than one reagent is used, or product produced, it is essential that product and reagent to which the yield figure refers is clearly stated.

Example 2.11

In the production of ethanol by the hydrolysis of ethylene, diethyl ether is produced as a by-product. A typical feed stream composition is: 55 per cent ethylene, 5 per cent inerts, 40 per cent water; and product stream: 52·26 per cent ethylene, 5·49 per cent ethanol, 0·16 per cent ether, 36·81 per cent water, 5·28 per cent inerts. Calculate the yield of ethanol and ether based on ethylene.

Solution

Reactions: $C_2H_4 + H_2O \rightarrow C_2H_5OH$ \qquad\qquad (a)

$2C_2H_4OH \rightarrow (C_2H_5)_2O + H_2O$ \qquad\qquad (b)

Basis: 100 mols feed (easier calculation than using the product stream)

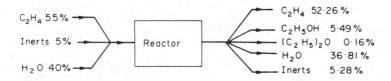

Note: the flow of inerts will be constant as they do not react, and it can be used to calculate the other flows from the compositions.

Feed stream	ethylene	55 mol
	inerts	5 mol
	water	40 mol

Product stream

$$\text{ethylene} \quad = \frac{52 \cdot 26}{5 \cdot 28} \times 5 \quad = 49 \cdot 49 \text{ mol}$$

$$\text{ethanol} \quad = \frac{5 \cdot 49}{5 \cdot 28} \times 5 \quad = 5 \cdot 20 \text{ mol}$$

$$\text{ether} \quad = \frac{0 \cdot 16}{5 \cdot 28} \times 5 \quad = 0 \cdot 15 \text{ mol}$$

Amount of ethylene reacted $= 55 \cdot 0 - 49.49 = 5 \cdot 51$ mol

$$\text{Yield of ethanol based on ethylene} = \frac{5 \cdot 2 \times 1}{5 \cdot 51} \times 100 = \underline{94 \cdot 4 \text{ per cent}}$$

As 1 mol of ethanol is produced per mol of ethylene the stoichiometric factor is 1.

$$\text{Yield of ether based on ethylene} = \frac{0 \cdot 15 \times 2}{5 \cdot 51} \times 100 = \underline{5 \cdot 44 \text{ per cent}}$$

The stoichiometric factor is 2, as 2 mol of ethylene produce 1 mol of ether.

Note: the conversion of ethylene, to all products, is given by:

$$\text{Conversion} = \frac{\text{mols fed} - \text{mols out}}{\text{mols fed}} = \frac{55 - 49 \cdot 49}{55} \times 100$$

$$= \underline{10 \text{ per cent}}$$

The yield based on water could also be calculated but is of no real interest as water is relatively inexpensive compared with ethylene. Water is clearly fed to the reactor in considerable excess.

Example 2.12

In the chlorination of ethylene to produce dichloroethane (DCE), the conversion of ethylene is reported as 99·0 per cent. If 94 mol of DCE are produced per 100 mol of ethylene fed, calculate the overall yield and the reactor (reaction) yield based on ethylene. The unreacted ethylene is not recovered.

Solution

$$\text{Reaction: } C_2H_4 + Cl_2 \rightarrow C_2H_4Cl_2$$

Stoichiometric factor 1.

$$\text{Overall yield (including physical losses)} = \frac{\text{mols DCE produced} \times 1}{\text{mols ethylene fed}} \times 100$$

$$= \frac{94}{100} \times 100 = \underline{94 \text{ per cent}}$$

$$\text{Chemical yield (reaction yield)} \quad = \frac{\text{mols DCE produced}}{\text{mols ethylene converted}} \times 100$$

$$= \frac{94}{99} \times 100 = \underline{94 \cdot 5 \text{ per cent}}$$

The principal by-product of this process is trichloroethane.

2.14. Recycle processes

Processes in which a flow stream is returned (recycled) to an earlier stage in the processing sequence are frequently used. If the conversion of a valuable reagent in a reaction process is appreciably less than 100 per cent, the unreacted material is usually separated and recycled. The return of reflux to the top of a distillation column is an example of a recycle process in which there is no reaction.

In mass balance calculations the presence of recycle streams makes the calculations more difficult.

Without recycle, the material balances on a series of processing steps can be carried out sequentially, taking each unit in turn; the calculated flows out of one unit become the feeds to the next. If a recycle stream is present, then at the point where the recycle is returned the flow will not be known as it will depend on downstream flows not yet calculated. Without knowing the recycle flow, the sequence of calculations cannot be continued to the point where the recycle flow can be determined.

Two approaches to the solution of recycle problems are possible:

1. The cut and try method. The recycle stream flows can be estimated and the calculations continued to the point where the recycle is calculated. The estimated flows are then compared with the calculated and a better estimate made. The procedure is continued until the difference between the estimated and the calculated flows are within acceptable limits.
2. The formal, algebraic, method. The presence of recycle implies that some of the mass balance equations will have to be solved simultaneously. The equations are set up with the recycle flows as unknowns and solved using standard methods for the solution of simultaneous equations.

With simple problems, with only one or two recycle loops, the calculation can often be simplified by the careful selection of the basis of calculation and the system boundaries. This is illustrated in Examples 2.4 and 2.13.

The solution of more complex material balance problems involving several recycle loops is discussed in Chapter 4.

Example 2.13

The block diagram shows the main steps in the balanced process for the production of vinyl chloride from ethylene. Each block represents a reactor and several other processing units. The main reactions are:

Block A, chlorination

$$C_2H_4 + Cl_2 \rightarrow C_2H_4Cl_2, \text{ yield on ethylene 98 per cent}$$

Block B, oxyhydrochlorination

$$C_2H_4 + 2HCl + \tfrac{1}{2}O_2 \rightarrow C_2H_4Cl_2 + H_2O, \text{ yields: on ethylene 95 per cent,}$$
$$\text{on HCl 90 per cent}$$

Block C, pyrolysis

$$C_2H_4Cl_2 \rightarrow C_2H_3Cl + HCl, \text{ yields: on DCE 99 per cent, on HCl 99·5 per cent}$$

The HCl from the pyrolysis step is recycled to the oxyhydrochlorination step. The flow of ethylene to the chlorination and oxyhydrochlorination reactors is adjusted so that the production of HCl is in balance with the requirement. The conversion in the pyrolysis reactor is limited to 55 per cent, and the unreacted dichloroethane (DCE) separated and recycled.

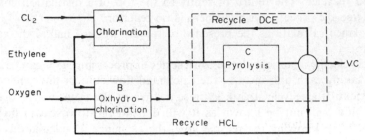

Using the yield figures given, and neglecting any other losses, calculate the flow of ethylene to each reactor and the flow of DCE to the pyrolysis reactor, for a production rate of 12,500 kg/h vinyl chloride (VC).

Solution

Molecular weights: vinyl chloride 62·5, DCE 99·0, HCl 36·5.

$$\text{VC per hour} = \frac{12,500}{62·5} = 200 \text{ kmol/h}$$

Draw a system boundary round each block, enclosing the DCE recycle within the boundary of step C.

Let flow of ethylene to block A be X and to block B be Y, and the HCl recycle be Z.

Then the total mols of DCE produced $= 0·98X + 0·95Y$, allowing for the yields, and the mols of HCl produced in block C

$$= (0·98X + 0·95Y)0·995 = Z \tag{a}$$

Consider the flows to and product from block B

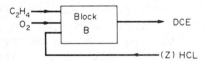

The yield of DCE based on HCl is 90 per cent, so the mols of DCE produced

$$= \frac{0·90Z}{2}$$

Note: the stoichiometric factor is 2 (2 mol HCl per mol DCE).

The yield of DCE based on ethylene is 95 per cent, so

$$\frac{0·9Z}{2} = 0·95Y$$

$$Z = \frac{0·95 \times 2Y}{0·9}$$

Substituting for Z into equation (a) gives

$$Y = (0.98X + 0.95Y)0.995 \times \frac{0.9}{2 \times 0.95}$$

$$Y = 0.837X \qquad\qquad (b)$$

Total VC produced $= 0.99 \times$ total DCE, so

$$0.99(0.98X + 0.95Y) = 200 \text{ kmol/h}$$

Substituting for Y from equation (b) gives $X = \underline{113.8 \text{ kmol/h}}$

and $\qquad\qquad Y = 0.837 \times 113.8 = \underline{95.3 \text{ kmol/h}}$

HCl recycle from equation (a)

$$Z = (0.98 \times 113.8 + 0.95 \times 95.3)0.995 = \underline{\underline{201.1 \text{ kmol/h}}}$$

Note: overall yield $= \dfrac{200}{(113.8 + 95.3)} \times 100 = \underline{\underline{96 \text{ per cent}}}$

2.15. Purge

It is usually necessary to bleed off a portion of a recycle stream to prevent the build-up of unwanted material. For example, if a reactor feed contains inert components that are not separated from the recycle stream in the separation units these inerts would accumulate in the recycle stream until the stream eventually consisted entirely of inerts. Some portion of the stream would have to be purged to keep the inert level within acceptable limits. A continuous purge would normally be used. Under steady-state conditions:

Loss of inert in the purge = Rate of feed of inerts into the system

The concentration of any component in the purge stream will be the same as that in the recycle stream at the point where the purge is taken off. So the required purge rate can be determined from the following relationship:

[Feed stream flow-rate] × [Feed stream inert concentration] =
[Purge stream flow-rate] × [Specified (desired) recycle inert concentration]

Example 2.14

In the production of ammonia from hydrogen and nitrogen the conversion, based on either raw material, is limited to 15 per cent. The ammonia produced is condensed from the reactor (convertor) product stream and the unreacted material recycled. If the feed contains 0.2 per cent argon (from the nitrogen separation process), calculate the purge rate required to hold the argon in the recycle stream below 5.0 per cent. Percentages are by volume.

Solution

Basis: 100 mols feed (purge rate will be expressed as mols per 100 mol feed, as the production rate is not given).

Process diagram

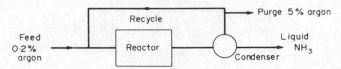

Volume percentages are taken as equivalent to mol per cent.

Argon entering system with feed = $100 \times 0.2/100 = 0.2$ mol.

Let purge rate per 100 mol feed be F.

Argon leaving system in purge = $F \times 5/100 = 0.05F$.

At the steady state, argon leaving = argon entering

$$0.05F = 0.2$$
$$F = 0.2/0.05 = \underline{\underline{4}}$$

Purge required: 4 mol per 100 mol feed.

2.16. By-pass

A flow stream may be divided and some part diverted (by-passed) around some units. This procedure is often used to control stream composition or temperature.

Material balance calculations on processes with by-pass streams are similar to those involving recycle, except that the stream is fed forward instead of backward. This usually makes the calculations easier than with recycle.

2.17. Unsteady-state calculations

All the previous material balance examples have been steady-state balances. The accumulation term was taken as zero, and the stream flow-rates and compositions did not vary with time. If these conditions are not met the calculations are more complex. Steady-state calculations are usually sufficient for the calculations of the process flow-sheet (Chapter 4). The unsteady-state behaviour of a process is important when considering the process start-up and shut-down, and the response to process upsets.

Batch processes are also examples of unsteady-state operation; though the total material requirements can be calculated by taking one batch as the basis for the calculation.

The procedure for the solution of unsteady-state balances is to set up balances over a small increment of time, which will give a series of differential equations describing the process. For simple problems these equations can be solved analytically. For more complex problems computer methods would be used.

The general approach to the solution of unsteady-state problems is illustrated in Example 2.15. Batch distillation is a further example of an unsteady-state material balance (see Volume 2, Chapter 11).

The behaviour of processes under non-steady-state conditions is a complex and specialised subject and beyond the scope of this book. It can be important in process design when assessing the behaviour of a process from the point of view of safety and control.

The use of material balances in the modelling of complex unsteady-state processes is discussed in the books by Myers and Seider (1976) and Henley and Rosen (1969).

Example 2.15

A hold tank is installed in an aqueous effluent-treatment process to smooth out fluctuations in concentration in the effluent stream. The effluent feed to the tank normally contains no more than 100 ppm of acetone. The maximum allowable concentration of acetone in the effluent discharge is set at 200 ppm. The surge tank working capacity is 500 m³ and it can be considered to be perfectly mixed. The effluent flow is 45,000 kg/h. If the acetone concentration in the feed suddenly rises to 1000 ppm, due to a spill in the process plant, and stays at that level for half an hour, will the limit of 200 ppm in the effluent discharge be exceeded?

Solution

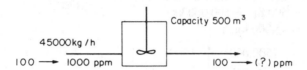

Basis: increment of time Δt.

To illustrate the general solution to this type of problem, the balance will be set up in terms of symbols for all the quantities and then actual values for this example substituted.

Let, Material in the tank $= M$,
      Flow-rate $= F$,
      Initial concentration in the tank $= C_0$,
      Concentration at time t after the feed concentration is increased $= C$,
      Concentration in the effluent feed $= C_1$,
      Change in concentration over time increment $\Delta t = \Delta C$,
      Average concentration in the tank during the time increment $= C_{av}$.

Then, as there is no generation in the system, the general material balance (Section 2.3) becomes:

$$\text{Input} - \text{Output} = \text{Accumulation}$$

Material balance on acetone.

Note: as the tank is considered to be perfectly mixed the outlet concentration will be the same as the concentration in the tank.

$$\text{Acetone in} - \text{Acetone out} = \text{Acetone accumulated in the tank}$$
$$FC_1 \Delta t - FC_{av} \Delta t = M(C + \Delta C) - MC$$

$$F(C_1 - C_{av}) = M \frac{\Delta C}{\Delta t}$$

Taking the limit, as $\Delta t \to 0$

$$\frac{\Delta C}{\Delta t} = \frac{dC}{dt}, \quad C_{av} = C$$

$$F(C_1 - C) = M\frac{dC}{dt}$$

Integrating

$$\int_0^t dt = \frac{M}{F}\int_{C_0}^C \frac{dC}{(C_1 - C)}$$

$$t = -\frac{M}{F}\ln\left[\frac{C_1 - C}{C_1 - C_0}\right]$$

Substituting the values for the example, noting that the maximum outlet concentration will occur at the end of the half-hour period of high inlet concentration.

$$
\begin{aligned}
t &= 0.5 \text{ h} \\
C_1 &= 1000 \text{ ppm} \\
C_0 &= 100 \text{ ppm (normal value)} \\
M &= 500 \text{ m}^3 = 500{,}000 \text{ kg} \\
F &= 45{,}000 \text{ kg/h}
\end{aligned}
$$

$$0.5 = -\frac{500{,}000}{45{,}000}\ln\left[\frac{1000 - C}{1000 - 100}\right]$$

$$0.045 = -\ln\left[\frac{1000 - C}{900}\right]$$

$$e^{-0.045} \times 900 = 1000 - C$$

$$C = \underline{\underline{140 \text{ ppm}}}$$

So the maximum allowable concentration will not be exceeded.

2.18. General procedure for material balance problems

The best way to tackle a problem will depend on the information given; the information required from the balance; and the constraints that arise from the nature of the problem. No all embracing, best method of solution can be given to cover all possible problems. The following step-by-step procedure is given as an aid to the efficient solution of material balance problems. The same general approach can be usefully employed to organise the solution of energy balance, and other design problems.

Procedure

Step 1. Draw a block diagram of the process.
Show each significant step as a block, linked by lines and arrows to show the stream connections and flow direction.

Step 2. List all the available data.

Show on the block diagram the known flows (or quantities) and stream compositions.

Step 3. List all the information required from the balance.

Step 4. Decide the system boundaries (see Section 2.6).

Step 5. Write out all the chemical reactions involved for the main products and by-products.

Step 6. Note any other constraints,

such as: specified stream compositions,

azeotropes,

phase equilibria,

tie substances (see Section 2.11).

The use of phase equilibrium relationships and other constraints in determining stream compositions and flows is discussed in more detail in Chapter 4.

Step 7. Note any stream compositions and flows that can be approximated.

Step 8. Check the number of conservation (and other) equations that can be written, and compare with the number of unknowns. Decide which variables are to be design variables; see Section 2.10.

This step would be used only for complex problems.

Step 9. Decide the basis of the calculation; see Section 2.7.

The order in which the steps are taken may be varied to suit the problem.

2.19. References (Further Reading)

Basic texts

FELDER, R. M. and ROUSSEAU, R. W. *Elementary Principles of Chemical Processes* (Wiley, 1978).
RUDD, D. F., POWERS, G. J. and SIIROLA, J. J. *Process Synthesis* (Prentice-Hall, 1973).
WHITWELL, J. C. and TONER, R. K. *Conservation of Mass and Energy* (McGraw-Hill, 1969).
WILLIAMS, E. T. and JACKSON, R. C. *Stoichiometry for Chemical Engineers* (McGraw-Hill, 1958).

Advanced texts

HENLEY, E. J. and ROSEN, E. M. (1969) *Material and Energy Balance Computations* (Wiley).
MYERS, A. L. and SEIDER, W. D. (1976) *Introduction to Chemical Engineering and Computer Calculations* (Prentice-Hall).

2.20. Nomenclature

		Dimensions in MLT
C	Concentration after time t, Example 2.15	—
C_{av}	Average concentration, Example 2.15	—
C_0	Initial concentration, Example 2.15	—
C_1	Concentration in feed to tank, Example 2.15	—
ΔC	Incremental change in concentration, Example 2.15	—
F	Flow-rate	MT^{-1}
F_n	Total flow in stream n	MT^{-1}
F_1	Water feed to reactor, Example 2.4	MT^{-1}
M	Quantity in hold tank, Example 2.15	M
N_c	Number of independent components	—

N_d	Number of variables to be specified	—
N_e	Number of independent balance equations	—
N_s	Number of streams	—
N_v	Number of variables	—
t	Time, Example 2.15	**T**
Δt	Incremental change in time, Example 2.15	**T**
X	Unknown flow, Examples 2.8, 2.10, 2.13	$\mathbf{MT}^{-1}$
$x_{n,m}$	Concentration of component m in stream n	—
Y	Unknown flow, Examples 2.8, 2.13	$\mathbf{MT}^{-1}$
Z	Unknown flow, Example 2.13	$\mathbf{MT}^{-1}$

CHAPTER 3

Fundamentals of Energy Balances

3.1. Introduction

As with mass, energy can be considered to be separately conserved in all but nuclear processes.

The conservation of energy, however, differs from that of mass in that energy can be generated (or consumed) in a chemical process. Material can change form, new molecular species can be formed by chemical reaction, but the total mass flow into a process unit must be equal to the flow out at the steady state. The same is not true of energy. The total enthalpy of the outlet streams will not equal that of the inlet streams if energy is generated or consumed in the processes; such as that due to heat of reaction.

Energy can exist in several forms: heat, mechanical energy, electrical energy, and it is the total energy that is conserved.

In process design, energy balances are made to determine the energy requirements of the process: the heating, cooling and power required. In plant operation, an energy balance (energy audit) on the plant will show the pattern of energy usage, and suggest areas for conservation and savings.

In this chapter the fundamentals of energy balances are reviewed briefly, and examples given to illustrate the use of energy balances in process design. The methods used for energy recovery and conservation are also discussed.

More detailed accounts of the principles and applications of energy balances are given in the texts covering material and energy-balance calculations which are cited at the end of Chapter 2.

3.2. Conservation of energy

As for material (Section 2.3), a general equation can be written for the conservation of energy:

$$\text{Energy out} = \text{Energy in} + \text{generation} - \text{consumption} - \text{accumulation}$$

This is a statement of the first law of thermodynamics.

An energy balance can be written for any process step.

Chemical reaction will evolve energy (exothermic) or consume energy (endothermic).

For steady-state processes the accumulation of both mass and energy will be zero.

Energy can exist in many forms and this, to some extent, makes an energy balance more complex than a material balance.

3.3. Forms of energy (per unit mass of material)

3.3.1. Potential energy

Energy due to position:

$$\text{Potential energy} = gz \tag{3.1}$$

where z = height above some arbitary datum, m,
g = gravitational acceleration (9·81 m/s^2).

3.3.2. Kinetic energy

Energy due to motion:

$$\text{Kinetic energy} = \frac{u^2}{2} \tag{3.2}$$

where u = velocity, m/s.

3.3.3. Internal energy

The energy associated with molecular motion. The temperature T of a material is a measure of its internal energy U:

$$U = \mathrm{f}(T) \tag{3.3}$$

3.3.4. Work

Work is done when a force acts through a distance:

$$W = \int_0^l F \, dx \tag{3.4}$$

where F = force, N,
x and l = distance, m.

Work done on a system by its surroundings is conventionally taken as negative; work done by the system on the surroundings as positive.

Where the work arises from a change in pressure or volume:

$$W = \int_1^2 P \, dv \tag{3.5}$$

where P = pressure, Pa (N/m^2),
v = volume per unit mass, m^3/kg.

To integrate this function the relationship between pressure and volume must be known. In process design an estimate of the work done in compressing or expanding a gas is often required. A rough estimate can be made by assuming either reversible adiabatic (isentropic) or isothermal expansion, depending on the nature of the process.

For isothermal expansion (expansion at constant temperature):

$$Pv = \text{constant}$$

For reversible adiabatic expansion (no heat exchange with the surroundings):

$$Pv^\gamma = \text{constant}$$

where γ = ratio of the specific heats, C_p/C_v.

The compression and expansion of gases is covered more fully in Section 3.13.

3.3.5. Heat

Energy is transferred either as heat or work. A system does not contain "heat", but the transfer of heat or work to a system changes its internal energy.

Heat taken in by a system from its surroundings is conventionally taken as positive and that given out as negative.

3.3.6. Electrical energy

Electrical, and the mechanical forms of energy, are included in the work term in an energy balance. Electrical energy will only be significant in energy balances on electrochemical processes.

3.4. The energy balance

Consider a steady-state process represented by Fig. 3.1. The conservation equation can be written to include the various forms of energy.

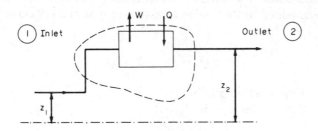

FIG. 3.1. General steady-state process

For unit mass of material:

$$U_1 + P_1 v_1 + u_1^2/2 + z_1 g + Q = U_2 + P_2 v_2 + u_2^2/2 + z_2 g + W \qquad (3.6)$$

The suffixes 1 and 2 represent the inlet and outlet points respectively. Q is the heat transferred across the system boundary; positive for heat entering the system, negative for heat leaving the system. W is the work done by the system; positive for work going from the system to the surroundings, and negative for work entering the system from the surroundings.

Equation 3.6 is a general equation for steady-state systems with flow.

In chemical processes, the kinetic and potential energy terms are usually small compared with the heat and work terms, and can normally be neglected.

It is convenient, and useful, to take the terms U and Pv together; defining the term enthalpy, usual symbol H, as:

$$H = U + Pv$$

Enthalpy is a function of temperature and pressure. Values for the more common substances have been determined experimentally and are given in the various handbooks (see Chapter 8).

Enthalpy can be calculated from specific and latent heat data; see Section 3.5.

If the kinetic and potential energy terms are neglected equation 3.6 simplifies to:

$$H_2 - H_1 = Q - W \qquad (3.7)$$

This simplified equation is usually sufficient for estimating the heating and cooling requirements of the various unit operations involved in chemical processes.

As the flow-dependent terms have been dropped, the simplified equation is applicable to both static (non-flow) systems and flow systems. It can be used to estimate the energy requirement for batch processes.

For many processes the work term will be zero, or negligibly small, and equation 3.7 reduces to the simple heat balance equation:

$$Q = H_2 - H_1 \qquad (3.8)$$

Where heat is generated in the system; for example, in a chemical reactor:

$$Q = Q_p + Q_s \qquad (3.9)$$

Q_s = heat generated in the system. If heat is evolved (exothermic processes) Q_s is taken as *positive*, and if heat is absorbed (endothermic processes) it is taken as *negative*.

Q_p = process heat added to the system to maintain required system temperature.

Hence:

$$Q_p = H_2 - H_1 - Q_s \qquad (3.10)$$

H_1 = enthalpy of the inlet stream,
H_2 = enthalpy of the outlet stream.

Example 3.1

Balance with no chemical reaction. Estimate the steam and the cooling water required for the distillation column shown in the figure.

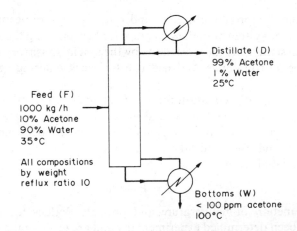

Feed (F)
1000 kg /h
10% Acetone
90% Water
35°C

All compositions
by weight
reflux ratio 10

Distillate (D)
99% Acetone
1% Water
25°C

Bottoms (W)
< 100 ppm acetone
100°C

Steam is available at 25 psig (276 kN/m²), dry saturated.

The rise in cooling water temperature is limited to 30°C.
Column operates at 1 bar.

Solution

Material balance

It is necessary to make a material balance to determine the top and bottoms product flow rates.

Balance on acetone, acetone loss in bottoms neglected.

$$1000 \times 0.1 = D \times 0.99$$

$$\text{Distillate, } D = 101 \text{ kg/h}$$

$$\text{Bottoms, } W = 1000 - 101 = 899 \text{ kg/h}$$

Energy balance

The kinetic and potential energy of the process streams will be small and can be neglected.

Take the first system boundary to include the reboiler and condenser.

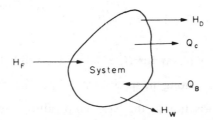

Inputs: reboiler heat input Q_B + feed sensible heat H_F.

Outputs: condenser cooling Q_c + top and bottom product sensible heats $H_D + H_W$.

The heat losses from the system will be small if the column and exchangers are properly lagged (typically less than 5 per cent) and will be neglected.

Basis 25°C, 1h.

Heat capacity data, from Volume 1, average values.

$$\text{Acetone:}\quad 25°C \text{ to } 35°C \quad 2.2 \text{ kJ/kg K}$$

$$\text{Water:}\quad 25°C \text{ to } 100°C \quad 4.2 \text{ kJ/kg K}$$

Heat capacities can be taken as additive.

$$\text{Feed, 10 per cent acetone} = 0.1 \times 2.2 + 0.9 \times 4.2 = 4.00 \text{ kJ/kg K}$$

$$\text{Tops, 99 per cent acetone, taken as acetone, } 2.2 \text{ kJ/kg K}$$

$$\text{Bottoms, as water, } 4.2 \text{ kJ/kg K}.$$

Q_c must be determined by taking a balance round the condenser and reboiler.

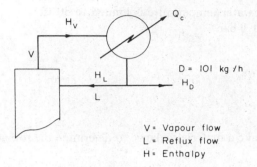

Reflux ratio (see Chapter 11)

$$R = L/D = 10$$
$$L = 10 \times 101 = 1010 \text{ kg/h}$$
$$V = L + D = 1111 \text{ kg/h}$$

From vapour–liquid equilibrium data:

boiling point of 99 per cent acetone/water = 56·5°C

At steady state:

input = output
$$H_V = H_D + H_L + Q_c,$$

Hence

$$Q_c = H_V - H_D - H_L$$

Assume complete condensation.

Enthalpy of vapour H_V = latent + sensible heat.

There are two ways of calculating the specific enthalpy of the vapour at its boiling point.

(1) Latent heat of vaporisation at the base temperature + sensible heat to heat the vapour to the boiling point.
(2) Latent heat of vaporisation at the boiling point + sensible heat to raise liquid to the boiling point.

Values of the latent heat of acetone and water as functions of temperature are given in Volume 1, so the second method will be used.

Latent heat acetone at 56·5°C (330 K) = 620 kJ/kg
Water at 56·5°C (330 K) = 2500 kJ/kg

Taking latent heats as additive:

$$H_V = 1111 \left[(0.01 \times 2500 + 0.99 \times 620) + (56.5 - 25)2.2 \right]$$
$$= 786,699 \text{ kJ/h}$$

The enthalpy of the top product and reflux are zero, as they are both at the base temperature. Both are liquid, and the reflux will be at the same temperature as the product.

Hence

$$Q_c = H_V = 786,699 \text{ kJ/h} \quad (218.5 \text{ kW})$$

Q_B is determined from a balance over complete system

Input   Output

$$Q_B + H_F = Q_c + H_D + H_W$$

$$H_F = 1000 \times 4{\cdot}00\,(35 - 25) = 40{,}000 \text{ kJ/h}$$

$$H_W = 899 \times 4{\cdot}2\,(100 - 25)\ = 283{,}185 \text{ kJ/h}$$

(boiling point of bottom product taken as $100°C$).

hence
$$\begin{aligned}
Q_B &= Q_c + H_W + H_D - H_F \\
&= 786{,}699 + 283{,}185 + 0 - 40{,}000 \\
&= \underline{\underline{1{,}029{,}884 \text{ kJ/h}}} \quad (286{\cdot}1 \text{ kW})
\end{aligned}$$

Q_B is supplied by condensing steam.

Latent heat of steam (Volume 1) = 2730 kJ/kg at $275{\cdot}8$ kN/m^2

$$\text{Steam required} = \frac{1{,}029{,}884}{2730} = \underline{\underline{377{\cdot}3 \text{ kg/h}}}$$

Q_c is removed by cooling water with a temperature rise of $30°C$

$$Q_c = \text{water flow} \times 30 \times 4{\cdot}2$$

$$\text{Water flow} = \frac{786{,}699}{4{\cdot}2 \times 30} = \underline{\underline{6244 \text{ kg/h}}}$$

3.5. Calculation of specific enthalpy

Tabulated values of enthalpy are available only for the more common materials. In the absence of published data the following expressions can be used to estimate the specific enthalpy (enthalpy per unit mass).

For pure materials, with no phase change:

$$H_T = \int_{T_d}^{T} C_p\,\mathrm{d}T \tag{3.11}$$

where H_T = specific enthalpy at temperature T,
$\quad C_p$ = specific heat capacity of the material, constant pressure,
$\quad T_d$ = the datum temperature.

If a phase transition takes place between the specified and datum temperatures, the latent heat of the phase transition is added to the sensible-heat change calculated by equation 3.11. The sensible-heat calculation is then split into two parts:

$$H_T = \int_{T_d}^{T_p} C_{p_1}\,\mathrm{d}T + \int_{T_p}^{T} C_{p_2}\,\mathrm{d}T \tag{3.12}$$

where T_p = phase transition temperature,
$\quad C_{p_1}$ = specific heat capacity first phase, below T_p,
$\quad C_{p_2}$ = specific heat capacity second phase, above T_p.

The specific heat at constant pressure will vary with temperature and to use equations 3.11 and 3.12, values of C_p must be available as a function of temperature. For solids and gases C_p is usually expressed as an empirical power series equation:

$$C_p = a + bT + cT^2 + dT^3 \qquad\qquad (3.13a)$$

$$\text{or} \qquad C_p = a + bT + cT^{-\frac{1}{2}} \qquad\qquad (3.13b)$$

Absolute (K) or relative (°C) temperature scales may be used when the relationship is in the form given in equation 3.13a. For equation 3.13b absolute temperatures must be used.

Example 3.2

Estimate the specific enthalpy of ethyl alcohol at 1 bar and 200° C, taking the datum temperature as 0° C.

$$\begin{aligned}
&C_p \text{ liquid } 0°C \;\; 24\cdot65 \text{ cal/mol°C} \\
&\qquad\qquad 100°C \;\; 37\cdot96 \text{ cal/mol°C} \\
&C_p \text{ gas } \quad (t°C) \;\; 14\cdot66 + 3\cdot758 \times 10^{-2}t - 2\cdot091 \times 10^{-5}t^2 \\
&\qquad\qquad\qquad\qquad + 4\cdot740 \times 10^{-9}t^3 \text{ cal/mol}
\end{aligned}$$

Boiling point of ethyl alcohol at 1 bar $= 78\cdot4°C$.
Latent heat of vaporisation $= 9\cdot22$ kcal/mol.

Solution

Note: as the data taken from the literature are given in cal/mol the calculation is carried out in these units and the result converted to SI units.

As no data are given on the exact variation of the C_p of the liquid with temperature, use an equation of the form $C_p = a + bt$, calculating a and b from the data given; this will be accurate enough over the range of temperature needed.

$$a = \text{value of } C_p \text{ at } 0°C, \; b = \frac{37\cdot96 - 24\cdot65}{100} = 0\cdot133$$

$$H_{200°C} = \int_0^{78\cdot4} (24\cdot65 + 0\cdot133t)\,dt + 9\cdot22 \times 10^3 + \int_{78\cdot4}^{200} (14\cdot66 + 3\cdot758 \times 10^{-2}t$$

$$- 2\cdot091 \times 10^{-5}t^2 + 4\cdot740 \times 10^{-9}t^3)\,dt$$

$$= \left[24\cdot65t + 0\cdot133t^2/2 \right]_0^{78\cdot4} + 9\cdot22 \times 10^3 + \left[14\cdot66t + 3\cdot758 \times 10^{-2}t^2/2 - 2\cdot091 \right.$$
$$\left. \times 10^{-5}t^3/3 + 4\cdot740 \times 10^{-9}t^4/4 \right]_{78\cdot4}^{200}$$

$$= 13\cdot95 \times 10^3 \text{ cal/mol}$$

$$= 13\cdot95 \times 10^3 \times 4\cdot18 = \underline{\underline{58\cdot31 \times 10^3 \text{ J/mol}}}$$

Specific enthalpy $= \underline{\underline{58\cdot31 \text{ kJ/mol}}}$.

Molecular weight of ethyl alcohol, C_2H_5OH, $= 46$

Specific enthalpy $= 58\cdot31 \times 10^3/46 = \underline{\underline{1268 \text{ kJ/kg}}}$

3.6. Mean heat capacities

The use of mean heat capacities often facilitates the calculation of sensible-heat changes; mean heat capacity over the temperature range t_1 to t_2 is defined by the following equation:

$$C_{p_m} = \int_{t_1}^{t_2} C_p \, dt \div \int_{t_1}^{t_2} dt \qquad (3.14)$$

Mean specific heat values are tabulated in various handbooks. If the values are calculated from some standard reference temperature, t_r, then the change in enthalpy between temperatures t_1 and t_2 is given by:

$$\Delta H = C_{p_{m,t_2}}(t_2 - t_r) - C_{p_{m,t_1}}(t_1 - t_r) \qquad (3.15)$$

where t_r is the reference temperature from which the values of C_{p_m} were calculated.

If C_p is expressed as a polynomial of the form: $C_p = a + bt + ct^2 + dt^3$, then the integrated form of equation 3.14 will be:

$$C_{p_m} = \frac{a(t - t_r) + \dfrac{b}{2}(t^2 - t_r^2) + \dfrac{c}{3}(t^3 - t_r^3) + \dfrac{d}{4}(t^4 - t_r^4)}{t - t_r} \qquad (3.16)$$

where t is the temperature at which C_{p_m} is required.

If the reference temperature is taken at 0°C, equation 3.16 reduces to:

$$C_{p_m} = a + \frac{bt}{2} + \frac{ct^2}{3} + \frac{dt^3}{4} \qquad (3.17)$$

and the enthalpy change from t_1 to t_2 becomes

$$\Delta H = C_{p_{m,t_2}} t_2 - C_{p_{m,t_1}} t_1 \qquad (3.18)$$

The use of mean heat capacities is illustrated in Example 3.3.

Example 3.3

The gas leaving a combustion chamber has the following composition: CO_2 7·8, CO 0·6, O_2 3·4, H_2O 15·6, N_2 72·6, all volume percentage. Calculate the heat removed if the gas is cooled from 800 to 200°C.

Solution

Mean heat capacities for the combustion gases are readily available in handbooks and texts on heat and material balances. The following values are taken from K. A. Kobe, *Thermochemistry of Petrochemicals*, reprint No. 44, Pet. Ref. 1958; converted to SI units, J/mol°C, reference temperature 0°C.

°C	N_2	O_2	CO_2	CO	H_2O
200	29·24	29·95	40·15	29·52	34·12
800	30·77	32·52	47·94	31·10	37·38

Heat extracted from the gas in cooling from 800 to 200°C, for each component:

$$= M_c(C_{p_{m,800}} \times 800 - C_{p_{m,200}} \times 200)$$

where M_c = mols of that component.

Basis 100 mol gas (as analysis is by volume), substitution gives:

CO_2	$7 \cdot 8(47 \cdot 94 \times 800 - 40 \cdot 15 \times 200) =$	$236 \cdot 51 \times 10^3$
CO	$0 \cdot 6(31 \cdot 10 \times 800 - 29 \cdot 52 \times 200) =$	$11 \cdot 39 \times 10^3$
O_2	$3 \cdot 4(32 \cdot 52 \times 800 - 29 \cdot 95 \times 200) =$	$68 \cdot 09 \times 10^3$
H_2O	$15 \cdot 6(37 \cdot 38 \times 800 - 34 \cdot 12 \times 200) =$	$360 \cdot 05 \times 10^3$
N_2	$72 \cdot 6(30 \cdot 77 \times 800 - 29 \cdot 24 \times 200) =$	$1362 \cdot 56 \times 10^3$

$$2038 \cdot 60 \text{ kJ}/100 \text{ mol}$$

$$= 20.38 \text{ kJ/mol}$$

3.7. The effect of pressure on heat capacity

The data on heat capacities given in the handbooks, and in Appendix A, are, usually for the ideal gas state. Equation 3.13a should be written as:

$$C_p^\circ = a + bT + cT^2 + dT^3 \qquad (3.19)$$

where the superscript $^\circ$ refers to the ideal gas state.

The ideal gas values can be used for the real gases at low pressures. At high pressures the effect of pressure on the specific heat may be appreciable.

Edmister (1948) published a generalised plot showing the isothermal pressure correction for real gases as a function of the reduced pressure and temperature. His chart, converted to SI units, is shown as Fig. 3.2. Edmister's chart was based on hydrocarbons, but can be used for other materials to give an indication of the likely error if the ideal gas specific heat values are used without corrections.

The method is illustrated in Example 3.4.

Example 3.4

The ideal state heat capacity of ethylene is given by the equation:

$$C_p^\circ = 3 \cdot 95 + 15 \cdot 6 \times 10^{-2} T - 8 \cdot 3 \times 10^{-5} T^2 + 17 \cdot 6 \times 10^{-9} T^3 \text{ J/mol K}$$

Estimate the value at 10 bar and 300 K.

Solution

Ethylene: critical pressure   50·5 bar

critical temperature 283 K

$$C_p^\circ = 3 \cdot 95 + 15 \cdot 6 \times 10^{-2} \times 300 - 8 \cdot 3 \times 10^{-5} \times 300^2 + 17 \cdot 6 \times 10^{-9} \times 300^3$$

$$= 43 \cdot 76 \text{ J/mol K}$$

$$P_r = \frac{10}{50 \cdot 5} = 0 \cdot 20$$

$$T_r = \frac{300}{283} = 1 \cdot 06$$

From Fig. 3.2:

$$C_p - C_p^\circ \simeq \underline{4 \text{ J/mol K}}$$

So

$$C_p = 43\cdot76 + 4 = \underline{47\cdot76 \text{ J/mol K}}$$

The error in C_p if the ideal gas value were used uncorrected would be approximately 10 per cent.

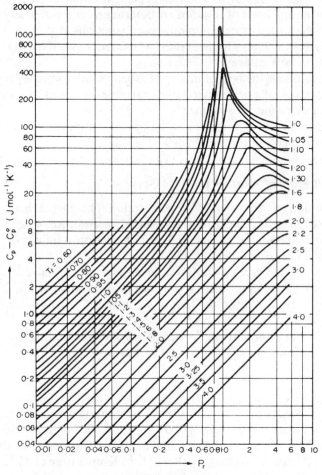

T_r = Reduced temperature
P_r = Reduced pressure

FIG. 3.2. Excess heat capacity chart (reproduced from Sterbacek *et al.* (1979), with permission)

3.8. Enthalpy of mixtures

For gases, the heats of mixing are usually negligible and the heat capacities and enthalpies can be taken as additive without introducing any significant error into design calculations; as was done in Example 3.3.

$$C_p \text{ (mixture)} = x_a C_{p_a} + x_b C_{p_b} + x_c C_{p_c} + \dots . \qquad (3.20)$$

where x_a, x_b, x_c, etc., are the mol fractions of the components a, b, c.

For mixtures of liquids and for solutions, the heat of mixing (heat of solution) may be significant, and so must be included when calculating the enthalpy of the mixture.

For binary mixtures, the specific enthalpy of the mixture at temperature t is given by:

$$H_{mixture,t} = x_a H_{a,t} + x_b H_{b,t} + \Delta H_{m,t} \qquad (3.21)$$

where $H_{a,t}$ and $H_{b,t}$ are the specific enthalpies of the components a and b and $-\Delta H_{m,t}$ is the heat of mixing when 1 mol of solution is formed, at temperature t.

Heats of mixing and heats of solution are determined experimentally and are available in the handbooks for the more commonly used solutions.

If no values are available, judgement must be used to decide if the heat of mixing for the system is likely to be significant.

For organic solutions the heat of mixing is usually small compared with the other heat quantities, and can usually be neglected when carrying out a heat balance to determine the process heating or cooling requirements.

The heats of solution of organic and inorganic compounds in water can be large, particularly for the strong mineral acids and alkalies.

3.8.1. Integral heats of solution

Heats of solution are dependent on concentration. The integral heat of solution at any given concentration is the cumulative heat released, or absorbed, in preparing the solution from pure solvent and solute. The integral heat of solution at infinite dilution is called the *standard integral heat of solution*.

Tables of the integral heat of solution over a range of concentration, and plots of the integral heat of solution as a function of concentration, are given in the handbooks for many of the materials for which the heat of solution is likely to be significant in process design calculations.

The integral heat of solution can be used to calculate the heating or cooling required in the preparation of solutions, as illustrated in Example 3.5.

Example 3.5

A solution of NaOH in water is prepared by diluting a concentrated solution in an agitated, jacketed, vessel. The strength of the concentrated solution is 50 per cent w/w and 2500 kg of 5 per cent w/w solution is required per batch. Calculate the heat removed by the cooling water if the solution is to be discharged at a temperature of 25°C. The temperature of the solutions fed to the vessel can be taken to be 25°C.

Solution

Integral heat of solution of $NaOH - H_2O$, at 25°C

mols H_2O/mol NaOH	$-\Delta H^{\circ}_{soln}$ kJ/mol NaOH
2	22·9
4	34·4
5	37·7
10	42·5
infinite	42·9

Conversion of weight per cents to mol/mol:

$$50 \text{ per cent w/w} = 50/18 \div 50/40 = 2{\cdot}22 \text{ mol } H_2O/\text{mol NaOH}$$
$$5 \text{ per cent w/w} = 95/18 \div 5/40 = 42{\cdot}2 \text{ mol } H_2O/\text{mol NaOH}$$

From a plot of the integral heats of solution versus concentration,

$$-\Delta H^{\circ}_{soln} \quad 2{\cdot}22 \text{ mol/mol} = 27{\cdot}0 \text{ kJ/mol NaOH}$$
$$42{\cdot}2 \text{ mol/mol} = 42{\cdot}9 \text{ kJ/mol NaOH}$$

Heat liberated in the dilution per mol NaOH

$$= 42{\cdot}9 - 27{\cdot}0 = \underline{\underline{15{\cdot}9 \text{ kJ}}}$$

Heat released per batch = mol NaOH per batch × 15·9

$$= \frac{2500 \times 10^3 \times 0{\cdot}05}{40} \times 15{\cdot}9 = \underline{\underline{49{\cdot}7 \times 10^3 \text{ kJ}}}$$

Heat transferred to cooling water, neglecting heat losses,

$$\underline{\underline{49{\cdot}7 \text{ MJ per batch}}}$$

In Example 3.5 the temperature of the feeds and final solution have been taken as the same as the standard temperature for the heat of solution, 25°C, to simplify the calculation. Heats of solution are analogous to heats of reaction, and examples of heat balances on processes where the temperatures are different from the standard temperature are given in the discussion of heats of reaction, Section 3.10.

3.9. Enthalpy-concentration diagrams

The variation of enthalpy for binary mixtures is conveniently represented on a diagram. An example is shown in Fig. 3.3. The diagram shows the enthalpy of mixtures of ammonia and water versus concentration; with pressure and temperature as parameters. It covers the phase changes from solid to liquid to vapour, and the enthalpy values given include the latent heats for the phase transitions.

The enthalpy is per kg of the mixture (ammonia + water)

Reference states: enthalpy ammonia at $-77°C$ = zero
enthalpy water at $0°C$ = zero

Enthalpy-concentration diagrams greatly facilitate the calculation of energy balances involving concentration and phase changes; this is illustrated in Example 3.6.

Example 3.6

Calculate the maximum temperature when liquid ammonia at 40°C is dissolved in water at 20°C to form a 10 per cent solution.

Solution

The maximum temperature will occur if there are no heat losses (adiabatic process). As no heat or material is removed the problem can be solved graphically on the enthalpy-

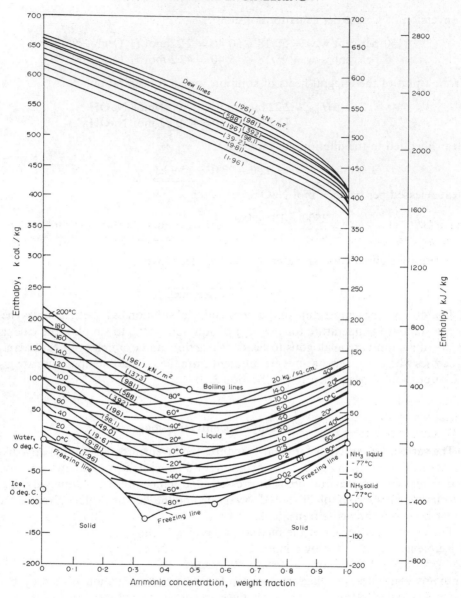

FIG. 3.3. Enthalpy-concentration diagram for aqueous ammonia. Reference states: enthalpies of liquid water at 0°C and liquid ammonia at −77°C are zero. (Bosniakovic, *Technische Thermodynamik*, T. Steinkopff, Leipzig, 1935)

concentration diagram (Fig. 3.3). The mixing operation is represented on the diagram by joining the point A representing pure ammonia at 40°C with the point B representing pure water at 20°C. The value of the enthalpy of the mixture lies on a vertical line at the required concentration, 0.1. The temperature of the mixture is given by the intersection of this vertical line with the line AB. This method is an application of the "lever rule" for phase diagrams. For a more detailed explanation of the method and further examples see Himmelblau (1967) or any of the general texts on material and energy balances listed at the

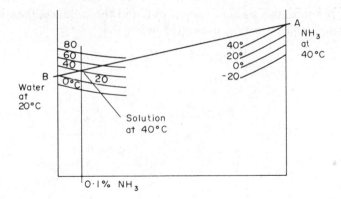

end of Chapter 2. The Ponchon–Savarit graphical method used in the design of distillation columns, described in Volume 2, Chapter 11, is a further example of the application of the lever rule, and the use of enthalpy-concentration diagrams.

3.10. Heats of reaction

If a process involves chemical reaction, heat will invariably have to be added or removed. The amount of heat given out in a chemical reaction depends on the conditions under which the reaction is carried out. The standard heat of reaction is the heat released when the reaction is carried out under standard conditions: pure components, pressure 1 atm (1·01325 bar), temperature usually, but not necessarily, 25°C.

Values for the standard heats of reactions are given in the literature, or may be calculated by the methods given in Sections 3.11 and 3.12.

When quoting heats of reaction the basis should be clearly stated. Either by giving the chemical equation, for example:

$$NO + \tfrac{1}{2}O_2 \rightarrow NO_2 \qquad \Delta H_r^\circ = -56\cdot68 \text{ kJ}$$

(The equation implies that the quantity of reactants and products are mols)

Or, by stating to which quantity the quoted value applies:

$$\Delta H_r^\circ = -56\cdot68 \text{ kJ per mol } NO_2$$

The reaction is exothermic and the enthalpy change ΔH_r° is therefore *negative*. The heat of reaction $-\Delta H_r^\circ$ is *positive*. The superscript ° denotes a value at *standard* conditions and the subscript r implies that a chemical reaction is involved.

The state of the reactants and products (gas, liquid or solid) should also be given, if the reaction conditions are such that they may exist in more than one state; for example:

$$H_2(g) + \tfrac{1}{2}O_2(g) \rightarrow H_2O(g), \ \Delta H_r^\circ = -241\cdot6 \text{ kJ}$$
$$H_2(g) + \tfrac{1}{2}O_2(g) \rightarrow H_2O(l), \ \Delta H_r^\circ = -285\cdot6 \text{ kJ}$$

The difference between the two heats of reaction is the latent heat of the water formed.

In process design calculations it is usually more convenient to express the heat of reaction in terms of the mols of product produced, for the conditions under which the reaction is carried out, kJ/mol product.

Standard heats of reaction can be converted to other reaction temperatures by making a heat balance over a hypothetical process, in which the reactants are brought to the standard temperature, the reaction carried out, and the products then brought to the required reaction temperature; as illustrated in Fig. 3.4.

$$\Delta H_{r,t} = \Delta H_r^\circ + \Delta H_{\text{prod.}} - \Delta H_{\text{react.}} \tag{3.22}$$

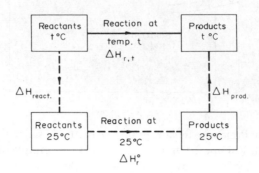

FIG. 3.4. ΔH_r at temperature t

where $-\Delta H_{r,t}$ = heat of reaction at temperature t,

$\Delta H_{\text{react.}}$ = enthalpy change to bring reactants to standard temperature,

$\Delta H_{\text{prod.}}$ = enthalpy change to bring products to reaction temperature, t.

For practical reactors, where the reactants and products may well be at temperatures different from the reaction temperature, it is best to carry out the heat balance over the actual reactor using the standard temperature (25°C) as the datum temperature; the standard heat of reaction can then be used without correction.

It must be emphasised that it is unnecessary to correct a heat of reaction to the reaction temperature for use in a reactor heat-balance calculation. To do so is to carry out two heat balances, whereas with a suitable choice of datum only one need be made. For a practical reactor, the heat added (or removed) Q_p to maintain the design reactor temperature will be given by (from equation 3.10):

$$Q_p = H_{\text{products}} - H_{\text{reactants}} - Q_r \tag{3.23}$$

where H_{products} is the *total* enthalpy of the product streams, including unreacted materials and by-products, evaluated from a datum temperature of 25°C;

$H_{\text{reactants}}$ is the total enthalpy of the feed streams, including excess reagent and inerts, evaluated from a datum of 25°C;

Q_r is the total heat generated by the reactions taking place, evaluated from the standard heats of reaction at 25°C (298 K).

$$Q_r = \sum -\Delta H_r^\circ \times (\text{mol of product formed}) \tag{3.24}$$

where $-\Delta H_r^\circ$ is the standard heat of reaction per mol of the particular product.

Note: A negative sign is necessary in equation 3.24 as Q_r is positive when heat is evolved by the reaction, whereas the standard enthalpy change will be negative for exothermic reactions. Q_p will be negative when cooling is required (see section 3.4).

3.10.1. Effect of pressure on heats of reaction

Equation 3.22 can be written in a more general form:

$$\Delta H_{r,P,T} = \Delta H_r^\circ + \int_1^P \left[\left(\frac{\partial H_{\text{prod.}}}{\partial P} \right)_T - \left(\frac{\partial H_{\text{react.}}}{\partial P} \right)_T \right] dP$$
$$+ \int_{298}^T \left[\left(\frac{\partial H_{\text{prod.}}}{\partial T} \right)_P - \left(\frac{\partial H_{\text{react.}}}{\partial T} \right)_P \right] dT \tag{3.25}$$

If the effect of pressure is likely to be significant, the change in enthalpy of the products and reactants, from the standard conditions, can be evaluated to include both the effects of temperature and pressure (for example, by using tabulated values of enthalpy) and the correction made in a similar way to that for temperature only.

Example 3.7

Illustrates the manual calculation of a reactor heat balance.

Vinyl chloride (VC) is manufactured by the pyrolysis of 1,2,dichloroethane (DCE). The reaction is endothermic. The flow-rates to produce 5000 kg/h at 55 per cent conversion are shown in the diagram (see Example 2.13).

The reactor is a pipe reactor heated with fuel gas, gross calorific value 33·5 MJ/m³. Estimate the quantity of fuel gas required.

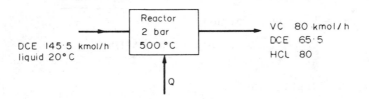

Solution

Reaction: $C_2H_4Cl_2(g) \rightarrow C_2H_3Cl(g) + HCl(g)$ $\Delta H_r^\circ = 70{,}224$ kJ/kmol.

The small quantity of impurities, less than 1 per cent, that would be present in the feed have been neglected for the purposes of this example. Also, the yield of VC has been taken as 100 per cent. It would be in the region of 99 per cent at 55 per cent conversion.

Heat capacity data, for vapour phase

$$C_p^\circ = a + bT + cT^2 + dT^3 \quad \text{kJ/kmol K}$$

	a	$b \times 10^2$	$c \times 10^5$	$d \times 10^9$
VC	5·94	20·16	−15·34	47·65
HCl	30·28	−0·761	1·325	−4·305
DCE	20·45	23·07	−14·36	33·83

for liquid phase: DCE at 20°C, $C_p = 116$ kJ/kmol K,

taken as constant over temperature rise from 20 to 25°C.

Latent heat of vaporisation of DCE 25°C = 34·3 MJ/kmol.

At 2 bar pressure the change in C_p with pressure will be small and will be neglected. Take base temperature as 25°C (298 K), the standard state for ΔH_r°.

Enthalpy of feed $= 145.5 \times 116\,(293 - 298) = -84{,}390 \text{ kJ/h} = -84.4 \text{ MJ/h}$

$$\text{Enthalpy of product stream} = \int_{298}^{773} \sum (n_i C_p)\,dT$$

Component	n_i (mol/h)	$n_i a$	$n_i b \times 10^2$	$n_i c \times 10^5$	$n_i d \times 10^9$
VC	80	475·2	1612·8	−1227·2	3812·0
HCl	80	2422·4	−60·88	106·0	−344·4
DCE	65·5	1339·5	1511·0	−940·6	2215·9
$\sum n_i C_p$		4237·1	3063·0	−2061·8	5683·5

$$\int_{298}^{773} \sum n_i C_p\,dT = \int_{298}^{773} (4237{\cdot}1 + 3063{\cdot}0 \times 10^{-2}\,T - 2061{\cdot}8 \times 10^{-5}\,T^2$$

$$+ 5683{\cdot}5 \times 10^{-9}T^3)\,dT$$

$$= 7307{\cdot}3 \text{ MJ/h}$$

Heat consumed in system by the endothermic reaction $= \Delta H_r^\circ \times$ mols produced

$$= 70{,}224 \times 80 = 5{,}617{,}920 \text{ kJ/h} = 5617{\cdot}9 \text{ MJ/h}$$

Heat to vaporise feed (gas phase reaction)

$$= 34{\cdot}3 \times 145{\cdot}5 = 4997{\cdot}9 \text{ MJ/h}$$

Heat balance:

$$\text{Output} = \text{Input} + \text{consumed} + Q$$

$$Q = H_{\text{product}} - H_{\text{feed}} + \text{consumed}$$

$$= 7307{\cdot}3 - (-84{\cdot}4) + (5617{\cdot}9 + 4997{\cdot}9) = 18{,}007{\cdot}5 \text{ MJ/h}$$

Taking the overall efficiency of the furnace as 70% the gas rate required

$$= \text{Heat input/calorific value} \times \text{efficiency}$$

$$= \frac{18{,}007{\cdot}5}{33{\cdot}5 \times 0{\cdot}7} = 768 \text{ m}^3/\text{h}$$

3.11. Standard heats of formation

The standard enthalpy of formation ΔH_f° of a compound is defined as the enthalpy change when one mol of the compound formed from its constituent elements in the standard state. The enthalpy of formation of the elements is taken as zero. The standard heat of any reaction can be calculated from the heats of formation $-\Delta H_f^\circ$ of the products and reactants; if these are available or can be estimated.

Conversely, the heats of formation of a compound can be calculated from the heats of reaction; for use in calculating the standard heat of reaction for other reactions.

The relationship between standard heats of reaction and formation is given by equation 3.26 and illustrated by Examples 3.8 and 3.9

$$\Delta H_r^\circ = \sum \Delta H_f^\circ, \text{ products} - \sum \Delta H_f^\circ, \text{ reactants} \tag{3.26}$$

A comprehensive list of enthalpies of formation is given in Appendix D.

As with heats of reaction, the state of the materials must be specified when quoting heats of formation.

Example 3.8

Calculate the standard heat of the following reaction, given the enthalpies of formation:

$$4NH_3(g) + 5O_2(g) \rightarrow 4NO(g) + 6H_2O(g)$$

Standard enthalpies of formation kJ/mol

$$
\begin{array}{ll}
NH_3(g) & -46\cdot2 \\
NO(g) & +90\cdot3 \\
H_2O(g) & -241\cdot6
\end{array}
$$

Solution

Note the enthalpy of formation of O_2 is zero.

$$\Delta H_r^\circ = \sum \Delta H_f^\circ, \text{ products} - \sum \Delta H_f^\circ, \text{ reactants}$$
$$= (4 \times 90\cdot3 + 6 \times (-241\cdot6)) - (4 \times (-46\cdot2))$$
$$= \underline{-903\cdot6 \text{ kJ/mol}}$$

Heat of reaction $-\Delta H_r^\circ = \underline{\underline{904 \text{ kJ/mol}}}$

3.12. Heats of combustion

The heat of combustion of a compound $-\Delta H_c^\circ$ is the standard heat of reaction for complete combustion of the compound with oxygen. Heats of combustion are relatively easy to determine experimentally. The heats of other reactions can be easily calculated from the heats of combustion of the reactants and products.

The general expression for the calculation of heats of reaction from heats of combustion is

$$\Delta H_r^\circ = \sum \Delta H_c^\circ, \text{ reactants} - \sum \Delta H_c^\circ, \text{ products} \tag{3.27}$$

Note: the product and reactant terms are the opposite way round to that in the expression for the calculation from heats of formation (equation 3.26).

Caution. Heats of combustion are large compared with heats of reaction. Do not round off the numbers before subtraction; round off the difference.

Two methods of calculating heats of reaction from heats of combustion are illustrated in Example 3.9.

Example 3.9

Calculate the standard heat of reaction for the following reaction: The hydrogenation of benzene to cyclohexane.

(1) $C_6H_6(g) + 3H_2(g) \rightarrow C_6H_{12}(g)$
(2) $C_6H_6(g) + 7\frac{1}{2}O_2(g) \rightarrow 6CO_2(g) + 3H_2O(l)$ $\Delta H_c^\circ = -3287\cdot4\,kJ$
(3) $C_6H_{12}(g) + 9O_2 \rightarrow 6CO_2(g) + 6H_2O(l)$ $\Delta H_c^\circ = -3949\cdot2\,kJ$
(4) $C(s) + O_2(g) \rightarrow CO_2(g)$ $\Delta H_c^\circ = -393\cdot12\,kJ$
(5) $H_2(g) + \frac{1}{2}O_2(g) \rightarrow H_2O(l)$ $\Delta H_c^\circ = -285\cdot58\,kJ$

Note: unlike heats of formation, the standard state of water for heats of combustion is liquid. Standard pressure and temperature are the same 25°C, 1 atm.

Solution

Method 1

Using the more general equation 3.26

$$\Delta H_r^\circ = \sum \Delta H_f^\circ, \text{ products} - \sum \Delta H_f^\circ \text{ reactants}$$

the enthalpy of formation of C_6H_6 and C_6H_{12} can be calculated, and from these values the heat of reaction (1).
From reaction (2)

$$\Delta H_c^\circ(C_6H_6) = 6 \times \Delta H_c^\circ(CO_2) + 3 \times \Delta H_c^\circ(H_2O) - \Delta H_f^\circ(C_6H_6)$$
$$3287\cdot4 = 6(-393\cdot12) + 3(-285\cdot58) - \Delta H_f^\circ(C_6H_6)$$

$$\Delta H_f^\circ(C_6H_6) = -3287\cdot4 - 3215\cdot52 = \underline{71\cdot88\,kJ/mol}$$

From reaction (3)

$$\Delta H_c^\circ(C_6H_{12}) = -3949\cdot2 = 6(-393\cdot12) + 6(-285\cdot58) - \Delta H_f^\circ(C_6H_{12})$$
$$\Delta H_f^\circ(C_6H_{12}) = 3949\cdot2 - 4072\cdot28 = \underline{-123\cdot06\,kJ/mol}$$
$$\Delta H_r^\circ = \Delta H_f^\circ(C_6H_{12}) - \Delta H_f^\circ(C_6H_6)$$
$$\Delta H_r^\circ = (-123\cdot06) - (71\cdot88) = \underline{-195\,kJ/mol}$$

Note: enthalpy of formation of H_2 is zero.

Method 2

Using equation 3.27

$$\Delta H_r^\circ = (\Delta H_c^\circ(C_6H_6) + 3 \times \Delta H_c^\circ(H_2) - \Delta H_c^\circ(C_6H_{12})$$
$$= (-3287\cdot4 + 3(-285\cdot88)) - (-3949\cdot2) = \underline{-196\,kJ/mol}$$

Heat of reaction $-\Delta H_r^\circ = \underline{196\,kJ/mol}$

3.13. Compression and expansion of gases

The work term in an energy balance is unlikely to be significant unless a gas is expanded or compressed as part of the process. To compute the pressure work term:

$$-W = \int_1^2 P\,dv \qquad\qquad \text{(equation 3.5)}$$

a relationship between pressure and volume during the expansion is needed.

If the compression or expansion is isothermal (at constant temperature) then for unit mass of an ideal gas:

$$Pv = \text{constant} \qquad (3.28)$$

and the work done, $\qquad -W = P_1 v_1 \ln P_2/P_1 = \dfrac{RT_1}{M} \ln P_2/P_1 \qquad (3.29)$

where $\quad P_1 =$ initial pressure,

$\qquad P_2 =$ final pressure,

$\qquad v_1 =$ initial volume.

In industrial compressors or expanders the compression or expansion path will be "polytropic", approximated by the expression:

$$Pv^n = \text{constant} \qquad (3.30)$$

The work produced (or required) is given by the general expression (see Volume 1, Chapter 6):

$$-W = P_1 v_1 \frac{n}{n-1}\left[\left(\frac{P_2}{P_1}\right)^{(n-1)/n} - 1\right] = Z\frac{RT_1}{M}\frac{n}{n-1}\left[\left(\frac{P_2}{P_1}\right)^{(n-1)/n} - 1\right] \qquad (3.31)$$

where $\quad Z =$ compressibility factor (1 for an ideal gas),

$\qquad R =$ universal gas constant, $8{\cdot}314\,\mathrm{J\,K^{-1}\,mol^{-1}}$,

$\qquad T_1 =$ inlet temperature, K,

$\qquad M =$ molecular mass (weight) of gas.

The value of n will depend on the design and operation of the machine.

The energy required to compress a gas, or the energy obtained from expansion, can be estimated by calculating the ideal work and applying a suitable efficiency value. For reciprocating compressors the isentropic work is normally used ($n = \gamma$) (see Fig. 3.7); and for centrifugal or axial machines the polytropic work (see Section 3.13.2).

3.13.1. Mollier diagrams

If a Mollier diagram (enthalpy–pressure–temperature–entropy) is available for the working fluid the isentropic work can be easily calculated.

$$W = H_1 - H_2 \qquad (3.32)$$

where $\quad H_1 \quad$ is the specific enthalpy at the pressure and temperature corresponding to point 1, the initial gas conditions,

$\qquad H_2 \quad$ is the specific enthalpy corresponding to point 2, the final gas condition.

Point 2 is found from point 1 by tracing a path (line) of constant entropy on the diagram.

The method is illustrated in Example 3.10.

Example 3.10

Methane is compressed from 1 bar and 290 K to 10 bar. If the isentropic efficiency is 0·85, calculate the energy required to compress 10,000 kg/h. Estimate the exit gas temperature.

Solution

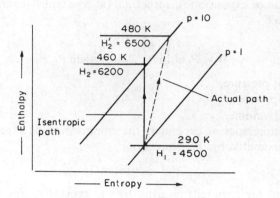

Fig. 3.5. Mollier diagram, methane

From the Mollier diagram, shown diagrammatically in Fig. 3.5

$$H_1 = 4500 \text{ cal/mol},$$
$$H_2 = 6200 \text{ cal/mol (isentropic path)},$$
$$\text{Isentropic work} = 6200 - 4500$$
$$= \underline{1700 \text{ cal/mol}}$$

For an isentropic efficiency of 0·85:

$$\text{Actual work done on gas} = \frac{1700}{0\cdot85} = \underline{2000 \text{ cal/mol}}$$

So, actual final enthalpy

$$H_2' = H_1 + 2000 = \underline{6500 \text{ cal/mol}}$$

From Mollier diagram, if all the extra work is taken as irreversible work done on the gas, the exit gas temperature = $\underline{480 \text{ K}}$

Molecular weight methane = 16

$$\text{Energy required} = (\text{mols per hour}) \times (\text{specific enthalpy change})$$
$$= \frac{10{,}000}{16} \times 2000 \times 10^3$$
$$= 1\cdot25 \times 10^9 \text{ cal/h}$$
$$= 1\cdot25 \times 10^9 \times 4\cdot187$$
$$= 5\cdot23 \times 10^9 \text{ J/h}$$
$$\text{Power} = \frac{5\cdot23 \times 10^9}{3600} = \underline{1\cdot45 \text{ MW}}$$

3.13.2. Polytropic compression and expansion

If no Mollier diagram is available, it is more difficult to estimate the ideal work in compression or expansion processes. Schultz (1962) gives a method for the calculation of the polytropic work, based on two generalised compressibility functions, X and Y; which supplement the familiar compressibility factor Z.

$$X = \frac{T}{V}\left(\frac{\partial V}{\partial T}\right)_P - 1 \tag{3.33}$$

$$Y = -\frac{P}{V}\left(\frac{\partial V}{\partial P}\right)_T \tag{3.34}$$

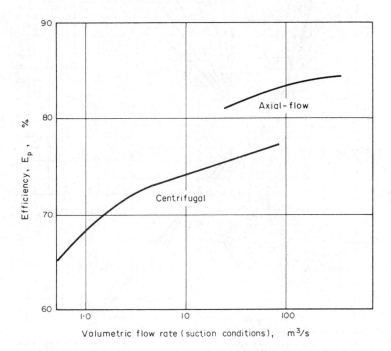

FIG. 3.6. Approximate polytropic efficiencies centrifugal and axial-flow compressors

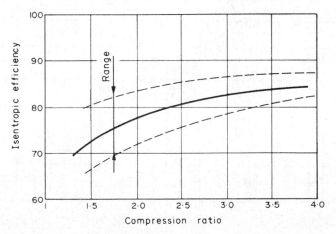

FIG. 3.7. Typical efficiencies for reciprocating compressors

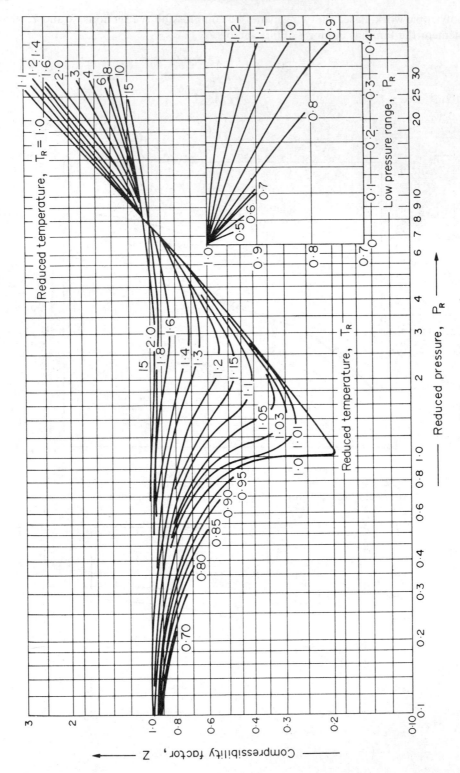

FIG. 3.8. Compressibility factors of gases and vapours

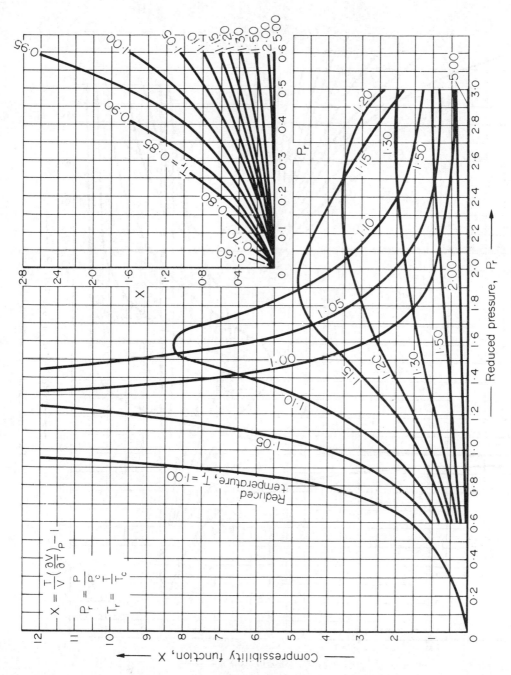

FIG. 3.9. Generalised compressibility function X

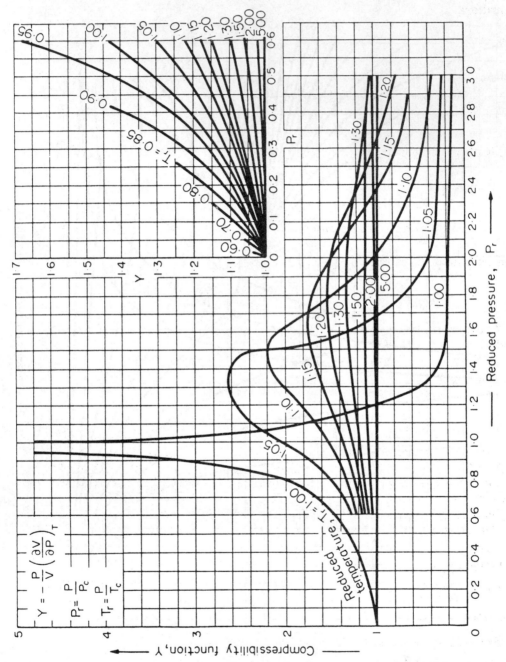

Fig. 3.10. Generalised compressibility function Y

His charts for X and Y as functions of reduced temperature and pressure are reproduced as Figs. 3.9 and 3.10. The functions are used to determine the polytropic exponent n for use in equation 3.31; and a polytropic temperature exponent m for use in the following equation:

$$T_2 = T_1 \left(\frac{P_2}{P_1} \right)^m \tag{3.35}$$

where

$$m = \frac{ZR}{C_p} \left(\frac{1}{E_p} + X \right) \quad \text{for compression,} \tag{3.36}$$

$$m = \frac{ZR}{C_p} (E_p + X) \quad \text{for expansion} \tag{3.37}$$

E_p is the polytropic efficiency, defined by:

$$\text{for compression } E_p = \frac{\text{polytropic work}}{\text{actual work required}}$$

$$\text{for expansion } E_p = \frac{\text{actual work obtained}}{\text{polytropic work}}$$

An estimate of E_p can be obtained from Fig. 3.6.

$$n = \frac{1}{Y - m(1 + X)} \tag{3.38}$$

At conditions well removed from the critical conditions equations 3.36, 3.37 and 3.38 reduce to:

$$m = \frac{(\gamma - 1)}{\gamma E_p} \tag{3.36a}$$

$$m = \frac{(\gamma - 1)E_p}{\gamma} \tag{3.37a}$$

$$n = \frac{1}{1 - m} \tag{3.38a}$$

These expressions can be used to calculate the polytropic work and outlet temperature by substitution in equations 3.31 and 3.35. They can also be used to make a first estimate of T_2 in order to estimate the mean reduced temperture for use with Figs. 3.9 and 3.10.

The use of Schultz's method is illustrated in Examples 3.11 and 3.15.

Example 3.11

Estimate the power required to compress 5000 kmol/h of HCl at 5 bar, 15°C, to 15 bar.

Solution

For HCl, $P_c = 82$ bar, $T_c = 324 \cdot 6$ K

$C_p^\circ = 30 \cdot 30 - 0 \cdot 72 \times 10^{-2}\, T + 12 \cdot 5 \times 10^{-6}\, T^2 - 3 \cdot 9 \times 10^{-9}\, T^3$ kJ/kmol K

Estimate T_2 from equations 3.35 and 3.36a.

For diatomic gases $\gamma \simeq 1 \cdot 4$.

Note: γ could be estimated from the relationship $\gamma = \dfrac{C_p}{C_v} = \dfrac{C_p}{C_p - R}$

At the inlet conditions, the flow rate in m^3/s

$$= \frac{5000}{3600} \times 22 \cdot 4 \times \frac{288}{273} \times \frac{1}{5} = 6 \cdot 56$$

From Fig. 3.6 $E_p = 0 \cdot 73$

$$m = \frac{1 \cdot 4 - 1}{1 \cdot 4 \times 0 \cdot 73} = 0 \cdot 39$$

$$T_2 = 288 \left(\frac{15}{5}\right)^{0 \cdot 39} = 442 \, K$$

$$T_{r \, (mean)} = \frac{442 + 228}{2 \times 324 \cdot 6} = 1 \cdot 03$$

$$P_{r \, (mean)} = \frac{5 + 15}{2 \times 82} = 0 \cdot 12$$

At $T_{(mean)} C_p^\circ = 29 \cdot 14 \, kJ/kmol \, K$

Correction for pressure from Fig. 3.2, 2 kJ/kmol K

$$C_p = 29 \cdot 14 + 2 \simeq 31 \, kJ/kmol \, K$$

From Figs. 3.8, 3.9 and 3.10 at mean conditions:

$$X = 0 \cdot 18, \; Y = 1 \cdot 04, \; Z = 0 \cdot 97$$
$$Z \text{ at inlet conditions} = 0 \cdot 98$$

From equations 3.36 and 3.38

$$m = \frac{0 \cdot 97 \times 8 \cdot 314}{31} \left(\frac{1}{0 \cdot 73} + 0 \cdot 18\right) = \underline{\underline{0 \cdot 40}}$$

$$n = \frac{1}{1 \cdot 04 - 0 \cdot 4(1 + 0 \cdot 18)} = \underline{\underline{1 \cdot 76}}$$

From equation 3.31

$$W \text{ polytropic} = 0 \cdot 98 \times 288 \times 8 \cdot 314 \times \frac{1 \cdot 76}{1 \cdot 76 - 1}\left(\left(\frac{15}{5}\right)^{\frac{1 \cdot 76 - 1}{1 \cdot 76}} - 1\right)$$

$$= \underline{\underline{3299 \, kJ/kmol}}$$

Actual work required $= \dfrac{\text{polytropic work}}{E_p}$

$$= \frac{3299}{0 \cdot 73} = \underline{\underline{4520 \, kJ/kmol}}$$

$$\text{Power} = \frac{4520}{3600} \times 5000 = 6275 \, kW$$

Say, $\underline{\underline{6 \cdot 3 \, MW}}$

$$T_2 = 288 \left(\frac{15}{5}\right)^{0 \cdot 4} = \underline{\underline{447 \, K}}$$

The first estimate was close enough. Note, for the conditions in this example the approximate equations 3.36a, 3.38a could have been used.

3.13.3. Electrical drives

The electrical power required to drive a compressor (or pump) can be calculated from a knowledge of the motor efficiency:

$$\text{Power} = \frac{-W \times \text{mass flow-rate}}{E_e} \tag{3.39}$$

where $-W$ = work of compression per unit mass (equation 3.31),
E_e = electric motor efficiency.

The efficiency of the drive motor will depend on the type, speed and size. The values given in Table 3.1 can be used to make a rough estimate of the power required.

TABLE 3.1. *Approximate efficiencies of electric motors*

Size (kW)	Efficiency (%)
5	80
15	85
75	90
200	92
750	95
>4000	97

3.14. A simple energy balance program

Manual energy-balance calculations, particularly those in which the specific heat capacities are expressed as polynomial equations (equation 3.13), are tedious and mistakes are easily made. It is worth while writing a short computer program for these problems. They can be solved using "desk-top" computers and programmable hand calculators. A typical program is listed in Table 3.2. This program can be used to calculate the heat input or cooling required for a process unit, where the stream enthalpies relative to the datum temperature can be calculated from the specific heat capacities of the components (equation 3.11).

The datum temperature in the program is 25°C (298 K), which is the standard for most heat of reaction data. Specific heats are represented by a cubic equation in temperature:

$$C_p = A + BT + CT^2 + DT^3$$

Any unspecified constants are typed in as zero.

If the process involves a reaction the heat generated or consumed is computed from the heat of reaction per kmol of product (at 25°C) and the kmols of product formed.

If any component undergoes a phase change in the unit the heat required is computed from the latent heat (at 25°C) and the quantity involved.

TABLE 3.2. *ENERGY 1, a simple energy balance program*

```
10 REM SHORT ENERGY BALANCE PROGRAM
20 REM CALCULATES HEAT INPUT OR COOLING REQUIRED
30 PRINT "HEAT BALANCE PROG, BASIS KMOL/H, TEMPS.K,DATUM 298K "
40 PRINT "INPUT NO. OF COMPONENTS, MAX. 10"
50 INPUT N1
60 PRINT "INPUT HEAT CAPACITY DATA, EQN. A+BT+CT↑2+DT↑3"
70 FOR I=1 TO N1
80 PRINT "FOR COMPONENT";I;"INPUT A,B,C,D"
90 INPUT A[I],B[I],C[I],D[I]
100 PRINT A[I],B[I],C[I],D[I]
110 NEXT I
120 H4=H5=H6=Q1=0
130 PRINT "INPUT THE NUMBER OF FEED STREAMS"
140 INPUT S1
150 FOR I=1 TO S1
160 PRINT "FOR FEED STREAM";I;"INPUT STREAM TEMP. AND NUMBER OF COMPONENTS"
170 INPUT T1,N2
180 GOSUB 540
190 PRINT "STREAM SENSIBLE HEAT =";H4;"KJ/H"
200 REM TOTAL SENSIBLE HEAT FEED STREAMS
210 H5=H5+H4
220 NEXT I
230 PRINT "INPUT THE NUMBER OF PRODUCT STREAMS"
240 INPUT S1
250 FOR I=1 TO S1
260 PRINT "FOR PRODUCT STREAM";I;"INPUT STREAM TEMP. AND NUMBER OF COMPONENTS"
270 INPUT T1,N2
280 GOSUB 540
290 PRINT "STREAM SENSIBLE HEAT =";H4;"KJ/H"
300 REM TOTAL SENSIBLE HEAT PRODUCT STREAMS
310 H6=H6+H4
320 NEXT I
330 PRINT "INPUT THE NUMBER OF REACTIONS AND PHASE CHANGES"
340 INPUT N4
345 IF N4=0 THEN 430
350 PRINT "FOR EACH REACTION OR PHASE CHANGE"
360 PRINT "INPUT REACTION/LATENT HEAT (KJ/MOL) & QUANTITY (MOL/H)"
365 PRINT " REMEMBER THAT HEAT EVOLVED IS TAKEN AS POSITIVE"
367 PRINT " AND HEAT ABSORBED AS NEGATIVE"
370 FOR I=1 TO N4
380 PRINT "NEXT REACTION/PHASE CHANGE"
390 INPUT R,F2
400 H7=F2*R
410 Q1=Q1+H7
420 NEXT I
430 REM HEAT BALANCE
440 Q=H6-H5-Q1
450 IF Q<0 THEN 480
460 PRINT "HEATING REQUIRED=";Q;"KJ/H"
470 GOTO 490
480 PRINT "COOLING REQUIRED=";-Q;"KJ/H"
490 PRINT "REPEAT CALCULATION WANTED ? , TYPE 1 FOR YES, 0 FOR NO"
500 INPUT P1
510 IF P1=0 THEN 530
520 GOTO 120
530 STOP
540 REM SUB-ROUTINE TO CALCULATE STREAM SENSIBLE HEATS
550 PRINT "FOR EACH COMPONENT INPUT THE COMP.NO.& FLOW RATE (KMOL/H)"
560 H4=0
570 FOR I1=1 TO N2
580 PRINT "NEXT COMPONENT"
590 INPUT J,F
600 PRINT "COMP.";J;",FLOW";F
610 REM HEAT CAPACITY EQN. SPLIT ONTO TWO LINES
620 H1=A[J]*(T1-298)+B[J]*(T1↑2-298↑2)/2
630 H2=C[J]*(T1↑3-298↑3/3+D[J]*(T1↑4-298↑4)/4
640 H3=F*(H1+H2)
650 H4=H4+H3
660 NEXT I1
670 RETURN
```

The component specific heat capacity coefficients A, B, C, D are stored as a matrix. If a heat balance is to be made on several units the coefficients for all the components can be typed in at the start, and the program rerun for each unit.

The program listing contains sufficient remark statements for the operation of the program to be easily followed. The program is written in BASIC for the Hewlett Packard model 9830 desk-top computer, and can be easily adapted for other machines.

The use of the program is illustrated in Example 3.12. It has also been used for other examples in this chapter and in the chapter on flow-sheeting, Chapter 4.

Example 3.12

Use of computer program ENERGY 1

A furnace burns a liquid coal tar fuel derived from coke-ovens. Calculate the heat transferred in the furnace if the combustion gases leave at 1500 K. The burners operate with 20 per cent excess air.

Take the fuel supply temperature as 50°C (323 K) and the air temperature as 15°C (288 K).

The properties of the fuel are:

Carbon	87·5 per cent w/w
Hydrogen	8·0
Oxygen	3·5
Nitrogen	1·0
Sulphur	trace
Ash	balance

Nett calorific value	39,540 kJ/kg
Latent heat of vaporisation	350 kJ/kg
Heat capacity	1·6 kJ/kg K

C_p° of gases, kJ/kmol K,

$$C_p = A + BT + CT^2 + DT^3$$

Component	A	B	C	D
1 CO_2	19·763	7·332E-2	−5·5E-5	17·125E-9
2 H_2O	32·190	19·207E-4	10·538E-6	−3·591E-9
3 O_2	28·06	−3·674E-6	17·431E-6	−10·634E-9
4 N_2	31·099	−1·354E-2	26·752E-6	−11·662E-9

Material balance

Basis: 100 kg (as analysis is by weight).
Assume complete combustion: maximum heat release.
Reactions: $C + O_2 \rightarrow CO_2$
$$H_2 + \tfrac{1}{2}O_2 \rightarrow H_2O$$

Element	kg	kmol	Stoichiometric O_2 kmol	kmol, products
C	87·5	7·29	7·29	7·29, CO_2
H_2	8·0	4·0	2·0	4·0, H_2O
O_2	3·5	0·11	—	0·11
N_2	1·0	0·04	—	0·04
Total		11·44	9·29	

O_2 required with 20 per cent excess $= 9·29 \times 1·2 = 11·15\,\text{kmol}$.

Unreacted O_2 from combustion air $= 11·15 - 9·29 = 1·86\,\text{kmol}$.

N_2 with combustion air $= 11·15 \times \dfrac{79}{21} = 41·94\,\text{kmol}$.

Composition of combustion gases:

$$CO_2 \qquad\qquad = 7·29\,\text{kmol}$$
$$H_2O \qquad\qquad = 4·0$$
$$O_2 \qquad 0·11 + 1·86 = 1·97$$
$$N_2 \qquad 0·04 + 41·94 = 41·98$$

Presentation of data to the program:
C_p of fuel (component 5), taken as constant,

$$A = 1·6, B = C = D = 0$$

Heat of reaction and latent heat, taken to be values at datum temperature of 298 K.

There is no need to convert to kJ/kmol, providing quantities are expressed in kg.
For the purposes of this example the dissociation of CO_2 and H_2O at 1500 K is ignored.

Computer print-out

Data inputs shown after the symbol (!)

```
HEAT BALANCE PROGRAM, BASIS KMOL/H, TEMPS. K, DATUM 298K
INPUT THE NUMBER OF COMPONENTS, MAX. 10
!5
INPUT HEAT CAPACITY DATA, EQN. A+BT+CT**2+DT**3
FOR COMPONENT 1 INPUT A, B, C, D
!19. 763, 0. 07332, -5. 518E-5, 1. 7125E-8
FOR COMPONENT 2 INPUT A, B, C, D
!32. 19, 1. 9207E-3, 1. 0538E-5, 3. 591E-09
FOR COMPONENT 3 INPUT A, B, C, D
!28. 06, -3. 67E-6, 1. 743E-5, -1. 0634E-8
FOR COMPONENT 4 INPUT A, B, C, D
!31. 099, -0. 01354, 2. 6752E-5, -1. 1662E-8
FOR COMPONENT 5 INPUT A, B, C, D
!1. 6, 0, 0, 0
INPUT THE NUMBER OF FEED STREAMS
!2
FOR FEED STREAM 1 INPUT STREAM TEMP. AND NUMBER OF COMPONENTS
!323, 1
```

```
FOR EACH COMPONENT, INPUT THE COMPONENT NO. & FLOW RATE, KMOL/H
NEXT COMPONENT
!5, 100
STREAM SENSIBLE HEAT = 4000 KJ/H
FOR FEED STREAM 2 INPUT STREAM TEMP. AND NUMBER OF COMPONENTS
!288, 2
FOR EACH COMPONENT, INPUT THE COMPONENT NO. & FLOW RATE, KMOL/H
NEXT COMPONENT
!3, 11. 15
NEXT COMPONENT
!4, 41. 94
STREAM SENSIBLE HEAT = -15484. 90158961 KJ/H
INPUT THE NUMBER OF PRODUCT STREAMS
!1
FOR PRODUCT STREAM 1 INPUT STREAM TEMP. AND NUMBER OF COMPONENTS
!1500, 4
FOR EACH COMPONENT, INPUT THE COMPONENT NO. & FLOW RATE, KMOL/H
NEXT COMPONENT
!1, 7. 29
NEXT COMPONENT
!2, 4. 0
NEXT COMPONENT
!3, 1. 97
NEXT COMPONENT
!4, 41. 98
STREAM SENSIBLE HEAT =2355987. 672535 KJ/H
INPUT THE NUMBER OF REACTIONS OR PHASE CHANGES
!2
FOR EACH REACTION OR PHASE CHANGE
INPUT THE REACTION OR LATENT HEAT (KJ/MOL) AND QUANTITY (MOL/H)
REMEMBER THAT HEAT EVOLVED IS TAKEN AS POSITIVE AND
HEAT ABSORBED NEGATIVE
NEXT REACTION OR PHASE CHANGE
!+39540, 100
NEXT REACTION OR PHASE CHANGE
!-350, 100
COOLING REQUIRED = 1551527. 425875 KJ/H
REPEAT CALCULATION WANTED ?, TYPE 1 FOR YES, 0 FOR NO
!0
```

Heat transferred (cooling required)

$$= \underline{\underline{1{,}550{,}000 \text{ kJ}/100 \text{ kg}}}$$

Note: though the program reports kJ/h, the basis used for this calculation was 100 kg fuel.

3.15 Unsteady state energy balances

All the examples of energy balances considered previously have been for steady-state processes; where the rate of energy generation or consumption did not vary with time and the accumulation term in the general energy balance equation was taken as zero.

If a batch process is being considered, or if the rate of energy generation or removal varies with time, it will be necessary to set up a differential energy balance, similar to the differential material balance considered in Chapter 2. For batch processes the total energy requirements can usually be estimated by taking as the time basis for the calculation 1 batch; but the maximum rate of heat generation will also have to be estimated to size any heat-transfer equipment needed.

The application of a differential energy balance is illustrated in Example 3.13.

Example 3.13

Differential energy balance

In the batch preparation of an aqueous solution the water is first heated to 80°C in a jacketed, agitated vessel; 1000 Imp. gal. (4545 kg) is heated from 15°C. If the jacket area is 300 ft² (27 m²) and the overall heat-transfer coefficient can be taken as 50 Btu ft⁻² h⁻¹ °F⁻¹ (285 W m⁻² K⁻¹), estimate the heating time. Steam is supplied at 25 psig (2·7 bar).

Solution

The rate of heat transfer from the jacket to the water will be given by the following expression (see Volume 1, Chapter 7):

$$\frac{dQ}{dt} = UA(t_s - t) \qquad (a)$$

where dQ is the increment of heat transferred in the time interval dt, and

U = the overall-heat transfer coefficient,
t_s = the steam-saturation temperature,
t = the water temperature.

The incremental increase in the water temperature dt is related to the heat transferred dQ by the energy-balance equation:

$$dQ = WC_p dt \qquad (b)$$

where WC_p is the heat capacity of the system.

Equating equations (a) and (b)

$$WC_p \frac{dt}{dt} = UA(t_s - t)$$

Integrating

$$\int_0^{t_B} dt = \frac{WC_p}{UA} \int_{t_1}^{t_2} \frac{dt}{(t_s - t)}$$

Batch heating time

$$t_B = -\frac{WC_p}{UA} \ln \frac{t_s - t_2}{t_s - t_1}$$

For this example $WC_p = 4·18 \times 4545 \times 10^3$ JK⁻¹

$$UA = 285 \times 27 \text{ WK}^{-1}$$

$$t_1 = 15°C, \ t_2 = 80°C$$

$$t_B = -\frac{4·18 \times 4545 \times 10^3}{285 \times 27} \ln \frac{116 - 80}{116 - 15}$$

$$= 2547 \, s = \underline{\underline{42·5 \text{ min}}}$$

In this example the heat capacity of the vessel and the heat losses have been neglected for simplicity. They would increase the heating time by 10 to 20 per cent.

3.16 Energy recovery

Process streams at high pressure or temperature, and those containing combustible material, contain energy that can be usefully recovered. Whether it is economic to recover the energy content of a particular stream will depend on the value of the energy that can be usefully extracted and the cost of recovery. The value of the energy will depend on the primary cost of energy at the site. It may be worth while recovering energy from a process stream at a site where energy costs are high but not where the primary energy costs are low. The cost of recovery will be the capital and operating cost of any additional equipment required. If the savings exceed the operating cost, including capital charges, then the energy recovery will usually be worth while. Maintenance costs should be included in the operating cost (see Chapter 6).

Some processes, such as air separation, depend on efficient energy recovery for economic operation, and in all processes the efficient utilization of energy recovery techniques will reduce product cost.

Some of the techniques used for energy recovery in chemical process plants are described briefly in the following sections. The references cited give fuller details of each technique. Miller (1968) gives a comprehensive review of process energy systems; including heat exchange, and power recover from high-pressure fluid streams.

3.16.1. Heat exchange

The most common energy-recovery technique is to utilise the heat in a high-temperature process stream to heat a colder stream: saving steam costs; and also cooling water, if the hot stream requires cooling. Conventional shell and tube exchangers are normally used. More total heat-transfer area will be needed, over that for steam heating and water cooling, as the overall driving forces will be smaller.

The cost of recovery will be reduced if the streams are located conveniently close.

The amount of energy that can be recovered will depend on the temperature, flow, heat capacity, and temperature change possible, in each stream. A reasonable temperature driving force must be maintained to keep the exchanger area to a practical size. The most efficient exchanger will be the one in which the shell and tube flows are truly countercurrent. Multiple tube pass exchangers are usually used for practical reasons. With multiple tube passes the flow will be part counter-current and part co-current and temperature crosses can occur, which will reduce the efficiency of heat recovery (see Chapter 12).

The hot process streams leaving a reactor or a distillation column are frequently used to preheat the feedstreams.

3.16.2. Heat-exchanger networks

In an industrial process there will be many hot and cold streams and there will be an optimum arrangement of the streams for energy recovery by heat exchange. The problem of synthesising a network of heat exchangers has been studied by many workers, particularly in respect of optimising heat recovery in crude petroleum distillation. An example of crude preheat train is shown in Fig. 3.11. The general problem of the synthesis

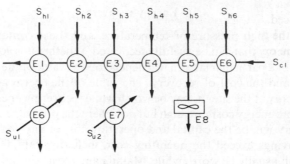

$$S_{h1} = \text{residue } (360°C)$$
$$S_{h2} = \text{reflux stream } (260°C)$$
$$S_{h3} = \text{heavy gas oil } (340°C)$$
$$S_{h4} = \text{light gas oil } (260°C)$$
$$S_{h5} = \text{reflux steam } (180°C)$$
$$S_{h6} = \text{reflux stream } (165°C)$$
$$S_{c1} = \text{crude oil } (15°C)$$
$$S_{u1} \text{ and } S_{u2} = \text{cooling water } (50°C)$$

FIG. 3.11. Typical heat-exchanger network

and optimisation of a network of heat exchangers has been defined by Masso and Rudd (1969).

Consider that there are M hot streams, $S_{hi} (i = 1, 2, 3, \ldots, M)$ to be cooled and N cold streams $S_{cj} (j = 1, 2, 3, \ldots, N)$ to be heated; each stream having an inlet temperature t_f, or an outlet temperature t_o, and a stream heat capacity W_i. There may also be $S_{uk} (k = 1, 2, 3, \ldots, L)$ auxiliary steam heated or water-cooled exchangers.

The problem is to create a minimum cost network of exchangers, that will also meet the design specifications on the required outlet temperature t_o of each stream. If the strictly mathematical approach is taken of setting up all possible arrangements and searching for the optimum, the problem, even for a small number of exchangers, would require an inordinate amount of computer time. Boland and Linnhoff (1979) point out that for a process with four cold and three hot streams, $2 \cdot 4 \times 10^{18}$ arrangements are possible. Most workers have taken a more pragmatic, "heuristic", approach to the problem, using "rules of thumb" to generate a limited number of feasible networks, which are then evaluated.

Porton and Donaldson (1974) suggest a simple procedure that involves the repeated matching of the hottest stream (highest t_f) against the cold stream with the highest required outlet temperature (highest t_o).

A general survey of computer and manual methods for optimising exchanger networks is given by Nishida et al. (1977); see also Siirola (1974).

3.16.3. Waste-heat boilers

If the process streams are at a sufficiently high temperature the heat recovered can be used to generate steam.

Waste-heat boilers are often used to recover heat from furnace flue gases and the process gas streams from high-temperature reactors. The pressure, and superheat temperature, of the stream generated will depend on the temperature of the hot stream and the approach temperature permissible at the boiler exit (see Chapter 12). As with any heat-transfer equipment, the area required will increase as the mean temperature driving force (log mean ΔT) is reduced. The permissible exit temperature may also be limited by process considerations. If the gas stream contains water vapour and soluble corrosive gases, such as HCl or SO_2, the exit gases temperature must be kept above the dew point.

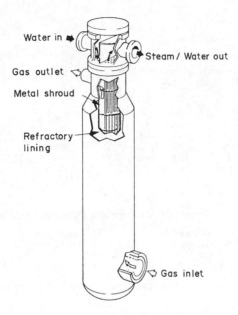

FIG. 3.12. Reformed gas waste-heat boiler arrangement of vertical U-tube water-tube boiler (Reprinted by permission of the Council of the Institution of Mechanical Engineers from the Proceedings of the Conference on Energy Recovery in the Process Industries, London, 1975.)

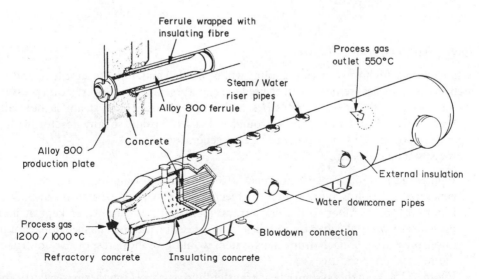

FIG. 3.13. Reformed gas waste-heat boiler, principal features of typical natural circulation fire-tube boilers (Reprinted by permission of the Council of the Institution of Mechanical Engineers from the Proceedings of the Conference on Energy Recovery in the Process Industries, London, 1975.)

Hinchley (1975) discusses the design and operation of waste heat boilers for chemical plant. Both fire tube and water tube boilers are used. A typical arrangement of a water tube boiler on a reformer furnace is shown in Fig. 3.12 and a fire tube boiler in Fig. 3.13. The application of a waste-heat boiler to recover energy from the reactor exit streams in a nitric acid plant is shown in Fig. 3.14.

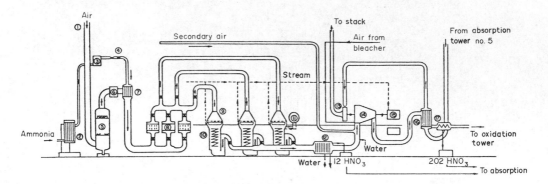

FIG. 3.14. Connections of a nitric acid plant, intermediate pressure type

1. Air entry	6. Air preheater	10. Lamont boilers	14. Compressor
2. Ammonia vaporizer	7. Gas mixer	11. Steam drum	15. Steam turbine
3. Ammonia filter	8. Gas filters	12. Gas cooler No. 1	16. Heat exchanger
4. Control valves	9. Converters	13. Exhaust turbine	17. Gas cooler No. 2
5. Air-scrubbing tower			

(From *Nitric Acid Manufacture*, Miles (1961), with permission)

The selection and operation of waste heat boilers for industrial furnaces is discussed in the *Efficient Use of Energy*, Dryden (1975).

3.16.4. High-temperature reactors

If a reaction is highly exothermic cooling will be needed and, if the reactor temperature is high enough, the heat removed can be used to generate steam. The lowest steam pressure normally used in the process industries is 2·7 bar (25 psig) and steam is normally distributed at a header pressure of around 8 bar (100 psig); so any reactor with a temperature above 200°C is a potential steam generator.

Three systems are used:

1. Figure 3.15a.   An arrangement similar to a conventional water-tube boiler. Steam is generated in cooling pipes within the reactor and separated in a steam drum.
2. Figure 3.15b.   Similar to the first arrangement but with the water kept at high pressure to prevent vaporisation. The high-pressure water is flashed to steam at lower pressure in a flash drum. This system would give more responsive control of the reactor temperature.
3. Figure 3.15c.   In this system a heat-transfer fluid, such as Dowtherm (see Perry and Chilton (1973) for details of heat-transfer fluids), is used to avoid the need for high-pressure tubes. The steam is raised in an external boiler.

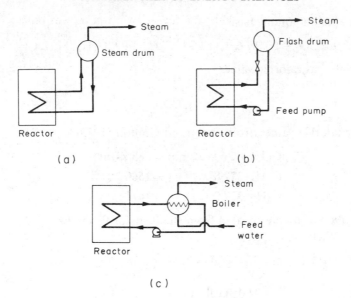

FIG. 3.15. Steam generation

3.16.5. Low-grade fuels

The waste products from any process (gases, liquids and solids) which contain significant quantities of combustible material can be used as low-grade fuels; for raising steam or direct process heating. Their use will only be economic if the intrinsic value of the fuel justifies the cost of special burners and other equipment needed to burn the waste. If the combustible content of the waste is too low to support combustion, the waste will have to be supplemented with higher calorific value primary fuels.

Reactor off-gases

The off-gases (vent gas) from reactors, and recycle stream purges are often of high enough calorific value to be used as fuels.

The calorific value of a gas can be calculated from the heats of combustion of its constituents; the method is illustrated in Example 3.14.

Other factors which, together with the calorific value, will determine the economic value of an off-gas as a fuel are the quantity available and the continuity of supply. Waste gases are best used for steam raising, rather than for direct process heating, as this decouples the source from the use and gives greater flexibility.

Example 3.14

Calculation of a waste-gas calorific value

The typical vent-gas analysis from the recycle stream in an oxyhydrochlorination process for the production of dichloroethane (DCE) (British patent BP 1,524, 449) is given below, percentages on volume basis.

$$O_2 \; 7.96, \; CO_2 + N_2 \; 87.6, \; CO \; 1.79, \; C_2H_4 \; 1.99, \; C_2H_6 \; 0.1, \; DCE \; 0.54$$

Estimate the vent gas calorific value.

Solution

Component calorific values, from Perry and Chilton (1973)

$$CO \quad 67.6 \text{ kcal/mol} = 283 \text{ kJ/mol}$$
$$C_2H_4 \; 372.8 \qquad\qquad = 1560.9$$
$$C_2H_6 \; 337.2 \qquad\qquad = 1411.9$$

The value for DCE can be estimated from the heats of formation.
Combustion reaction:

$$C_2H_4Cl_2(g) + 2\tfrac{1}{2}O_2(g) \rightarrow 2CO_2(g) + H_2O(g) + 2HCl(g)$$

ΔH_f°

$$CO_2 \quad -94.05 \text{ kcal/mol} = -393.8 \text{ kJ/mol}$$
$$H_2O \quad -57.8 \qquad\qquad = -242.0$$
$$HCl \quad -22.06 \qquad\qquad = -92.4$$
$$DCE \quad -31.05 \qquad\qquad = -130.0$$
$$\Delta H_c^\circ = \sum \Delta H_f^\circ \text{ products} - \sum \Delta H_f^\circ \text{ reactants}$$
$$[2(-393.8) - 242.0 + 2(-92.4)] - [-130.0]$$
$$= -1084.4 \text{ kJ}$$

Estimation of vent gas c.v., basis 100 mols.

Component	mols/100 mols		Calorific value (kJ/mol)		Heating value
CO	1.79	×	283.0	=	506.6
C$_2$H$_4$	1.99		1560.9		3106.2
C$_2$H$_6$	0.1		1411.9		141.2
DCE	0.54		1084.4		585.7
				Total	4339.7

Calorific value = 4339.7/100 = 43.4 kJ/mol
of vent gas

$$= \frac{43.4}{22.4} \times 10^3 = 1938 \text{ kJ/m}^3 \quad (52 \text{ Btu/ft}^3) \text{ at 1 bar, } 0°C$$

Barely worth recovery, but if the gas has to be burnt to avoid pollution it could be used in an incinerator such as that shown in Fig. 3.16, giving a useful steam production to offset the cost of disposal.

Liquid and solid wastes

Combustible liquid and solid waste can be disposed of by burning, which is usually preferred to dumping. Incorporating a steam boiler in the incinerator design will enable an otherwise unproductive, but necessary operation, to save energy. If the combustion products are corrosive, corrosion-resistant materials will be needed, and the flue gases scrubbed to reduce air pollution. An incinerator designed to handle chlorinated and other liquid and solid wastes is shown in Fig. 3.16. This incinerator incorporates a steam boiler and a flue-gas scrubber. The disposal of chlorinated wastes is discussed by Santoleri (1973).

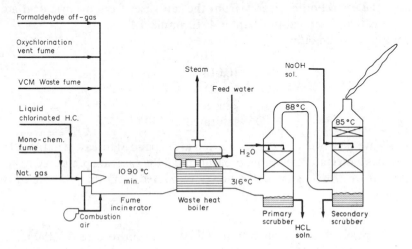

Fig. 3.16. Typical incinerator-heat recovery-scrubber system for vinyl-chloride-monomer process waste
(Courtesy of John Thurley Ltd.)

Dunn and Tomkins (1975) discuss the design and operation of incinerators for process wastes. They give particular attention to the need to comply with the current clean-air legislation, and the problem of corrosion and erosion of refractories and heat-exchange surfaces.

3.16.6. *High-pressure process streams*

Where high-pressure gas or liquid process streams are throttled to lower pressures, energy can be recovered by carrying out the expansion in a suitable turbine.

Gas streams

The economic operation of processes which involve the compression and expansion of large quantities of gases, such as ammonia synthesis, nitric acid production and air separation, depends on the efficient recovery of the energy of compression. The energy recovered by expansion is often used to drive the compressors directly; as shown in Fig. 3.14. If the gas contains condensible components it may be advisable to consider heating the gas by heat exchange with a higher temperature process stream before expansion. The gas can then be expanded to a lower pressure without condensation and the power generated increased.

An interesting process incorporating an expansion turbine is described by Barlow (1975) who discusses energy recovery in an organic acids plant (acetic and propionic). In this process a thirteen-stage turbo-expander is used to recover energy from the off-gases. The pressure range is deliberately chosen to reduce the off-gases to a low temperature at the expander outlet ($-60°C$), for use for low-temperature cooling, saving refrigeration.

The energy recoverable from the expansion of a gas can be estimated by assuming polytropic expansion; see Section 3.13.2 and Example 3.15.

Example 3.15

Consider the extraction of energy from the tail gases from a nitric acid adsorption tower, such as that described in Chapter 4, Example 4.4.

Gas composition, kmol/h:

$$
\begin{array}{ll}
O_2 & 371\cdot5 \\
N_2 & 10,014\cdot7 \\
NO & 21\cdot9 \\
NO_2 & \text{Trace} \\
H_2O & \text{saturated at } 250°C
\end{array}
$$

If the gases leave the tower at 6 atm, 25°C, and are expanded to, say, 1.5 atm, calculate the turbine exit gas temperatures without preheat, and if the gases are preheated to 400°C with the reactor off-gas. Also, estimate the power recovered from the preheated gases.

Solution

For the purposes of this calculation it will be sufficient to consider the tail gas as all nitrogen, flow 10410 kmol/h.

$$P_c = 33\cdot5 \text{ atm}, \ T_c = 126\cdot2 \text{ K}$$

Figure 3.6 can be used to estimate the turbine efficiency.

$$\text{Exit gas volumetric flow-rate} = \frac{10,410}{3600} \times 22\cdot4 \times \frac{1}{1\cdot5}$$

$$\simeq 43 \text{ m}^3/\text{s}$$

from Fig. 3.6 $E_P = 0\cdot75$

$$P_r \text{, inlet} = \frac{6}{33\cdot5} = 0\cdot18$$

$$T_r \text{, inlet} = \frac{298}{126\cdot2} = 2\cdot4$$

For these values the simplified equations can be used, equations 3.37a and 3.38a.
For N_2 $\gamma = 1\cdot4$

$$m = \frac{1\cdot4 - 1}{1\cdot4} \times 0\cdot75 = 0\cdot21$$

$$n = \frac{1}{1 - m} = \frac{1}{1 - 0\cdot21} = 1\cdot27$$

without preheat $T_2 = 298\left(\dfrac{1\cdot5}{6\cdot0}\right)^{0\cdot21} = 223\ K$

$= \underline{-50°C}$ (acidic water would condense out)

with preheat $T_2 = 673\left(\dfrac{1\cdot5}{6\cdot0}\right)^{0\cdot21} = 503\ K$

$= \underline{230°C}$

From equation 3.31, work done by gases as a result of polytropic expansion

$$= -1 \times 673 \times 8\cdot314 \times \frac{1\cdot27}{1\cdot27-1}\left\{\left(\frac{1\cdot5}{6\cdot0}\right)^{\frac{1\cdot27-1}{1\cdot27}} - 1\right\}$$

$$= 6718\ kJ/kmol$$

Actual work = polytropic work $\times E_p$

$$= 6718 \times 0\cdot75 = \underline{5039\ kJ/kmol}$$

Power output = work/kmol $\times$ kmol/s $= 5039 \times \dfrac{10{,}410}{3600}$

$$= 14{,}571\ kJ/s = \underline{14\cdot6\ MW}$$

Liquid streams

As liquids are essentially incompressible, less energy is stored in a compressed liquid than a gas. However, it is worth considering power recovery from high-pressure liquid streams (> 15 bar) as the equipment required is relatively simple and inexpensive. Centrifugal pumps are used as expanders and are often coupled directly to pumps. The design, operation and cost of energy recovery from high-pressure liquid streams is discussed fully in Perry and Chilton (1973) and by Jenett (1968).

3.17. References

BARLOW, J. A. (1975) Inst. Mech. Eng. Conference on Energy Recovery in the Process Industries, London. Energy recovery in a petro-chemical plant: advantages and disadvantages.

BOLAND, D. and LINNHOFF, B. (1979) *Chem. Engr, London* No. 343 (April) 222. The preliminary design of networks for heat exchangers by systematic methods.

DRYDEN, I. (Ed.) (1975) *The Efficient Use of Energy* (IPC Science and Technology Press).

DUNN, K. S. and TOMKINS, A. G. (1975) *Inst. Mech. Eng. Conference on Energy Recovery in the Process Industries,* London. Waste heat recovery from the incineration of process wastes.

EDMISTER, W. C. (1948) *Pet. Ref.* **27** (Nov.) 129 (609). Applications of thermodynamics to hydrocarbon processing, part XIII—heat capacities.

HINCHLEY, P. (1975) *Inst. Mech. Eng. Conference on Energy Recovery in the Process Industries,* London. Waste heat boilers in the chemical industry.

JENETT, E. (1968) *Chem. Eng., Albany* **75** (April 8th) 159, (June 17th) 257 (in two parts). Hydraulic power recovery systems.

MASSO, A. H. and RUDD, D. F. (1969) *AIChEJl* **15**, 10. The synthesis of system design: heuristic structures.

MILES, F. D. (1961) *Nitric Acid Manufacture and Uses* (Oxford U.P.)

MILLER, R. (1968) *Chem. Eng., Albany* **75** (May 20th) 130. Process energy systems.

NISHIDA, N., LIU, Y. A. and LAPIDUS, L. (1977) *AIChEJl* **23**, 77. Studies in chemical process design and synthesis.

PERRY, R. H. and CHILTON, C. H. (Eds.) (1973) *Chemical Engineers Handbook*, 5th ed. (McGraw-Hill).

PORTON, J. W. and DONALDSON, R. A. B. (1974) *Chem. Eng. Sci.* **29**, 2375. A fast method for the synthesis of optimal heat exchanger networks.

SANTOLERI, J. J. (1973) *Chem. Eng. Prog.* **69** (Jan.) 69. Chlorinated hydrocarbon waste disposal and recovery systems.

SIIROLA, J. J. (1974) AIChE 76th National Meeting, Tulsa, Oklahoma. Studies of heat exchanger network synthesis.

STERBACEK, Z., BISKUP, B. and TAUSK, P. (1979) *Calculation of Properties Using Corresponding-state Methods* (Elsevier).

SHULTZ, J. M. (1962) *Trans. ASME* **84** (*Journal of Engineering for Power*) (Jan.) 69, (April) 222 (in two parts). The polytropic analysis of centrifugal compressors.

3.18. Nomenclature

		Dimensions in MLTθ
a	Constant in specific heat equation (equation 3.13)	$L^2T^{-2}\theta^{-1}$
b	Constant in specific heat equation (equation 3.13)	$L^2T^{-2}\theta^{-2}$
C_p	Specific heat at constant pressure	$L^2T^{-2}\theta^{-1}$
C_{p_a}	Specific heat component a	$L^2T^{-2}\theta^{-1}$
C_{p_b}	Specific heat component b	$L^2T^{-2}\theta^{-1}$
C_{p_c}	Specific heat component c	$L^2T^{-2}\theta^{-1}$
C_{p_m}	Mean specific heat	$L^2T^{-2}\theta^{-1}$
C_{p_1}	Specific heat first phase	$L^2T^{-2}\theta^{-1}$
C_{p_2}	Specific heat second phase	$L^2T^{-2}\theta^{-1}$
C_v	Specific heat at constant volume	$L^2T^{-2}\theta^{-1}$
C_p°	Ideal gas state specific heat	$L^2T^{-2}\theta^{-1}$
c	Constant in specific heat equation (equation 3.13)	$L^2T^{-2}\theta^{-3}$ or $L^2T^{-2}\theta^{-1/2}$
E_e	Efficiency, electric motors	—
E_p	Polytropic efficiency, compressors and turbines	—
F	Force	MLT^{-2}
g	Gravitational acceleration	LT^{-2}
H	Enthalpy	ML^2T^{-2}
H_a	Specific enthalpy of component a	L^2T^{-2}
H_b	Specific enthalpy of component b	L^2T^{-2}
H_d	Enthalpy top product stream (Example 3.1)	ML^2T^{-3}
H_f	Enthalpy feed stream (Example 3.1)	ML^2T^{-3}
H_T	Specific enthalpy at temperature T	L^2T^{-2}
H_w	Enthalpy bottom product stream (Example 3.1)	ML^2T^{-3}
ΔH	Change in enthalpy	ML^2T^{-2}
$-\Delta H_{m,t}$	Heat of mixing at temperature t	L^2T^{-2}
$-\Delta H_{r,t}$	Heat of reaction at temperature t	L^2T^{-2}
$-\Delta H_c^\circ$	Standard heat of combustion	L^2T^{-2}
ΔH_f°	Standard enthalpy of formation	L^2T^{-2}
$-\Delta H_m^\circ$	Standard heat of mixing	L^2T^{-2}
$-\Delta H_r^\circ$	Standard heat of reaction	L^2T^{-2}
L	Number of auxiliary streams, heat exchanger networks	—
l	Distance	L
M	Number of hot streams, heat-exchanger networks	—
M	Molecular mass (weight)	—
m	Polytropic temperature exponent	—
N	Number of cold streams, heat-exchanger networks	—
n	Expansion or compression index (equation 3.30)	—
P	Pressure	$ML^{-1}T^{-2}$
P_r	Reduced pressure	—
Q	Heat transferred across system boundary	ML^2T^{-2} or ML^2T^{-3}
Q_b	Reboiler heat load (Example 3.1)	ML^2T^{-3}
Q_c	Condenser heat load (Example 3.1)	ML^2T^{-3}
Q_p	Heat added (or subtracted) from a system	ML^2T^{-2} or ML^2T^{-3}

Q_r	Heat from reaction	ML^2T^{-2} or ML^2T^{-3}
Q_s	Heat generated in the system	ML^2T^{-2} or ML^2T^{-3}
R	Universal gas constant	$L^2T^{-2}\theta^{-1}$
S_{cj}	Cold streams, heat-exchanger networks	—
S_{hi}	Hot streams, heat-exchanger networks	—
S_{uk}	Auxiliary streams, heat-exchanger networks	—
T	Temperature, absolute	θ
T_d	Datum temperature for enthalpy calculations	θ
T_p	Phase-transition temperature	θ
T_r	Reduced temperature	—
t	Temperature, relative scale	θ
t	Time	**T**
t_r	Reference temperature, mean specific heat	θ
t_f	Inlet-stream temperatures, heat-exchanger networks	θ
t_o	Outlet-stream temperatures, heat-exchanger networks	θ
U	Internal energy per unit mass	L^2T^{-2}
u	Velocity	LT^{-1}
V_1	Initial volume	L^3
V_2	Final volume	L^3
v	Volume per unit mass	$M^{-1}L^3$
X	Compressibility function defined by equation 3.33	—
x	Distance	**L**
x_a	Mol fraction component a in a mixture	—
x_b	Mol fraction component b in a mixture	—
x_c	Mol fraction component c in a mixture	—
Y	Compressibility function defined by equation 3.34	—
W	Work per unit mass	L^2T^{-2}
W_i	Heat capacity of streams in a heat-exchanger network	$ML^2T^{-3}\theta^{-1}$
Z	Compressibility factor	—
z	Height above datum	**L**

CHAPTER 4

Flow-sheeting

4.1. Introduction

This chapter covers the preparation and presentation of the process flow-sheet. The flow-sheet is the key document in process design. It shows the arrangement of the equipment selected to carry out the process; the stream connections; stream flow-rates and compositions; and the operating conditions. It is a diagrammatic model of the process.

The flow-sheet will be used by the specialist design groups as the basis for their designs. This will include piping, instrumentation, equipment design and plant layout. It will also be used by operating personnel for the preparation of operating manuals and operator training. During plant start-up and subsequent operation, the flow-sheet forms a basis for comparison of operating performance with design.

The flow-sheet is drawn up from material balances made over the complete process and each individual unit. Energy balances are also made to determine the energy flows and the service requirements.

Manual flow-sheeting calculations can be tedious and time consuming when the process is large or complex, and computer-aided flow-sheeting programs are being increasingly used to facilitate this stage of process design. Their use enables the designer to consider different processes, and more alterative processing schemes, in his search for the best process and optimum process conditions. Some of the proprietary flow-sheeting programs available are discussed in this chapter. A simple linear flow-sheeting program is presented in detail and listed in the appendices.

In this chapter the calculation procedures used in flow-sheeting have for convenience been divided into manual calculation procedures and computer-aided procedures.

The next step in process design after the flow-sheet is the preparation of Piping and Instrumentation diagrams (abbreviated to P & I diagrams) often also called the Engineering Flow-sheet or Mechanical Flow-sheet. The P & I diagrams, as the name implies, show the engineering details of the process, and are based on the process flow-sheet. The preparation and presentation of P & I diagrams is discussed in Chapter 5.

4.2. Flow-sheet presentation

As the process flow-sheet is the definitive document on the process, the presentation must be clear, comprehensive, accurate and complete. The various types of flow-sheet are discussed below.

4.2.1. Block diagrams

A block diagram is the simplest form of presentation. Each block can represent a single piece of equipment or a complete stage in the process. Block diagrams were used to illustrate the examples in Chapters 2 and 3. They are useful for showing simple processes. With complex processes, their use is limited to showing the overall process, broken down into its principal stages; as in Example 2.13 (Vinyl Chloride). In that example each block represented the equipment for a complete reaction stage: the reactor, separators and distillation columns.

Block diagrams are useful for representing a process in a simplified form in reports and textbooks, but have only a limited use as engineering documents.

The stream flow-rates and compositions can be shown on the diagram adjacent to the stream lines, when only a small amount of information is to be shown, or tabulated separately.

The blocks can be of any shape, but it is usually convenient to use a mixture of squares and circles, drawn with a template.

4.2.2. Pictorial representation

On the detailed flow-sheets used for design and operation, the equipment is normally drawn in a stylised pictorial form. For tender documents or company brochures, actual scale drawings of the equipment are sometimes used, but it is more usual to use a simplified representation. The symbols given in British Standard, BS 1553 (1977) "Graphical Symbols for General Engineering" Part 1, 'Piping Systems and Plant' are recommended; though most design offices use their own standard symbols. A selection of symbols from BS 1553 is given in Appendix A. The American National Standards Institute (ANSI) has also published a set of symbols for use on flow-sheets. Austin (1979) has compared the British Standard, ANSI, and some proprietary flow-sheet symbols.

4.2.3. Presentation of stream flow-rates

The data on the flow-rate of each individual component, on the total stream flow-rate, and the percentage composition, can be shown on the flow-sheet in various ways. The simplest method, suitable for simple processes with few equipment pieces, is to tabulate the data in blocks alongside the process stream lines, as shown in Fig. 4.1. Only a limited amount of information can be shown in this way, and it is difficult to make neat alterations or to add additional data.

A better method for the presentation of data on flow-sheets is shown in Fig. 4.2. In this method each stream line is numbered and the data tabulated at the bottom of the sheet. Alterations and additions can be easily made. This is the method generally used by professional design offices. A typical commercial flow-sheet is shown in Fig. 4.3. Guide rules for the layout of this type of flow-sheet presentation are given in Section 4.2.5.

4.2.4. Information to be included

The amount of information shown on a flow-sheet will depend on the custom and practice of the particular design office. The list given below has therefore been divided into essential items and optional items. The essential items must always be shown, the optional items add to the usefulness of the flow-sheet but are not always included.

Essential information

1. Stream composition, either:
   (i) the flow-rate of each individual component, kg/h, which is preferred, or
   (ii) the stream composition as a weight fraction.
2. Total stream flow-rate, kg/h.
3. Stream temperature, degrees Celsius preferred.
4. Nominal operating pressure (the required operating pressure).

Optional information

1. Molar percentages composition.
2. Physical property data, mean values for the stream, such as:
   (i) density, kg/m^3,
   (ii) viscosity, mN s/m^2.
3. Stream name, a brief, one or two-word, description of the nature of the stream, for example "ACETONE COLUMN BOTTOMS".
4. Stream enthalpy, kJ/h.

The stream physical properties are best estimated by the process engineer responsible for the flow-sheet. If they are then shown on the flow-sheet, they are available for use by the specialist design groups responsible for the subsequent detailed design. It is best that each group use the same estimates, rather than each decide its own values.

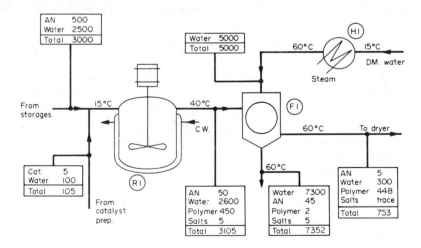

FIG. 4.1. Flow-sheet: polymer production

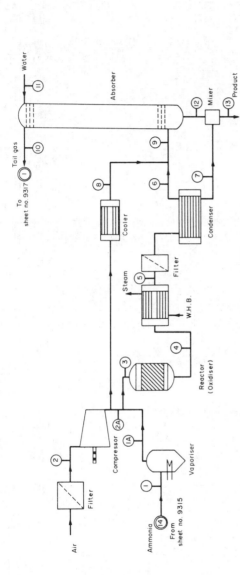

Flows kg/h   Pressures nominal

Line no. Stream	1 Ammonia feed	1A Ammonia vapour	2 Filtered air	2A Oxidiser air	3 Oxidiser feed	4 Oxidiser outlet	5 W.H.B. outlet	6 Condenser gas	7 Condenser acid	8 Secondary air	9 Absorber feed	10 Tail (2) gas	11 Water feed	12 Absorber acid	13 Product acid
Component															
NH_3	731.0	731.0	—	—	731.0	Nil	(—)[1]	—	Trace	408.7	683.9	371.5	—	Trace	Trace
O_2	—	—	3036.9	2628.2	2628.2	935.7	(935.7)[1]	275.2	Trace	1346.1	10,014.7	10,014.7	—	Trace	Trace
N_2	—	—	9990.8	8644.7	8644.7	8668.8	8668.8	8668.8	—	—	202.5	21.9	—	Trace	Trace
NO	—	—	—	—	—	1238.4	(1238.4)[1]	202.5	—	—	967.2	(Trace)[1]	—	Trace	Trace
NO_2	—	—	—	—	—	Trace	(?)[1]	967.2	850.6	—	29.4	26.3	—	17040	2554.6
HNO_3	—	—	—	—	—	Nil	Nil	—	850.6	—	—	—	—	1704.0	2554.6
H_2O	—	—	Trace	—	—	1161.0	1161.0	29.4	1010.1	—	29.4	26.3	1376.9	1136.0	2146.0
Total	731.0	731.0	13,027.7	11,272.9	12,003.9	12,003.9	12,003.9	10,143.1	1860.7	1754.8	11,897.7	10,434.4	1376.9	2840.0	4700.6
Press bar	8	8	1	8	8	8	8	8	1	8	8	1	8	1	1
Temp. °C	15	20	15	230	204	907	234	40	40	40	40	25	25	40	43

C & R Construction Inc

Nitric acid 60 per cent
100,000 t/y
Client BOP Chemicals
SLIGO
Sheet no. 9316

Dwg by	Date
Checked	25/7/1980

Fig. 4.2. Flow-sheet: simplified nitric acid process (Example 4.2) (1) **See example**

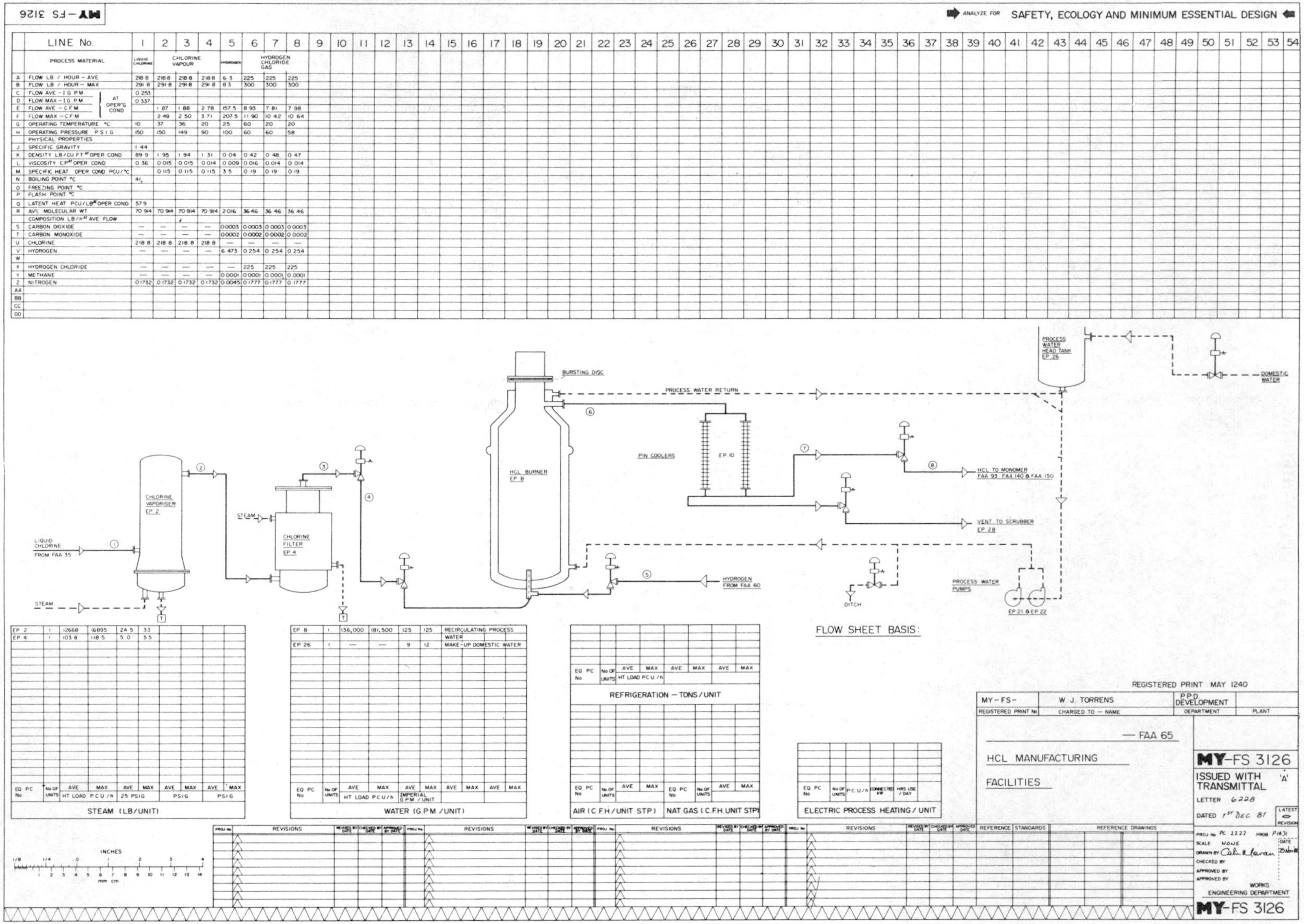

FIG. 4.3. A typical flow-sheet

4.2.5. Layout

The sequence of the main equipment items shown symbolically on the flow-sheet follows that of the proposed plant layout. Some licence must be exercised in the placing of ancillary items, such as heat exchangers and pumps, or the layout will be too congested. But the aim should be to show the flow of material from stage to stage as it will occur, and to give a general impression of the layout of the actual process plant.

The equipment should be drawn approximately to scale. Again, some licence is allowed for the sake of clarity, but the principal equipment items should be drawn roughly in the correct proportion. Ancillary items can be drawn out of proportion. For a complex process, with many process units, several sheets may be needed, and the continuation of the process streams from one sheet to another must be clearly shown. One method of indicating a line continuation is shown in Fig. 4.2; those lines which are continued over to another are indicated by a double concentric circle round the line number and the continuation sheet number written below.

The table of stream flows and other data can be placed above or below the equipment layout. Normal practice is to place it below. The components should be listed down the left-hand side of the table, as in Fig. 4.2. For a long table it is good practice to repeat the list at the right-hand side, so the components can be traced across from either side.

The stream line numbers should follow consecutively from left to right of the layout, as far as is practicable; so that when reading the flow-sheet it is easy to locate a particular line and the associated column containing the data.

All the process stream lines shown on the flow-sheet should be numbered and the data for the stream given. There is always a temptation to leave out the data on a process stream if it is clearly just formed by the addition of two other streams, as at a junction, or if the composition is unchanged when flowing through a process unit, such as a heat exchanger; this should be avoided. What may be clear to the process designer is not necessarily clear to the others who will use the flow-sheet. Complete, unambiguous, information on all streams should be given, even if this involves some repetition. The purpose of the flow-sheet is to show the function of each process unit; even to show when it has no function.

4.2.6. Precision of data

The total stream and individual component flows do not normally need to be shown to a high precision on the process flow-sheet; at most one decimal place is all that is usually justified by the accuracy of the flow-sheet calculations, and is sufficient. The flows should, however, balance to within the precision shown. If a stream or component flow is so small that it is less than the precision used for the larger flows, it can be shown to a greater number of places, if its accuracy justifies this and the information is required. Imprecise small flows are best shown as "TRACE". If the composition of a trace component is specified as a process constraint, as, say, for an effluent stream or product quality specification, it can be shown in parts per million, ppm.

A trace quantity should not be shown as zero, or the space in the tabulation left blank, unless the process designer *is sure* that it has no significance. Trace quantities can be important. Only a trace of an impurity is needed to poison a catalyst, and trace quantities can determine the selection of the materials of construction; see Chapter 7. If the space in the data table is left blank opposite a particular component the quantity may be assumed to be zero by the specialist design groups who take their information from the flow-sheet.

4.2.7. Basis of the calculation

It is good practice to show on the flow-sheet the basis used for the flow-sheet calculations. This would include: the operating hours per year; the reaction and physical yields; and the datum temperature used for energy balances. It is also helpful to include a list of the principal assumptions used in the calculations. This alerts the user to any limitations that may have to be placed on the flow-sheet information.

4.2.8. Batch processes

Flow-sheets drawn up for batch processes normally show the quantities required to produce one batch. If a batch process forms part of an otherwise continuous process, it can be shown on the same flow-sheet, providing a clear break is made when tabulating the data between the continuous and batch sections; the change from kg/h to kg/batch. A continuous process may include batch make-up of minor reagents, such as the catalyst for a polymerisation process.

4.2.9. Services (utilities)

To avoid cluttering up the flow-sheet, it is not normal practice to show the service headers and lines on the process flow-sheet. The service connections required on each piece of equipment should be shown and labelled. The service requirements for each piece of equipment can be tabulated on the flow-sheet.

4.2.10. Equipment identification

Each piece of equipment shown on the flow-sheet must be identified with a code number and name. The identification number (usually a letter and some digits) will normally be that assigned to a particular piece of equipment as part of the general project control procedures, and will be used to identify it in all the project documents.

If the flow-sheet is not part of the documentation for a project, then a simple, but consistent, identification code should be devised. The easiest code is to use an initial letter to identify the type of equipment, followed by digits to identify the particular piece. For example, H – heat exchangers, C – columns, R – reactors. The key to the code should be shown on the flow-sheet.

4.3. Manual flow-sheet calculations

This section is a general discussion of the techniques used for the preparation of flow-sheets from manual calculations. The stream flows and compositions are calculated from material balances; combined with the design equations that arise from the process and equipment design constraints.

As discussed in Chapter 1, there will be two kinds of design constraints:

External constraints: not directly under the control of the designer, and which cannot normally be relaxed. Examples of this kind of constraint are:

(i) Product specifications, possibly set by customer requirements.
(ii) Major safety considerations, such as flammability limits.
(iii) Effluent specifications, set by government agencies.

Internal constraints: determined by the nature of the process and the equipment functions. These would include:

(i) The process stoichiometry, reactor conversions and yields.
(ii) Chemical equilibria.
(iii) Physical equilibria, involved in liquid–liquid and gas/vapour–liquid separations.
(iv) Azeotropes and other fixed compositions.
(v) Energy-balance constraints. Where the energy and material balance interact, as for example in flash distillation.
(vi) Any general limitations on equipment design.

The flow-sheet is usually drawn up at an early stage in the development of the project. A preliminary flow-sheet will help clarify the designer's concept of the process; and serve as basis for discussions with other members of the design team.

The extent to which the flow-sheet can be drawn up before any work is done on the detailed design of the equipment will depend on the complexity of the process and the information available. If the design is largely a duplication of an existing process, though possibly for a different capacity, the equipment performance will be known and the stream flows and compositions can be readily calculated. For new processes, and for major modifications of existing processes, it will only be possible to calculate some of the flows independently of the equipment design considerations; other stream flows and compositions will be dependent on the equipment design and performance. To draw up the flow-sheet the designer must use his judgement in deciding which flows can be calculated directly; which are only weakly dependent on the equipment design; and which are determined by the equipment design.

By weakly dependent is meant those streams associated with equipment whose performance can be assumed, or approximated, without introducing significant errors in the flow-sheet. The detailed design of these items can be carried out later, to match the performance then specified by the flow-sheet. These will be items which in the designer's estimation do not introduce any serious cost penalty if not designed for their optimum performance. For example, in a phase separator, such as a decanter, if equilibrium between the phases is assumed the outlet stream compositions can be often calculated directly, independent of the separator design. The separator would be designed later, to give sufficient residence time for the streams to approach the equilibrium condition assumed in the flow-sheet calculation.

Strong interaction will occur where the stream flows and compositions are principally determined by the equipment design and performance. For example, the optimum conversion in a reactor system with recycle of the unreacted reagents will be determined by the performance of the separation stage, and reactor material balance cannot be made without considering the design of the separation equipment. To determine the stream flows and compositions it would be necessary to set up a mathematical model of the reactor–separator system, including costing.

To handle the manual calculations arising from complex processes, with strong interactions between the material balance calculations and the equipment design, and where physical recycle streams are present, it will be necessary to sub-divide the process into manageable sub-systems. With judgement, the designer can isolate those systems with strong interactions, or recycle, and calculate the flows sequentially, from sub-system to sub-system, making approximations as and where required. Each sub-system can be considered separately, if necessary, and the calculations repeatedly revised till a satisfactory flow-sheet for the complete process is obtained. To attempt to model a complex process without subdivision and approximation would involve too many variables and design equations to be handled manually. Computer flow-sheeting programs should be used if available.

When sub-dividing the process and approximating equipment performance to produce a flow-sheet, the designer must appreciate that the resulting design for the complete process, as defined by the flow-sheet, will be an approximation to the optimum design. He must continually be aware of, and check, the effect of his approximations on the performance of the complete process.

4.3.1. Basis for the flow-sheet calculations

Time basis

No plant will operate continuously without shut-down. Planned shut-down periods will be necessary for maintenance, inspection, equipment cleaning, and the renewal of catalysts and column packing. The frequency of shut-downs, and the consequent loss of production time, will depend on the nature of the process. For most chemical and petrochemical processes the plant attainment will typically be between 90 to 95 per cent of the total hours in a year (8760). Unless the process is known to require longer shut-down periods, a value of 8000 hours per year can be used for flow-sheet preparation.

Scaling factor

It is usually easiest to carry out the sequence of flow-sheet calculations in the same order as the process steps; starting with the raw-material feeds and progressing stage by stage, where possible, through the process to the final product flow. The required production rate will usually be specified in terms of the product, not the raw-material feeds, so it will be necessary to select an arbitrary basis for the calculations, say 100 kmol/h of the principal raw material. The actual flows required can then be calculated by multiplying each flow by a scaling factor determined from the actual production rate required.

$$\text{Scaling factor} = \frac{\text{mols product per hour specified}}{\text{mols product produced per 100 kmol}\atop\text{of the principal raw material}}$$

4.3.2. Flow-sheet calculations on individual units

Some examples of how design constraints can be used to determine stream flows and compositions are given below.

1. Reactors

(i) Reactor yield and conversion specified.

The reactor performance may be specified independently of the detailed design of the reactor. The conditions for the optimum, or near optimum, performance may be known from the operation of existing plant or from pilot plant studies.

For processes that are well established, estimates of the reactor performance can often be obtained from the general and patent literature; for example, the production of nitric and sulphuric acids.

If the yields and conversions are known, the stream flows and compositions can be calculated from a material balance; see Example 2.13.

(ii) Chemical equilibrium.

With fast reactions, the reaction products can often be assumed to have reached equilibrium. The product compositions can then be calculated from the equilibrium data for the reaction, at the chosen reactor temperature and pressure; see Example 4.1.

2. Equilibrium stage

In a separation or mixing unit, the anticipated equipment performance may be such that it is reasonable to consider the outlet streams as being in equilibrium; the approach to equilibrium being in practice close enough that no significant inaccuracy is introduced by assuming that equilibrium is reached. The stream compositions can then be calculated from the phase equilibrium data for the components. This approximation can often be made for single-stage gas–liquid and liquid–liquid separators, such as quench towers, partial condensers and decanters. It is particularly useful if one component is essentially non-condensable and can be used as a tie substance (see Section 2.11). Some examples of the use of this process constraint are given in Examples 4.2 and 4.4.

3. Fixed stream compositions

If the composition (or flow-rate) of one stream is fixed by "internal" or "external" constraints, this may fix the composition and flows of other process streams. In Chapter 1, the relationship between the process variables, the design variables and design equations was discussed. If sufficient design variables are fixed by external constraints, or by the designer, then the other stream flow round a unit will be uniquely determined. For example, if the composition of one product stream from a distillation column is fixed by a product specification, or if an azeotrope is formed, then the other stream composition can be calculated directly from the feed compositions; see section 2.10. The feed composition would be fixed by the outlet composition of the preceding unit.

4. Combined heat and material balances

It is often possible to make a material balance round a unit independently of the heat balance. The process temperatures may be set by other process considerations, and the energy balance can then be made separately to determine the energy requirements to maintain the specified temperatures. For other processes the energy input will determine the process stream flows and compositions, and the two balances must be made simultaneously; for instance, in flash distillation or partial condensation; see also Example 4.1.

Example 4.1

An example illustrating the calculation of stream composition from reaction equilibria, and also an example of a combined heat and material balance.

In the production of hydrogen by the stream reforming of hydrocarbons, the classic water–gas reaction is used to convert CO in the gases leaving the reforming furnace to hydrogen, in a shift convertor.

$$CO(g) + H_2O(g) \rightarrow CO_2(g) + H_2(g)$$
$$\Delta H^\circ_{298} \; -41{,}197 \text{ kJ/kmol}$$

In this example the exit gas stream composition from a convertor will be determined for a given inlet gas composition and stream ratio; by assuming that in the outlet stream the gases reach chemical equilibrium. In practice the reaction is carried out over a catalyst, and the assumption that the outlet composition approaches the equilibrium composition is valid. Equilibrium constants for the reaction are readily available in the literature.

A typical gases composition obtained by steam reforming methane is:

$$CO_2 \; 8\cdot5, \quad\quad CO \; 11\cdot0, \quad\quad H_2 \; 76\cdot5 \text{ mol per cent dry gas}$$

If this is fed to a shift converter at 500°K, with a steam ratio of 3 mol H_2O to 1 mol CO, estimate the outlet composition and temperature.

Solution

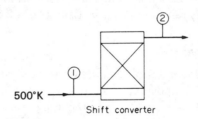

Shift converter

Basis: 100 mol/h dry feed gas.

$$H_2O \text{ in feed stream} = 3\cdot0 \times 11\cdot0 = 33 \text{ mol.}$$

Let fractional conversion of CO to H_2 be C. Then mols of CO reacted $= 11\cdot0 \times C$. From the stoichiometric equation and feed composition, the exit gas composition will be:

$$CO = 11\cdot0(1 - C)$$
$$CO_2 = 8\cdot5 + 11\cdot0 \times C$$
$$H_2O = 33 - 11\cdot0 \times C$$
$$H_2 = 76\cdot5 + 11\cdot0 \times C$$

At equilibrium
$$K_p = \frac{P_{CO} \times P_{H_2O}}{P_{CO_2} \times P_{H_2}}$$

The temperature is high enough for the gases to be considered ideal, so the equilibrium constant is written in terms of partial pressure rather than fugacity, and the constant will not be affected by pressure. Mol fraction can be substituted for partial pressure. As the total mols in and out is constant, the equilibrium relationship can be written directly in mols of the components.

$$K_p = \frac{11(1-C)(33-11C)}{(8\cdot5+11C)(76\cdot5+11C)}$$

Expanding and rearranging

$$(K_p 121 - 121)C^2 + (K_p 935 + 484)C + (K_p 650 - 363) = 0 \tag{1}$$

K_p is a function of temperature.

For illustration, take T out $= 700\,°K$, at which $K_p = 1\cdot11 \times 10^{-1}$

$$-107\cdot6C^2 + 587\cdot8C - 290\cdot85 = 0$$
$$C = 0\cdot57$$

The reaction is exothermic and the operation can be taken as adiabatic, as no cooling is provided and the heat losses will be small.

The gas exit temperature will be function of the conversion. The exit temperature must satisfy the adiabatic heat balance and the equilibrium relationship.

A heat balance was carried over a range of values for the conversion C, using the program Energy 1, Chapter 3. The value for which the program gives zero heat input or output required (adiabatic) is the value that satisfies the conditions above. For a datum temperature of $25\,°C$.

Data for energy-balance program

| | Component | Stream (mol) | | C_p°(kJ/kmol) | | | |
		1	2	a	b	c	d
1	CO_2	8·5	$8\cdot5+11C$	19·80	7·34 E-2	$-5\cdot6$ E-5	17·15 E-9
2	CO	11·0	$11(1-C)$	30·87	$-1\cdot29$ E-2	27·9 E-6	$-12\cdot72$ E-9
3	H_2O	33·0	$33-11C$	32·24	19·24 E-4	10·56 E-6	$-3\cdot60$ E-9
4	H_2	76·5	$76\cdot5+11C$	27·14	9·29 E-3	$-13\cdot81$ E-6	7·65 E-9

Results

| Outlet temp. (K) | K_p | C | Mols converted | Outlet composition, mol | | | | Heat required Q |
				CO	CO_2	H_2O	H_2	
550	$1\cdot86 \times 10^{-2}$	0·88	9·68	1·32	18·18	23·32	86·18	$-175{,}268$
600	$3\cdot69 \times 10^{-2}$	0·79	8·69	2·31	17·19	24·31	85·19	76,462
650	$6\cdot61 \times 10^{-2}$	0·68	7·48	3·52	15·98	25·52	83·98	337,638

The values for the equilibrium constant K_p were taken from *Technical Data on Fuel*, Spiers.

The outlet temperature at which $Q = 0$ was found by plotting temperature versus Q to be 580 K.

At 580 K, $K_p = 2.82 \times 10^{-2}$.
From equation (1)

$$-117.6C^2 + 510.4 + -344.7 = 0,$$

$$C = 0.83$$

Outlet gas composition

$$CO_2 = 8.5 + 11 \times 0.83 = 17.6$$

$$CO = 11(1 - 0.83) = 1.9$$

$$H_2O = 33.0 - 11 \times 0.83 = 23.9$$

$$H_2 = 76.5 + 11 \times 0.83 = \underline{85.6}$$

$$129.0 \text{ mol}$$

In this example the outlet exit gas composition has been calculated for an arbitrarily chosen steam: CO ratio of 3. In practice the calculation would be repeated for different steam ratios, and inlet temperatures, to optimise the design of the converter system. Two converters in series are normally used, with gas cooling between the stages. For large units a waste-heat boiler could be incorporated between the stages. The first stage conversion is normally around 80 per cent.

Example 4.2

This example illustrates the use of phase equilibrium relationships (vapour–liquid) in material balance calculations.

In the production of dichloroethane (EDC) by oxyhydrochlorination of ethylene, the products from the reaction are quenched by direct contact with dilute HCl in a quench tower. The gaseous stream from this quench tower is fed to a condenser and the uncondensed vapours recycled to the reactor. A typical composition for this stream is shown in the diagram below; operating pressure 4 bar. Calculate the outlet stream compositions leaving the condenser.

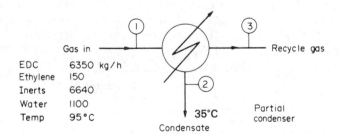

The EDC flow includes some organic impurities and a trace of HCl. The inerts are mainly N_2, CO, O_2—non-condensable.

Solution

In order to calculate the outlet stream composition it is reasonable, for a condenser, to assume that the gas and liquid streams are in equilibrium at the outlet liquid temperature of 35°C.

The vapour pressures of the pure liquids can be calculated from the Antoine equation (see Chapter 8):

At 35°C (308 K)
EDC   0·16 bar
Ethylene   70·7
H_2O 0·055

From the vapour pressures it can be seen that the EDC and water will be essentially totally condensed, and that the ethylene remains as vapour. Ethylene will, however, tend to be dissolved in the condensed EDC. As a first trial, assume all the ethylene stays in the gas phase.

Convert flows to mol/h.

	Mol wt.	kmol/h	
EDC	99	64	
C_2H_4	28	5·4	
Inerts	32 (estimated)	208	} 213·4
H_2O	18	61	

Take the "non-condensables" (ethylene and inerts) as the tie substance. Treat gas phase as ideal, and condensed EDC–water as immiscible.

Partial pressure   = (total pressure) − (vapour pressure of EDC + vapour
of non-condensables   pressure of water)
                     = $4 - 0·16 - 0·055 = \underline{3·79\ \text{bar}}$

Flow of EDC in vapour $= \dfrac{\text{vapour press. EDC}}{\text{partial press. non-condensables}} \times \text{flow non-condensables}$

$$= \frac{0·16}{3·79} \times 213·4 = \underline{\underline{9\ \text{kmol/h}}}$$

Similarly, flow of $= \dfrac{0·055}{3·79} \times 213·4 = \underline{\underline{3·1\ \text{kmol/h}}}$
H_2O in vapour

So composition of gas streams is

	kmol/h	Per cent mol	kg/h
EDC	9	4·0	891
H_2O	3·1	1·4	56
Inerts	208	92·3	6640
C_2H_4	5·4	2·3	150

Check on dissolved ethylene

Partial pressure of ethylene = total pressure × mol fraction

$$= 4 \times \frac{2·3}{100} = 0·092\ \text{bar}$$

By assuming EDC and C_2H_4 form an ideal solution, the mol fraction of ethylene dissolved in the liquid can be estimated, from Raoults Law (see Chapter 8).

$$y_A = \frac{x_A P_A^\circ}{P}$$

y_A = gas phase mol fraction,
x_A = liquid phase mol fraction,
P_A° = sat. vapour pressure,
P = total pressure,

Substituting

$$\frac{2 \cdot 3}{100} = \frac{x_A \, 70 \cdot 7}{4}$$

$$x_A = 1 \cdot 3 \times 10^{-3}$$

hence quantity of ethylene in liquid = kmol EDC $\times x_A$

$$= (64 - 9) \times 1 \cdot 3 \times 10^{-3} = 0 \cdot 07 \text{ kmol/h}$$

so kmol ethylene in gas phase $= 5 \cdot 4 - 0 \cdot 07 = \underline{\underline{5 \cdot 33 \text{ kmol/h}}}$

This is little different from calculated value and shows that initial assumption that no ethylene was condensed or dissolved was reasonable; so report ethylene in liquid as "trace".

Stream no.:	Material balance 1	Flows (kg/h) 2	 3
Title	Condenser feed	Condensate	Recycle gas
EDC	6350	5459	891
H_2O	1100	1044	56
Ethylene	150	Trace	150
Inerts	6640	—	6640
Total	14,240	6503	7737
Temp.°C	95	35	35
Pressure bar:	4	4	4

Example 4.3

This example illustrates the use of liquid–liquid phase equilibria in material balance calculations. The condensate stream from the condenser described in Example 4.2 is fed to a decanter to separate the condensed water and dichloroethane (EDC). Calculate the decanter outlet stream compositions.

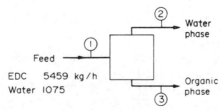

EDC   5459 kg/h
Water 1075

Solution

Assume outlet phases are in equilibrium.
The solubilities of the components at 20°C are:

EDC in water   0·86 kg/100 kg
Water in EDC   0·16 kg/100 kg

Note the water will contain a trace of HCl, but as data on the solubility of EDC in dilute HCl are not available, the solubility in water will be used.

As the concentrations of dissolved water and EDC are small, the best approach to this problem is by successive approximation; rather than by setting up and solving equations for the unknown concentrations.

As a first approximation take organic stream flow = EDC flow in.

Then water in EDC $= \dfrac{0·16}{100} \times 5459 = 8·73 \text{ kg/h}$

So water flow out $= 1075 - 8·73 = 1066·3 \text{ kg/h}$

and EDC dissolved in the water stream $= \dfrac{1066·3}{100} \times 0·86 = 9·2 \text{ kg/h}$

so, revised organic stream flow $= 5459 - 9·2 = \underline{\underline{5449·8 \text{ kg/h}}}$

and quantity of water dissolved in the stream $= \dfrac{5449·8}{100} \times 0·16 = \underline{\underline{8·72 \text{ kg/h}}}$

Which is not significantly lower than the first approximation. So stream flows, kg/h, will be:

Stream no.	1	2	3
Title	Decanter feed	Organic phase	Aqueous phase
EDC	5459	5449·8	9·2
H$_2$O	1075	8·7	1066·3
Total	6534	5458·5	1075·5

Example 4.4

This example illustrates the manual calculation of a material and energy balance for a process involving several processing units.

Draw up a preliminary flow-sheet for the manufacture of 20,000t/y nitric acid (basis 100 per cent HNO$_3$) from anhydrous ammonia, concentration of acid required 50 to 60 per cent.

The technology of nitric acid manufacture is well established and has been reported in several articles:

1. R. M. Stephenson: *Introduction to the Chemical Process Industries* (Reinhold, 1966).
2. C. H. Chilton: *The Manufacture of Nitric Acid by the Oxidation of Ammonia* (American Institute of Chemical Engineers).

3. S. Strelzoff: *Chem. Eng. Albany* **63** (5), 170 (1956).
4. F. D. Miles: *Nitric Acid Manufacture and Uses* (Oxford University Press, 1961).

Three processes are used:

1. Oxidation and absorption at atmospheric pressure.
2. Oxidation and absorption at high pressure (approx. 8 atm).
3. Oxidation at atmospheric pressure and absorption at high pressure.

The relative merits of the three processes are discussed by Chilton (2), and Strelzoff (3).
For the purposes of this example the high-pressure process has been selected. A typical process is shown in the block diagram.

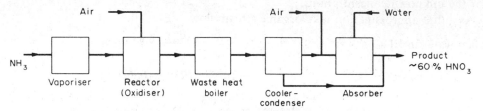

Schematic (block) diagram; production of nitric acid by oxidation of ammonia

The principal reactions in the reactor (oxidiser) are:

Reaction 1. $NH_3(g) + \frac{5}{4}O_2(g) \rightarrow NO(g) + \frac{3}{2}H_2O(g)$
$$\Delta H^\circ_{298} = -226,334 \text{ kJ/kmol}$$

Reaction 2. $NH_3(g) + \frac{3}{4}O_2(g) \rightarrow \frac{1}{2}N_2(g) + \frac{3}{2}H_2O(g)$
$$\Delta H^\circ_{298} = -316,776 \text{ kJ/kmol}$$

The nitric oxide formed can also react with ammonia:

Reaction 3. $NH_3(g) + \frac{3}{2}NO(g) \rightarrow \frac{5}{4}N_2(g) + \frac{3}{2}H_2O(g)$
$$\Delta H^\circ_{298} = -452,435 \text{ kJ/kmol}$$

The oxidation is carried out over layers of platinum–rhodium catalyst; and the reaction conditions are selected to favour reaction 1. Yields for the oxidation step are reported to be 95 to 96 per cent.

Solution

Basis of the flow-sheet calculations

Typical values, taken from the literature cited:

1. 8000 operating hours per year.
2. Overall plant yield on ammonia 94 per cent.
3. Oxidiser (reactor) chemical yield 96 per cent.
4. Acid concentration produced 58 per cent w/w HNO_3.
5. Tail gas composition 0·2 per cent v/v NO.

Material balances
Basis: 100 kmol NH_3 feed to reactor.

Oxidiser

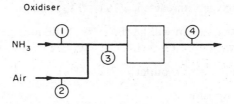

From reaction 1, at 96 per cent yield,

$$\text{NO produced} \quad = 100 \times \frac{96}{100} = 96\,\text{kmol}$$

$$\text{oxygen required} \quad = 96 \times 5/4 \quad = 120\,\text{kmol}$$

$$\text{water produced} \quad = 96 \times 3/2 \quad = 144\,\text{kmol}$$

The remaining 4 per cent ammonia reacts to produce nitrogen; production of 1 mol of N_2 requires 3/2 mol of O_2, by either reaction 2 or 1 and 3 combined.

$$\text{nitrogen produced} = 4/2 = 2\,\text{kmol}$$

$$\text{oxygen required} \quad = 2 \times 3/2 = 3\,\text{kmol}$$

All the oxygen involved in these reactions produces water,

$$\text{water produced} = 3 \times 2 = 6\,\text{kmol}$$

So, total oxygen required and water produced;

$$\text{water} = 144 + 6 = 150\,\text{kmol}$$

$$\text{oxygen (stoichiometric)} = 120 + 3 = 123\,\text{kmol}$$

Excess air is supplied to the oxidiser to keep the ammonia concentration below the explosive limit (see Chapter 9), reported to be 12 to 13 per cent (Chilton), and to provide oxygen for the oxidation of NO to NO_2.

Reaction 4. $NO(g) + \frac{1}{2}O_2 \rightarrow NO_2(g)$ $\Delta H^\circ_{298} = 57{,}120\,\text{kJ/kmol}$

The inlet concentration of ammonia will be taken as 11 per cent v/v.

$$\text{So, air supplied} = \frac{100}{11} \times 100 = 909\,\text{kmol}$$

Composition of air: 79 per cent N_2, 21 per cent O_2, v/v.

So, oxygen and nitrogen flows to oxidiser:

$$\text{oxygen} \quad = 909 \times \frac{21}{100} \quad = 191\,\text{kmol}$$

$$\text{nitrogen} = 909 \times \frac{79}{100} = 718\,\text{kmol}$$

And the oxygen unreacted (oxygen in the outlet stream) will be given by:

$$\text{oxygen unreacted} = 191 - 123 = 68 \text{ kmol}$$

The nitrogen in the outlet stream will be the sum of the nitrogen from the air and that produced from ammonia:

$$\text{nitrogen in outlet} = 718 + 2 = 720 \text{ kmol}$$

Summary, stream compositions:

	Feed (3)		Outlet (4)	
	kmol	kg	kmol	kg
NH_3	100	1700	nil	
NO	nil		96	2880
H_2O	trace		150	2700
O_2	191	6112	68	2176
N_2	718	20,104	720	20,016
Total		27,916		27,916

Notes

(1) The small amount of water in the inlet air is neglected.
(2) Some NO_2 will be present in the outlet gases, but at the oxidiser temperature used, 1100 to 1200 K, the amount will be small, typically < 1 per cent.
(3) It is good practice always to check the balance across a unit by calculating the totals; total flow in must equal total flow out.

Waste-heat boiler (WHB) and cooler-condenser

The temperature of the gases leaving the oxidiser is reduced in a waste-heat boiler and cooler-condenser. There will be no separation of material in the WHB but the composition will change, as NO is oxidised to NO_2 as the temperature falls. The amount oxidised will depend on the residence time and temperature (see Stephenson). The oxidation is essentially complete at the cooler-condenser outlet. The water in the gas condenses in the cooler-condenser to form dilute nitric acid, 40 to 50 per cent w/w.

Balance on cooler-condenser

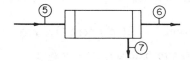

The inlet stream (5) will be taken as having the same composition as the reactor outlet stream (4).

Let the cooler-condenser outlet temperature be 40°C. The maximum temperature of the cooling water will be about 30°C, so this gives a 10°C approach temperature.

If the composition of the acid leaving the unit is taken as 45 per cent w/w (a typical value) the composition of the gas phase can be estimated by assuming that the gas and condensed liquid are in equilibrium at the outlet temperature.

At 40°C the vapour pressure of water over 45 per cent HNO_3 is 29 mm Hg (Perry's *Chemical Engineers Handbook*, 5th ed., pp. 3–65). Take the total pressure as 8 atm. The mol fraction of water in the outlet gas stream will be given by the ratio of the vapour pressure to the total pressure:

$$\text{mol fraction water} = \frac{29}{760 \times 8} = 4.77 \times 10^{-3}$$

As a first trial, assume that all the water in the inlet stream is condensed, then:

$$\text{water condensed} = 150 \text{ kmol} = 2700 \text{ kg}$$

NO_2 combines with this water to produce a 45 per cent solution:

$$\text{Reaction 5.} \quad 3NO_2 + H_2O \rightarrow 2HNO_3 + NO$$

For convenience, take as a subsidiary basis for this calculation 100 kmol of HNO_3 (100 per cent basis) in the condensate.

From reaction 5, the mols of water required to form 100 kmol HNO_3 will be:

$$50 \text{ kmol} \qquad\qquad\qquad = 900 \text{ kg}$$

$$\text{mass of 100 kmol } HNO_3 \qquad = 100 \times 63 = 6300 \text{ kg}$$

$$\text{water to dilute this to 45 per cent} = \frac{6300 \times 55}{45} = 7700 \text{ kg}$$

So, total water to form dilute acid = 900 + 7700 = 8600 kg .

Changing back to the original basis of 100 kmol NH_3 feed:

$$HNO_3 \text{ formed} = 100 \times \frac{\text{Water condensed per 100 kmol } NH_3 \text{ feed}}{\text{Total water to form 45 per cent acid, per 100 kmol } HNO_3}$$

$$= 100 \times \frac{2700}{8600} = 31.4 \text{ kmol}$$

$$NO_2 \text{ consumed (from reaction 5)} = 31.4 \times 3/2 = 47.1 \text{ kmol}$$

$$NO \text{ formed} = 31.4 \times 1/2 = 15.7 \text{ kmol}$$

$$H_2O \text{ reacted} = 15.7 \text{ kmol}$$

Condensed water not reacted with NO_2 = 150 − 15.7 = 134.3 kmol .

The quantity of unoxidised NO in the gases leaving the cooler-condenser will depend on the residence time and the concentration of NO and NO_2 in the inlet stream. For simplicity in this preliminary balance the quantity of NO in the outlet gas will be taken as equal to the quantity formed from the absorption of NO_2 in the condensate to form nitric acid:

$$\text{NO in outlet gas} = 15.7 \text{ kmol}$$

The unreacted oxygen in the outlet stream can be calculated by making a balance over the unit on the nitric oxides, and on oxygen.

Balance on oxides

$$\text{Total } (NO + NO_2) \text{ entering} = NO \text{ in stream } 4 = 96 \text{ kmol}$$

Of this, 31·4 kmol leaves as nitric acid, so $(NO + NO_2)$ left in the gas stream $= 96 - 31·4 = 64·6$ kmol.

Of this, 15·7 kmol is assumed to be NO, so NO_2 in exit gas $= 64·6 - 15·7$
$$= 48·9 \text{ kmol}.$$

Balance on oxygen

Let unreacted O_2 be x kmol. Then oxygen out of the unit will be given by:

$$\left[\frac{NO}{2} + NO_2 + x\right]_{\substack{\text{gas} \\ \text{stream (6)}}} + \left[\frac{3}{2} HNO_3 + \frac{H_2O}{2}\right]_{\substack{\text{acid} \\ \text{stream (7)}}}$$

$$= \left(\frac{15·7}{2} + 48·9 + x\right) + \left(\frac{3}{2} \times 31·4 + \frac{134·3}{2}\right) = (171 + x) \text{ kmol}$$

$$\text{Oxygen into the unit} = \left[\frac{NO}{2} + O_2 + H_2O\right]_{\text{stream (5)}}$$

$$= \frac{96}{2} + 68 + \frac{150}{2} = 191 \text{ kmol}$$

Equating O_2 in and out:

$$\text{unreacted } O_2, x, = 191 - 171 = 20·0 \text{ kmol}$$

As a first trial, all the water vapour was assumed to condense; this assumption will now be checked.

The quantity of water in the gas stream will be given by:

$$\text{mol fraction} \times \text{total flow.}$$

The total flow of gas (neglecting water) $= 804·6$ kmol, and the mol fraction of water was estimated to be $4·77 \times 10^{-3}$.

$$\text{So, water vapour} = 4·77 \times 10^{-3} \times 804·6 = 3·8 \text{ kmol}$$

And, mols of water condensed $= 134·3 - 3·8 = 130·5$ kmol.

The calculations could be repeated using this adjusted value for the quantity of water condensed, to get a better approximation, but the change in the acid, nitric oxides, oxygen and water flows will be small. So, the only change that will be made to the original estimates will be to reduce the quantity of condensed water by that estimated to be in the gas stream:

$$\text{Water in stream (6) } 3·8 \text{ kmol} = 68·4 \text{ kg}$$

So, water in stream (7) $= 134·3 - 3·8 = 130·5$ kmol $= 2349$ kg.

Summary, stream compositions:

	Gas (6)		Acid (7)	
	kmol	kg	kmol	kg
NO	15·7	471·0	Trace	
NO_2	48·9	2249·4	Trace	
O_2	20·0	640	—	
N_2	720	20,160		
HNO_3	—	—	31·4	1978·2
H_2O	3·8	68·4	130·5	2349·0
Total		23,588·4		4327·2

Total, stream $(6)+(7) = 23,588\cdot4 + 4327\cdot2 = 27,915\cdot6$ kg, checks with inlet stream (4) total of 27,915.

Absorber

In the absorber the NO_2 in the gas stream is absorbed in water to produce acid of about 60 per cent w/w. Sufficient oxygen must be present in the inlet gases to oxidise the NO formed to NO_2. The rate of oxidation will be dependent on the concentration of oxygen, so an excess is used. For satisfactory operation the tail gases from absorber should contain about 3 per cent O_2 (Miles).

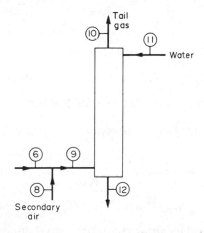

From stream (6) composition:

NO in inlet stream to absorber $= 15\cdot7$ kmol and $O_2 = 20\cdot0$ kmol

Note: Though the NO/NO_2 ratio in this stream is not known exactly, this will not affect the calculation of the oxygen required; the oxygen is present in the stream either as free, uncombined, oxygen or combined in the NO_2.

So, O_2 required to oxidise the NO in the inlet to stream to NO_2, from reaction 4,

$$= \frac{15\cdot7}{2} = 7\cdot85 \text{ kmol .}$$

Hence, the "free" oxygen in the inlet stream $= 20\cdot0 - 7\cdot85 = 12\cdot15$ kmol .

Combining reactions (4) and (5) gives the overall reaction for the absorption of NO_2 to produce HNO_3.

Reaction 6. $4NO_2 + 2H_2O + O_2 \rightarrow 4HNO_3$

Using this reaction, the oxygen required to oxidise the NO formed in the absorber can be calculated:

$$O_2 \text{ required to oxidise NO formed} = \{(NO + NO_2) \text{ in stream (6)}\} \times \tfrac{1}{4}$$
$$= (48.9 + 15.7) \times \tfrac{1}{4} = 16.15 \text{ kmol}$$

So O_2 required for complete oxidation, in addition to that in inlet gas

$$= 16.15 - 12.15 = 4 \text{ kmol}$$

Let the secondary air flow be y kmol. Then the O_2 in the secondary air will be $= 0.21\,y$ kmol. Of this, 4 kmol react with NO in the absorber, so the free O_2 in the tail gases will be $= 0.21\,y - 4$ kmol.

N_2 passes through the absorber unchanged, so the N_2 in the tail gases $=$ the N_2 entering the absorber from the cooler-condenser and the secondary air. Hence:

$$N_2 \text{ in tail gas} = 720 + 0.79\,y \text{ kmol.}$$

The tail gases are essentially all N_2 and O_2 (the quantity of other constituents is negligible) so the percentage O_2 in the tail gas will be given by:

$$O_2 \text{ per cent} = 3 = \frac{(0.21\,y - 4)100}{(720 + 0.79\,y) + (0.21\,y - 4)}$$

from which

$$y = 141.6 \text{ kmol}$$

and the O_2 in the tail gases $= 141.6 \times 0.21 - 4 = 25.7 \text{ kmol}$

and the N_2 in the tail gases $= 720 + 111.8 = 831.8 \text{ kmol}$.

Tail gas composition, the tail gases will contain from 0.2 to 0.3 per cent NO, say 0.2 per cent, then:

$$NO \text{ in tail gas} = \text{total flow} \times \frac{0.2}{100} = (N_2 + O_2)\text{flow} \times 0.002$$

$$= (831.8 + 25.7)0.002 = \underline{\underline{1.7 \text{ kmol}}}$$

The quantity of the secondary air was based on the assumption that all the nitric oxides were absorbed. This figure will not be changed as it was calculated from an assumed (approximate) value for the concentration of the O_2 in the tail gases. The figure for O_2 in the tail gases must, however, be adjusted to maintain the balance.

The unreacted O_2 can be calculated from Reactions (4) and (6). 1.7 kmol of NO are not oxidised or absorbed, so the adjusted O_2 in tail gases $= 25.7 + 1.7(\tfrac{1}{4} + \tfrac{1}{2}) = \underline{\underline{27.0 \text{ kmol}}}$.

The tail gases will be saturated with water at the inlet water temperature, say 25°C. Partial pressure of water at 25°C $= 0.032$ atm. The absorber pressure will be approximately 8 atm, so mol fraction water $= 0.032/8 = 4 \times 10^{-3}$ and H_2O in tail gas $= 857.5 \times 4 \times 10^{-3} = \underline{\underline{3.4 \text{ kmol.}}}$

Water required, stream (11).

The nitrogen oxides absorbed, allowing for the NO in the tail gases, will equal the HNO_3 formed

$$= (48·9 + 15·7) - 1·7 = 62·9 \text{ kmol} = 3962·7 \text{ kg}$$

Stoichiometric H_2O required, from reaction 6

$$= \frac{62·9}{4} \times 2 = 31·5 \text{ kmol}$$

The acid strength leaving the absorber will be taken as 60 per cent w/w. Then, water required for dilution

$$= \frac{3962·7}{0·6} \times 0·4 = 2641·8 \text{ kg} = 146·8 \text{ kmol}$$

So, total water required, allowing for the water vapour in the inlet stream (6), but neglecting the small amount in the secondary air

$$= 31·5 + 146·8 + 3·4 - 3·8 = 177·9 \text{ kmol}$$

Summary, stream compositions:

Stream	Secondary air (8)		Inlet (9)		Acid (12)		Tail gas (10)		Water feed (11)	
	kmol	kg	kmol	kg	kmol	kg	kmol	kg	kmol	kg
NO	—	—	15·7	471·0	—	—	1·7	51·0	—	—
NO_2	—	—	48·9	2249·4	trace	—	—	—	—	—
O_2	29·7	950·4	49·7	1590·4	—	—	27·0	864	—	—
N_2	111·8	3130·4	831·8	23,290·0	—	—	831·8	23,290·4	—	—
HNO_3	—	—	—	—	62·9	3962·7	—	—	—	—
H_2O	trace	—	3·8	68·4	146·8	2641·8	3·4	61·2	177·9	3202·2
Total		4080·8		27,669·2		6604·5		24,266·6		3202·6

Check on totals: Stream (6) + (8) = (9)? 4080·8 + 23,588·4 = 27,669·2
27,669·2 = 27,669·2 checks

Stream (9) + (11) = (10) + (12)? 27,669·2 + 3203·2 = 24,266·6 + 6604·5
30,871·4 = 30,871·1 near enough.

Acid produced

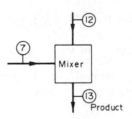

From cooler-condenser	HNO_3	=	31·4 kmol	=	1978·2 kg
	H_2O	=	130·5 kmol	=	2349·0 kg
From absorber	HNO_3	=	62·9 kmol	=	3962·7 kg
	H_2O	=	146·8 kmol	=	2641·8 kg
Totals	HNO_3	=	1978·2 + 3962·7	=	5940·9 kg
	H_2O	=	2349·0 + 2641·8	=	4990·8 kg
					10,931·7 kg

So, concentration of mixed acids $= \dfrac{5940·9}{10,931·7} \times 100 = 54$ per cent.

Summary, stream composition:

	Acid product (13)	
Stream	kmol	kg
HNO_3	94·3	5940·3
H_2O	277·3	4990·8
		10,931·7

Overall plant yield

The overall yield can be calculated by making a balance on the combined nitrogen:

$$\text{Yield} = \frac{\text{mols } N_2 \text{ in } HNO_3 \text{ produced}}{\text{mols } N_2 \text{ in } NH_3 \text{ feed}} = \frac{94·3/2}{100/2} = 94·3 \text{ per cent}$$

Note: the acid from the cooler-condenser could be added to the acid flow in the absorber, on the appropriate tray, to produce a more concentrated final acid. The secondary air flow is often passed through the acid mixer to strip out dissolved NO.

Scale-up to the required production rate

Production rate, 20,000 t/y HNO_3 (as 100 per cent acid).
With 8000 operating hours per year

$$\text{kg/h} = \frac{20,000 \times 10^3}{8000} = 2500 \text{ kg/h}$$

From calculations on previous basis:
100 kmol NH_3 produces 5940·9 kg HNO_3.

$$\text{So, scale-up factor} = \frac{2500}{5940·9} = 0·4208$$

To allow for unaccounted physical yield losses, round off to 0·43

All the stream flows, tabulated, were multiplied by this factor and are shown on the flowsheet, Fig. 4.2. A sample calculation is given below:
Stream (6) gas from condenser

	Mass 100 kmol NH_3 basis (kg)		Mass flow for 20,000 t/y (kg/h)
NO	471		202·5
NO_2	2249·4		967·2
O_2	640·0	$\times 0·43 =$	275·2
N_2	20,160·0		8668·0
H_2O	68·4		29·4
Total	23,588·8		10,143·1

Energy balance

Basis 1 hour.

Compressor

Calculation of the compressor power and energy requirements (see Chapter 3).

$$\text{Inlet flow rate, from flow sheet} = \frac{13{,}027{\cdot}7}{29 \times 3600} = 0{\cdot}125 \text{ kmol/s}$$

Volumetric flow rate

$$\text{at inlet conditions, 15°C, 1 bar} = 0{\cdot}125 \times 22{\cdot}4 \times \frac{288}{273} = 2{\cdot}95 \text{ m}^3/\text{s}$$

From Fig. 3.6, for this flow rate a centrifugal compressor would be used, $E_p = 74$ per cent.

(3.31) $$\text{Work (per kmol)} = Z_1 T_1 \mathbf{R} \frac{n}{n-1} \left[\left(\frac{P_2}{P_1} \right)^{\frac{n-1}{n}} - 1 \right]$$

(3.35) $$\text{Outlet temperature, } T_2 = T_1 \left(\frac{P_2}{P_1} \right)^m$$

As the conditions are well away from the critical conditions for air, equations (3.36*a*) and (3.38*a*) can be used

(3.36*a*) $$m = \frac{(\gamma - 1)}{\gamma E_p}$$

(3.38*a*) $$n = \frac{1}{1 - m}$$

γ for air can be taken as 1·4

$$m = \frac{1{\cdot}4 - 1}{1{\cdot}4 \times 0{\cdot}74} = 0{\cdot}39$$

$$n = \frac{1}{1 - 0{\cdot}39} = 1{\cdot}64$$

The inlet air will be at the ambient temperature, take as 15°C. With no intercooling

$$T_2 = 288 \times 8^{0{\cdot}39} = \underline{\underline{648 \text{ K}}}$$

This is clearly too high and intercooling will be needed. Assume compressor is divided into two sections, with approximately equal work in each section. Take the intercooler gas outlet temperature as 60°C (which gives a reasonable approach to the normal cooling water temperature of 30°C).

For equal work in each section the interstage pressure

$$= \sqrt{\frac{P_{\text{out}}}{P_{\text{in}}}} = \sqrt{8} = 2{\cdot}83$$

Taking the interstage pressure as 2.83 atm will not give exactly equal work in each section, as the inlet temperatures are different; however, it will be near enough for the purposes of this example.

First section work, inlet 15°C $= 1 \times 288 \times 8 \cdot 314 \times \dfrac{1 \cdot 64}{1 \cdot 64 - 1} \left[(2 \cdot 83)^{\frac{1 \cdot 64 - 1}{1 \cdot 64}} - 1 \right]$

$$= 3072 \cdot 9 \text{ kJ/kmol}$$

Second section work, inlet 60°C $= 1 \times 333 \times 8 \cdot 314 \times \dfrac{1 \cdot 64}{1 \cdot 64 - 1} \left[(2 \cdot 83)^{\frac{1 \cdot 64 - 1}{1 \cdot 64}} - 1 \right]$

$$= 3552 \cdot 6 \text{ kJ/kmol}$$

Total work $= 3072 \cdot 9 + 3552 \cdot 6 = 6625 \cdot 5 \text{ kJ/kmol}$

Compressor power $= \dfrac{\text{work/mol} \times \text{mol/s}}{\text{efficiency}} = \dfrac{6625 \cdot 5 \times 0 \cdot 125 \times 10^3}{0 \cdot 74}$

$$= 1119 \text{ kJ/s} = 1 \cdot 12 \text{ MW}$$

Energy required per hour $= 1 \cdot 12 \times 3600 = 4032 \text{ MJ}$

Compressor outlet temperature $= 333(2 \cdot 83)^{0 \cdot 39} = 500 \text{ K}$

say, 230°C

This temperature will be high enough for no preheating of the reactor feed to be needed (Strelzoff).

Ammonia vaporiser

The ammonia will be stored under pressure as a liquid. The saturation temperature at 8 atm is 20°C. Assume the feed to the vaporiser is at ambient temperature, 15°C.

Specific heat at 8 bar = 4·5 kJ/kgK
Latent heat at 8 bar   = 1186 kJ/kg
Flow to vaporiser   = 731·0 kg/h

Heat input required to raise to 20°C and vaporise

$$= 731 \cdot 0 \, [4 \cdot 5 \, (20 - 15) + 1186] = 883{,}413 \cdot 5 \text{ kJ/h}$$

add 10 per cent for heat losses $= 1 \cdot 1 \times 883{,}413 \cdot 5 = 971{,}754 \cdot 9 \text{ kJ/h}$

say, 972 MJ

Mixing tee

air   11,272·9 kg/h
230°C

t_3

NH$_3$ vapour 731·0 kg/h
20°C

C_p air $= 1$ kJ/kgK,
C_p ammonia vapour $2 \cdot 2$ kJ/kgK.

Note: as the temperature of the air is only an estimate, there is no point in using other than average values for the specific heats at the inlet temperatures.

Energy balance around mixing tee, taking as the datum temperature the inlet temperature to the oxidiser, t_3.

$$11,272 \cdot 9 \times 1(230 - t_3) + 731 \times 2 \cdot 2(20 - t_3) = 0$$
$$t_3 = 204°C$$

Oxidiser

The program ENERGY 1 (see Chapter 3) was used to make the balance over on the oxidiser. Adiabatic operation was assumed (negligible heat losses) and the outlet temperature found by making a series of balances with different outlet temperatures to find the value that reduced the computed cooling required to zero (adiabatic operation). The data used in the program are listed below:

$$\Delta H_r^\circ \text{ reaction } 1 = -226,334 \text{ kJ/kmol (per kmol } NH_3 \text{ reacted)}$$
$$\Delta H_r^\circ \text{ reaction } 2 = -316,776 \text{ kJ/kmol (per kmol } NH_3 \text{ reacted)}$$

All the reaction yield losses were taken as caused by reaction 2.
NH_3 reacted, by reaction 1

$$\text{Flow of } NH_3 \text{ to oxidiser} \times \text{reactor yield} = \frac{731 \cdot 0 \times 0 \cdot 96}{17} = 41 \cdot 3 \text{ kmol/h}$$

$$\text{balance by reaction } 2 = \frac{731 \cdot 0 \times 0 \cdot 04}{17} = 1 \cdot 7 \text{ kmol/h}$$

Summary, flows and heat capacity data:

Stream component	Feed (3) kmol/h	Product (4) kmol/h	C_p° kJ/kmol K			
			a	b	c	d
NH_3	43	—	27·32	23·83E-3	17·07E-6	−11·85E-9
O_2	82·1	29·2	28·11	−3·68E-6	17·46E-6	−10·65E-9
N_2	308·7	309·6	31·15	−1·36E-2	26·80E-6	−11·68E-9
NO	—	41·3	29·35	−0·94E-3	9·75E-6	−4·19E-9
H_2O	—	64·5	32·24	19·24E-4	10·5E-6	−3·60E-9
Temp. K	477	T_4				

The outlet temperature T_4 was found to be 1180 K = $\underline{907°C}$.

Waste-heat Boiler

As the amount of NO oxidised to NO_2 in this unit has not been estimated, it is not possible to make an exact energy balance over the unit. However, the maximum possible quantity steam generated can be estimated by assuming that all the NO is oxidised; and the minimum quantity by assuming that none is. The plant steam pressure would be typically 150 to 200 psig $\approx$ 11 bar, saturation temperature 184°C. Taking the approach temperature of the outlet gases (difference between gas and steam temperature) to be 50°C, the gas outlet temperature will be = $184 + 50 = 234°C$ (507 K).

From the flow-sheet, NO entering WHB $= \dfrac{1238\cdot4}{30} = 41\cdot3$ kmol

$\quad$ O_2 entering $\qquad\qquad\qquad = \dfrac{935\cdot7}{32} = 29\cdot2$ kmol/h

If all the NO is oxidised, reaction 4, the oxygen leaving the WHB will be reduced to

$$29\cdot2 - \frac{41\cdot3}{2} = 8\cdot6 \text{ kmol/h}$$

$$\Delta H_r^\circ = -57{,}120 \text{ kJ/kmol, NO oxidised}$$

If no NO is oxidised the composition of the outlet gas will be the same as the inlet. The inlet gas has the same composition as the reactor outlet, which is summarised above. Summarised below are the flow changes if the NO is oxidised:

| | (kmol/h) | C_p° (kJ/kmol K) | | | |
		a	b	c	d
O_2	7·46		as above		
NO_2	41·3	24·23	4·84 E-2	$-20\cdot81$ E-2	0·29 E-9
Temp.	507K				

Using the program ENERGY 1, the following values were calculated for the heat transferred to the steam:

$$\text{no NO oxidised} \quad 9\cdot88 \text{ GJ/h}$$

$$\text{all NO oxidised} \quad 12\cdot29 \text{ GJ/h}$$

Steam generated; take feed water temperature as 20°C,

$\qquad$ enthalpy of saturated steam at 11 bar = 2781 kJ/kg
$\qquad$ enthalpy of water at 20°C = 84 kJ/kg
$\qquad$ heat to form 1 kg steam = 2781 − 84 = 2697 kJ

$$\text{steam generated} = \frac{\text{heat transferred}}{\text{enthalpy change per kg}}$$

so, minimum quantity generated $= \dfrac{9{,}880{,}000}{2697} = 3662$ kg/h

$\qquad$ maximum $\qquad\qquad\qquad = \dfrac{12{,}290{,}000}{2697} = 4555$ kg/h

Note: in practice superheated steam would probably be generated, for use in a turbine driving the air compressor.

Cooler-condenser

The sources of heat to be considered in the balance on this unit are:

1. Sensible heat: cooling the gases from the inlet temperature of 234°C to the required outlet temperature (the absorber inlet temperature) 40°C.

2. Latent heat of the water condensed.
3. Exothermic oxidation of NO to NO_2.
4. Exothermic formation of nitric acid.
5. Heat of dilution of the nitric acid formed, to 40 per cent w/w.
6. Sensible heat of the outlet gas and acid streams.

So that the magnitude of each source can be compared, each will be calculated separately. Take the datum temperature as 25°C.

1. Gas sensible heat

The programme ENERGY 1 was used to calculate the sensible heat in the inlet and outlet gas streams. The composition of the inlet stream and the heat capacity data will be the same as that for the WHB outlet given above. Outlet stream flows from flow-sheet, converted to kmol/h:

Condenser outlet (6)	
	kmol/h
O_2	8·6
N_2	309·6
NO	6·75
NO_2	21·03
H_2O	1·63
Temp. 313K	

Sensible heat inlet stream (5) = 2·81 GJ/h,
              outlet stream (6) = 0·15 GJ/h.

2. Condensation of water

Water condensed = (inlet H_2O − outlet H_2O) = (1161 − 29) = 1131·6 kg/h

Latent heat of water at the inlet temperature, 230°C = 1812 kJ/kg

The steam is considered to condense at the inlet temperature and the condensate then cooled to the datum temperature.

Heat from condensation = 1131·6 × 1812 = $2·05 \times 10^6$ kJ/h

Sensible heat to cool condensate = 1131·6 × 4·18 (230 − 25)
$$= 0·97 \times 10^6 \text{ kJ/h}$$

Total, condensation and cooling $= (2·05 + 0·97) 10^6$ kJ/h
$$= 3·02 \text{ GJ/h}$$

3. Oxidation of NO

The greatest heat load will occur if all the oxidation occurs in cooler-condenser (i.e. none in the WHB) which gives the worst condition for the cooler-condenser design.

Mols of NO oxidised = mols in − mols out = 41·3 − 6·75 = 34·55 kmol/h

From reaction 4, heat generated $= 34·55 \times 57,120$
$$= 1·97 \times 10^6 \text{ kJ/h} = 1·97 \text{ GJ/h}$$

4. Formation of nitric acid

$$\text{HNO}_3 \text{ formed, from flow sheet,} = \frac{850 \cdot 6}{63} = 13 \cdot 50 \text{ kmol/h}$$

The enthalpy changes in the various reactions involved in the formation of aqueous nitric acid are set out below (Miles):

(6a) $2\text{NO}_2(\text{g}) \rightarrow \text{N}_2\text{O}_4(\text{g})$ $\Delta H = -57 \cdot 32 \text{ kJ}$

(6b) $\text{N}_2\text{O}_4(\text{g}) + \text{H}_2\text{O}(\text{l}) + \frac{1}{2}\text{O}_2(\text{g}) \rightarrow 2\text{HNO}_3(\text{g})$ $\Delta H = + 9 \cdot 00 \text{ kJ}$

(7) $\text{HNO}_3(\text{g}) \rightarrow \text{HNO}_3(\text{l})$ $\Delta H = -39 \cdot 48 \text{ kJ}$

Combining reactions 6a, 6b and 7.

Reaction 8 $2\text{NO}_2(\text{g}) + \text{H}_2\text{O}(\text{l}) + \frac{1}{2}\text{O}_2 \rightarrow 2\text{HNO}_3(\text{l})$

overall enthalpy change $= -57 \cdot 32 + 9 \cdot 00 + 2(-39 \cdot 48)$

$$= -127 \cdot 28 \text{ kJ}$$

$$\text{heat generated per kmol of HNO}_3(\text{l}) \text{ formed} = \frac{127 \cdot 28}{2} \times 10^3$$

$$= 63,640 \text{ kJ}$$

heat generated $= 13 \cdot 50 \times 63,640$ $= 0 \cdot 86 \times 10^6 \text{ kJ/h}$

$$= 0 \cdot 86 \text{ GJ/h}$$

Note, the formation of N_2O_4 and the part played by N_2O_4 in the formation of nitric acid was not considered when preparing the flow-sheet, as this does not affect the calculation of the components flow-rates.

5. Heat of dilution of HNO₃

The heat of dilution was calculated from an enthalpy – concentration diagram given in Perry's *Chemical Engineers Handbook*, 5th ed., p. 3.205, Fig. 3.42.

The reference temperature for this diagram is 32°F (0°C). From the diagram:

enthalpy of 100 per cent $\text{HNO}_3 = 0$

enthalpy of 45 per cent $\text{HNO}_3 = -80 \text{ Btu/lb solution}$

specific heat 45 per cent $\text{HNO}_3 = 0 \cdot 67$

So, heat released on dilution, at $32°\text{F} = 80 \times \dfrac{4 \cdot 186}{1 \cdot 8} = 186 \text{ kJ/kg soln.}$

Heat to raise solution to calculation datum temperature of 25°C $= 0 \cdot 67(25 - 0)4 \cdot 186$

$$= 70 \cdot 1 \text{ kJ/kg.}$$

So, heat generated on dilution at 25°C $= 186 - 70 \cdot 1 = 115 \cdot 9 \text{ kJ/kg soln.}$

Quantity of solution produced by dilution of 1 kmol 100 per cent $\text{HNO}_3 = \dfrac{63}{45} \times 100$

$$= 140 \text{ kg,}$$

so, heat generated on dilution of 1 kmol $= 140 \times 115 \cdot 9 = 16,226 \text{ kJ,}$

so, total heat generated $= 13 \cdot 5 \times 16,226 = 219,051 \text{ kJ/h}$

$$= 0 \cdot 22 \text{ GJ/h.}$$

6. *Sensible heat of acid*

Acid outlet temperature was taken as 40°C, which is above the datum temperature.
Sensible heat of acid $= 0.67 \times 4.186(40 - 25) \times 1860.7 = 78,278$ kJ/h $= 0.08$ GJ/h

Heat balance (GJ/h)

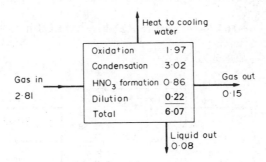

Heat transferred to cooling water $= 2.81 + 6.07 - 0.15 - 0.08$
$$= 8.65 \text{ GJ/h}$$

Air cooler

The secondary air from the compressor must be cooled before mixing with the process
gas stream at the absorber inlet; to keep the absorber inlet temperature as low as possible.
Take the outlet temperature as the same as exit gases from the cooler condenser, 40°C.

Secondary air flow, from flow-sheet, 1754·8 kg/h
Specific heat of air 1 kJ/kgK
Heat removed from secondary air $= 1754.8 \times 1 \times (230 - 40)$
$$= 333,412 \text{ kJ/h} = 0.33 \text{ GJ/h}$$

Absorber

The sources of heat in the absorber will be the same as the cooler-condenser and the
same calculation methods have been used. The results are summarised below:
Sensible heat in inlet gases from cooler-condenser $= 0.15$ GJ/h

Sensible heat in secondary air $= 1754.8 \times 1.0(40 - 25) = 0.018$ GJ/h
Sensible heat in tail gases (at datum) $= 0$
Sensible heat in water feed (at datum) $= 0$

NO oxidised $= \dfrac{202.5 - 21.9}{30} = 6.02$ kmol/h

Heat generated $= 6.02 \times 57,120 = 0.34$ GJ/h

HNO_3 formed $= \dfrac{1704}{63} = 27.05$ kmol/h

Heat generated $= 27.05 \times 63,640 = 1.72$ GJ/h

Heat of dilution to 60 per cent at 25°C $= 27.05 \times 14,207 = 0.38$ GJ/h

Water condensed $= 29\cdot4 - 26\cdot3 = 3\cdot1$ kg/h

Latent heat at $40°C = 2405$ kJ/h

Sensible heat above datum temperature $= 4\cdot18(40 - 25) = 63$ kJ/kg

Heat released $= 3\cdot1(2405 + 63) = 7\cdot6 \times 10^{-3}$ GJ/h (negligible)

Sensible heat in acid out, specific heat 0·64, take temperature out as same as gas inlet, 40°C

$$= 0\cdot64(40 - 25)4\cdot18 \times 2840 = 0\cdot11 \text{ GJ/h}$$

Heat balance (GJ/h)

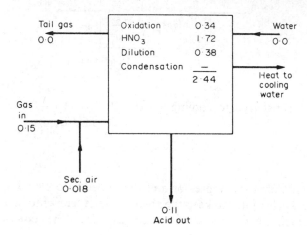

Heat transferred to cooling water $= 0\cdot15 + 0\cdot018 + 2\cdot44 - 0\cdot11 = \underline{\underline{2\cdot5 \text{ GJ/h}}}$

Mixer

Calculation of mixed acid temperature.

Taking the datum as 0°C for this calculation, so the enthalpy-concentration diagram can be used directly.

From diagram:

enthalpy 45 per cent acid at $0°C = -186$ kJ/kg

specific heat $= 0\cdot67$ kcal/kg°C

enthalpy 60 per cent acid at $0°C = -202$ kJ/kg

specific heat $= 0\cdot64$ kcal/kg°C

So, enthalpy 45 per cent acid at $40°C = -186 + 0\cdot67 \times 4\cdot186(40) = -73\cdot8$ kJ/kg

and enthalpy 60 per cent acid at $40°C = -202 + 0\cdot64 \times 4\cdot186(40) = -94\cdot8$ kJ/kg

Enthalpy of mixed acid $= \dfrac{(-73\cdot8 \times 1860\cdot7) + (-94\cdot8 \times 2840\cdot0)}{(1860\cdot7 + 2840\cdot0)}$

$$= -86\cdot5 \text{ kJ/kg}$$

From enthalpy-concentration diagram, enthalpy of mixed acid

(54 per cent) at $0°C = -202$ kJ/kg; specific heat $= 0\cdot65$ kcal/kg°C

so, "sensible" heat in mixed acid above datum of $0°C$

$$= -86·5 - (-202) = 115·5 \text{ kJ/kg}$$

and, mixed acid temperature $= \dfrac{115·5}{0·65 \times 4·186} = 43°C$

Energy recovery

In an actual nitric acid plant, the energy in the tail gases would normally be recovered by expansion through a turbine coupled to the air compressor. The tail gases would be preheated before expansion, by heat exchange with the process gas leaving the WHB.

4.4. Computer-aided flow-sheeting

The computer programs available for flow-sheeting in process design can be classified into two basic types:

1. Full simulation programs, which require powerful computing facilities.
2. Simple material balance programs requiring only a relatively small core size.

The full simulation programs are capable of carrying out rigorous simultaneous heat and material balances, and preliminary equipment design: producing accurate and detailed flow-sheets. In the early stages of a project the use of a full simulation package is not justified and a simple material balance program is more suitable. These are an aid to manual calculations and enable preliminary flow-sheets to be quickly, and cheaply, produced.

4.5. Full steady-state simulation programs

Complex flow-sheeting programs, that simulate the steady-state operation of the process being designed, have been developed by the central design departments of many of the major manufacturing companies and contracting organisations, and by university departments. Flow-sheeting packages and computing facilities are also available from several commercial computing bureaux. A survey of the programs available has been published by Flower and Whitehead (1973). Brief details of the programs available to university departments are given in Table 4.1. Detailed discussion of these programs is beyond the scope of this book; for a general review of the requirements and development of full simulation programs the reader is referred to Crowe *et al.* (1971) and Westerburg *et al.* (1979); see also Benedek (1980). The structure of a typical program is shown in Fig. 4.4.

The program consists of:

1. A main executive program; which controls and keeps track of the flow-sheet calculations and the flow of information to and from the sub-routines.
2. A library of equipment performance sub-routines (modules); which simulate the equipment and enable the output streams to be calculated from information on the inlet streams.
3. A data bank of physical properties. To a large extent the utility of a sophisticated flow-sheeting program will depend on the comprehensiveness of the physical property data bank. The collection of the physical property data required for the design of a particular process, and its transformation into a form suitable for a particular flow-sheeting program can be very time-consuming.

4. Sub-programs for thermodynamic routines; such as the calculation of vapour–liquid equilibria and stream enthalpies.

5. Sub-programs and data banks for costing; the estimation of equipment capital costs and operating costs. Full simulation flow-sheeting programs enable the designer to consider alternative processing schemes, and the cost routines allow quick economic comparisons to be made. Some programs include optimisation routines. To make use of a costing routine, the program must be capable of producing at least approximate equipment designs.

TABLE 4.1. *Simulation packages available: 1980*

Acronym	Full name	Source	Reference	Availability
GEMCS	General Engineering Management Computation System	McMasters University	GEMCS Users manual Johnson (1972)	Open
FLOWPACK	—	Imperial Chemical Industries Ltd., UK	FLOWPACK Users manual. Davies (1971)	FLOWPACK 1 available to UK universities
FLOWTRAN	—	Monsanto Company and CACHE (Computer Aids in Chemical Engineering)	Seader *et al.* (1977) Clark (1977)	US universities
CONCEPT	Computation On-line of Networks of Chemical Engineering Process Technology	CAD Centre Cambridge, UK	CONCEPT Users manual. Hutchinson and Leesley (1973)	Commercial bureaux
PROCESS	Process simulation program	Simulation Science Inc., USA	Process Users, and Applications manuals	Commercial bureux. Available to UK universities

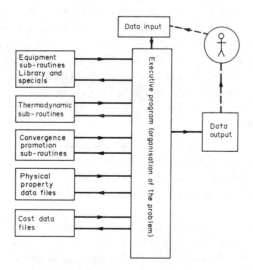

FIG. 4.4. A typical simulation program

General simulation flow-sheeting programs normally use an iterative method for the solution of the material and energy balances. Units are taken sequentially until a recycle loop is encountered. The loop is then torn, suitable values assigned to the unknown variables, and the calculation carried out iteratively until satisfactory convergence is obtained. The program then reverts to sequential calculation until another loop is encountered. For complex processes, involving several recycle loops, efficient ordering of the calculations is essential to economise on computing time and ensure successful completion of the calculations. The executive program normally includes procedures for determining the optimum order of calculation and routines to promote convergence. Kehat and Shacham (1973) has discussed and compared the techniques that have been developed for determining the order of calculation in process flow-sheet calculations.

4.5.1. Information flow diagrams

To present the problem to the computer, the basic process flow diagram, which shows the sequence of unit operations and stream connections, must be transformed into an information flow diagram, such as that shown in Fig. 4.5b. Each block represents a calculation module in the simulation program; usually a process unit or part of a unit. Units in which no change of composition, or temperature or pressure, occurs are omitted from the information flow diagram. But other operations not shown on the process flow diagram as actual pieces of equipment, but which cause changes in the stream compositions, such as mixing tees, must be shown.

The lines and arrows connecting the blocks show the flow of information from one sub-program to the next. An information flow diagram is a form of directed graph (a diagraph).

The calculation topology defined by the information diagram is transformed into a numerical form suitable for input into the computer, usually as a matrix.

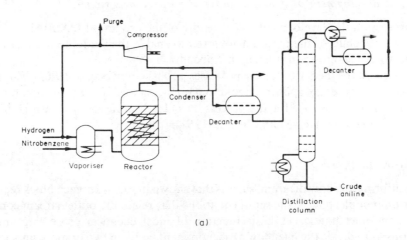

(a)

FIG. 4.5a. Process flow diagram: hydrogenation of nitrobenzene to aniline

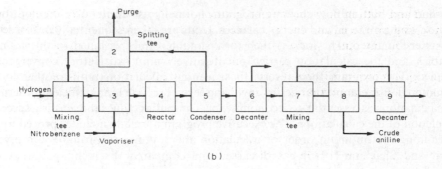

FIG. 4.5b Information flow diagram hydrogenation of nitrobenzene to aniline (Fig. 4.5a)

Note: (1) Modules have been added to represent mixing and separation tees.
(2) The compressor is omitted.
(3) The distillation module includes the condenser and reboiler.

4.6. Simple material balance programs

In the initial stages of the process design and evaluation, when only a rough, approximate, material balance is required, the use of a full simulation program is not justified. Simpler programs, which calculate only the material balance, have been developed and these can be used as an aid to manual flow-sheeting calculations. They will be particularly useful if the process involves several recycle streams.

Some of the full simulation flow-sheeting packages, such as FLOWPACK, can also be used to calculate the material balance without simultaneous solution of the energy balance, or use of the equipment design routines. They should be used in this mode for the initial, scouting, flow-sheet calculations, to economise on computing costs.

Simple material balance programs need only a small core size, and can be run on desk-top computers, or in a small part of a larger machine.

4.6.1. The development of a simple material balance program

In this section the development and structure of the program MASSBAL is described, and sufficient details of the program are given to enable the reader to use it as an aid to flow-sheeting. The program is listed in Appendix B.

It is based on the theory of recycle processes published by Nagiev (1964). This method, which uses the concept of split-fractions to set up the set of simultaneous equations which define the material balance for the process, has also been used by Rosen (1962) and is described in detail by Henley and Rosen (1969).

The split-fraction concept

In an information flow diagram, such as that shown in Fig. 4.5b, each block represents a calculation module; that is, the set of equations that relate the outlet stream component flows to the inlet flows. The basic function of most chemical processing units (unit operations) is to divide the inlet flow of a component between two or more outlet streams; for example, a distillation column divides the components in the feed between the overhead and bottom product streams, and any side streams. It is therefore convenient, when setting up the equations describing a unit operation, to express the flow of any component in any outlet stream as a fraction of the flow of that component in the inlet stream.

The block shown in Fig. 4.6 represents any unit in an information flow diagram, and shows the nomenclature that will be used in setting up the material balance equations.

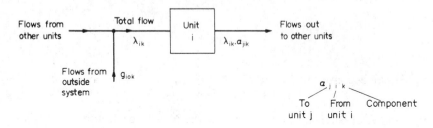

FIG. 4.6

i = the unit number,

$\lambda_{i,k}$ = the total flow into the unit i of the component k,

$\alpha_{j,i,k}$ = the *fraction* of the total flow of component k entering unit i that leaves in the outlet stream connected to the unit j; the "split-fraction coefficient",

$g_{i,0,k}$ = any fresh feed of component k into unit i; flow from outside the system (from unit 0).

The flow of any component from unit i to unit j will equal the flow into unit i multiplied by the split-fraction coefficient.

$$= \lambda_{i,k} \times \alpha_{j,i,k}$$

The value of the split-fraction coefficient will depend on the nature of the unit and the inlet stream composition.

The outlet streams from a unit can feed forward to other units, or backward (recycle).

An information flow diagram for a process consisting of three units, with two recycle streams is shown in Fig. 4.7. The nomenclature defined in Fig. 4.6 is used to show the stream flows.

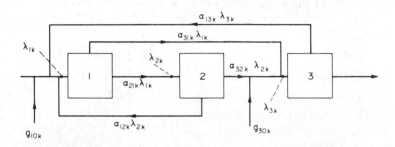

FIG. 4.7

Consider the streams entering unit 1.

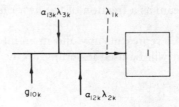

FIG. 4.8

A material balance gives:

$$g_{10k} + \alpha_{13k}\lambda_{3k} + \alpha_{12k}\lambda_{2k} = \lambda_{1k} \tag{4.1}$$

A similar material balance can be written at the inlet to each unit:

$$\text{unit 2:} \quad \alpha_{21k}\lambda_{1k} = \lambda_{2k} \tag{4.2}$$

$$\text{unit 3:} \quad \alpha_{32k}\lambda_{2k} + g_{30k} + \alpha_{31k}\lambda_{1k} = \lambda_{3k} \tag{4.3}$$

Rearranging each equation

$$\lambda_{1k} - \alpha_{12k}\lambda_{2k} - \alpha_{13k}\lambda_{3k} = g_{10k} \tag{4.1a}$$

$$-\alpha_{21k}\lambda_{1k} + \lambda_{2k} = 0 \tag{4.2b}$$

$$-\alpha_{31k}\lambda_{1k} - \alpha_{32k}\lambda_{2k} + \lambda_{3k} = g_{30k} \tag{4.3c}$$

This is simply a set of three simultaneous equations in the unknown flows λ_{1k}, λ_{2k}, λ_{3k}.
These equations are written in matrix form:

$$
j \quad
\begin{array}{c}
 \\ 1 \\ 2 \\ 3
\end{array}
\overset{\displaystyle i}{
\begin{array}{ccc}
1 & 2 & 3 \\
\end{array}}
\begin{bmatrix}
1 & -\alpha_{12k} & -\alpha_{13k} \\
-\alpha_{21k} & 1 & 0 \\
-\alpha_{31k} & -\alpha_{32k} & 1
\end{bmatrix}
\times
\begin{bmatrix}
\lambda_{1k} \\
\lambda_{2k} \\
\lambda_{3k}
\end{bmatrix}
=
\begin{bmatrix}
g_{10} \\
0 \\
g_{30}
\end{bmatrix}
$$

There will be a set of such equations for each component.

This procedure for deriving the set of material balance equations is quite general. For a process with n units there will be a set of n equations for each component.

The matrix form of the n equations will be as shown in Fig. 4.9.

$$
\begin{bmatrix}
(1-\alpha_{11k})-\alpha_{12k} & -\alpha_{13k} & \cdots & -\alpha_{1nk} \\
-\alpha_{21k} & (1-\alpha_{22k})-\alpha_{23k} & \cdots & -\alpha_{2nk} \\
 & & & \\
-\alpha_{n1k} \cdots & \cdots\cdots\cdots\cdots & \cdots & (1-\alpha)_{nnk}
\end{bmatrix}
\times
\begin{bmatrix}
\lambda_{1k} \\
\lambda_{2k} \\
 \\
\lambda_{nk}
\end{bmatrix}
=
\begin{bmatrix}
g_{10k} \\
g_{20k} \\
 \\
g_{n0k}
\end{bmatrix}
$$

FIG. 4.9. Matrix form of equations for n units

For practical processes most of the split-fraction coefficients are zero and the matrix is sparse.

In general, the equations will be non-linear, as the split-fractions coefficients (α's) will be functions of the inlet flows, as well as the unit function. However, many of the coefficients will be fixed by the process constraints, and the remainder can usually be taken as independent of the inlet flows (λ's) as a first approximation.

The fresh feeds will be known from the process specification; so if the split-fraction coefficients can be estimated, the equations can be solved to determine the flows of each component to each unit. Where the split-fractions are strongly dependent on the inlet flows, the values can be adjusted and the calculation repeated until a satisfactory convergence between the estimated values and those required by the calculated inlet flows is reached.

Processes with reaction

In a chemical reactor, components in the inlet streams are consumed and new components, not necessarily in the inlet streams, are formed. The components formed cannot be shown as split-fractions of the inlet flows and must therefore be shown as pseudo-fresh feeds.

A reactor is represented as two units (Fig. 4.10). The split-fractions for the first unit are chosen to account for the loss of material by reaction. The second unit divides the reactor output between the streams connected to the other units. If the reactor has only one outlet stream (one connection to another unit), the second unit forming the reactor can be omitted.

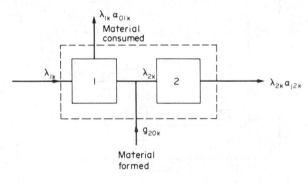

FIG. 4.10

4.6.2. Illustration of the method

The procedure for setting up the equations and assigning suitable values to the split-fraction coefficients is best illustrated by considering a short problem: the manufacture of acetone from isopropyl alcohol.

Process description

Reaction:
$$C_3H_7OH \xrightarrow[\text{cat.}]{\text{heat}} (CH_3)_2CO + H_2$$

Isopropyl alcohol is vaporised, heated and fed to a reactor, where it undergoes catalytic dehydrogenation to acetone. The reactor exit gases (acetone, water, hydrogen and unreacted isopropyl alcohol) pass to a condenser where most of the acetone, water and alcohol condense out. The final traces of acetone and alcohol are removed in a water scrubber. The effluent from the scrubber is combined with the condensate from the condenser, and distilled in a column to produce "pure" acetone and an effluent consisting of water and alcohol. This effluent is distilled in a second column to separate the excess water. The product from the second column is an azeotrope of water and isopropyl alcohol containing approximately 91 per cent alcohol. This is recycled to the reactor. Zinc oxide or copper is used as the catalyst, and the reaction carried out at 400 to 500°C and 40 to 50 psig pressure (4.5 bar). The yield to acetone is around 98 per cent, and the conversion of isopropyl alcohol per pass through the reactor is 85 to 90 per cent.

The process flow diagram is shown in Fig. 4.11. This diagram is simplified and drawn as an information flow diagram in Fig. 4.12. Only those process units in which there is a

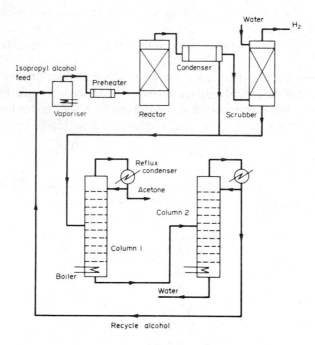

FIG. 4.11. Process flow diagram

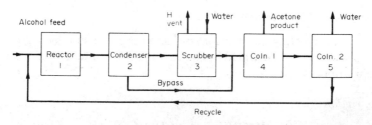

FIG. 4.12. Information flow diagram

difference in composition between the inlet and outlet streams are shown. The preheater and vaporiser are not shown, as there is no change in composition in these units and no division of the inlet stream into two or more outlet streams.

Figure 4.12 is redrawn in Fig. 4.13, showing the fresh feeds, split-fraction coefficients and component flows. Note that the fresh feed g_{20k} represents the acetone and hydrogen generated in the reactor. There are 5 units so there will be 5 simultaneous equations. The equations can be written out in matrix form (Fig. 4.14) by inspection of Fig. 4.13. The fresh feed vector contains three terms.

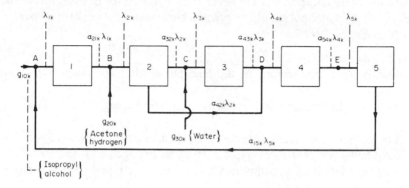

FIG. 4.13. Split-fractions and fresh feeds

$$
\begin{array}{c}
\begin{array}{ccccc} 1 & 2 & 3 & 4 & 5 \end{array} \\
\begin{array}{c} 1 \\ 2 \\ 3 \\ 4 \\ 5 \end{array}
\begin{bmatrix}
1 & 0 & 0 & 0 & -\alpha_{15k} \\
-\alpha_{21k} & 1 & 0 & 0 & 0 \\
0 & -\alpha_{32k} & 1 & 0 & 0 \\
0 & -\alpha_{42k} & -\alpha_{43k} & 1 & 0 \\
0 & 0 & 0 & -\alpha_{54k} & 1
\end{bmatrix}
\times
\begin{bmatrix}
\lambda_{1k} \\ \lambda_{2k} \\ \lambda_{3k} \\ \lambda_{4k} \\ \lambda_{5k}
\end{bmatrix}
=
\begin{bmatrix}
g_{10k} \\ g_{20k} \\ g_{30k} \\ 0 \\ 0
\end{bmatrix}
\end{array}
$$

FIG. 4.14. The set of equations

Estimation of the split-fraction coefficients

The values of the split-fraction coefficients will depend on the function of the processing unit and the constraints on the stream flow-rates and compositions. Listed below are suggested first trial values, and the basis for selecting the particular value for each component.

Component 1, isopropyl alcohol ($k = 1$)

Unit 1, Reactor. The conversion per pass is given as 90 per cent, so for each mol entering only 10 per cent leave, hence α_{211} is fixed at 0·1. For this example it is assumed that the conversion is independent of the feed stream composition.

Unit 2, Condenser. Most of the alcohol will condense as its boiling point is 82°C. Assume 90 per cent condensed, $\alpha_{421} = 0·9$ (liquid out) and $\alpha_{321} = 0·1$ (vapour out). The actual amounts will depend on the condenser design.

Unit 3, Scrubber. To give a high plant yield, the scrubber would be designed to recover most of the alcohol in the vent stream. Assume 99 per cent recovery, allowing for the small loss that must theoretically occur, $\alpha_{431} = 0.99$.

Unit 4, First column. The fraction of alcohol in the overheads would be fixed by the amount allowed in the acetone product specification. Assume 1 per cent loss to the acetone is acceptable, which will give less than 1 per cent alcohol in the product; fraction in the bottoms 99 per cent, $\alpha_{541} = 0.99$.

Unit 5, Second column. No distillation column can be designed to give complete separation of the components. However, the volatilities for this system are such that a high recovery of alcohol should be practicable. Assume 99 per cent recovery, alcohol recycled, $\alpha_{151} = 0.99$.

Component 2, Acetone ($k = 2$)

Unit 1. Assume that any acetone in the feed passes through the reactor unchanged, $\alpha_{212} = 1$.

Unit 2. Most of the acetone will condense (b.p. 56°C) say 80 per cent, $\alpha_{322} = 0.2$, $\alpha_{422} = 0.8$.

Unit 3. As for alcohol, assume 99 per cent absorbed, allows for a small loss, $\alpha_{432} = 0.99$.

Unit 4. Assume 99 per cent recovery of acetone as product, $\alpha_{542} = 0.01$.

Unit 5. Because of its high volatility in water all but a few ppm of the acetone will go overhead, put $\alpha_{152} = 0.01$.

Component 3, Hydrogen ($k = 3$)

Unit 1. Passes through unreacted, $\alpha_{213} = 1$.

Unit 2. Non-condensable, $\alpha_{323} = 1$, $\alpha_{423} = 0$.

Unit 3. None absorbed, $\alpha_{433} = 0$.

Unit 4. Any present in the feed would go out with the overheads, $\alpha_{543} = 1$.

Unit 5. As for unit 4, $\alpha_{153} = 1$.

Component 4, Water ($k = 4$)

Unit 1. Passes through unreacted, $\alpha_{214} = 1$.

Unit 2. A greater fraction of the water will condense than the alcohol or acetone (b.p. 100°C) assume 95 per cent condensed, $\alpha_{324} = 0.05$, $\alpha_{423} = 0.95$.

Unit 3. There will be a small loss of water in the vent gas stream, assume 1 per cent lost, $\alpha_{434} = 0.99$.

Unit 4. Some water will appear in the acetone product; as for the alcohol this will be fixed by the acetone product specification. Putting $\alpha_{544} = 0.99$ will give less than 1 per cent water in the product.

Unit 5. The overhead composition will be close to the azeotropic composition, approximately 9 per cent water. The value of α_{154} (recycle to the reactor) must be selected so that the overheads from this unit approximate to the azeotropic composition, as a first try put $\alpha_{154} = 0.05$

Estimation of fresh feeds

1. Isopropyl alcohol, take the basis of the flow sheet as 100 mol feed, $g_{101} = 100$.
2. Acetone formed in the reaction. The overall yield to acetone is approximately 98 per cent, so acetone formed $= 100 \times \dfrac{98}{10} = 98$ mol, $g_{202} = 98$ mol.

3. Hydrogen, it is formed in equimolar proportion to acetone, so $g_{203} = 98$ mol.

4. Water, the feed of water to the scrubber will be dependent on the scrubber design. A typical design value for mG_m/L_m for a scrubber is 0·7 (see Volume 2, Chapter 4). For the acetone absorption this would require a value of L_m of 200 mol, $g_{304} = 200$ mol.

Matrices

Substituting the values for alcohol ($k = 1$) into the matrix (Fig. 4.14) gives the following set of equations for the flow of alcohol into each unit;

$$
\begin{bmatrix}
1 & 0 & 0 & 0 & -0.99 \\
-0.1 & 1 & 0 & 0 & 0 \\
0 & -0.1 & 1 & 0 & 0 \\
0 & -0.9 & -0.99 & 1 & 0 \\
0 & 0 & 0 & -0.99 & 1
\end{bmatrix}
\times
\begin{bmatrix}
\lambda_{11} \\
\lambda_{21} \\
\lambda_{31} \\
\lambda_{41} \\
\lambda_{51}
\end{bmatrix}
=
\begin{bmatrix}
100 \\
0 \\
0 \\
0 \\
0
\end{bmatrix}
$$

Substitution of the values of the split-fraction coefficients for the other components will give the sets of equations for the component flows to each unit. The values of the split-fraction coefficients and fresh feeds are summarised in Table 4.2.

TABLE 4.2. *Split-fraction coefficients and feeds*

α	1	2	3	4
$21k$	−0·1	−1	−1	−1·0
$32k$	−0·1	−0·2	−1	−0.05
$42k$	−0·9	−0·8	0	−0·95
$43k$	−0·99	−0·99	0	−0·99
$54k$	−0·99	−0·01	−1	−0·99
$15k$	−0·99	−0·01	−1	−0·05
	g_{101}	g_{202}	g_{203}	g_{304}
Mol	100	98	98	200

Solution of the equations

The equations for each components can be solved using any of the standard procedures for linear simultaneous equations; see Volume 3, Chapter 4. The sub-routine available in the Hewlett Packard model 9820A, desk computer, MATHS PACK, was used for the solution of this example:

TABLE 4.3. *Solution of equations, feeds to units*

Unit	Component	1	2	3	4	Total
1	λ_{1k}	110·85	0·01	0·0	10·31	121·17
2	λ_{2k}	11·09	98·01	98·0	10·31	217·41
3	λ_{3k}	1·11	19·6	98·0	200·51	319·22
4	λ_{4k}	11·07	97·81	0·0	208·3	317·19
5	λ_{5k}	10·96	0·98	0·0	206·22	218·16

Comments on the first trial solutions

Table 4.3 shows the feed of each component and the total flow to each unit. The composition of any other stream of interest can be calculated from these values and the split-fraction coefficients. The compositions and flows should be checked for compliance with the process constraints, the split-fraction values adjusted, and the calculation repeated, as necessary, until a satisfactory fit is obtained. Some of the constraints to check in this example are discussed below.

Recycle flow from the second column

This should approximate to the azeotropic composition (9 per cent alcohol, 91 per cent water). The flow of any component in this stream is given by multiplying the feed to the column (λ_{5k}) by the split-fraction coefficient for the recycle stream (α_{15k}). The calculated flows for each component are shown in Table 4.4.

TABLE 4.4. *Calculation of recycle stream flow*

Component	1	2	3	4	Total
λ_{5k}	10·96	0·98	0·0	206·22	
α_{15k}	0·99	0·01	1	0·05	
Flow					
$\alpha_{15k}\lambda_{5k}$	10·85	0·01	0	10·31	21.17
Per cent	51·3	0·05	0	48·7	

Calculated percentage alcohol = 51·3 per cent, required value 91 per cent. Clearly the initial value selected for α_{154} was too high; too much recycle. Try $\alpha_{154} = 0.02$ for the second trial.

Reactor conversion and yield

$$\text{Conversion} = \frac{\text{alcohol in} - \text{alcohol out}}{\text{alcohol in}} = \frac{\lambda_{11} - \lambda_{21}}{\lambda_{11}} = \frac{110.85 - 11.09}{110.85}$$

= 90 per cent, which is the value given

$$\text{Yield} = \frac{\text{acetone out}}{\text{alcohol in} - \text{alcohol out}} = \frac{\lambda_{22}}{\lambda_{11} - \lambda_{21}} = \frac{98.01}{110.85 - 11.09}$$

= 98·3 per cent, near enough.

Condenser vapour and liquid composition.

The liquid and vapour streams from the partial condenser should be approximately in equilibrium.

The component flows in the vapour stream = $\alpha_{32k}\lambda_{2k}$ and in the liquid stream = $\alpha_{42k}\lambda_{2k}$. The calculation is shown in Table 4.5.

These compositions should be checked against the vapour–liquid equilibrium data for acetone–water and the values of the split-fraction coefficients adjusted, as necessary.

TABLE 4.5. *Condenser vapour and liquid compositions*

Component k	1	2	3	4	Total
λ_{2k}	11·09	98·01	98·0	10·31	
α_{32k}	0·1	0·2	1	0·05	
Vapour flow					
$\quad\alpha_{32k}\lambda_{2k}$	1·11	19·6	98·0	0·52	119·23
Per cent	0·9	16·4	82·2	0·4	
α_{42k}	0·9	0·8	0	0·95	
Liquid flow					
$\quad\alpha_{42k}\lambda_{2k}$	9·98	78·41	0	9·79	98·18
Per cent	10·2	79·9	0	10·0	

4.6.3. *Guide Rules for estimating split-fraction coefficients*

The split-fraction coefficients can be estimated by considering the function of the process unit, and by making use of any constraints on the stream flows and compositions that arise from considerations of product quality, safety, phase equilibria, other thermodynamic relationships; and general process and mechanical design considerations. The procedure is similar to the techniques used for the manual calculation of material balances discussed in Section 4.3.

Suggested techniques for use in estimating the split-fraction coefficients for some of the more common unit operations are given below.

1. Reactors

The split-fractions for the reactants can be calculated directly from the percentage conversion. The conversion may be dependent on the relative flows of the reactants (feed composition) and, if so, iteration may be necessary to determine values that satisfy the feed condition.

Conversion is not usually very dependent on the concentration of any inert components.

The pseudo fresh feeds of the products formed in the reactor can be calculated from the specified, or estimated, yields for the process.

2. Mixers

For a unit that simply combines several inlet streams into one outlet stream, the split-fraction coefficients for each component will be equal to 1. $\alpha_{j,i,k} = 1$.

3. Stream dividers

If the unit simply divides the inlet stream into two or more outlet streams, each with the same composition as the inlet stream, then the split-fraction coefficient for each component will have the same value as the fractional division of the total stream. A purge stream is an example of this simple division of a process stream into two streams: the main stream and the purge. For example, for a purge rate of 10 per cent the split-fraction coefficients for the purge stream would be 0·1.

4. Absorption or stripping columns

The amount of a component absorbed or stripped in a column is dependent on the column design (the number of stages), the component solubility, and the gas and liquid rates. The fraction absorbed can be estimated using the absorption factor method, attributed to Kremser (1930) (see Volume 2, Chapter 12). If the concentration of solute in the solvent feed to the column is zero, or can be neglected, then for the solute component the fraction absorbed =

$$\frac{(L_m/mG_m)^{s+1} - L_m/mG_m}{(L_m/mG_m)^{s+1} - 1}$$

and for a stripping column, the fraction stripped =

$$\frac{(mG_m/L_m)^{s+1} - (mG_m/L_m)}{(mG_m/L_m)^{s+1} - 1}$$

where G_m = gas flow rate, kmol m^{-2} h^{-1},
  L_m = liquid flow rate, kmol m^{-2} h^{-1},
  m = slope of the equilibrium curve,
  s = the number of stages.

For a packed column the chart by Colburn (1939) can be used (see Volume 2, Chapter 11). This gives the ratio of the inlet and outlet concentrations, y_1/y_2, in terms of the number of transfer units and mG_m/L_m.

The same general approach can be used for solvent extraction processes.

5. Distillation columns

A distillation column divides the feed stream components between the top and bottom streams, and any side streams. The product compositions are often known; they may be specified, or fixed by process constraints, such as product specifications, effluent limits or an azeotropic composition. For a particular stream, "s", the split-fraction coefficient is given by:

$$\frac{x_{sk} r_s}{x_{fk}}$$

where x_{sk} = the concentration of the component k in the stream, s,
  x_{fk} = the concentration component k in the feed stream,
  r_s = the fraction of the total feed that goes to the stream, s.

If the feed composition is fixed, or can be estimated, the value of r_s can be calculated from a mass balance.

The split-fraction coefficients are not very dependent on the feed composition, providing the reflux flow-rate is adjusted so that the ratio of reflux to feed flow is held constant; Vela (1961), Hachmuth (1952).

It is not necessary to specify the reflux when calculating a preliminary material balance; the system boundary can be drawn to include the reflux condenser.

For a column with no side streams the fraction of the total feed flow going to the overheads is given by:

$$r_{\text{overheads}} = \frac{x_{fk} - x_{wk}}{x_{dk} - x_{wk}}$$

where x is the component composition and the suffixes f, d, w refer to feed, overheads and bottoms respectively.

6. Equilibrium separators

This is a stream divider with two outlet streams, a and b, which may be considered to be in equilibrium.

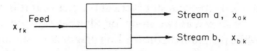

where x_{ak} = concentration of component k in stream a,
x_{bk} = concentration of component k in stream b,
x_{fk} = concentration of component k in the feed stream.

If the equilibrium relationship can be expressed by a simple equilibrium constant, K_k, such that:

$$x_{ak} = K_k x_{bk}$$

Then the split-fraction coefficients can be calculated from a material balance.

Split fraction for stream $a = \dfrac{K_k}{K_k - 1} \dfrac{(x_{fk} - x_{bk})}{x_{fk}}$.

4.6.4. MASSBAL, a mass balance program

A simple material balance program, based on the split-fraction concept, is listed in Appendix B. This program can be used to calculate material balances for processes with up to fifty units and twenty components. It will be found to be particularly useful for processes that contain several recycle loops. The procedure for using the program is similar to that illustrated in Section 4.6.2. The process flow diagram is reduced to an information flow diagram showing all the connections between the units, and the values of the component split fractions and any fresh feeds estimated for each unit. These values are typed in from a terminal and a program calculates and prints out the component flows to each unit.

The program includes a routine to enable the initial estimates of the split-fraction coefficients to be easily changed, and can be run in an interactive manner to find the values that satisfy the design constraints (process specifications and equipment parameters).

The program is written in BASIC. The version listed was run on a PRIME 750 machine. The program has also been run on an ICL 1904S, and in a modified form on a Hewlett Packard 9830A desk-top computer. It can be readily adapted for other computer systems. Sufficient comments (REM statements) are included in the listing for the structure and logic of the program to be readily followed.

MASSBAL consists of three separate BASIC programs:

MM1 – a program to set up the coefficient matrix and fresh feed vector, and file the values;

MM2 – a program to enable the filed values to be altered, as required;

MM3 – a program to solve the set of equations for each component, sum the values, and print out the flows and percentage compositions.

A full set of operating instructions is included in the program listing.

Self-recycle

The program assumes that there are no self-recycle streams, recycle round a single unit (Fig. 4.15). These are unlikely to occur in practical problems. If it is necessary to include a self-recycle loop, it can be shown as two units; or, alternatively, the value of 1 that will be automatically set up by the program on the leading diagonal (assuming no self-recycle) can be changed to $(1 - \alpha_{iik})$ by typing in this value in the same manner as for the other non-zero coefficients. (*Note*, putting $\alpha_{iik} = 1$ implies total recycle, and there will be no unique solution to the set of equations.)

Dummy units

The program only calculates and prints out the inlet stream flows and composition. Though this will give sufficient information for the flows and compositions of all other process streams to be calculated by hand, a direct print-out of any non-inlet stream can be obtained by inserting a "dummy" unit in the line wanted, with the split-fraction coefficients for the dummy unit set at 1.

The stream flows will be printed out as the inlet stream to the dummy unit (Fig. 4.16). Dummy units can also be used to obtain directly the flow and composition of any streams that leave the system, such as a vent or product stream. There will be no outlet streams from these units.

FIG. 4.15. Self-recycle   FIG. 4.16. Dummy units

Equation solution routine

For all practical material balance problems, the matrix of split-fraction coefficients will be very sparse, as the number of connections between units will only be a fraction of the total possible.

For a process with n units the total number for possible connections will be equal to the dimensions of the matrix, $n \times n$, but the actual number will be between $2n$ and $3n$.

To make the most efficient use of computer storage, and to give a quick response time, the efficient sparse matrix solution algorithm developed by D. J. Gunn (1977) is used in program MM3.

In Gunn's procedure the matrix of split-fraction coefficients is represented by three vectors: a vector D containing the non-zero coefficients, in column order within consecutive rows; an integer vector Z, of the same dimensions as D, containing the column address of each non-zero element; and an integer vector L giving the position in the other vectors of the first element in each row. The program MM3 contains a sub-routine that automatically reads the values from the data file into these vectors for the calculation procedure.

4.7. References

AUSTIN, D. G. (1979) *Chemical Engineering Drawing Symbols* (George Godwin).

BENEDEK, P. (Ed.) (1980) *Steady-state Flow-sheeting of Chemical Plants* (Elsevier).

CLARK, A. P. (1977) *Exercises in Process Simulation Using FLOWTRAN* (CACHE Corporation).

COLBURN, A. P. (1939) *Trans. Am. Inst. Chem. Eng.* **35**, 211. The simplified calculation of diffusional processes, general considerations of two-film resistances.

CROWE, C. M., HAMIELEE, A. E., HOFFMAN, T. N., JOHNSON, A. I., SHANNON, P. T. and WOODS, D. R. (1971) *Chemical Plant Simulation* (Prentice-Hall).

DAVIES, C. (1971) *Chem. Engr. London.* No. 248 (April) 149. Applications of systems engineering techniques to projects in the chemical process industry.

FLOWER, J. R. and WHITEHEAD, B. D. (1973) *Chem. Engr. London*, No. 272 (April) 208, No. 273 (May) 271 (in two parts). A survey of flow-sheeting programs.

GUNN, D. J. (1977) *Inst. Chem. Eng.*, 4th Annual Research Meeting, Swansea, April. A sparse matrix technique for the calculation of linear reactor-separator simulations of chemical plant.

HACHMUTH, K. H. (1952) *Chem. Eng. Prog.* **48** (Oct.) 523, (Nov.) 570, (Dec.) 570 (in three parts). Industrial viewpoints on separation processes.

HENLEY, E. J. and ROSEN, E. M. (1969) *Material and Energy Balance Computations* (Wiley).

HUTCHINSON, H. P. and LEESLEY, M. E. (1973) *Computer Aided Design* **5**, 228. A balanced approach to process design by computer.

JOHNSON, A. I. (1972) *Brit. Chem. Eng. & Proc. Tech.* **17**, 28. Computer-aided process analysis and design – a modular approach.

KEHAT, E. and SHACHAM, M. (1973) *Process Design and Development* (formerly *Brit. Chem. Eng.*) **18**, No. 1/2, 35; No. 3, 115 (in two parts). Chemical process simulation programs.

KREMSER, A. (1930) *Nat. Petroleum News* **22** (21 May) 43. Theoretical analysis of absorption columns.

NAGIEV, M. F. (1964) *The Theory of Recycle Processes in Chemical Engineering* (Pergamon).

ROSEN, E. M. (1962) *Chem. Eng. Prog.* **58** (Oct.) 69. A machine computation method for performing material balances.

SEADER, J. D., SEIDER, W. D. and PAULS, A. C. (1977) *FLOWTRAN Simulation – an introduction*, 2nd ed. (CACHE Corporation).

VELA, M. A. (1961) *Pet. Ref.* **40** (May) 247, (June) 189 (in two parts). Use of fractions for recycle balances.

WESTERBURG, A. W., HUTCHINSON, H. P., MOTARD, R. L. and WINTER, P. (1979) *Process Flow-sheeting* (Cambridge U.P.).

British Standards
BS 1553: ----   Specification for graphical symbols for general engineering
           Part 1: 1977 Piping systems and plant.

4.8. Nomenclature

		Dimension in **MLT**
G_m	Molar flow-rate of gas per unit area	$ML^{-2}T^{-1}$
g_{iok}	Fresh feed to unit i of component k	MT^{-1}
K_k	Equilibrium constant for component k	—
L_m	Liquid flow-rate per unit area	$ML^{-2}T^{-1}$
m	Slope of equilibrium line	—
r_s	Fraction of total feed that goes to stream s	—
s	Number of stages	—
x_{ak}	Concentration of component k in stream a	—
x_{bk}	Concentration of component k in stream b	—
x_{dk}	Concentration of component k in distillate	—
x_{fk}	Concentration of component k in feed	—
x_{wk}	Concentration of component k in bottom product	—
λ_{ik}	Total flow of component k to unit i	MT^{-1}
α_{jik}	Split-fraction coefficient: fraction of component k flowing from unit i to unit j	—

CHAPTER 5

Piping and Instrumentation

5.1. Introduction

The process flow-sheet shows the arrangement of the major pieces of equipment and their interconnection. It is a description of the nature of the process.

The Piping and Instrument diagram (P & I diagram) shows the engineering details of the equipment, instruments, piping, valves and fittings; and their arrangement. It is often called the Engineering Flow-sheet or Engineering Line Diagram.

This chapter covers the preparation of the preliminary P & I diagrams at the process design stage of the project.

The design of piping systems, and the specification of the process instrumentation and control systems, is usually done by specialist design groups, and a detailed discussion of piping design and control systems is beyond the scope of this book. Only general guide rules are given. The piping handbook edited by Holmes (1973) is particularly recommended for the guidance on the detailed design of piping systems and process instrumentation and control. The references cited in the text and listed at the end of the chapter should also be consulted.

5.2. The P & I diagram

The P & I diagram shows the arrangement of the process equipment, piping, pumps, instruments, valves and other fittings. It should include:

1. All process equipment identified by an equipment number. The equipment should be drawn roughly in proportion, and the location of nozzles shown.
2. All pipes, identified by a line number. The pipe size and material of construction should be shown. The material may be included as part of the line identification number.
3. All valves, control and block valves, with an identification number. The type and size should be shown. The type may be shown by the symbol used for the valve or included in the code used for the valve number.
4. Ancillary fittings that are part of the piping system, such as inline sight-glasses, strainers and steam traps; with an identification number.
5. Pumps, identified by a suitable code number.
6. All control loops and instruments, with an identification number.

For simple processes, the utility (service) lines can be shown on the P & I diagram. For complex processes, separate diagrams should be used to show the service lines, so the information can be shown clearly, without cluttering up the diagram. The service connections to each unit should, however, be shown on the P & I diagram.

148

The P & I diagram will resemble the process flow-sheet, but the process information is not shown. The same equipment identification numbers should be used on both diagrams.

5.2.1. Symbols and layout

The symbols used to show the equipment, valves, instruments and control loops will depend on the practice of the particular design office. The equipment symbols are usually more detailed than those used for the process flow-sheet. A typical example of a P & I diagram is shown in Fig. 5.17.

Standard symbols for instruments, controllers and valves are given in the British Standard BS 1646.

Austin (1979) gives a comprehensive summary of the British Standard symbols, and also shows the American standard symbols (ANSI) and examples of those used by some process plant contracting companies.

When laying out the diagram, it is only necessary to show the relative elevation of the process connections to the equipment where these affect the process operation; for example, the net positive suction head (NPSH) of pumps, barometric legs, syphons and the operation of thermosyphon reboilers.

5.2.2. Basic symbols

The symbols illustrated below are those given in BS 1646.

Control valve

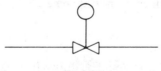

FIG. 5.1

This symbol is used to represent all types of control valve, and both pneumatic and electric actuators.

Failure mode

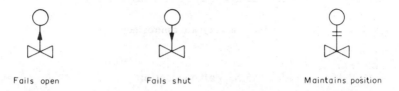

Fails open   Fails shut   Maintains position

FIG. 5.2

The direction of the arrow shows the position of the valve on failure of the power supply.

Instruments and controllers

FIG. 5.3

Locally mounted means that the controller and display is located out on the plant near to the sensing instrument location. *Main panel* means that they are located on a panel in the control room. Except on small plants, most controllers would be mounted in the control room.

Type of instrument

This is indicated on the circle representing the instrument-controller by a letter code (see Table 5.1).

The first letter indicates the property measured; for example, F = flow. Subsequent letters indicate the function; for example,

$$I = \text{indicating}$$

$$RC = \text{recorder controller}$$

The suffixes E and A can be added to indicate emergency action and/or alarm functions.

Lines

The instrument connecting lines should be drawn in a manner to distinguish them from the main process lines. Dotted or cross-hatched lines are normally used.

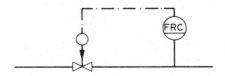

FIG. 5.4. A typical control loop

5.3. Valve selection

The valves used for chemical process plant can be divided into two broad classes, depending on their primary function:

1. Shut-off valves (block valves), whose purpose is to close off the flow.
2. Control valves, both manual and automatic, used to regulate flow.

TABLE 5.1. *Letter code*
(Based on BS 1646: 1979)

1	2	3	4	5	6	7	8	9	10	11
		Sequence of additional letters denoting instrument function								
Property measured	First letter	Indicating only	Recording only	Indicating and Integrating	Recording and Integrating	Controlling only	Indicating and Controlling	Recording and Controlling	Indicating Controlling and Integrating	Recording Controlling and Integrating
		(I)	(R)	(IS)	(RS)	(C)	(IC)	(RC)	(ICS)	(RCS)
Flow rate (volumetric or gravimetric)	F	FI	FR	FIS	FRS	FC	FIC	FRC	FICS	FRCS
Level (surface height, depth, contents)	L	LI	LR			LC	LIC	LRC		
Movement, displacement or dimensions of solid elements	U	UI	UR			C	UIC	URC		
Pressure	P	PI	PR			PC	PIC	PRC		
Quality, analysis or concentration*	Q	QI	QR			QC	QIC	QRC		
Radiation	R	RI	RR	RIS	RRS	RC	RIC	RRC	RICS	RRCS
Speed (linear or rotary)	S	SI	SR	SIS	SRS	SC	SIC	SRC	SICS	SRCS
Temperature	T	TI	TR			TC	TIC	TRC		
Weight, mass or load	W	WI	WR			WC	WIC	WRC		
Any other property*	X	XI	XR			XC	XIC	XRC		
Combinations of different properties†	D	DI	DR			DC	DIC	DRC		

NOTE 1. Where applicable the following suffixes may be added to the first letter:
d—denoting differential, e.g. Td, Pd,
r—denoting ratio, e.g. Fr, Sr.

NOTE 2. The letter E and/or A may be added as suffixes to any of the above letter codes to indicate an emergency action and/or alarm function respectively.

NOTE 3. If the first letter is F the suffix "q" may be added to denote quantity transfer.

NOTE 4. If the first letter is R, the suffixes α, β, γ, n (neutrons) may be added to indicate the form of radiation followed by the suffix "q" when radiant energy transfer is to be denoted.

* A note shall be added to specify the property measured.

† The letter D shall only be used where measurements normally represented by different letters are being combined in one instrument (e.g. data logger). In all such cases the normal symbols for the separate measurement shall also be shown individually at the points of measurement.

The main types of valves used are:

Gate   Fig. 5.5a
Plug   Fig. 5.5b
Ball   Fig. 5.5c
Globe   Fig. 5.5d
Diaphragm   Fig. 5.5e
Butterfly   Fig. 5.5f

A valve selected for shut-off purposes should give a positive seal in the closed position and minimum resistance to flow when open. Gate, plug and ball valves are most frequently used for this purpose.

If flow control is required, the valve should be capable of giving smooth control over the full range of flow, from fully open to closed. Globe valves are normally used, though the other types can be used. Butterfly valves are often used for the control of gas and

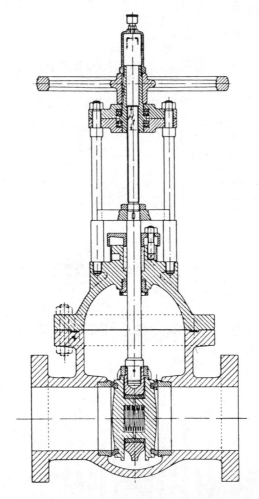

Fig. 5.5a. Gate valve (slide valve)

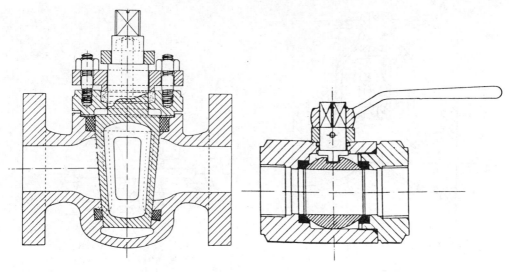

FIG. 5.5b. Plug valve   FIG. 5.5c. Ball valve

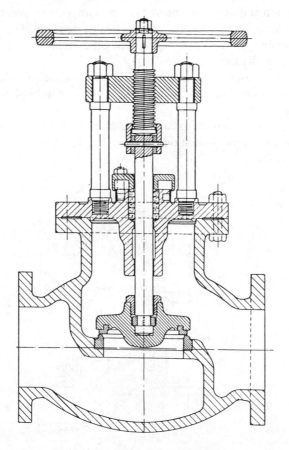

FIG. 5.5d. Globe valve

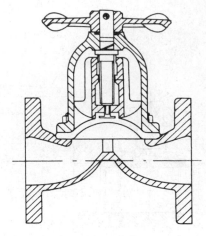

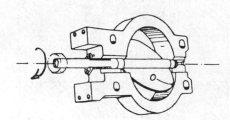

FIG. 5.5e. Diaphragm valve   FIG. 5.5f. Butterfly valve

vapour flows. Automatic control valves are basically globe valves with special trim designs (see Volume 3, Chapter 3).

The careful selection and design of control valves is important; good flow control must be achieved, whilst keeping the pressure drop as low as possible. The valve must also be sized to avoid the flashing of hot liquids and the super-critical flow of gases and vapours. Control valve sizing is discussed by Chaflin (1974).

Non-return valves are used to prevent back-flow of fluid in a process line. They do not normally give an absolute shut-off of the reverse flow. A typical design is shown in Fig. 5.5g.

Details of valve types and standards can be found in the Technical Data Manual of the British Valve Manufacturers Association (1972). Valve design is covered in detail by Pearson (1978).

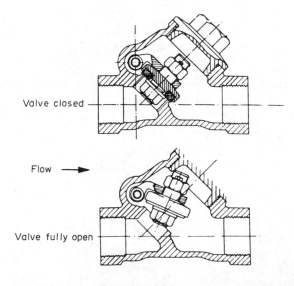

FIG. 5.5g. Non-return valve, check valve, hinged disc type

5.4. Pumps

The pumping of liquids is covered in Volume 1, Chapter 6. Reference should be made to that chapter for a discussion of the principles of pump design and illustrations of the more commonly used pumps.

Pumps can be classified into two general types:

1. Dynamic pumps, such as centrifugal pumps.
2. Positive displacement pumps, such as reciprocating and diaphragm pumps.

The single-stage, horizontal, overhung, centrifugal pump is by far the most commonly used type in the chemical process industry. Other types are used where a high head or other special process considerations are specified.

Pump selection is made on the flow rate and head required, together with other process considerations, such as corrosion or the presence of solids in the fluid.

The chart shown in Fig. 5.6 can be used to determine the type of pump required for a particular head and flow rate. This figure is based on one published by Doolin (1977).

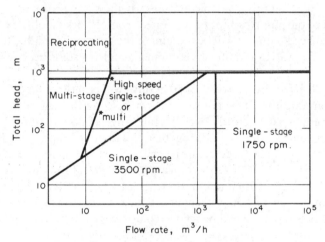

FIG. 5.6. Centrifugal pump selection guide.
* Single-stage > 1750 rpm, multi-stage 1750 rpm

Centrifugal pumps are characterised by their specific speed (see Volume 1, Chapter 6). In the dimensionless form, specific speed is given by:

$$N_s = \frac{NQ^{1/2}}{(gh)^{3/4}} \tag{5.1}$$

where N = revolutions per second,
Q = flow, m^3/s,
h = head, m,
g = gravitational acceleration m/s^2.

Pump manufacturers do not generally use the dimensionless specific speed, but define it by the equation:

$$N'_s = \frac{NQ^{1/2}}{h^{3/4}} \tag{5.2}$$

where N'_s = revolutions per minute (rpm),
          Q = flow, US gal/min,
          h = head, ft.

Values of the non-dimensional specific speed, as defined by equation 5.1, can be converted to the form defined by equation 5.2 by multiplying by 1.73×10^4.

The specific speed for centrifugal pumps (equation 5.2) usually lies between 400 and 10,000, depending on the type of impeller. Generally, pump impellers are classified as radial for specific speeds between 400 and 1000, mixed flow between 1500 and 7000, and axial above 7000. Doolin (1977) states that below a specific speed of 1000 the efficiency of single-stage centrifugal pumps is low and multi-stage pumps should be considered.

For a detailed discussion of the factors governing the selection of the best centrifugal pump for a given duty the reader should refer to the articles by De Santis (1976), Neerkin (1974), or Jacobs (1965).

Positive displacement, reciprocating, pumps are normally used where a high head is required at a low flow-rate. Holland and Chapman (1966) review the various types of positive displacement pumps available and discuss their applications.

The selection of the pump cannot be separated from the design of the complete piping system. The total head required will be the sum of the dynamic head due to friction losses in the piping, fittings, valves and process equipment, and any static head due to differences in elevation.

The pressure drop required across a control valve will be a function of the valve design. Sufficient pressure drop must be allowed for when sizing the pump to ensure that the control valve operates satisfactorily over the full range of flow required. If possible, the control valve and pump should be sized together, as a unit, to ensure that the optimum size is selected for both. As a rough guide, if the characteristics are not specified, the control valve pressure drop should be taken as at least 30 per cent of the total dynamic pressure drop through the system, with a minimum value of 50 kPa (7 psi). The valve should be sized for a maximum flow rate 30 per cent above the normal stream flow-rate. Some of the pressure drop across the valve will be recovered down stream, the amount depending on the type of valve used.

Methods for the calculation of pressure drop through pipes and fittings are given in Volume 1, Chapter 3. It is important that a proper analysis is made of the system and the use of a calculation form (work sheet) to standardise pump-head calculations is recommended. A standard calculation form ensures that a systematic method of calculation is used, and provides a check list to ensure that all the usual factors have been considered. It is also a permanent record of the calculation. Example 5.4 has been set out to illustrate the use of a typical calculation form. The calculation should include a check on the net positive suction head (NPSH) available (see Volume 1, Chapter 6). It is important that the pump suction pressure does not fall below the vapour pressure of the liquid being pumped, or vapour bubbles may form; which can lead to a reduction in flow or cavitation in the pump. The available suction head is determined by the suction piping design and the vapour pressure of the fluid; and can be calculated using the methods given in Volume 1.

The NPSH required is a function of the pump design and is specified by the manufacturer. As a general guide, the NPSH should be above 3 m for pump capacities up to a flow rate of 100 m^3/h, and 6 m above this capacity. Special impeller designs can be used to overcome problems of low suction head (Doolin, 1977).

Kern (1975) discusses the practical design of pump suction piping, in a series of articles on the practical aspects of piping system design published in the journal *Chemical Engineering* from December 1973 through to November 1975. A detailed presentation of pipe-sizing techniques is also given by Simpson (1968), who covers liquid, gas and two-phase systems. Line sizing and pump selection is also covered in a comprehensive article by Ludwig (1960).

5.5. Mechanical design of piping systems

5.5.1. Wall thickness: pipe schedule

The pipe wall thickness is selected to resist the internal pressure, with an allowance for corrosion. Processes pipes can normally be considered as thin cylinders; only high-pressure pipes, such as high-pressure steam lines, are likely to be classified as thick cylinders and must be given special consideration (see Chapter 13).

The British Standard BS 3351 gives the following formula for pipe thickness:

$$t = \frac{Pd}{20\sigma_d + P} \qquad (5.3)$$

where P = internal pressure, bar,
d = pipe od, mm,
σ_d = design stress at working temperature, N/mm^2.

Pipes are often specified by a schedule number (based on the thin cylinder formula). The schedule number is defined by:

$$\text{Schedule number} = \frac{P_s \times 1000}{\sigma_s} \qquad (5.4)$$

P_s = safe working pressure, lb/in^2 (or N/mm^2),
σ_s = safe working stress, lb/in^2 (or N/mm^2).

Schedule 40 pipe is commonly used for general purposes.

Full details of the preferred dimensions for pipes can be found in the appropriate Handbook and Standards. The main United Kingdom codes for pipes and piping systems are the British Standards BS 3351 and BS 1600.

The UK pipe schedule numbers are the same as the American (US). A summary of the US standards is given in Perry and Chilton (1973).

Example 5.1

Estimate the safe working pressure for a 4 in. (100 mm) dia., schedule 40 pipe, carbon steel, butt welded, working temperature 100°C. The safe working stress for butt welded steel pipe up to 120°C is 6000 lb/in^2 (41·4 N/mm^2).

Solution

$$P_s = \frac{(\text{schedule no.}) \times \sigma_s}{1000} = \frac{40 \times 6000}{1000} = \underline{\underline{240 \text{ lb/in}^2}} = \underline{\underline{1656 \text{ kN/m}^2}}$$

5.5.2. Pipe supports

Over long runs, between buildings and equipment, pipes are usually carried on pipe racks. These carry the main process and service pipes, and are laid out to allow easy access to the equipment.

Various designs of pipe hanger and support are used to support individual pipes. Details of typical supports can be found in Perry and Chilton (1973) and Holmes (1973). Pipe supports frequently incorporate provision for thermal expansion.

5.5.3. Pipe fittings

Pipe runs are normally made up from lengths of pipe, incorporating standard fittings for joints, bends and tees. Joints are usually welded but small sizes may be screwed. Flanged joints are used where this is a more convenient method of assembly, or if the joint will have to be frequently broken for maintenance. Flanged joints are normally used for the final connection to the process equipment, valves and ancillary equipment.

Details of the standard fittings, welded, screwed and flanged, can be found in the Handbooks and in the appropriate British and American standards. The standards for metal pipes and fittings are discussed by Masek (1968).

5.5.4. Pipe stressing

Piping systems must be designed so as not to impose unacceptable stresses on the equipment to which they are connected.

Loads will arise from:

1. Thermal expansion of the pipes and equipment.
2. The weight of the pipes, their contents, insulation and any ancillary equipment.
3. The reaction to the fluid pressure drop.
4. Loads imposed by the operation of ancillary equipment, such as relief valves.
5. Vibration.

Thermal expansion is a major factor to be considered in the design of piping systems. The reaction load due to pressure drop will normally be negligible. The dead-weight loads can be carried by properly designed supports.

Flexibility is incorporated into piping systems to absorb the thermal expansion. A piping system will have a certain amount of flexibility due to the bends and loops required by the layout. If necessary, expansion loops, bellows and other special expansion devices can be used to take up expansion.

A discussion of the methods used for the calculation of piping flexibility and stress analysis are beyond the scope of this book. Manual calculation techniques, and the application of computers in piping stress analysis, are discussed in Chapter 12 of the handbook edited by Holmes (1973). Other texts which give methods for the flexibility analysis of piping systems are those by King (1967) and the M. W. Kellog Co. (1964).

5.6. Pipe size selection

If the motive power to drive the fluid through the pipe is available free, for instance when pressure is let down from one vessel to another or if there is sufficient head for gravity flow, the smallest pipe diameter that gives the required flow-rate would normally be used.

If the fluid has to be pumped through the pipe, the size should be selected to give the least annual operating cost.

Typical pipe velocities and allowable pressure drops, which can be used to estimate pipe sizes, are given below:

	Velocity m/s	ΔP kPa/m
Liquids, pumped (not viscous)	1–3	0·5
Liquids, gravity flow	—	0·05
Gases and vapours	15–30	0·02 per cent of line pressure
High-pressure steam, > 8 bar	30–60	—

Rase (1953) gives expressions for design velocities in terms of the pipe diameter. His expressions, converted to SI units, are:

Pump discharge	$0·06d + 0·4$ m/s
Pump suction	$0·02d + 0·1$ m/s
Steam or vapour	$0·2d$ m/s

where d is the internal diameter in mm.

Simpson (1968) gives values for the optimum velocity in terms of the fluid density. His values, converted to SI units and rounded, are:

Fluid density kg/m^3	Velocity m/s
1600	2·4
800	3·0
160	4·9
16	9·4
0·16	18·0
0·016	34·0

The maximum velocity should be kept below that at which erosion is likely to occur. For gases and vapours the velocity cannot exceed the critical velocity (sonic velocity) (see Volume 1, Chapter 4) and would normally be limited to 30 per cent of the critical velocity.

Economic pipe diameter

The capital cost of a pipe run increases with diameter, whereas the pumping costs decrease with increasing diameter. The most economic pipe diameter will be the one which gives the lowest annual operating cost. Several authors have published formulae and nomographs for the estimation of the economic pipe diameter, Genereaux (1937), Peters and Timmerhaus (1968) and Nolte (1978). Most apply to American practice and costs, but the method used by Peters and Timmerhaus has been modified to take account of UK prices (Anon, 1971).

The formulae developed in this section are presented as an illustration of a simple optimisation problem in design, and to provide an estimate of economic pipe diameter that is based on UK costs and in SI units. The method used is essentially that first published by Genereaux (1937).

The cost equations can be developed by considering a 1 metre length of pipe.

The purchase cost will be roughly proportional to the diameter raised to some power.

$$\text{Purchase cost} = Kd^n \; \pounds/\text{m} \qquad (5.5)$$

The value of the constant A and the index n depend on the pipe material and schedule.

The installed cost can be calculated by using the factorial method of costing discussed in Chapter 6.

$$\text{Installed cost} = Kd^n(1+F) \qquad (5.6)$$

where the factor F includes the cost of valves, fittings and erection, for a typical run of the pipe.

The capital cost can be included in the operating cost as an annual capital charge. There will also be an annual charge for maintenance, based on the capital cost.

$$Cp = Kd^n(1+F)(a+b) \qquad (5.7)$$

where Cp = capital cost portion of the annual operating cost, $\pounds$,

a = capital charge, per cent/100,

b = maintenance costs, per cent/100.

The power required for pumping is given by:

$$\text{Power} = \text{volumetric flow-rate} \times \text{pressure drop.}$$

Only the friction pressure drop need be considered, as any static head is not a function of the pipe diameter.

To calculate the pressure drop the pipe friction factor needs to be known. This is a function of Reynolds number, which is in turn a function of the pipe diameter. Several expressions have been proposed for relating friction factor to Reynolds number (see Volume 1, Chapter 3). For simplicity the relationship proposed by Genereaux (1937) for turbulent flow in clean commercial steel pipes will be used.

$$f = 0.04 \, Re^{-0.16} \qquad (5.8)$$

where f is the Fanning friction factor $= 2 \dfrac{R}{\rho u^2}$.

Substituting this into the Fanning pressure drop equation gives:

$$\Delta P = 4.07 \times 10^{10} G^{1.84} \mu^{0.16} \rho^{-1} d^{-4.84} \qquad (5.9)$$

where ΔP = pressure drop, kN/m^2 (kPa),

G = flow rate, kg/s,

ρ = density, kg/m^3,

d = pipe id, mm.

The annual pumping costs will be given by:

$$Cf = \frac{Hp}{E} \Delta P \frac{G}{\rho} \qquad (5.10)$$

where H = plant attainment, hours/year,

p = cost of power, $\pounds$/kWh,

E = pump efficiency, per cent/100.

Substituting from equation 5.9

$$Cf = \frac{Hp}{E} 4 \cdot 07 \times 10^{10} G^{2 \cdot 84} \mu^{0 \cdot 16} \rho^{-2} d^{-4 \cdot 84} \qquad (5.11)$$

The total annual operating cost $Ct = Cp + Cf$.

Adding equations 5.7 and 5.11, differentiating, and equating to zero to find the pipe diameter to give the minimum cost gives:

$$d, \text{optimum} = \left[\frac{19 \cdot 80 \times 10^{10} \times HpG^{2 \cdot 84} \mu^{0 \cdot 16} \rho^{-2}}{EnK(1+F)(a+b)} \right]^{\frac{1}{(4 \cdot 84 + n)}} \qquad (5.12)$$

Equation 5.12 is a general equation and can be used to estimate the economic pipe for any particular situation.

The equation can be simplified by substituting typical values for the constants.

H The normal attainment for a chemical plant will be 90–95 per cent, so take operating hours as 8000 hours/year.

E Pump and compressor efficiencies will be between 50–70 per cent, put $E = 0 \cdot 6$.

P Use current cost of power, $0 \cdot 016 \pounds/\text{kWh}$ (mid-1976).

F This is the most difficult factor to estimate. Other authors have used values ranging from $1 \cdot 5$ (Peters and Timmerhaus, 1968) to $6 \cdot 75$ (Nolte, 1978).

K, n Can be estimated from the current cost of piping. Most authors have used a value of n of $1 \cdot 5$, first proposed by Genereaux.

F, K, n have been estimated from the cost data published by the Institution of Chemical Engineers (1977) (p. 48, table 7.7). This includes the cost of fittings and erection. A log-log plot of the data gives the following expressions for the installed cost (equation 5.6):

Carbon steel, 50 to 250 mm	$3 \cdot 9d^{0 \cdot 6}$	$\pounds/\text{m}$
Stainless steel, 40 to 150 mm	$1 \cdot 6d^{0 \cdot 9}$	$\pounds/\text{m}$

The date of these costs is given as April 1976.

The equations should remain valid with time; the cost of steel is dependent on the cost of power and the two costs appear in the equations as a ratio raised to a small fractional exponent. Substitution in equation 5.12 gives, for carbon steel:

$$d, \text{optimum} = 352 \cdot 8 G^{0 \cdot 52} \mu^{0 \cdot 03} \rho^{-0 \cdot 37} \qquad (5.13)$$

Because the exponent of the viscosity term is small, its value will change very little over a wide range of viscosity

at $\mu = 10^{-5} \, \text{N s/m}^2 \, (0 \cdot 01 \text{ cp}), \mu^{0 \cdot 03} = 0 \cdot 71$
                    $\mu = 10^{-2} \, \text{N s/m}^2 \, (10 \text{ cp}), \mu^{0 \cdot 03} \quad = 0 \cdot 88$

Taking a mean value of $0 \cdot 8$, the approximate optimum diameter for carbon steel is given by:

$$d, \text{optimum} = 282 \, G^{0 \cdot 52} \rho^{-0 \cdot 37} \qquad (5.14)$$

For stainless steel the expression is:

$$d, \text{optimum} = 226 G^{0 \cdot 50} \rho^{-0 \cdot 35} \qquad (5.15)$$

Equations 5.14 and 5.15 can be used to make an approximate estimate of the economic pipe diameter for normal pipe runs. For a more accurate estimate, or if the fluid or pipe run, is unusual, the method used to develop equation 5.12 can be used, taking into account the special features of the particular pipe run.

Equations for the optimum pipe diameter with laminar flow can be developed by using a suitable equation for pressure drop in the equation for pumping costs (equation 5.10).

The approximate equations should not be used for steam, as the quality of steam depends on its pressure, and hence the pressure drop.

Nolte (1978) gives detailed methods for the selection of economic pipe diameters, taking into account all the factors involved. He gives equations for liquids, gases, steam and two-phase systems. He includes in his method an allowance for the pressure drop due to fittings and valves, which was neglected in the development of equation 5.12, and by most other authors.

The use of equations 5.14 and 5.15 are illustrated in Examples 5.2 and 5.3, and the results compared with those obtained by other authors. Peters and Timmerhaus's formula give larger values for the economic pipe diameters, which is probably due to their low value for the installation cost factor, F.

Example 5.2

Estimate the optimum pipe diameter for a water flow rate of 10 kg/s, at 20°C. Carbon steel pipe will be used. Density of water 1000 kg/m^3.

Solution

(5.14)
$$d \text{ optimum} = 282 \times (10)^{0.52} \, 1000^{-0.37}$$
$$= \underline{\underline{72.5 \text{ mm}}}$$

use 80-mm pipe.

Viscosity of water at 20°C = 1.1×10^{-3} N s/m^2,

$$Re = \frac{4G}{\pi \mu d} = \frac{4 \times 10}{\pi \times 1.1 \times 10^{-3} \times 80 \times 10^{-3}} = 1.45 \times 10^5$$

> 4000, so flow is turbulent.
Comparison of methods:

	Economic diameter
Equation 5.14	80 mm
Peters and Timmerhaus (1968)	4 in. (100 mm)
Nolte (1978)	80 mm

Example 5.3

Estimate the optimum pipe diameter for a flow of HCl of 7000 kg/h at 5 bar, 15°C, stainless steel pipe. Molar volume 22.4 m^3/kmol, at 1 bar, 0°C.

Solution

Molecular weight HCl = 36·5.
Density at operating conditions

$$= \frac{36·5}{22·4} \times \frac{5}{1} \times \frac{273}{288} = \underline{7·72\,kg/m^3}$$

(5.15)   Optimum diameter $= 226\left(\frac{7000}{3600}\right)^{0·5} 7·72^{-0·35}$

$$= \underline{154\,mm}$$

use 150-mm pipe.
Viscosity of HCl 0·013 m N s/m²

$$Re = \frac{4}{\pi} \times \frac{7000}{3600} \times \frac{1}{0·013 \times 10^{-3} \times 150 \times 10^{-3}} = \underline{1·27 \times 10^6},\ turbulent$$

Comparison of methods:

	Economic diameter
Equation 5.15	150 mm
Peters and Timmerhaus (1968)	9 in. (220 mm) carbon steel
Nolte (1978)	7 in. (180 mm) carbon steel

Example 5.4

Calculate the line size and specify the pump required for the line shown in Fig. 5.7; material ortho-dichlorobenzene (ODCB), flow-rate 10,000 kg/h, temperature 20°C, pipe material carbon steel.

Solution

ODCB density at 20°C = 1306 kg/m³.
Viscosity: 0·9 mNs/m² (0·9 cp).
Estimation of pipe diameter required:

typical velocity for liquid 2 m/s
mass flow = 10,000/3600 = 2·78 kg/s
volumetric flow = 2·78/1306 = 2·13 × 10⁻³ m³/s

$$area\ of\ pipe = \frac{volumetric\ flow}{velocity} = \frac{2·13 \times 10^{-3}}{2} = 1·06 \times 10^{-3}\,m^2$$

$$diameter\ of\ pipe = \sqrt{\left(1·06 \times 10^{-3} \times \frac{4}{\pi}\right)} = 0·037\,m$$

$$= 37\,mm$$

Or, use economic pipe diameter formula:

(5.14) d, optimum $= 282 \times 2·78^{0·52} \times 1306^{-0·37}$

$$= 34\,mm$$

Take diameter as $\underline{40\,mm}$

$$cross\text{-}sectional\ area = \frac{\pi 40^2}{4} = 1257\,mm^2$$

$$= 1·26 \times 10^{-3}\,m^2$$

Pressure-drop calculation:

$$\text{fluid velocity} = \frac{2 \cdot 13 \times 10^{-3}}{1 \cdot 26 \times 10^{-3}} = 1 \cdot 70 \text{ m/s}$$

friction loss (Genereaux's formula):

(5.9) $\Delta P = 4 \cdot 07 \times 10^{10} (2 \cdot 78)^{1 \cdot 84} (0 \cdot 9 \times 10^{-3})^{0 \cdot 16} 1306^{-1} \times 40^{-4 \cdot 84} = 1 \cdot 17 \text{ kPa/m}$

The friction loss can also be calculated using the formula and friction factor chart given in Volume 1. This will be done to give a check on the above value.

$$Re = \frac{4G}{\pi \mu d} = \frac{4 \times 2 \cdot 8}{\pi 0 \cdot 9 \times 10^{-3} \times 40 \times 10^{-3}} = 98,324$$

The absolute roughness for commercial steel pipe, $e = 0 \cdot 046$ mm (see table in Volume 1, Chapter 3).

$$\text{relative roughness } e/d = \frac{0 \cdot 046}{40} = 1 \cdot 15 \times 10^{-3}$$

From the friction factor chart in Volume 1, Chapter 3,

$$\text{at } Re = 98,324, \frac{R}{\rho u^2} = 0 \cdot 0027$$

$$\text{Friction loss per metre} = 4 \left(\frac{R}{\rho u^2} \right) \frac{\rho u^2}{d}$$

$$= 4 \times 0 \cdot 0027 \times 1306 \times \frac{1 \cdot 7^2}{40 \times 10^{-3}} = 1019 \text{ N/m}^2$$

$$= \underline{1 \cdot 02 \text{ kPa}}$$

Take the higher value, and design for a maximum flow rate of 20 per cent above the normal (average) flow:

$$\text{friction loss at maximum flow} = 1 \cdot 02 \times 1 \cdot 2^2$$
$$= \underline{1 \cdot 5 \text{ kPa/m}}$$

The loss through the bends and block valves can be included in line pressure-loss calculation as an "equivalent length of pipe" (see Volume 1, Chapter 3):

All the bends will be taken as 90° elbows of standard radius, equivalent length = 30d, and the valves as plug valves, fully open, equivalent length = 18d. Line to pump suction:

$$\begin{aligned} \text{length} &= 1 \cdot 5 \text{ m} \\ \text{bend, } 1 \times 30 \times 40 \times 10^{-3} &= 1 \cdot 2 \text{ m} \\ \text{valve, } 1 \times 18 \times 40 \times 10^{-3} &= \frac{0 \cdot 7}{3 \cdot 4} \text{ m} \end{aligned}$$

$$\text{entry loss} = \frac{\rho u^2}{2} \quad \text{(see Volume 1, Chapter 3)}$$

$$\text{at maximum design velocity} = \frac{1306 (1 \cdot 7 \times 1 \cdot 2)^2}{2 \times 10^3} = 2 \cdot 7 \text{ kPa}$$

Control valve pressure drop, allow normal 140 kPa

$(\times 1.2^2)$ maximum 200 kPa

Heat exchanger, allow normal 70 kPa

$(\times 1.2^2)$ maximum 100 kPa

Orifice, allow normal 15 kPa

$(\times 1.2^2)$ maximum 22 kPa

Line from pump discharge:

$$\text{length} = 4 + 5.5 + 20 + 5 + 0.5 + 1 + 6.5 + 2 = 44.5 \text{ m}$$
$$\text{bends, } 6 \times 30 \times 40 \times 10^{-3} = 7.2 \text{ m} \qquad = 7.2 \text{ m}$$
$$\text{valves, } 3 \times 18 \times 40 \times 10^{-3} = 2.2 \text{ m} \qquad = \underline{2.2 \text{ m}}$$
$$\underline{54.0 \text{ m}}$$

The line pressure-drop calculation is set out on the calculation sheet shown in Table 5.2.

Pump selection:

$$\text{flow-rate} = 2.13 \times 10^{-3} \times 3600 = 7.7 \text{ m}^3/\text{h}$$
$$\text{differential head, maximum, } \underline{\underline{44 \text{ m}}}$$

select single-stage centrifugal (Table 5.3)

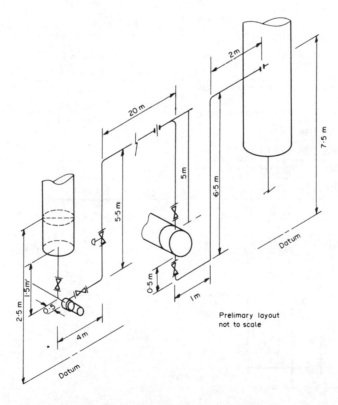

FIG. 5.7. Piping isometric drawing (Example 5.4)

TABLE 5.2. *Line Calculation form (Example 5.4)*

Pump and Line Calculation Sheet					
Job no.	**Sheet no.**	**By** RKS, 7/7/79		**Checked**	
4415A	1				

Fluid		ODCB		**DISCHARGE CALCULATION**			
Temperature °C		20		Line size mm		40	
Density kg/m³		1306		Flow	Norm.	Max.	Units
Viscosity mNs/m²		0·9	u_2	Velocity	1·7	2·0	m/s
Normal Flow kg/s		2·78	Δf_2	Friction Loss	1·0	1·5	kPa/m
Design Max. Flow kg/s		3·34	L_2	Line Length	54	—	m
			$\Delta f_2 L_2$	Line Loss	54		kPa
SUCTION CALCULATION				Orifice	15	22	kPa
Line size mm		40		30%	Control Valve	140	200	kPa
	Flow	Norm.	Max.	Units		*Equipment*		
u_1	Velocity	1·7	2·0	m/s		(a) Heat Ex.	70	100	kPa
Δf_1	Friction Loss	1·0	1·5	kPa/m		(b)	—	—	kPa
L_1	Line Length	3·4	—	m		(c)	—	—	kPa
$\Delta f_1 L_1$	Line Loss	3·4	5·1	kPa		(6) Dynamic Loss	279	403	kPa
$\rho u_1^2/2$	Entrance	1·9	2·7	kPa					
(40 kPa)	Strainer	—	—	kPa	z_2	Static Head	6·5	—	m
	(1) Sub-total	5·3	7·8	kPa	$\rho g z_2$		85	85	kPa
						Equip. Press (Max)	200	200	kPa
z_1	Static Head	1·5	1·5	m		Contingency	None	None	kPa
$\rho g z_1$		19·6	19·6	kPa		(7) Sub-total	285	285	kPa
	Equip. Press	100	100	kPa	(7) + (6)	Discharge Press	564	685	kPa
	(2) Sub-total	119·6	119·6	kPa	(3)	Suction Press	114·3	111·8	kPa
(2) − (1)	(3) Suction Press	114·3	111·8	kPa		(8) Diff. Press	450	576	kPa
	(4) VAP. PRESS.	0·1	0·1	kPa	(8)/ρg		34	44	m
(3) − (4)	(5) NPSH	114·2	111·7	kPa					
(5)/ρg			8·7	8·6	m	Valve/(6)	Control Valve % Dyn. Loss	50%		

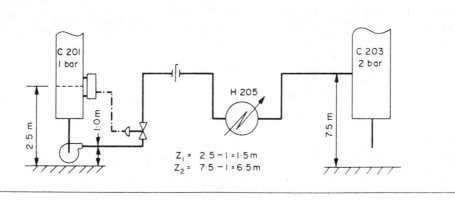

C 201
1 bar

C 203
2 bar

H 205

2·5 m

1·0 m

7·5 m

$Z_1 = 2·5 - 1 = 1·5$ m
$Z_2 = 7·5 - 1 = 6·5$ m

TABLE 5.3. *Pump Specification Sheet* (Example 5.4)

	Pump Specification
Type:	Centrifugal
No. stages:	1
Single/Double suction:	Single
Vertical/Horizontal mounting:	Horizontal
Impeller type:	Closed
Casing design press.:	600 kPa
design temp.:	20°C
Driver:	Electric, 440 V, 50 c/s 3-phase.
Seal type:	Mechanical, external flush
Max. flow:	7·7 m^3/h
Diff. press.:	600 kPa (47 m, water)

5.7. Control and instrumentation

5.7.1. Instruments

Instruments are provided to monitor the key process variables during plant operation. They may be incorporated in automatic control loops, or used for the manual monitoring of the process operation. They may also be part of an automatic computer data logging system. Instruments monitoring critical process variables will be fitted with automatic alarms to alert the operators to critical and hazardous situations.

The main types of instruments used for chemical process plants are described in Volumes 1 and 3; flow and pressure measurement is covered in detail in Volume 1, Chapter 5, and temperature and level measurement, briefly, in Volume 3, Chapter 3. Comprehensive reviews of process instruments and control equipment are published periodically in the journal *Chemical Engineering*. These reviews give details of all the instruments and control hardware available commercially, including those for the on-line analysis of stream compositions, (Anon., 1969). Details of process instruments and control equipment can also be found in various handbooks, Perry and Chilton (1973) and Considine (1957).

It is desirable that the process variable to be monitored be measured directly; often, however, this is impractical and some dependent variable, that is easier to measure, is monitored in its place. For example, in the control of distillation columns the continuous, on-line, analysis of the overhead product is desirable but difficult and expensive to achieve reliably, so temperature is often monitored as an indication of composition. The temperature instrument may form part of a control loop controlling, say, reflux flow; with the composition of the overheads checked frequently by sampling and laboratory analysis.

5.7.2. Instrumentation and control objectives

The primary objectives of the designer when specifying instrumentation and control schemes are:

1. Safe plant operation:
   (a) To keep the process variables within known safe operating limits.
   (b) To detect dangerous situations as they develop and to provide alarms and automatic shut-down systems.

(c) To provide interlocks and alarms to prevent dangerous operating procedures.
2. Production rate:
   To achieve the design product output.
3. Product quality:
   To maintain the product composition within the specified quality standards.
4. Cost:
   To operate at the lowest production cost, commensurate with the other objectives.

These are not separate objectives and must be considered together. The order in which they are listed is not meant to imply the precedence of any objective over another, other than that of putting safety first. Product quality, production rate and the cost of production will be dependent on sales requirements. For example, it may be a better strategy to produce a better-quality product at a higher cost.

In a typical chemical processing plant these objectives are achieved by a combination of automatic control, manual monitoring and laboratory analysis.

5.7.3. Automatic-control schemes

The detailed design and specification of the automatic control schemes for a large project is usually done by specialists. The basic theory underlying the design and specification of automatic control systems is covered in Volume 3, Chapter 3. A bibliography of the numerous specialised texts that have been published on the subject is given at the end of that chapter. The books by Buckley (1964) and Shinskey (1979a), cited in this chapter, cover many of the more practical aspects of process control system design, and are recommended.

In this chapter only the first step in the specification of the control systems for a process will be considered: the preparation of a preliminary scheme of instrumentation and control, developed from the process flow-sheet. This can be drawn up by the process designer based on his experience with similar plant and his critical assessment of the process requirements. Many of the control loops will be conventional and a detailed analysis of the system behaviour will not be needed, nor justified. Judgement, based on experience, must be used to decide which systems are critical and need detailed analysis and design.

Some examples of typical (conventional) control systems used for the control of specific process variables and unit operations are given in the next section, and can be used as a guide in preparing preliminary instrumentation and control schemes.

Guide rules

The following procedure can be used when drawing up preliminary P & I diagrams:

1. Identify and draw in those control loops that are obviously needed for steady plant operation, such as:
   (a) level controls,
   (b) flow controls,
   (c) pressure controls,
   (d) temperature controls.

2. Identify the key process variables that need to be controlled to achieve the specified product quality. Include control loops using direct measurement of the controlled variable, where possible; if not practicable, select a suitable dependent variable.
3. Identify and include those additional control loops required for safe operation, not already covered in steps 1 and 2.
4. Decide and show those ancillary instruments needed for the monitoring of the plant operation by the operators; and for trouble-shooting and plant development. It is well worthwhile including additional connections for instruments which may be needed for future trouble-shooting and development, even if the instruments are not installed permanently. This would include: extra thermowells, pressure tappings, orifice flanges, and extra sample points.
5. Decide on the location of sample points.
6. Decide on the need for recorders and the location of the readout points, local or control room. This step would be done in conjunction with steps 1 to 4.
7. Decide on the alarms and interlocks needed; this would be done in conjunction with step 3 (see Chapter 9).

5.8. Typical control systems

5.8.1. Level control

In any equipment where an interface exists between two phases (e.g. liquid–vapour), some means of maintaining the interface at the required level must be provided. This may be incorporated in the design of the equipment, as is usually done for decanters, or by automatic control of the flow from the equipment. Figure 5.8 shows a typical arrangement for the level control at the base of a column. The control valve should be placed on the discharge line from the pump.

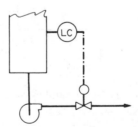

FIG. 5.8. Level control

5.8.2. Pressure control

Pressure control will be necessary for most systems handling vapour or gas. The method of control will depend on the nature of the process. Typical schemes are shown in Figs. 5.9a,b,c,d. The scheme shown in Fig. 5.9a would not be used where the vented gas was toxic, or valuable. In these circumstances the vent should be taken to a vent recovery system, such as a scrubber.

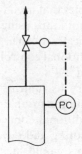

FIG. 5.9a Pressure control by direct venting

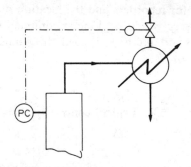

FIG. 5.9b. Venting of non-condensables after a condenser

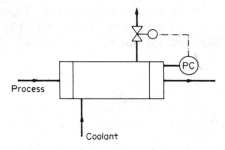

Process

Coolant

FIG. 5.9c. Condenser pressure control by controlling coolant flow

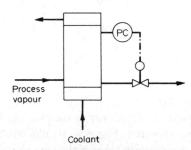

Process vapour

Coolant

FIG. 5.9d. Pressure control of a condenser by varying the heat-transfer area, area dependent on liquid level

5.8.3. *Flow control*

Flow control is usually associated with inventory control in a storage tank or other equipment. There must be a reservoir to take up the changes in flow-rate.

To provide flow control on a compressor or pump running at a fixed speed and supplying a near constant volume output, a by-pass control would be used, as shown in Figs. 5.10a,b.

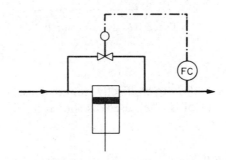

FIG. 5.10a. Flow control for a reciprocating pump

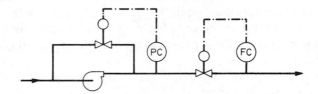

FIG. 5.10b. Alternative scheme for a centrifugal compressor or pump

5.8.4. *Heat exchangers*

Figure 5.11a shows the simplest arrangement, the temperature being controlled by varying the flow of the cooling or heating medium.

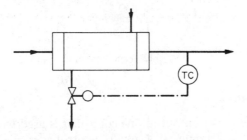

FIG. 5.11a. Control of one fluid stream

If the exchange is between two process streams whose flows are fixed, by-pass control will have to be used, as shown in Fig. 5.11b.

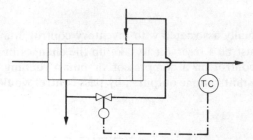

FIG. 5.11b. By-pass control

Condenser control

Temperature control is unlikely to be effective for condensers, unless the liquid stream is sub-cooled. Pressure control is often used, as shown in Fig. 5.9d, or control can be based on the outlet coolant temperature.

Reboiler and vaporiser control

As with condensers, temperature control is not effective, as the saturated vapour temperature is constant at constant pressure. Level control is often used for vaporisers; the controller controlling the steam supply to the heating surface, with the liquid feed to the vaporiser on flow control, as shown in Fig. 5.12. An increase in the feed results in an automatic increase in steam to the vaporiser to vaporise the increased flow and maintain the level constant.

Reboiler control systems are selected as part of the general control system for the column and are discussed in Section 5.8.7.

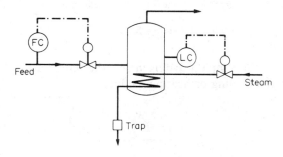

FIG. 5.12. Vaporiser control

5.8.5. Cascade control

With this arrangement, the output of one controller is used to adjust the set point of another (see Volume 3, Chapter 3). Cascade control can give smoother control in situations where direct control of the variable would lead to unstable operation. The "slave" controller can be used to compensate for any short-term variations in, say, a service stream flow, which would upset the controlled variable; the primary (master) controller controlling long-term variations. Typical examples are shown in Figs. 5.14e and 5.15.

5.8.6. Ratio control

Ratio control can be used where it is desired to maintain two flows at a constant ratio; for example, reactor feeds and distillation column reflux. A typical scheme for ratio control is shown in Fig. 5.13. In Fig. 5.13 the controller on stream A controls the flow of that stream and provides a signal to the ratio relay, which controls the set point of the controller on stream B; the set point is automatically adjusted to maintain a fixed, preset, ratio between the two stream flows.

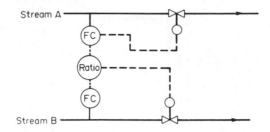

FIG. 5.13. Ratio control

5.8.7. Distillation column control

The primary objective of distillation column control is to maintain the specified composition of the top and bottom products, and any side streams; correcting for the effects of disturbances in:

1. Feed flow-rate, composition and temperature.
2. Steam supply pressure.
3. Cooling water pressure and header temperature.
4. Ambient conditions, which cause changes in internal reflux (see Chapter 11).

The compositions are controlled by regulating reflux flow and boil-up. The column overall material balance must also be controlled; distillation columns have little surge capacity (hold-up) and the flow of distillate and bottom product (and side-streams) must match the feed flows.

Shinskey (1979b) has shown that there are 120 ways of connecting the five main pairs of measured and controlled variables, in single loops. A variety of control schemes has been devised for distillation column control. Some typical schemes are shown in Figs. 5.14a,b,c,d,e; ancillary control loops and instruments are not shown.

Distillation column control is discussed in detail by Parkins (1959), Bertrand and Jones (1961) and Shinskey (1979b).

Column pressure is normally controlled at a constant value. The use of variable pressure control to conserve energy has been discussed by Shinskey (1976).

The feed flow-rate is often set by the level controller on a preceding column. It can be independently controlled if the column is fed from a storage or surge tank.

Feed temperature is not normally controlled, unless a feed preheater is used.

Temperature is often used as an indication of composition. The temperature sensor should be located at the position in the column where the rate of change of temperature with change in composition of the key component is a maximum; see Parkins, 1959. Near the top and bottom of the column the change is usually small. With multicomponent systems, temperature is not a unique function of composition.

Top temperatures are usually controlled by varying the reflux ratio, and bottom temperatures by varying the boil-up rate. If reliable on-line analysers are available they can be incorporated in the control loop, but more complex control equipment will be needed.

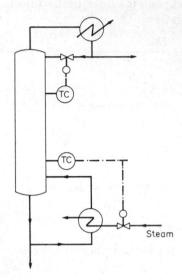

FIG. 5.14a. Temperature pattern control. With this arrangement interaction can occur between the top and bottom temperature controllers

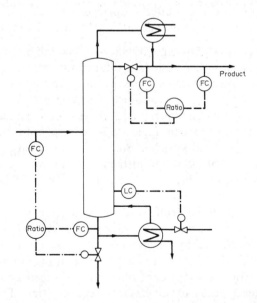

FIG. 5.14b. Composition control. Reflux ratio controlled by a ratio controller, or splitter box, and the bottom product as a fixed ratio of the feed flow

Differential pressure control is often used on packed columns to ensure that the packing operates at the correct loading; see Fig. 5.14 d.

Additional temperature indicating or recording points should be included up the column for monitoring column performance and for trouble shooting.

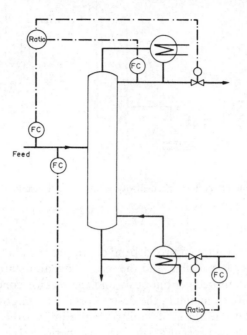

FIG. 5.14c. Composition control. Top product take-off and boil-up controlled by feed

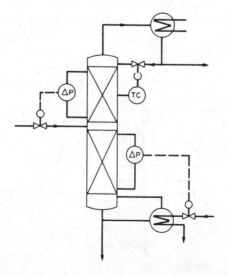

FIG. 5.14d. Packed column, differential pressure control. Eckert (1964) discusses the control of packed columns

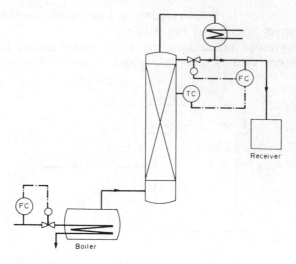

FIG. 5.14e. Batch distillation, reflux flow cascaded with temperature to maintain constant top composition

5.8.8. Reactor control

The schemes used for reactor control depend on the process and the type of reactor. If a reliable on-line analyser is available, and the reactor dynamics are suitable, the product composition can be monitored continuously and the reactor conditions and feed flows controlled automatically to maintain the desired product composition and yield. More often, the operator is the final link in the control loop, adjusting the controller set points to maintain the product within specification, based on periodic laboratory analyses.

Reactor temperature will normally be controlled by regulating the flow of the heating or cooling medium. Pressure is usually held constant. Material balance control will· be necessary to maintain the correct flow of reactants to the reactor and the flow of products and unreacted materials from the reactor. A typical reactor control scheme is shown in Fig. 5.15.

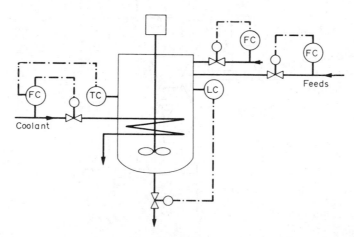

FIG. 5.15. A typical stirred tank reactor control scheme, temperature: cascade control, and reagent: flow control

5.9. Alarms and safety trips, and interlocks

Alarms are used to alert operators of serious, and potentially hazardous, deviations in process conditions. Key instruments are fitted with switches and relays to operate audible and visual alarms on the control panels and annunciator panels. Where delay, or lack of response, by the operator is likely to lead to the rapid development of a hazardous situation, the instrument would be fitted with a trip system to take action automatically to avert the hazard; such as shutting down pumps, closing valves, operating emergency systems.

The basic components of an automatic trip system are:

1. A sensor to monitor the control variable and provide an output signal when a preset value is exceeded (the instrument).
2. A link to transfer the signal to the actuator, usually consisting of a system of pneumatic or electric relays.
3. An actuator to carry out the required action; close or open a valve, switch off a motor.

A description of some of the equipment (hardware) used is given by Rasmussen (1975).

A safety trip can be incorporated in a control loop; as shown in Fig. 5.16a. In this system the high-temperature alarm operates a solenoid valve, releasing the air on the pneumatic activator, closing the valve on high temperature. However, the safe operation of such a system will be dependent on the reliability of the control equipment, and for potentially hazardous situations it is better practice to specify a separate trip system; such as that shown in Fig. 5.16b. Provision must be made for the periodic checking of the trip system to ensure that the system operates when needed.

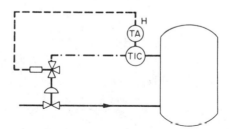

FIG. 5.16a. Trip as part of control system

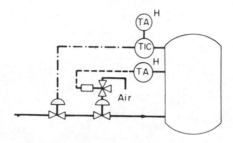

FIG. 5.16b. Separate shut-down trip

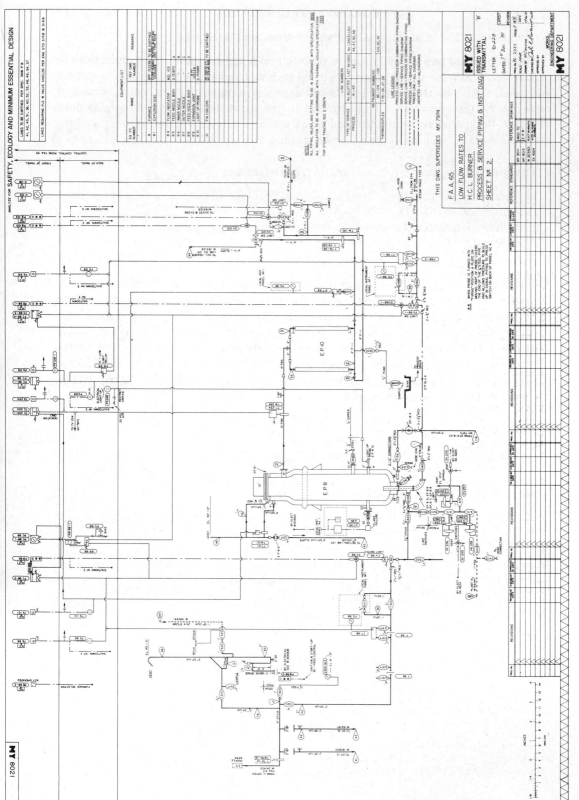

Fig. 5.17. Piping and instrumentation diagram

Interlocks

Where it is necessary to follow a fixed sequence of operations—for example, during a plant start-up and shut-down, or in batch operations—interlocks are included to prevent operators departing from the required sequence. They may be incorporated in the control system design, as pneumatic or electric relays, or may be mechanical interlocks. Various proprietary special lock and key systems are available.

5.10. References

ANON. (1969) *Chem. Eng., Albany* **76** (June 2nd) 136. Process instrument elements.
ANON. (1971) *Brit. Chem. Eng.* **16**, 313. Optimum pipeline diameters by nomograph.
AUSTIN, D. G. (1979) *Chemical Engineering Drawing Symbols* (George Godwin).
BERTRAND, L. and JONES, J. B. (1961) *Chem. Eng., Albany* **68** (Feb. 20th) 139. Controlling distillation columns.
BRITISH VALVE MANUFACTURERS ASSOC. (1972) *Technical Reference Book on Valves for the Control of Fluids*, 3rd ed. (British Valve Manufacturers Association).
BUCKLEY, P. S. (1964) *Techniques of Process Control* (Wiley).
CHAFLIN, S. (1974) *Chem. Eng., Albany* **81** (Oct. 14th) 105. Specifying control valves.
CONSIDINE, D. M. (1957) *Process Instruments and Control Handbook* (McGraw-Hill).
DE SANTIS, G. J. (1976) *Chem. Eng., Albany* **83** (Nov. 22nd) 163. How to select a centrifugal pump.
DOOLIN, J. H. (1977) *Chem. Eng., Albany* **84** (Jan. 17th) 137. Select pumps to cut energy cost.
ECKERT, J. S. (1964) *Chem. Eng., Albany* **71** (Mar. 30th) 79. Controlling packed-column stills.
GENEREAUX, R. P. (1937) *Ind. Eng. Chem.* **29**, 385. Fluid-flow design methods.
HOLLAND, F. A. and CHAPMAN, F. S. (1966) *Chem. Eng., Albany* **73** (Feb. 14th) 129. Positive displacement pumps.
HOLMES, E. (Ed.) (1973) *Handbook of Industrial Pipework Engineering* (McGraw-Hill).
ICHEME (1977) *A New Guide to Capital Cost Estimation* (Institution of Chemical Engineers, London).
JACOBS, J. K. (1965) *Hydrocarbon Proc.* **44** (June) 122. How to select and specify process pumps.
KERN, R. (1975) *Chem. Eng., Albany* **82** (April 28th) 119. How to design piping for pump suction conditions.
KING, R. C. (Ed.) (1967) *Piping Handbook*, 5th ed. (McGraw-Hill).
LUDWIG, E. E. (1960) *Chem. Eng., Albany* **67** (June 13th) 162. Flow of fluids.
M. W. KELLOG CO. (1964) *Design of Piping Systems* (Wiley).
MASEK, J. A. (1968) *Chem. Eng., Albany* **75** (June 17th) 215. Metallic piping.
NEERKIN, R. F. (1974) *Chem. Eng., Albany* **81** (Feb. 18) 104. Pump selection for chemical engineers.
NOLTE, C. B. (1978) *Optimum Pipe Size Selection* (Trans. Tech. Publications).
PARKINS, R. (1959) *Chem. Eng. Prog.* **55** (July) 60. Continuous distillation plant controls.
PEARSON, G. H. (1978) *Valve Design* (Mechanical Engineering Publications).
PERRY, R. H. and CHILTON, C. H. (Eds) (1973) *Chemical Engineers Handbook*, 5th ed. (McGraw-Hill).
PETERS, M. S. and TIMMERHAUS, K. D. (1968) *Plant Design and Economics for Chemical Engineers*, 2nd. ed. (McGraw-Hill).
RASE, H. F. (1953) *Petroleum Refiner* **32** (Aug.) 14. Take another look at economic pipe sizing.
RASMUSSEN, E. J. (1975) *Chem. Eng., Albany* **82** (May 12th) 74. Alarm and shut down devices protect process equipment.
SIMPSON, L. L. (1968) *Chem. Eng., Albany* **75** (June 17th) 192. Sizing piping for process plants.
SHINSKEY, F. G. (1976) *Chem. Eng. Prog.* **72** (May) 73. Energy-conserving control systems for distillation units.
SHINSKEY, F. G. (1979a) *Process Control Systems*, 2nd ed. (McGraw-Hill).
SHINSKEY, F. G. (1979b) *Distillation Control* (McGraw-Hill).

British Standards
BS 806: 1975   Ferrous pipes and piping for and in connection with land boilers.
BS 1600:   Dimension of steel pipes for the petroleum industry.
    Part 1: 1970 Imperial units.
    Part 2: 1970 Metric units.
BS 1646: 1979   Graphical symbols for process measurement and control functions.
BS 3351: 1971   Piping systems for petroleum refineries and petrochemical plant.

American Standards
USAS B31.1.0: The ASME standard code for pressure piping.
ASA B31.3.0: The ASME code for petroleum refinery piping.

Useful references, not cited in the text
WEAVER, R. (1973) *Process Piping Design*, Volumes 1 and 2 (Gulf Publishing Co.).
EVANS, F. L. (1980) *Equipment Design Handbook for Refineries and Chemical Plant*, Volume 2, 2nd ed. (Gulf).
LAMIT, L. C. (1981) *Piping Systems, Drafting and Design* (Prentice-Hall).

5.11. Nomenclature

		Dimensions in MLT
a	Capital charges factor, piping	—
b	Maintenance cost factor, piping	—
Cf	Annual pumping cost, piping	$£L^{-1}T^{-1}$
Cp	Capital cost, piping	$£L^{-1}$
Ct	Total annual cost, piping	$£L^{-1}T^{-1}$
d	Pipe diameter	L
E	Pump efficiency	—
F	Installed cost factor, piping	—
f	Fanning friction factor	—
G	Mass flow rate	MT^{-1}
g	Gravitational acceleration	LT^{-2}
H	Plant attainment (hours operated per year)	—
h	Pump head	L
K	Purchased cost factor, pipes	$£L^{-1}$
N	Pump speed, revolutions per unit time	T^{-1}
N_s	Pump specific speed	—
n	Index relating pipe cost to diameter	—
P	Pressure	$ML^{-1}T^{-2}$
P_s	Safe working pressure	$ML^{-1}T^{-2}$
ΔP	Pressure drop[†]	$ML^{-1}T^{-2}$
p	Cost of power, pumping	—
Q	Volumetric flow rate	L^3T^{-1}
R	Shear stress on surface, pipes	$ML^{-1}T^{-2}$
t	Pipe wall thickness	L
u	Mean velocity	LT^{-1}
z	Height above datum	L
ρ	Fluid density	ML^{-3}
μ	Viscosity of fluid	$ML^{-1}T^{-1}$
σ_d	Design stress	$ML^{-1}T^{-2}$
σ_s	Safe working stress	$ML^{-1}T^{-2}$
Re	Reynolds number	—

† *Note*: In Volumes 1, 2 and 3 this symbol is used for pressure difference, and pressure drop (negative pressure gradient) indicated by a minus sign. In this chapter, as the symbol is only used for pressure drop, the minus sign is omitted for convenience.

CHAPTER 6

Costing and Project Evaluation

6.1. Introduction

Cost estimation is a specialised subject and a profession in its own right. The design engineer, however, needs to be able to make quick, rough, cost estimates to decide between alternative designs and for project evaluation. Chemical plants are built to make a profit, and an estimate of the investment required and the cost of production are needed before the profitability of a project can be assessed.

In this chapter the various components that make up the capital cost of a plant and the components of the operating costs are discussed, and the techniques used for estimating reviewed briefly. Simple costing methods and some cost data are given, which can be used to make preliminary estimates of capital and operating costs at the flow-sheet stage. They can also be used to cost out alternative processing schemes and equipment.

For a more detailed treatment of the subject the reader should refer to the numerous specialised texts that have been published on cost estimation. The following books are particularly recommended: Aries and Newton (1955), Happle and Jordan (1975) and Guthrie (1974).

6.2. Accuracy and purpose of capital cost estimates

The accuracy of an estimate depends on the amount of design detail available; the accuracy of the cost data available; and the time spent on preparing the estimate. In the early stages of a project only an approximate estimate will be required, and justified, by the amount of information by then developed.

Capital cost estimates can be broadly classified into three types according to their accuracy and purpose:

1. Preliminary (approximate) estimates, accuracy typically ± 30 per cent, which are used in initial feasibility studies and to make coarse choices between design alternatives. They are based on limited cost data and design detail.
2. Authorisation (Budgeting) estimates, accuracy typically ± 10–15 per cent. These are used for the authorisation of funds to proceed with the design to the point where an accurate and more detailed estimate can be made. Authorisation may also include funds to cover cancellation charges on any long delivery equipment ordered at this stage of the design to avoid delay in the project. In a contracting organisation this type of estimate could be used with a large contingency factor to obtain a price for tendering. Normally, however, an accuracy of about ± 5 per cent would be needed and a more detailed estimate would be made, if time permitted.

   With experience, and where a company has cost data available from similar projects, estimates of acceptable accuracy can be made at the flow-sheet stage of the

181

project. A rough P & I diagram and the approximate sizes of the major items of equipment would also be needed.

3. Detailed (Quotation) estimates, accuracy ±5–10 per cent, which are used for project cost control and estimates for fixed price contracts. These are based on the completed (or near complete) process design, firm quotations for equipment, and a detailed breakdown and estimation of the construction cost.

The cost of preparing an estimate increases from about 0·1 per cent of the total project cost for ± 30 per cent accuracy, to about 2 per cent for a detailed estimate with an accuracy of ± 5 per cent.

6.3   Fixed and working capital

Fixed capital is the total cost of the plant ready for start-up. It is the cost paid to the contractors.

It includes the cost of:

1. Design, and other engineering and construction supervision.
2. All items of equipment and their installation.
3. All piping, instrumentation and control systems.
4. Buildings and structures.
5. Auxiliary facilities, such as utilities, land and civil engineering work.

It is a once-only cost that is not recovered at the end of the project life, other than the scrap value.

Working capital is the additional investment needed, over and above the fixed capital, to start the plant up and operate it to the point when income is earned.

It includes the cost of:

1. Start-up.
2. Initial catalyst charges.
3. Raw materials and intermediates in the process.
4. Finished product inventories.
5. Funds to cover outstanding accounts from customers.

Most of the working capital is recovered at the end of the project. The total investment needed for a project is the sum of the fixed and working capital.

Working capital can vary from as low as 5 per cent of the fixed capital for a simple, single-product, process, with little or no finished product storage; to as high as 30 per cent for a process producing a diverse range of product grades for a sophisticated market, such as synthetic fibres. A typical figure for petrochemical plants is 15 per cent of the fixed capital.

Methods for estimating the working capital requirements are given by Bechtel (1960) and Lyda (1972).

6.4.   Cost escalation (inflation)

The cost of materials and labour has been subject to inflation since Elizabethan times. All cost-estimating methods use historical data, and are themselves forecasts of future costs. Some method has to be used to update old cost data for use in estimating at the design stage, and to forecast the future construction cost of the plant.

The method usually used to update historical cost data makes use of published cost indices. These relate present costs to past costs, and are based on data for labour, material and energy costs published in government statistical digests; such as *Trade and Industry Monthly* and the *Employment and Productivity Gazette*.

$$\text{Cost in year A} = \text{Cost in year B} \times \frac{\text{Cost index in year A}}{\text{Cost index in year B}} \qquad (6.1)$$

To get the best estimate, each job should be broken down into its components and separate indices used for labour and materials. It is often more convenient to use the composite indices published for various industries in the trade journals. These produce a weighted average index combining the various components in proportions considered typical for the particular industry. Such an index for the chemical industry in the United Kingdom is published in the journal *Process Engineering* (previously entitled *Chemical and Process Engineering*), see Cran (1973) (1979). The composition of this index is:

$$I = 0.37\,Im + 0.081\,Ie + 0.10\,Ic + 0.19\,Is + 0.26\,Io$$

where: I = the composite index
         Im = mechanical engineering index
         Ie = electrical engineering index
         Ic = civil engineering index
         Is = site engineering index
         Io = overheads engineering index

The base year currently used for the index is 1975, and the index was first published in 1965. The indices used to calculate the composite index are taken from the *Monthly Digest of Statistics*, published by the Department of Industry and Employment, Central Statistical Office. *The Process Engineering Index*, together with composite indices for some other countries, over a ten-year period, is shown in Fig. 6.1.

A composite index for the United States process plant industry is published monthly in the journal *Chemical Engineering*, the CPE plant cost index. This journal also publishes the Marshall and Stevens index (M & S equipment cost index), base year 1926.

All cost indices should be used with caution and judgement. They do not necessarily relate the true make-up of costs for any particular piece of equipment or plant; nor the effect of supply and demand on prices. The longer the period over which the correlation is made the more unreliable the estimate.

Since 1970 prices have risen dramatically and this is reflected in the UK index which, as can be seen from Fig. 6.1, rose from 45 to 160 over the period from 1969 to 1978, a factor of 3.6. The use of the index to update costs over this period can only give an approximate indication of the true cost; to be used only when up-to-date cost data are not available.

To estimate the future cost of a plant some prediction has to be made of the future annual rate of inflation. This can be based on an extrapolation of one of the published indices, tempered with the engineer's own assessment of what the future may hold. Prior to 1970 costs were escalating at about 7 per cent per year and this figure was often used to predict future costs.

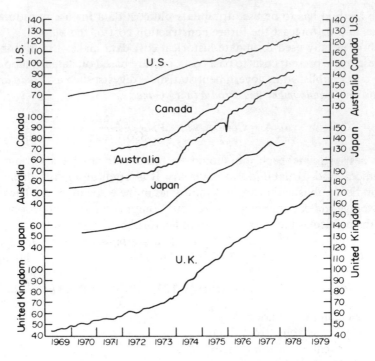

FIG. 6.1. Plant cost indices (from *Process Engineering*, with permission)

Example 6.1

The purchased cost of a tubular heat exchanger, carbon steel shell, stainless tubes, area 500 m², was £10,000 in December 1972; estimate the present-day cost (mid-1979).

Solution

Use the Process Engineering index, Fig. 6.1:

$$\text{value of index, Dec. 1972} = 64$$
$$\text{mid-1979} = 170$$
$$1979 \text{ cost} = 10,000 \times \frac{170}{64} = £26,563, \text{ say } £\underline{\underline{27,000}}$$

6.5. Rapid capital cost estimating methods

An approximate estimate of the capital cost of a project can be obtained from a knowledge of the cost of earlier projects using the same manufacturing process. This method can be used prior to the preparation of the flow-sheets to get a quick estimate of the investment likely to be required.

The capital cost of a project is related to capacity by the equation

$$C_2 = C_1 \left(\frac{S_2}{S_1}\right)^n \tag{6.2}$$

where C_2 = capital cost of the project with capacity S_2,
   C_1 = capital cost of the project with capacity S_1.

The value of the index n is traditionally taken as 0·6; the well-known six-tenths rule. This value can be used to get a rough estimate of the capital cost if there are not sufficient data available to calculate the index for the particular process. Estrup (1972) gives a critical review of the six-tenths rule. Equation 6.2 is only an approximation, and if sufficient data are available the relationship is best represented on a log-log plot. Guthrie (1970) has published investment-capacity plots for fifty-four processes. Capital cost data for several processes have been presented in a different form by Winfield and Dryden (1962).

Example 6.2

Obtain a rough estimate of the capital cost of a plant to produce 750 tonnes per day of ammonia.

Use the costs given by Guthrie (1970), reproduced in Fig. 6.2.

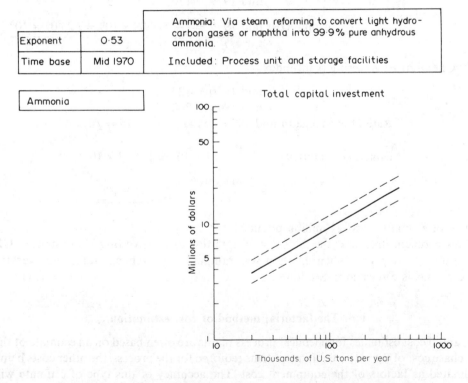

FIG. 6.2. Investment vs. capacity (from Guthrie (1970), with permission)

Solution

Guthrie's units are US dollars and tons, and refer to mid-1970.

$$1 \text{ US ton} = 2000 \text{ lb} = 0.91 \text{ tonne } (1000 \text{ kg})$$

$$\text{so, } 750 \text{ tonnes/day} = \frac{750}{0.91} \times 365 = 301{,}000 \text{ US tons/y}$$

From Fig. 6.2 the fixed capital cost for this capacity is

$$17 \times 10^6 \text{ US dollars}$$

There are two possible ways of converting this cost to a UK cost:

1. Convert at the 1970 exchange rate and update using a UK inflation index.
2. Update to 1979 using a US index and convert at the 1979 exchange rate.

1. In mid-1970 US and UK costs were roughly equivalent and the rate of exchange was $\$2.4 = £1$.

$$1970 \text{ cost} = \frac{17 \times 10^6}{2.4} = £7.08 \times 10^6$$

Updating to 1979 using the index published in *Process Engineering*:

$$\text{Index mid-1970} = 50$$
$$\text{mid-1979} = 170$$
$$\text{capital cost mid 1979} = \frac{170}{50} \times 7.08 \times 10^6 = \underline{\underline{£24.07 \times 10^6}}$$

2. Chemical Engineering plant index

$$\text{mid-1970} = 123$$
$$\text{mid-1979} = 260$$

Rate of exchange in mid-1979 was around $\$2.15 = £1$.

$$\text{capital cost mid-1979} = \frac{260}{123} \times 17 \times 10^6 = \$35.93 \times 10^6$$

$$= \frac{35.93 \times 10^6}{2.15} = \underline{\underline{£16.71 \times 10^6}}$$

Range of estimate 17 to 24 million pounds.

This problem illustrates the difficulty of updating and converting US costs to a UK situation over a period when inflation was rampant and exchange rates were varying widely. This is discussed in Section 6.7.

6.6. The factorial method of cost estimation

Capital cost estimates for chemical process plants are often based on an estimate of the purchase cost of the major equipment items required for the process, the other costs being estimated as factors of the equipment cost. The accuracy of this type of estimate will depend on what stage the design has reached at the time the estimate is made, and on the reliability of the data available on equipment costs. In the later stages of the project design, when detailed equipment specifications are available and firm quotations have been obtained, an accurate estimation of the capital cost of the project can be made.

6.6.1. Lang factors

The factorial method of cost estimation is often attributed to Lang (1948). The fixed capital cost of the project is given as a function of the total purchase equipment cost by the equation:

$$Cf = f_L Ce \qquad (6.3)$$

where Cf = fixed capital cost,

Ce = the total delivered cost of all the major equipment items: storage tanks, reaction vessels, columns, heat exchangers, etc.,

f_L = the "Lang factor", which depends on the type of process.

f_L = 3·1 for predominantly solids processing plant
f_L = 4·7 for predominantly fluids processing plant
f_L = 3·6 for a mixed fluids–solids processing plant

The values given above should be used as a guide; the factor is best derived from an organisation's own cost files.

Equation 6.3 can be used to make a quick estimate of capital cost in the early stages of project design, when the preliminary flow-sheets have been drawn up and the main items of equipment roughly sized.

6.6.2. Detailed factorial estimates

To make a more accurate estimate, the cost factors that are compounded into the "Lang factor" are considered individually. The direct-cost items that are incurred in the construction of a plant, in addition to the cost of equipment are:

1. Equipment erection, including foundations and minor structural work.
2. Piping, including insulation and painting.
3. Electrical, power and lighting.
4. Instruments, local and control room.
5. Process buildings and structures.
6. Ancillary buildings, offices, laboratory buildings, workshops.
7. Storages, raw materials and finished product.
8. Utilities (Services), provision of plant for steam, water, air, firefighting services (if not costed separately).
9. Site, and site preparation.

The contribution of each of these items to the total capital cost is calculated by multiplying the total purchased equipment by an appropriate factor. As with the basic "Lang factor", these factors are best derived from historical cost data for similar processes. Typical values for the factors are given in several references, Aries and Newton (1955), Happle and Jordan (1975). Guthrie (1974), splits the costs into the material and labour portions and gives separate factors for each. In a booklet published by the Institution of Chemical Engineers, *IChemE* (1977), the factors are shown as a function of plant size and complexity.

The accuracy and reliability of an estimate can be improved by dividing the process into sub-units and using factors that depend on the function of the sub-units; see Guthrie (1969). In Guthrie's detailed method of cost estimation the installation, piping and instrumentation costs for each piece of equipment are costed separately. Detailed costing is only justified if the cost data available are reliable and the design has been taken to the point where all the cost items can be identified and included.

Typical factors for the components of the capital cost are given in Table 6.1. These can be used to make an approximate estimate of capital cost using equipment cost data published in the literature.

TABLE 6.1. *Typical factors for estimation of project fixed capital cost*

Item	Process type		
	Fluids	Fluids–solids	Solids
1. MAJOR EQUIPMENT, TOTAL PURCHASE COST	PCE	PCE	PCE
f_1 Equipment erection	0·4	0·45	0·50
f_2 Piping	0·70	0·45	0·20
f_3 Instrumentation	0·20	0·15	0·10
f_4 Electrical	0·10	0·10	0·10
f_5 Buildings, process	0·15	0·10	0·05
*f_6 Utilities	0·50	0·45	0·25
*f_7 Storages	0·15	0·20	0·25
*f_8 Site development	0·05	0·05	0·05
*f_9 Ancillary buildings	0·15	0·20	0·30
2. TOTAL PHYSICAL PLANT COST (PPC) $\quad$ PPC = PCE $(1 + f_1 \ldots + f_9)$ $\qquad\qquad\qquad$ = PCE ×	3·40	3·15	2·80
f_{10} Design and Engineering	0·30	0·25	0·20
f_{11} Contractor's fee	0·05	0·05	0·05
f_{12} Contingency	0·10	0·10	0·10
FIXED CAPITAL = PPC$(1 + f_{10} + f_{11} + f_{12})$ $\qquad\qquad\qquad$ = PPC ×	1·45	1·40	1·35

* Omitted for minor extensions or additions to existing sites.

In addition to the direct cost of the purchase and installation of equipment, the capital cost of a project will include the indirect costs listed below. These can be estimated as a function of the direct costs.

Indirect costs

1. Design and engineering costs, which cover the cost of design and the cost of "engineering" the plant: purchasing, procurement and construction supervision. Typically 20 per cent to 30 per cent of the direct capital costs.
2. Contractor's fees, if a contractor is employed his fees (profit) would be added to the total capital cost and would range from 5 per cent to 10 per cent of the direct costs.
3. Contingency allowance, this is an allowance built into the capital cost estimate to cover for *unforeseen* circumstances (labour disputes, design errors, adverse weather). Typically 5 per cent to 10 per cent of the direct costs.

The indirect cost factors are included in Table 6.1.

The capital cost required for the provision of utilities and other plant services will depend on whether a new (green field) site is being developed, or if the plant is to be built on an existing site and will make use of some of the existing facilities. The term "battery limits" is used to define a contractor's responsibility. The main processing plant, within the battery limits, would normally be built by one contractor. The utilities and other ancillary equipment would often be the responsibility of other contractors and would be said to be outside the battery limits. They are often also referred to as "off-sites".

6.7. Estimation of purchased equipment costs

The cost of the purchased equipment is used as the basis of the factorial method of cost estimation and must be determined as accurately as possible. It should preferably be based on recent prices paid for similar equipment.

The relationship between size and cost given in equation 6.1 can also be used for equipment, but the relationship is best represented by a log-log plot if the size range is wide. A wealth of data has been published on equipment costs; see Aries and Newton (1955), Chilton (1960), Chemical Engineering (1970, 1979), Guthrie (1969, 1974), Winfield

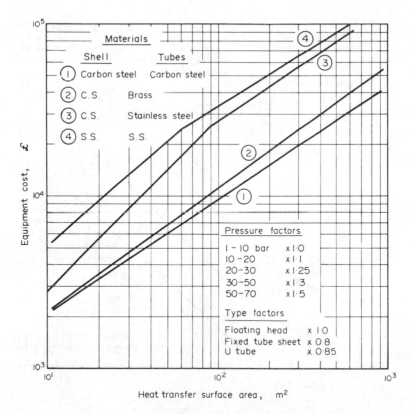

FIG. 6.3. Shell and tube heat exchangers. Time base mid-1979
Purchased cost = (bare cost from figure) × Type factor × Pressure factor

and Dryden (1962), IChemE (1977) and Karbanda (1978). Articles giving the cost of process equipment are frequently published in the journals *Chemical Engineering* and *Hydrocarbon Processing*.

The cost of specialised equipment, which cannot be found in the literature, can usually be estimated from the cost of the components that make up the equipment. For example, a reactor design is usually unique for a particular process but the design can be broken down into standard components (vessels, heat-exchange surfaces, spargers, agitators) the cost of which can be found in the literature and used to build up an estimate of the reactor cost.

Pikulik and Diaz (1977) give a method of costing major equipment items from cost data on the basic components: shells, heads, nozzles, and internal fittings.

Almost all the information on costs available in the open literature is in American journals and refers to dollar prices in the US. Some UK equipment prices were published in the journals *British Chemical Engineering* and *Chemical and Process Engineering* before they ceased publication. The only comprehensive collection of UK prices available is given in the Institution of Chemical Engineers booklet, IChemE (1977).

Up to 1970 US and UK prices for equipment could be taken as roughly equivalent, converting from dollars to pounds using the rate of exchange ruling on the date the prices were quoted. Since 1970 the rate of inflation in the US has been significantly lower than in

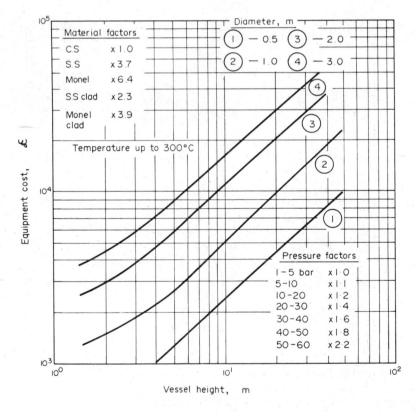

FIG. 6.4. Vertical pressure vessels. Time base mid-1979
Purchased cost = (bare cost from figure) × Material factor × Pressure factor

the UK, and rates of exchange have fluctuated since the pound was floated in 1972. Miller (1979), in a paper discussing the conversion of construction costs from one country to another, suggests conversion rates for equipment costs, US dollars to pounds sterling, of:

1976	2·10
1977	2·05
1978	1·95

These are roughly equivalent to the rates of exchange in these years.

If it can be assumed that world market forces will level out the prices of equipment, the UK price can be estimated from the US price by bringing the cost up to date using a suitable US price index, converting to pounds sterling at the current rate of exchange, and adding an allowance for freight and duty.

If an estimate is being made to compare two processes, the costing can be done in dollars and any conclusion drawn from the comparison should still be valid for the United Kingdom.

The cost data given in Figs. 6.3 to 6.6, and Table 6.2 have been compiled from various sources. They can be used to make preliminary estimates. The base date is mid-1979, and the prices are thought to be accurate to within ± 25 per cent. To use Table 6.2, substitute

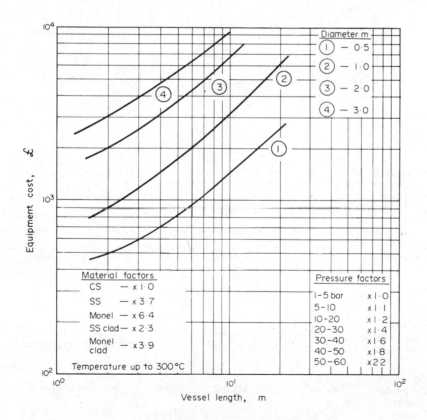

FIG. 6.5. Horizontal pressure vessels. Time base mid-1979

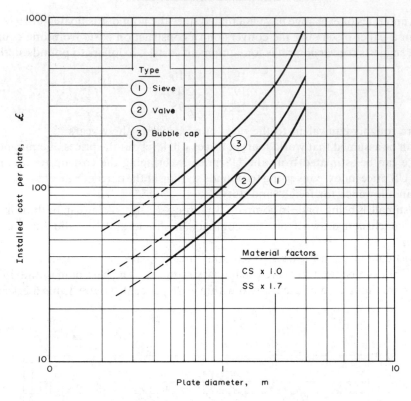

FIG. 6.6. Column plates. Time base mid-1979 (for column costs see Fig. 6.4)
Installed cost = (cost from figure) × Material factor

TABLE 6.2. *Purchase cost of miscellaneous equipment, cost factors for use in equation 6.4*

Equipment	Size unit S	Size range	Constant C £	Index n	Comment
Agitators					
Propellor	driver power, kW	5–25 kW	400	0·50	carbon steel
Turbine			820	0·30	
Boilers					
Packaged					
up to 10 bar	kg/h, steam	20–50 × 10³	700	0·50	oil or
up to 60 bar					gas fired
10 to 60 bar			900	0·50	
Centrifuges					
Horizontal basket	dia., m	0·5–1·0	16,000	1·3	
Vertical basket			16,000	1·0	

TABLE 6.2. (Cont.)

Equipment	Size unit S	Size range	Constant C £	Index n	Comment
Compressors					
Centrifugal	driver power, kW	20–200	700	0·8	electric drive, max. press. 50 bar
Conveyors					
Belt					
0·5 m wide	length, m	2–40	1000	0·65	
1·0 m wide			1500	0·65	
Crushers					
Cone	te/h	20–200	800	0·85	
pulverisers	kg/h	—	700	0·35	
Dryers					
Rotary vac.	area, m^2	5–30	6000	0·45	carbon steel
pan		2–10	4000	0·35	
Evaporators					
Vertical tube	area, m^2	10–1000	4000	0·53	carbon steel
Filters					
Plate and frame					
	area, m^2	5–50	1000	0·60	cast iron
Vacuum drum		1–10	6000	0·60	carbon steel
Reactors					
Jacketed, agitated	capacity, m^3	3–30	6000	0·40	carbon steel
			12,000	0·45	glass lined
Tanks					
Process					
Vertical	capacity, m^3	1–50	500	0·59	atmos. press.
Horizontal		10–100	600	0·60	carbon steel ×2·5 for stainless
Tanks					
Storage	capacity, m^3				
Floating roof		50–8000	650	0·65	carbon steel
Fixed cone roof			600	0·65	
Furnaces					
Process	heat absorbed, kW	10^3–10^4	76	0·77	carbon steel tubes, for stainless × 2·5
Packing			£/m^3		
Ceramic saddles	size, mm	25–75	980	−0·4	
Plastic pall rings		25–75	8300	−1·0	
Stainless-steel pall rings		25–100	8750	−1·0	

the values given for the particular type of equipment into the equation:

$$Ce = CS^n \tag{6.4}$$

where Ce = purchased equipment cost, £,
$\quad\quad S$ = characteristic size parameter, in the units given in Table 6.2,
$\quad\quad C$ = cost constant from Table 6.2,
$\quad\quad n$ = index for that type of equipment.

6.8. Summary of the factorial method

Many variations on the factorial method are used. The method outlined below can be used with the data given in this chapter to make a quick, approximate, estimate of the investment need for a project.

Procedure

1. Prepare material and energy balances, draw up preliminary flow-sheets, size major equipment items and select materials of construction.
2. Estimate the purchase cost of the major equipment items. Use Figs. 6.3 to 6.6 and Table 6.2, or the general literature
3. Calculate the total physical plant cost (PPC), using the factors given in Table 6.1

$$PPC = PCE\ (1 + f_1 + \ldots f_9) \tag{6.5}$$

4. Calculate the indirect costs from the direct costs using the factors given in Table 6.1.
5. The direct plus indirect costs give the total fixed capital.
6. Estimate the working capital as a percentage of the fixed capital; 10 to 20 per cent.
7. Add the fixed and working capital to get the total investment required.

6.9. Operating costs

An estimate of the operating costs, the cost of producing the product, is needed to judge the viability of a project, and to make choices between possible alternative processing schemes. These costs can be estimated from the flow-sheet, which gives the raw material and service requirements, and the capital cost estimate.

The cost of producing a chemical product will include the items listed below. They are divided into two groups.

1. Fixed operating costs: costs that do not vary with production rate. These are the bills that have to be paid whatever the quantity produced.
2. Variable operating costs: costs that are dependent on the amount of product produced.

Fixed costs

1. Maintenance (labour and materials).
2. Operating labour.
3. Laboratory costs.

4. Plant overheads.
5. Capital charges.
6. Rates (and any other local taxes).
7. Insurance.
8. Licence fees and royalty payments.

Variable costs

1. Raw materials.
2. Miscellaneous operating materials.
3. Utilities (Services).
4. Shipping and packaging.

The division into fixed and variable costs is somewhat arbitrary. Certain items can be classified without question, but the classification of other items will depend on the accounting practice of the particular organisation.

The items may also be classified differently in cost sheets and cost standards prepared to monitor the performance of the operating plant. For this purpose the fixed-cost items should be those over which the plant supervision has no control, and the variable items those for which they can be held accountable.

The costs listed above are the direct costs of producing the product at the plant site. In addition to these costs the site will have to carry its share of the Company's general operating expenses. These will include:

1. General overheads.
2. Research and development costs.
3. Sales expense.
4. Reserves.

How these costs are apportioned will depend on the Company's accounting methods. They would add about 20 to 30 per cent to direct production costs at the site.

6.9.1. Estimation of operating costs

In this section the components of the fixed and variable costs are discussed and methods given for their estimation.

It is usually convenient to do the costing on an annual basis.

Raw materials

These are the major (essential) materials required to manufacture the product. The quantities can be obtained from the flow-sheet and multiplied by the operating hours per year to get the annual requirements.

The price of each material is best obtained by getting quotations from potential suppliers, but in the preliminary stages of a project prices can be taken from the literature. The current prices of some of the major chemical products and raw materials are published in the journal *European Chemical News*, monthly.

Miscellaneous materials (plant supplies)

Under this heading are included all the miscellaneous materials required to operate the plant that are not covered under the headings raw materials or maintenance materials; such as, safety clothing, instrument charts, gaskets.

An accurate estimate can be made by detailing and costing all the items needed, based on experience with similar plants. As a rough guide the cost of miscellaneous materials can be taken as 10 per cent of the total maintenance cost.

Utilities (Services)

This term includes, power, steam, compressed air, cooling and process water, and effluent treatment; unless costed separately. The quantities required can be obtained from the energy balances and the flow-sheets. The prices should be taken from Company records, if available. They will depend on the primary energy sources and the plant location. The figures given in Table 6.3 can be used to make preliminary estimates. The current cost of utilities supplied by the public corporations, electricity, gas and water, can be obtained from their local area offices.

TABLE 6.3. *Cost of utilities, UK, mid-1979*

Mains water	11–16p/m^3	50–70p/1000 gal
Natural gas	190p/GJ	20p/therm
Electricity	760p/GJ	2·8p/kWh
Fuel oil	170p/GJ	31p/gal
Cooling water (cooling tower)	0·3p/m^3	1·4p/1000 gal
Demineralised water	20p/m^3	91p/1000 gal
Steam (from direct fired boilers)	0·62p/kg	£2·8 /1000 lb

(£1 = 100p)

Shipping and packaging

This cost will depend on the nature of the product. For liquids collected at the site in the customer's own tankers the cost to the product would be small; whereas the cost of packaging and transporting synthetic fibres or polymers to a central distribution warehouse would add significantly to the product cost.

Maintenance

This item will include the cost of maintenance labour, which can be as high as the operating labour cost, and the materials (including equipment spares) needed for the maintenance of the plant. The annual maintenance costs for chemical plants are high, typically 5 to 15 per cent of the installed capital costs. They should be estimated from a knowledge of the maintenance costs on similar plant. As a first estimate the annual maintenance cost can be taken as 10 per cent of the fixed capital cost; the cost can be considered to be divided evenly between labour and materials.

Operating labour

This is the manpower needed to operate the plant: that directly involved with running the process.

The costs should be calculated from an estimate of the number of shift and day personnel needed, based on experience with similar processes. It should be remembered that to operate three shifts per day, at least four shift crews will be needed. The figures used for the cost of each man should include an allowance for holidays, shift allowances, national insurance, pension contributions and any other overheads. The current wage rates per hour in the UK chemical industry (mid-1979) are £2 to £2·50, to which must be added up to 50 per cent for the various allowances and overheads mentioned above.

Chemical plants do not normally employ many people and the cost of operating labour would not normally exceed 15 per cent of the total operating cost. The direct overhead charges would add 20 to 30 per cent to this figure.

Wessel (1952) gives a method of estimating the number of man-hours required based on the plant capacity and the number of discreet operating steps.

Supervision

This heading covers the direct operating supervision: the management directly associated with running the plant. The number required will depend on the size of the plant and the nature of the process. The site would normally be broken down into a number of manageable units. A typical management team for a unit would consist of four to five shift foremen, a general foreman, and an area supervisor (manager) and his assistant. The cost of supervision should be calculated from an estimate of the total number required and the current salary levels, including the direct overhead costs. On average, one "supervisor" would be needed for each four to five operators. Typical salaries, mid-1979, are £5000 to £10,000, depending on seniority. An idea of current salaries can be obtained from the salary reviews published periodically by the Institution of Chemical Engineers and the Council of Engineering Institutions.

Laboratory costs

The annual cost of the laboratory analyses required for process monitoring and quality control is a significant item in most modern chemical plants. The costs should be calculated from an estimate of the number of analyses required and the standard charge for each analysis, based on experience with similar processes.

As a rough estimate the cost can be taken as 20 to 30 per cent of the operating labour cost, or 2 to 4 per cent of the total production cost.

Plant overheads

Included under this heading are all the general costs associated with operating the plant not included under the other headings; such as, general management, plant security, medical, canteen, general clerical staff and safety. It would also normally include the plant technical personnel not directly associated with and charged to a particular operating area. This group may be included in the cost of supervision, depending on the organisation's practice.

The plant overhead cost is usually estimated from the total labour costs: operating, maintenance and supervision. A typical range would be 50 to 100 per cent of the labour costs; depending on the size of the plant and whether the plant was on a new site, or an extension of an existing site.

Capital charges

The investment required for the project is recovered as a charge on the project. How this charge is shown on an organisation's books will depend on its accounting practices. Capital is often recovered as a depreciation charge, which sets aside a given sum each year to repay the cost of the plant. If the plant is considered to "depreciate" at a fixed rate over its predicted operating life, the annual sum to be included in the operating cost can be easily calculated. The operating life of a chemical plant is usually taken as 10 years, which gives a depreciation rate of 10 per cent per annum. The plant is not necessarily replaced at the end of the depreciation period. The depreciation sum is really an internal transfer to the organisation's fund for future investment. If the money for the investment is borrowed, the sum set aside would be used to repay the loan. Interest would also be payable on the loan at the current market rates. Normally the capital to finance a particular project is not taken as a direct loan from the market but comes from the Company's own reserves. Any interest charged would, like depreciation, be an internal (book) transfer of cash to reflect the cost of the capital used.

Rather than consider the cost of capital as depreciation or interest, or any other of the accounting terms used, which will depend on the accounting practice of the particular organisation and the current tax laws, it is easier to take the cost as a straight, unspecified, capital charge on the operating cost. This would be typically 10 to 20 per cent of the fixed capital, annually, depending on the cost of money. As an approximate estimate the "capital charge" can be taken as 2 per cent above the current minimum lending rate. For a full discussion on the nature of depreciation and the cost of capital see Happle and Jordan (1975) or Holland *et al.* (1974).

Rates

This term covers local authority taxes, which are calculated on the assessed rateable value of the site. A typical figure would be 1 to 2 per cent of the fixed capital.

Insurance

The cost of the site and plant insurance: the annual insurance premium paid to the insurers; usually about 1 to 2 per cent of the fixed capital.

Royalties and licence fees

If the process used has not been developed exclusively by the operating company, royalties and licence fees may be payable. These may be paid as a lump sum, included in the fixed capital, or as an annual fee; or payments based on the amount of product sold.

The cost would add about 1 per cent to 5 per cent to the sales price.

Summary of production costs

The various components of the operating costs are summarised in Table 6.4. The typical values given in this table can be used to make an approximate estimate of production costs.

TABLE 6.4. *Summary of production costs*

Variable costs	Typical values
1. Raw materials	from flow-sheets
2. Miscellaneous materials	10 per cent of item (5)
3. Utilities	from flow-sheet
4. Shipping and packaging	usually negligible
Sub-total A	. . . .
Fixed costs	
5. Maintenance	5–10 per cent of fixed capital
6. Operating labour	from manning estimates
7. Supervision	20 per cent of item (6)
8. Plant overheads	50 per cent of item (6)
9. Capital charges	15 per cent of the fixed capital
10. Insurance	1 per cent of the fixed capital
11. Rates	2 per cent of the fixed capital
12. Royalties	1 per cent of the fixed capital
Sub-total B	. . . .
Direct production costs A + B	. . . .
13. Sales expense	20–30 per cent of the direct
14. General overheads	production cost
15. Research and development	
Sub-total C	. . . .
Annual production cost = A + B + C =	. . . .

$$\text{Production cost } \pounds/\text{kg} = \frac{\text{Annual production cost}}{\text{Annual production rate}}$$

Example 6.3

Preliminary design work has been done on a process to recover a valuable product from an effluent gas stream. The gas will be scrubbed with a solvent in a packed column; the recovered product and solvent separated by distillation; and the solvent cooled and recycled. The major items of equipment that will be required are detailed below.

1. Absorption column: diameter 1 m, vessel overall height 15 m, packed height 12 m, packing 38 mm stainless-steel Pall rings, vessel carbon steel, operating pressure 5 bar.
2. Recovery column: diameter 1 m, vessel overall height 20 m, 35 sieve plates, vessel and plates stainless steel, operating pressure 1 bar.
3. Reboiler: forced convection type, fixed tube sheets, area 18·6 m^2, carbon steel shell, stainless-steel tubes, operating pressure 1 bar.
4. Condenser: fixed tube sheets, area 25·3 m^2, carbon steel shell and tubes, operating pressure 1 bar.

5. Recycle solvent cooler: U-tubes, area $10 \cdot 1$ m^2, carbon steel shell and tubes, operating pressure 5 bar.

6. Solvent and product storage tanks: fixed cone roof, capacity 35 m^3, carbon steel.

Estimated service requirements:

Steam	200 kg/h
Cooling water	5000 kg/h
Electrical power	100 kWh/d (360 MJ/d)

Estimated solvent loss 10 kg/d. Price: £400/t.

Plant attainment 95 per cent.

Estimate the capital investment required for this project, and the annual operating cost; date mid-1979.

Solution

Purchased cost of the major equipment items:

Absorption column:

Bare vessel cost (Fig. 6.4) £7500; material factor $1 \cdot 0$, pressure factor $1 \cdot 1$.

Vessel cost $= 7500 \times 1 \cdot 0 \times 1 \cdot 1 = £8250$

Packing cost per m^3 (Table 6.2) $= 8750 \times 38^{-1} = £230$

Volume of packing $= \pi/4 \times 12 = 9 \cdot 4$ m^3

Cost of column packing $= 9 \cdot 4 \times 230 = £2168$

Total cost of column $= 2168 + 8250 = £10,418$ say, £11,000

Recovery column:

Bare vessel cost (Fig. 6.4) £10,000; material factor $3 \cdot 7$, pressure factor $1 \cdot 0$.

Vessel cost $= 10,000 \times 3 \cdot 7 \times 1 \cdot 0 = £37,000$

Cost of a plate (Fig. 6.6), material factor $1 \cdot 7$, $= 67 \times 1 \cdot 7 = £114$

Total cost of plates $= 35 \times 114 = £3990$

Total cost of column $= 37,000 + 3990 = £40,990$ say, £41,000

Reboiler:

Bare cost (Fig. 6.3) £5200; type factor $0 \cdot 8$, pressure factor $1 \cdot 0$.

Purchased cost $= 5200 \times 0 \cdot 8 \times 1 \cdot 0 = £4160$

Condenser:

Bare cost (Fig. 6.3) £3800; type factor $0 \cdot 8$, pressure factor $1 \cdot 0$.

Purchased cost $= 3800 \times 0 \cdot 8 \times 1 \cdot 0 = £3040$

Cooler:

Bare cost (Fig. 6.3) £2100; type factor $0 \cdot 85$, pressure factor $1 \cdot 0$.

Purchased cost $= 2100 \times 0 \cdot 85 \times 1 \cdot 0 = £1785$

Solvent tank:

Purchase cost (Table 6.2) $= 600 \times (35)^{0 \cdot 65} = £6000$

Product tank:

Same as solvent tank, cost $= £6000$

Total purchase cost of major equipment items (PCE)

	Absorption column	11,000
	Recovery column	41,000
	Reboiler	4160
	Condenser	3040
	Cooler	1785
	Solvent tank	6000
	Product tank	6000

Total £72,985   round up to £73,000

Estimation of fixed capital cost, reference Table 6.1, fluids processing plant:

PCE £73,000

f_1	Equipment erection	0·40
f_2	Piping	0·70
f_3	Instrumentation	0·20
f_4	Electrical	0·10
f_5	Buildings	none required
f_6	Utilities	not applicable
f_7	Storages	provided for in PCE
f_8	Site development	not applicable
f_9	Ancillary buildings	none required

Total Physical plant cost (PPC) $= 73,000(1 + 0·4 + 0·7 + 0·2 + 0·1) = 175,200$

f_{10}	Design and Engineering	0·30
f_{11}	Contractor's fee	none (contractors unlikely to be employed for such a small project)
f_{12}	Contingencies	0·10

Fixed capital $= 175,000 (1 + 0·3 + 0·1) = £245,280$ round up to £250,000

Working capital, allow 5 per cent of fixed capital to cover cost
of the initial solvent charge, $= 250,000 \times 5/100 = £12,500$

Total investment required for project $= £262,500$ say £260,000

Annual operating costs, reference Table 6.4:
Operating time $= 365 \times 95/100 = 345$ d/y $= 8280$ h/y.

Variable costs:
1. Raw material, solvent make-up, $= 10 \times 345 \times 400/1000$ $= £1380$
2. Miscellaneous materials, 10 per cent of maintenance cost (item 5) $= £1250$
3. Utilities, costs from Table 6.3:

Steam, at 0·62p/kg, $= 200 \times 8280 \times 0·62/100$ $= £10,267$

Cooling water, at 0·3p/m^3, $= \dfrac{5000}{1000} \times 8280 \times \dfrac{0·3}{100}$ $= £124$

Power, at 2·8p/kWh, $= 100 \times 345 \times 2·8/100$ $= £966$

Total   £11,357   round to £12,000

4. Shipping and packaging—not applicable

$$\text{Variable costs} = £14,630$$

Fixed costs:

5. Maintenance, take as 5 per cent of fixed capital, $= 5/100 \times 250,000$ $= £12,500$
6. Operating labour: allow one man per shift; it is unlikely that one extra man per shift would be needed to operate this small plant, but this is the minimum possible increment.

   Four men needed at, say, £6000 per year, $= £24,000$
7. Supervision, no additional supervision would be needed   —
8. Plant overheads, take as 50 per cent of operating labour $= £12,000$
9. Capital charges, 15 per cent of fixed capital $= £37,500$
10. Insurance, 1 per cent of fixed capital $= £2500$
11. Rates—neglect   —
12. Royalties—not applicable   —

                            Fixed costs $= £88,500$

Direct production costs $= 14,630 + 88,500$ $= £103,130$

13. Sales expense   ⎫
14. General overheads   ⎬ not applicable
15. Research and development   ⎭

   Annual operating cost, rounded $= £103,000$

6.10. Economic evaluation of projects

As the purpose of investing money in chemical plant is to earn money, some means of comparing the economic performance of projects is needed.

For small projects, and for simple choices between alternative processing schemes and equipment, the decisions can usually be made by comparing the capital and operating costs. More sophisticated evaluation techniques and economic criteria are needed when decisions have to be made between large, complex projects, particularly when the projects differ widely in scope, time scale and type of product. Some of the more commonly used techniques of economic evaluation and the criteria used to judge economic performance are outlined in this section. For a full discussion of the subject one of the many specialist texts that have been published should be consulted; ICI (1968), Merrett and Sykes (1963), Alfred and Evans (1967). The booklet published by the Institution of Chemical Engineers, Allen (1972), is particularly recommended to students.

Making major investment decisions in the face of the uncertainties that will undoubtedly exist about plant performance, costs, the market, government policy, and the world economic situation, is a difficult and complex task (if not an impossible task) and in a large design organisation the evaluation would be done by a specialist group.

6.10.1. Cash flow and cash-flow diagrams

The flow of cash is the life blood of any commercial organisation. The cash flows in a manufacturing company can be likened to the material flows in a process plant. The inputs are the cash needed to pay for research and development; plant design and construction; and plant operation. The outputs are goods for sale; and cash returns, is recycled, to the organisation from the profits earned. The "net cash flow" at any time is the difference between the earnings and expenditure. A cash-flow diagram, such as that shown in Fig. 6.7, shows the forecast cumulative net cash flow over the life of a project. The cash flows are based on the best estimates of investment, operating costs, sales volume and sales price, that can be made for the project. A cash-flow diagram gives a clear picture of the resources required for a project and the timing of the earnings. The diagram can be divided into the following characteristic regions:

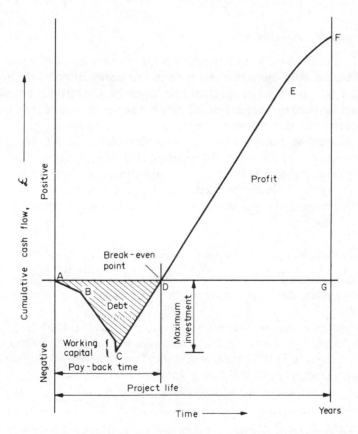

FIG. 6.7. Project cash-flow diagram

A–B   The investment required to design the plant.

B–C   The heavy flow of capital to build the plant, and provide funds for start-up.

C–D   The cash-flow curve turns up at C, as the process comes on stream and income is generated from sales. The net cash flow is now positive but the cumulative amount remains negative until the investment is paid off, at point D.

Point D is known as the *break-even point* and the time to reach the break-even point is called the *pay-back time*. In a different context, the term "break-even point" is used for the percentage of plant capacity at which the income equals the cost for production.

D–E   In this region the cumulative cash flow is positive. The project is earning a return on the investment.

E–F   Toward the end of project life the rate of cash flow may tend to fall off, due to increased operating costs and falling sale volume and price, and the slope of the curve changes.

The point F gives the final cumulative net cash flow at the end of the project life.

Net cash flow is a relatively simple and easily understood concept, and forms the basis for the calculation of other, more complex, measures of profitability.

6.10.2. Tax and depreciation

In calculating cash flows, as in Example 6.5, the project is usually considered as an isolated system, and taxes on profits and the effect of depreciation of the investment are not considered; tax rates are not constant and depend on government policy. In recent years, profit tax has been running at around 50 per cent and this figure can be used to make an estimate of the cash flow after tax. Depreciation rates depend on government policy, and on the accounting practices of the particular company. At times, it has been government practice to allow higher depreciation rates for tax purposes in development areas; or to pay capital grants to encourage investment in these areas. The effect of government policy must clearly be taken into account at some stage when evaluating projects, particularly when considering projects in different countries.

6.10.3. Discounted cash flow (*time value of money*)

In Fig. 6.7 the net cash flow is shown at its value in the year in which it occurred. So the figures on the ordinate show the "future worth" of the project: the cumulative "net future worth" (NFW).

The money earned in any year can be put to work (reinvested) as soon as it is available and start to earn a return. So money earned in the early years of the project is more valuable than that earned in later years. This "time value of money" can be allowed for by using a variation of the familiar compound interest formula. The net cash flow in each year of the project is brought to its "present worth" at the start of the project by discounting it at some chosen compound interest rate.

$$\frac{\text{Net present worth (NPW)}}{\text{of cash flow in year } n} = \frac{\text{Estimated net cash flow in year } n\,(\text{NFW})}{(1+r)^n} \qquad (6.6)$$

Where r is the discount rate (interest rate) per cent/100 and

$$\text{Total NPW of project} = \sum_{n=1}^{n=t} \frac{\text{NFW}}{(1+r)^n} \qquad (6.7)$$

t = life of project, years.

The discount rate is chosen to reflect the earning power of money. It would be roughly equivalent to the current interest rate that the money could earn if invested.

The total NPW will be less than the total NFW, and reflects the time value of money and the pattern of earnings over the life of the project; see Example 6.5.

6.10.4. Rate of return calculations

Cash-flow figures do not show how well the capital invested is being used; two projects with widely different capital costs may give similar cumulative cash-flow figures. Some way of measuring the performance of the capital invested is needed. Rate of return (ROR), which is the ratio of annual profit to investment, is a simple index of the performance of the money invested. Though basically a simple concept, the calculation of the ROR is complicated by the fact that the annual profit (net cash flow) will not be constant over the life of the project. The simplest method is to base the ROR on the average income over the life of the project and the original investment.

$$\text{ROR} = \frac{\text{Cumulative net cash flow at end of project}}{\text{Life of project} \times \text{original investment}} \times 100 \text{ per cent} \qquad (6.8)$$

From Fig. 6.7.

$$\begin{aligned}
\text{Cumulative income} &= F - C, \; \pounds \\
\text{Investment} &= C \quad , \pounds \\
\text{Life of project} &= G \quad , y
\end{aligned}$$

$$\text{then, ROR} = \frac{F - C}{C \times G} \times 100 \text{ per cent}$$

The rate of return is often calculated for the anticipated best year of the project: the year in which the net cash flow is greatest. It can also be based on the book value of the investment, the investment after allowing for depreciation. Simple rate of return calculations take no account of the time value of money.

6.10.5. Discounted cash-flow rate of return (DCFRR)

Discounted cash-flow analysis, used to calculate the present worth of future earnings (Section 6.10.3), is sensitive to the interest rate assumed. By calculating the NPW for various interest rates, it is possible to find an interest rate at which the cumulative net present worth at the end of the project is zero. This particular rate is called the "discounted cash-flow rate of return" (DCFRR) and is a measure of the maximum rate that the project could pay and still break even by the end of the project life.

$$\sum_{n=1}^{n=t} \frac{\text{NFW}}{(1+r')^n} = 0 \qquad (6.9)$$

where r' = the discounted cash-flow rate of return (per cent/100),

NFW = the future worth of the net cash flow in year n, $\pounds$,

t = the life of the project, years.

The value of r' is found by trial-and-error calculations. Finding the discount rate that just pays off the project investment over the project's life is analogous to paying off a mortgage.

The more profitable the project, the higher the DCFRR that it can afford to pay.

DCFRR provides a useful way of comparing the performance of capital for different projects; independent of the amount of capital used and the life of the plant, or the actual interest rates prevailing at any time.

Other names for DCFRR are interest rate of return and internal rate of return.

6.10.6. Pay-back time

Pay-back time is the time required after the start of the project to pay off the initial investment from income; point D on Fig. 6.7. Pay-back time is a useful criterion for judging projects that have a short life, or when the capital is only available for a short time.

It is often used to judge small improvement projects on operating plant. Typically, a pay-back time of 3 to 5 years would be expected from such projects.

Pay-back time as a criterion of investment performance does not, by definition, consider the performance of the project after the pay-back period.

6.10.7. Allowing for inflation

Inflation depreciates money in a manner similar to, but different from, the idea of discounting to allow for the time value of money. The effect of inflation on the net cash flow in future years can be allowed for in a similar manner to the net present worth calculation given by equation 6.6, using an inflation rate in place of, or added to, the discount rate r. However, the difficulty is to decide what the inflation rate is likely to be in future years. Also, inflation may well affect the sales price, operating costs and raw material prices differently. One approach is to argue that a decision between alternative projects made without formally considering the effect of inflation on future earnings will still be correct, as inflation is likely to affect the predictions made for both projects in a similar way.

6.10.8. Sensitivity analysis

The economic analysis of a project can only be based on the best estimates that can be made of the investment required and the cash flows. The actual cash flows achieved in any year will be affected by any changes in raw-materials costs, and other operating costs; and will be very dependent on the sales volume and price. A sensitivity analysis is a way of examining the effects of uncertainties in the forecasts on the viability of a project. To carry out the analysis the investment and cash flows are first calculated using what are considered the most probable values for the various factors; this establishes the base case for analysis. The cash flows, and whatever criteria of performance are to be used, are then calculated assuming a range of error for each of the factors in turn; for example, an error of, say, $\pm$ 10 per cent on the sales price might be assumed. This will show how sensitive the cash flows and economic criteria are to errors in the forecast figures. It gives some idea of the degree of risk involved in making judgements on the forecast performance of the project.

6.10.9. Summary

The investment criteria discussed in this section are set out in Table 6.5, which shows the main advantage and disadvantage of each criterion.

TABLE 6.5. *Investment criteria*

Criterion	Abbreviation	Units	Main advantage	Main shortcoming
Investment	—	£	Shows financial resources needed	No indication of project performance
Net future worth	NFW	£	Simple. When plotted as cash-flow diagram, shows timing of investment and income	Takes no account of the time value of money
Pay-back time	—	years	Shows how soon investment will be recovered	No information on later years
Net present worth	NPW	£	As for NFW but accounts for timing of cash flows	Dependent on discount rate used
Rate of return	ROR	%	Measures performance of capital	Takes no account of timing of cash flows. Dependent on definition of income (profit) and investment
Discounted cash-flow rate of return	DCFRR	%	Measures performance of capital allowing for timing of cash flows	No indication of the resources needed

There is no one best criterion on which to judge an investment opportunity. A company will develop its own methods of economic evaluation, using the techniques discussed in this section, and will have a "target" figure of what to expect for the criterion used, based on their experience with previous successful, and unsuccessful, projects.

A figure of 20 to 30 per cent for the return on investment (ROR) can be used as a rough guide for judging small projects, and when decisions have to be made on whether to install additional equipment to reduce operating costs. This is equivalent to saying that for a project to be viable the investment needed should not be greater than about 4 to 5 times the annual savings achieved.

As well as economic performance, many other factors have to be considered when evaluating projects; such as those listed below:

1. Safety.
2. Environmental problems (waste disposal).
3. Political considerations (government policies).
4. Location of customers.
5. Availability of labour.
6. Availability of supporting services.
7. Company experience in the particular technology.

Example 6.4

A plant is producing 10,000 t/y of a product. The overall yield is 70 per cent, on a mass basis (kg of product per kg raw material). The raw material costs £10/t, and the product sells for £35/t. A process modification has been devised that will increase the yield to 75 per cent. The additional investment required is £35,000, and the additional operating costs are negligible. Is the modification worth making?

Solution

There are two ways of looking at the earnings to be gained from the modification:

1. If the additional production given by the yield increase can be sold at the current price, the earnings on each additional ton of production will equal the sales price less the raw material cost.
2. If the additional production cannot be readily sold, the modification results in a reduction in raw material requirements, rather than increased sales, and the earnings (savings) are from the reduction in annual raw material costs.

The second way gives the lowest figures and is the safest basis for making the evaluation.

At 10,000 t/y production

$$\text{Raw material requirements at 70 per cent yield} = \frac{10,000}{0\cdot7} = 14,286$$

$$\text{at 75 per cent yield} = \frac{10,000}{0\cdot75} = \underline{13,333}$$

$$\text{savings} \quad \underline{\underline{953 \text{ t/y}}}$$

Cost savings, at £10/t, = $953 \times 10 = £9530$ per year

$$\text{ROR} = \frac{9530}{35,000} \times 100 = \underline{\underline{27 \text{ per cent}}}$$

Pay-back time (as the annual savings are constant, the pay-back time will be the reciprocal of the ROR)

$$= \frac{100}{27} = \underline{\underline{3\cdot7 \text{ years}}}$$

On these figures the modification would be considered worth while.

Example 6.5

It is proposed to build a plant to produce a new product. The estimated investment required is 12·5 million pounds and the timing of the investment will be:

year 1	1·0 million (design costs)
year 2	5·0 million (construction costs)
year 3	5·0 million „ „
year 4	1·5 million (working capital)

The plant will start up in year 4.

The forecast sales price, sales volume, and raw material costs are shown in Table 6.6. The fixed operating costs are estimated to be:

£400,000 per year up to year 9
£500,000 per year from year 9 to 13
£550,000 per year from year 13

TABLE 6.6. *Summary of data and results for example 6.5*

End of year	Forecast sales 10³ t	Forecast selling Price £/t	Raw Material costs £/t product	During year — Sale income Less operating costs 10⁶ £	During year — Net cash flow 10⁶ £	At year end — Cumulative cash flow 10⁶ £ (Project NFW)	At year end — Discounted cash flow at 15 per cent 10⁶ £	At commencement of project — Cumulative DCF (Project NPW) 10⁶ £	Project NPW at 25 per cent discount rate	Project NPW at 35 per cent discount rate	Project NPW at 37 per cent discount rate
1	0	—	—	0	−1·0	−1·00	−0·87	−0·87	−0·80	−0·74	0·73
2	0	—	—	0	−5·0	−6·00	−3·78	−4·65	−4·00	−3·48	−3·39
3	0	—	—	0	−5·0	−11·00	−3·29	−7·94	−6·56	−5·52	−5·34
4	100	150	90	4·6	3·10	−7·90	1·77	−6·17	−5·29	−4·58	−4·46
5	105	150	90	4·85	4·85	−3·05	2·41	−4·03	−3·70	−3·50	−3·45
6	110	150	90	5·10	5·10	2·05	2·20	−1·83	−2·36	−2·66	−2·68
7	120	150	90	5·60	5·60	7·65	2·11	0·28	−1·19	−1·97	−2·06
8	130	150	90	6·10	6·10	13·75	1·99	2·27	−0·17	−1·42	−1·57
9	140	150	90	6·50	6·50	20·25	1·85	4·12	0·70	−0·98	−1·19
10	150	145	85	7·00	7·00	27·25	1·73	5·85	1·45	−0·64	−0·89
11	165	140	85	6·93	6·93	34·18	1·49	7·34	2·05	−0·38	−0·67
12	180	140	85	7·60	7·60	41·78	1·42	8·76	2·57	−0·17	−0·50
13	190	140	85	8·05	8·05	49·83	1·31	10·07	3·01	−0·01	−0·36
14	200	135	80	8·05	8·05	57·88	1·14	11·21	3·36	0·11	−0·27
15	190	130	75	7·62	7·62	65·50	0·94	12·15	3·63	0·19	−0·20
16	180	120	75	7·19	7·19	72·69	0·77	12·92	3·83	0·25	−0·15
17	170	115	70	5·06	5·06	77·75	0·47	13·39	3·95	0·28	−0·13
18	160	110	70	3·93	3·93	81·68	0·32	13·71	4·02	0·30	−0·12
19	150	100	70	2·15	2·15	83·83	0·15	13·86	4·05	0·31	−0·11

The variable operating costs are estimated to be:

£10 per ton of product up to year 13
£13 per ton of product from year 13

Calculate:

1. The net cash flow in each year.
2. The future worth of the project, NFW.
3. The present worth, NPW, at a discount rate of 15 per cent.
4. The discounted cash-flow rate of return, DCFRR.
5. The pay-back time.

No account needed to be taken of tax in this exercise; or the scrap value of the equipment and value of the site at the end of the project life. For the discounting calculation, cash flows can be assumed to occur at the end of the year in which they actually occur.

Solution

The cash-flow calculations are summarised in Table 6.6. Sample calculations to illustrate the methods used are given below.

For year 4

Investment (negative cash flow)	$= £1.5 \times 10^6$
Sales income $= 100 \times 10^3 \times 150$	$= £15.0 \times 10^6$
Raw material costs $= 100 \times 10^3 \times 90$	$= £9.0 \times 10^6$
Fixed operating costs	$= £0.4 \times 10^6$
Variable operating costs $= 100 \times 10^3 \times 10$	$= £1.0 \times 10^6$

Net cash flow = sales income − costs − investment
$$= 15.0 - 10.4 - 1.5 = 3.1 \text{ million pounds}$$

Discounted cash flow (at 15 per cent) $= \dfrac{3.1}{(1+0.15)^4}$ $\qquad = £1.77 \times 10^6$

For year 8

Investment	nil
Sales income $= 130 \times 10^3 \times 150$	$= £19.5 \times 10^6$
Raw material costs $= 130 \times 10^3 \times 90$	$= £11.7 \times 10^6$
Fixed operating costs	$= £0.4 \times 10^6$
Variable operating costs $= 130 \times 10^3 \times 10$	$= £1.3 \times 10^6$

Net cash flow $= 19.5 - 13.4 = 6.10$ million pounds

$$DCF = \frac{6.1}{(1.15)^8} = 1.99$$

DCFRR

This is found by trial-and-error calculations. The present worth has been calcuated at discount rates of 25, 35 and 37 per cent. From the results shown in Table 6.6 it will be seen that the rate to give zero present worth will be around 36 per cent. This is the discounted cash-flow rate of return for the project.

6.11. References

ALFRED, A. M. and EVANS, J. B. (1967) *Appraisal of Investment Projects by DCF* (Chapman & Hall).
ALLEN, D. H. (1972) *A Guide to Capital Cost Estimation* (Institution of Chemical Engineers, London).
ARIES, R. S. and NEWTON, R. D. (1955) *Cost Estimation* (McGraw-Hill).
BECHTEL, L, B. (1960) *Chem. Eng., Albany* **67** (Feb. 22nd) 127. Estimate working capital needs.
CHEM. ENG. (1970) *Modern Cost Estimating Techniques* (McGraw-Hill).
CHEM. ENG. (1977) *Modern Cost Engineering* (McGraw-Hill).
CHILTON, C. H. (1960) *Cost Engineering in the Process Industries* (McGraw-Hill).
CRAN, J. (1973) *Process Engineering* (Jan.) 18. Process engineering indices help estimate the cost of new plant.
CRAN, J. (1979) *Process Engineering* (June) 10. Plant cost indices change with time.
ESTRUP, C. (1972) *Brit. Chem. Eng. Proc. Tech.* **17,** 213. The history of the six-tenths rule in capital cost estimation.
GUTHRIE, K. M. (1969) *Chem. Eng., Albany* **76** (March 24th) 114. Capital cost estimating.
GUTHRIE, K. M. (1970) *Chem. Eng., Albany* **77** (June 15th) 140. Capital and operating costs for 54 processes.(*Note*: correction Dec. 14th, 7)
GUTHRIE, K. M. (1974) *Process Plant Estimating, Evaluation, and Control* (Craftsman books).
HAPPLE, J. and JORDAN, D. G. (1975) *Chemical Process Economics*, 2nd ed. (Marcel Dekker).
HOLLAND, F. A., WATSON, F. A. and WILKINSON, J. K. (1974) *Introduction to Process Economics* (Wiley).
ICI (1968) *Assessing Projects – a programme for learning* (Methuen).
ICHEME (1977) *A New Guide to Capital Cost Estimation* (Institution of Chemical Engineers, London).

KARBANDA, O. P. (1978) *Process Plant and Equipment Cost Estimating* (Sevak Publications, Bombay).
LANG, H. J. (1948) *Chem. Eng., Albany* **55** (June) 112. Simplified approach to preliminary cost estimates.
LYDA, T. B. (1972) *Chem. Eng., Albany* **79** (Sept. 18th) 182. How much working capital will the new project need?
MERRETT, A. J. and SYKES, A. (1963) *The Finance and Analysis of Capital Projects* (Longmans & Green).
MILLER, C. A. (1979) *Chem. Eng., Albany* **86** (July 2nd) 89. Converting construction costs from one country to another.
PIKULIK, A. and DIAZ, H. E. (1977) *Chem. Eng., Albany* **84** (Oct. 10th) 106. Cost estimating for major process equipment.
WESSEL, H. E. (1952) *Chem. Eng., Albany* (July) 209. New graph correlates operating labor data for chemical processes.
WINFIELD, M. D. and DRYDEN, C. E. (1962) *Chem. Eng., Albany* (Dec. 24th) 100. Chart gives equipment, plant costs.

6.12. Nomenclature

		Dimensions in MT
A	Year in which cost is known (equation 6.1)	**T**
B	Year in which cost is to be estimated (equation 6.1)	**T**
C	Cost constant in equation 6.4	*
Ce	Purchased equipment cost	£
Cf	Fixed capital cost	£
C_1	Capital cost of plant 1	£
C_2	Capital cost of plant 2	£
f_L	Lang factors (equation 6.3)	—
$f_1 \ldots f_9$	Capital cost factors (Table 6.1)	—
n	Capital cost index in equation 6.4	—
S	Equipment size unit in equation 6.4	*
S_1	Capacity of plant 1	**MT^{-1}**
S_2	Capacity of plant 2	**MT^{-1}**

Asterisk (*) indicates that these dimensions are dependent on the type of equipment.

CHAPTER 7

Materials of Construction

7.1. Introduction

This chapter covers the selection of materials of construction for process equipment and piping.

Many factors have to be considered when selecting engineering materials, but for chemical process plant the overriding consideration is usually the ability to resist corrosion. The process designer will be responsible for recommending materials that will be suitable for the process conditions. He must also consider the requirements of the mechanical design engineer; the material selected must have sufficient strength and be easily worked. The most economical material that satisfies both process and mechanical requirements should be selected; this will be the material that gives the lowest cost over the working life of the plant, allowing for maintenance and replacement. Other factors, such as product contamination and process safety, must also be considered. The mechanical properties that are important in the selection of materials are discussed briefly in this chapter. Several books have been published on the properties of materials, and the metal-working processes used in equipment fabrication, and a selection suitable for further study is given in the list of references at the end of this chapter. The mechanical design of process equipment is discussed in Chapter 13.

A detailed discussion of the theoretical aspects of corrosion is not given in this chapter, as this subject is comprehensively covered in several books; those by Evans (1963a), Uhlig (1963) and Fontana and Greene (1978) are recommended. The book by Stewart and Tulloch (1968) gives a condensed but comprehensive account of the subject. Corrosion and corrosion prevention are also the subject of one of the design guides published by the Design Council, Ross (1977).

7.2. Material properties

The most important characteristics to be considered when selecting a material of construction are:

1. Mechanical properties
   (a) Strength—tensile strength
   (b) Stiffness—elastic modulus (Young's modulus)
   (c) Toughness—fracture resistance
   (d) Hardness—wear resistance
   (e) Fatigue resistance
   (f) Creep resistance
2. The effect of high and low temperatures on the mechanical properties
3. Corrosion resistance

4. Any special properties required; such as, thermal conductivity, electrical resistance, magnetic properties
5. Ease of fabrication—forming, welding, casting (see Table 7.1)
6. Availability in standard sizes—plates, sections, tubes
7. Cost

TABLE 7.1. *A Guide to the Fabrication Properties of Common Metals and Alloys*

	Machining	Cold working	Hot working	Casting	Welding	Annealing temp. °C
Mild steel	S	S	S	D	S	750
Low alloy steel	S	D	S	D	S	750
Cast iron	S	U	U	S	D/U	—
Stainless steel (18Cr, 8Ni)	S	S	S	D	S	1050
Nickel	S	S	S	S	S	1150
Monel	S	S	S	S	S	1100
Copper (deoxidised)	D	S	S	S	D	800
Brass	S	D	S	S	S	700
Aluminium	S	S	S	D	S	550
Dural	S	S	S	—	S	350
Lead	—	S	—	—	S	—
Titanium	S	S	U	U	D	—

S – Satisfactory, D – Difficult, special techniques needed.
U – Unsatisfactory.

7.3. Mechanical properties

Typical values of the mechanical properties of the more common materials used in the construction of chemical process equipment are given in Table 7.2.

7.3.1. Tensile strength

The tensile strength (tensile stress) is a measure of the basic strength of a material. It is the maximum stress that the material will withstand, measured by a standard tensile test. The older name for this property, which is more descriptive of the property, was Ultimate Tensile Strength (UTS).

The design stress for a material, the value used in any design calculations, is based on the tensile strength, or on the yield or proof stress (see Chapter 13).

Proof stress is the stress to cause a specified permanent extension, usually 0·1 per cent.

The tensile testing of materials is covered by BS 18.

7.3.2. Stiffness

Stiffness is the ability to resist bending and buckling. It is a function of the elastic modulus of the material and the shape of the cross-section of the member (the second moment of area).

TABLE 7.2. *Mechanical properties of common metals and alloys (typical values at room temperature)*

	Tensile strength (N/mm^2)	0·1 per cent proof stress (N/mm^2)	Modulus of elasticity (kN/mm^2)	Hardness Brinell	Specific gravity
Mild steel	430	220	210	100–200	7·9
Low alloy steel	420–660	230–460	210	130–200	7·9
Cast iron	140–170	—	140	150–250	7·2
Stainless steel (18 Cr, 8 Ni)	> 540	200	210	160	8·0
Nickel (> 99 per cent Ni)	500	130	210	80–150	8·9
Monel	650	170	170	120–250	8·8
Copper (deoxidised)	200	60	110	30–100	8·9
Brass (Admiralty)	400–600	130	115	100–200	8·6
Aluminium (> 99 per cent)	80–150	—	70	30	2·7
Dural	400	150	70	100	2·7
Lead	30	—	15	5	11·3
Titanium	500	350	110	150	4·5

7.3.3. Toughness

Toughness is associated with tensile strength, and is a measure of the material's resistance to crack propagation. The crystal structure of ductile materials, such as steel, aluminium and copper, is such that they stop the propagation of a crack by local yielding at the crack tip. In other materials, such as the cast irons and glass, the structure is such that local yielding does not occur and the materials are brittle. Brittle materials are weak in tension but strong in compression. Under compression any incipient cracks present are closed up. Various techniques have been developed to allow the use of brittle materials in situations where tensile stress would normally occur. For example, the use of prestressed concrete, and glass-fibre-reinforced plastics in pressure vessels construction.

A detailed discussion of the factors that determine the fracture toughness of materials can be found in the books by Institute of Metallurgists (1960) and Boyd (1970). Gordon (1976) gives an elementary, but very readable, account of the strength of materials in terms of their macroscopic and microscopic structure.

7.3.4. Hardness

The surface hardness, as measured in a standard test, is an indication of a material's ability to resist wear. Hardness testing is covered by British Standards: BS 240, 4175, 427 and 860. This will be an important property if the equipment is being designed to handle abrasive solids, or liquids containing suspended solids which are likely to cause erosion.

7.3.5. Fatigue

Fatigue failure is likely to occur in equipment subject to cyclic loading; for example, rotating equipment, such as pumps and compressors, and equipment subjected to pressure cycling. A comprehensive treatment of this subject is given by Harris (1976).

7.3.6. Creep

Creep is the gradual extension of a material under a steady tensile stress, over a prolonged period of time. It is usually only important at high temperatures; for instance, with steam and gas turbine blades. For a few materials, notably lead, the rate of creep is significant at moderate temperatures. Lead will creep under its own weight at room temperature and lead linings must be supported at frequent intervals.

The creep strength of a material is usually reported as the stress to cause rupture in 100,000 hours, at the test temperature.

7.3.7. Effect of temperature on the mechanical properties

The tensile strength and elastic modulus of metals decrease with increasing temperature. For example, the tensile strength of mild steel (low carbon steel, $C < 0.25$ per cent) is $450 \, N/mm^2$ at $25°C$ falling to 210 at $500°C$ and the value of Young's modulus $200,000 \, N/mm^2$ at $25°C$ falling to $150,000 \, mm^2$ at $500°C$. If equipment is being designed to operate at high temperatures, materials that retain their strength must be selected. The stainless steels are superior in this respect to plain carbon steels.

Creep resistance will be important if the material is subjected to high stresses at elevated temperatures. Special alloys, such as Inconel (International Nickel Co.), are used for high temperature equipment such as furnace tubes.

The selection of materials for high-temperature applications is discussed by Day (1979).

At low temperatures, less than $10°C$, metals that are normally ductile can fail in a brittle manner. Serious disasters have occurred through the failure of welded carbon steel vessels at low temperatures. The phenomenon of brittle failure is associated with the crystalline structure of metals. Metals with a body-centered-cubic (bcc) lattice are more liable to brittle failure than those with a face-centred-cubic (fcc) or hexagonal lattice. For low-temperature equipment, such as cryogenic plant and liquified-gas storages, austenitic stainless steel (fcc) or aluminium alloys (hex) should be specified; see Wigley (1978).

V-notch impact tests, such as the Charpy test, are used to test the susceptibility of materials to brittle failure: see Wells (1968) and BS 131.

The brittle fracture of welded structures is a complex phenomenon and is dependent on plate thickness and the residual stresses present after fabrication; as well as the operating temperature. A comprehensive discussion of brittle fracture in steel structures is given by Boyd (1970).

7.4. Corrosion resistance

The conditions that cause corrosion can arise in a variety of ways. For this brief discussion on the selection of materials it is convenient to classify corrosion into the following categories:

1. General wastage of material—uniform corrosion.
2. Galvanic corrosion—dissimilar metals in contact.
3. Pitting—localised attack.
4. Intergranular corrosion.
5. Stress corrosion.
6. Erosion–corrosion.

7. Corrosion fatigue.
8. High temperature oxidation.
9. Hydrogen embrittlement.

Metallic corrosion is essentially an electrochemical process. Four components are necessary to set up an electrochemical cell:

1. Anode—the corroding electrode.
2. Cathode—the passive, non-corroding electrode.
3. The conducting medium—the electrolyte—the corroding fluid.
4. Completion of the electrical circuit—through the material.

Cathodic areas can arise in many ways:

  (i) Dissimilar metals.
 (ii) Corrosion products.
(iii) Inclusions in the metal, such as slag.
(iv) Less well-aerated areas.
 (v) Areas of differential concentration.
(vi) Differentially strained areas.

7.4.1. Uniform corrosion

This term describes the more or less uniform wastage of material by corrosion, with no pitting or other forms of local attack. If the corrosion of a material can be considered to be uniform the life of the material in service can be predicted from experimentally determined corrosion rates.

Corrosion rates are usually expressed as a penetration rate in inches per year, or mills per year (mpy) (where a mill $= 10^{-3}$ inches). They are also expressed as a weight loss in milligrams per square decimetre per day (mdd). In corrosion testing, the corrosion rate is measured by the reduction in weight of a specimen of known area over a fixed period of time.

$$\text{ipy} = \frac{12w}{tA\rho} \qquad (7.1)$$

where w = mass loss in time t, lb,
    t = time, years,
    A = surface area, ft^2,
    ρ = density of material, lb/ft^3,

as most of the published data on corrosion rates are in imperial units. In SI units 1 ipy = 25 mm per year.

When judging corrosion rates expressed in mdd it must be remembered that the penetration rate depends on the density of the material. For ferrous metals 100 mdd = 0·02 ipy.

What can be considered as an acceptable rate of attack will depend on the cost of the material; the duty, particularly as regards to safety; and the economic life of the plant. For the more commonly used inexpensive materials, such as the carbon and low alloy steels, a guide to what is considered acceptable is given in Table 7.3. For the more expensive alloys,

TABLE 7.3. *Acceptable corrosion rates*

	Corrosion rate	
	ipy	mm/y
Completely satisfactory	< 0·01	0·25
Use with caution	< 0·03	0·75
Use only for short exposures	< 0·06	1·5
Completely unsatisfactory	> 0·06	1·5

such as the high alloy steels, the brasses and aluminium, the figures given in Table 7.3 should be divided by 2.

The corrosion rate will be dependent on the temperature and concentration of the corrosive fluid. An increase in temperature usually results in an increased rate of corrosion; though not always. The rate will depend on other factors that are affected by temperature, such as oxygen solubility.

The effect of concentration can also be complex. For example, the corrosion of mild steel in sulphuric acid, where the rate is unacceptably high in dilute acid and at concentrations above 70 per cent, but is acceptable at intermediate concentrations.

7.4.2. Galvanic corrosion

If dissimilar metals are placed in contact, in an electrolyte, the corrosion rate of the anodic metal will be increased, as the metal lower in the electrochemical series will readily act as a cathode. The galvanic series in sea water for some of the more commonly used metals is shown in Table 7.4. Some metals under certain conditions form a natural protective film; for example, stainless steel in oxidising environments. This state is denoted by "passive" in the series shown in Table 7.4; active indicates the absence of the protective film. Minor shifts in position in the series can be expected in other electrolytes, but the series for sea water is a good indication of the combinations of metals to be avoided. If

TABLE 7.4. *Galvanic series in sea water*

Noble end (protected end)	18/8 stainless steel (passive)
	Monel
	Inconel (passive)
	Nickel (passive)
	Copper
	Aluminium bronze (Cu 92 per cent, Al 8 per cent)
	Admiralty brass (Cu 71 per cent, Zn 28 per cent, Sn 1 per cent)
	Nickel (active)
	Inconel (active)
	Lead
	18/8 stainless steel (active)
	Cast iron
	Mild steel
	Aluminium
	Galvanised steel
	Zinc
	Magnesium

metals which are widely separated in the galvanic series have to be used together, they should be insulated from each other, breaking the conducting circuit. Alternatively, if sacrificial loss of the anodic material can be accepted, the thickness of this material can be increased to allow for the increased rate of corrosion. The corrosion rate will depend on the relative areas of the anodic and cathodic metals. A high cathode to anode area should be avoided. Sacrificial anodes are used to protect underground steel pipes.

7.4.3. Pitting

Pitting is the term given to very localised corrosion that forms pits in the metal surface. If a material is liable to pitting, penetration can occur prematurely and corrosion rate data are not a reliable guide to the equipment life.

Pitting can be caused by a variety of circumstances; any situation that causes a localised increase in corrosion rate may result in the formation of a pit. In an aerated medium the oxygen concentration will be lower at the bottom of a pit, and the bottom will be anodic to the surrounding metal, causing increased corrosion and deepening of the pit. A good surface finish will reduce this type of attack. Pitting can also occur if the composition of the metal is not uniform; for example, the presence of slag inclusions in welds. The impingement of bubbles can also cause pitting; the effect of cavitation in pumps, which is an example of erosion–corrosion.

7.4.4. Intergranular corrosion

Intergranular corrosion is the preferential corrosion of material at the grain (crystal) boundaries. Though the loss of material will be small, intergranular corrosion can cause the catastrophic failure of equipment. Intergranular corrosion is a common form of attack on alloys but occurs rarely with pure metals. The attack is usually caused by a differential couple being set up between impurities existing at the grain boundary. Impurities will tend to accumulate at the grain boundaries after heat treatment. The classic example of intergranular corrosion in chemical plant is the weld decay of unstabilised stainless steel. This is caused by the precipitation of chromium carbides at the grain boundaries in a zone adjacent to the weld, where the temperature has been between 500–800°C during welding. Weld decay can be avoided by annealing after welding, if practical; or by using low carbon grades (< 0.3 per cent C); or grades stabilised by the addition of titanium or niobium. A test for the susceptibility of stainless steels to weld decay is given in BS 1501.

7.4.5. Effect of stress

Corrosion rate and the form of attack can be changed if the material is under stress. Generally, the rate of attack will not change significantly within normal design stress values. However, for some combinations of metal, corrosive media and temperature, the phenomenon called stress cracking can occur. This is the general name given to a form of attack in which cracks are produced that grow rapidly, and can cause premature, brittle failure, of the metal. The conditions necessary for stress corrosion cracking to occur are:

1. Simultaneous stress and corrosion.
2. A specific corrosive substance; in particular the presence of Cl^-, OH^-, NO_3^-, or NH_4^+ ions.

Mild stress can cause cracking; the residual stresses from fabrication and welding are sufficient.

For a general discussion of the mechanism of stress corrosion cracking see Fontana and Greene (1978).

Some classic examples of stress corrosion cracking are:

The season cracking of brass cartridge cases.
Caustic embrittlement of steel boilers.
The stress corrosion cracking of stainless steels in the presence of chloride ions.

Stress corrosion cracking can be avoided by selecting materials that are not susceptible in the specific corrosion environment; or, less certainly, by stress relieving by annealing after fabrication and welding.

Comprehensive tables of materials susceptible to stress corrosion cracking in specific chemicals are given by Moore (1979). Moore's tables are taken from the corrosion data survey published by NACE (1974).

The term corrosion fatigue is used to describe the premature failure of materials in corrosive environments caused by cyclic stresses. Even mildly corrosive conditions can markedly reduce the fatigue life of a component. Unlike stress corrosion cracking, corrosion fatigue can occur in any corrosive environment and does not depend on a specific combination of corrosive substance and metal. Materials with a high resistance to corrosion must be specified for critical components subjected to cyclic stresses.

7.4.6. Erosion–corrosion

The term erosion–corrosion is used to describe the increased rate of attack caused by a combination of erosion and corrosion. If a fluid stream contains suspended particles, or where there is high velocity or turbulence, erosion will tend to remove the products of corrosion and any protective film, and the rate of attack will be markedly increased. If erosion is likely to occur, more resistant materials must be specified, or the material surface protected in some way. For example, plastics inserts are used to prevent erosion–corrosion at the inlet to heat-exchanger tubes.

7.4.7. High-temperature oxidation

Corrosion is normally associated with aqueous solutions but oxidation can occur in dry conditions. Carbon and low alloy steels will oxidise rapidly at high temperatures and their use is limited to temperatures below 500°C.

Chromium is the most effective alloying element to give resistance to oxidation, forming a tenacious oxide film. Chromium alloys should be specified for equipment subject to temperatures above 500°C in oxidising atmospheres.

7.4.8. Hydrogen embrittlement

Hydrogen embrittlement is the name given to the loss of ductility caused by the absorption (and reaction) of hydrogen in a metal. It is of particular importance when specifying steels for use in hydrogen reforming plant. Alloy steels have a greater resistance to hydrogen embrittlement than the plain carbon steels. A chart showing the suitability of various alloy steels for use in hydrogen atmospheres, as a function of hydrogen partial pressure and temperature, is given in the NACE (1974) corrosion data survey. Below 500°C plain carbon steel can be used.

7.5. Selection for corrosion resistance

In order to select the correct material of construction, the process environment to which the material will be exposed must be clearly defined. Additional to the main corrosive chemicals present, the following factors must be considered:

1. Temperature—affects corrosion rate and mechanical properties.
2. Pressure.
3. pH.
4. Presence of trace impurities—stress corrosion.
5. The amount of aeration—differential oxidation cells.
6. Stream velocity and agitation—erosion–corrosion.
7. Heat-transfer rates—differential temperatures.

The conditions that may arise during abnormal operation, such as at start-up and shut-down, must be considered, in addition to normal, steady state, operation.

Corrosion charts

The resistance of some commonly used materials to a range of chemicals is shown in Appendix C. More comprehensive corrosion data, covering most of the materials used in the construction of process plant, in a wide range of corrosive media, are given by, Rabald (1968), NACE (1974), Hamner (1974), Perry and Chilton (1973) and Schweitzer (1976).

These corrosion guides can be used for the preliminary screening of materials that are likely to be suitable, but the fact that published data indicate that a material is suitable cannot be taken as a guarantee that it will be suitable for the process environment being considered. Slight changes in the process conditions, or the presence of unsuspected trace impurities, can markedly change the rate of attack or the nature of the corrosion. The guides will, however, show clearly those materials that are manifestly unsuitable. Judgement, based on experience with the materials in similar processes environments, must be used when assessing published corrosion data.

Pilot plant tests, and laboratory corrosion tests under simulated plant conditions, will help in the selection of suitable materials if actual plant experience is not available. Care is needed in the interpretation of laboratory tests. Corrosion test procedures are described by Ailor (1971) and Champion (1967).

The advice of the technical service department of the company supplying the materials should also be sought.

7.6. Material costs

An indication of the cost of some commonly used metals is given in Table 7.5.

The quantity of a material used will depend on the material density and strength (design stress) and these must be taken into account when comparing material costs. Moore

TABLE 7.5. *Basic cost of metals (April 1976)*

	£/tonne
Carbon steel	140
Low alloy steels (Cr-Mo)	280
Nickel steel (9 per cent)	650
Austentic stainless steels	
304	840
321	950
316	1140
Copper	800
Aluminium	1140
Monel	3500
Titanium	3700

(1970) compares costs by calculating a cost rating factor defined by the equation:

$$\text{Cost rating} = \frac{C \times \rho}{\sigma_d} \qquad (7.2)$$

where C = cost per unit mass, £/kg,
ρ = density, kg/m^3,
σ_d = design stress, N/mm^2.

His calculated cost ratings, relative to the rating for mild steel (low carbon), are shown in Table 7.6. Materials with a relatively high design stress, such as stainless and low alloy steels, can be used more efficiently than carbon steel.

TABLE 7.6. *Relative cost ratings for metals*

		Design stress (N/mm^2)
Carbon steel	1	100
Al-alloys (Mg)	4	70
Stainless steel 18/8 (Ti)	5	130
Inconel	12	140
Brass	10–15	76
Al-bronzes	16	87
Aluminium	18	14
Monel	19	120
Copper	27	46
Nickel	35	70

Note: the design stress figures are shown for the purposes of illustration only and should not be used as design values.

The relative cost of equipment made from different materials will depend on the cost of fabrication, as well as the basic cost of the material. Unless a particular material requires special fabrication techniques, the relative cost of the finished equipment will be lower than the relative bare material cost. For example; the purchased cost of a stainless-steel storage tank will be 2 to 3 times the cost of the same tank in carbon steel, whereas the relative cost of the metals is between 5 to 8.

If the corrosion rate is uniform, then the optimum material can be selected by calculating the annual costs for the possible candidate materials. The annual cost will depend on the predicted life, calculated from the corrosion rate, and the purchased cost of the equipment. In a given situation, it may prove more economic to install a cheaper material with a high corrosion rate and replace it frequently; rather than select a more resistant but more expensive material. This strategy would only be considered for relatively simple equipment with low fabrication costs, and where premature failure would not cause a serious hazard. For example, carbon steel could be specified for an aqueous effluent line in place of stainless steel, accepting the probable need for replacement. The pipe wall thickness would be monitored *in situ* frequently to determine when replacement was needed.

The more expensive, corrosion-resistant, alloys are frequently used as a cladding on carbon steel. If a thick plate is needed for structural strength, as for pressure vessels, the use of clad materials can substantially reduce the cost.

7.7. Contamination

With some processes, the prevention of the contamination of a process stream, or a product, by certain metals, or the products of corrosion, overrides any other considerations when selecting suitable materials. For instance, in textile processes, stainless steel or aluminium is often used in preference to carbon steel, which would be quite suitable except that any slight rusting will mark the textiles (iron staining).

With processes that use catalysts, care must be taken to select materials that will not cause contamination and poisoning of the catalyst.

Some other examples that illustrate the need to consider the effect of contamination by trace quantities of other materials are:

1. For equipment handling acetylene the pure metals, or alloys containing copper, silver, mercury, gold, must be avoided to prevent the formation of explosive acetylides.
2. The presence of trace quantities of mercury in a process stream can cause the catastrophic failure of brass heat-exchanger tubes, from the formation of a mercury–copper amalgam. Incidents have occurred where the contamination has come from unsuspected sources, such as the failure of mercury-in-steel thermometers.
3. In the Flixborough disaster (see Chapter 9), there was evidence that the stress corrosion cracking of a stainless-steel pipe had been caused by zinc contamination from galvanised-wire supporting lagging.

7.8. Commonly used materials of construction

The general mechanical properties, corrosion resistance, and typical areas of use of some of the materials commonly used in the construction of chemical plant are given in this section. The values given are for a typical, representative, grade of the material or alloy. The multitude of alloys used in chemical plant construction is known by a variety of trade names, and code numbers designated in the various national standards. With the exception of the stainless steels, no attempt has been made in this book to classify the alloys discussed by using one or other of the national standards; the commonly used, generic, names for the alloys have been used. For the full details of the properties and compositions of the grades available in a particular class of alloy, and the designated code numbers, reference should be made to the appropriate national code, to the various handbooks, or to manufacturers' literature. For the United Kingdom standards, the British Standards Institute year book should be consulted.

The US trade names and codes are given by Perry and Chilton (1973). A comprehensive review of the engineering materials used for chemical and process plant can be found in the books by Evans (1974), Hepner (1962) and Rumford (1954). Hepner's book is a collection of articles previously published in the journal *Chemical and Process Engineering*, in the period 1960 to 1961. The articles cover the complete range of materials used for process plant.

7.8.1. Iron and steel

Low carbon steel (mild steel) is the most commonly used engineering material. It is cheap; is available in a wide range of standard forms and sizes; and can be easily worked and welded. It has good tensile strength and ductility.

The carbon steels and iron are not resistant to corrosion, except in certain specific environments, such as concentrated sulphuric acid and the caustic alkalies. They are suitable for use with most organic solvents, except chlorinated solvents; but traces of corrosion products may cause discoloration.

Mild steel is susceptible to stress-corrosion cracking in certain environments.

The corrosion resistance of the low alloy steels (less than 5 per cent of alloying elements) where the alloying elements are added to improve the mechanical strength and not for corrosion resistance is not significantly different from that of the plain carbon steels.

The use of carbon steel in the construction of chemical plant is discussed by Clark (1970).

The high silicon irons (14 to 15 per cent Si) have a high resistance to mineral acids, except hydrofluoric acid. They are particularly suitable for use with sulphuric acid at all concentrations and temperatures. They are, however, very brittle.

7.8.2. Stainless steel

The stainless steels are the most frequently used corrosion resistant materials in the chemical industry.

To impart corrosion resistance the chromium content must be above 12 per cent, and the higher the chromium content, the more resistant is the alloy to corrosion in oxidising conditions. Nickel is added to improve the corrosion resistance in non-oxidising environments.

Types

A wide range of stainless steels is available, with compositions tailored to give the properties required for specific applications. They can be divided into three broad classes according to their microstructure:

1. Ferritic: 13–20 per cent Cr, < 0·1 per cent C, with no nickel
2. Austenitic: 18–20 per cent Cr, > 7 per cent Ni
3. Martensitic: 12–10 per cent Cr, 0·2 to 0·4 per cent C, up to 2 per cent Ni

The uniform structure of Austenite (fcc, with the carbides in solution) is the structure desired for corrosion resistance, and it is these grades that are widely used in the chemical industry. The composition of the main grades of austenitic steels, and the US, and equivalent UK designations are shown in Table 7.7. Their properties are discussed below.

TABLE 7.7. *Commonly used grades of austenitic stainless steel*

Specification no.		Composition per cent							
BS 1501	AISI	C max	Si max	Mn max	Cr range	Ni range	Mo range	Ti	Nb
801B	304	0·08	—	2·00	17·5 20·0	8·0 11·0	—	—	—
810 C	304 ELC	0·03	1·00	2·00	17·5 20·0	10 min	—	—	—
801 Ti	321	0·12	1·00	2·00	17·00 20·0	7·5 min	—	4 × C	—
801 Nb	347	0·08	1·00	2·00	17·0 20·0	9 min	—	—	10 × C
821 Ti	—	0·12	1·00	2·00	17·0 20·0	25 min	—	4 × C	—
845 B	316	0·08	1·00	2·00	16·5 18·5	10 min	2·25 3·00		—
845 Ti	—	0·08	0·06	2·00	16·5 18·5	10 min	2·25 3·00	4 × C	—
846	—	0·08	1·00	2·00	18·0 20·0	11·0 14·0	3·0 4·0		

S and P   0·045 per cent all grades.
AISI   American Iron and Steel Institute.

Type 304 (the so-called 18/8 stainless steels): the most generally used stainless steel. It contains the minimum Cr and Ni that give a stable austenitic structure. The carbon content is low enough for heat treatment not to be normally needed with thin sections to prevent weld decay (see Section 7.4.4).

Type 304L: low carbon version of type 304 (< 0·03 per cent C) used for thicker welded sections, where carbide precipitation would occur with type 304.

Type 321: a stabilised version of 304, stabilised with titanium to prevent carbide precipitation during welding. It has a slightly higher strength than 304L, and is more suitable for high-temperature use.

Type 347: stabilised with niobium.

Type 316: in this alloy, molybdenum is added to improve the corrosion resistance in reducing conditions, such as in dilute sulphuric acid, and, in particular, to solutions containing chlorides.

Type 316L: a low carbon version of type 316, which should be specified if welding or heat treatment is liable to cause carbide precipitation in type 316.

Types 309/310: alloys with a high chromium content, to give greater resistance to oxidation at high temperatures. Alloys with greater than 25 per cent Cr are susceptible to embrittlement due to sigma phase formation at temperatures above 500°C. Sigma phase is an intermetallic compound, FeCr. The formation of the sigma phase in austenitic stainless steels is discussed by Hills and Harries (1960).

Mechanical properties

The austenitic stainless steels have greater strength than the plain carbon steels, particularly at elevated temperatures (see Table 7.8).

TABLE 7.8. *Comparative strength of stainless steel*

Temperature°C		300	400	500	600
Typical design stress N/mm^2	mild steel	77	62	31	—
	stainless 18/8	108	100	92	62

As was mentioned in Section 7.3.7, the austenitic stainless steels, unlike the plain carbon steels, do not become brittle at low temperatures. It should be noted that the thermal conductivity of stainless steel is significantly lower than that of mild steel.

Typical at 100°C values are, type 304 (18/8)   0·16 W/cm°C
mild steel   0·60 W/cm°C

Austenitic stainless steels are non-magnetic in the annealed condition.

General corrosion resistance

The higher the alloying content, the better the corrosion resistance over a wide range of conditions, strongly oxidising to reducing, but the higher the cost. A ranking in order of increasing corrosion resistance, taking type 304 as 1, is given below:

304	304L	321	316	316L	310
1·0	1·1	1·1	1·25	1·3	1·6

Intergranular corrosion (weld decay) and stress corrosion cracking are problems associated with the use of stainless steels, and must be considered when selecting types suitable for use in a particular environment. Stress corrosion cracking in stainless steels can be caused by a few ppm of chloride ions (see Section 7.4.5). In general, stainless steels are used for corrosion resistance when oxidising conditions exist. Special types, or other high nickel alloys, should be specified if reducing conditions are likely to occur. The properties, corrosion resistance, and uses of the various grades of stainless steel are discussed fully by Peckner and Bernstein (1977). A comprehensive discussion of the corrosion resistance of stainless steels is given in Sedriks (1979).

7.8.3. Nickel

Nickel has good mechanical properties and is easily worked. The pure metal (> 99 per cent) is not generally used for chemical plant, its alloys being preferred for most applications. The main use is for equipment handling caustic alkalies at temperatures above that at which carbon steel could be used; above 70°C. Nickel is not subject to corrosion cracking like stainless steel.

7.8.4. Monel

Monel, the classic nickel–copper alloy with the metals in the ratio 2 : 1, is probably, after the stainless steels, the most commonly used alloy for chemical plant. It is easily worked and has good mechanical properties up to 500°C. It is more expensive than stainless steel but is not susceptible to stress-corrosion cracking in chloride solutions. Monel has good resistance to dilute mineral acids and can be used in reducing conditions, where the stainless steels would be unsuitable. It may be used for equipment handling, alkalies, organic acids and salts, and sea water.

7.8.5. Inconel

Inconel (typically 76 per cent Ni, 7 per cent Fe, 15 per cent Cr) is used primarily for acid resistance at high temperatures. It maintains its strength at elevated temperature and is resistant to furnace gases, if sulphur free.

7.8.6. The Hastelloys

The trade name Hastelloy covers a range of nickel, chromium, molybdenum, iron alloys that were developed for corrosion resistance to strong mineral acids, particularly HCl. The corrosion resistance, and use, of the two main grades, Hastelloy B (65 per cent Ni, 28 per cent Mo, 6 per cent Fe) and Hastelloy C (54 per cent Ni, 17 per cent Mo, 15 per cent Cr, 5 per cent Fe), are discussed in papers by Weisert (1952a,b).

7.8.7. Copper and copper alloys

Pure copper is not widely used for chemical equipment. It has been used traditionally in the food industry, particularly in brewing. Copper is a relatively soft, very easily worked metal, and is used extensively for small-bore pipes and tubes.

The main alloys of copper are the brasses, alloyed with zinc, and the bronzes, alloyed with tin. Other, so-called bronzes are the aluminium bronzes and the silicon bronzes.

Copper is attacked by mineral acids, except cold, dilute, unaerated sulphuric acid. It is resistant to caustic alkalies, except ammonia, and to many organic acids and salts. The brasses and bronzes have a similar corrosion resistance to the pure metal. Their main use in the chemical industry is for valves and other small fittings, and for heat-exchanger tubes and tube sheets. If brass is used, a grade must be selected that is resistant to dezincification.

The cupro-nickel alloys (70 per cent Cu) have a good resistance to corrosion–erosion and are used for heat-exchanger tubes, particularly where sea water is used as a coolant.

7.8.8. Aluminium and its alloys

Pure aluminium lacks mechanical strength but has higher resistance to corrosion than its alloys. The main structural alloys used are the Duralumin (Dural) range of aluminium–copper alloys (typical composition 4 per cent Cu, with 0·5 per cent Mg) which have a tensile strength equivalent to that of mild steel. The pure metal can be used as a cladding on Dural plates, to combine the corrosion resistance of the pure metal with the strength of the alloy. The corrosion resistance of aluminium is due to the formation of a thin oxide film (as with the stainless steels). It is therefore most suitable for use in strong oxidising conditions. It is attacked by mineral acids, and by alkalies; but is suitable for concentrated nitric acid, greater than 80 per cent. It is widely used in the textile and food industries, where the use of mild steel would cause contamination. It is also used for the storage and distribution of demineralised water.

7.8.9. Lead

Lead was one of the traditional materials of construction for chemical plant but has now, due to its price, been largely replaced by other materials, particularly plastics. It is a soft, ductile material, and is mainly used in the form of sheets (as linings) or pipe. It has a good resistance to acids, particularly sulphuric.

7.8.10. Titanium

Titanium is now used quite widely in the chemical industry, mainly for its resistance to chloride solutions, including sea water and wet chlorine. It is rapidly attacked by dry chlorine, but the presence of as low a concentration of moisture as 0.01 per cent will prevent attack. Like the stainless steels, titanium depends for its resistance on the formation of an oxide film.

Alloying with palladium (0.15 per cent) significantly improves the corrosion resistance, particularly to HCl. Titanium is being increasingly used for heat exchangers, for both shell and tube, and plate exchangers; replacing cupro-nickel for use with sea water.

7.8.11. Tantalum

The corrosion resistance of tantalum is similar to that of glass, and it has been called a metallic glass. It is expensive, and is used for special applications, where glass or a glass lining would not be suitable. Tantalum plugs are used to repair glass-lined equipment.

7.9. Plastics as materials of construction for chemical plant

Plastics are being increasingly used as corrosion-resistant materials for chemical plant construction. They can be divided into two broad classes:

1. Thermoplastic materials, which soften with increasing temperature; for example, polyvinyl chloride (PVC) and polyethylene.
2. Thermosetting materials, which have a rigid, cross-linked structure; for example, the polyester and epoxy resins.

Details of the chemical composition and properties of the wide range of plastics used as engineering material can be found in the books by Evans (1974) and Hepner (1962).

The biggest use of plastics is for piping; sheets are also used for lining vessels and for fabricated ducting and fan casings. Mouldings are used for small items; such as, pump impellers, valve parts and pipe fittings.

The mechanical strength of plastics is low compared with that of metals. The mechanical strength, and other properties, can be modified by the addition of fillers and plasticisers. When reinforced with glass or carbon fibres thermosetting plastics can have a strength equivalent to mild steel, and are used for pressure vessels and pressure piping. Unlike metals, plastics are flammable. Plastics can be considered to complement metals as corrosion-resistant materials of construction. They generally have good resistance to dilute acids and inorganic salts, but suffer degradation in organic solvents that would not attack metals. Unlike metals, plastics can absorb solvents, causing swelling and softening. The properties and typical areas of use of the main plastics used for chemical plant are reviewed briefly in the following sections. A comprehensive discussion of the use of plastics as corrosion-resistant materials is given in the book by Evans (1966). The mechanical properties and relative cost of plastics are given in Table 7.9.

TABLE 7.9. *Mechanical properties and relative cost of polymers*

Material	Tensile strength (N/mm^2)	Elastic modulus (kN/mm^2)	Density (kg/m^3)	Relative cost
PVC	55	3·5	1400	1·5
Polyethylene (low density)	12	0·2	900	1·0
Polypropylene	35	1·5	900	1·5
PTFE	21	1·0	2100	30·0
GRP polyester	100	7·0	1500	3·0
GRP epoxy	250	14·0	1800	5·0

Approximate cost relative to polyethylene, volumetric basis.

7.9.1. Poly-vinyl chloride (PVC)

PVC is probably the most commonly used thermoplastic material in the chemical industry. Of the available grades, rigid (unplasticised) PVC is the most widely used. It is resistant to most inorganic acids, except strong sulphuric and nitric, and inorganic salt solutions. It is unsuitable, due to swelling, for use with most organic solvents. The use of PVC as a material of construction in chemical engineering is discussed in a series of articles by Mottram and Lever (1957).

7.9.2. Polyolefines

Low-density polyethylene (polythene) is a relatively cheap, tough, flexible plastic. It has a low softening point and is not suitable for use above about 60°C. The higher density polymer (950 kg/m³) is stiffer, and can be used at higher temperatures. Polypropylene is a stronger material than the polyethylenes and can be used at temperatures up to 120°C.

The chemical resistance of the polyolefines is similar to that of PVC.

7.9.3. Polytetrafluroethylene (PTFE)

PTFE, known under the trade names Teflon and Fluon, is resistant to all chemicals, except molten alkalies and fluorine, and can be used at temperatures up to 250°C. It is a relatively weak material, but its mechanical strength can be improved by the addition of fillers (glass and carbon fibres). It is expensive and difficult to fabricate. PTFE is used extensively for gaskets and gland packings. As a coating, it is used to confer non-stick properties to surfaces, such as filter plates. It can also be used as a liner for vessels.

7.9.4. Glass-fibre reinforced plastics (GRP)

The polyester resins, reinforced with glass fibre, are the most common thermosetting plastics used for chemical plant. Complex shapes can be easily formed using the techniques developed for working with reinforced plastics. Glass-reinforced plastics are relatively strong and have a good resistance to a wide range of chemicals. The mechanical strength depends on the resin used; the form of the reinforcement (chopped mat or cloth); and the ratio of resin to glass.

By using special techniques, in which the reinforcing glass fibres are wound on in the form of a continuous filament, high strength can be obtained, and this method is used to produce pressure vessels.

The polyester resins are resistant to dilute mineral acids, inorganic salts and many solvents. They are less resistant to alkalies.

Glass-fibre-reinforced epoxy resins are also used for chemical plant but are more expensive than the polyster resins. In general they are resistant to the same range of chemicals as the polyesters, but are more resistant to alkalies.

The chemical resistance of GRP is dependent on the amount of glass reinforcement used. High ratios of glass to resin give higher mechanical strength but generally lower resistance to some chemicals. The design of chemical plant equipment in GRP is the subject of a book by Malleson (1969); see also Shaddock (1971).

7.9.5. Rubber

Rubber, particularly in the form of linings for tanks and pipes, has been extensively used in the chemical industry for many years. Natural rubber is most commonly used, because of its good resistance to acids (except concentrated nitric) and alkalies. It is unsuitable for use with most organic solvents.

Synthetic rubbers are also used for particular applications. Hypalon (trademark, E. I. du Pont de Nemours) has a good resistance to strongly oxidising chemicals and can be used with nitric acid. It is unsuitable for use with chlorinated solvents. Viton (trademark, E. I. du Pont de Nemours) has a better resistance to solvents, including chlorinated solvents, than other rubbers. Both Hypalon and Viton are expensive, compared with other synthetic, and natural, rubbers.

The use of natural rubber lining is discussed by Saxman (1965), and the chemical resistance of synthetic rubbers by Evans (1963b). Rubber and other linings for chemical plant are covered by the British Standard code of practice CP 3003.

7.10. Ceramic materials (silicate materials)

Ceramics are compounds of non-metallic elements and include the following materials used for chemical plant:

Glass, the borosilicate glasses (hard glass).
Stoneware.
Acid-resistant bricks and tiles.
Refractory materials.
Cements and concrete.

Ceramic materials have a cross-linked structure and are therefore brittle.

7.10.1. Glass

Borosilicate glass (known by several trade names, including Pyrex) is used for chemical plant as it is stronger than the soda glass used for general purposes; it is more resistant to thermal shock and chemical attack.

Glass equipment is available from several specialist manufacturers. Pipes and fittings are produced in a range of sizes, up to 0·5 m. Special equipment, such as heat exchangers, is available and, together with the larger sizes of pipe, is used to construct distillation and absorption columns. Teflon gaskets are normally used for jointing glass equipment and pipe.

Where failure of the glass could cause injury, pipes and equipment should be protected by external shielding or wrapping with plastic tape.

Glass linings, also known as glass enamel, have been used on steel and iron vessels for many years. Borosilicate glass is used, and the thickness of the lining is about 1 mm. The techniques used for glass lining, and the precautions to be taken in the design and fabrication of vessels to ensure a satisfactory lining, are discussed by Landels and Stout (1970). Glass linings are also covered by the British Standard Code of practice CP 3003. Borosilicate glass is resistant to acids, salts and organic chemicals. It is attacked by the caustic alkalies and fluorine.

7.10.2. Stoneware

Chemical stoneware is similar to the domestic variety, but of higher quality; stronger and with a better glaze. It is available in a variety of shapes for pipe runs and columns. As for glass, it is resistant to most chemicals, except alkalies and fluorine. The composition and properties of chemical stoneware are discussed by Holdridge (1961). Stoneware and porcelain shapes are used for packing absorption and distillation columns (see Chapter 11).

7.10.3. Acid-resistant bricks and tiles

High-quality bricks and tiles are used for lining vessels, ditches and to cover floors. The linings are usually backed with a corrosion-resistant membrane of rubber or plastic, placed behind the titles, and special acid-resistant cements are used for the joints. Brick and tile linings are covered in part 10 of BS CP 3003.

7.10.4. Refractory materials (refractories)

Refractory bricks and cements are needed for equipment operating at high temperatures; such as, fired heaters, high-temperature reactors and boilers.

The refractory bricks in common use are composed of mixtures of silica (SiO_2) and alumina (Al_2O_3). The quality of the bricks is largely determined by the relative amounts of these materials and the firing temperature. Mixtures of silica and alumina form an eutectic (94·5 per cent SiO_2, 1545 °C) and for a high refractoriness under load (the ability to resist distortion at high temperature) the composition must be well removed from the eutectic composition. The highest quality refractory bricks, for use in load-bearing structures at high temperatures, contain high proportions of silica or alumina. "Silica bricks", containing greater than 98 per cent SiO_2, are used for general furnace construction. High alumina bricks, 60 per cent Al_2O_3, are used for special furnaces where resistance to attack by alkalies is important; such as lime and cement kilns. Fire bricks, typical composition 50 per cent SiO_2, 40 per cent Al_2O_3, balance CaO and Fe_2O_3, are used for general furnace construction. Silica can exist in a variety of allotropic forms, and bricks containing a high proportion of silica undergo reversible expansion when heated up to working temperature. The higher the silica content the greater the expansion, and this must be allowed for in furnace design and operation.

Ordinary fire bricks, fire bricks with a high porosity, and special bricks composed of diatomaceous earths are used for insulating walls.

Full details of the refractory materials used for process and metallurgical furnaces can be found in the books by Norton (1968) and Lyle (1947).

7.11. Carbon

Impervious carbon, impregnated with chemically resistant resins, is used for specialised equipment; particularly heat exchangers. It has a high conductivity and a good resistance to most chemicals, except oxidising acids, of concentrations greater than 30 per cent. Carbon tubes can be used in conventional shell and tube exchanger arrangements; or proprietary designs can be used, in which the fluid channels are formed in blocks of carbon; see Hilland (1960).

7.12. Protective coatings

A wide range of paints and other organic coatings is used for the protection of mild steel structures. Paints are used mainly for protection from atmospheric corrosion. Special chemically resistant paints have been developed for use on chemical process equipment. Chlorinated rubber paints and epoxy-based paints are used. In the application of paints and other coatings, good surface preparation is essential to ensure good adhesion of the paint film or coating.

7.13. Design for corrosion resistance

The life of equipment subjected to corrosive environments can be increased by proper attention to design details. Equipment should be designed to drain freely and completely. The internal surfaces should be smooth and free from crevasses where corrosion products and other solids can accumulate. Butt joints should be used in preference to lap joints. The use of dissimilar metals in contact should be avoided, or care taken to ensure that they are effectively insulated to avoid galvanic corrosion. Fluid velocities and turbulence should be high enough to avoid the deposition of solids, but not so high as to cause erosion–corrosion.

7.14. References

AILOR, W. H. (Ed.) (1971) *Handbook of Corrosion Testing and Evaluation* (Wiley).

BOYD, G. M. (1970) *Brittle Fracture of Steel Structures* (Butterworths).

CHAMPION, F. A. (1967) *Corrosion Testing Procedures* 3rd ed. (Chapman & Hall).

CLARK, E. E. (1970) *Chem. Engr. London* No. 242 (Oct.) 312. Carbon Steels for the construction of chemical and allied plant.

DAY, M. F. (1979) *Materials for High Temperature Use, Engineering Design Guide* No. 28 (Oxford U.P.).

EVANS, U. R. (1963a) *An Introduction to Metallic Corrosion* (Arnold).

EVANS, L. S. (1963b) *Rubber and Plastics Age* **44**, 1349. The chemical resistance of rubber and plastics.

EVANS, L. S. (1974) *Selecting Engineering Materials for Chemical and Process Plant* (Business Books); see also 2nd ed. (Hutchinson, 1980).

EVANS, V. (1966) *Plastics as Corrosion Resistant Materials* (Pergamon).

FONTANA, M. G. and GREENE, N. D. (1978) *Corrosion Engineering*, 2nd ed. (McGraw-Hill).

GORDON, J. E. (1976) *The New Science of Strong Materials*, 2nd ed. (Penguin Books).

HAMNER, N. E. (1974) *Corrosion Data Survey*, 5th ed. (National Association of Corrosion Engineers).

HARRIS, W. J. (1976) *The Significance of Fatigue* (Oxford U.P.).

HEPNER, I. L. (Ed.) (1962) *Materials of Construction for Chemical Plant* (Leonard Hill).

HILLAND, A. (1960) *Chem. & Proc. Eng.* **41**, 416. Graphite for heat exchangers.

HILLS, R. F. and HARRIES, D. P. (1960) *Chem. & Proc. Eng.* **41**, 391. Sigma phase in austenitic stainless steel.

HOLDRIDGE, D. A. (1961) *Chem. & Proc. Eng.* **42**, 405. Ceramics.

INSTITUTE OF METALLURGISTS (1960) *Toughness and Brittleness of metals* (Iliffe).

LANDELS, H. H. and STOUT, E. (1970) *Brit. Chem. Eng.* **15**, 1289. Glassed steel equipment: a guide to current technology.

LYLE, O. (1947) *Efficient Use of Steam* (HMSO).

MALLESON, J. H. (1969) *Chemical Plant Design with Reinforced Plastics* (McGraw-Hill).

MOORE, D. C. (1970) *Chem. Engr. London* No. 242 (Oct.) 326. Copper.

MOORE, R. E. (1979) *Chem. Eng., Albany* **86** (July 30th) 91. Selecting materials to resist corrosive conditions.

MOTTRAM, S. and LEVER, D. A. (1957) *The Ind. Chem.* **33**, 62, 123, 177 (in three parts). Unplasticized P.V.C. as a constructional material in chemical engineering.

NACE (1974) *Standard TM-01-69 Laboratory Corrosion Testing of Metals for the Process Industries* (National Association of Corrosion Engineers).

NORTON, F. H. (1968) *Refractories*, 4th ed. (McGraw-Hill).

PECKNER, D. and BERNSTEIN, I. M. (1977) *Handbook of Stainless Steels* (McGraw-Hill).

PERRY, R. H. and CHILTON, C. H. (Eds.) (1973) *Chemical Engineer's Handbook*, 5th ed. (McGraw-Hill).

RABALD, E. (1968) *Corrosion Guide*, 2nd ed. (Elsevier).

ROSS, T. K. (1977) *Metal Corrosion* (Oxford U.P.).

RUMFORD, F. (1954) *Chemical Engineering Materials* (Constable).

SAXMAN, T. E. (1965) *Materials Protection* **4** (Oct.) 43. Natural rubber tank linings.

SCHWEITZER, P. A. (1976) *Corrosion Resistance Tables* (Dekker).

SEDRIKS, A. J. (1979) *Corrosion Resistance of Stainless Steel* (Wiley).

SHADDOCK, A. K. (1971) *Chem. Eng., Albany* **78** (Aug. 9th) 116. Designing for reinforced plastics.

SOAR, D. G. (1962) *Chem. & Proc. Eng.* **43**, 81. Paints.

STEWART, D. and TULLOCH, D. S. (1968) *Principles of Corrosion Protection* (Macmillan).

UHLIG, H. H. (1963) *Corrosion and Corrosion Control* (Wiley); see also 2nd ed., 1971.

WEISERT, E. D. (1952a) *Chem. Eng., Albany* **59** (June) 267. Hastelloy alloy C.

WEISERT, E. D. (1952b) *Chem. Eng., Albany* **59** (July) 314. Hastelloy alloy B.

WELLS, A. A. (1968) *British Welding Journal* **15**, 221. Fracture control of thick steels for pressure vessels.

WIGLEY, D. A. (1978) *Materials for Low Temperatures, Engineering Design Guide* No. 28 (Oxford U.P.).

British Standards

BS 18:	-----	Method for tensile testing of metals.
	Part 1: 1970	Non-ferrous metals.
	Part 2: 1971	Steel (general).
BS 131:	-----	Methods for notched bar tests.
	Part 1: 1961	The Izod impact test on metals.
	Part 2: 1972	The Charpy V-notch impact test on metals.
	Part 3: 1972	The Charpy U-notch impact test on metals.
	Part 4: 1972	Calibration of impact testing machines for metals.
	Part 5: 1965	Determination of crystallinity.

BS 240:   -----   Method for Brinell hardness testing.
           Part 1: 1962   Testing of metals.
           Part 2: 1964   Verification of testing machine.
BS 427:   -----   Method for Vickers hardness test.
           Part 1: 1961   Testing of metals.
           Part 2: 1962   Verification of the testing machine.
BS 860:   1967   Tables for comparison of hardness scales.
BS 970:   -----   Wrought steel in the form of blooms, billet, bars, and forgings.
           Part 1: 1972   Carbon and carbon manganese steels including free cutting steels.
           Part 2: 1970   Direct hardening alloy steels, including steels capable of surface hardening by nitriding.
           Part 3: 1971   Steels for case harding.
           Part 4: 1970   Stainless, heat resisting, and valve steels.
           Part 5: 1972   Carbon and alloy spring steels.
           Part 6: 1973   SI metric values (for use with BS 970 parts 1–5).
BS 1501:   -----   Steels for fired and unfired pressure vessels.
           PLATES
           Part 1: 1964   Carbon and Carbon manganese steels.
           Part 2: 1970   Alloy steels.
           Part 3: 1973   Corrosion and heat resisting steel.
BS 1501–6: 1958   Steels for use in the chemical, petroleum and allied industries.
BS 1502:   1968   Steels for fired and unfired pressure vessels. Sections and bars.
BS 1503:   1980   Steels for fired and unfired pressure vessels.
           FORGINGS
BS 1504:   1976   Specification for steels for pressure purposes.
BS 1510:   1958   Steels for use in the chemical, petroleum and allied industries (Low temperature supplementary requirements to BS 1501–6).
BS 4175:   -----   Method for Rockwell superficial hardness test (N and T scales).
           Part 1: 1967   Testing of metals.
           Part 2: 1970   Verification of testing machine.

Codes of Practice
CP 3003:   -----   Lining of vessels and equipment for chemical processes.
           Part 1: 1967   Rubber, natural and synthetic.
           Part 2: 1966   Glass enamel.
           Part 3: 1965   Lead.
           Part 4: 1965   Plasticized PVC.
           Part 5: 1966   Epoxide resins.
           Part 6: 1966   Phenolic resins.
           Part 7: 1970   Corrosion and heat resistant metals, stainless steel and nickel alloys.
           Part 8: 1970   Precious metals, gold, silver, platinum.
           Part 9: 1970   Titanium.
           Part 10: 1970   Brick and tile.

7.15. Nomenclature

		Dimensions in **MLT**
A	Area	L^2
C	Cost of material	$£/M$
t	Time	T
w	Mass loss	M
ρ	Density	ML^{-3}
σ_d	Design stress	$ML^{-1}T^{-2}$

CHAPTER 8

Design Information and Data

8.1. Introduction

Information on manufacturing processes, equipment parameters, materials of construction, costs and the physical properties of process materials are needed at all stages of design; from the initial screening of possible processes, to the plant start-up and production.

Sources of data on costs were discussed in Chapter 6 and materials of construction in Chapter 7. This chapter covers sources of information on manufacturing processes and physical properties; and the estimation of physical property data. Information on the types of equipment (unit operations) used in chemical process plants is given in Volumes 2 and 3, and in the Chapters concerned with equipment selection and design in this Volume, Chapters 10, 11 and 12.

When a project is largely a repeat of a previous project, the data and information required for the design will be available in the Company's process files, if proper detailed records are kept. For a new project or process, the design data will have to be obtained from the literature, or by experiment (research laboratory and pilot plant), or purchased from other companies. The information on manufacturing processes available in the general literature can be of use in the initial stages of process design, for screening potential process; but is usually mainly descriptive, and too superficial to be of much use for detailed design and evaluation.

The literature on the physical properties of elements and compounds is extensive, and reliable values for common materials can usually be found. The principal sources of physical property data are listed in the references at the end of this chapter.

Where values cannot be found, the data required will have to be measured experimentally or estimated. Methods of estimating (predicting) the more important physical properties required for design are given in this chapter. A physical property data bank is given in Appendix D.

Readers who are unfamiliar with the sources of information, and the techniques used for searching the literature, should consult one of the many guides to the technical literature that have been published; such as those by Antony (1979), Burman (1965) and Mount (1976).

8.2. Sources of information on manufacturing processes

In this section the sources of information available in the open literature on commercial processes for the production of chemicals and related products are reviewed.

The chemical process industries are competitive, and the information that is published on commercial processes is restricted. The articles on particular processes published in the

234

technical literature and in textbooks invariably give only a superficial account of the chemistry and unit operations used. They lack the detailed information needed on reaction kinetics, process conditions, equipment parameters, and physical properties needed for process design. The information that can be found in the general literature is, however, useful in the early stages of a project, when searching for possible process routes. It is often sufficient for a flow-sheet of the process to be drawn up and a rough estimate of the capital and production costs made.

The most comprehensive collection of information on manufacturing processes is probably the *Encyclopedia of Chemical Technology* edited by Kirk and Othmer (1966, 1977), which covers the whole range of chemical and associated products. Another encyclopedia covering manufacturing processes is that edited by McKetta (1977). Several books have also been published which give brief summaries of the production processes used for the commercial chemicals and chemical products. The most well known of these is probably Shreve's book on the chemical process industries, now updated by Brink, Brink and Shreve (1977). Others worth consulting are those by Faith *et al.* (1965), Groggins (1958), Stephenson (1966) and Weissermal and Arpe (1978).

Specialised texts have been published on some of the more important bulk industrial chemicals, such as that by Miller (1969) on ethylene and its derivatives; these are too numerous to list but should be available in the larger reference libraries and can be found by reference to the library catalogue.

Books quickly become outdated, and many of the processes described are obsolete, or at best obsolescent. More up-to-date descriptions of the processes in current use can be found in the technical journals. The journal *Hydrocarbon Processing* publishes an annual review of petrochemical processes, which was entitled *Petrochemical Handbook Issue* but is now called *Petrocarbon Developments*; this gives flow diagrams and brief process descriptions of new process developments. Patents are a useful source of information; but it should be remembered that the patentee will try to write the patent in a way that protects his invention, whilst disclosing the least amount of useful information to his competitors. The examples given in a patent to support the claims often give an indication of the process conditions used; though they are frequently examples of laboratory preparations, rather than of the full-scale manufacturing processes. Several short guides have been written to help engineers understand the use of patents for the protection of inventions, and as sources of information; such as those by Capsey (1963), Lieberry (1972) and HMSO (1970, 1971).

8.3. General sources of physical properties

In this section those references that contain comprehensive compilations of physical property data are reviewed. Sources of data on specific physical properties are given in the remaining sections of the chapter.

International Critical Tables (1933) is still probably the most comprehensive compilation of physical properties, and is available in most reference libraries. Though it was first published in 1933, physical properties do not change, except in as much as experimental techniques improve, and ICT is still a useful source of engineering data.

Tables and graphs of physical properties are given in many handbooks and textbooks on Chemical Engineering and related subjects. Many of the data given are duplicated from book to book, but the various handbooks do provide quick, easy access to data on the more commonly used substances.

An extensive compilation of thermophysical data has been published by Plenum Press, Touloukian (1970–77). This multiple-volume work covers conductivity, specific heat, thermal expansion, viscosity and radiative properties (emittance, reflectance, absorptance and transmittance).

The Engineering Sciences Data Unit (ESDU) was set up to provide authenticated data for engineering design. Its publications include some physical property data, and other design data and methods of interest to chemical engineering designers. They also cover data and methods of use in the mechanical design of equipment.

Caution should be exercised when taking data from the literature, as typographical errors often occur. If a value looks doubtful it should be cross-checked in an independent reference, or by estimation.

The values of some properties will be dependent on the method of measurement; for example, surface tension and flash point, and the method used should be checked, by reference to the original paper if necessary, if an accurate value is required.

The results of research work on physical properties are reported in the general engineering and scientific literature. The *Journal of Chemical Engineering Data* specialises in publishing physical property data for use in chemical engineering design. A quick search of the literature for data can be made by using the abstracting journals; such as *Chemical Abstracts* (American Chemical Society) and *Engineering Index* (Engineering Index Inc., New York).

Computerised physical property data banks have been set up by various organisations to provide a service to the design engineer. They can be incorporated into computer-aided design programs and will be increasingly used to provide reliable, authenticated, design data. An example of such a data bank is the Physical Property Data Service (PPDS) available from the Institution of Chemical Engineers.

8.4. Accuracy required of engineering data

The accuracy needed depends on the use to which the data will be put. Before spending time and money searching for the most accurate value, or arranging for special measurements to be made, the designer must decide what accuracy is required; this will depend on several factors:

1. The level of design; less accuracy is obviously needed for rough scouting calculations, made to sort out possible alternative designs, than in the final stages of design; when money will be committed to purchase equipment, and for construction.
2. The reliability of the design methods; if there is some uncertainty in the techniques to be used, it is clearly a waste of time to search out highly accurate physical property data that will add little or nothing to the reliability of the final design.
3. The sensitivity to the particular property: how much will a small error in the property affect the design calculation. For example, it was shown in Chapter 4 that the estimation of the optimum pipe diameter is insensitive to viscosity. The sensitivity of a design method to errors in physical properties, and other data, can be checked by repeating the calculation using slightly altered values.

It is often sufficient to estimate a value for a property (sometimes even to make an intelligent guess) if the value has little effect on the final outcome of the design calculation.

For example, in calculating the heat load for a reboiler or vaporiser an accurate value of the liquid specific heat is seldom needed, as the latent heat load is usually many times the sensible heat load and a small error in the sensible heat calculation will have little effect on the design. The designer must, however, exercise caution when deciding to use less reliable data, and to be sure that they are sufficiently accurate for his purpose. For example, it would be correct to use an approximate value for density when calculating the pressure drop in a pipe system where a small error could be tolerated, considering the other probable uncertainties in the design; but it would be quite unacceptable in the design of a decanter, where the operation depends on small differences in density.

Consider the accuracy of the equilibrium data required to calculate the number of equilibrium stages needed for the separation of a mixture of acetone and water by distillation (see Chapter 11, Example 11.2). Several investigators have published vapour–liquid equilibrium data for this system: Othmer et al. (1952), York and Holmes (1942), Kojima et al. (1968), Reinders and De Minjer (1947).

If the purity of the acetone product required is less than 95 per cent, inaccuracies in the v–l–e plot will have little effect on the estimate of the number of stages required, as the relative volatility is very high. If a high purity is wanted, say > 99 per cent, then reliable data are needed in this region as the equilibrium line approaches the operating line (a pinch point occurs). Of the references cited, none gives values in the region above 95 per cent, and only two give values above 90 per cent; more experimental values are needed to design with confidence. There is a possibility that the system forms an azeotrope in this region. An azeotrope does form at higher pressure, Othmer et al. (1952).

8.5. Prediction of physical properties

Whenever possible, experimentally determined values of physical properties should be used. If reliable values cannot be found in the literature and if time, or facilities, are not available for their determination, then in order to proceed with the design the designer must resort to estimation. Techniques are available for the prediction of most physical properties with sufficient accuracy for use in process and equipment design. A detailed review of all the different methods available is beyond the scope of this book; selected methods are given for the more commonly needed properties. The criterion used for selecting a particular method for presentation in this chapter was to choose the most easily used, simplest, method that had sufficient accuracy for general use. If highly accurate values are required, then specialised texts on physical property estimation should be consulted; such as those by: Reid et al. (1977), Bretsznajder (1971) and Sterbacek et al. (1979).

A quick check on the probable accuracy of a particular method can be made by using it to estimate the property for an analogous compound, for which experimental values are available.

The techniques used for prediction are also useful for the correlation, and extrapolation and interpolation, of experimental values.

Group contribution techniques; which are based on the concept that a particular physical property of a compound can be considered to be made up of contributions from the constituent atoms, groups, and bonds, the contributions being determined from experimental data; provide the designer with simple, convenient, methods for physical property estimation; requiring only a knowledge of the structural formula of the compound.

Also useful, and convenient, are the prediction methods based on the use of reduced conditions: temperature, pressure and volume, providing values for the critical constants are available, or can be estimated.

8.6. Density

8.6.1. Liquids

Values for the density of pure liquids can usually be found in the handbooks. It should be noted that the density of most organic liquids, other than those containing a halogen or other "heavy atom", usually lies between 800 and 1000 kg/m^3. Liquid densities are given in Appendix D.

An approximate estimate of the density at the normal boiling point can be obtained from the molar volume (see Table 8.6)

$$\rho_b = \frac{M}{V_m} \tag{8.1}$$

where ρ_b = density, kg/m^3,
         M = molecular weight,
         V_m = molar volume, m^3/kmol.

For mixtures, it is usually sufficient to take the density of the components as additive; even for non-ideal solutions, as is illustrated by Example 8.1.

The densities of many aqueous solutions are given by Perry and Chilton (1973).

Example 8.1

Calculate the density of a mixture of methanol and water at 20°C, composition 40 per cent w/w methanol.

$$\begin{array}{ll}
\text{Density of water at } 20°\text{C} & 998\cdot2 \text{ kg/m}^3 \\
\text{Density of methanol at } 20°\text{C} & 791\cdot2 \text{ kg/m}^3
\end{array}$$

Solution

Basis: 1000 kg

$$\text{Volume of water} \quad = \frac{0\cdot6 \times 1000}{998\cdot2} \quad = 0\cdot601 \text{ m}^3$$

$$\text{Volume of methanol} \quad = \frac{0\cdot4 \times 1000}{791\cdot2} \quad = \underline{0\cdot506 \text{ m}^3}$$

$$\text{Total} \quad 1\cdot107 \text{ m}^3$$

$$\text{Density of mixture} \quad = \frac{1000}{1\cdot107} = \underline{\underline{903\cdot3 \text{ kg/m}^3}}$$

$$\text{Experimental value} \quad = 934\cdot5 \text{ kg/m}^3$$

$$\text{Error} = \frac{934\cdot5 - 903\cdot3}{903\cdot3} = 3 \text{ per cent, which would be acceptable for most engineering purposes}$$

If data on the variation of density with temperature cannot be found, they can be approximated for non-polar liquids from Smith's equation for thermal expansion (Smith et al., 1954).

$$\beta = \frac{0 \cdot 04314}{(T_c - T)^{0 \cdot 641}}$$ (8.2)

where β = coefficient of thermal expansion, K^{-1},
 T_c = critical temperature, K,
 T = temperature, K.

8.6.2. Gas and vapour density (specific volume)

For general engineering purposes it is often sufficient to consider that real gases, and vapours, behave ideally, and to use the gas law:

$$PV = nRT$$ (8.3)

where P = absolute pressure N/m^2 (Pa),
 V = volume m^3,
 n = mols of gas
 T = absolute temperature, K,
 R = universal gas constant, $8 \cdot 314 \, J \, K^{-1} \, mol^{-1}$ (or $kJ \, K^{-1} \, kmol^{-1}$).

$$\text{Specific volume} = \frac{RT}{P}$$ (8.4)

These equations will be sufficiently accurate up to moderate pressures, in circumstances where the value is not critical. If greater accuracy is needed, the simplest method is to modify equation 8.3 by including the compressibility factor z:

$$PV = znRT$$ (8.5)

The compressibility factor can be estimated from a generalised compressibility plot, which gives z as a function of reduced pressure and temperature (Chapter 3, Fig. 3.8); see also Volume 1, Chapter 2.

For mixtures, the pseudocritical properties of the mixture should be used to obtain the compressibility factor.

$$P_{c,m} = P_{c,a} y_a + P_{c,b} y_b + \ldots$$ (8.6)

$$T_{c,m} = T_{c,a} y_a + T_{c,b} y_b + \ldots$$ (8.7)

where P_c = critical pressure,
 T_c = critical temperature,
 y = mol fraction,

suffixes
 m = mixture
 a, b, etc. = components

8.7. Viscosity

Viscosity values will be needed for any design calculations involving the transport of fluids or heat. Values for pure substances can usually be found in the literature. Liquid viscosities are given in Appendix D. Methods for the estimation of viscosity are given below.

8.7.1. Liquids

A rough estimate of the viscosity of a pure liquid at its boiling point can be obtained from the modified Arrhenius equation:

$$\mu_b = 0.01 \, \rho_b^{0.5} \qquad (8.8)$$

where μ_b = viscosity, mN s/m^2,

ρ_b = density at boiling point, kg/m^3.

A more accurate value can be obtained if reliable values of density are available, or can be estimated with sufficient accuracy, from Souders' equation, Souders (1938):

$$\log(\log 10\mu) = \frac{I}{M}\rho \times 10^{-3} - 2.9 \qquad (8.9)$$

where μ = viscosity, mN s/m^2,

M = molecular weight,

I = Souders' index, estimated from the group contributions given in Table 8.1,

ρ = density at the required temperature, kg/m^3.

TABLE 8.1. *Contributions for calculating the Viscosity constant I in Souders' equation*

Atom	H	O	C	N	Cl	Br	I
Contribution	+2.7	+29.7	+50.2	+37.0	+60	+79	+110

Contributions of groups and bonds			
Double bond	−15.5	H—C—R $\quad \overset{\parallel}{O}$	+10
Five-member ring	−24		
Six-member ring	−21		
		—CH=CH—CH$_2$—X†	+4
Side groups on a six-member ring:		$\underset{R}{\overset{R}{\diagdown}}$CH—X†	+6
Molecular weight < 17	−9		
Molecular weight > 16	−17		
Ortho or para position	+3		
Meta position	−1		
		OH	+57.1
		COO	+90
$\underset{R}{\overset{R}{\diagdown}}$CH—CH$\underset{R}{\overset{R}{\diagup}}$	+8	COOH	+104.4
		NO$_2$	+80
R—$\overset{\overset{\displaystyle R}{\vert}}{\underset{\underset{\displaystyle R}{\vert}}{C}}$—R	+10		
—CH$_2$—	+55.6		

† X is a negative group.

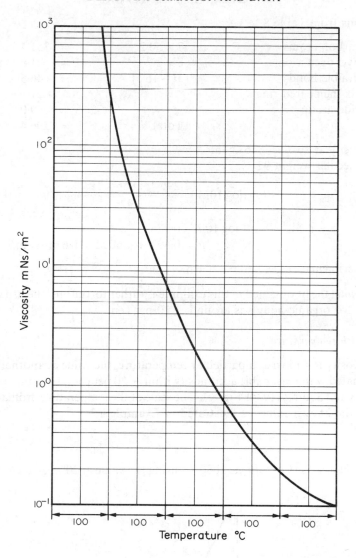

Fig. 8.1. Generalised viscosity vs. temperature curve for liquids

Example 8.2

Estimate the viscosity of toluene at 20°C.

Solution

Toluene   ⌬—CH$_3$

Contributions from Table 8.1:

7 carbon atoms	$7 \times 50\cdot2$	$= \quad 351\cdot4$
8 hydrogen atoms	$8 \times 2\cdot7$	$= \quad 21\cdot6$
3 double bonds	$3(-15\cdot5)$	$= -46\cdot5$
1 six-membered ring		$-21\cdot1$
1 side group		$- \ 9\cdot0$
	Total, I	$= \quad 296\cdot4$

Density at $20°C = 866 \, kg/m^3$
Molecular weight 92

$$\log(\log 10\mu) = \frac{296\cdot4 \times 866 \times 10^{-3}}{92} - 2\cdot9 = -0\cdot11$$

$$\log 10\mu = 0\cdot776$$
$$\mu = 0\cdot597, \text{ rounded} = 0\cdot6 \, mN \, s/m^2$$

experimental value, $0\cdot6 \, cp = 0\cdot6 \, mN \, s/m^2$

Author's note: the fit obtained in this example is rather fortuitous, the usual accuracy of the method for organic liquids is around ± 10 per cent.

Variation with temperature

If the viscosity is known at a particular temperature, the value at another temperature can be estimated with reasonable accuracy (within ± 20 per cent) by using the generalised plot of Lewis and Squires (1934), Fig. 8.1. The scale of the temperature ordinate is obtained by plotting the known value, as illustrated in Example 8.3.

Example 8.3

Estimate the viscosity of toluene at $80°C$, using the value at $20°C$ given in Example 8.2.

Solution

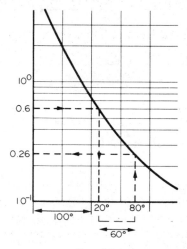

FIG. 8.1a

Temperature increment $80 - 20 = 60°C$.
From Fig. 8.1a, viscosity at $80°C = 0.26 \, mN \, s/m^2$.

Effect of pressure

The viscosity of a liquid is dependent on pressure as well as temperature, but the effect is not significant except at very high pressures. A rise in pressure of 300 bar is roughly equivalent to a decrease in temperature of $1°C$.

Mixtures

It is difficult to predict the viscosity of mixtures of liquids. Viscosities are rarely additive, and the shape of the viscosity–concentration curve can be complex. The viscosity of the mixture may be lower or, occasionally, higher than that of the pure components. A rough check on the magnitude of the likely error in a design calculation, arising from uncertainty in the viscosity of a mixture, can be made by using the smallest and largest values of the pure components in the calculation, and noting the result.

As an approximation, the variation can be assumed to be linear, if the range of viscosity is not very wide, and a weighted average viscosity calculated. For organic liquid mixtures a modified form of Souders' equation can be used; using a mol fraction weighted average value for the viscosity constant for the mixture I_m, and the average molecular weight.

For a binary mixture equation 8.9 becomes:

$$\log(\log 10 \, \mu_m) = \rho_m \left[\frac{x_1 I_1 + x_2 I_2}{x_1 M_1 + x_2 M_2} \right] \times 10^{-3} - 2.9 \tag{8.10}$$

where μ_m = viscosity of mixture,
$\quad \rho_m$ = density of mixture,
$\quad x_1, x_2$ = mol fraction of components,
$\quad M_1, M_2$ = molecular weights of components.

Bretsznajder (1971) gives a detailed review of the methods that have been developed for estimating the viscosity of mixtures, including methods for aqueous solutions and dispersions.

For heat-transfer calculations, Kern (1950) gives a rough rule of thumb for organic liquid mixtures:

$$\frac{1}{\mu_m} = \frac{w_1}{\mu_1} + \frac{w_2}{\mu_2} \tag{8.11}$$

where w_1, w_2 = mass fractions of the components 1 and 2,
$\quad \mu_1, \mu_2$ = viscosities of components 1 and 2.

8.7.2. Gases

Reliable methods for the prediction of gas viscosities, and the effect of temperature and pressure, are given by Bretsznajder (1971) and Reid et al. (1977).

Where an estimate of the viscosity is needed to calculate Prandtl numbers (see Volume 1, Chapter 7) the methods developed for the direct estimation of Prandtl numbers should be used.

For gases at low pressure Bromley (1952) has suggested the following values:

	Prandtl number
Monotomic gases (e.g. Ar, He)	0.67 ± 5 per cent
Non-polar, linear molecules (e.g. O_2, Cl_2)	0.73 ± 15 per cent
Non-polar, non-linear molecules (e.g. CH_4, C_6H_6)	0.79 ± 15 per cent
Strongly polar molecules (e.g. CH_3OH, SO_2, HCl)	0.86 ± 8 per cent

The Prandtl number for gases varies only slightly with temperature.

8.8. Thermal conductivity

The experimental methods used for the determination of thermal conductivity are described by Tsederberg (1965), who also lists values for many substances. Ho *et al.* (1972) give values for the thermal conductivity of the elements.

8.8.1. Solids

The thermal conductivity of a solid is determined by its form and structure, as well as composition. Values for the commonly used engineering materials are given in various handbooks.

8.8.2. Liquids

The data available in the literature up to 1973 have been reviewed by Jamieson *et al.* (1965). The Weber equation (Weber, 1880) can be used to make a rough estimate of the thermal conductivity of organic liquids, for use in heat-transfer calculations.

$$k = 3.56 \times 10^{-5} C_p \left(\frac{\rho^4}{M} \right)^{1/3} \tag{8.12}$$

where k = thermal conductivity. $W/m°C$,
   M = molecular weight,
   C_p = specific heat capacity, $kJ/kg°C$,
   ρ = density, kg/m^3.
Bretsznajder (1971) gives a group contribution method for estimating the thermal conductivity of liquids.

Example 8.4

Estimate the thermal conductivity of benzene at 30°C.

Solution

Density at 30°C = 875 kg/m^3
Molecular weight = 78
Specific heat capacity = 1.75 $kJ/kg°C$

(8.12) $k = 3.56 \times 10^{-5} \times 1.75 \left(\frac{875^4}{78} \right)^{1/3} = \underline{\underline{0.12 \ W/m°C}}$

Experimental value, 0.16 $W/m°C$

8.8.3. Gases

Approximate values for the thermal conductivity of pure gases, up to moderate pressures, can be estimated from values of the gas viscosity, using Eucken's equation, Eucken (1911):

$$k = \mu\left(C_p + \frac{10\cdot4}{M}\right) \tag{8.13}$$

where μ = viscosity, mN s/m^2,

$\quad C_p$ = specific heat capacity, kJ/kg°C,

$\quad M$ = molecular weight.

Example 8.5

Estimate the thermal conductivity of ethane at 1 bar and 450°C.

Solution

$\qquad$ Viscosity = 0·0134 mN s/m^2

$\qquad$ Specific heat capacity = 2·47 kJ/kg°C

(8.13) $\qquad k = 0\cdot0134\left(2\cdot47 + \dfrac{10\cdot4}{30}\right) = \underline{\underline{0\cdot038 \text{ W/m°C}}}$

Experimental value, 0·043 W/m°C, error 12 per cent.

8.8.4. Mixtures

In general, the thermal conductivities of liquid mixtures, and gas mixtures, are not simple functions of composition and the thermal conductivity of the components. Bretsznajder (1971) discusses the methods that are available for estimating the thermal conductivities of mixtures from a knowledge of the thermal conductivity of the components.

If the components are all non-polar a simple weighted average is usually sufficiently accurate for design purposes.

$$k_m = k_1 w_1 + k_2 w_2 + \ldots \tag{8.14}$$

where k_m = thermal conductivity of mixture,

$\quad k_1, k_2$ = thermal conductivity of components,

$\quad w_1, w_2$ = components mass fractions.

8.9. Specific heat capacity

The specific heats of the most common organic and inorganic materials can usually be found in the handbooks.

8.9.1. Solids and liquids

Approximate values can be calculated for solids, and liquids, by using a modified form of Kopp's law, which is given by Werner (1941). The heat capacity of a compound is taken as the sum of the heat capacities of the individual elements of which it is composed. The values attributed to each element, for liquids and solids, at room temperature, are given in Table 8.2; the method illustrated in Example 8.6.

TABLE 8.2. *Heat capacities of the elements, J/mol °C*

Element	Solids	Liquids
C	7·5	11·7
H	9·6	18·0
B	11·3	19·7
Si	15·9	24·3
O	16·7	25·1
F	20·9	29·3
P and S	22·6	31·0
all others	26·0	33·5

Example 8.6

Estimate the specific heat capacity of urea, CH_4N_2O.

Solution

Element	mol. wt.	Heat capacity		
C	12	7·5	=	7·5
H	4	4 × 9·6	=	38·4
N	28	2 × 26·0	=	52·0
O	16	16·7	=	16·7
	60			114·6 J/mol°C

$$\text{Specific heat capacity} = \frac{114\cdot6}{60} = \underline{\underline{1\cdot91 \text{ J/g°C (kJ/kg°C)}}}$$

Experimental value 1·34 kJ/kg°C.

Kopp's rule does not take into account the arrangement of the atoms in the molecule, and, at best, gives only very approximate, "ball-park" values.

For organic liquids, the group contribution method proposed by Chueh and Swanson (1973a,b) will give accurate predictions. The contributions to be assigned to each molecular group are given in Table 8.3 and the method illustrated in Examples 8.7 and 8.8.

Liquid specific heats do not vary much with temperature, at temperatures well below the critical temperature (reduced temperature < 0·7).

The specific heats of liquid mixtures can be estimated, with sufficient accuracy for most technical calculations, by taking heat capacities of the components as additive.

For dilute aqueous solutions it is usually sufficient to take the specific heat of the solution as that of water.

Example 8.7

Using Chueh and Swanson's method, estimate the specific heat capacity of ethyl bromide at 20°C.

TABLE 8.3. *Group contributions for liquid heat capacities at 20°C, kJ/kmol°C (Chueh and Swanson, 1973a,b)*

Group	Value	Group	Value
Alkane		O‖	
—CH₃	36·84	—C—O—	60·71
—CH₂—	30·40	—CH₂OH	73·27
—CH—	20·93	—CHOH	76·20
—C—	7·37	—COH	111·37
Olefin		—OH	44·80
=CH₂	21·77	—ONO₂	119·32
=C—H	21·35	**Halogen**	
=C—	15·91	—Cl (first or second on a carbon)	36·01
Alkyne		—Cl (third or fourth on a carbon)	25·12
—C≡H	24·70	—Br	37·68
—C≡	24·70	—F	16·75
In a ring		—I	36·01
—CH=	18·42	**Nitrogen**	
—C= or —C—	12·14	H—N— (with H above)	58·62
—CH=	22·19	—N— (with H above)	43·96
—CH₂—	25·96	—N—	31·40
Oxygen		—N= (in a ring)	18·84
—O—	35·17	—C≡N	58·70
>C=O	53·00	**Sulphur**	
—C—O	53·00	—SH	44·80
—C—O with H	79·97	—S—	33·49
O‖ —C—OH		**Hydrogen**	
		H— (for formic acid, formates, hydrogen cyanide, etc.)	14·65

Add 18·84 for any carbon group which fulfils the following criterion: a carbon group which is joined by a single bond to a carbon group connected by a double or triple bond with a third carbon group. In some cases a carbon group fulfils the above criterion in more ways than one; 18·84 should be added each time the group fulfil the criterion.

Exceptions to the above 18·84 rule:

1. No such extra 18·84 additions for —CH₃ groups.
2. For a —CH₂— group fulfilling the 18·84 addition criterion add 10·47 instead of 18·84. However, when the —CH₂— group fulfils the addition criterion in more ways than one, the addition should be 10·47 the first time and 18·84 for each subsequent addition.
3. No such extra addition for any carbon group in a ring.

Solution

Ethyl bromide CH_3CH_2Br

Group	Contribution	No. of		
—CH_3	36·84	1	=	36·84
—CH_2—	30·40	1	=	30·40
—Br	37·68	1	=	37·68
		Total		104·92 kJ/kmol°C

mol. wt. $= 109$

$$\text{Specific heat capacity} = \frac{104\cdot92}{109} = 0\cdot96 \text{ kJ/kg°C}$$

Experimental value, 0·90 kJ/kg°C

Example 8.8

Estimate the specific heat capacity of chlorobutadiene at 20°C, using Chueh and Swanson's method.

Solution

Structural formula $CH_2 = C - CH = CH_2$, mol. wt. 88·5
$\qquad\qquad\qquad\qquad\quad |$
$\qquad\qquad\qquad\qquad\ Cl$

Group	Contribution	No. of	Addition rule		Total
=CH_2	21·77	2	—	=	43·54
=C—	15·91	1	18·84	=	34·75
=CH	21·35	1	18·84	=	40·19
—Cl	36·01	1	—	=	36·01
					154·49 kJ/kmol°C

$$\text{Specific heat capacity} = \frac{154\cdot49}{88\cdot5} = 1\cdot75 \text{ kJ/kg°C}$$

8.9.2. Gases

The dependence of gas specific heats on temperature was discussed in Chapter 3, Section 3.5. For a gas in the ideal state the specific heat capacity at constant pressure is given by:

$$C_p^\circ = a + bT + cT^2 + dT^3 \qquad\qquad \text{(equation 3.19)}$$

Values for the constants in this equation for the more common gases can be found in the handbooks, and in Appendix D.

Several group contribution methods have been developed for the estimation of the constants, such as that by Rihani and Doraiswamy (1965) for organic compounds. Their values for each molecular group are given in Table 8.4, and the method illustrated in Example 8.9. The values should not be used for acetylenic compounds.

The correction of the ideal gas heat capacity to account for real conditions of temperature and pressure was discussed in Chapter 3, Section 3.7.

Example 8.9

Estimate the specific heat capacity of isopropyl alcohol at 500 K.

Solution

Structural formula

$$CH_3-CH-OH$$

with CH_3 attached above the CH.

Group	No. of	a	$b \times 10^2$	$c \times 10^4$	$d \times 10^6$
—CH$_3$	2	5·0970	17·9480	−0·7134	0·0095
—CH	1	−14·7516	14·3020	−1·1791	0·03356
—OH	1	27·2691	−0·5640	0·1733	−0·0068
Total		17·6145	31·6860	−1·7190	0·0363

$$C_p^\circ = 17\cdot6145 + 31\cdot6860 \times 10^{-2}\,T - 1\cdot7192 \times 10^{-4}\,T^2 + 0\cdot0363 \times 10^{-6}\,T^3.$$

At 500 K, substitution gives:

$$C_p = \underline{\underline{137\cdot6\ \text{kJ/kmol}^\circ\text{C}}}$$

Experimental value, 31·78 cal/mol°C = 132·8 kJ/kmol°C, error 4 per cent.

8.10. Enthalpy of vaporisation (latent heat)

The latent heats of vaporisation of the more commonly used materials can be found in the handbooks and in Appendix D.

A very rough estimate can be obtained from Trouton's rule (Trouton, 1884), one of the oldest prediction methods.

$$\frac{L_v}{T_b} = \text{constant} \tag{8.15}$$

where L_v = latent heat of vaporisation, kJ/kmol,

T_b = normal boiling point, K.

For organic liquids the constant can be taken as 100.

More accurate estimates, suitable for most engineering purposes, can be made from a knowledge of the vapour pressure–temperature relationship for the substance. Several correlations have been proposed; see Reid *et al.* (1977).

The equation presented here, due to Haggenmacher (1946), is derived from the Antoine vapour pressure equation (see Section 8.11).

$$L_v = \frac{8\cdot32\,BT^2\Delta z}{(T+C)^2} \tag{8.16}$$

TABLE 8.4. *Group contributions to ideal gas heat capacities, kJ/kmol °C (Rihani and Doraiswamy, 1965)*

Group	a	$b \times 10^2$	$c \times 10^4$	$d \times 10^6$
Aliphatic hydrocarbon groups				
—CH_3	2·5485	8·9740	−0·3567	0·004752
—CH_2—	1·6518	8·9447	−0·5012	0·0187
=CH_2	2·2048	7·6857	−0·3994	0·008264
—C—H	−14·7516	14·3020	−1·1791	0·03356
—C—	−24·4131	18·6493	−1·7619	0·05288
$\overset{H}{>}C=CH_2$	1·1610	14·4786	−0·8031	0·01792
$>C=CH_2$	−1·7472	16·2694	−1·1652	0·03083
$\overset{H}{>}C=C\overset{H}{<}$	−13·0676	15·9356	−0·9877	0·02305
$\overset{H}{>}C=C\underset{H}{<}$	3·9261	12·5208	−0·7323	0·01641
$>C=C\overset{H}{<}$	−6·161	14·1696	−0·9927	0·02594
$>C=C<$	1·9829	14·7304	−1·3188	0·03854
$\overset{H}{>}C=C=CH_2$	9·3784	17·9597	−1·07433	0·02474
$>C=C=CH_2$	11·0146	17·4414	−1·1912	0·03047
$\overset{H}{>}C=C=C\overset{H}{<}$	−13·0833	20·8878	−1·8018	0·05447
Aromatic hydrocarbon groups				
HC	−6·1010	8·0165	−0·5162	0·01250
—C	−5·8125	6·3468	−0·4476	0·01113
↔C	0·5104	5·0953	−0·3580	0·00888
Contributions due to ring formation				
Three-membered ring	−14·7878	−0·1256	0·3129	−0·02309
Four-membered ring	−36·2368	4·5134	0·1779	−0·00105
Five-membered ring:				
Pentane	−51·4348	7·7913	−0·4342	0·00898
Pentene	−28·8106	3·2732	−0·1445	0·00247

TABLE 8.4. (*cont.*)

Group	a	$b \times 10^2$	$c \times 10^4$	$d \times 10^6$
Six-membered ring:				
Hexane	−56·0709	8·9564	−0·1796	−0·00781
Hexene	−33·5941	9·3110	−0·80118	0·02291
Oxygen-containing groups				
—OH	27·2691	−0·5640	0·1733	−0·00680
—O—	11·9161	−0·04187	0·1901	−0·01142
—C=O (with H)	14·7308	3·9511	0·2571	−0·02922
>C=O	4·1935	8·6931	−0·6850	0·01882
—C—O—H (C=O)	5·8846	14·4997	−1·0706	0·02883
—C(=O)O—	11·4509	4·5012	0·2793	−0·03864
O<	−15·6352	5·7472	−0·5296	0·01586
Nitrogen-containing groups				
—C≡N	18·8841	2·2864	0·1126	−0·01587
—N≡C	21·2941	1·4620	0·1084	−0·01020
—NH$_2$	17·4937	3·0890	0·2843	−0·03061
>NH	−5·2461	9·1825	−0·6716	0·01774
>N—	−14·5186	12·3230	−1·1191	0·03277
N<	10·2401	1·4386	0·07159	−0·01138
—NO$_2$	4·5638	11·0536	−0·7834	0·01989
Sulphur-containing groups				
—SH	10·7170	5·5881	−0·4978	0·01599
—S—	17·6917	0·4719	−0·0109	−0·00030
S<	17·0922	−0·1260	0·3061	−0·02546
—SO$_3$H	28·9802	10·3561	0·7436	−0·09397
Halogen-containing groups				
—F	6·0215	1·4453	−0·0444	−0·00014
—Cl	12·8373	0·8885	−0·0536	0·00116
—Br	11·5577	1·9808	−0·1905	0·0060
—I	13·6703	2·0520	−0·2257	0·00746

where L_v = latent heat at the required temperature, kJ/kmol,
         T = temperature, K,
      B, C = coefficients in the Antoine equation (equation 8.20),
        $\Delta z = z_{gas} - z_{liquid}$ (where z is the compressibility constant), calculated
              from the equation:

$$\Delta z = \left[1 - \frac{P_r}{T_r^3} \right]^{0.5} \tag{8.17}$$

         P_r = reduced pressure,
         T_r = reduced temperature.

If an experimental value of the latent heat at the boiling point is known, the Watson
equation (Watson, 1943), can be used to estimate the latent heat at other temperatures.

$$L_v = L_{v,b} \left[\frac{T_c - T}{T_c - T_b} \right]^{0.38} \tag{8.18}$$

where L_v = latent heat at temperature T, kJ/kmol,
      $L_{v,b}$ = latent heat at the normal boiling point, kJ/kmol,
      T_b = boiling point, K,
      T_c = critical temperature, K,
      T = temperature, K.

Over a limited range of temperature, up to 100°C, the variation of latent heat with
temperature can usually be taken as linear.

8.10.1. Mixtures

For design purposes it is usually sufficiently accurate to take the latent heats of the
components of a mixture as additive:

$$L_v \text{ mixture} = L_{v1} x_1 + L_{v2} x_2 + \ldots . \tag{8.19}$$

where L_{v1}, L_{v2} = latent heats of the components kJ/kmol,
      x_1, x_2 = mol fractions of components.

Example 8.10

Estimate the latent heat of vaporisation of acetic anhydride, $C_4H_6O_3$, at its boiling
point, 139·6°C (412·7 K), and at 200°C (473 K).

Solution

For acetic anhydride $T_c = 569·1$ K, $P_c = 46$ bar,

                    Antoine constants $A = 16·3982$
                                        $B = 3287·56$
                                        $C = -75·11$

Experimental value at the boiling point 41,242 kJ/kmol.
From Trouton's rule:

$$L_{v,b} = 100 \times 412·7 = \underline{\underline{41,270 \text{ kJ/kmol}}}$$

Note: the close approximation to the experimental value is fortuitous, the rule normally gives only a very approximate estimate.

From Haggenmacher's equation:

at the b.p. $\quad P_r = 1/46 = 0.02124$

$$T_r = 412.7/569.1 = 0.7252$$

$$\Delta z = \left[1 - \frac{0.02124}{0.7252^3} \right]^{0.5} = 0.972$$

$$L_{v,b} = \frac{8.32 \times 3287.6 \times (412.7)^2 \times 0.972}{(412.7 - 75.11)^2} = \underline{39,733 \text{ kJ/mol}}$$

At 200°C, the vapour pressure must first be estimated, from the Antoine equation:

$$\ln P = A - \frac{B}{T + C}$$

$$\ln P = 16.3982 - \frac{3287.56}{473 - 75.11} = 8.14$$

$$P = 3421.35 \text{ mm Hg} = 4.5 \text{ bar}$$

$$P_c = 4.5/46 = 0.098$$

$$T_z = \frac{473}{569.1} = 0.831$$

$$\Delta z = \left[1 - \frac{0.098}{0.831^3} \right]^{0.5} = 0.911$$

$$L_v = \frac{8.32 \times 3287.6 \times (473)^2 \times 0.911}{(473 - 75.11)^2} = \underline{35,211 \text{ kJ/kmol}}$$

Using Watson's equation and the experimental value at the b.p.

$$L_v - 41,242 \left[\frac{569.1 - 473}{569.1 - 412.7} \right]^{0.38} = \underline{34,260 \text{ kJ/kmol}}$$

8.11. Vapour pressure

If the normal boiling point (vapour pressure = 1 atm) and the critical temperature and pressure are known, then a straight line drawn through these two points on a plot of log-pressure versus reciprocal absolute temperature can be used to make a rough estimation of the vapour pressure at intermediate temperatures.

Several equations have been developed to express vapour pressure as a function of temperature. One of the most commonly used is the three-term Antoine equation, Antoine (1888);

$$\ln P = A - \frac{B}{T + C} \tag{8.20}$$

where P = vapour pressure, mm Hg,
   A, B, C = the Antoine coefficients,
      T = temperature, K.

Vapour pressure data, in the form of the constants in the Antoine equation, are given in several references; the compilations by Ohe (1976), Dreisbach (1952), Hala et al. (1968) and Hirata (1975) give values for several thousand compounds. Antoine vapour pressure coefficients for the elements are given by Nesmeyanov (1963). Care must be taken when using Antoine coefficients taken from the literature in equation 8.20, as the equation is often written in different and ambiguous forms; the logarithm of the pressure may be to the base 10, instead of the natural logarithm, and the temperature may be degrees Centigrade, not absolute temperature. Also, occasionally, the minus sign shown in equation 8.20 is included in the constant B and the equation written with a plus sign. The pressure may also be in units other than mm Hg. Always check the actual form of the equation used in the particular reference. Antoine constants for use in equation 8.20 are given in Appendix D.

8.12. Diffusion coefficients (Diffusivities)

Diffusion coefficients are needed in the design of mass transfer processes; such as gas absorption, distillation and liquid–liquid extraction.

Experimental values for the more common systems can be often found in the literature, but for most design work the values will have to be estimated. Methods for the prediction of gas and liquid diffusivities are given in Volume 1, Chapter 8; some experimental values are also given.

8.12.1. Gases

The equation developed by Fuller et al. (1966) is as easy to apply and gives more reliable predictions than the Gilliland equation given in Volume 1, Chapter 8:

$$D_v = \frac{1 \cdot 013 \times 10^{-7} \, T^{1 \cdot 75} (1/M_a + 1/M_b)^{1/2}}{P\left[\left(\sum_a v_i\right)^{1/3} + \left(\sum_b v_i\right)^{1/3}\right]^2} \tag{8.21}$$

where D_v = diffusivity, m^2/s,
      T = temperature, K,
   M_a, M_b = molecular weights of components a and b,
      P = total pressure, bar,

$\sum_a v_i, \sum_b v_i$ = the summation of the special diffusion volume coefficients for components a and b, given in Table 8.5.
The method is illustrated in Example 8.11.

Example 8.11

Estimate the diffusivity of methanol in air at atmospheric pressure and 25°C.

TABLE 8.5. *Special atomic diffusion volumes* (*Fuller* et al., 1966)

Atomic and structural diffusion volume increments

C	16·5	Cl	19·5*
H	1·98	S	17·0*
O	5·48	Aromatic or hetrocyclic rings	−20·0
N	5·69*		

Diffusion volumes of simple molecules

H_2	7·07	CO	18·9
D_2	6·70	CO_2	26·9
He	2·88	N_2O	35·9
N_2	17·9	NH_3	14·9
O_2	16·6	H_2	12·7
Air	20·1	CCL_2F_2	114·8*
Ne	5·59	SF_6	69·7*
Ar	16·1	Cl_2	37·7*
Kr	22·8	Br_2	67·2*
Xe	37·9*	SO_2	41·1*

* Value based on only a few data points.

Solution

Diffusion volumes from Table 8.5; methanol:

Element	v_i	No. of	
C	16·50 ×	1	= 16·50
H	1·98 ×	4	= 7·92
O	5·48 ×	1	= 5·48
		$\sum_a v_i$	29·90

Diffusion volume for air = 20·1.
1 standard atmosphere = 1·013 bar.
molecular weight $CH_3OH = 32$, air = 29.

(8.21)
$$D_v = \frac{1·013 \times 10^{-7} \times 298^{1·75}(1/32 + 1/29)^{1/2}}{1·013\left[(29·90)^{1/3} + (20·1)^{1/3}\right]^2}$$

$$= \underline{16·2 \times 10^{-6} \text{ m}^2/\text{s}}$$

Experimental value, $15·9 \times 10^{-6}$ m²/s.

8.12.2. Liquids

The equation given in Volume 1, Chapter 8, for estimating liquid diffusivity is based on a correlation in graphical form published by Wilke (1949). A better correlation was published later by Wilke and Chang (1955):

$$D_L = \frac{1·173 \times 10^{-13}(\phi M)^{0·5}T}{\mu V_m^{0·6}} \qquad (8.22)$$

where D_L = liquid diffusivity, m^2/s,

ϕ = an association factor for the solvent,

= 2·6 for water (some workers recommend 2·26),

= 1·9 for methanol,

= 1·5 for ethanol,

= 1·0 for unassociated solvents,

M = molecular weight of solvent,

μ = viscosity of solvent, $mN\,s/m^2$,

T = temperature, K,

V_m = molar volume of the solute at its boiling point, $m^3/kmol$. This can be estimated from the group contributions given in Table 8.6.

The method is illustrated in Example 8.12.

TABLE 8.6 *Structural contributions to molar volumes, $m^3/kmol$ (Gambil, 1958)*

		Molecular volumes				
Air	0·0299	CO_2	0·0340	H_2S	0·0329	NO	0·0236
Br_2	0·0532	COS	0·0515	I_2	0·0715	N_2O	0·0364
Cl_2	0·0484	H_2	0·0143	N_2	0·0312	O_2	0·0256
CO	0·0307	H_2O	0·0189	NH_3	0·0258	SO_2	0·0448

		Atomic volumes				
As	0·0305	F	0·0087	P	0·0270	Sn	0·0423
Bi	0·0480	Ge	0·0345	Pb	0·0480	Ti	0·0357
Br	0·0270	H	0·0037	S	0·0256	V	0·0320
C	0·0148	Hg	0·0190	Sb	0·0342	Zn	0·0204
Cr	0·0274	I	0·037	Si	0·0320		

Cl, terminal, as in RCl	0·0216	in higher esters, ethers	0·0110
medial, as in R—CHCl—R	0·0246	in acids	0·0120
Nitrogen, double-bonded	0·0156	in union with S, P, N	0·0083
triply bonded, as in nitriles	0·0162	three-membered ring	−0·0060
in primary amines, RNH_2	0·0105	four-membered ring	−0·0085
in secondary amines, R_2NH	0·012	five-membered ring	−0·0115
in tertiary amines, R_3N	0·0108	six-membered ring as in benzene, cyclohexane, pyridine	−0·0150
Oxygen, except as noted below	0·0074		
in methyl esters	0·0091	Naphthalene ring	−0·0300
in methyl ethers	0·0099	Anthracene ring	−0·0475

The Wilke–Chang correlation is shown graphically in Fig. 8.2. This figure can be used to determine the association constant for a solvent from experimental values for D_L in the solvent.

The Wilke–Chang equation gives satisfactory predictions for the diffusivity of organic compounds in water but not for water in organic solvents.

Example 8.12

Estimate the diffusivity of phenol in ethanol at 20°C (293 K).

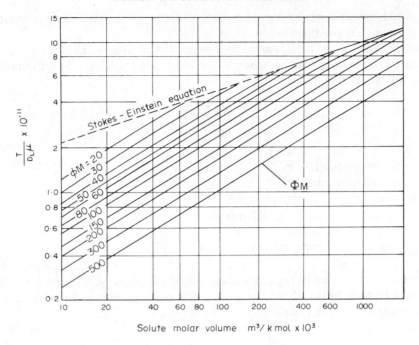

FIG. 8.2 The Wilke–Chang correlation

Solution

Viscosity of ethanol at 20°C, $1.2 \, \text{mN s/m}^2$.

Molecular weight, 46.

Molar volume of phenol ⬡—OH , from Table 8.6:

Atom	Vol.		No. of		
C	0.0148	×	6	=	0.0888
H	0.0037	×	6	=	0.0222
O	0.0074	×	1	=	0.0074
ring	−0.015	×	1	=	−0.015
					$0.1034 \, \text{m}^3/\text{k mol}$

$$(8.22) \qquad D_L = \frac{1.173 \times 10^{-13}(1.5 \times 46)^{0.5} \, 293}{1.2 \times 0.1034^{0.6}} = \underline{\underline{9.28 \times 10^{-10} \, \text{m}^2/\text{s}}}$$

Experimental value, $8 \times 10^{-10} \, \text{m}^2/\text{s}$

8.13. Surface tension

It is usually difficult to find experimental values for surface tension for any but the more commonly used liquids. A useful compilation of experimental values is that by Jasper

(1972), which covers over 2000 pure liquids. Othmer *et al.* (1968) give a nomograph covering about 100 compounds.

If reliable values of the liquid and vapour density are available, the surface tension can be estimated from the Sugden parachor; which can be estimated by a group contribution method, Sugden (1924).

$$\sigma = \left[\frac{P_{ch}(\rho_L - \rho_v)}{M} \right]^4 \times 10^{-12} \tag{8.23}$$

where σ = surface tension, mJ/m^2 (dyne/cm),
     P_{ch} = Sugden's parachor,
     ρ_L = liquid density, kg/m^3,
     ρ_v = density of the saturated vapour, kg/m^3,
     M = molecular weight.
σ, ρ_L, ρ_v evaluated at the system temperature.

The vapour density can be neglected when it is small compared with the liquid density.

The parachor can be calculated using the group contributions given in Table 8.7. The method is illustrated in Example 8.13.

TABLE 8.7. *Contributions to Sugden's parachor for organic compounds (Sugden, 1924)*

Atom, group or bond	Contribution	Atom, group or bond		Contribution
C	4·8	Si		25·0
H	17·1	Al		38·6
H in (OH)	11·3	Sn		57·9
O	20·0	As		50·1
O_2 in esters, acids	60·0	Double bond:	terminal	
N	12·5		2,3-position	23·2
S	48·2		3,4-position	
P	37·7	Triple bond		46·6
F	25·7	Rings		
Cl	54·3		3-membered	16·7
Br	68·0		4-membered	11·6
I	91·0		5-membered	8·5
Se	62·5		6-membered	6·1

8.13.1. Mixtures

The surface tension of a mixture is rarely a simple function of composition. However, for hydrocarbons a rough value can be calculated by assuming a linear relationship.

$$\sigma_m = \sigma_1 x_1 + \sigma_2 x_2 \ldots \tag{8.24}$$

where σ_m = surface tension of mixture,
     σ_1, σ_2 = surface tension of components,
     x_1, x_2 = component mol fractions.

Example 8.13

Estimate the surface tension of pure methanol at 20°C, density 791·7 kg/m³, molecular weight 32·04.

Solution

Calculation of parachor, CH_3OH, Table 8.7.

Group	Contribution	No.			
C	4·8	×	1	=	4·8
H–O	11·3	×	1	=	11·3
H–C	17·1	×	3	=	51·3
O	20·0	×	1	=	20·0
					87·4

$$(8.23) \qquad \sigma = \left[\frac{87 \cdot 4 \times 791 \cdot 7}{32 \cdot 04} \right]^4 \times 10^{-12} = \underline{\underline{21 \cdot 8 \ mJ/m^2}}$$

Experimental value 22·5 mJ/m².

8.14. Critical constants

Values of the critical temperature and pressure will be needed for prediction methods that correlate physical properties with the reduced conditions. Experimental values for many substances can be found in various handbooks; and in Appendix D. Critical reviews of the literature on critical constants, and summaries of selected values, have been published by Kudchadker *et al.* (1968), for organic compounds, and by Mathews (1972), for inorganic compounds. An earlier review was published by Kobe and Lynn (1953).

If reliable experimental values cannot be found, techniques are available for estimating the critical constants with sufficient accuracy for most design purposes. For organic compounds Lydersen's method is normally used, Lydersen (1955):

$$T_c = T_b / [0 \cdot 567 + \sum \Delta T - (\sum \Delta T)^2] \qquad (8.25)$$

$$P_c = M / (0 \cdot 34 + \sum \Delta P)^2 \qquad (8.26)$$

$$V_c = 0 \cdot 04 + \sum \Delta V \qquad (8.27)$$

where T_c = critical temperature, K,

P_c = critical pressure, atm (1·0133 bar),

V_c = molar volume at the critical conditions, m³/kmol,

T_b = normal boiling point, K,

M = molecular weight,

ΔT = critical temperature increments, Table 8.8,

ΔP = critical pressure increments, Table 8.8,

ΔV = molar volume increments, Table 8.8.

TABLE 8.8. *Critical constant increments (Lydersen, 1955)*

Non-ring increments	ΔT	ΔP	ΔV	Non-ring increments	ΔT	ΔP	ΔV
—CH₃	0·020	0·227	0·055	=C—	0·0	0·198	0·036
—CH₂	0·020	0·227	0·055	=C=	0·0	0·198	0·036
—CH	0·012	0·210	0·051	≡CH	0·005	0·153	0·036*
				≡C—	0·005	0·153	0·036*
				H	0	0	0
—C—	0·00	0·210	0·041				
=CH₂	0·018	0·198	0·045				
=CH	0·018	0·198	0·045				

Ring increments							
—CH₂—	0·013	0·184	0·0445	=CH	0·011	0·154	0·037
—CH	0·012	0·192	0·046	=C—	0·011	0·154	0·036
				=C=	0·011	0·154	0·036
—C—	−0·007*	0·154*	0·031*				

Halogen increments							
—F	0·018	0·224	0·018	—Br	0·010	0·50*	0·070*
—Cl	0·017	0·320	0·049	—I	0·012	0·83*	0·095*

Oxygen increments							
—OH (alcohols)	0·082	0·06	0·018*	—CO (ring)	0·033*	0·2*	0·050*
—OH (phenols)	0·031	−0·02*	0·030*				
—O— (non-ring)	0·021	0·16	0·020	HC=O (aldehyde)	0·048	0·33	0·073
—O— (ring)	0·014*	0·12*	0·080*	—COOH (acid)	0·085	0·4*	0·080
				—COO— (ester)	0·047	0·47	0·080
—C=O (non-ring)	0·040	0·29	0·060	=O (except for combinations above)	0·02*	0·12*	0·011*

Nitrogen increments							
—NH₂	0·031	0·095	0·028	—N— (ring)	0·007*	0·013*	0·032*
—NH (non-ring)	0·031	0·135	0·037*	—CN	0·060*	0·36*	0·080*
				—NO₂	0·055*	0·42*	0·078*
—NH (ring)	0·024*	0·09*	0·027*				
—N— (non-ring)	0·014	0·17	0·042*				

TABLE 8.8 (cont.)

	ΔT	ΔP	ΔV		ΔT	ΔP	ΔV
Sulphur Increments							
—SH	0·015	0·27	0·055	—S— (ring)	0·008*	0·24*	0·045*
—S— (non-ring)	0·015	0·27	0·055	S	0·003*	0·24*	0·047*
Miscellaneous							
—Si—	0·03	0·54*		—B—	0·03*		

Dashes represent bonds with atoms other than hydrogen.
Values marked with an asterisk are based on too few experimental points to be reliable.

Example 8.14

Estimate the critical constants for diphenylmethane using Lydersen's method; normal boiling point 537·5 K, molecular weight 168·2, structural formula:

Solution

Group	No. of	Total contribution		
		ΔT	ΔP	ΔV
H—C—(ring)	10	0·11	1·54	0·37
=C—(ring)	2	0·022	0·308	0·072
—CH$_2$—	1	0·02	0·227	0·055
$\sum$		0·152	2·075	0·497

$$T_c = 537·5/(0·567 + 0·152 - 0·152^2) = \underline{\underline{772\,k}}$$
experimental value 767 K,

$$P_c = 168·2/(0·34 + 2·075)^2 = \underline{\underline{28·8\,atm}}$$
experimental value 28·2 atm,

$$V_c = 0·04 + 0·497 = \underline{\underline{0·537\,m^3/kmol}}$$

8.15. Enthalpy of reaction and enthalpy of formation

Enthalpies of reaction (heats of reaction) for the reactions used in the production of commercial chemicals can usually be found in the literature. Stephenson (1966) gives values for most of the production processes he describes in his book.

Heats of reaction can be calculated from the heats of formation of the reactants and products, as described in Chapter 3, Section 3.11. Values of the standard heats of formation for the more common chemicals are given in various handbooks; see also appendix D. A useful source of data on heats of formation, and combustion, is the critical review of the literature by Domalski (1972).

Benson has developed a detailed group contribution method for the estimation of heats of formation; see Benson (1976) and Benson et al. (1968). He estimates the accuracy of the method to be from ± 2.0 kJ/mol for simple compounds, to about ± 12 kJ/mol for highly substituted compounds. Benson's method, and other group contribution methods for the estimation of heats of formation, are described by Reid et al. (1977).

8.16. Phase equilibrium data

Phase equilibrium data are needed for the design of all separation processes that depend on differences in concentration between phases.

8.16.1. Vapour–liquid equilibrium data (vle)

Distillation is probably the most important separation and purification process used in the chemical industry, and reliable vle data are required for the estimation of the number of theoretical stages required in distillation column design.

Experimental data have been published for several thousand binary and many multicomponent systems. The books by Chu et al. (1956) and Hala et al. (1968, 1973) cover most of the published experimental data. In a book by Hirata et al. (1975) the published data for several hundred binary systems have been correlated and plotted with the aid of a computer; low- and high-pressure systems are covered.

The number of possible combinations of components of interest in chemical process design will always far exceed the published data, and the designer is unlikely to find all the vle data he needs. Where data cannot be found they will have to be measured or estimated. Some of the techniques that have been developed for the extrapolation, interpolation and prediction of vle data are discussed briefly in the following sections. Whenever possible estimates should be supported by experimentally determined values.

Many equations have been developed to correlate the vapour–liquid equilibria data; those considered the most useful in design work are discussed below. Only a limited discussion of the subject can be given and one of the numerous specialised texts on phase equilibria should be consulted for a full discussion of the thermodynamic basis of the equations, and their derivation. The books by Null (1970), Prausnitz (1969) and Prausnitz et al. (1967) are recommended.

For ideal mixtures the relationship between the liquid and vapour compositions is given by Raoult's law (see Volume 2, Chapter 11):

$$y_i = \frac{P_i^\circ x_i}{P}$$

(8.28)

where y_i = mol fraction of component i in the vapour,
x_i = mol fraction of component i in the liquid,
P_i° = vapour pressure of pure component i,
P = total pressure.

It is often sufficient to use Raoult's law for real mixtures when the deviations from ideality are likely to be small; for instance, for mixtures comprised of members of homologous series, such as the *benzene–toluene* system used as an example in Volume 2. The departures from Raoult's law with real mixtures are due to deviations in the liquid phase from the ideal mixture laws; and deviations in the gas phase from the ideal mixture laws and from the ideal gas laws (equation 8.3). Deviations from the ideal mixture laws can be accounted for by the use of activity coefficients γ and deviations from the gas laws in the vapour phase by the fugacity coefficient ϕ.

In chemical engineering applications the relationship between the liquid and gas phase compositions is often expressed in the form of equilibrium K-values, particularly for hydrocarbon systems.

$$K_i = y_i/x_i \tag{8.29}$$

where K_i = K-value for component i.

The relative volatility of two components α_{ij} (see Volume 2, Chapter 11) can be expressed as the ratio of their K-values:

$$\alpha_{ij} = K_i/K_j \tag{8.30}$$

For ideal mixtures (obeying Raoult's law):

$$K_i^\circ = P_i^\circ/P \tag{8.31}$$

and

$$\alpha_{ij} = K_i^\circ/K_j^\circ = P_i^\circ/P_j^\circ \tag{8.32}$$

where K_i° and K_j° are the *ideal* K-values for the components i and j.

For real mixtures:

$$K_i = K_i^\circ \frac{\gamma_i}{\phi_i} \tag{8.33}$$

where γ_i = the liquid phase activity coefficient for component i,
ϕ_i = the gas phase fugacity coefficient for component i.

At low pressures ϕ_i can be taken as 1, and K_i° as given by equation 8.31. Equation 8.33 then simplifies to:

$$K_i = \gamma_i \frac{P_i^\circ}{P} \tag{8.34}$$

and

$$y_i = \gamma_i \frac{P_i^\circ}{P} x_i. \tag{8.35}$$

The validity of these equations in the design of columns operating at more than a few bars pressure will depend on the nature of the components in the mixture and the difficulty of the separation. In the gas phase, the deviations from the ideal gas laws can be judged by estimating the compressibility factor z (see Section 8.6.2). The difficulty of separation will depend on the closeness of the component K-values. Equation 8.35 is suitable for the design of relatively easy separations, say less than 50 stages, but should not be used for difficult separations.

At moderate pressures, up to about 20 bar, the deviation from the ideal mixture laws in the gas phase will still be negligible, and deviations from the ideal gas laws can be allowed for by introducing the second viral coefficients B_i:

$$K_i = y_i/x_i = \frac{\gamma_i P_i^\circ}{P} \exp\left[\frac{(V_{m,i} - B_i)(P - P_i^\circ)}{RT} \right] \qquad (8.36)$$

where $V_{m,i}$ = molar volume (liquid state) component i,
          B_i = second viral coefficient in the viral equation of state, for component i,
and

$$z_i = 1 + \frac{B_i P}{RT} \qquad (8.37)$$

8.16.2. Correlations for liquid phase activity coefficients

The liquid phase activity coefficient is a function of pressure, temperature and liquid composition, but at conditions remote from the critical conditions for the mixture the activity coefficient is virtually independent of pressure and, in the range of temperature encountered in most distillation processes, can be taken as independent of temperature. The many equations that have been developed to represent the dependence on liquid composition are discussed by Reid et al. (1977) and Null (1970).

The Wilson equation (Wilson, 1964) is probably the most convenient for use in process design:

$$\ln \gamma_k = 1 \cdot 0 - \ln\left[\sum_{j=1}^{n} (x_j A_{kj}) \right] - \sum_{i=1}^{n} \left[\frac{x_i A_{ik}}{\sum\limits_{j=1}^{n} (x_j A_{ij})} \right] \qquad (8.38)$$

where γ_k = activity coefficient for component k,
       A_{ij}, A_{ji} = Wilson coefficients (A values) for the binary pair i, j,
              n = number of components.

Wilson's equation is superior to the older activity coefficient equations, such as the Margules and Van–Laar equations, for systems that are very non-ideal but which do not form two phases. Like the other three suffix equations, it cannot be used to represent systems that form two phases in the concentration range of interest. A significant advantage of the Wilson equation is that it can be used to calculate the equilibrium compositions for multicomponent systems using only the Wilson coefficients obtained from the binary systems that comprise multicomponent mixture. The Wilson coefficients for several hundred binary pairs are given by Hirata (1975) who also discusses methods for calculating the Wilson coefficients from experimental vle data.

The Wilson equation is best solved using a short computer program with the Wilson coefficients in a matrix form. A suitable program is given in Table 8.9 and its use illustrated in Example 8.15. The program language is BASIC and it is intended for interactive use. It can be extended for use with any number of components by changing the value of the constant N in the first data statement and including the appropriate Wilson coefficients (Wilson A values) in the other data statements. The program can easily be modified for use as a sub-routine for bubble-point and other vapour composition programs.

TABLE 8.9. *Program for Wilson equation*
(Example 8.15)

```
100 REM WILSON EQUATION
110 REM CALCULATES ACTIVITY COEFFICIENTS FOR MULTICOMPONENT SYSTEMS
120 PRINT" DATA STATEMENTS LINES 370 TO 490"
130 READ N
140 MAT READ A
150 PRINT" TYPE IN LIQUID COMPOSITION, ONE COMPONENT AT A TIME"
160 FOR P=1 TO N
170 PRINT" X";P;"?"
180 INPUT X(P)
190 NEXT P
200 FOR K=1 TO N
210 Q1=0
220 FOR J=1 TO N
230 Q1=Q1+X(J)*A(K,J)
240 NEXT J
250 Q2=0
260 FOR I=1 TO N
270 Q3=0
280 FOR J=1 TO N
290 Q3=Q3+X(J)*A(I,J)
300 NEXT J
310 Q2=Q2+(X(I)*A(I,K))/Q3
320 NEXT I
330 G(K) = EXP(1-LOG(Q1)-Q2)
340 PRINT"GAMMA";K;"=";G(K)
350 NEXT K
360 DIM A(4,4),X(4),G(4)
370 DATA 4
380 DATA 1,2.3357,2.7385,0.4180
390 DATA 0.1924,1,1.6500,0.1108
400 DATA 0.2419,0.5343,1,0.0465
410 DATA 0.9699,0.9560,0.7795,1
500 END
```

Example 8.15

Using the Wilson equation, calculate the activity coefficients for isopropyl alcohol (IPA) and water in a mixture of IPA, methanol, water, and ethanol; composition, all mol fraction:

IPA	Water	Methanol	Ethanol
0·18	0·72	0·05	0·05

Solution

Use the binary Wilson A values given by Hirata (1975). The program "WILSON", Table 8.9, is used to solve this example.

The Wilson A-values for the binary pairs are $A_{i,j}$

$$
j
$$

	1	2	3	4
1	1	2·3357	2·7385	0·4180
2	0·1924	1	1·6500	0·1108
3	0·2419	0·5343	1	0·0465
4	0·9699	0·9560	0·7795	1

i

Component 1 = MeOH
      2 = EtOH
      3 = IPA
      4 = H_2O

The output from the program for the concentrations given was:

$$\gamma_3 = 2·11, \qquad \gamma_4 = 1·25$$

Experimental values from Hirata (1975)

$$\gamma_3 = 2·1, \qquad \gamma_4 = 1·3$$

8.16.3. Prediction of vapour–liquid equilibria

The designer will often be confronted with the problem of how to proceed with the design of a separation process without adequate experimentally determined equilibrium data. Some techniques are available for the prediction of vle data and for the extrapolation of experimental values. Caution must be used in the application of these techniques in design and the predictions should be supported with experimentally determined values whenever practicable. The same confidence cannot be placed on the prediction of equilibrium data as that for many of the prediction techniques for other physical properties given in this chapter. Some of the techniques most useful in design are given in the following paragraphs.

Estimation of activity coefficients from azeotropic data

If a binary system forms an azeotrope, the activity coefficients can be calculated from a knowledge of the composition of the azeotrope and the azeotropic temperature. At the azeotropic point the composition of the liquid and vapour are the same, so from equation 8.35:

$$\gamma_i = P/P_i^\circ \qquad\qquad (8.38)$$

where P_i° is determined at the azeotropic temperature.

The values of the activity coefficients determined at the azeotropic composition can be used to calculate the coefficients in the Wilson equation (or any other of the three-suffix equations) and the equation used to estimate the activity coefficients at other compositions.

Horsley (1973) gives an extensive collection of data on azeotropes.

Activity coefficients at infinite dilution

The constants in any of the activity coefficient equations can be readily calculated from experimental values of the activity coefficients at infinite dilution. For the Wilson equation:

$$\ln \gamma_1^\infty = -\ln A_{12} - A_{21} + 1 \qquad (8.39a)$$

$$\ln \gamma_2^\infty = -\ln A_{21} - A_{12} + 1 \qquad (8.39b)$$

where $\gamma_1^\infty, \gamma_2^\infty$ = the activity coefficients at infinite dilution
for components 1 and 2, respectively,

A_{12} = the Wilson A-value for component 1 in component 2,

A_{21} = the Wilson A-value for component 2 in component 1.

Relatively simple experimental techniques, using ebulliometry and chromatography, are available for the determination of the activity coefficients at infinite dilution. The methods used are described by Null (1970) and Conder and Young (1979).

Pieratti *et al.* (1955) have developed correlations for the prediction of the activity coefficients at infinite dilution for systems containing water, hydrocarbons and some other organic compounds. Their method, and the data needed for predictions, is described by Treybal (1963) and Reid *et al.* (1977).

Calculation of activity coefficients from mutual solubility data

For systems that are only partially miscible in the liquid state, the activity coefficient in the homogeneous region can be calculated from experimental values of the mutual solubility limits. The methods used are described by Reid *et al.* (1977), Treybal (1963), Brian (1965) and Null (1970). Treybal (1963) has shown that the Van–Laar equation should be used for predicting activity coefficients from mutual solubility limits.

Group contribution methods

Group contribution methods have been developed for the prediction of liquid-phase activity coefficients. The objective has been to enable the prediction of phase equilibrium data for the tens of thousands of possible mixtures of interest to the process designer to be made from the contributions of the relatively few functional groups which made up the compounds. The UNIFAC method, Fredenslund *et al.* (1977a), is probably the most useful for process design. Its use is described in detail in a book by Fredenslund *et al.* (1977b), which includes computer programs and data for the use of the UNIFAC method in the design of distillation columns.

8.16.4. K-values for hydrocarbons

A useful source of K-values for light hydrocarbons is the well-known "De Priester charts", Dabyburjor (1978), which are reproduced as Figs. 8.3a and b. These charts give the K-values over a wide range of temperature and pressure.

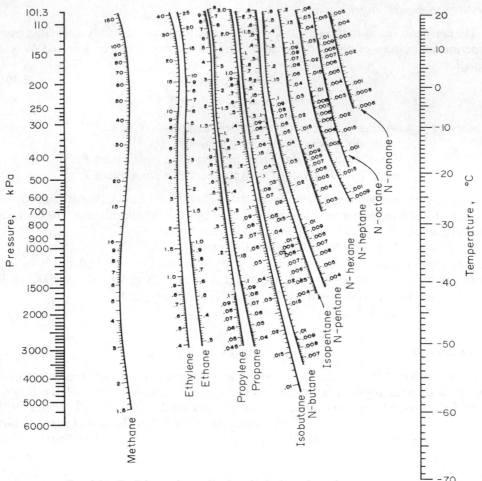

Fig. 8.3a. De Priester chart—K-values for hydrocarbons, low temperature

8.16.5. Vapour–liquid equilibria at high pressures

At pressures above a few atmospheres, the deviations from ideal behaviour in the gas phase will be significant and must be taken into account in process design. The effect of pressure on the liquid-phase activity coefficient must also be considered. A discussion of the methods used to correlate and estimate vapour–liquid equilibrium data at high pressures is beyond the scope of this book. The reader should refer to the texts by Null (1970) or Prausnitz and Chueh (1968).

Prausnitz and Chueh also discuss phase equilibria in systems containing components above their critical temperature (super-critical components).

8.16.6. Liquid–liquid equilibria

Experimental data, or predictions, that give the distribution of components between the two solvent phases, are needed for the design of liquid–liquid extraction processes; and mutual solubility limits will be needed for the design of decanters, and other liquid–liquid separators.

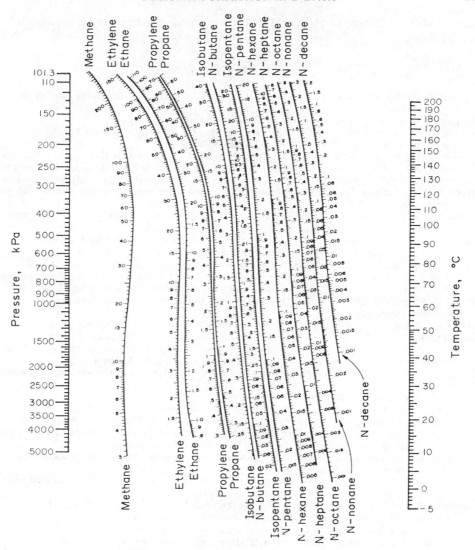

FIG. 8.3b. De Priester chart—K-values for hydrocarbons, high temperature

Perry and Chilton (1973) give a useful summary of solubility data. Liquid–liquid equilibrium compositions can be predicted from vapour–liquid equilibrium data, but the predictions are seldom accurate enough for use in the design of liquid–liquid extraction processes.

Null (1970) gives a computer program for the calculation of ternary diagrams from vle data, using the Van–Laar equation.

8.16.7. Gas solubilities

At low pressures, most gases are only sparingly soluble in liquids, and at dilute concentrations the systems obey Henry's law (see Volume 2, Chapter 11). Markham and

Kobe (1941) and Battino and Clever (1966) give comprehensive reviews of the literature on gas solubilities.

8.17. References

ANTOINE, C. (1888) *Compte rend.* **107,** 681 and 836. Tensions des vapeurs: nouvelle relation entre les tensions et les températures.

ANTONY, A. (1979) *Guide to Basic Information Sources in Chemistry* (Wiley).

BATTINO, R. and CLEVER, H. L. (1966) *Chem. Rev.* **66,** 395. The solubility of gases in liquids.

BENSON, S. W. (1976) *Thermochemical Kinetics*, 2nd ed. (Wiley).

BENSON, S. W., CRUICKSHANK, F. R., GOLDEN, D. M., HAUGEN, G. R., O'NEAL, H. E., ROGERS, A. S., SHAW, R. and WALSH, R. (1969) *Chem. Rev.* **69,** 279. Activity rules for the estimation of thermochemical properties.

BRETSZNAJDER, S. (1971) *Prediction of Transport and other Physical Properties of Fluids* (Pergamon Press).

BRIAN, P. L. T. (1965) *Ind. Eng. Chem. Fundamentals* **4,** 100. Predicting activity coefficients from liquid phase solubility limits.

BROMLEY, L. A. (1952) *Thermal Conductivity of Gases at Moderate Pressure*, University of California Radiation Laboratory Report UCRL-1852 (University of California, Berkeley).

BRINK, J. A. and SHREVE, R. N. (1977) *The Chemical Process Industries*, 4th ed. (McGraw-Hill).

BURMAN, C. R. (1965) *How to find out in Chemistry* (Pergamon Press).

CAPSEY, S. R. (1963) *Patents, an Introduction for Engineers and Scientists* (Newnes–Butterworths).

CHUEH, C. F. and SWANSON, A. C. (1973a) *Can. J. Chem. Eng.* **51,** 576. Estimation of liquid heat capacity.

CHUEH, C. F. and SWANSON, A. C. (1973b) *Chem. Eng. Prog.* **69** (July) 83. Estimating liquid heat capacity.

CHU, J. C., WANG, S. L., LEVY, S. L. and PAUL, R. (1956) *Vapour–liquid Equilibrium Data* (J. W. Edwards Inc., Ann Arbor, Michigan).

CONDER, J. R. and YOUNG, C. L. (1979) *Physicochemical Measurement by Gas Chromatography* (Wiley).

DABYBURJOR, D. B. (1978) *Chem. Eng. Prog.* **74** (April) 85. SI units for distribution coefficients.

DOMALSKI, E. S. (1972) *J. Phys. Chem. Ref. Data* **1,** 221. Selected values of heats of combustion and heats of formation of organic compounds containing the elements C, H, N, O, P, and S.

DREISBACH, R. R. (1952) *Pressure–volume–temperature Relationships of Organic Compounds*, 3rd ed. (Handbook Publishers).

EUCKEN, A. (1911) *Phys. Z.* **12,** 1101.

FAITH, W. L., KEYES, W. L. and CLARK, R. L. (1965) *Industrial Chemicals*, 3rd ed. (Wiley).

FREDENSLUND, A., GMEHLING, J., MICHELSEN, M. L., RASMUSSEN, P. and PRAUSNITZ, J. M. (1977a) *Ind. Eng. Chem. Proc. Des. & Dev.* **16,** 450. Computerized design of multicomponent distillation columns using the UNIFAC group contribution method for calculation of activity coefficients.

FREDENSLUND, A., GMEHLING, J. and RASMUSSEN, P. (1977b) *Vapour–liquid Equilibria using UNIFAC: a Group Contribution Method* (Elsevier).

FULLER, E. N., SCHETTLER, P. D. and GIDDINGS, J. C. (1966) *Ind. Eng. Chem.* **58** (May) 19. A new method for the prediction of gas-phase diffusion coefficients.

GAMBILL, W. R. (1958) *Chem. Eng., Albany* **65** (June 2nd) 125. Predict diffusion coefficient, D.

GROGGINS, P. (1958) *Unit Processes in Organic Synthesis*, 5th ed. (McGraw-Hill).

HAGGENMACHER, J. E. (1946) *J. Am. Chem. Soc.* **68,** 1633. Heat of vaporisation as a function of temperature.

HALA, E., WICHTERLE, I. POLAK, J. and BOUBLIK, T. (1968) *Vapour–liquid Equilibrium Data at Normal Pressure* (Pergamon).

HALA, E., WICHTERLE, I. and LINEK, J. (1973) *Vapour–liquid Equilibrium Data Bibliography* (Elsevier). Supplement 1, 1976.

HIRATA, M., OHE, S. and NAGAHAMA, K. (1975) *Computer Aided Data Book of Vapour–liquid Equilibria* (Elsevier).

HMSO (1970) *Searching British Patent Literature*.

HMSO (1971) *About Patents*—patents as a source of technical information.

HO, C. Y., POWELL, R. W. and LILEY, P. E. (1972) *J. Phys. Chem. Ref. Data* **1,** 279. Thermal conductivity of the elements.

HORSLEY, L. H. (1973) *Azeotropic Data III* (American Chemical Society).

JAMIESON, D. T., IRVING, J. B. and TUDHOPE, J. S. (1965) *Thermal Conductivity of Gases and Liquids* (Arnold).

JASPER, J. J. (1972) *J. Phys. Chem. Ref. Data* **1,** 841. The surface tension of pure liquids.

KERN, D. Q. (1950) *Process Heat Transfer* (McGraw-Hill).

KIRK, R. E. and OTHMER, D. F. (Eds.) (1966) *Encyclopedia of Chemical Technology*, 2nd ed. (Wiley).

KIRK, R. E. and OTHMER, D. F. (Eds.) (1977) *Encyclopedia of Chemical Technology*, 3rd ed. (Wiley).

KOBE, K. A. and LYNN, R. E. (1953) *Chem. Rev.* **52,** 177. The critical properties of elements and compounds.

KOJIMA, K., TOCHIGI, K., SEKI, H. and WATASE, K. (1968) *Kagaku Kogaku* **32,** 149. Determination of vapour–liquid equilibrium from boiling point curve.

KUDCHADKER, A. P., ALANI, G. H. and ZWOLINSK, B. J. (1968) *Chem. Rev.* **68,** 659. The critical constants of organic substances.

LEWIS, W. K. and SQUIRES, L. (1934) *Oil & Gas J.* (Nov. 15th) 92. The mechanism of oil viscosity as related to the structure of liquids.

LIEBERRY, F. (Ed.) (1972) *Mainly on Patents; the use of Industrial Property and its Literature* (Butterworths).

LYDERSEN, A. L. (1955) *Estimation of Critical Properties of Organic Compounds*, University of Wisconsin Coll. Eng. Exp. Stn. Report 3 (University of Wisconsin).

MARKHAM, A. E. and KOBE, K. A. (1941) *Chem. Rev.* **28,** 519. The solubility of gases in liquids.

MATHEWS, J. F. (1972) *Chem. Rev.* **72,** 71. The critical constants of inorganic substances.

McKETTA, J. J. (Ed.) (1977) *Encyclopedia of Chemical Processes and Design* (Marcel Dekker).

MILLER, S. A. (1969) *Ethylene and its Industrial Derivatives* (Benn).

MOUNT, E. (1976) *Guide to Basic Information Sources in Engineering* (Wiley).

NESMEYANOV, A. N. (1963) *Vapour Pressure of Elements* (Infosearch Ltd., London).

NULL, H. R. (1970) *Phase Equilibrium in Process Design* (Wiley).

OHE, S. (1976) *Computer Aided Data Book of Vapour Pressure* (Data Book Publishing Co., Japan).

OTHMER, D. F., CHUDGAR, M. M. and LEVY, S. L. (1952) *Ind. Eng. Chem.* **44,** 1872. Binary and ternary systems of acetone, methyl ethyl ketone and water.

OTHMER, D. F., JOSEFOWITZ, S. and SCHMUTZLER, A. F. (1968) *Ind. Eng. Chem.* **40,** 886. Correlating surface tensions of liquids.

PERRY, R. H. and CHILTON, C. H. (Eds.) (1973) *Chemical Engineers Handbook*, 5th ed. (McGraw-Hill).

PIERATTI, G. J., DEAL, C. H. and DERR, E. L. (1955) *Ind. Eng. Chem.* **51,** 95. Activity coefficients and molecular structure.

PRAUSNITZ, J. M., ECKERT, C. A., ORYE, R. V. and O'CONNELL, J. P. (1967) *Computer Calculation of Multicomponent Vapour–liquid Equilibria* (Prentice-Hall).

PRAUSNITZ, J. M. and CHUEH, P. L. (1968) *Computer Calculations for High-pressure Vapour–liquid–equilibria* (Prentice-Hall).

PRAUSNITZ, J. M. (1969) *Molecular Thermodynamics of Fluid-phase Equilibria* (Prentice-Hall).

REID, R. C., PRAUSNITZ, J. M. and SHERWOOD, T. K. (1977) *The Properties of Gases and Liquids*, 3rd ed. (McGraw-Hill).

REINDERS, W. and DE MINJER, C. H. (1947) *Trav. Chim. Pays-Bas* **66,** 573. Vapour–liquid equilibria in ternary systems VI. The system water–acetone–chloroform.

RIHANI, D. N. and DORAISWAMY, L. K. (1965) *Ind. Eng. Chem. Fundamentals* **4,** 17. Estimation of heat capacity of organic compounds from group contributions.

SMITH, W. T., GREENBAUM, S. and RUTLEDGE, G. P. (1954) *J. Phys. Chem.* **58,** 443. Correlation of critical temperature with thermal expansion coefficients of organic liquids.

SOUDERS, M. (1938) *J. Am. Chem. Soc.* **60,** 154. Viscosity and chemical constitution.

STEPHENSON, R. M. (1966) *Introduction to the Chemical Process Industries* (Reinhold).

STERBACEK, Z., BISKUP, B. and TAUSK, P. (1979) *Calculation of Properties using Corresponding-state Methods* (Elsevier).

SUGDEN, S. (1924) *J. Chem. Soc.* **125,** 1177. A relation between surface tension, density, and chemical composition.

TOULOUKIAN, Y. S. (Ed.) (1970–77) *Thermophysical Properties of Matter, TPRC Data Services* (Plenum Press).

TREYBAL, R. E. (1963) *Liquid Extraction*, 2nd ed. (McGraw-Hill).

TROUTON, F. T. (1884) *Phil. Mag.* **18,** 54. On molecular latent heat.

TSEDERBERG, N. V. (1965) *Thermal Conductivity of Gases and Liquids* (Arnold).

WATSON, K. M. (1943) *Ind. Eng. Chem.* **35,** 398. Thermodynamics of the liquid state: generalized prediction of properties.

WEBER, H. F. (1980) *Ann Phy. Chem.* **10,** 103. Untersuchungen über die wärmeleitung in flüssigkeiten.

WEISSERMAL, K. and ARPE, H. (1978) *Industrial Organic Chemistry* (Verlag Chemie).

WERNER, R. R. (1941) *Thermochemical Calculations* (McGraw-Hill).

WILKE, C. R. (1949) *Chem. Eng. Prog.* **45,** 218. Estimation of liquid diffusion coefficients.

WILKE, C. R. and CHANG, P. (1955) *A.I.Ch.E. Jl.* **1,** 264. Correlation of diffusion coefficients in dilute solutions.

WILSON, G. M. (1964) *J. Am. Chem. Soc.* **86,** 127. A new expression for excess energy of mixing.

YORK, R. and HOLMES, R. C. (1942) *Ind. Eng. Chem.* **34,** 345. Vapor–liquid equilibria of the system acetone–acetic acid–water.

Bibliography: general sources of physical properties

DREISBACH, R. R. (1955–61) *Physical Properties of Chemical Compounds*, Vols. I, II, III (American Chemical Society).

DREISBACH, R. R. (1952) *Pressure–volume–temperature Relationships of Organic Compounds*, 3rd ed. (Handbook Publishers).

FENSKE, M., BRAUN, W. G. and THOMPSON, W. H. (1966) *Technical Data Book – Petroleum Refining* (American Petroleum Institute).

GALLANT, R. W. (1968) (1970) *Physical Properties of Hydrocarbons*, Vols. 1 and 2 (Gulf).

LANGE, N. A. (Ed.) (1961) *Handbook of Chemistry*, 10th ed. (McGraw-Hill).

MAXWELL, J. B. (1950) *Data Book on Hydrocarbons* (Van Nostrand).

NATIONAL BUREAU OF STANDARDS (1951) *Selected Values of Thermodynamic Properties*, Circular C500 (US Government Printing Office).

PERRY, R. H. and CHILTON, C. H. (Eds.) (1973) *Chemical Engineers Handbook*, 5th ed. (McGraw-Hill).

ROSS, T. K. and FRESHWATER, D. C. (1962) *Chemical Engineers Data Book* (Leonard Hill).

ROSSINI, F. D. (1953) *Selected Values of Physical and Thermodynamic Properties of Hydrocarbons and Related Compounds* (American Chemical Society).

SEIDELL, A. (1952) *Solubilities of Inorganic and Organic Compounds*, 3rd ed. (Van Nostrand).

SPIERS, H. M. (Ed.) (1961) *Technical Data on Fuel*, 6th ed. (British National Committee, Conference on World Power).

STEPHEN, T. and STEPHEN, H. (1963) *Solubilities of Inorganic and Organic Compounds*, 2 vols. (Macmillan).

TIMMERMANNS, J. (1950) *Physico–chemical Constants of Pure Organic Compounds* (Elsevier).

TIMMERMANNS, J. (1959) *Physico–chemical Constants of Binary Systems*, 4 vols. (Interscience).

WEAST, R. C. (Ed.) (1972) *Handbook of Chemistry and Physics*, 53rd ed. (the Chemical Rubber Co.).

WASHBURN, E. W. (Ed.) (1933) *International Critical Tables of Numerical Data, Physics, Chemistry, and Technology*, 8 vols. (McGraw-Hill).

YAWS, C. L. (1977) *Physical Properties* (McGraw-Hill).

8.18. Nomenclature

		Dimensions in $\mathbf{MLT}\theta$
A	Coefficient in the Antoine equation	—
$A_{1,2}$	Coefficients in the Wilson equation for the binary pair 1, 2	—
B	Coefficient in the Antoine equation	θ
B_i	Second viral coefficient for component i	$M^{-1}L^3$
C	Coefficient in the Antoine equation	θ
C_p	Specific heat capacity at constant pressure	$L^2T^{-2}\theta^{-1}$
D_L	Liquid diffusivity	L^2T^{-1}
D_v	Gas diffusivity	L^2T^{-1}
I	Souders' index (equation 8.9)	$M^{-1}L^3$
K	Equilibrium constant (ratio)	—
K^0	Equilibrium constant for an ideal mixture	—
k	Thermal conductivity	$MLT^{-3}\theta^{-1}$
k_m	Thermal conductivity of a mixture	$MLT^{-3}\theta^{-1}$
L_v	Latent heat of vaporisation	L^2T^{-2}
$L_{v,b}$	Latent heat at normal boiling point	L^2T^{-2}
M	Molecular weights (mass)	M
n	Number of components	—
P	Pressure	$ML^{-1}T^{-2}$ or L
P_c	Critical pressure	$ML^{-1}T^{-2}$
P_{ch}	Sugden's parachor (equation 8.23)	—
P_i^0	Vapour pressure of component i	$ML^{-1}T^{-2}$ or L
P_k	Vapour pressure of component k	$ML^{-1}T^{-2}$ or L
P_r	Reduced pressure	—
ΔP_c	Critical constant increment in Lydersen equation (equation 8.26)	$M^{-1/2}L^{1/2}T$
$\mathbf{R}$	Universal gas constant	$L^2T^{-2}\theta^{-1}$
T	Temperature, absolute scale	θ
T_b	Normal boiling point, absolute scale	θ
T_c	Critical temperature	θ
T_r	Reduced temperature	—
ΔT_c	Critical constant increment in Lydersen equation (equation 8.25)	—
t	Temperature, relative scale	θ
V_c	Critical volume	$M^{-1}L^3$
V_m	Molar volume at normal boiling point	$M^{-1}L^3$
ΔV_c	Critical constant increment in Lydersen equation (equation 8.27)	$M^{-1}L^3$
v_i	Special diffusion volume coefficient for component i (Table 8.5)	L^3

w	Mass fraction (weight fraction)	—
x	Mol fraction, liquid phase	—
y	Mol fraction, vapour phase	—
z	Compressibility factor	—
α	Relative volatility	—
β	Coefficient of thermal expansion	θ^{-1}
γ	Liquid activity coefficient	—
γ^{∞}	Activity coefficient at infinite dilution	—
μ	Dynamic viscosity	$\mathbf{ML^{-1}T^{-1}}$
μ_b	Viscosity at boiling point	$\mathbf{ML^{-1}T^{-1}}$
μ_m	Viscosity of a mixture	$\mathbf{ML^{-1}T^{-1}}$
ρ	Density	$\mathbf{ML^{-3}}$
ρ_L	Liquid density	$\mathbf{ML^{-3}}$
ρ_v	Vapour (gas) density	$\mathbf{ML^{-3}}$
ρ_b	Density at normal boiling point	$\mathbf{ML^{-3}}$
σ	Surface tension	$\mathbf{MT^{-2}}$
σ_m	Surface tension of a mixture	$\mathbf{MT^{-2}}$
ϕ	Fugacity coefficient	—

Suffixes

$\left.\begin{array}{l} a,\ b \\ i,\ j,\ k \\ 1,\ 2 \end{array}\right\}$ Components

CHAPTER 9

Safety and Loss Prevention

9.1. Introduction

Any organisation has a legal and moral obligation to safeguard the health and welfare of its employees and the general public. Safety is also good business; the good management practices needed to ensure safe operation will also ensure efficient operation.

The term "loss prevention" is an insurance term, the loss being the financial loss caused by an accident. This loss will not only be the cost of replacing damaged plant and third party claims, but also the loss of earnings from lost production and lost sales opportunity.

All manufacturing processes are to some extent hazardous, but in chemical processes there are additional, special, hazards associated with the chemicals used and the process conditions. The designer must be aware of these hazards, and ensure, through the application of sound engineering practice, that the risks are reduced to acceptable levels.

In this book only the particular hazards associated with chemical and allied processes will be considered. The more general, normal, hazards present in all manufacturing process such as, the dangers from rotating machinery, falls, falling objects, use of machine tools, and of electrocution will not be considered. General industrial safety and hygiene are covered in several books, King and Magid (1979), Hadley (1969).

Safety and loss prevention in process design can be considered under the following broad headings:

1. Identification and assessment of the hazards.
2. Control of the hazards: for example, by containment of flammable and toxic materials.
3. Control of the process. Prevention of hazardous deviations in process variables (pressure, temperature, flow), by provision of automatic control systems, interlocks, alarms, trips; together with good operating practices and management.
4. Limitation of the loss. The damage and injury caused if an incident occurs: pressure relief, plant layout, provision of fire-fighting equipment.

In this chapter the discussion of safety in process design will of necessity be limited. A more complete treatment of the subject can be found in the books by Wells (1980) and Lees (1980) and in the general literature. The proceedings of the symposia on the subject, organised regularly by the American Institute of Chemical Engineers, the Institution of Chemical Engineers and the European Federation of Chemical Engineering, contain many articles of interest on general safety philosophy, techniques and organisation; and the hazards associated with specific processes and equipment. The Institution of Chemical Engineers has also published two booklets on safety for Chemical Engineering students, IChemE (1976, 1977).

9.2. Intrinsic and extrinsic safety

Processes can be divided into those that are intrinsically safe, and those for which the safety has to be engineered in. An intrinsically safe process is one in which safe operation is inherent in the nature of the process; a process which causes no danger, or negligible danger, under all foreseeable circumstances (all possible deviations from the design operating conditions). Clearly, the designer should always select a process that is intrinsically safe whenever it is practical, and economic, to do so. However, most chemical manufacturing processes are, to a greater or lesser extent, inherently unsafe, and dangerous situations can develop if the process conditions deviate from the design values.

The safe operation of such processes depends on the design and provision of engineered safety devices, and on good operating practices, to prevent a dangerous situation developing, and to minimise the consequences of any incident that arises from the failure of these safeguards.

The term "engineered safety" covers the provision in the design of control systems, alarms, trips, pressure-relief devices, automatic shut-down systems, duplication of key equipment services; and fire-fighting equipment, sprinkler systems and blast walls, to contain any fire or explosion.

9.3. The hazards

In this section the special hazards of chemicals are reviewed (toxicity, flammability and corrosivity); together with the other hazards of chemical plant operation.

9.3.1. Toxicity

Most of the materials used in the manufacture of chemicals are poisonous, to some extent. The potential hazard will depend on the inherent toxicity of the material and the frequency and duration of any exposure. It is usual to distinguish between the short-term effects (acute) and the long-term effects (chronic). A highly toxic material that causes immediate injury, such as phosgene or chlorine, would be classified as a safety hazard. Whereas a material whose effect was only apparent after long exposure at low concentrations, for instance, carcinogenic materials, such as vinyl chloride, would be classified as industrial health and hygiene hazards. The permissible limits and the precautions to be taken to ensure the limits are met will be very different for these two classes of toxic materials. Industrial hygiene is as much a matter of good operating practice and control as of good design.

TABLE 9.1. Some LD_{50} values

Compound	mg/kg
Potassium cyanide	10
Tetraethyl lead	35
Lead	100
DDT	150
Aspirin	1500
Table salt	3000

Source: Lowrance (1976)

The inherent toxicity of a material is measured by tests on animals. It is usually expressed as the lethal dose at which 50 per cent of the test animals are killed, the LD_{50} (lethal dose fifty) value. The dose is expressed as the quantity in milligrams of the toxic substance per kilogram of body weight of the test animal.

Some values for tests on rats are given in Table 9.1. Estimates of the LD_{50} for man are made based on tests on animals. The LD_{50} measures the acute effects; it gives only a crude indication of the possible chronic effects.

There is no generally accepted definition of what can be considered toxic and non-toxic. Kusnetz (1974) gives two examples of attempts to set limits based on LD_{50} values:

$$LD_{50} \quad < 1 \text{ mg/kg—extremely toxic}$$
$$> 15 \text{ mg/kg—relatively non-toxic}$$
$$\text{or } LD_{50} \quad < 5 \text{ mg/kg—supertoxic}$$
$$> 15 \text{ mg/kg—relatively non-toxic}$$

These definitions apply only to the short-term (acute) effects. In fixing permissible limits on concentration for the long-term exposure of workers to toxic materials, the exposure time must be considered together with the inherent toxicity of the material. The "Threshold Limit Value" (TLV) is the most commonly used guide for controlling the long-term exposure of workers to contaminated air. The TLV is defined as the concentration to which it is believed the average worker could be exposed to, day by day, for 8 hours a day, 5 days a week, without suffering harm. It is expressed in ppm for vapours and gases, and in mg/m^3 (or $grains/ft^3$) for dusts and liquid mists. A comprehensive source of data on the toxicity of industrial materials is Sax's handbook, Sax (1975); which also gives guidance on the interpretation and use of the data. Toxicity data on solvents is given by Browning (1965). Recommended TLV values are published in bulletins by the United States Occupational Safety and Health Administration and the United Kingdom Health and Safety Executive.

Control of toxic materials

Points to consider at the design stage for the control of toxic materials are:

Containment: sound design of equipment and piping; specify welded joints in preference to flanges (liable to leak).
Disposal: provision of effective vent stacks to disperse material vented from pressure-relief devices; or use vent scrubbers.
Ventilation: use open structures, or provide adequate ventilation systems.
Emergency equipment: escape routes, rescue equipment, safety showers, eye baths.

In addition, operating practices should include:

Regular medical check-ups for employees, to check for chronic effects.
Regular monitoring of the environment to check exposure levels; consider installation of permanent instruments fitted with alarms.
Good general hygiene; washing facilities, operating instructions.

9.3.2. Flammability

The term "flammable" is now more commonly used in the technical literature than "inflammable" to describe materials that will burn, and will be used in this book. The hazard caused by a flammable material depends on a number of factors:

1. The flash-point of the material.
2. The autoignition temperature of the material.
3. The flammability limits of the material.
4. The energy released in combustion.

Flash-point

The flash-point is a measure of the ease of ignition of the liquid. It is the lowest temperature at which the material will ignite from an open flame. The flash-point is a function of the vapour pressure and the flammability limits of the material. It is measured in standard apparatus, following standard procedures (BS 2839 and 4688). Both open- and closed-cup apparatus is used. Closed-cup flash-points are lower than open cup, and the type of apparatus used should be stated clearly when reporting measurements. Flash-points are given in Sax's handbook, Sax (1975). The flash-points of many volatile materials are below normal ambient temperature; for example, ether $-45°C$, petrol (gasoline) $-43°C$ (open cup).

Autoignition temperature

The autoignition temperature of a substance is the temperature at which it will ignite spontaneously in air, without any external source of ignition. It is an indication of the maximum temperature to which a material can be heated in air; for example, in drying operations.

Flammability limits

The flammability limits of a material are the lowest and highest concentrations in air, at normal pressure and temperature, at which a flame will propagate through the mixture. They show the range of concentration over which the material will burn in air, if ignited.

Flammability limits are characteristic of the particular material, and differ widely for different materials. For example, for hydrogen the lower limit is 4·1 per cent v/v and the upper 74·2 per cent v/v; for methane the range is from 3·1 per cent to 32 per cent v/v; but for petrol (gasoline) the range is only 1·4 per cent to 7·6 per cent v/v.

Flammability limits are given in Sax's handbook, Sax (1975).

A flammable mixture may exist in the space above the liquid surface in a storage tank. The vapour space above highly flammable liquids is usually purged with inert gas (nitrogen) or floating-head tanks are used. In a floating-head tank a "piston" floats on top of the liquid, eliminating the vapour space.

Flame traps

Flame arresters are fitted in the vent lines of equipment that contains flammable material to prevent the propagation of flame through the vents. Various types of proprietary flame arresters are used. In general, they work on the principle of providing a heat sink, usually expanded metal grids or plates, to dissipate the heat of the flame; the main types are described in Rogowski (1980).

Traps should also be installed in plant ditches to prevent the spread of flame. These are normally liquid U-legs, which block the spread of flammable liquid along ditches.

Fire precautions

Recommendations on the fire precautions to be taken in the design of chemical plant are given in the British Standard code of practice CP 3013.

9.3.3. Explosions

An explosion is the sudden, catastrophic, release of energy, causing a pressure wave (blast wave). An explosion can occur without fire, such as the failure through over-pressure of a steam boiler or an air receiver.

When discussing the explosion of a flammable mixture it is necessary to distinguish between detonation and deflagration. If a mixture detonates the reaction zone propagates at supersonic velocity (approximately 300 m/s) and the principal heating mechanism in the mixture is shock compression. In a deflagration the combustion process is the same as in the normal burning of a gas mixture; the combustion zone propagates at subsonic velocity, and the pressure build-up is slow. Whether detonation or deflagration occurs in a gas–air mixture depends on a number of factors; including the concentration of the mixture and the source of ignition. Unless confined or ignited by a high-intensity source (a detonator) most materials will not detonate. However, the pressure wave (blast wave) caused by a deflagration can still cause considerable damage.

Certain materials, for example, acetylene, can decompose explosively in the absence of oxygen; such materials are particularly hazardous.

Unconfined vapour cloud explosions

This type of explosion results from the release of a considerable quantity of flammable gas, or vapour, into the atmosphere, and its subsequent ignition. Such an explosion can cause extensive damage, such as occurred at Flixborough, HMSO (1975). Unconfined vapour explosions are discussed by Munday (1976) and Gugan (1979).

Dust explosions

Finely divided combustible solids, if intimately mixed with air, can explode. Several disastrous explosions have occurred in grain silos.

Dust explosions usually occur in two stages: a primary explosion which disturbs deposited dust; followed by the second, severe, explosion of the dust thrown into the atmosphere. Any finely divided combustible solid is a potential explosion hazard. Particular care must be taken in the design of dryers, conveyors, cyclones, and storage hoppers for polymers and other combustible products or intermediates. The hazard of dust explosions and their prevention is discussed fully by Palmer (1973).

9.3.4. Sources of ignition

Though precautions are normally taken to eliminate sources of ignition on chemical plants, it is best to work on the principle that a leak of flammable material will ultimately find an ignition source.

Electrical equipment

The sparking of electrical equipment, such as motors, is a major potential source of ignition, and flame-proof equipment is normally specified.

In all areas where flammable gases are likely to be present in flammable concentrations under normal operating conditions, intrinsically safe equipment should be specified, or the equipment enclosed in a purged, gas-tight, chamber.

In areas where a flammable mixture will only be present under abnormal circumstances, non-sparking equipment can be specified: equipment that does not normally spark but could spark if a fault develops. Some risk is involved, but the coincident failure of the electrical equipment and a leak of flammable gas would be required to cause a fire or explosion.

The use of electrical equipment in hazardous areas is covered by a British Standards Institute code of practice (CP 1003) and several standards (BS 229, BS 1259 and BS 4683).

Static electricity

The movement of any non-conducting material, powder, liquid or gas, can generate static electricity, producing sparks. Precautions must be taken to ensure that all piping is properly earthed (grounded) and that electrical continuity is maintained around flanges. Escaping steam, or other vapours and gases, can generate a static charge. Gases escaping from a ruptured vessel can self-ignite from a static spark. For a review of the dangers of static electricity in the process industries, see Napier (1971) and Napier and Russell (1974).

Process flames

Open flames from process furnaces and incinerators are obvious sources of ignition and must be sited well away from plant containing flammable materials.

Miscellaneous sources

It is the usual practice on plants handling flammable materials to control the entry on to the site of obvious sources of ignition; such as matches, cigarette lighters and battery-operated equipment. The use of portable electrical equipment, welding, spark-producing tools and the movement of petrol-driven vehicles would also be subject to strict control.

9.3.5. Ionising radiation

The radiation emitted by radioactive materials is harmful to living matter. Small quantities of radioactive isotopes are used in the process industry for various purposes; for example, in level and density-measuring instruments, and for the non-destructive testing of equipment.

The use of radioactive isotopes in industry is covered by government legislation, HMSO (1968, 1969).

A discussion of the particular hazards that arise in the chemical processing of nuclear fuels is outside the scope of this book.

9.3.6. Pressure

Over-pressure, a pressure exceeding the system design pressure, is one of the most serious hazards in chemical plant operation. Failure of a vessel, or the associated piping, can precipitate a sequence of events that culminate in a disaster.

Pressure vessels are invariably fitted with some form of pressure-relief device, set at the design pressure, so that (in theory) potential over-pressure is relieved in a controlled manner.

Three basically different types of relief device are commonly used:

Directly actuated valves: weight or spring-loaded valves that open at a predetermined pressure, and which normally close after the pressure has been relieved. The system pressure provides the motive power to operate the valve.

Indirectly actuated valves: pneumatically or electrically operated valves, which are activated by pressure-sensing instruments.

Bursting discs: thin discs of material that are designed and manufactured to fail at a predetermined pressure, giving a full bore opening for flow.

Relief valves are normally used to regulate minor excursions of pressure; and bursting discs as safety devices to relieve major over-pressure. Bursting discs are often used in conjunction with relief valves to protect the valve from corrosive process fluids during normal operation. The design and selection of relief valves is discussed by Connison (1960) and Issacs (1971) and is also covered in the pressure vessel standards, BS 5500 (see Chapter 13). Bursting discs are discussed by Kayser (1972) and Fitzsimmons and Cockram (1979). In the UK the use of bursting discs is covered by BS 2915. The discs are manufactured in a variety of materials for use in corrosive conditions; such as, impervious carbon, gold and silver; and suitable discs can be found for use with all process fluids.

Bursting discs and relief valves are proprietary items and the vendors should be consulted when selecting suitable types and sizes.

The factors to be considered in the design of relief systems are set out in a comprehensive paper by Parkinson (1979).

Vent piping

When designing relief venting systems it is important to ensure that flammable or toxic gases are vented to a safe location. This will normally mean venting at a sufficient height to ensure that the gases are dispersed without creating a hazard. For highly toxic materials it may be necessary to provide a scrubber to absorb and "kill" the material; for instance, the provision of caustic scrubbers for chlorine and hydrochloric acid gases. If flammable materials has to be vented at frequent intervals; as, for example, in some refinery operations, flare stacks are used.

The rate at which material can be vented will be determined by the design of the complete venting system: the relief device and the associated piping. The maximum venting rate will be limited by the critical (sonic) velocity, whatever the pressure drop (see Volume 1, Chapter 4). The design of venting systems to give adequate protection against over-pressure is a complex and difficult subject, particularly if two-phase flow is likely to occur. For complete protection the venting system must be capable of venting at the same rate as the vapour is being generated. For reactors, the maximum rate of vapour generation resulting from a loss of control can usually be estimated. Vessels must also be protected against over-pressure caused by external fires. In these circumstances the maximum rate of vapour generation will depend on the rate of heating. Standard formulae are available for the estimation of the maximum rates of heat input; see ROSPA (1971).

For some vessels, particularly where complex vent piping systems are needed, it may be impractical to the size of the vent to give complete protection against the worst possible situation.

For a comprehensive discussion of the problem of vent system design, and the design methods available, see the papers by Duxbury (1976, 1979).

Under-pressure (Vacuum)

Unless designed to withstand external pressure (see Chapter 13) a vessel must be protected against the hazard of under-pressure, as well as over-pressure. Under-pressure will normally mean vacuum on the inside with atmospheric pressure on the outside. It requires only a slight drop in pressure below atmospheric pressure to collapse a storage tank. Though the pressure differential may be small, the force on the tank roof will be considerable. For example, if the pressure in a 10-m diameter tank falls to 10 millibars below the external pressure, the total load on the tank roof will be around 80,000 N (8 tonne). It is not an uncommon occurrence for a storage tank to be sucked in (collapsed) by the suction pulled by the discharge pump, due to the tank vents having become blocked. Where practical, vacuum breakers (valves that open to atmosphere when the internal pressure drops below atmospheric) should be fitted.

9.3.7. Temperature deviations

Excessively high temperature, over and above that for which the equipment was designed, can cause structural failure and initiate a disaster. High temperatures can arise from loss of control of reactors and heaters; and, externally, from open fires. In the design of processes where high temperatures are a hazard, protection against high temperatures is provided by:

1. Provision of high temperature alarms and interlocks to shut down reactor feeds, or heating systems, if the temperature exceeds critical limits.
2. Provision of emergency cooling systems for reactors, where heat continues to be generated after shut-down; for instance, in some polymerisation systems.
3. Structural design of equipment to withstand the worst possible temperature excursion.
4. The selection of intrinsically safe heating systems for hazardous materials.

Steam, and other vapour heating systems, are intrinsically safe; as the temperature cannot exceed the saturation temperature at the supply pressure. Other heating systems rely on control of the heating rate to limit the maximum process temperature. Electrical heating systems can be particularly hazardous.

Fire protection

To protect against structural failure, water-deluge systems are usually installed to keep vessels and structural steelwork cool in a fire.

The lower section of structural steel columns are also often lagged with concrete or other suitable materials.

9.3.8. Noise

Excessive noise is a hazard to health and safety. Long exposure to high noise levels can cause permanent damage to hearing. At lower levels, noise is a distraction and causes fatigue.

The unit of sound measurement is the decibel, defined by the expression:

$$\text{Sound level} = 20 \log_{10}\left[\frac{\text{RMS sound pressure (Pa)}}{2 \times 10^{-5}}\right], \text{dB} \tag{9.1}$$

The subjective effect of sound depends on frequency as well as intensity.

Industrial sound meters include a filter network to give the meter a response that corresponds roughly to that of the human ear. This is termed the "A" weighting network and the readings are reported as dB(A); see BS 3489 and BS 4197.

Permanent damage to hearing can be caused at sound levels above about 90 dB(A), and it is normal practice to provide ear protection in areas where the level is above 80 dB(A).

Excessive plant noise can lead to complaints from neighbouring factories and local residents. Due attention should be given to noise levels when specifying, and when laying out, equipment that is likely to be excessively noisy; such as, compressors, fans, burners and steam relief valves.

Several books are available on the general subject of industrial noise control (Warring, 1974; Sharland, 1972) and on noise control in the process industries (PPA, 1973; Lipcombe and Taylor, 1978).

9.4. Dow fire and explosion index

The safety and loss prevention guide developed by the Dow Chemical Company, and published by the American Institute of Chemical Engineers, Dow (1973), gives a method for evaluating the potential hazards of a process and assessing the safety and loss prevention measures needed. A numerical "Fire and Explosion Index" is calculated, based on the nature of the process and the properties of the materials. The larger the value of the index, the more hazardous is the process. When used to evaluate the design of a new plant, the index is normally calculated after the Piping and Instrumentation diagrams and equipment layout have been prepared, and is used as a guide to the selection and design of the preventive and protection equipment needed for safe plant operation. It may be calculated at an early stage in the process design, after the preliminary flow-sheets have been prepared, and will indicate whether alternative, less hazardous, processes should be considered.

The Dow index applies only to the main process units; it does not cover process auxiliaries, such as, warehouses, tank farms, utilities and control rooms. Only the fire and explosion hazard is considered; toxicity and corrosion hazards are not covered. Nor does it deal with the special requirements of plants manufacturing explosives.

Only a brief discussion of the Dow Safety and Loss Prevention Guide will be given in this section; sufficient to show how the Fire and Explosion Index is calculated and used. The full guide should be studied before applying the technique to a particular process design. Judgement, based on experience with similar processes, is needed to calculate the index, and when using it to decide what preventive and protective measures are needed.

9.4.1. Calculation of the Dow F & E index

The basis of the F & E index is a material factor (MF), which is normally determined from the heat of combustion of the main process material. This primary material factor is multiplied by factors to allow for special material hazards; and for general and special process hazards.

The process is divided into units and the index calculated for each unit. A unit is defined as a part of the process that can be considered as a separate entity. It may be a section that is separated from the remainder of the plant by a physical barrier, or by distance; or it may be a section of the plant in which a particular hazard occurs.

Material factor

The material factor is a number from 0 to 60 that indicates the magnitude of the energy released in a fire or explosion.

For non-combustible materials the factor is zero; examples: water, carbon tetrachloride.

For combustible materials the factor is calculated from the following equation:

$$MF = -\Delta H_c \times 10^{-3} \qquad (9.2)$$

where $-\Delta H_c$ = heat of combustion, Btu/lb.
Converted to SI units, equation 9.2 becomes:

$$MF = -\Delta H_c^\circ \times \frac{4\cdot3 \times 10^{-4}}{\text{mol wt}} \qquad (9.3)$$

where $-\Delta H_c^\circ$ is now the standard heat of combustion at 25°C, kJ/kmol.

For combinations of highly reactive materials, such as mixtures of oxidising and reducing agents, the heat of combustion may not be an adequate indication of the potential energy release, and the heat of reaction (or decomposition) is used, if numerically larger than the heat of combustion.

The Dow guide includes a list of the material factors for commonly used chemicals.

The material factor should be evaluated for all process materials present in the unit in sufficient quantity to constitute a hazard, to decide the dominant material.

Example 9.1

Calculate the material factor for methane, standard heat of combustion 801,700 kJ/kmol.

From equation 9.3
$$MF = 801,700 \times \frac{4\cdot3 \times 10^{-4}}{16} = \underline{\underline{21\cdot6}}$$

Special material hazards

These factors are included to take account of any special hazards associated with the materials present in the unit. The primary material factor is increased by a percentage for

Note: In the latest edition of the Dow guide (5th ed., 1981), the material factor is based on the flammability and reactivity of the material.

each of the hazards listed below, if applicable to any material present in a significant quantity in the unit. The percentages shown in Fig. 9.1 are given as a guide; the values to be used will depend on judgement: the designer's assessment of the hazard.

A. Oxidising materials: materials that in a fire will release oxygen.
B. Reaction with water: materials that produce a combustible gas on reaction with water; example: calcium carbide.
C. Spontaneous heating: materials that are subject to spontaneous heating, or are pyrophoric; example: coal.
D. Spontaneous polymerisation: materials that are liable to polymerise when heated; examples: chloroprene, butadiene.
E. Explosive decomposition: materials that are liable to decompose, accompanied by explosion; example: acetylene.
F. Detonation: materials that could detonate under the process conditions, if the protective control systems fail.
G. Others: any other unusual hazards associated with the materials, that are considered appropriate.

General process hazards

These factors are intended to allow for the general process hazards associated with the unit being considered. The material factor, after adjustment for the special material factors, is increased by a percentage for each of the hazards listed below that are applicable to the process. The values given in Fig. 9.1 will serve as a guide.

A. Handling and physical changes only: processes that do not involve chemical reactions. The larger percentage factor is used if the process involves disconnecting and connecting pipes handling flammable liquids.
B. Continuous reactions: the lower factor is used if a runaway reaction is not likely to cause an excessive temperature rise; for other, exothermic, reactions use 50 per cent.
C. Batch reactions: a percentage added, in addition to the reaction factor B, to take account of any special hazard associated with batch operations.
D. Multiplicity of reactions: a factor added to allow for possible contamination from one reaction to another, for processes carried out in the same equipment, if this is likely to constitute a hazard.

Special process hazards

These factors allow for any of the special process hazards given below. The percentages shown in Fig. 9.1 are used as a guide; the percentage to be used depending on the magnitude of the hazard.

A. Low pressure: a factor to be added if inleakage of air is possible and likely to cause a hazard.
B. Operation near the explosive range: this factor is added if a concentration within the explosive (flammable) range is likely to occur in normal operation, or where the process relies on instrumentation to avoid explosive concentrations. The larger factor (150 per cent) is added for processes that always operate within the explosive range.

FIRE AND EXPLOSION INDEX CALCULATION SHEET	UNIT:		
1. MATERIAL FACTOR FOR:		⟶	
2. SPECIAL MATERIAL HAZARDS	% FACTOR SUGGESTED	% FACTOR USED	
A. OXIDISING MATERIALS	0–20		
B. REACTS WITH WATER TO PRODUCE A COMBUSTIBLE GAS	0–30		
C. SUBJECT TO SPONTANEOUS HEATING	30		
D. SUBJECT TO RAPID SPONTANEOUS POLYMERISATION	50–75		
E. SUBJECT TO EXPLOSIVE DECOMPOSITION	125		
F. SUBJECT TO DETONATION	150		
G. OTHER	0–150		
ADD PERCENTAGES A–G FOR SPECIAL MATERIAL HAZARD (S.M.H) TOTAL			
$((100 + \text{S.M.H. TOTAL})/100) \times (\text{MATERIAL FACTOR})$ = SUB-TOTAL No. 2 ⟶			
3. GENERAL PROCESS HAZARDS			
A. HANDLING AND PHYSICAL CHANGES ONLY	0–50		
B. CONTINUOUS REACTIONS	25–50		
C. BATCH REACTIONS	25–60		
D. MULTIPLICITY OF REACTIONS IN SAME EQUIPMENT	0–50		
ADD PERCENTAGES A–D FOR GENERAL PROCESS (G.P.H.) TOTAL			
$((100 + \text{G.P.H. TOTAL})/100) \times (\text{SUB-TOTAL No. 2})$ = SUB-TOTAL No. 3 ⟶			
4. SPECIAL PROCESS HAZARDS			
A. LOW PRESSURE (BELOW 1 BAR)	0–100		
B. OPERATION IN OR NEAR EXPLOSION RANGE	0–150		
C. LOW TEMPERATURE: 1. (CARBON STEELS 10 to −30°C)	15		
2. (BELOW −30°C)	25		
D. HIGH TEMPERATURE (USE ONE ONLY)			
1. (ABOVE FLASH POINT)	10–20		
2. (ABOVE BOILING POINT)	25		
3. (ABOVE AUTOIGNITION POINT)	35		
E. HIGH PRESSURE: 1. (15–200 BAR)	30		
2. (ABOVE 200 BAR)	60		
F. PROCESSES OR REACTIONS DIFFICULT TO CONTROL	50 100		
G. DUST OR MIST HAZARD	30–60		
H. GREATER THAN AVERAGE EXPLOSION HAZARD	60–100		
I. LARGE QUANTITIES OF COMBUSTIBLE LIQUIDS			
(USE ONE ONLY)			
1. 10–25 m³	40–55		
2. 25–75 m³	55–75		
3. 75–200 m³	75–100		
4. ABOVE 200 m³	100+		
J. OTHER	0–20		
ADD PERCENTAGES A–J FOR SPECIAL PROCESS (S.P.H.) TOTAL			
$((100 + \text{S.P.H. TOTAL})/100) \times (\text{SUB-TOTAL No. 3}) = \text{FIRE \& EXPLOSION INDEX}$ ⟶			

FIG. 9.1. Form for calculation of Dow index.

C. Low temperature: a factor to allow for the possible low-temperature brittleness of structural materials (see Chapter 7).
D. High temperature: a factor to account for the hazard of operating at temperatures above the boiling point, flash-point, or autoignition temperature of the process materials. Use the percentages given in Fig. 9.1.
E. High pressure: a factor to allow for the hazard of operating at high pressures, greater than 15 bar. Use the factors in Fig. 9.1.
F. Reactions difficult to control: applies to exothermic reactions for which, because of the nature of the process, there is a strong possibility of the reaction going out of control.
G. Dust or mist hazards: applies to processes in which the malfunction of equipment could lead to a dust or mist explosion.
H. Greater than average explosion hazard: applies to processes where process conditions are such that the hazard of an explosion due to the malfunction of equipment is greater than usual; such as processes containing flammable liquids at temperatures and pressures such that release would result in rapid vaporisation and the formation of an explosive vapour cloud.
I. Large quantities of combustible liquids: a factor to allow for the increased fire hazard associated with the storage of large quantities of flammable liquids. The percentage to be added depends on the quantity of material; see Fig. 9.1 for suggested values.
J. Others: a factor added for processes with unusual hazards, particularly when the F & E index is such as to expose nearby units to increased risk.

9.4.2. Selection of preventive and protective measures

In the Dow Safety and Loss Prevention Guide, the F & E index is used as an aid to determining the equipment and facilities needed to control the hazards and reduce the losses from any incident that may occur; the preventive and protective measures (P & P). The preventive and protective measures to be taken are divided into three categories:

1. The basic P & P features, which must be provided for all processes, regardless of the F & E index.
2. The recommended minimum features, which depend on the value of the F & E index.
3. Specific preventive features; measures that provide specific protection for the hazards considered in evaluating the index.

1. Basic preventive and protective measures

The basic safety and fire protective measures that should be included in all chemical process designs are listed below. This list is based on that given in the Dow Guide, with some minor amendments.

1. Adequate, and secure, water supplies for fire fighting.
2. Correct structural design of vessels, piping, steel work.
3. Pressure-relief devices.
4. Corrosion-resistant materials, and/or adequate corrosion allowances.
5. Segregation of reactive materials.

6. Earthing of electrical equipment.
7. Safe location of auxiliary electrical equipment, transformers, switch gear.
8. Provision of back-up utility supplies and services.
9. Compliance with national codes and standards.
10. Fail-safe instrumentation.
11. Provision for access of emergency vehicles and the evacuation of personnel.
12. Adequate drainage for spills and fire-fighting water.
13. Insulation of hot surfaces.
14. No glass equipment used for flammable or hazardous materials, unless no suitable alternative is available.
15. Adequate separation of hazardous equipment.
16. Protection of pipe racks and cable trays from fire.
17. Provision of block valves on lines to main processing areas.
18. Protection of fired equipment (heaters, furnaces) against accidental explosion and fire.
19. Safe design and location of control rooms.

Note: the design and location of control rooms, particularly as regards protection against an unconfined vapour explosion, is covered in a publication of the Chemical Industries Association, CIA (1979a).

2. *Recommended minimum preventive and protective measures*

This category covers measures which should be considered in addition to the basic measures, but which may, or may not, be specified, depending on the designer's assessment of the risks and the magnitude of the resulting fire and explosion. In the Dow Guide the F & E index is used as a guide to the degree of hazard; see Table 9.2.

This rating can be used as a guide when selecting the minimum preventive and protective measures, Table 9.3. Judgement, based on experience with similar processes, must be used when selecting the minimum measures required from Table 9.3; where any doubt exists, the measure should always be included. The Dow guide should be consulted for a full discussion of the factors to be considered when selecting the minimum protective and preventive features.

In addition to the features listed in Table 9.3, the need for building ventilation, dust-explosion control, and building explosion relief, should be considered.

3. *Specific preventive features*

This category includes measures for the specific protection from the hazards considered when evaluating the F & E index under the headings of special material hazards, and general and special process hazards. Listed below is a selection of the main features in this category from the Dow guide. The guide should be consulted for the full list.

Special material hazards

(a) Oxidising materials: separate from combustible materials, store in a fire-proof area.

(b) Reacts with water: protect from water, ventilate, remove sources of ignition.
(c) Spontaneous heating: provide emergency cooling.
(d) Spontaneous polymerisation: use inhibitors, provide cooling, pressure relief, high-temperature alarms.
(e) Explosive decomposition: design equipment to contain, or safely relieve, the explosion.
(f) Detonation: as for explosive decomposition.

General process hazards

(a) Handling and physical change: excess-flow valves, purge procedures, ventilation, remotely-operated valves.
(b) Continuous reactions: tight process control, instrumentation for detection and protection against over-pressure, over-temperature.
(c) Batch reactions: as (b), and process interlocks to prevent cross contamination.
(d) Multiplicity of reactions: as (b).

Special process hazards

(a) Low pressure: trips and alarms.
(b) Operation near explosive range: design to contain or safely relieve explosions; explosion-suppression systems; instrumentation to control composition; purges or inert dilution systems.
(c) Low temperature: specify suitable materials of construction.
(d) High temperature: special ventilation and dump systems, combustible gas monitors, automatic deluge systems; control systems to minimise the flow of flammable materials.
(e) High pressures: as for high temperature, (d).
(f) Reactions and processes difficult to control: containment, safe venting, dump systems, quench systems.
(g) Dust or mist hazards: as (b).
(h) Above average explosion hazards: as (b).
(i) Large quantities of flammable materials: remotely operated valves to minimise flow; combustible gas monitors linked to automatic deluge systems; adequate drainage.

TABLE 9.2. *Assessment of hazard*

Degree of hazard	Fire and explosion index range
Mild	0–20
Light	20–40
Moderate	40–60
Moderately heavy	60–75
Heavy	75–90
Extreme	> 90

TABLE 9.3. *Minimum preventive and protective measures*

Legend

Feature Optional	1
Feature Suggested	2
Feature Recommended	3
Feature Required	4

	Fire and Explosion Index					
	0–20	20–40	40–60	60–75	75–90	> 90
Fireproofing	1	2	2	3	4	4
Water spray						
(a) directional	1	2	3	3	4	4
(b) area	1	2	3	3	4	4
(c) curtain	1	1	2	2	2	4
Special instrumentation						
(a) temperature	1	2	3	3	4	4
(b) pressure	1	2	3	3	3	4
(c) flow control	1	2	3	4	4	4
Dump, blowdown spill control	1	1	2	3	3	4
Internal explosion protection	1	2	3	3	4	4
Combustible gas monitors						
(a) signal alarm	1	1	2	3	3	4
(b) actuate equipment	1	1	2	2	3	4
Remote operation	1	1	2	3	3	4
Dykeing	1	4	4	4	4	4
Blast and barrier walls separation	1	1	2	3	4	4

9.4.3. *Mond–Dow index*

The principles and general approach used in the Dow method of hazard evaluation have been further developed by ICI Mond Division. Their revised, Mond, fire, explosion and toxicity index is discussed in a series of papers by Lewis (1979a, 1979b).

The main developments made to the Dow index in the Mond index are:

1. It covers a wider range of process and storage installations.
2. It covers the processing of chemicals with explosive properties.
3. A calculation procedure is included for the evaluation of a toxicity hazards index.
4. A procedure is included to allow for the off-setting effects of good design, and control and safety instrumentation.
5. The procedure has been extended to cover plant layout.
6. Separate indices are calculated to assess the hazards of fire, internal explosion and aerial explosion.

The procedure followed in calculating the Mond method is, briefly:

1. An initial assessment of the hazards of each unit, in a similar manner to that used for the Dow index. At this stage no account is taken of the off-setting effect of any protective and preventive features.
2. An analysis of the different types of potential hazard: fire, explosion and toxicity. Comparison of these hazard levels with standards of acceptable risk.

3. Review of the hazard factors used in the assessment; for example, the general and special process hazards, to see if the risks can be reduced by design changes.
4. Application of the appropriate off-setting factors to allow for the preventive features included in the design; calculation of the final hazard indices.

The Mond technique of hazard evaluation is fully explained in the paper by Lewis (1979a), which includes a Technical Manual setting out the calculation procedures. The calculations are made on a calculation sheet similar to that given in Fig. 9.1 extended to include the additional features in the Mond index.

9.4.4. Summary

The Dow and Mond indexes are useful techniques, which can be used in the early stages of a project design to evaluate the hazards and risks of the proposed process.

Calculation of the indexes for the various sections of the process will highlight any particularly hazardous sections and indicate where a detailed study is needed to reduce the hazards.

Example 9.2

Evaluate the Dow fire and explosion index for the nitric acid plant described in Example 4.4.

Solution

The calculation is set out on the special form shown in Fig. 9.1. Notes on the decisions taken and the factors used are given below:

Unit: consider the total plant, there are no separate areas.

Material factor: main combustible material is ammonia and the factor is calculated from the heat of combustion for the reaction:

$$NH_3(g) + 5/4O_2(g) \rightarrow NO(g) + 3/2H_2O(g), \Delta H = -226,334 \, kJ/kmol$$

From equation 9.3

$$\text{Material factor} = \frac{4\cdot3 \times 10^{-4}}{17} \times 226,334 = \underline{\underline{5\cdot7}}$$

Note: Hydrogen is present, and has a large material factor (51.6), but the concentration is too small for it to be considered the dominant material.

Special material hazards: none.

General process hazards

B. oxidation reaction: 50 per cent

Special process hazards

B. Operation near explosive range, relies on instrumentation to prevent an explosive mixture entering the reactor: 100 per cent.
C. Low temperature storage of NH_3: 15 per cent.

FIRE AND EXPLOSION INDEX CALCULATION SHEET	UNIT: Complete plant		
1. MATERIAL FACTOR FOR:　　　Ammonia	———————————▶		5·7

2. SPECIAL MATERIAL HAZARDS	% FACTOR SUGGESTED	% FACTOR USED
A. OXIDISING MATERIALS	0–20	
B. REACTS WITH WATER TO PRODUCE A COMBUSTIBLE GAS	0–30	
C. SUBJECT TO SPONTANEOUS HEATING	30	
D. SUBJECT TO RAPID SPONTANEOUS POLYMERISATION	50–75	
E. SUBJECT TO EXPLOSIVE DECOMPOSITION	125	
F. SUBJECT TO DETONATION	150	
G. OTHER	0–150	
ADD PERCENTAGES A–G FOR SPECIAL MATERIAL HAZARD (S.M.H)　　　TOTAL		0·0

$((100 + \text{S.M.H. TOTAL})/100) \times (\text{MATERIAL FACTOR})$ = SUB-TOTAL No. 2 ——————▶　　　5·7

3. GENERAL PROCESS HAZARDS		
A. HANDLING AND PHYSICAL CHANGES ONLY	0–50	
B. CONTINUOUS REACTIONS	25–50	50
C. BATCH REACTIONS	25–60	
D. MULTIPLICITY OF REACTIONS IN SAME EQUIPMENT	0–50	
ADD PERCENTAGES A–D FOR GENERAL PROCESS (G.P.H.)　　　TOTAL		50

$((100 + \text{G.P.H. TOTAL})/100) \times (\text{SUB-TOTAL No. 2})$ = SUB-TOTAL No. 3 ——————▶　　　8·6

4. SPECIAL PROCESS HAZARDS		
A. LOW PRESSURE (BELOW 1 BAR)	0–100	
B. OPERATION IN OR NEAR EXPLOSION RANGE	0–150	100
C. LOW TEMPERATURE: 1. (CARBON STEELS 10 to −30°C)	15	15
2. (BELOW −30°C)	25	
D. HIGH TEMPERATURE (USE ONE ONLY)		
1. (ABOVE FLASH POINT)	10–20	
2. (ABOVE BOILING POINT)	25	
3. (ABOVE AUTOIGNITION POINT)	35	
E. HIGH PRESSURE: 　1. (15–200 BAR)	30	
2. (ABOVE 200 BAR)	60	
F. PROCESSES OR REACTIONS DIFFICULT TO CONTROL	50–100	
G. DUST OR MIST HAZARD	30–60	
H. GREATER THAN AVERAGE EXPLOSION HAZARD	60–100	
I. LARGE QUANTITIES OF COMBUSTIBLE LIQUIDS		
(USE ONE ONLY)		
1. 10–25 m³	40–55	
2. 25–75 m³	55–75	
3. 75–200 m³	75–100	
4. ABOVE 200 m³	100+	
J. OTHER	0–20	
ADD PERCENTAGES A–J FOR SPECIAL PROCESS (S.P.H.)　　　TOTAL		115

$((100 + \text{S.P.H. TOTAL})/100) \times (\text{SUB-TOTAL No. 3})$ = FIRE & EXPLOSION INDEX ——————▶　　　18·5

Fig. 9.1a (Example 9.2)

The index works out at 18·5: classified as "Mild". Ammonia would not normally be considered a dangerously flammable material; the danger of an internal explosion in the reactor is the main process hazard. The toxicity of ammonia and the corrosiveness of nitric acid would also need to be considered in a full hazard evaluation.

9.5. Hazard and operability studies

A Hazard and Operability Study is a procedure for the systematic, critical, examination of the operability of a process. When applied to a process design or an operating plant, it indicates potential hazards that may arise from deviations from the intended design conditions.

The technique was developed by the Petrochemicals Division of Imperial Chemical Industries, see Lawley (1974), and is now in general use in the chemical and process industries.

The term "Operability Study" should more properly be used for this type of study, though it is usually referred to as a Hazard and Operability study, or HAZOP study. This can cause confusion with the term "Hazard Analysis", which is a technique for the quantitative assessment of a hazard, after it has been identified by an operability study, or similar technique. The Chemical Industries Association has published a *Guide to Hazard and Operability Studies*, CIA (1979b), which gives a comprehensive description of the technique and examples of its applications. Further examples are given by Lawley (1974), Wells (1980) and Austin and Jeffreys (1979).

A brief outline of the technique is given in this section to illustrate its use in process design. It can be used to make a preliminary examination of the design at the flow-sheet stage; and for a detailed study at a later stage, when a full process description, final flow-sheets, P & I diagrams, and equipment details are available.

9.5.1. Basic principles

A formal operability study is the systematic study of the design, vessel by vessel, and line by line, using "Guide Words" to help generate thought about the way deviations from the intended operating conditions can cause hazardous situations.

The seven guide words recommended in the CIA booklet are given in Table 9.4. In addition to these words, the following words are also used in a special way, and have the precise meanings given below:

Intention: the intention defines how the particular part of the process was intended to operate; the intention of the designer.
Deviations: these are departures from the designer's intention which are detected by the systematic application of the guide words.
Causes: reasons why, and how, the deviations could occur. Only if a deviation can be shown to have a realistic cause is it treated as meaningful.
Consequences: the results that follow from the occurrence of a meaningful deviation.
Hazards: consequences that can cause damage (loss) or injury.

The use of the guide words can be illustrated by considering a simple example. Figure 9.2 shows a chlorine vaporiser, which supplies chlorine at 2 bar to a chlorination reactor. The vaporiser is heated by condensing steam.

TABLE 9.4. *A list of guide words*

Guide words	Meanings	Comments
No or Not	The complete negation of these intentions	No part of the intentions is achieved but nothing else happens
More	Quantitative increases or decreases	These refer to quantities and properties such as flow rates and temperatures, as well as activities like "Heat" and "React"
Less		
As well as	A qualitative increase	All the design and operating intentions are achieved together with some additional activity
Part of	A qualitative decrease	Only some of the intentions are achieved; some are not
Reverse	The logical opposite of the intention	This is mostly applicable to activities, for example reverse flow or chemical reaction. It can also be applied to substances, e.g. "Poison" instead of "Antidote" or "D" instead of "L" optical isomers
Other than	Complete substitution	No part of the original intention is achieved. Something quite different happens

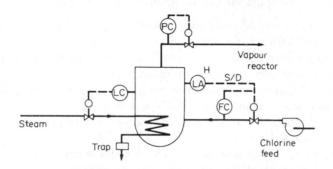

Fig. 9.2. Chlorine vaporiser instrumentation

Consider the steam supply line and associated control instrumentation. The designer's intention is that steam shall be supplied at a pressure and flow rate to match the required chlorine demand.

Apply the guide word No:

Possible deviation—no steam flow.

Possible causes—blockage, valve failure (mechanical or power), failure of steam supply (fracture of main, boiler shut-down).

Clearly this is a meaningful deviation, with several plausible causes.

Consequences—the main consequence is loss of chlorine flow to the chlorination reactor. The effect of this on the reactor operation would have to be considered. This would be brought out in the operability study on the reactor; it would be a possible cause of no chlorine flow.

Apply the guide word MORE:

Possible deviation—more steam flow.

Possible cause—valve stuck open.

Consequences—low level in vaporiser (this should activate the low level alarm), higher rate
 of flow to the reactor.
 Note: to some extent the level will be self-regulating, as the level falls the heating
 surface is uncovered.

Hazard—depends on the possible effect of high flow on the reactor.

Possible deviation—more steam pressure (increase in mains pressure).

Possible causes—failure of pressure-regulating valves.

Consequences—increase in vaporisation rate. Need to consider the consequences of the
 heating coil reaching the maximum possible steam system pressure.

Hazard—rupture of lines (unlikely), effect of sudden increase in chlorine flow on reactor.

9.5.2. Explanation of guide words

The basic meaning of the guide words in Table 9.4. The meaning of the words NO/NOT,
MORE and LESS are easily understood; the NO/NOT, MORE and LESS could, for example,
refer to flow, pressure, temperature, level and viscosity. All circumstances leading to NO
flow should be considered, including reverse flow.

The other words need some further explanation:

AS WELL AS: something in addition to the design intention; such as, impurities, side-
 reactions, ingress of air, extra phases present.

PART OF: something missing, only part of the intention realised; such as, the change in
 composition of a stream, a missing component.

REVERSE: the reverse of, or opposite to, the design intention. This could mean reverse flow
 if the intention was to transfer material. For a reaction, it could mean the reverse
 reaction. In heat transfer, it could mean the transfer of heat in the opposite direction
 to what was intended.

OTHER THAN: an important and far-reaching guide word, but consequently more vague in
 its application. It covers all conceivable situations other than that intended; such as,
 start-up, shut-down, maintenance, catalyst regeneration and charging, failure of plant
 services.

When referring to time, the guide words SOONER THAN and LATER THAN can also be used.

9.5.3. Procedure

An operability study would normally be carried out by a team of experienced people,
who have complementary skills and knowledge; led by a team leader who is experienced in
the technique.

The team examines the process vessel by vessel, and line by line, using the guide words to
detect any hazards.

The information required for the study will depend on the extent of the investigation.

A preliminary study can be made from a description of the process and the process flow-
sheets. For a detailed, final, study of the design, the flow-sheets, piping and instrument

diagrams, equipment specifications and layout drawings would be needed. For a batch process information on the sequence of operation will also be required, such as that given in operating instructions, logic diagrams and flow charts.

A typical sequence of events is shown in Fig. 9.3. After each line has been studied it is marked on the flow-sheet as checked.

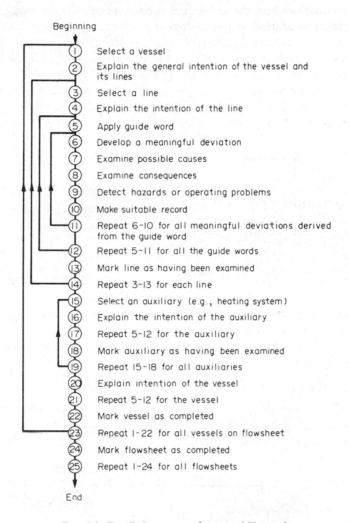

Beginning

1. Select a vessel
2. Explain the general intention of the vessel and its lines
3. Select a line
4. Explain the intention of the line
5. Apply guide word
6. Develop a meaningful deviation
7. Examine possible causes
8. Examine consequences
9. Detect hazards or operating problems
10. Make suitable record
11. Repeat 6-10 for all meaningful deviations derived from the guide word
12. Repeat 5-11 for all the guide words
13. Mark line as having been examined
14. Repeat 3-13 for each line
15. Select an auxiliary (e.g., heating system)
16. Explain the intention of the auxiliary
17. Repeat 5-12 for the auxiliary
18. Mark auxiliary as having been examined
19. Repeat 15-18 for all auxiliaries
20. Explain intention of the vessel
21. Repeat 5-12 for the vessel
22. Mark vessel as completed
23. Repeat 1-22 for all vessels on flowsheet
24. Mark flowsheet as completed
25. Repeat 1-24 for all flowsheets

End

FIG. 9.3. Detailed sequence of an operability study.

A written record is not normally made of each step in the study, only those deviations that lead to a potential hazard are recorded. If possible, the action needed to remove the hazard is decided by the team and recorded. If more information, or time, is needed to decide the best action, the matter is referred to the design group for action, or taken up at another meeting of the study team.

When using the operability study technique to vet a process design, the action to be taken to deal with a potential hazard will often be modifications to the control systems and instrumentation: the inclusion of additional alarms, trips, or interlocks. If major hazards are identified, major design changes may be necessary; alternative processes, materials or equipment.

Example 9.3

This example illustrates how the techniques used in an operability study can be used to decide the instrumentation required for safe operation. Figure 9.4a shows the

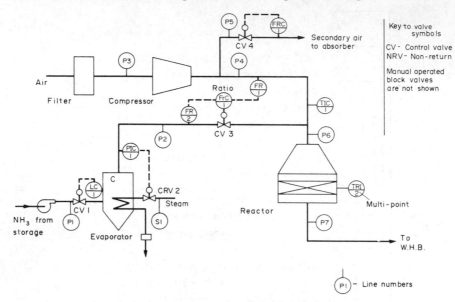

FIG. 9.4a. Nitric acid plant, reactor section basic instrumentation

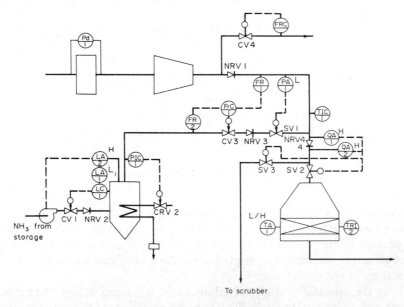

FIG. 9.4b. Nitric acid plant, reactor section full instrumentation

basic instrumentation and control systems required for the steady-state operation of the reactor section of the nitric acid process considered in Example 4.4. Figure 9.4*b* shows the additional instrumentation and safety trips added after making the operability study set out below. The instrument symbols used are explained in Chapter 5.

The most significant hazard of this process is the probability of an explosion if the concentration of ammonia in the reactor is inadvertently allowed to reach the explosive range, > 14 per cent.

Operability study

The sequence of steps shown in Fig. 9.3 is followed. Only deviations leading to action, and those having consequences of interest, are recorded.

Vessel – **Air Filter**
Intention – to remove particles that would foul the reactor catalyst

Guide word	Deviation	Cause	Consequences and action

Line No. P3
Intention – transfers clear air at atmospheric pressure and ambient temperature to compressor

LESS OF	Flow	Partially blocked filter	Possible dangerous increase in NH_3 concentration: measure and log pressure differential
AS WELL AS	Composition	Filter damaged, incorrectly installed	Impurities, possible poisoning of catalyst: proper maintenance

Vessel – **Compressor**
Intention – to supply air at 8 bar, 12,000 kg/h, 250°C, to the mixing tee

Line No. P4
Intention – transfers air to reactor (mixing tee)

NO/NONE	Flow	Compressor failure	Possible dangerous NH_3 conc.: low flow pressure alarm (PA1) interlocked to shut-down NH_3 flow
MORE	Flow	Failure of compressor controls	High rate of reaction, high reactor temperature: high-temperature alarms (TA1)
REVERSE	Flow	Fall in line press. (compressor fails) high pressure at reactor	NH_3 in compressor – explosion hazard: fit non-return valve (NRV1); hot wet acid gas – corrosion; fit second valve (NRV4)

Line No. P5
Intention – transfer secondary air to absorber

No	Flow	Compressor failure CV4 failure	Incomplete oxidation, air pollution from absorber vent: operating procedures
LESS	Flow	CV4 pluggage FRC1 failure	As no flow

Vessel – **Ammonia vaporiser**
Intention – evaporate liquid ammonia at 8 bar, 25°C, 731 kg/h

Line No. P1
Intention – transfer liquid NH_3 from storage

No	Flow	Pump failure CV1 fails	Level falls in vaporiser: fit low-level alarm (LA1)
LESS	Flow	Partial failure pump/valve	(LA1) alarms

Guide word	Deviation	Cause	Consequences and action
MORE	Flow	CV1 sticking, LC1 fails	Vaporiser floods, liquid to reactor: fit high-level alarm (LA2) with automatic pump shut-down
AS WELL AS	Water brine	Leakage into storages from refrigeration	Concentration of NH_4OH in vaporiser: routine analysis, maintenance
REVERSE	Flow	Pump fails, vaporiser press. higher than delivery	Flow of vapour into storages: (LA1) alarms; fit non-return valve (NRV2)

Line No. P2
Intention – transfers vapour to mixing tee

No	Flow	Failure of steam flow, CV3 fails closed	(LA1) alarms, reaction ceases: considered low flow alarm, rejected – needs resetting at each rate
LESS	Flow	Partial failure or blockage CV3	As no flow
	Level	LC1 fails	LA2 alarms
MORE	Flow	FR2/ratio control mis-operation	Danger of high ammonia concentration: fit alarm, fit analysers (duplicate) with high alarm 12 per cent NH_3 (QA1, QA2)
	Level	LC1 fails	LA2 alarms
REVERSE	Flow	Steam failure	Hot, acid gases from reactor – corrosion: fit non-return valve (NRV3)
Line S1 (auxiliary)		CRV2 fails, trap frozen	High level in vaporiser: LA2 actuated

*Vessel – **Reactor***
Intention – oxidises NH_3 with air, 8 bar, 900°C

Line No. P6
Intention – transfers mixture to reactor, 250°C

No	Flow	NRV4 stuck closed	Fall in reaction rate: fit low temp. alarm (TA1)
LESS	Flow	NRV4 partially closed	As No
	NH_3 concentration	Failure of ratio control	Temperatures fall: TA1 alarms (consider low conc. alarm on QA1, 2)
MORE	NH_3 conc.	Failure of ratio control, air flow restricted	High reactor temp.: TA1 alarms 14 per cent explosive mixture enters reactor – disaster: include automatic shut-down/by-pass actuated by QA1, 2, SV2, SV3
	Flow	Control systems failure	High reactor temp.: TA1 alarms

Line No. P7
Intention – transfers reactor products to waste-heat boiler

AS WELL AS	Composition	Refractory particles from reactor	Possible pluggage of boiler tubes: install filter up-stream of boiler

9.6. Hazard analysis

An operability study will identify potential hazards, but gives no guidance on the likelihood of an incident occurring, or the loss suffered; this is left to the intuition of the team members. Incidents usually occur through the coincident failure of two or more items; failure of equipment, control systems and instruments, and mis-operation. The sequence of events that leads to a hazardous incident can be shown as a fault tree (logic tree), such as that shown in Fig. 9.5. This figure shows the set of circumstances that would result in the flooding of the chlorine vaporiser shown in Fig. 9.2. The AND symbol is used where coincident inputs are necessary before the system fails, and the OR symbol where

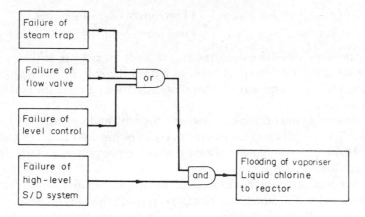

FIG. 9.5. Simple fault chart (logic diagram)

failure of any input, by itself, would cause failure of the system. A fault tree is analogous to the type of logic diagram used to represent computer operations, and the symbols are analogous to logic AND and OR gates.

The fault trees for even a simple process unit will be complex, with many branches. Fault trees are used to make a quantitive assessment of the likelihood of failure of a system, using data on the reliability of the individual components of the system. For example, if the following figures represent an estimate the probability of the events shown in Fig. 9.5 happening, the probability of failure of the total system by this route can be calculated.

	Probability of failure $\times 10^3$
Steam trap	1
Flow control valve	0·1
Level control, sub-system	0·5
High level shut-down, sub-system	0·04

The probabilities are added for OR gates, and multiplied for AND gates; so the probability of flooding the vaporiser is given by:

$$(1 + 0·1 + 0·5)\,10^{-3} \times 0·04 \times 10^{-3} = 0·06 \times 10^{-6}$$

The data on probabilities given in this example are for illustration only, and do not represent actual data for these components. Some quantitive data on the reliability of instruments and control systems is given by Lees (1976). Examples of the application of quantitive hazard analysis techniques in chemical plant design are given by Lawley (1974), Kletz (1971), Wells (1980), Gibson (1976) and Prugh (1980). Much of the work on the development of hazard analysis techniques, and the reliability of equipment, has been done in connection with the development of the nuclear energy programmes in the USA (USAEC, 1975) and the UK.

9.7. Acceptable risk and safety priorities

If the consequences of an incident can be predicted quantitatively (property loss and the possible number of fatalities), then a quantitive assessment can be made of the risk.

$$\begin{matrix} \text{Quantitive assessment} \\ \text{of risk} \end{matrix} = \begin{Bmatrix} \text{Frequency of} \\ \text{incident} \end{Bmatrix} \times \begin{Bmatrix} \text{loss per} \\ \text{incident} \end{Bmatrix}$$

If the loss can be measured in money, the cash value of the risk can be compared with the cost of safety equipment or design changes to reduce the risk. In this way, decisions on safety can be made in the same way as other design decisions: to give the best return of the money invested.

Hazards invariably endanger life as well as property, and any attempt to make cost comparisons will be difficult and controversial. It can be argued that no risk to life should be accepted. However, resources are always limited and some way of establishing safety priorities is needed.

One approach is to compare the risks, calculated from a hazard analysis, with risks that are generally considered acceptable; such as, the average risks in the particular industry, and the kind of risks that people accept voluntarily. One measure of the risk to life is the "Fatal Accident Frequency Rate" (FAFR), defined as the number of deaths per 10^8 working hours. This is equivalent to the number of deaths in a group of 1000 men over their working lives. The FAFR can be calculated from statistical data for various industries and activities; some of the published values are shown in Tables 9.5 and 9.6. Table 9.5 shows the relative position of the chemical industry compared with other industries; Table 9.6 gives values for some of the risks that people accept voluntarily.

In the chemical process industries it is generally accepted that risks with an FAFR greater than 0·4 (one-tenth of the average for the industry) should be eliminated as a matter of priority, the elimination of lesser risks depending on the resources available; see Kletz (1977a). This criterion is for risks to employees; for risks to the general public (undertaken involuntarily) a lower criterion must be used. The level of risk to which the

TABLE 9.5. *FAFR for some UK industries*

	FAFR
Manufacturing (1971–74)	2
Chemical (1971–74)	4
Construction (1971–74)	9
Mining and quarrying (1971–73)	10
Air crews in flight (1964–73)	200
Professional boxers in the ring (1963–74)	20,000

TABLE 9.6. *FAFR for some non-industrial activities*

Travelling by train (1963–72)	3
Staying at home (1972)	4
Travelling by bus (1963–73)	4
Travelling by car (1963–73)	48
Travelling by air (1964–73)	190
Canoeing (1962–72)	670
Gliding (1964–73)	1000
Motor cycling (1963–73)	1040

Source: Gibson (1976).

public outside the factory gate should be exposed by the operations will always be a matter of debate and controversy. Kletz (1977b) suggests that a hazard can be considered acceptable if the average risk is less than one in 10 million, per person, per year. This is equivalent to a FAFR of 0·001; about the same as deaths from the bites of venomous creatures in the UK, or the chance of being struck by lightning.

9.8. Safety check lists

Check lists are useful aids to memory. A check list that has been drawn up by experienced engineers can be a useful guide for the less experienced. However, too great a reliance should never be put on the use of check lists, to the exclusion of all other considerations and techniques. No check list can be completely comprehensive, covering all the factors to be considered for any particular process or operation.

A short safety check list, covering the main items which should be considered in process design, is given below.

More detailed lists have been published by the Institution of Chemical Engineers, IChemE (1976), Wells (1980).

Balemans (1974) gives a comprehensive list of guidelines for the safe design of chemical process plants, drawn up in the form of a check list.

The lists in the Dow Safety Guide, the general and specific hazards, and the preventive and protective features, can also be used as a check list of the factors to be considered in design. The guide also includes a detailed "Engineer's check list".

Design safety check list

Materials
    (a) flash-point
    (b) flammability range
    (c) autoignition temperature
    (d) composition
    (e) stability (shock sensitive?)
    (f) toxicity, TLV
    (g) corrosion
    (h) physical properties (unusual?)
    (i) heat of combustion/reaction

Process
1. Reactors
    (a) exothermic—heat of reaction
    (b) temperature control—emergency systems
    (c) side reactions—dangerous?
    (d) effect of contamination
    (e) effect of unusual concentrations (including catalyst)
    (f) corrosion
2. Pressure systems
    (a) need?
    (b) design to current codes (BS 5500)

(c) materials of construction—adequate?
(d) pressure relief—adequate?
(e) safe venting systems
(f) flame arresters

Control systems
(a) fail safe
(b) back-up power supplies
(c) high/low alarms and trips on critical variables
   (i) temperature
   (ii) pressure
   (iii) flow
   (iv) level
   (v) composition
(d) back-up/duplicate systems on critical variables
(e) remote operation of valves
(f) block valves on critical lines
(g) excess-flow valves
(h) interlock systems to prevent mis-operation
(i) automatic shut-down systems

Storages
(a) limit quantity
(b) inert purging/blanketing
(c) floating roof tanks
(d) dykeing
(e) loading/unloading facilities—safety
(f) earthing
(g) ignition sources—vehicles

General
(a) inert purging systems needed
(b) compliance with electrical codes
(c) adequate lighting
(d) lightning protection
(e) sewers and drains adequate, flame traps
(f) dust-explosion hazards
(g) build-up of dangerous impurities—purges
(h) plant layout
   (i) separation of units
   (ii) access
   (iii) siting of control rooms and offices
   (iv) services
(i) safety showers, eye baths

Fire protection
(a) emergency water supplies
(b) fire mains and hydrants
(c) foam systems

(d) sprinklers and deluge systems

(e) insulation and protection of structures

(f) access to buildings

(g) fire-fighting equipment

The check list is intended to promote thought; to raise questions such as: is it needed, what are the alternatives, has provision been made for, check for, has it been provided?

9.9. References

AUSTIN, D. G. and JEFFERYS, G. V. (1979) *The Manufacture of Methyl Ethyl Ketone from 2-Butanol* (IChemE/Godwin).

BALEMANS, A. W. M. (1974) Check-lists: guide lines for safe design of process plants. *Loss Prevention and Safety Promotion in the Process Industries*, C. H. Bushmann (Ed.) (Elsevier).

BROWNING, E. (1965) *Toxicity and Metabolism of Industrial Solvents* (Elsevier).

CIA (1979a) *Process Plant Hazards and Control Building Design* (Chemical Industries Association, London).

CIA (1979b) *A Guide to Hazard and Operability Studies* (Chemical Industries Association, London).

CONNISON, J. (1960) *Chem. Eng., Albany* **67** (July 25th) 113. How to design a pressure relief system.

DOW CHEMICAL CO. (1973) The Dow Safety Guide, a reprint from *Chemical Engineering Progress* (AIChE).

DUXBURY, H. A. (1976) *Loss Prevention* No. 10 (AIChE) 147. Gas vent sizing methods.

DUXBURY, H. A. (1979) *Chem. Engr. London* No. 350 (Nov.) 783. Relief line sizing for gases.

FITZSIMMONS, P. E. and COCKRAM, M. D. (1979) *Northwestern Branch Paper No. 2* (IChemE, London). The safe venting of chemical reactors.

GIBSON, S. B. (1976) *Inst. Chem. Eng. Sym. Ser.* No. 47, 135. The design of new chemical plants using hazard analysis.

GUGAN, K. (1979) *Unconfined Vapour Cloud Explosions* (IChemE/Godwin).

HADLEY, W. (Ed.) (1969) *Industrial Safety Handbook* (McGraw-Hill).

HMSO (1968) *The Ionising Radiation (unsealed radioactive substances) Regulations* (Stationery Office).

HMSO (1969) *The Ionising Radiation (sealed sources) Regulations* (Stationery Office).

HMSO (1975) *The Flixborough Disaster, Report of the Court of Enquiry* (Stationery Office).

IChemE (1976) *Flowsheeting for Safety* (Institution of Chemical Engineers, London).

IChemE (1977) *A First Guide to Loss Prevention* (Institution of Chemical Engineers, London).

ISSACS, M. (1971) *Chem. Eng., Albany* **78** (Feb. 22nd) 113. Pressure relief systems.

KAYSER, D. S. (1972) *Loss Prevention* No. 6 (AIChE) 82. Rupture disc selection.

KING, R. and MAGID, J. (1979) *Industrial Hazard and Safety Handbook* (Newnes–Butterworths).

KLETZ, T. A. (1971) *Inst. Chem. Eng. Sym. Ser.* No. 34, 75. Hazard analysis – a quantitative approach to safety.

KLETZ, T. A. (1977a) *New Scientist* (May 12th) 320. What risks should we run.

KLETZ, T A. (1977b) *Hyd. Proc.* **56** (May) 207. Evaluate risk in plant design.

KUSNETZ, H. L. (1974) *Loss Prevention* No. 8 (AIChE) 20. Industrial hygiene factors in design and operating practice.

LAWLEY, H. G. (1974) *Loss Prevention* No. 8 (AIChE) 105. Operability studies and hazard analysis.

LEES, F. P. (1976) *Inst. Chem. Eng. Sym. Ser.* No. 47, 73. A review of instrument failure data.

LEES, F. P. (1980) *Loss Prevention in the Process Industries*, 2 vols. (Butterworths).

LEWIS, D. J. (1979a) *AIChE Loss Prevention Symposium, Houston, April.* The Mond fire, explosion and toxicity index: a development of the Dow index.

LEWIS, D. J. (1979b) *Loss Prevention* No. 13 (AIChE) 20. The Mond fire, explosion and toxicity index applied to plant layout and spacing.

LIPCOMBE, D. M. and TAYLOR, A. C. (1978) *Noise Control Handbook of Principles and Practice* (Van Nostrand).

LOWRANCE, W. W. (1976) *Of Acceptable Risk* (W. Kaufmann, USA).

MUNDAY, G. (1976) *Chem. Engr. London* No. 308 (April) 278. Unconfined vapour explosions.

NAPIER, D. H. (1971) *Inst. Chem. Eng. Sym. Ser.* **34**, 170. Static electrification in the process industries.

NAPIER, D. H. and RUSSELL, D. A. (1974) *Proc. First Int. Sym. on Loss Prevention* (Elsevier). Hazard assessment and critical parameters relating to static electrification in the process industries.

PALMER, K. N. (1973) *Dust Explosions and Fires* (Chapman & Hall).

PARKINSON, J. S. (1979) *Inst. Chem. Eng. Sym. Design 79*, K1. Assessment of plant pressure relief systems.

PPA (1973) *Noise Control* (Process Plant Association, London).

PRUGH, R. N. (1980) *Chem. Eng. Prog.* **76** (July) 59. Applications of fault tree analysis.

ROGOWSKI, Z. W. (1980) *Inst. Chem. Eng. Sym. Ser.* No. 58, 53. Flame arresters in industry.

ROSPA (1971) *Liquid Flammable Gases; Storage and Handling* (Royal Society for the Prevention of Accidents, London).

SAX, N. I. (1975) *Dangerous Properties of Industrial Materials*, 4th ed. (Reinhold).

SHARLAND, I. (1972) *Woods Practical Guide to Noise Control* (Woods Acoustics Ltd., England).

USAEC (1975) *Reactor Safety Study*, WASH-1400 (United States Atomic Energy Commission).

WARRING, R. H. (1974) *Handbook of Noise and Vibration Control*, 2nd ed. (Trade and Technical Publications).

WELLS, G. L. (1980) *Safety in Process Plant Design* (IChemE/Godwin).

British Standards

BS 229: 1957 Flameproof enclosure of electrical apparatus.

BS 1259: 1958 Intrinsically safe electrical apparatus and circuits for use in explosive atmospheres.

BS 2839: 1979 Method for determination of flashpoint of petroleum products by Pensky–Martens closed tester.

BS 2915: 1974 Bursting discs and bursting disc assemblies.

BS 3489: 1962 Sound level meters (industrial grade).

BS 4197: 1967 A precision sound level meter.

BS 4683: ---- Electrical apparatus for explosive atmospheres.
    Part 1: 1971 Classification of maximum surface temperature.
    Part 2: 1971 The construction and testing of flameproof enclosures of electrical apparatus.
    Part 3: 1972 Type of protection N.
    Part 4: 1973 Type of protection "e".

BS 4688: 1971 Method for determination of flash point (open) and fire point of petroleum products by the Pensky–Martens apparatus.

British Standard Codes of Practice

CP 1003: ---- Electrical apparatus and associated equipment for use in explosive atmosphere of gas or vapour other than mining applications.
    Part 1: 1964 Choice, installation and maintenance of flameproof and instrinsically safe equipment.
    Part 2: 1966 Methods of meeting the explosion hazard other than by the use of flameproof or intrinsically-safe equipment.
    Part 3: 1967 Division 2 areas.

CP 3013: 1974 Fire precautions in chemical plant.

BS 5345: ----- Code of practice for the selection, installation and maintenance of electrical apparatus for use in potentially explosive atmospheres (other than mining applications or explosive processing and manufacture).
    Part 1: 1976 Basic requirements for all parts of the code.
    Part 3: 1979 Installation and maintenance requirements for electrical apparatus with type of protection "d". Flameproof enclosure.
    Part 4: 1977 Installation and maintenance requirements for electrical apparatus with type of protection "i". Intrinsically safe electrical apparatus and systems.
    Part 6: 1978 Installation and maintenance requirements for electrical appliances apparatus with type of protection "e". Increased safety.

9.10. Nomenclature

Dimensions in MLT

		Dimensions in **MLT**
$-\Delta H_c^\circ$	Standard heat of combustion	L^2T^{-2}

CHAPTER 10

Equipment Selection, Specification and Design

10.1 Introduction

The first chapters of this book covered process design: the synthesis of the complete process as an assembly of units; each carrying out a specific process operation. In this and the following chapters, the selection, specification and design of the equipment required to carry out the function of these process units (unit operations) is considered in more detail. The equipment used in the chemical processes industries can be divided into two classes: proprietary and non-proprietary. Proprietary equipment, such as pumps, compressors, filters, centrifuges and dryers, is designed and manufactured by specialist firms. Non-proprietary equipment is designed as special, one-off, items for particular processes; for example, reactors, distillation columns and heat exchangers.

Unless employed by one of the specialist equipment manufacturers, the chemical engineer is not normally involved in the detailed design of proprietary equipment. His job will be to select and specify the equipment needed for a particular duty; consulting with the vendors to ensure that the equipment supplied is suitable. He may be involved with the vendor's designers in modifying standard equipment for particular applications; for example, a standard tunnel dryer designed to handle particulate solids may be adapted to dry synthetic fibres.

As was pointed out in Chapter 1, the use of standard equipment, whenever possible, will reduce costs.

Reactors, columns and other vessels are usually designed as special items for a given project. In particular, reactor designs are usually unique, except where more or less standard equipment is used; such as an agitated, jacketed, vessel.

Distillation columns, vessels and tubular heat exchangers, though non-proprietary items, will be designed to conform to recognised standards and codes; this reduces the amount of design work involved.

The chemical engineer's part in the design of "non-proprietary" equipment is usually limited to selecting and "sizing" the equipment. For example, in the design of a distillation column his work will typically be to determine the number of plates; the type and design of plate; diameter of the column; and the position of the inlet, outlet and instrument nozzles. This information would then be transmitted, in the form of sketches and specification sheets, to the specialist mechanical design group, or the fabricator's design team, for detailed design.

In this chapter the emphasis is put on equipment selection, rather than equipment design; as most of the equipment described is proprietary equipment. Design methods are given for some miscellaneous non-proprietary items. A brief discussion of reactor design is included to supplement that given in Volume 3. The design of two important classes of

305

equipment, columns and heat exchangers, is covered separately in Chapters 11 and 12. A great variety of equipment is used in the process industries, and it is only possible to give very brief descriptions of the main types in this volume. Further details are given in Volume 2; and descriptions and illustrations of most of the equipment used can be found in various handbooks (Perry and Chilton, 1973; Schweitzer, 1979; Mead, 1964; and Henglein, 1969). Equipment manufacturers' advertisements in the technical press should also be studied; and it is worthwhile building up a personal file of vendors' catalogues to supplement those that may be held in a firm's library. Two commercial organisations in the United Kingdom, EDACS Data Ltd. and Technical Data Ltd., publish indexes to equipment vendors' catalogues, and provide an information service.

The scientific principles and theory that underlie the design of and operation of processing equipment is covered in Volume 2.

10.2. Separation processes

As was discussed in Chapter 1, chemical processes consist essentially of reaction stages followed by separation stages in which the products are separated and purified.

TABLE 10.1. *Separation processes*

Numbers refer to the sections in this chapter. Processes in brackets are used for separating dissolved components (solutions). The terms major and minor component only apply where different phases are to be separated; i.e. not to those on the diagonal.

		MINOR COMPONENT				
		SOLID		LIQUID		GAS/VAPOUR
MAJOR COMPONENT	SOLID	Sorting Screening Hydrocyclones Classifiers Jigs Tables Centrifuges Dense media Flotation Magnetic Electrostatic	10.3 10.3.1 10.3.2 10.3.3 10.3.4 10.3.5 10.3.6 10.3.7 10.3.8 10.3.9 10.3.10	Pressing Drying	10.4.5 10.4.6	Crushing 10.10 Heating —
	LIQUID	Thickeners Clarifiers Hydrocyclones Filtration Centrifuges (Crystallisers) (Evaporators)	10.4.1 10.4.1 10.4.4 10.4.2 10.4.3 10.5.2 10.5.1	Decanters Coalescers (Solvent extraction) (Distillation) (Adsorption) (Ion exchange)	10.6.1 10.6.3 10.7.1 Chapter 11 Volume 3 Volume 3	(Stripping) Volume 2

	GAS/VAPOUR	Gravity
		  settlers
		Impingement
		  settlers
		Cyclones
		Filters
		Wet scrubbers
		Electrostatic
		  precipitators
			10.8.1
			
			10.8.2
			10.8.3
			10.8.4
			10.8.5
			
			10.8.6
				Separating
				  vessels
				Demisting pads
				Cyclones
				Wet scrubbers
				Electrostatic
				  precipitators
					10.9
					10.9
					10.8.3
					10.8.5
					
					10.8.6
						(Adsorption) Volume 3
						(Absorption) Volume 2

The main techniques used to separate phases, and the components within phases, are listed in Table 10.1 and discussed in Sections 10.3 to 10.9.

10.3. Solid–solid separations

Processes and equipment are required to separate valuable solids from unwanted material, and for size grading (classifying) solid raw materials and products.

The equipment used for solid–solid separation processes was developed primarily for the minerals processing and metallurgical industries for the beneficiation (up-grading) of ores. The techniques used depend on differences in physical, rather than chemical, properties, though chemical additives may be used to enhance separation. The principal techniques used are shown in Fig. 10.1; which can be used to select the type of processes likely to be suitable for a particular material and size range.

Sorting material by appearance, by hand, is now rarely used due to the high cost of labour.

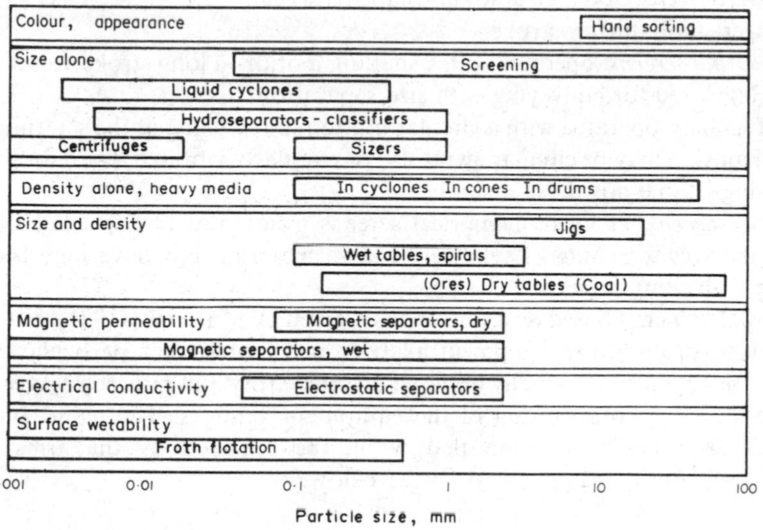

FIG. 10.1. A particle size selection guide to solid–solid separation techniques and equipment (after Roberts *et al.* 1971)

10.3.1. Screening (sieving)

The methods used for laboratory particle size analysis are discussed in detail in Volume 2, Chapter 1.

Screens separate particles on the basis of size. Their main application is in grading raw materials and products into size ranges, but they are also used for the removal of trash (over-and under-sized contaminants) and for dewatering. Industrial screening equipment is used over a wide range of particle sizes, from fine powders to large rocks. For small particles woven cloth or wire screens are used, and for larger sizes, perforated metal plates or grids.

Screen sizes are defined in two ways: by a mesh size number for small sizes and by the actual size of opening in the screen for the larger sizes. There are several different standards in use for mesh size, and it is important to quote the particular standard used when specifying particle size ranges by mesh size. In the UK the appropriate British Standards should be used; BS 410 and BS 1796. A comparison of the various international standard sieve mesh sizes is given in Volume 2, Chapter 1.

The simplest industrial screening equipment are stationary screens, over which the material to be screened flows. Typical of this type are "Grizzly" screens, which consist of rows of equally spaced parallel bars, and which are used to "scalp" off over-sized rocks in the feed to crushers.

Dynamic screening equipment can be categorised according to the type of motion used to shake-up and transport the material on the screen. The principal types used in the chemical process industries are described briefly below.

Vibrating screens: horizontal and inclined screening surfaces vibrated at high frequencies (1000 to 7000 Hz). High capacity units, with good separating efficiency, which are used for a wide range of particle sizes.

Oscillating screens: operated at lower frequencies than vibrating screens (100–400 Hz) with a longer, more linear, stroke.

Reciprocating screens: operated with a shaking motion, a long stroke at low frequency (20–200 Hz). Used for conveying with size separation.

Shifting screens: operated with a circular motion in the plane of the screening surface. The actual motion may be circular, gyratory, or circularly vibrated. Used for the wet and dry screening of fine powders.

Revolving screens: inclined, cylindrical screens, rotated at low speeds (10–20 rpm). Used for the wet screening of relatively coarse material, but have now been largely replaced by vibrating screens.

Figure 10.2, which is based on a similar chart given by Matthews (1971), can be used to select the type of screening equipment likely to be suitable for a particular size range. Equipment selection will normally be based on laboratory and pilot scale screening tests, conducted with the co-operation of the equipment vendors. The main factors to be considered, and the information that would be required by the firms supplying proprietary screening equipment, are listed below:

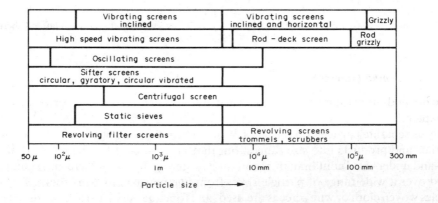

FIG. 10.2. Screen selection by particle size range

1. Rate, throughput required.
2. Size range (test screen analysis).
3. Characteristics of the material: free-flowing or sticky, bulk density, abrasiveness.
4. Hazards: flammability, toxicity, dust explosion.
5. Wet or dry screening to be used.

10.3.2. Liquid–solid cyclones

Cyclones can be used for the classification of solids, as well as for liquid–solid, and liquid–liquid separations. The design and application of liquid cyclones (hydrocyclones) is discussed in Section 10.4.4. A typical unit is shown in Fig. 10.3.

Liquid cyclones can be used for the classification of solids particles over a size range from 5 to 100 μm. Commercial units are available in a wide range of materials of construction and sizes; from as small as 10 mm to up to 30 m diameter. The separating efficiency of liquid cyclones depends on the particle size and density, and the density and viscosity of the liquid medium.

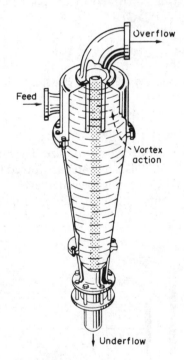

FIG. 10.3. Liquid–solid cyclone (hydrocyclone)

10.3.3. Hydroseparators and sizers (classifiers)

Classifiers that depend on the difference in the settling rates of different size particles in water are frequently used for separating fine particles, in the 50 to 300 μm range. Various designs are used. The principal ones used in the chemical process industries are described below.

Thickeners: thickeners are primarily used for liquid-solid separation (see Section 10.4). When used for classification, the feed rate is such that the overflow rate is greater than the settling rate of the slurry, and the finer particles remain in the overflow stream.

Rake classifiers: are inclined, shallow, rectangular troughs, fitted with mechanical rakes at the bottom to rake the deposited solids to the top of the incline (Fig. 10.4). Several rake classifiers can be used in series to separate the feed into different size ranges.

Bowl classifiers: are shallow bowls with concave bottoms, fitted with rakes. Their operation is similar to that of thickeners.

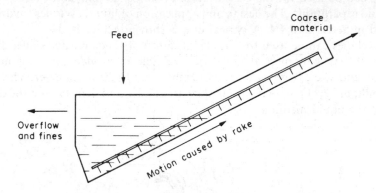

FIG. 10.4. Rake classifier

10.3.4. Hydraulic jigs

Jigs separate solids by difference in density and size. The material is immersed in water, supported on a screen (Fig. 10.5). Pulses of water are forced through the bed of material, either by moving the screen or by pulsating the water level. The flow of water fluidises the bed and causes the solids to stratify with the lighter material at the top and the heavier at the bottom.

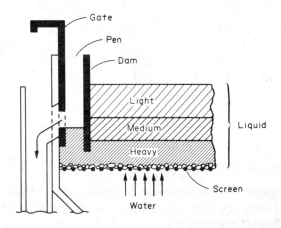

FIG. 10.5. A hydraulic jig

10.3.5. Tables

Tables are used wet and dry. The separating action of a wet table resembles that of the traditional miner's pan. Riffled tables (Fig. 10.6) are basically rectangular decks, inclined at a shallow angle to the horizontal (2 to 5°), with shallow slats (riffles) fitted to the surface. The table is mechanically shaken, with a slow stroke in the forward direction and a faster backward stroke. The particles are separated into different size ranges under the combined action of the vibration, water flow, and the resistance to flow over the riffles.

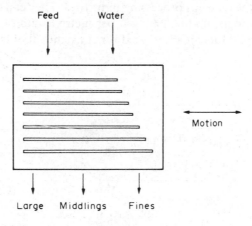

FIG. 10.6. Wilfley riffled table

10.3.6. Classifying centrifuges

Centrifuges are used for the classification of particles in size ranges below 10 μm. Two types are used: solid bowl centrifuges, usually with a cylindrical, conical bowl, rotated about a horizontal axis; and "nozzle" bowl machines, fitted with discs.

These types are described in Section 10.4.3.

10.3.7. Dense-medium separators (sink and float processes)

Solids of different densities can be separated by immersing them in a fluid of intermediate density. The heavier solids sink to the bottom and the lighter float to the surface. Water suspensions of fine particles are often used as the dense liquid (heavy-medium). The technique is used extensively for the benefication (concentration) of mineral ores.

10.3.8. Flotation separators (froth-flotation)

Froth-flotation processes are used extensively for the separation of finely divided solids. Separation depends on differences in the surface properties of the materials. The particles are suspended in an aerated liquid (usually water), and air bubbles adhere preferentially to the particles of one component and bring them to the surface. Frothing agents are used so that the separated material is held on the surface as a froth and can be removed.

Froth-flotation is an extensively used separation technique, having a wide range of applications in the minerals processing industries and other industries. It can be used for particles in the size range from 50 to 400 μm.

10.3.9. Magnetic separators

Magnetic separators can be used for materials that are affected by magnetic fields; the principle is illustrated in Fig. 10.7. Rotating-drum magnetic separators are used for a wide range of materials in the minerals processing industries. They can be designed to handle relatively high throughputs, up to 3000 kg/h per metre length of drum.

Simple magnetic separators are often used for the removal of iron from the feed to a crusher.

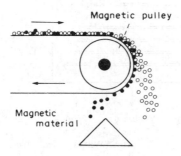

FIG. 10.7. Magnetic separator

10.3.10. Electrostatic separators

Electrostatic separation depends on differences in the electrical properties (conductivity) of the materials to be treated. In a typical process the material particles pass through a high-voltage electric field as it is fed on to a revolving drum, which is at earth potential (Fig. 10.8). Those particles that acquire a charge adhere to the drum surface and are carried further around the drum before being discharged.

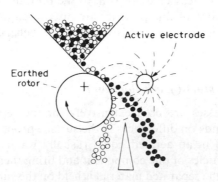

FIG. 10.8. Electrostatic separator

10.4. Liquid–solid (solid–liquid) separators

The need to separate solid and liquid phases is probably the most common phase separation requirement in the process industries, and a variety of techniques is used (Fig. 10.9). Separation is effected by either the difference in density between the liquid and

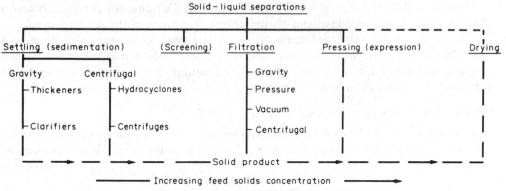

FIG. 10.9. Solid–liquid separation techniques

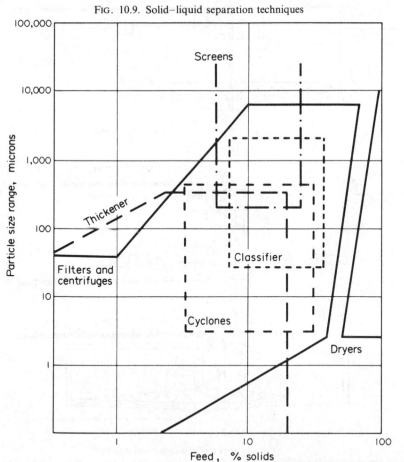

FIG. 10.10. Solid–liquid separation techniques (after Dahlstrom and Cornell, 1971)

solids, using either gravity or centrifugal force, or, for filtration, depends on the particle size and shape. The most suitable technique to use will depend on the solids concentration and feed rate, as well as the size and nature of the solid particles. The range of application of various techniques and equipment, as a function of slurry concentration and particle size, is shown in Fig. 10.10.

The choice of equipment will also depend on whether the prime objective is to obtain a clear liquid or a solid product, and on the degree of dryness of the solid required.

The design, construction and application of thickeners, centrifuges and filters is a specialised subject, and firms who have expertise in these fields should be consulted when selecting and specifying equipment for new applications. The theory of sedimentation processes is covered in Volume 2, Chapter 5.

10.4.1. Thickeners and clarifiers

Thickening and clarification are sedimentation processes, and the equipment used for the two techniques are similar. The primary purpose of thickening is to increase the

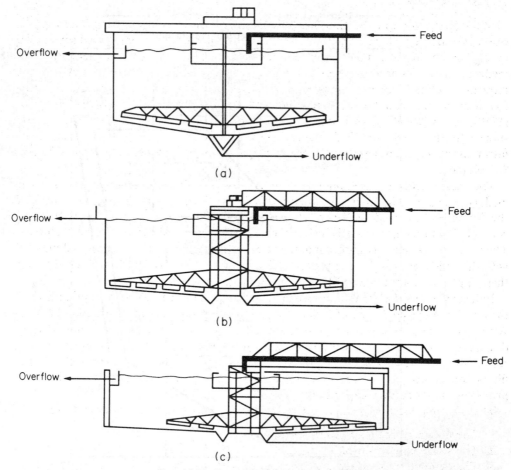

FIG. 10.11. Types of thickener and clarifier
(a) Bridge supported (up to < 40 m dia.) (b) Centre column supported (< 30 m dia.) (c) Traction driven (< 60 m dia.)

concentration of a relatively large quantity of suspended solids; whereas that of clarifying, as the name implies, is to remove a small quantity of fine solids to produce a clear liquid effluent. Thickening and clarification are relatively cheap processes when used for the treatment of large volumes of liquid.

A thickener, or clarifier, consists essentially of a large circular tank with a rotating rake at the base. Rectangular tanks are also used, but the circular design is preferred. They can be classified according to the way the rake is supported and driven. The three basic designs are shown in Fig. 10.11. Various designs of rake are used, depending on the nature of the solids.

The design and construction of thickeners and clarifiers is described by Dahlstrom and Cornell (1971).

Flocculating agents are often added to promote the separating performance of thickeners.

10.4.2. Filtration

In filtration processes the solids are separated from the liquid by passing (filtering) the slurry through some form of porous filter medium. Filtration is a widely used separation process in the chemical and other process industries. Many types of equipment and filter media are used; designed to meet the needs of particular applications. Descriptions of the filtration equipment used in the process industries and their fields of application can be found in various handbooks; Perry and Chilton (1973), Schweitzer (1979), Henglein (1969), Mead (1964), Cremer and Davies (1956), and in several specialist texts on the subject, Suttle (1969), Orr (1977), Purchas (1967), Wakeman (1975). A short discussion of filtration theory and descriptions of the principal types of equipment is given in Chapter 9, Volume 2.

The most commonly used filter medium is woven cloth, but a great variety of other media is also used. The main types are listed in Table 10.2. A comprehensive discussion of the factors to be considered when selecting filter media is given by Purchas (1971) and Mais (1971). Filter aids are often used to increase the rate of filtration of difficult slurries. They are either applied as a precoat to the filter cloth or added to the slurry, and deposited with the solids, assisting in the formation of a porous cake. The various filter aids and their application are discussed by Purchas (1967).

Industrial filters use vacuum, pressure, or centrifugal force to drive the liquid (filtrate) through the deposited cake of solids. Filtration is essentially a discontinuous process. With batch filters, such as plate and frame presses, the equipment has to be shut down to discharge the cake; and even with those filters designed for continuous operation, such as rotating-drum filters, periodic stoppages are necessary to change the filter cloths. Batch filters can be coupled to continuous plant by using several units in parallel, or by providing buffer storage capacity for the feed and product.

The principal factors to be considered when selecting filtration equipment are:

1. The nature of the slurry and the cake formed.
2. The solids concentration in the feed.
3. The throughput required.
4. The nature and physical properties of the liquid: viscosity, flammability, toxicity, corrosiveness.

5. Whether cake washing is required.
6. The cake dryness required.
7. Whether contamination of the solid by a filter aid is acceptable.
8. Whether the valuable product is the solid or the liquid, or both.

The overriding factor will be the filtration characteristics of the slurry; whether it is fast filtering (low specific cake resistance) or slow filtering (high specific cake resistance). The filtration characteristics can be determined by laboratory or pilot plant tests. A guide to filter selection by the slurry characteristics is given in Table 10.3; which is based on a similar selection chart given by Porter *et al.* (1971).

The principal types of industrial scale filter used are described briefly below.

TABLE 10.2. *Filter media*

Type	Examples	Minimum size particle trapped (μm)
1. Solid fabrications	Scalloped washers Wire–wound tubes	 5
2. Rigid porous media	Ceramics, stoneware Sintered metal	1 3
3. Metal	Perforated sheets Woven wire	100 5
4. Porous plastics	Pads, sheets Membranes	3 0·005
5. Woven fabrics	Natural and synthetic fibre cloths	 10
6. Non-woven sheets	Felts, lap Paper, cellulose	10 5
7. Cartridges	Yarn-wound spools, graded fibres	 2
8. Loose solids	Fibres, asbestos, cellulose	sub-micron

Nutsche (gravity and vacuum operation)

This is the simplest type of batch filter. It consists of a tank with a perforated base, which supports the filter medium.

Plate and frame press (pressure operation) (Fig. 10.12)

The oldest and most commonly used batch filter. Versatile equipment, made in a variety of materials, and capable of handling viscous liquids and cakes with a high specific resistance.

TABLE 10.3 *Guide to filter selection*

Slurry characteristics	Fast filtering	Medium filtering	Slow filtering	Dilute	Very dilute
Cake formation rate	cm/s	mm/s	0·02–0·12 mm/s	0·02 mm/s	No cake
Normal concentration	>20%	10–20%	1–10%	<5%	<0·1%
Settling rate	Very fast	Fast	Slow	Slow	—
Leaf test rate, kg/h m²	>2500	250–2500	25–250	<25	—
Filtrate rate, m³/h m²	>10	5–10	0·02–0·05	0·02–5	0·02–5

Filter application
Continuous vacuum filters
Multicompartment drum
Single compartment drum
Top feed drum
Scroll discharge drum
Tilting pan
Belt
Disc

Batch vacuum leaf
Batch nutsche
Batch pressure filters
Plate and frame
Vertical leaf
Horizontal plate
Cartridge edge

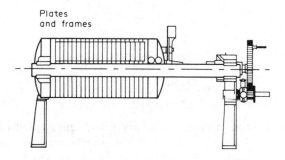

FIG. 10.12. Plate and frame filter press

Leaf filters (pressure and vacuum operation)

Various types of leaf filter are used, with the leaves arranged in horizontal or vertical rows. The leaves consist of metal frames over which filter cloths are draped. The cake is removed either mechanically or by sluicing it off with jets of water. Leaf filters are used for similar applications as plate and frame presses, but generally have lower operating costs.

Rotary drum filters (usually vacuum operation) (Fig. 10.13)

A drum filter consists essentially of a large hollow drum round which the filter medium is fitted. The drum is partially submerged in a trough of slurry, and the filtrate sucked through the filter medium by vacuum inside the drum. Wash water can be sprayed on to the drum surface and multicompartment drums are used so that the wash water can be kept separate from the filtrate. A variety of methods is used to remove the cake from the drum: knives, strings, air jets and wires. Rotating drum filters are essentially continuous in operation. They can handle large throughputs, and are widely used for free filtering slurries.

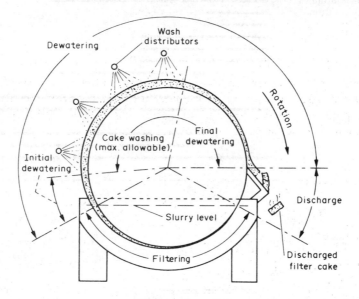

FIG. 10.13. Drum filter

Disc filters (pressure and vacuum operation)

Disc filters are similar in principle to rotary filters, but consist of several thin discs mounted on a shaft, in place of the drum. This gives a larger effective filtering area on a given floor area, and vacuum disc filters are used in preference to drum filters where space is restricted. At sizes above approximately 25 m² filtration area, disc filters are cheaper; but their applications are more restricted, as they are not as suitable for the application of wash water, or precoating.

Belt filters (vacuum operation) (Fig. 10.14)

A belt filter consists of an endless reinforced rubber belt, with drainage hole along its centre, which supports the filter medium. The belt passes over a stationary suction box, into which the filtrate is sucked. Slurry and wash water are sprayed on to the top of the belt.

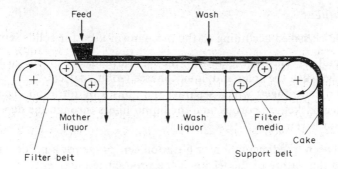

FIG. 10.14. Belt filter

Horizontal pan filters (vacuum operation) (Fig. 10.15)

This type is similar in operation to a vacuum nutsche filter. It consists of shallow pans with perforated bases, which support the filter medium. By arranging a series of pans around the circumference of a rotating wheel, the operation of filtering, washing, drying and discharging can be made automatic.

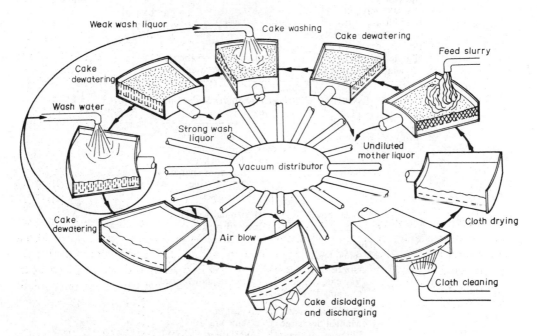

FIG. 10.15. Pan filters

Centrifugal filters

Centrifugal filters use centrifugal force to drive the filtrate through the filter cake. The equipment used is described in the next section.

10.4.3. Centrifuges

Centrifuges are classified according to the mechanism used for solids separation:

(a) Sedimentation centrifuges: in which the separation is dependent on a difference in density between the solid and liquid phases (solid heavier).

(b) Filtration centrifuges: which separate the phases by filtration. The walls of the centrifuge basket are porous, and the liquid filters through the deposited cake of solids and is removed.

The choice between a sedimentation or filtration centrifuge for a particular application will depend on the nature of the feed and the product requirements.

The main factors to be considered are summarised in Table 10.4. As a general rule, sedimentation centrifuges are used when it is required to produce a clarified liquid, and filtration centrifuges to produce a pure, dry, solid.

TABLE 10.4. *Selection of sedimentation or filter centrifuge*

Factor	Sedimentation	Filtration
Solids size, fine		x
Solids size, > 150 μm	x	
Compressible cakes	x	
Open cakes		x
Dry cake required		x
High filtrate clarity	x	
Crystal breakage problems		x
Pressure operation		
High-temperature operation	will depend on the type of centrifuge used	

TABLE 10.5. *Centrifuge types (after Sutherland, 1970)*

Sedimentation	Filtration–fixed bed

Laboratory
  Bottle
  Ultra
	Vertical basket
	  Manual discharge
	  Bag discharge
	  Knife discharge

Tubular bowl	Horizontal basket Inclined basket

Disc
  Batch bowl
  Nozzle discharge
  Valve discharge
  Opening bowl
	Filtration–moving bed

Imperforate basket
  Manual discharge
  Skimmer discharge
	Conical bowl
	  Wide angle
	  Vibrating
	  Torsional
	  Tumbling

Scroll discharge
  Horizontal
  Cantilevered
  Vertical
  Screen bowl
	Scroll discharge
	
	Cylindrical bowl
	  Scroll discharge
	  Pusher

A variety of centrifugal filter and sedimenter designs is used. The main types are listed in Table 10.5. They can be classified by a number of design and operating features, such as:

1. Mode of operation—batch or continuous.
2. Orientation of the bowl/basket—horizontal or vertical.
3. Position of the suspension and drive—overhung or underhung.
4. Type of bowl—solid, perforated basket, disc bowl.
5. Method of solids cake removal.
6. Method of liquid removal.

Detailed descriptions of the various types of centrifuge used in the process industries and their applications can be found in various handbooks, and in articles by Morris (1966), Bradley (1965a) and Ambler (1971).

The fields of application of each type, classified by the size range of the solid particles separated, are given in Fig. 10.16.

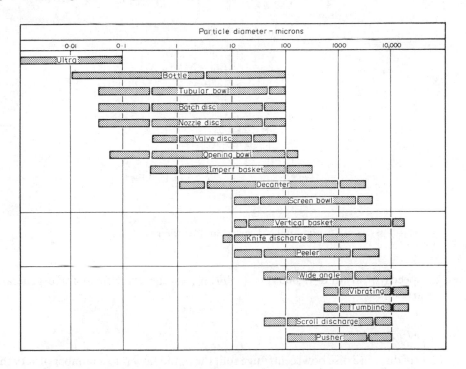

FIG. 10.16. Classification of centrifuges by particle size (after Sutherland, 1970)

Sedimentation centrifuges

There are four main types of sedimentation centrifuge:

1. Tubular bowl (Fig. 10.17)

High-speed, vertical axis, tubular bowl centrifuges are used for the separation of immiscible liquids, such as water and oil, and for the separation of fine solids. The bowl is

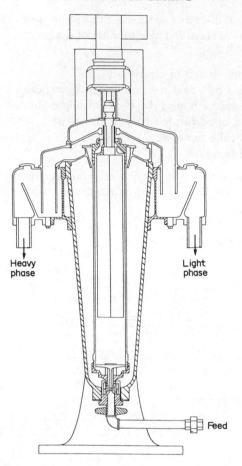

FIG. 10.17. Tubular Bowl centrifuge

driven at speeds of around 15,000 rpm (250 Hz) and the centrifugal force generated exceeds 130,000 N.

2. Disc bowl (Fig. 10.18)

The conical discs in a disc bowl centrifuge split the liquid flow into a number of very thin layers, which greatly increases the separating efficiency. Disc bowl centrifuges are used for separating liquids and fine solids, and for solids classification.

3. Scroll discharge

In this type of machine the solids deposited on the wall of the bowl are removed by a scroll (a helical screw conveyer) which revolves at a slightly different speed from the bowl. Scroll discharge centrifuges can be designed so that solids can be washed and relatively dry solids be discharged.

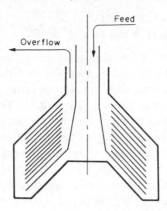

FIG. 10.18. Disc bowl centrifuge

4. Solid bowl batch centrifuge

The simplest type; similar to the tubular bowl machine type but with a smaller bowl length to diameter ratio (less than 0·75). The tubular bowl type is rarely used for solids concentrations above 1 per cent by volume. For concentrations between 1 to 15 per cent, any of the other three types can be used. Above 15 per cent, either the scroll discharge type or the batch type may be used, depending on whether continuous or intermittent operation is required.

Sigma theory for sedimentation centrifuges

The basic equations describing sedimentation in a centrifugal field have been developed in Volume 2, Chapter 5. In that discussion the term *sigma* (Σ) is introduced, which can be used to define the performance of a centrifuge independently of the physical properties of the solid–fluid system. The sigma value of a centrifuge, normally expressed in cm^2, is equal to the cross-sectional area of a gravity settling tank having the same clarifying capacity.

This approach to describing centrifuge performance has become known as the "Sigma theory". It provides a means for comparing the performance of sedimentation centrifuges and for scaling up from laboratory and pilot scale tests; see Ambler (1952) and Trowbridge (1962).

In the general case, it can be shown that:

$$Q = 2u_g \Sigma \tag{10.1}$$

and (where Stokes' law applies)
$$u_g = \frac{\Delta \rho d_s^2 g}{18\mu} \tag{10.2}$$

where Q = volumetric flow of liquid through the centrifuge, cm^3/s,

u_g = terminal velocity of the solid particle settling under gravity through the liquid, cm/s,

Σ = sigma value of the centrifuge, cm^2,

$\Delta\rho$ = density difference between solid and liquid, g/cm^3,

d_s = The diameter of the solid particle; the "cut-off" size, 50 per cent of the particles of this size will be removed in passage through the centrifuge, cm (μm $\times 10^{-4}$),

g = gravitational acceleration, 981 cm/s²,

μ = viscosity of the liquid, poise (N s/m² × 10).

Morris (1966) gives a method for the selection of the appropriate type of sedimentation centrifuge for a particular application based on the ratio of the liquid overflow to sigma value (Q/Σ). His values for the operating range of each type, and their approximate efficiency rating, are given in Table 10.6. The efficiency term is used to account for the different amounts by which the various designs differ from the theoretical sigma values given by equation 10.1. Sigma values depend solely on the geometrical configuration and speed of the centrifuge. Details of the calculation for various types are given by Ambler (1952). To use Table 10.6, it is necessary to know the feed rate of slurry (and hence the liquid overflow Q), the density of the liquid and solid, the liquid viscosity; and the diameter of the particle for, say, a 98 per cent size removal. The use of Table 10.6 is illustrated in Example 10.1.

TABLE 10.6. *Selection of sedimentation centrifuges*

Type	Approximate efficiency (%)	Normal operating range Q, cm³/s at Q/Σ, cm/s
Tubular bowl	90	100 at 5×10^{-6} to 1000 at $3\cdot5 \times 10^{-5}$
Disc	45	25 at 7×10^{-6} to 30,000 at $4\cdot5 \times 10^{-5}$
Solid bowl scroll discharge	60	200 at $1\cdot5 \times 10^{-4}$ to 40,000 at $1\cdot5 \times 10^{-3}$
Solid bowl batch basket	75	100 at 5×10^{-4} to 1000 at $1\cdot5 \times 10^{-2}$

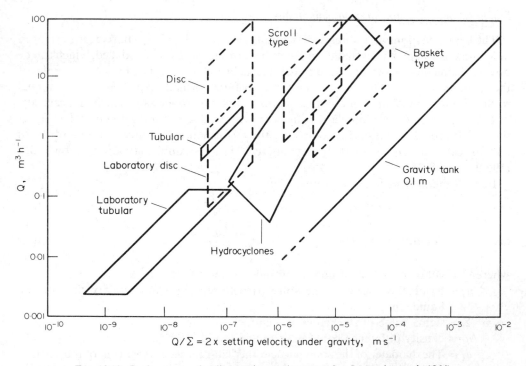

FIG. 10.19. Performance of sedimentation equipment (after Lavanchy *et al.*, 1964)

A selection guide for sedimentation centrifuges by Lavanchy *et al.* (1964), which includes other types of solid–liquid separators, is shown in Fig. 10.19, adapted to SI units. (Note that the units for Q in that figure are m^3/h and for sigma, m^2.)

Example 10.1

A precipitate is to be continuously separated from a slurry. The solids concentration is 5 per cent and the slurry feed rate $5.5 \, m^3/h$. The relevant physical properties at the system operating temperature are:

liquid density $1.5 \, g/cm^3$, viscosity $4 \, cp$ $(mN \, s/m^2)$,
solid density $2.3 \, g/cm^3$, 'cut-off' particle size $50 \, \mu m = 50 \times 10^{-4} \, cm$.

Solution

$$\text{Overflow rate, } Q = 0.95 \times 5.5 \times \frac{10^6}{3600} = 1451 \, cm^3/s$$

$$\Delta\rho = 2.3 - 1.5 = 0.8 \, g/cm^3$$

From equations 10.1 and 10.2

$$Q/\Sigma = 2 \times \frac{0.8(50 \times 10^{-4})^2}{18 \times 4 \times 10^{-2}} \times 981 = 5.45 \times 10^{-2}$$

From Table 10.6, for a Q of $1451 \, cm^3/s$ at a Q/Σ of 5.45×10^{-2} a scroll discharge centrifuge should be used.

To obtain an idea of the size of machine needed the sigma value can be calculated using the efficiency value from Table 10.6.

From Equation 10.1:

$$\Sigma = \frac{Q}{\text{Eff.} \times 2u_g} = \frac{1451}{0.6 \times 5.45 \times 10^{-2}}$$

$$= \underline{\underline{4.4 \times 10^4 \, cm^2}}$$

Filtration centrifuges (centrifugal filters)

It is convenient to classify centrifugal filters into two broad classes, depending on how the solids are removed: fixed bed or moving bed.

In the fixed-bed type, the cake of solids remains on the walls of the bowl until removed manually, or automatically by means of a knife mechanism. It is essentially cyclic in operation. In the moving-bed type, the mass of solids is moved along the bowl by the action of a scroll (similar to the solid-bowl sedimentation type); or by a ram (pusher type); or by a vibration mechanism; or by the bowl angle. Washing and drying zones can be incorporated into the moving bed type.

Bradley (1965a) has grouped the various types into the family tree shown in Fig. 10.20.

Schematic diagrams of the various types are shown in Fig. 10.21. The simplest machines are the basket types (Figs. 10.21*a,b,c*), and these form the basic design from which the other types have been developed (Figs. 10.21*d* to *o*).

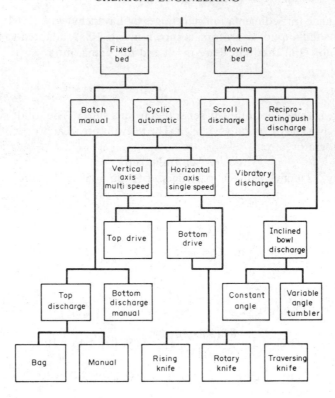

Fig. 10.20. Filtration centrifuge family tree (after Bradley, 1965a)

The various arrangements of knife mechanisms used for automatic removal of the cake are shown in Figs. 10.21*d* to *h*. The bottom discharge-type machines (Figs. 10.21*d, e*) can be designed for variable speed, automatic discharge; and are suitable for use with fragile, or plate or needle-shaped crystals, where it is desirable to avoid breakage or compaction of the bed. They can be loaded and discharged at low speeds, which reduces breakage and compaction of the cake. The single-speed machines (Figs. 10.21*f,g,h*) are used where cakes are thin, and short cycle times are required. They can be designed for high-temperature and pressure operation. When continuous operation is required, the scroll, pusher, or other self-discharge types are used (Figs. 10.21*i* to *o*). The scroll discharge centrifuge is a low-cost, flexible machine, capable of a wide range of applications; but is not suitable for handling fragile materials. It is normally used for coarse particles, where some contamination of the filtrate with fines can be tolerated.

The capacity of filtration centrifuges is very dependent on the solids concentration in the feed. For example, at 10 per cent feed slurry concentration 9 kg of liquid will be centrifuged for every 1 kg of solids separated; whereas with a 50 per cent solids concentration the quantity will be less than 1 kg. For dilute slurries it is well worth considering using some form of pre-concentration; such as gravity sedimentation or a hydrocyclone.

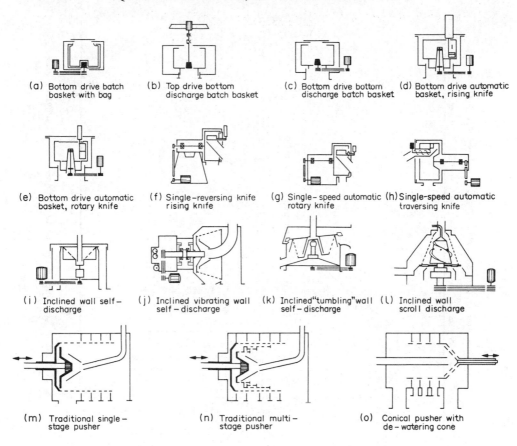

(a) Bottom drive batch basket with bag

(b) Top drive bottom discharge batch basket

(c) Bottom drive bottom discharge batch basket

(d) Bottom drive automatic basket, rising knife

(e) Bottom drive automatic basket, rotary knife

(f) Single-reversing knife rising knife

(g) Single-speed automatic rotary knife

(h) Single-speed automatic traversing knife

(i) Inclined wall self-discharge

(j) Inclined vibrating wall self-discharge

(k) Inclined "tumbling" wall self-discharge

(l) Inclined wall scroll discharge

(m) Traditional single-stage pusher

(n) Traditional multi-stage pusher

(o) Conical pusher with de-watering cone

FIG. 10.21. Schematic diagrams of filtration centrifuge types (Bradley, 1965a)

10.4.4. Hydrocyclones (liquid-cyclones)

Hydrocyclones are used for solid–liquid separations; as well as for solids classification, and liquid–liquid separation. It is a centrifugal device with a stationary wall, the centrifugal force being generated by the liquid motion. The operating principle is basically the same as that of the gas cyclone described in Section 10.8.3, and in Volume 2, Chapter 8. Hydrocyclones are simple, robust, separating devices, which can be used over the particle size range from 4 to 500 μm. The design and application of hydrocyclones is discussed fully in a book by Bradley (1965b). Design methods and charts are also given by Zanker (1977), and by Day and Grichar (1979).

The nomographs given by Zanker (Figs. 10.22 and 10.23) can be used to estimate the size of cyclone needed for a particular application. The method is outlined below and illustrated in Example 10.2. Figure 10.23 is based on an empirical equation by Bradley (1960):

$$d_{50} = 4 \cdot 5 \left[\frac{D_c^3 \mu}{L^{1 \cdot 2}(\rho_s - \rho_L)} \right] \tag{10.3}$$

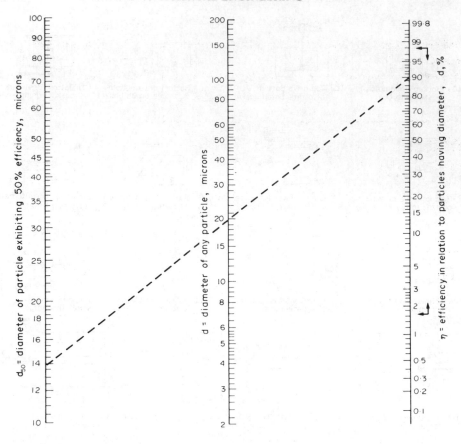

Fig. 10.22. Determination of d_{50} from the desired particle separation (Equation 10.3, Zanker, 1977) (Example 10.2)

where d_{50} = the particle diameter for which the cyclone is 50 per cent efficient, μm,

D_c = diameter of the cyclone chamber, cm,

μ = liquid viscosity, centipoise (mN s/m^2),

L = feed flow rate, l/min,

ρ_L = density of the liquid, g/cm^3,

ρ_s = density of the solid, g/cm^3.

The equation gives the chamber diameter required to separate the so-called d_{50} particle diameter, as a function of the slurry flow rate and the liquid and solid physical properties. The d_{50} particle diameter is the diameter of the particle, 50 per cent of which will appear in the overflow, and 50 per cent in the underflow. The separating efficiency for other particles is related to the d_{50} diameter by Fig 10.22, which is based on a formula by Bennett (1936).

$$\eta = 100\left[1 - e^{-(d/d_{50} - 0.115)^3} \right] \qquad (10.4)$$

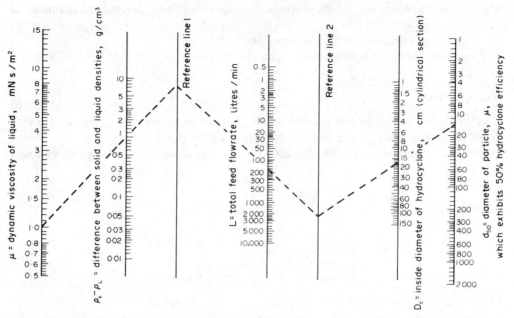

FIG. 10.23. Chamber dia. D_c from flow-rate, physical properties, and d_{50} particle size (Equation 10.4, Zanker, 1977) (Example 10.2)

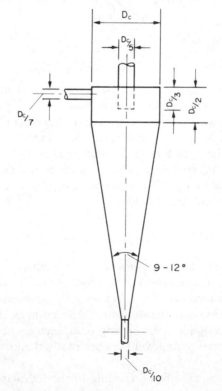

FIG. 10.24. Hydrocyclone–typical proportions

where η = the efficiency of the cyclone in separating any particle of diameter d, per cent,

   d = the selected particle diameter, μm.

The method applies to hydrocyclones with the proportions shown in Fig. 10.24.

Example 10.2

Estimate the size of hydrocyclone needed to separate 90 per cent of particles with a diameter greater than 20 μm, from 10 m^3/h of a dilute slurry.

Physical properties: solid density 2000 kg/m^3, liquid density 1000 kg/m^3, viscosity 1 mN s/m^2

Solution

$$\text{Flow-rate} = \frac{10 \times 10^3}{60} \quad = 166 \cdot 7 \, \text{l/min}$$

$$(\rho_s - \rho_L) = 2 \cdot 0 - 1 \cdot 0 \quad = 1 \cdot 0 \, \text{g/cm}^3$$

From Fig. 10.22, for 90 per cent removal of particles above 20 μm

$$d_{50} = 14 \, \mu\text{m}$$

From Fig. 10.23, for $\mu = 1 \, \text{mN s/m}^2$, $(\rho_s - \rho_L) = 1 \cdot 0 \, \text{g/cm}^3$, $L = 167/\text{min}$

$$D_c = \underline{16 \, \text{cm}}$$

10.4.5. Pressing (expression)

Pressing, in which the liquid is squeezed (expressed) from a mass of solids by compression, is used for certain specialised applications. Pressing consumes a great deal of energy, and should not be used unless no other separating technique is suitable. However, in some applications dewatering by pressing can be competitive with drying.

Presses are of two basic types: hydraulic batch presses and screw presses. Hydraulic presses are used for extracting fruit juices, and screw presses for dewatering materials; such as paper pulp, rubbish and manure. The equipment used is described in the handbooks; Perry and Chilton (1973) and Mead (1964).

10.4.6. Solids drying

Drying is the removal of water, or other volatile liquids, by evaporation. Most solid materials require drying at some stage in their production. The choice of suitable drying equipment cannot be separated from the selection of the upstream equipment feeding the drying stage. The overriding consideration in the selection of drying equipment is the nature and concentration of the feed. Drying is an energy-intensive process, and the removal of liquid by thermal drying will be more costly than by mechanical separation techniques.

Drying equipment can be classified according to the following design and operating features:

TABLE 10.7. *Dryer selection*

Mode of operation	Generic type	Feed condition 1	2	3	Specific Dryer types	Jacketed	Suitable for heat-sensitive materials	Suitable for vacuum service	Retention or cycle time	Heat transfer method	Capacity	Typical evaporation capacity
Batch	Stationary				1. Shelf 2. Cabinet 3. Compartment	Yes	Yes	Yes	6–48 h	Radiant and conduction	Limited	0·15–1·0
					Truck	No	Yes	No	6–48 h	Convection	Limited	0·15–1·0
					1. Kettle 2. Pan	Yes	No	Yes	3–12 h	Conduction	Limited	1·5–15
					Rotary shell	Yes	Yes	Yes	4–48 h	Conduction	Limited	0·5–12
					Rotary internal	Yes	Yes	Yes	4–48 h	Conduction	Limited	0·5–12
					Double cone	Yes	Yes	Yes	3–12 h	Conduction	Limited	0·5–12
Continuous	Drum				1. Single drum 2. Double drum 3. Twin drum	No	Yes	Yes	Very short	Conduction	Medium	5–50
	Rotary				Rotary, direct heat	No	No	No	Long	Convection	High	3–110
					Rotary, indirect heat	No	No	No	Long	Conduction	Medium	15–200
					Rotary, steam tube	No	Depends on material	No	Long	Conduction	High	15–200
					Rotary, direct-indirect heat	No	No	No	Long	Conduction Convection	High	50–150
					Louver	No	Depends on material	No	Long	Convection	High	5–240
	Conveyor				Tunnel: belt, screen	No	Yes	No	Long	Convection	Medium	1·5–35
					Rotary shelf	Yes	Depends on material	No	Medium	Conduction Convection	Medium	0·5–10
					Trough	Yes	Depends on material	Yes	Varies	Conduction	Medium	0·5–15
					Vibrating	Yes	Depends on material	No	Medium	Convection Conduction	Medium	0·5–100
					Turbo	No	Depends on material	No	Medium	Convection	Medium	1–10
	Suspended particle				Spray	No	Yes	No	Short	Convection	High	1·5–50
					Flash	No	Yes	No	Short	Convection	High	—
					Fluid bed	No	Yes	No	Short	Convection	Medium	—

Batch: kg/hm^2 heat transfer area

Continuous (Rotary): kg/hm^3 dryer volume

Conveyor: kg/hm^2

Suspended particle: kg/hm^3

←·····→ = applicable to feed conditions noted

Key to feed conditions:
1. Solutions, colloidal suspensions and emulsions, pumpable solids suspensions, pastes and sludges.
2. Free-flowing powders, granular, crystalline or fibrous solids that can withstand mechanical handling.
3. Solids incapable of withstanding mechanical handling.

1. Batch or continuous.
2. Physical state of the feed: liquid, slurry, wet solid.
3. Method of conveyance of the solid: belt, rotary, fluidised.
4. Heating system: conduction, convection, radiation.

Except for a few specialised applications, hot air is used as the heating and mass transfer medium in industrial dryers. The air may be directly heated by the products of combustion of the fuel used (oil, gas or coal) or indirectly heated, usually by banks of steam-heated finned tubes. The heated air is usually propelled through the dryer by electrically driven fans.

TABLE 10.8. *Dryer applications*

Dryer type	System	Feed form	Typical products
Batch ovens	Forced convection	Paste, granules, extrude cake	Pigment dyestuffs, pharmaceuticals, fibres
	Vacuum	Extrude cake	Pharmaceuticals
„ pan (agitated)	Atmospheric and vacuum	Crystals, granules, powders	Fine chemicals, food products
„ rotary	Vacuum	Crystals, granules solvent recovery	Pharmaceuticals
„ fluid bed	Forced convection	Granular, crystals	Fine chemicals, pharmaceuticals, plastics
„ infra-red	Radiant	Components sheets	Metal products, plastics
Continuous rotary	Convection Direct/indirect Direct Indirect Conduction	Crystals, coarse powders, extrudes, preformed cake lumps, granular paste and fillers, cakes back-mixed with dry product	Chemical ores, food products, clays, pigments, chemicals Carbon black
„ film drum	Conduction	Liquids, suspensions	Foodstuffs, pigment
„ trough	Conduction		Ceramics, adhesives
„ spray	Convection	Liquids, suspensions	Foodstuffs, pharmaceuticals, ceramics, fine chemicals, detergents, organic extracts
„ band	Convection	Preformed solids	Foodstuffs, pigments, chemicals, rubber, clays, ores, textiles
„ fluid bed	Convection	Preformed solids granules, crystals	Ores, coal, clays, chemicals
„ pneumatic	Convection	Preformed pastes, granules, crystals, coarse products	Chemicals, starch, flour, resins, wood-products, food products
„ infra-red	Radiant	Components sheets	Metal products, moulded fibre articles, painted surfaces

Table 10.7, adapted from a similar selection guide by Parker (1963a), shows the basic features of the various types of solids dryer used in the process industries; and Table 10.8, by Williams-Gardner (1965), shows typical applications.

Batch dryers are normally used for small-scale production and where the drying cycle is likely to be long. Continuous dryers require less labour, less floor space; and produce a more uniform quality product.

When the feed is solids, it is important to present the material to the dryer in a form that will produce a bed of solids with an open, porous, structure.

For pastes and slurries, some form of pretreatment equipment will normally be needed, such as extrusion or granulation.

The main factors to be considered when selecting a dryer are:

1. Feed condition: solid, liquid, paste, powder, crystals.
2. Feed concentration, the initial liquid content.
3. Product specification: dryness required, physical form.
4. Throughput required.
5. Heat sensitivity of the product.
6. Nature of the vapour: toxicity, flammability.
7. Nature of the solid: flammability (dust explosion hazard), toxicity.

The drying characteristics of the material can be investigated by laboratory and pilot plant tests; which are best carried out in consultation with the equipment vendors.

The theory of drying processes is discussed in Volume 2, Chapter 11. Full descriptions of the various types of equipment used and their applications are given in that chapter, and in several specialist books; Keey (1972, 1978); Nonhebel and Moss (1971); Williams–Gardner (1971). Only brief descriptions of the principal types will be given in this section.

The basic types used in the chemical process industries are: tray, band, rotary, fluidised, pneumatic, drum and spray dryers.

Tray dryers (Fig. 10.25)

Batch tray dryers are used for drying small quantities of solids, and are used for a wide range of materials.

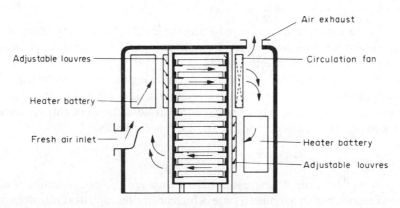

FIG. 10.25. Tray dryer

The material to be dried is placed in solid bottomed trays over which hot air is blown; or perforated bottom trays through which the air passes.

Batch dryers have high labour requirements, but close control can be maintained over the drying conditions and the product inventory, and they are suitable for drying valuable products.

Conveyor dryers (continuous circulation band dryers) (Fig. 10.26)

In this type, the solids are fed on to an endless, perforated, conveyor belt, through which hot air is forced. The belt is housed in a long rectangular cabinet, which is divided up into zones, so that the flow pattern and temperature of the drying air can be controlled. The relative movement through the dryer of the solids and drying air can be parallel or, more usually, counter-current.

This type of dryer is clearly only suitable for materials that form a bed with an open structure. High drying rates can be achieved, with good product-quality control. Thermal efficiencies are high and, with steam heating, steam usage can be as low as 1·5 kg per kg of water evaporated. The disadvantages of this type of dryer are high initial cost and, due to the mechanical belt, high maintenance costs.

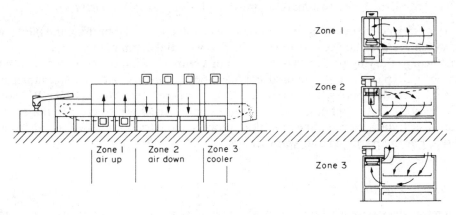

FIG. 10.26. Conveyor dryer

Rotary dryer (Fig. 10.27)

In rotary dryers the solids are conveyed along the inside of a rotating, inclined, cylinder and are heated and dried by direct contact with hot air gases flowing through the cylinder. In some, the cylinders are indirectly heated.

Rotating dryers are suitable for drying free-flowing granular materials. They are suitable for continuous operation at high throughputs; have a high thermal efficiency and relatively low capital cost and labour costs. Some disadvantages of this type are: a non-uniform residence time, dust generation and high noise levels.

Fluidised bed dryers (Fig. 10.28)

In this type of dryer, the drying gas is passed through the bed of solids at a velocity sufficient to keep the bed in a fluidised state; which promotes high heat transfer and drying rates.

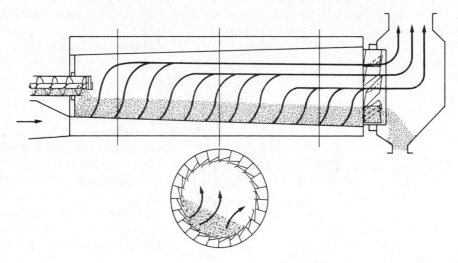

FIG. 10.27. Rotary dryer

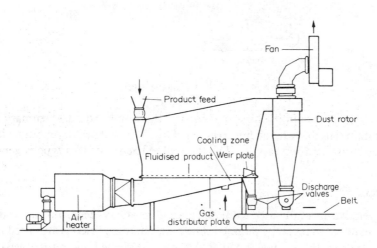

FIG. 10.28. Fluidised bed dryer

Fluidised bed dryers are suitable for granular and crystalline materials within the particle size range 1 to 3 mm. They are designed for continuous and batch operation.

The main advantages of fluidised dryers are: rapid and uniform heat transfer; short drying times, with good control of the drying conditions; and low floor area requirements. The power requirements are high compared with other types.

Pneumatic dryers (Fig. 10.29)

Pneumatic dryers, also called flash dryers, are similar in their operating principle to spray dryers. The product to be dried is dispersed into an upward-flowing stream of hot gas by a suitable feeder. The equipment acts as a pneumatic conveyor and dryer. Contact

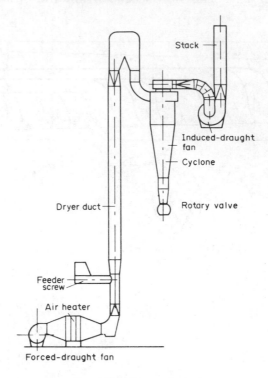

FIG. 10.29. Pneumatic dryer

times are short, and this limits the size of particle that can be dried. Pneumatic dryers are suitable for materials that are too fine to be dried in a fluidised bed dryer but which are heat sensitive and must be dried rapidly. The thermal efficiency of this type is generally low.

Spray dryers (Fig. 10.30)

Spray dryers are normally used for liquid and dilute slurry feeds, but can be designed to handle any material that can be pumped. The material to be dried is atomised in a nozzle, or by a disc-type atomiser, positioned at the top of a vertical cylindrical vessel. Hot air flows up the vessel (in some designs downward) and conveys and dries the droplets. The liquid vaporises rapidly from the droplet surface and open, porous particles are formed. The dried particles are removed in a cyclone separator or bag filter.

The main advantages of spray drying are the short contact time, making it suitable for drying heat-sensitive materials, and good control of the product particle size, bulk density, and form. Because the solids concentration in the feed is low the heating requirements will be high. Spray drying is discussed in a book by Masters (1972).

Rotary drum dryers (Fig. 10.31)

Drum dryers are used for liquid and dilute slurry feeds. They are an alternative choice to spray dryers when the material to be dried will form a film on a heated surface, and is not heat sensitive.

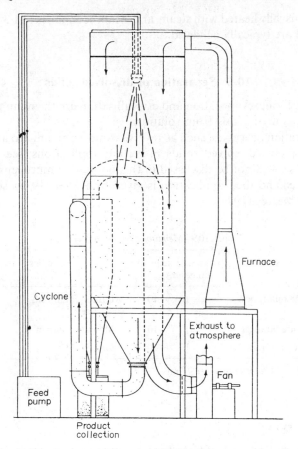

FIG. 10.30. Spray dryer

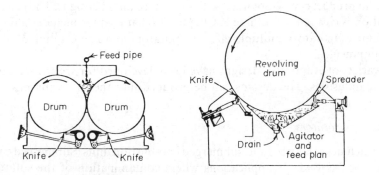

FIG. 10.31. Rotary drum dryers

They consist essentially of a revolving, internally heated, drum, on which a film of the solids is deposited and dried. The film is formed either by immersing part of the drum in a trough of the liquid or by spraying, or splashing, the feed on to the drum surface; double drums are also used in which the feed is fed into the "nip" formed between the drums.

The drums are usually heated with steam, and steam economies of 1·3 kg steam per kg of water evaporated are typically achieved.

10.5. Separation of dissolved solids

On an industrial scale, evaporation and crystallisation are the main processes used for the recovery of dissolved solids from solutions.

Membrane filtration processes, such as reverse osmosis, and micro and ultra filtration, are used to "filter out" dissolved solids in certain applications; see Table 10.9. These specialised processes will not be discussed in this book. A comprehensive description of the techniques used and their application is given by Porter (1979); see also Lacey and Loeb (1972) and Meares (1976).

TABLE 10.9. *Membrane filtration processes*

Process	Approximate size range (m)	Examples
Microfiltration	10^{-8} –10^{-4}	Pollen, blood cells, bacteria
Ultrafiltration	10^{-9} –10^{-8}	Albumin, vitamin B_{12}
Reverse osmosis	10^{-10}–10^{-9}	Ions, Na^+Cl^-

(*Note*: 1 angstrom = 10^{-10} m)

10.5.1 Evaporators

Evaporation is the removal of a solvent by vaporisation, from solids that are not volatile. It is normally used to produce a concentrated liquid, often prior to crystallisation, but a dry solid product can be obtained with some specialised designs. The general subject of evaporation is covered in Volume 2, Chapter 9. That chapter includes a discussion of heat transfer in evaporators, multiple-effect evaporators, and a description of the principal types of equipment.

A great variety of evaporator designs have been developed for specialised applications in particular industries. The designs can be grouped into the following basic types.

Direct-heated evaporators

This type includes solar pans and submerged combustion units. Submerged combustion evaporators can be used for applications where contamination of the solution by the products of combustion is acceptable.

Long-tube evaporators (Fig. 10.32)

In this type the liquid flows as a thin film on the walls of a long, vertical, heated, tube. Both falling film and rising film types are used. They are high capacity units; suitable for low viscosity solutions.

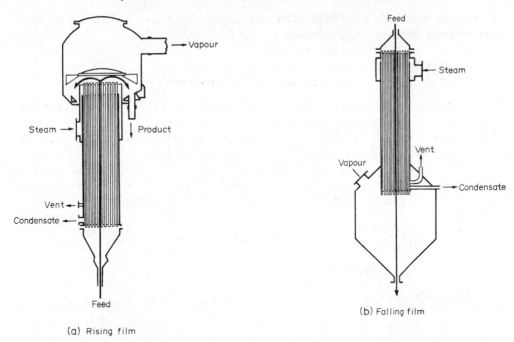

FIG. 10.32. Long-tube evaporators

Forced-circulation evaporators (Fig. 10.33)

In forced circulation evaporators the liquid is pumped through the tubes. They are suitable for use with materials which tend to foul the heat transfer surfaces, and where crystallisation can occur in the evaporator.

Agitated thin-film evaporators (Fig. 10.34)

In this design a thin layer of solution is spread on the heating surface by mechanical means. Wiped-film evaporators are used for very viscous materials and for producing solid products. The design and applications of this type of evaporator are discussed by Mutzenburg (1965), Parker (1965) and Fischer (1965).

Short-tube evaporators

Short-tube evaporators, also called callandria evaporators, are used in the sugar industry; see Volume 2.

Evaporator selection

The selection of the most suitable evaporator type for a particular application will depend on the following factors:

1. The throughput required.
2. The viscosity of the feed and the increase in viscosity during evaporation.
3. The nature of the product required; solid, slurry, or concentrated solution.

4. The heat sensitivity of the product.
5. Whether the materials are fouling or non-fouling.
6. Whether the solution is likely to foam.
7. Whether direct heating can be used.

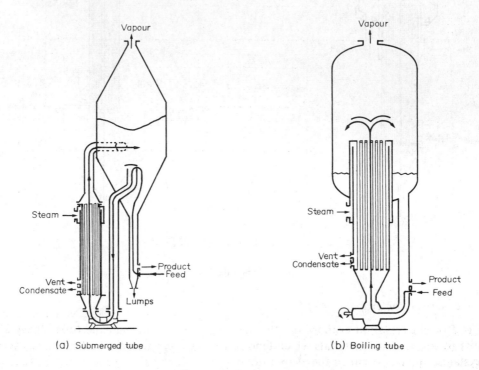

FIG. 10.33. Forced-circulation evaporators

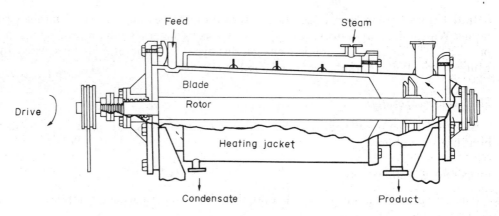

FIG. 10.34. Horizontal wiped-film evaporator

A selection guide based on these factors is given in Fig. 10.35; see also Parker (1963b).

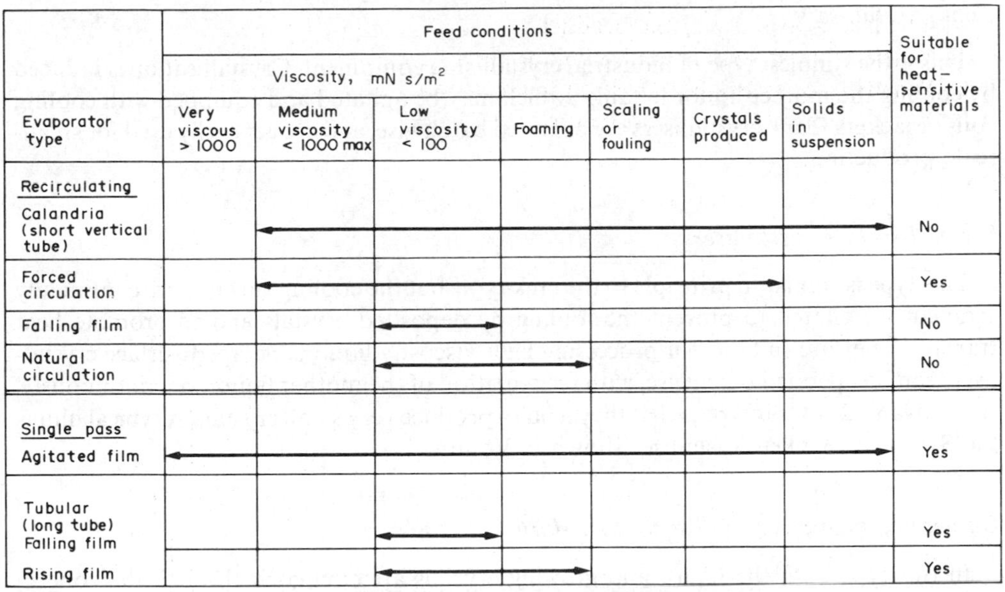

| Evaporator type | Feed conditions | | | | | | | Suitable for heat-sensitive materials |
| | Viscosity, mN s/m^2 | | | | | | | |
	Very viscous > 1000	Medium viscosity < 1000 max	Low viscosity < 100	Foaming	Scaling or fouling	Crystals produced	Solids in suspension	
Recirculating								
Calandria (short vertical tube)	←———————————————————→							No
Forced circulation	←——————————————→							Yes
Falling film			←——→					No
Natural circulation			←————→					No
Single pass								
Agitated film	←———————————————————→							Yes
Tubular (long tube) Falling film			←——→					Yes
Rising film			←————→					Yes

FIG. 10.35. Evaporator selection guide

Auxiliary equipment

Condensers and vacuum pumps will be needed for evaporators operated under vacuum. For aqueous solutions, steam ejectors and jet condensers are normally used. Jet condensers are direct-contact condensers, where the vapour is condensed by contact with jets of cooling water. Indirect, surface condensers, are used where it is necessary to keep the condensed vapour and cooling water effluent separate.

10.5.2. Crystallisation

Crystallisation is used for the production, purification and recovery of solids. Crystalline products have an attractive appearance, are free flowing, and easily handled and packaged. The process is used in a wide range of industries: from the small-scale production of specialised chemicals, such as pharmaceutical products, to the tonnage production of products such as sugar, common salt and fertilisers.

Crystallisation theory is covered in Volume 2, Chapter 15, and in several other texts; Mullin (1972), Van Hook (1961), Nývlt (1971). Descriptions and illustrations of the many types of commercial crystalliser used and their applications can be found in these texts and in various handbooks; Perry and Chilton (1973), Schweitzer (1979). Bamforth (1965) includes details of the ancillary equipment required: vacuum pumps, circulation pumps, valves; and filters, centrifuges and dryers.

Crystallisation equipment can be classified by the method used to obtain supersaturation of the liquor, and also by the method used to suspend the growing crystals. Supersaturation is obtained by cooling or evaporation. There are four basic types of crystalliser; these are described briefly below.

Tank crystallisers

This is the simplest type of industrial crystallising equipment. Crystallisation is induced by cooling the mother liquor in tanks; which may be agitated and equipped with cooling coils or jackets. Tank crystallisers are operated batchwise, and are generally used for small-scale production.

Scraped-surface crystallisers

This type is similar in principle to the tank type, but the cooling surfaces are continually scraped or agitated to prevent the fouling by deposited crystals and to promote heat transfer. They are suitable for processing high-viscosity liquors. Scraped-surface crystallisers can be operated batchwise, with recirculation of the mother liquor, or continuously. A disadvantage of this type is that they tend to produce very small crystals. A typical unit is the Swenson–Walker crystalliser shown in Volume 2.

Circulating magma crystallisers (Fig. 10.36)

In this type both the liquor and growing crystals are circulated through the zone in which supersaturation occurs. Circulating magma crystallisers are probably the most important type of large-scale crystallisers used in the chemical process industry. Designs are available in which supersaturation is achieved by direct cooling, evaporation or evaporative cooling under vacuum.

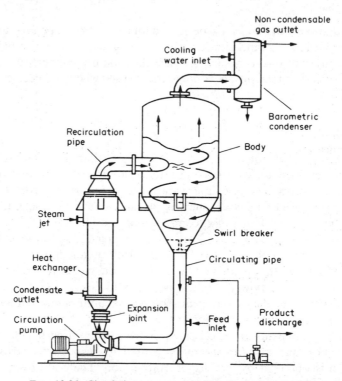

Fig. 10.36. Circulating magma crystalliser (evaporative type)

Circulating liquor crystallisers (Fig. 10.37)

In this type only the liquor is circulated through the heating or cooling equipment; the crystals are retained in suspension in the crystallising zone by the up-flow of liquor. Circulating liquor crystallisers produce crystals of regular size. The basic design consists of three components: a vessel in which the crystals are suspended and grow and are removed; a means of producing supersaturation, by cooling or evaporation; and a means of circulating the liquor. The Oslo crystalliser (Fig. 10.37) is the archetypical design for this type of crystallising equipment.

Circulating liquor crystallisers and circulating magma crystallisers are used for the large-scale production of a wide range of crystal products.

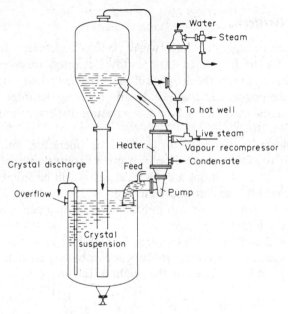

FIG. 10.37. Oslo evaporative crystalliser

Typical applications of the main types of crystalliser are summarised in Table 10.10; see also Larson (1978).

TABLE 10.10. *Selection of crystallisers*

Crystalliser type	Applications	Typical uses
Tank	Batch operation, small-scale production	Fatty acids, vegetable oils, sugars
Scraped surface	Organic compounds, where fouling is a problem, viscous materials	Chlorobenzenes, organic acids, paraffin waxes, napthalene, urea
Circulating magma	Production of large-sized crystals. High throughputs	Ammonium and other inorganic salts, sodium and potassium chlorides
Circulating liquor	Production of uniform crystals (smaller size than circulating magma). High throughputs.	Gypsum, inorganic salts, sodium and potassium nitrates, silver nitrates

10.6. Liquid–liquid separation

Separation of two liquid phases, immiscible or partially miscible liquids, is a common requirement in the process industries. For example, in the unit operation of liquid–liquid extraction the liquid contacting step must be followed by a separation stage (Volume 2, Chapter 13). It is also frequently necessary to separate small quantities of entrained water from process streams. The simplest form of equipment used to separate liquid phases is the gravity settling tank, the decanter. Various proprietary equipment is also used to promote coalescence and improve separation in difficult systems, or where emulsions are likely to form. Centrifugal separators are also used.

10.6.1. Decanters (settlers)

Decanters are used to separate liquids where there is a sufficient difference in density between the liquids for the droplets to settle readily. Decanters are essentially tanks which give sufficient residence time for the droplets of the dispersed phase to rise (or settle) to the interface between the phases and coalesce. In an operating decanter there will be three distinct zones or bands: clear heavy liquid; separating dispersed liquid (the dispersion zone); and clear light liquid.

Decanters are normally designed for continuous operation, but the same design principles will apply to batch operated units. A great variety of vessel shapes is used for decanters, but for most applications a cylindrical vessel will be suitable, and will be the cheapest shape. Typical designs are shown in Figs. 10.38 and 10.39. The position of the interface can be controlled, with or without the use of instruments, by use of a syphon take-off for the heavy liquid, Fig. 10.38.

The height of the take-off can be determined by making a pressure balance. Neglecting friction loss in the pipes, the pressure exerted by the combined height of the heavy and light liquid in the vessel must be balanced by the height of the heavy liquid in the take-off leg, Fig. 10.38.

$$(z_1 - z_3)\rho_1 g + z_3 \rho_2 g = z_2 \rho_2 g$$

hence
$$z_2 = \frac{(z_1 - z_3)\rho_1}{\rho_2} + z_3 \tag{10.5}$$

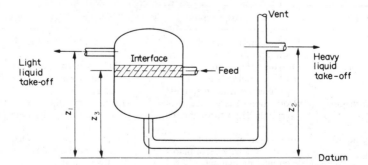

FIG. 10.38. Vertical decanter

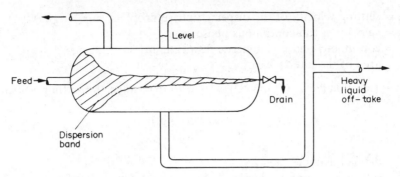

FIG. 10.39. Horizontal decanter

where ρ_1 = density of the light liquid, kg/m^3,
   ρ_2 = density of the heavy liquid, kg/m^3,
   z_1 = height from datum to light liquid overflow, m,
   z_2 = height from datum to heavy liquid overflow, m,
   z_3 = height from datum to the interface, m.

The height of the liquid interface should be measured accurately when the liquid densities are close, when one component is present only in small quantities, or when the throughput is very small. A typical scheme for the automatic control of the interface, using a level instrument that can detect the position of the interface, is shown in Fig. 10.40. Where one phase is present only in small amounts it is often recycled to the decanter feed to give more stable operation.

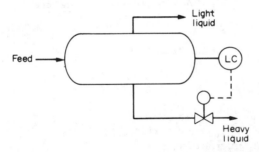

FIG. 10.40. Automatic control, level controller detecting interface

Decanter design

A rough estimate of the decanter volume required can be made by taking a hold-up time of 5 to 10 min, which is usually sufficient where emulsions are not likely to form. Methods for the design of decanters are given by Hooper and Jacobs (1979) and Signales (1975). The general approach taken is outlined below and illustrated by Example 10.3.

The decanter vessel is sized on the basis that the velocity of the continuous phase must be less than settling velocity of the droplets of the dispersed phase. Plug flow is assumed, and the velocity of the continuous phase calculated using the area of the interface:

$$u_c = L_c/A_i < u_d \tag{10.6}$$

where u_d = settling velocity of the dispersed phase droplets, m/s,
   u_c = velocity of the continuous phase, m/s,
   L_c = continuous phase volumetric flow rate, m³/s,
   A_i = area of the interface, m².

Stokes' law (see Volume 2, Chapter 3) is used to determine the settling velocity of the droplets:

$$u_d = \frac{d_d^2 g (\rho_d - \rho_c)}{18 \, \mu_c} \tag{10.7}$$

where d_d = droplet diameter, m,
   u_d = settling (terminal) velocity of the dispersed phase droplets
      with diameter d, m/s,
   ρ_c = density of the continuous phase, kg/m³,
   ρ_d = density of the dispersed phase, kg/m³,
   μ_c = viscosity of the continuous phase, N s/m²,
   g = gravitational acceleration, 9·81 m/s².

   Equation 10.7 is used to calculate the settling velocity with an assumed droplet size of 150 μm, which is well below the droplet sizes normally found in decanter feeds. If the calculated settling velocity is greater than 4×10^{-3} m/s, then a figure of 4×10^{-3} m/s is used.

   For a horizontal, cylindrical, decanter vessel, the interfacial area will depend on the position of the interface.

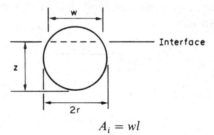

$$A_i = wl$$

and

$$w = 2(2rz - z^2)^{1/2}$$

where w = width of the interface, m,
   z = height of the interface from the base of the vessel, m,
   l = length of the cylinder, m,
   r = radius of the cylinder, m.

For a vertical, cylindrical decanter:

$$A_i = \pi r^2$$

The position of the interface should be such that the band of droplets that collect at the interface waiting to coalesce and cross the interface does not extend to the bottom (or top) of the vessel. Ryon *et al.* (1959) and Mizrahi and Barnea (1973) have shown that the depth of the dispersion band is a function of the liquid flow rate and the interfacial area. A value of 10 per cent of the decanter height is usually taken for design purposes. If the performance of the decanter is likely to be critical the design can be investigated using scale models. The model should be scaled to operate at the same Reynolds number as the proposed design, so that the effect of turbulence can be investigated; see Hooper (1975).

Example 10.3

Design a decanter to separate a light oil from water.
The oil is the dispersed phase.
Oil, flow rate 1000 kg/h, density 900 kg/m^3, viscosity 3 mN s/m^2.
Water, flow rate 5000 kg/h, density 1000 kg/m^3, viscosity 1 mN s/m^2.

Solution

Take $d_d = 150 \ \mu m$

(10.7)
$$u_d = \frac{(150 \times 10^{-6})^2 \, 9 \cdot 81 \, (900-1000)}{18 \times 1 \times 10^{-3}}$$

$$= -0 \cdot 0012 \text{ m/s}, \ -1 \cdot 2 \text{ mm/s (rising)}$$

As the flow rate is small, use a vertical, cylindrical vessel.

$$L_c = \frac{5000}{1000} \times \frac{1}{3600} = 1 \cdot 39 \times 10^{-3} \text{ m}^3/\text{s}$$

$$u_c \not> u_d, \quad \text{and} \quad u_c = \frac{L_c}{A_i}$$

hence
$$A_i = \frac{1 \cdot 39 \times 10^{-3}}{0 \cdot 0012} = 1 \cdot 16 \text{ m}^2$$

$$r = \sqrt{\frac{1 \cdot 16}{\pi}} = 0 \cdot 61 \text{ m}$$

diameter = 1·2 m

Take the height as twice the diameter, a reasonable value for a cylinder:

height = 2·4 m

Take the dispersion band as 10 per cent of the height = 0·24 m

Check the residence time of the droplets in the dispersion band

$$= \frac{0 \cdot 24}{u_d} = \frac{0 \cdot 24}{0 \cdot 0012} = 200 \text{ s} \ (\sim 3 \text{ min})$$

This is satisfactory, a time of 2 to 5 min is normally recommended. Check the size of the water (continuous, heavy phase) droplets that could be entrained with the oil (light phase).

$$\text{Velocity of oil phase} = \frac{1000}{900} \times \frac{1}{3600} \times \frac{1}{1 \cdot 16}$$

$$= 2 \cdot 7 \times 10^{-4} \text{ m/s} \ (0 \cdot 27 \text{ mm/s})$$

From equation 10.7

$$d_d = \left[\frac{u_d \, 18\mu_c}{g(\rho_d - \rho_c)} \right]^{1/2}$$

so the entrained droplet size will

$$= \left[\frac{2{\cdot}7 \times 10^{-4} \times 18 \times 3 \times 10^{-3}}{9{\cdot}81 \, (1000\text{--}900)} \right]^{1/2}$$

$$= \underline{1{\cdot}2 \times 10^{-4} \text{ m}} \; = 120 \; \mu m$$

which is satisfactory; below 150 μm.

Piping arrangement

To minimise entrainment by the jet of liquid entering the vessel, the inlet velocity for a decanter should keep below 1 m/s.

$$\text{Flow-rate} = \left[\frac{1000}{900} + \frac{5000}{1000} \right] \frac{1}{3600} = 1{\cdot}7 \times 10^{-3} \text{ m}^3/\text{s}$$

$$\text{Area of pipe} = \frac{1{\cdot}7 \times 10^{-3}}{1} = 1{\cdot}7 \times 10^{-3} \text{ m}^2$$

$$\text{Pipe diameter} = \sqrt{\frac{1{\cdot}7 \times 10^{-3} \times 4}{\pi}} = 0{\cdot}047 \text{ m, say } \underline{\underline{50 \text{ mm}}}$$

Take the position of the interface as half-way up the vessel and the light liquid off-take as at 90 per cent of the vessel height, then

$$z_1 = 0{\cdot}9 \times 2{\cdot}4 = 2{\cdot}16 \text{ m}$$

$$z_3 = 0{\cdot}5 \times 2{\cdot}4 = 1{\cdot}2 \text{ m}$$

(10.5)

$$z_2 = \frac{(2{\cdot}16 - 1{\cdot}2)}{1000} \times 900 + 1{\cdot}2 = \underline{\underline{2{\cdot}06 \text{ m}}}$$

$$\text{say } \underline{\underline{2{\cdot}0 \text{ m}}}$$

Proposed design

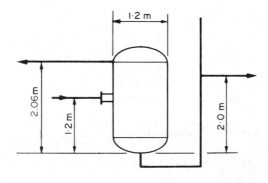

Drain valves should be fitted at the interface so that any tendency for an emulsion to form can be checked; and the emulsion accumulating at the interface drained off periodically as necessary.

10.6.2. Plate separators

Stacks of horizontal, parallel, plates are used in some proprietary decanter designs to increase the interfacial area per unit volume and to reduce turbulence. They, in effect, convert the decanter volume into several smaller separators connected in parallel.

10.6.3. Coalescers

Proprietary equipment, in which the dispersion is forced through some form of coalescing medium, is often used for the coalescence and separation of finely dispersed droplets. A medium is chosen that is preferentially wetted by the dispersed phase; knitted wire or plastic mesh, beds of fibrous material, or special membranes are used. The coalescing medium works by holding up the dispersed droplets long enough for them to form globlets of sufficient size to settle. A typical unit is shown in Fig. 10.41; see Redmon (1963). Coalescing filters are suitable for separating small quantities of dispersed liquids from large throughputs.

Electrical coalescers, in which a high voltage field is used to break down the stabilising film surrounding the suspended droplets, are used for desalting crude oils and for similar applications; see Waterman (1965).

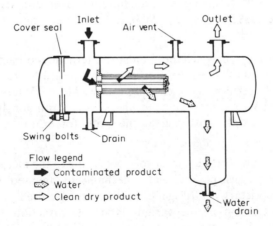

FIG. 10.41. Typical coalescer design

10.6.4. Centrifugal separators

Sedimentation centrifuges

For difficult separations, where simple gravity settling is not satisfactory, sedimentation centrifuges should be considered. Centrifuging will give a cleaner separation than that obtainable by gravity settling. Centrifuges can be used where the difference in gravity between the liquids is very small, as low as $100 \, kg/m^3$, and they can handle high throughputs, up to around $100 \, m^3/h$. Also, centrifuging will usually break any emulsion that may form. Bowl or disc centrifuges are normally used (see Section 10.4.3).

Hydrocyclones

Hydrocyclones are used for some liquid–liquid separations, but are not so effective in this application as in separating solids from liquids.

10.7. Separation of dissolved liquids

The most commonly used techniques for the separation and purification of miscible liquids are distillation, solvent extraction and, for a few special applications, adsorption. Distillation is probably the most widely used separation technique in the chemical process industries, and is covered in Chapter 11 of this Volume, and Chapter 11 of Volume 2. Solvent extraction and the associated technique, leaching (solid–liquid extraction) are covered in Volume 2, Chapters 10 and 13. Adsorption processes are covered in Volume 3, Chapter 7.

In this section, the discussion is restricted to a brief description of the equipment used in solvent-extraction processes.

10.7.1. Solvent extraction (liquid–liquid extraction)

The basic principles of liquid–liquid extraction are covered in several specialist texts; Hanson (1971), Alders (1955), Treybal (1963). Extraction equipment can be divided into two broad groups:

1. Stage-wise contactors, in which the liquids are alternately contacted (mixed) and then separated in a series of stages. Examples of this type are "mixer–settler" contactors, which consist of a series of agitated tanks followed by decanters; and plate columns.
2. Differential contactors, in which the phases are continuously in contact in the extractor and are only separated at the exits; for example, in packed column extractors.

Extraction columns can be further sub-divided according to the method used to promote contact between the phases: packed, plate, mechanically agitated, or pulsed columns. Various types of proprietary centrifugal extractors are also used.

The fields of application of the various types of extraction equipment are well summarised in Volume 2. The following factors need to be taken into consideration when selecting an extractor for a particular application:

1. The number of stages required.
2. The throughputs.
3. The settling characteristics of the phases.
4. The available floor area and head room.

Hanson (1968) has given a selection guide based on these factors, which can be used to select the type of equipment most likely to be suitable; Fig. 10.42.

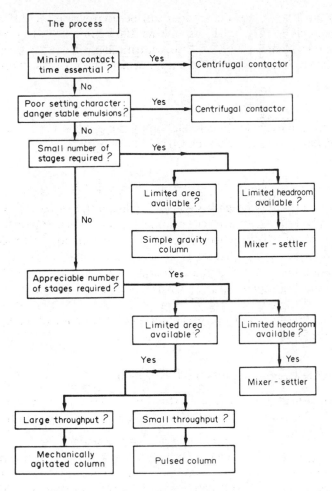

FIG. 10.42. Selection guide for Liquid–liquid contactors (after Hanson, 1968)

10.8. Gas–solids separations (Gas cleaning)

The primary need for gas–solid separation processes is for gas cleaning: the removal of dispersed finely divided solids (dust) and liquid mists from gas streams. Process gas streams must often be cleaned up to prevent contamination of catalysts or products, and to avoid damage to equipment, such as compressors. Also, effluent gas streams must be cleaned to comply with air-pollution regulations and for reasons of hygiene, to remove toxic and other hazardous materials.

There is also often a need for clean, filtered, air for process using air as a raw material, and where clean working atmospheres are needed: for instance, in the pharmaceutical and electronics industries.

The particles to be removed may range in size from large molecules, measuring a few hundredths of a micrometre, to the coarse dusts arising from the attrition of catalysts or the fly ash from the combustion of pulverised fuels.

TABLE 10.11. *Gas-cleaning equipment*

Type of equipment	Minimum particle size (μm)	Minimum loading (mg/m³)	Approx. efficiency (%)	Typical gas velocity (m/s)	Maximum capacity (m³/s)	Gas pressure drop (mm H_2O)	Liquid rate (m³/10³ m³ gas)	Space required (relative)
Dry collectors								
Settling chamber	50	12,000	50	1.5–3	none	5	—	Large
Baffle chamber	50	12,000	50	5–10	none	3–12	—	Medium
Louvre	20	2500	80	10–20	15	10–50	—	Small
Cyclone	10	2500	85	10–20	25	10–70	—	Medium
Multiple cyclone	5	2500	95	10–20	100	50–150	—	Small
Impingement	10	2500	90	15–30	none	25–50	—	Small
Wet scrubbers								
Gravity spray	10	2500	70	0.5–1	50	25	0.05–0.3	Medium
Centrifugal	5	2500	90	10–20	50	50–150	0.1–1.0	Medium
Impingement	5	2500	95	15–30	50	50–200	0.1–0.7	Medium
Packed	5	250	90	0.5–1	25	25–250	0.7–2.0	Medium
Jet	0.5 to 5 (range)	250	90	10–100	50	none	7–14	Small
Venturi	0.5	250	99	50–200	50	250–750	0.4–1.4	Small
Others								
Fabric filters	0.2	250	99	0.01–0.1	100	50–150	—	Large
Electrostatic precipitators	2	250	99	5–30	1000	5–25	—	Large

A variety of equipment has been developed for gas cleaning. The principal types used in the process industries are listed in Table 10.11, which is adapted from a selection guide given by Sargent (1971). Table 10.11 shows the general field of application of each type in terms of the particle size separated, the expected separation efficiency, and the throughput. It can be used to make a preliminary selection of the type of equipment likely to be suitable for a particular application. Descriptions of the equipment shown in Table 10.11 can be found in various handbooks; Perry and Chilton (1973), Schweitzer (1979); and in several specialist texts, Nonhebel (1972), Strauss (1966), Dorman (1974) and Rose and Wood (1966). Gas cleaning is also covered in Volume 2, Chapter 8.

Gas-cleaning equipment can be classified according to the mechanism employed to separate the particles: gravity settling, impingement, centrifugal force, filtering, washing and electrostatic precipitation.

10.8.1. Gravity settlers (settling chambers)

Settling chambers are the simplest form of industrial gas-cleaning equipment, but have only a limited use; they are suitable for coarse dusts, particles larger than 40 μm. They are essentially long, horizontal, rectangular chambers; through which the gas flows. The solids settle under gravity and are removed from the bottom of the chamber. Horizontal plates or vertical baffles are used in some designs to improve the separation. Settling chambers offer little resistance to the gas flow, and can be designed for operation at high temperature and high pressure, and for use in corrosive atmospheres.

The length of chamber required to settle a given particle size can be estimated from the settling velocity (calculated using Stokes' law) and the gas velocity. A design procedure is given by Maas (1979).

10.8.2. Impingement separators

Impingement separators employ baffles to achieve the separation. The gas stream flows easily round the baffles, whereas the solid particles, due to their higher momentum, tend to continue in their line of flight, strike the baffles and are collected. A variety of baffle designs is used in commercial equipment; a typical example is shown in Fig. 10.43. Impingement separators cause a higher pressure drop than settling chambers, but are capable of separating smaller particle sizes, 10–20 μm.

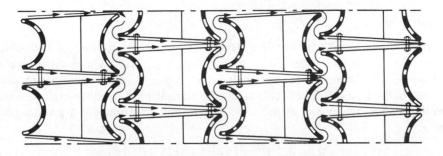

FIG. 10.43. Impingement separator (section showing gas flow)

10.8.3. Centrifugal separators (cyclones)

Cyclones are the principal type of gas–solids separator employing centrifugal force, and are widely used. They are basically simple constructions; can be made from a wide range of materials; and can be designed for high temperature and pressure operation.

Cyclones are suitable for separating particles above about 5 μm diameter; smaller particles, down to about 0·5 μm, can be separated where agglomeration occurs.

The most commonly used design is the reverse-flow cyclone, Fig. 10.44; other configurations are used for special purposes. In a reverse-flow cyclone the gas enters the top chamber tangentially and spirals down to the apex of the conical section; it then moves upward in a second, smaller diameter, spiral, and exits at the top through a central vertical pipe. The solids move radially to the walls, slide down the walls, and are collected at the bottom. Design procedures for cyclones are given by Strauss (1966), Koch and Licht (1977) and Stairmand (1951). The theoretical concepts and experimental work on which the design methods are based are discussed in Volume 2, Chapter 8. Stairmand's method is outlined below and illustrated in Example 10.4.

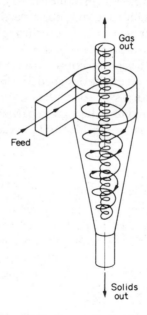

FIG. 10.44. Reverse-flow cyclone

Cyclone design

Stairmand developed two standard designs for gas–solid cyclones: a high-efficiency cyclone, Fig. 10.45a, and a high throughput design, Fig. 10.45b. The performance curves for these designs, obtained experimentally under standard test conditions, are shown in Figs. 10.46a and 10.46b. These curves can be transformed to other cyclone sizes and operating conditions by use of the following scaling equation, for a given separating efficiency:

$$d_2 = d_1 \left[\left(\frac{D_{c_2}}{D_{c_1}} \right)^3 \times \frac{Q_1}{Q_2} \times \frac{\Delta \rho_1}{\Delta \rho_2} \times \frac{\mu_2}{\mu_1} \right]^{1/2} \tag{10.8}$$

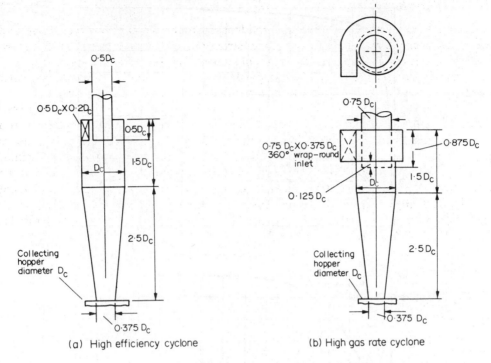

FIG. 10.45. Standard cyclone dimension

where d_1 = mean diameter of particle separated at the standard conditions, at the chosen separating efficiency, Figs. 10.46a or 10.46b,

d_2 = mean diameter of the particle separated in the proposed design, at the same separating efficiency,

D_{c_1} = diameter of the standard cyclone = 8 inches (203 mm),

D_{c_2} = diameter of proposed cyclone, mm,

Q_1 = standard flow rate:

for high efficiency design = 223 m³/h,

for high throughput design = 669 m³/h,

Q_2 = proposed flow rate, m³/h,

$\Delta\rho_1$ = solid–fluid density difference in standard conditions = 2000 kg/m³,

$\Delta\rho_2$ = density difference, proposed design,

μ_1 = test fluid viscosity (air at 1 atm, 20°C)

= 0·018 mN s/m²,

μ_2 = viscosity, proposed fluid.

A performance curve for the proposed design can be drawn up from Fig. 10.46a or 10.46b by multiplying the grade diameter at, say, each 10 per cent increment of efficiency, by the scaling factor given by equation 10.8; as shown in Fig. 10.47.

An alternative method of using the scaling factor, that does not require redrawing the performance curve, is used in Example 10.4. The cyclone should be designed to give an inlet velocity of between 9 and 27 m/s (30 to 90 ft/s); the optimum inlet velocity has been found to be 15 m/s (50 ft/s).

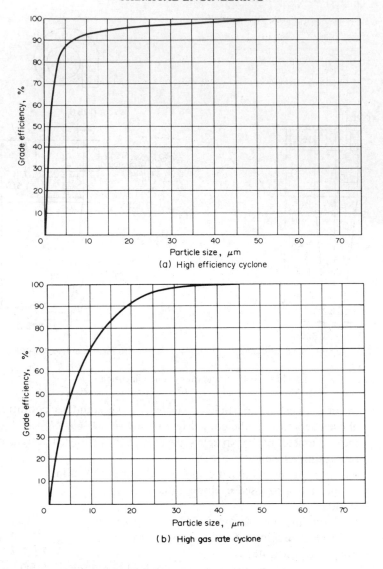

FIG. 10.46. Performance curves, standard conditions

Pressure drop

The pressure drop in a cyclone will be due to the entry and exit losses, and friction and kinetic energy losses in the cyclone. The empirical equation given by Stairmand (1949) can be used to estimate the pressure drop:

$$\Delta P = \frac{\rho_f}{203} \left\{ u_1^2 \left[1 + 2\phi^2 \left(\frac{2r_t}{r_e} - 1 \right) \right] + 2u_2^2 \right\} \qquad (10.9)$$

where ΔP = cyclone pressure drop, millibars,
   ρ_f = gas density, kg/m³,
   u_1 = inlet duct velocity, m/s,
   u_2 = exit duct velocity, m/s,
   r_t = radius of circle to which the centre line of the inlet is tangential, m,
   r_e = radius of exit pipe, m,
   ϕ = factor from Fig. 10.48,

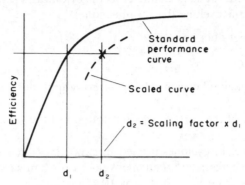

FIG. 10.47. Scaled performance curve

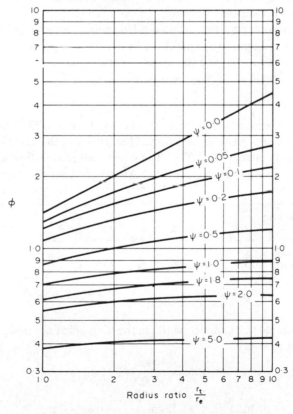

FIG. 10.48. Cyclone pressure drop factor

ψ = parameter in Fig. 10.48, given by:

$$\psi = f_c \frac{A_s}{A_1}$$

f_c = friction factor, taken as 0·005 for gases,

A_s = surface area of cyclone exposed to the spinning fluid, m^2.

For design purposes this can be taken as equal to the surface area of a cylinder with the same diameter as the cylone and length equal to the total height of the cyclone (barrel plus cone).

A_1 = area of inlet duct, m^2.

General design procedure

1. Select either the high-efficiency or high-throughput design, depending on the performance required.
2. Obtain an estimate of the particle size distribution of the solids in the stream to be treated.
3. Estimate the number of cyclones needed in parallel.
4. Calculate the cyclone diameter for an inlet velocity of 15 m/s (50 ft/s). Scale the other cyclone dimensions from Figs. 10.45a or 10.45b.
5. Calculate the scale-up factor for the transposition of Figs. 10.46a or 10.46b.
6. Calculate the cyclone performance and overall efficiency (recovery of solids). If unsatisfactory try a smaller diameter.
7. Calculate the cyclone pressure drop and, if required, select a suitable blower.
8. Cost the system and optimise to make the best use of the pressure drop available, or, if a blower is required, to give the lowest operating cost.

Example 10.4

Design a cyclone to recover solids from a process gas stream. The anticipated particle size distribution in the inlet gas is given below. The density of the particles is 2500 kg/m^3, and the gas is essentially nitrogen at 150°C. The stream volumetric flow-rate is 4000 m^3/h, and the operation is at atmospheric pressure. An 80 per cent recovery of the solids is required.

Particle size (μm)	50	40	30	20	10	5	2
Percentage by weight less than	90	86	80	70	45	25	10

Solution

As 45 per cent of the particles are below 10 μm the high-efficiency design will be required to give the specified recovery.

$$\text{Flow-rate} = \frac{4000}{3600} = 1\cdot11 \text{ m}^3/\text{s}$$

$$\text{Area of inlet duct, at 15 m/s} = \frac{1\cdot11}{15} = 0\cdot07 \text{ m}^2$$

From Fig. 10.45a, duct area $= 0.5 D_c \times 0.2 D_c$

$$\text{so, } D_c = 0.84$$

This is clearly too large compared with the standard design diameter of 0.203 m. Try four cyclones in parallel, $D_c = 0.42$ m.

$$\text{Flow-rate per cyclone } = 1000 \, \text{m}^3/\text{h}$$

$$\text{Density of gas at } 150°C = \frac{28}{22.4} \times \frac{273}{423} = 0.81 \, \text{kg/m}^2,$$

negligible compared with the solids density

$$\text{Viscosity of N}_2 \text{ at } 150°C = 0.023 \, \text{cp(mN s/m}^2\text{)}$$

From equation 10.8,

$$\text{scaling factor} = \left[\left(\frac{0.42}{0.203} \right)^3 \times \frac{223}{1000} \times \frac{2000}{2500} \times \frac{0.023}{0.018} \right]^{1/2} = \underline{\underline{1.42}}$$

The performance calculations, using this scaling factor and Fig. 10.46a, are set out in the table below:

Calculated performance of cyclone design, Example 10.4

1	2	3	4	5	6	7
Particle size (μm)	Per cent in range	Mean particle size × scaling factor	Efficiency at scaled size (%) (Fig. 10.46a)	Collected (2) × (4)	Grading at exit (2)−(5)	Per cent at exit
50	10	71	100	10	0	0
40–50	4	64	99.5	4.0	0	0
40–30	6	50	99	5.9	0.1	0.6
30–20	10	36	98	9.8	0.2	1.1
20–10	25	21	95	24.0	1.0	5.6
10–5	20	11	92	18.4	1.6	8.9
5–2	15	5	60	9.0	6.0	33.5
0–2	10	1.5	10	1.0	9.0	50.3
	100		overall collection efficiency	$\underline{\underline{82.0}}$	17.9	100.0

The collection efficiencies shown in column 4 of the table were read from Fig. 10.46a at the scaled particle size, column 3. The overall collection efficiency satisfies the specified solids recovery. The proposed design with dimension in the proportions given in Fig. 10.45a is shown in Fig. 10.49.

Pressure-drop calculation

$$\text{Area of inlet duct, } A_1, = 210 \times 80 = 16,800 \, \text{mm}^2$$

$$\text{Cyclone surface area, } A_s = \pi 420 \times (630 + 1050)$$

$$= 2.218 \times 10^6 \, \text{mm}^2$$

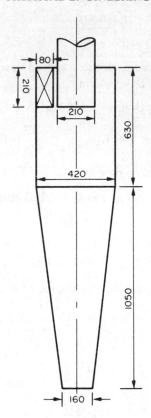

FIG. 10.49. Proposed cyclone design, all dimensions mm (Example 10.4)

f_c taken as 0·005

$$\psi = \frac{f_c, A_s}{A_1} = \frac{0·005 \times 2·218 \times 10^6}{16,800} = 0·66$$

$$\frac{r_t}{r_e} = \frac{(420 - (80/2))}{210} = 1·81$$

From Fig. 10.48, $\phi = 0·9$.

$$u_1 = \frac{1000}{3600} \times \frac{10^6}{16,800} = 16·5 \text{ m/s}$$

Area of exit pipe $= \dfrac{\pi \times 210^2}{4} = 34,636 \text{ mm}^2$

$$u_2 = \frac{1000}{3600} \times \frac{10^6}{34,636} = 8·0 \text{ m/s}$$

From equation 10.6

$$\Delta P = \frac{0·81}{203} [16·5^2[1 + 2 \times 0·9^2(2 \times 1·81 - 1)] + 2 \times 8·0^2]$$

$$= \underline{6·4 \text{ millibar}} \text{ (67 mm } H_2O)$$

This pressure drop looks reasonable.

10.8.4. Filters

The filters used for gas cleaning separate the solid particles by a combination of impingement and filtration; the pore sizes in the filter media used are too large simply to filter out the particles. The separating action relies on the precoating of the filter medium by the first particles separated; which are separated by impingement on the filter medium fibres. Woven or felted cloths of cotton and various synthetic fibres are commonly used as the filter media. Glass-fibre mats and paper filter elements are also used.

A typical example of this type of separator is the bag filter, which consists of a number of bags supported on a frame and housed in a large rectangular chamber, Fig. 10.50. The deposited solids are removed by mechanically vibrating the bag, or by periodically reversing the gas flow. Bag filters can be used to separate small particles, down to around 1 μm, with a high separating efficiency. Commercial units are available to suit most applications and should be selected in consultation with the vendors.

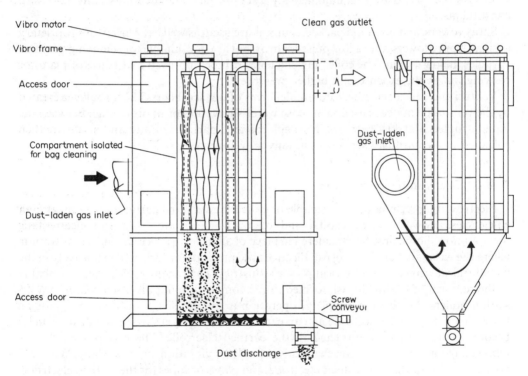

FIG. 10.50. Multi-compartment vibro bag filter

Air filters

Dust-free air is required for many process applications. The requirements of air filtration differ from those of process gas filtration mainly in that the quantity of dust to be removed will be lower, typically less than 10 mg/m^3 ($\sim$ 5 grains per 1000 ft^3); and also in that there is no requirement to recover the material collected.

Three basic types of air filter are used: viscous, dry and continuous. Viscous and dry units are similar in construction, but the filter medium of the viscous type is coated with a viscous material, such as a mineral oil, to retain the dust. The filters are made up from standard, preformed, sections, supported on a frame in a filter housing. The sections are removed periodically for cleaning or replacement. Various designs of continuous filtration equipment are also available, employing either viscous or dry filter elements, but in which the filter is cleaned continuously. A comprehensive description of air-filtration equipment is given by Strauss (1966).

10.8.5. Wet scrubbers (washing)

In wet scrubbing the dust is removed by counter-current washing with a liquid, usually water, and the solids are removed as a slurry. The principal mechanism involved is the impact (impingement) of the dust particles and the water droplets. Particle sizes down to 0·5 μm can be removed in suitably designed scrubbers. In addition to removing solids, wet scrubbers can be used to simultaneously cool the gas and neutralise any corrosive constituents.

Spray towers, and plate and packed columns are used, as well as a variety of proprietary designs. Spray towers have a low pressure drop but are not suitable for removing very fine particles, below 10 μm. The collecting efficiency can be improved by the use of plates or packing but at the expense of a higher pressure drop.

Venturi and orifice scrubbers are simple forms of wet scrubbers. The turbulence created by the venturi or orifice is used to atomise water sprays and promote contact between the liquid droplets and dust particles. The agglomerated particles of dust and liquid are then collected in a centrifugal separator, usually a cyclone.

10.8.6. Electrostatic precipitators

Electrostatic precipitators are capable of collecting very fine particles, $< 2\mu$m, at high efficiencies. However, their capital and operating costs are high, and electrostatic precipitation should only be considered in place of alternative processes, such as filtration, where the gases are hot or corrosive. Electrostatic precipitators are used extensively in the metallurgical, cement and electrical power industries. Their main application is probably in the removal of the fine fly ash formed in the combustion of pulverised coal in power-station boilers. The basic principle of operation is simple. The gas is ionised in passing between a high-voltage electrode and an earthed (grounded) electrode; the dust particles become charged and are attracted to the earthed electrode. The precipitated dust is removed from the electrodes mechanically, usually by vibration, or by washing. Wires are normally used for the high-voltage electrode, and plates or tubes for the earthed electrode. A typical design is shown in Fig. 10.51. A full description of the construction, design and application of electrostatic precipitators is given by Schneider et al. (1975) and by Rose and Wood (1966).

10.9. Gas–liquid separators

The separation of liquid droplets and mists from gas and vapour streams is analogous to the separation of solid particles and, with the possible exception of filtration, the same techniques and equipment can be used.

Where some carry-over of fine droplets can be tolerated, it is often sufficient to rely on gravity settling in vertical or horizontal separating vessels (knock-out pots).

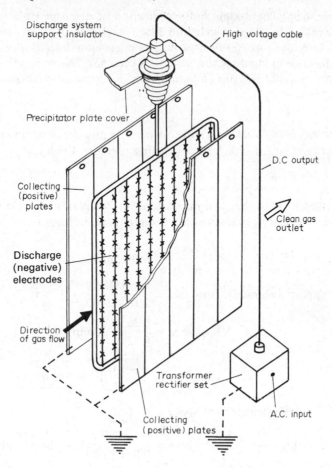

Discharge system
support insulator

High voltage cable

Precipitator plate cover

D.C output

Collecting
(positive)
plates

Clean gas
outlet

Discharge
(negative)
electrodes

Direction
of gas flow

Transformer
rectifier set

A.C. input

Collecting
(positive) plates

FIG. 10.51. Electrostatic precipitator

Equation 10.10 can be used to estimate the design gas velocity for the approximate sizing of vertical separating vessels; see Example 10.5

$$\hat{u}_v = 0.035 \sqrt{\rho_L/\rho_v} \qquad (10.10)$$

where $\hat{u}_v$ = maximum design vapour velocity, m/s,
   ρ_v = vapour density, kg/m^3,
   ρ_L = liquid density, kg/m^3.

A disengagement height equal to the vessel diameter should be provided above the liquid level. The liquid level will depend on the hold-up time required for smooth operation and control; typically, 10 minutes would be allowed.

   The separating efficiency can be improved by fitting baffles or plates in the disengagement space.

   Detailed design methods for vapour–liquid separation vessels are given by Evans (1980).

   Knitted mesh demisting pads are frequently used to improve the performance of separating vessels where the droplets are likely to be small, down to 1 μm, and where high separating efficiencies are required. Proprietary demister pads are available in a wide range of materials, metals and plastics. Separating efficiencies above 99 per cent can be obtained, with low pressure drop. The design and specification of demister pads for gas–liquid separations is discussed by York (1954), and Pryce Bayley and Davies (1973).

Cyclone separators are also frequently used for gas–liquid separation, and can be designed using the same methods as for gas–solids cyclones. The inlet velocity should be kept below 30 m/s to avoid pick-up of liquid from the cyclone surfaces.

Example 10.5

Make a preliminary design of a separator to separate a mixture of steam and water; flow rates: water 1000 kg/h, steam 2000 kg/h; operating pressure 4 bar.

Solution

From steam tables, at 4 bar, steam saturation temperature = 143·6°C, density = 2·16 kg/m^3, water density = 926·4 kg/m^3. From equation 10.10:

$$\hat{u}_v = 0\cdot035 \sqrt{\frac{926\cdot4}{2\cdot16}} = 0\cdot72\,\text{m/s}$$

$$\text{Vapour volumetric flow-rate} = \frac{2000}{2\cdot16 \times 3600} = 0\cdot26\,\text{m}^3/\text{s}$$

$$\text{Vessel area} = \frac{0\cdot26}{0\cdot72} = 0\cdot36\,\text{m}^2$$

$$\text{Diameter} = \sqrt{\frac{4 \times 0\cdot36}{\pi}} = \underline{\underline{0\cdot68\,\text{m}}}$$

$$\text{Disengagement space} = \underline{\underline{0\cdot68\,\text{m}}}$$

$$\text{Liquid volumetric flow} = \frac{1000}{926\cdot4 \times 3600} = 3 \times 10^{-4}\,\text{m}^3/\text{s}$$

$$\text{Volume for 10 min hold-up} = 3 \times 10^{-4} \times 600 = 0\cdot18\,\text{m}^3$$

$$\text{Liquid depth} = \frac{0\cdot18}{0\cdot36} = \underline{\underline{0\cdot5\,\text{m}}}$$

Proposed design

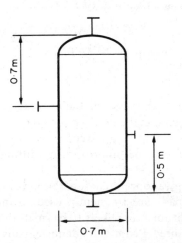

0·7 m

0·5 m

0·7 m

10.10. Crushing and grinding (comminution) equipment

Crushing is the first step in the process of size reduction; reducing large lumps to manageable sized pieces. For some processes crushing is sufficient, but for chemical processes it is usually followed by grinding to produce a fine-sized powder. Though many articles have been published on comminution, and Marshall (1974) mentions over 4000, the subject remains essentially empirical. The designer must rely on experience, and the advice of the equipment manufacturers, when selecting and sizing crushing and grinding equipment; and to estimate the power requirements. Several models have been proposed for the calculation of the energy consumed in size reduction; some of which are discussed in Volume 2, Chapter 2. For a fuller treatment of the subject the reader should refer to the book by Lowrison (1974).

The main factors to be considered when selecting equipment for crushing and grinding are:

1. The size of the feed.
2. The size reduction ratio.
3. The required particle size distribution of the product.
4. The throughput.
5. The properties of the material: hardness, abrasiveness, stickiness, density, toxicity, flammability.
6. Whether wet grinding is permissible.

The selection guides given by Lowrison (1974) and Marshall (1974), which are reproduced in Tables 10.12 and 10.13, can be used to make a preliminary selection based on particle size and material hardness. Descriptions of most of the equipment listed in these tables are given in Volume 2, or can be found in the literature; Perry and Chilton (1973), Hiorns (1970), Lowrison (1974). The most commonly used equipment for coarse size reduction are jaw crushers and rotary crushers; and for grinding, ball mills or their variants: pebble, roll and tube mills.

TABLE 10.12. *Selection of comminution equipment (after Lowrison, 1974)*

TABLE 10.13. Selection of comminution equipment for various materials (after Marshall, 1974)
Note: Moh's scale of hardness is given in Table 10.12.

Material class no	Material classification	Typical materials in class	Suitable equipment for product size classes			Remarks
			Down to 5 mesh	Between 5 and 300 mesh	Less than 300 mesh	
1	Hard and tough	Mica Scrap and powdered metals	Jaw crushers Gyratory crushers Cone crushers Autogeneous mills	Ball, pebble, rod and cone mills Tube mills Vibration mills	Ball, pebble and cone mills Tube mills Vibration and vibro-energy mills Fluid-energy mills	Moh's Hardness 5–10, but includes other tough materials of lower hardness
2	Hard, abrasive and brittle	Coke, quartz, granite	Jaw crushers Gyratory and cone crushers Roll crushers	Ball, pebble, rod and cone mills Vibration mills Roller mills	Ball, pebble and cone mills Tube mills Vibration and vibro-energy mills Fluid-energy mills	Moh's Hardness 5–10 High wear rate/contamination in high-speed machinery Use machines with abrasion resistant linings
3	Intermediate hard, and friable	Barytes, fluor-spar, limestone	Jaw crushers Gyratory crushers Roll crushers Edge runner mills Impact breakers Autogeneous mills Cone crushers	Ball, pebble, rod and cone mills Tube mills Ring roll mills Ring ball mills Roller mills Peg and disc mills Cage mills Impact breakers Vibration mills	Ball, pebble and cone mills Tube mills Perl mills Vibration and vibro-energy mills Fluid-energy mills	Moh's Hardness 3–5

4	Fibrous, low abrasion and possibly tough	Wood, asbestos	Cone crushers Roll crushers Edge runner mills Autogeneous mills Impact breakers	Ball, pebble, rod and cone mills Tube mills Roller mills Peg and disc mills Cage mills Impact breakers Vibration mills Rotary cutters and dicers	Ball, pebble and cone mills Tube mills Sand mills Perl mills Vibration and vibro-energy mills Colloid mills	Wide range of hardness Low-temperature, liquid nitrogen, useful to embrittle soft but tough materials
5	Soft and friable	Sulphur, gypsum rock salt	Cone crushers Roll crushers Edge runner mills Impact breakers Autogeneous mills	Ball, pebble and cone mills Tube mills Ring roll mills Ring ball mills Roller mills Peg and disc mills Cage mills Impact breakers Vibration mills	Ball, pebble and cone mills Tube mills Sand mills Perl mills Vibration and vibro-energy mills Colloid mills Fluid-energy mills Peg and disc mills	Moh's hardness 1-3
6	Sticky	Clays, certain organic pigments	Roll crushers Impact breakers Edge runner mills	Ball, pebble, rod and cone mills* Tube mills* Peg and disc mills Cage mills Ring roll mills	Ball, pebble and cone mills* Tube mills* Sand mills Perl mills Vibration and vibro-energy mills Colloid mills	Wide range of Moh's Hardness although mainly less than 3 Tends to clog *Wet grinding employed except for certain exceptional cases

* All ball, pebble, rod and cone mills, edge runner mills, tube mills, vibration mills and some ring ball mills may be used wet or dry except where stated. The perl mills, sand mills and colloid mills may be used for wet milling only.

10.11. Mixing equipment

The preparation of mixtures of solids, liquids and gases is an essential part of most production processes in the chemical and allied industries; covering all processing stages, from the preparation of reagents through to the final blending of products. The equipment used depends on the nature of the materials and the degree of mixing required. Mixing is often associated with other operations, such as reaction and heat transfer. Liquid and solids mixing operations are frequently carried out as batch processes.

In this section, mixing processes will be considered under three separate headings: gases, liquids and solids.

10.11.1. Gas mixing

Specialised equipment is seldom needed for mixing gases, which because of their low viscosities mix easily. The mixing given by turbulent flow in a length of pipe is usually sufficient for most purposes. Turbulence promoters, such as orifices or baffles, can be used to increase the rate of mixing. The piping arrangements used for inline mixing are discussed in the section on liquid mixing.

10.11.2. Liquid mixing

The following factors must be taken into account when choosing equipment for mixing liquids:

1. Batch of continuous operation.
2. Nature of the process: miscible liquids, preparation of solutions, or dispersion of immiscible liquids.
3. Degree of mixing required.
4. Physical properties of the liquids, particularly the viscosity.
5. Whether the mixing is associated with other operations: reaction, heat transfer.

For the continuous mixing of low viscosity fluids inline mixers can be used. For other mixing operations stirred vessels or proprietary mixing equipment will be required.

Inline mixing

Static devices which promote turbulent mixing in pipelines provide an inexpensive way of continuously mixing fluids. Some typical designs are shown in Figs. 10.52a,b,c. A simple mixing tee, Fig. 10.52a, followed by a length of pipe equal to 10 to 20 pipe diameters, is suitable for mixing low viscosity fluids ($\leqslant 50 \, \text{mN s/m}^2$) providing the flow is turbulent, and the densities and flow-rates of the fluids are similar.

With injection mixers (Figs. 10.52b,c), in which the one fluid is introduced into the flowing stream of the other through a concentric pipe or an annular array of jets, mixing will take place by entrainment and turbulent diffusion. Such devices should be used where one flow is much lower than the other, and will give a satisfactory blend in about 80 pipe diameters. The inclusion of baffles or other flow restrictions will reduce the mixing length required.

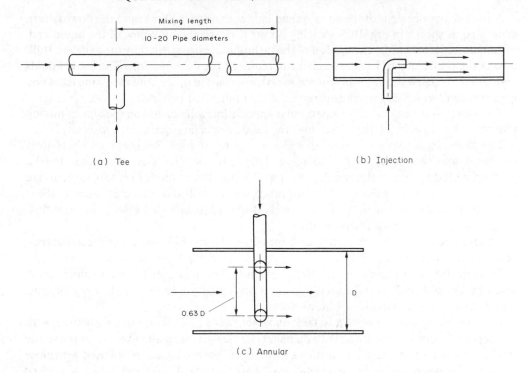

(a) Tee (b) Injection

(c) Annular

FIG. 10.52. Inline mixers

The static inline mixer shown in Fig. 10.53 is effective in both laminar and turbulent flow, and can be used to mix viscous mixtures. The division and rotation of the fluid at each element causes rapid radial mixing; see Rosenzweig (1977). The dispersion and mixing of liquids in pipes is discussed by Clayton *et al.* (1968) and Lee and Brodskey (1964).

Centrifugal pumps are effective inline mixers for blending and dispersing liquids. Various proprietary motor-driven inline mixers are also used for special applications; see Perry and Chilton (1973).

FIG. 10.53. Static mixer (Kenics Corporation)

Stirred tanks

Mixing vessels fitted with some form of agitator are the most commonly used type of equipment for blending liquids and preparing solutions.

Liquid mixing in stirred tanks is covered in Volume 2, Chapter 13, and in several textbooks; Uhl and Gray (1967), Holland and Chapman (1966a), Sterbacek and Tausk (1965), Nagata (1975).

A typical arrangement of the agitator and baffles in a stirred tank, and the flow pattern generated, is shown in Fig. 10.54. Mixing occurs through the bulk flow of the liquid and, on a microscopic scale, by the motion of the turbulent eddies created by the agitator. Bulk flow is the predominant mixing mechanism required for the blending of miscible liquids and for solids suspension. Turbulent mixing is important in operations involving mass and heat transfer; which can be considered as shear controlled processes.

The most suitable agitator for a particular application will depend on the type of mixing required, the capacity of the vessel, and the fluid properties, mainly the viscosity.

The three basic types of impeller which are used at high Reynolds numbers (low viscosity) are shown in Figs. 10.55a,b,c. They can be classified according to the predominant direction of flow leaving the impeller. The flat-bladed (Rushton) turbines are essentially radial-flow devices, suitable for processes controlled by turbulent mixing (shear controlled processes). The propeller and pitched-bladed turbines are essentially axial-flow devices, suitable for bulk fluid mixing.

Paddle, anchor and helical ribbon agitators (Figs. 10.56a,b,c), and other special shapes, are used for more viscous fluids.

The selection chart given in Fig. 10.57, which has been adapted from a similar chart given by Penney (1970), can be used to make a preliminary selection of the agitator type, based on the liquid viscosity and tank volume.

For turbine agitators, impeller to tank diameter ratios of up to about 0·6 are used, with the depth of liquid equal to the tank diameter. Baffles are normally used, to improve the mixing and reduce problems from vortex formation. Anchor agitators are used with close clearance between the blades and vessel wall, anchor to tank diameter ratios of 0·95 or higher. The selection of agitators for dispersing gases in liquids is discussed by Hicks (1976).

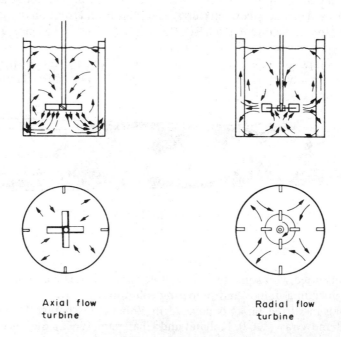

Axial flow
turbine

Radial flow
turbine

FIG. 10.54. Agitator arrangements and flow patterns

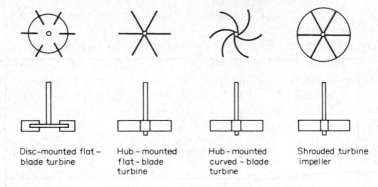

Disc-mounted flat-blade turbine   Hub-mounted flat-blade turbine   Hub-mounted curved-blade turbine   Shrouded turbine impeller

(a) Turbine impeller

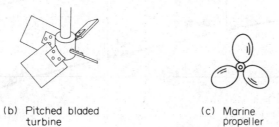

(b) Pitched bladed turbine

(c) Marine propeller

FIG. 10.55. Basic impeller types

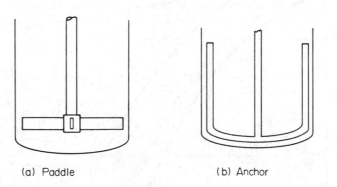

(a) Paddle   (b) Anchor

(c) Helical ribbon

FIG. 10.56. Low-speed agitators

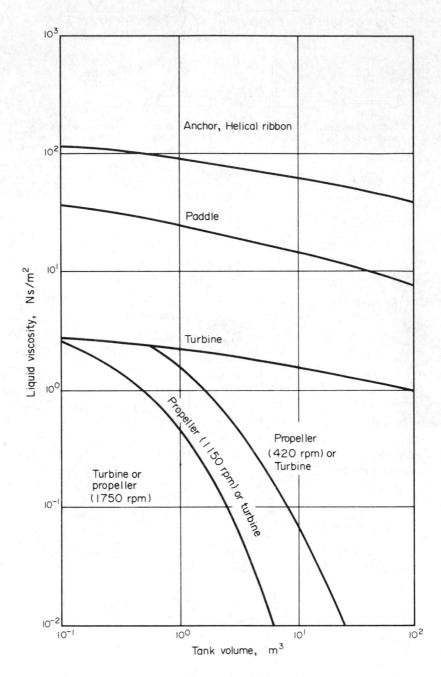

Fig. 10.57. Agitator selection guide

Agitator power consumption

The shaft power required to drive an agitator can be estimated using the following generalised dimensionless equation, the derivation of which is given in Volume 2, Chapter 13.

$$N_p = K \, Re^b \, Fr^c \tag{10.11}$$

where N_p = power number = $\dfrac{P}{D^5 N^3 \rho}$,

Re = Reynolds number = $\dfrac{D^2 N \rho}{\mu}$,

Fr = Froude number = $\dfrac{D N^2}{g}$,

P = shaft power, W,

K = a constant, dependent on the agitator type, size, and the agitator-tank geometry,

ρ = fluid density, kg/m^3,
μ = fluid viscosity, $N \, s/m^2$,
N = agitator speed, s^{-1} (rps),
D = agitator diameter, m,
g = gravitational acceleration, $9\cdot81 \ m/s^2$.

Values for the constant K and the indices b and c for various types of agitator, tank–agitator geometries, and dimensions, can be found in the literature; Rushton *et al.* (1950), Nagata (1975). A useful review of the published correlations for agitator power consumption and heat transfer in agitated vessels is given by Wilkinson and Edwards (1972); they include correlations for non-Newtonian fluids. Typical power curves for propeller and turbine agitators are given in Figs. 10.58 and 10.59. In the laminar flow region the index "b" = 1; and at high Reynolds number the power number is independent of the Froude number; index "c" = 0.

Side-entering agitators

Side-entering agitators are used for blending low viscosity liquids in large tanks, where it is impractical to use conventional agitators supported from the top of the tank; see Oldshue *et al.* (1956).

Where they are used with flammable liquids, particular care must be taken in the design and maintenance of the shaft seals, as any leakage may cause a fire.

For blending flammable liquids, the use of liquid jets should be considered as an "intrinsically" safer option; see Fossett and Prosser (1949).

10.11.3. Solids and pastes

A great variety of specialised equipment has been developed for mixing dry solids and pastes (wet solids). The principal types of equipment and their fields of application are

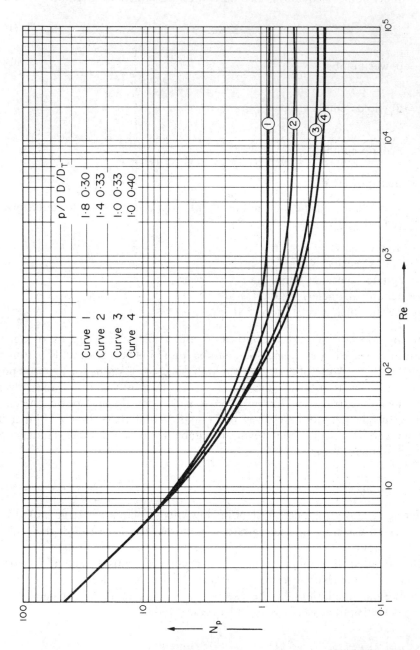

Fig. 10.58. Power correlation for single three-bladed propellers baffled, (from Uhl and Gray (1967) with permission). p = blade pitch, D = impeller diameter, D_T = tank diameter

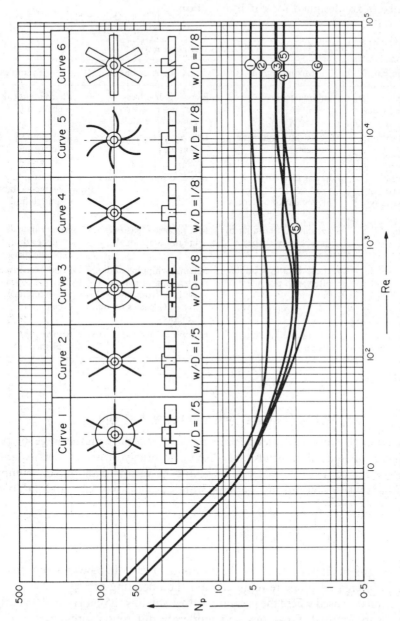

FIG. 10.59. Power correlations for baffled turbine impellers, for tank with 4 baffles (From Uhl and Gray (1967) with permission). w = impeller width, D = impeller diameter

given in Table 10.14. Descriptions of the equipment can be found in the literature; Perry and Chilton (1973), Reid (1979) and Mead (1964). Cone blenders are used for free-flowing solids. Ribbon blenders can be used for dry solids and for blending liquids with solids. Z-blade mixers and pan mixers are used for kneading heavy pastes and doughs. Most solid and paste mixers are designed for batch operation.

TABLE 10.14. *Solids and paste mixers*

Type of equipment	Mixing action	Applications	Examples
Rotating: cone, double cone, drum	Tumbling action	Blending dry, free-flowing powders, granules, crystals	Pharmaceuticals, food, chemicals
Air blast fluidisation	Air blast lifts and mixes particles	Dry powders and granules	Milk powder; detergents, chemicals
Horizontal trough mixer, with ribbon blades, paddles or beaters	Rotating element produces contra-flow movement of materials	Dry and moist powders	Chemicals, food, pigments, tablet granulation
Z-blade mixers	Shearing and kneading by the specially shaped blades	Mixing heavy pastes, creams and doughs	Bakery industry, rubber doughs, plastic dispersions
Pan mixers	Vertical, rotating paddles, often with planetary motion	Mixing, whipping and kneading of materials ranging from low viscosity pastes to stiff doughs	Food, pharmaceuticals and chemicals, printing inks and ceramics
Cylinder mixers, single and double	Shearing and kneading action	Compounding of rubbers and plastics	Rubbers, plastics, and pigment dispersion

10.12. Transport and storage of materials

In this section the principal means used for the transport and storage of process materials: gases, liquids and solids are discussed briefly. Further details and full descriptions of the equipment used can be found in various handbooks. Pumps and compressors are also discussed in Chapters 3 and 5 of this volume, and in Volume 1, Chapter 6.

10.12.1. Gases

The type of equipment best suited for the pumping of gases in pipelines depends on the flow-rate, the differential pressure required, and the operating pressure.

In general, fans are used where the pressure drop is small, $< 35\,\text{cm}\,H_2O$ (0·03 bar); axial flow compressors for high flow-rates and moderate differential pressures; centrifugal compressors for high flow-rates and, by staging, high differential pressures. Reciprocating compressors can be used over a wide range of pressures and capacities, but are normally only specified in preference to centrifugal compressors where high pressures are required at relatively low flow-rates.

Reciprocating, centrifugal and axial flow compressors are the principal types used in the chemical process industries, and the range of application of each type is shown in Fig. 10.60 which has been adapted from a similar diagram by Dimoplon (1978). A more comprehensive selection guide is given in Table 10.15. Diagrammatic sketches of the compressors listed are given in Fig. 10.61.

Several textbooks are available on compressor design, selection and operation; Stepanoff (1955), Chlumsky (1965), Kovat (1964).

Vacuum production

The production of vacuum (sub-atmospheric pressure) is required for many chemical engineering processes; for example, vacuum distillation, drying and filtration. The type of vacuum pump needed will depend on the degree of vacuum required, the capacity of the system and the rate of air inleakage.

Reciprocating and rotary positive displacement pumps are commonly used where moderately low vacuum is required, about 10 mm Hg (0·013 bar), at moderate to high flow rates; such as in vacuum filtration.

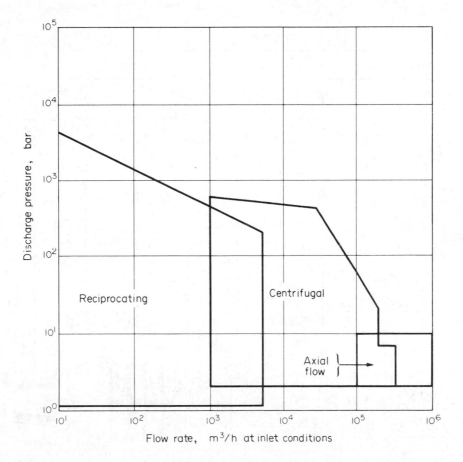

FIG. 10.60. Compressor operating ranges

TABLE 10.15. *Operating range of compressors and blowers (after Begg, 1966)*

Type of compressor	Normal maximum speed	Normal maximum capacity	Normal maximum pressure (differential) (bar)	
	(rpm)	(m³/h)	Single stage	Multiple stage
Displacement				
1. Reciprocating	300	85,000	3·5	5000
2. Sliding vane	300	3400	3·5	8
3. Liquid ring	200	2550	0·7	1·7
4. Rootes	250	4250	0·35	1·7
5. Screw	10,000	12,750	3·5	17
Dynamic				
6. Centrifugal fan	1000	170,000		0·2
7. Turbo blower	3000	8500	0·35	1·7
8. Turbo compressor	10,000	136,000	3·5	100
9. Axial flow fan	1000	170,000	0·35	2·0
10. Axial flow blower	3000	170,000	3·5	10

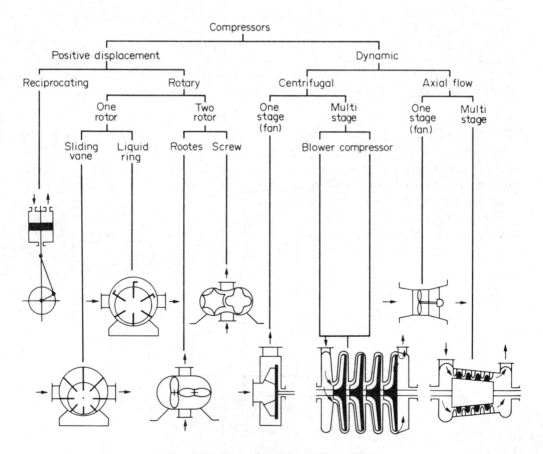

FIG. 10.61. Type of compressor (Begg, 1966)

Steam-jet ejectors are versatile and economic vacuum pumps and are frequently used, particularly in vacuum distillation. They can handle high vapour flow rates and, by using several ejectors in series, can produce low pressures, down to about 0·1 mm Hg (0·13 mbar).

The operating principle of steam-jet ejectors is explained in Volume 1, Chapter 6. Their specification, sizing and operation are covered in a comprehensive series of papers by Power (1964). Diffusion pumps are used where very low pressures are required (hard vacuum) for processes such as molecular distillation.

Storage

Gases are stored at low pressure in gas holders similar to those used for towns gas, which are a familiar sight in any town. The liquid sealed type are most commonly used. These consist of a number of telescopic sections (lifts) which rise and fall as gas is added to or withdrawn from the holder. The dry sealed type is used where the gas must be kept dry. In this type the gas is contained by a piston moving in a large vertical cylindrical vessel. Water seal holders are intrinsically safer for use with flammable gases than the dry seal type; as any leakage through the piston seal may form an explosive mixture in the closed space between the piston and the vessel roof. Details of the construction of gas holders can be found in text books on Gas Engineering; Meade (1921), Smith (1945).

Gases are stored at high pressures where this is a process requirement and to reduce the storage volume. For some gases the volume can be further reduced by liquefying the gas by pressure or refrigeration. Cylindrical and spherical vessels (Horton spheres) are used. The design of pressure vessels is discussed in Chapter 13.

10.12.2. *Liquids*

The selection of pumps for liquids is discussed in Chapter 5. Descriptions of most of the types of pumps used in the chemical process industries are given in Volume 1, Chapter 6. Several textbooks and handbooks have also been published on this subject; Holland and Chapman (1966b), Pollak (1980), Warring (1979).

The principal types used and their operating pressures and capacity ranges are summarised in Table 10.16. Centrifugal pumps will normally be the first choice for

TABLE 10.16. *Normal operating range of pumps*

Type	Capacity range (m³/h)	Typical head (m of water)
Centrifugal	$0·25–10^3$	10–50 300 (multistage)
Reciprocating	0·5 –500	50–200
Diaphragm	0·05–50	5–60
Rotary gear and similar	0·05–500	60–200
Rotary sliding vane or similar	0·25–500	7–70

pumping process fluids, the other types only being used for special applications; such as the use of reciprocating and gear pumps for metering.

Pump shaft power

The power required for pumping an incompressible fluid is given by:

$$\text{Power} = \frac{\Delta P Q_p}{\eta_p} \times 100 \tag{10.12}$$

where ΔP = pressure differential across the pump, N/m^2,
$\quad Q_p$ = flow rate, m^3/s,
$\quad \eta_p$ = pump efficiency, per cent.

The efficiency of centrifugal pumps depends on their size. The values given in Fig. 10.62 can be used to estimate the power and energy requirements for preliminary design purposes. The efficiency of reciprocating pumps is usually around 90 per cent.

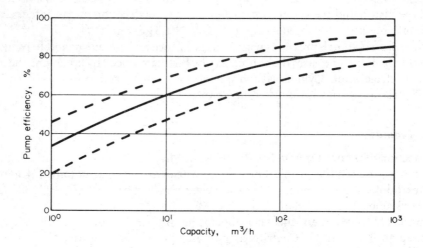

FIG. 10.62. Efficiencies of centrifugal pumps

Storage

Liquids are usually stored in bulk in vertical cylindrical steel tanks. Fixed and floating-roof tanks are used. In a floating-roof tank a movable piston floats on the surface of the liquid and is sealed to the tank walls. Floating-roof tanks are used to eliminate evaporation losses and, for flammable liquids, to obviate the need for inert gas blanketing to prevent an explosive mixture forming above the liquid, as would be the situation with a fixed-roof tank.

Horizontal cylindrical tanks and rectangular tanks are also used for storing liquids, usually for relatively small quantities.

10.12.3. Solids

The movement and storage of solids is usually more expensive than the movement of liquids and gases, which can be easily pumped down a pipeline. The best equipment to use will depend on a number of factors:

1. The throughput.
2. Length of travel.
3. Change in elevation.
4. Nature of the solids: size, bulk density, angle of repose, abrasiveness, corrosiveness, wet or dry.

Belt conveyors are the most commonly used type of equipment for the continuous transport of solids. They can carry a wide range of materials economically over long and short distances; both horizontally or at an appreciable angle, depending on the angle of repose of the solids. A belt conveyor consists of an endless belt of a flexible material, supported on rollers (idlers), and passing over larger rollers at each end, one of which is driven. The belt material is usually fabric-reinforced rubber or plastics; segmental metal belts are also used. Belts can be specified to withstand abrasive and corrosive materials; see BS 490.

Screw conveyors, also called worm conveyors, are used for materials that are free flowing. The basic principle of the screw conveyor has been known since the time of Archimedes. The modern conveyor consists of helical screw rotating in a U-shaped trough. They can be used horizontally or, with some loss of capacity, at an incline to lift materials. Screw conveyors are less efficient than belt conveyors, due to the friction between the solids and the flights of the screw and the trough, but are cheaper and easier to maintain. They are used to convey solids over short distances, and when some elevation (lift) is required. They can also be used for delivering a metered flow of solids.

The most widely used equipment where a vertical lift is required is the bucket elevator. This consists of buckets fitted to an endless chain or belt, which passes over a driven roller or sprocket at the top end. Bucket elevators can handle a wide range of solids, from heavy lumps to fine powders, and are suitable for use with wet solids and slurries.

Pneumatic and hydraulic conveying, in which the solid particles are transported along a pipeline in suspension in a fluid, are discussed in Volume 2, Chapter 7.

Storage

The simplest way to store solids is to pile them on the ground in the open air. This is satisfactory for the long-term storage of materials that do not deteriorate on exposure to the elements; for example, the seasonal stock piling of coal at collieries and power stations. For large stockpiles, permanent facilities are usually installed for distributing and reclaiming the material; travelling gantry cranes, grabs and drag scrapers feeding belt conveyors are used. For small, temporary, storages mechanical shovels and trunks can be used. Where the cost of recovery from the stockpile is large compared with the value of the stock held, storage in silos or bunkers should be considered.

Overhead bunkers, also called bins or hoppers, are normally used for the short-term storage of materials that must be readily available for the process. They are arranged so that the material can be withdrawn at a steady rate from the base of the bunker on to a suitable conveyor. Bunkers must be carefully designed to ensure the free flow of material within the bunker, to avoid packing and bridging. Jenike (1967) and Jenike and Johnson (1970), has studied the flow of solids in containers and developed design methods. All aspects of the design of bins and hoppers, including feeding and discharge systems, are covered in books by Reisner and Rothe (1971), and Stepanoff (1969).

10.13. Reactors

The reactor is the heart of a chemical process. It is the only place in the process where raw materials are converted into products, and reactor design is a vital step in the overall design of the process.

The fundamentals of reactor design are covered comprehensively in Volume 3: Chapter 1 – general principles, Chapter 2 – catalytic reactors and Chapter 5 – biochemical reactors. Many other textbooks have also been published on this subject; Smith (1970), Levenspiel (1972), Carberry (1976). The volumes by Rase (1977) cover the practical aspects of reactor design and include case studies of some industrial reactor designs. The design of electrochemical reactors is covered by Pickett (1979).

The treatment of reactor design in this section will be restricted to a discussion of the selection of the appropriate reactor type for a particular process, and an outline of the steps to be followed in the design of a reactor.

The design of an industrial chemical reactor must satisfy the following requirements:

1. The chemical factors: the kinetics of the reaction. The design must provide sufficient residence time for the desired reaction to proceed to the required degree of conversion.

2. The mass transfer factors: with heterogeneous reactions the reaction rate may be controlled by the rates of diffusion of the reacting species; rather than the chemical kinetics.

3. The heat transfer factors: the removal, or addition, of the heat of reaction.

4. The safety factors: the confinement of hazardous reactants and products, and the control of the reaction and the process conditions.

The need to satisfy these interrelated, and often contradictory factors, makes reactor design a complex and difficult task. However, in many instances one of the factors will predominate and will determine the choice of reactor type and the design method.

10.13.1. Principal types of reactor

The following characteristics are normally used to classify reactor designs:

1. Mode of operation: batch or continuous.
2. Phases present: homogeneous or heterogeneous.
3. Reactor geometry: flow pattern and manner of contacting the phases
    (i) stirred tank reactor;
    (ii) tubular reactor;
    (iii) packed bed, fixed and moving;
    (iv) fluidised bed.

Batch or continuous processing

In a batch process all the reagents are added at the commencement; the reaction proceeds, the compositions changing with time, and the reaction is stopped and the product withdrawn when the required conversion has been reached. Batch processes are suitable for small-scale production and for processes where a range of different products, or grades, is to be produced in the same equipment; for instance, pigments, dyestuffs and polymers.

In continuous processes the reactants are fed to the reactor and the products withdrawn continuously; the reactor operates under steady-state conditions. Continuous production will normally give lower production costs than batch production, but lacks the flexibility of batch production. Continuous reactors will usually be selected for large-scale production. Processes that do not fit the definition of batch or continuous are often referred to as semi-continuous or semi-batch. In a semi-batch reactor some of the reactants may be added, or some of the products withdrawn, as the reaction proceeds. A semi-continuous process can be one which is interrupted periodically for some purpose; for instance, for the regeneration of catalyst.

Homogeneous and heterogeneous reactions

Homogeneous reactions are those in which the reactants, products, and any catalyst used form one continuous phase: gaseous or liquid.

Homogeneous gas phase reactors will always be operated continuously; whereas liquid phase reactors may be batch or continuous. Tubular (pipe-line) reactors are normally used for homogeneous gas-phase reactions; for example, in the thermal cracking of petroleum crude oil fractions to ethylene, and the thermal decomposition of dichloroethane to vinyl chloride. Both tubular and stirred tank reactors are used for homogeneous liquid-phase reactions.

In a heterogeneous reaction two or more phases exist, and the overriding problem in the reactor design is to promote mass transfer between the phases. The possible combination of phases are:

1. Liquid–liquid: immiscible liquid phases; reactions such as the nitration of toluene or benzene with mixed acids, and emulsion polymerisations.
2. Liquid–solid: with one, or more, liquid phases in contact with a solid. The solid may be a reactant or catalyst.
3. Liquid–solid–gas: where the solid is normally a catalyst; such as in the hydrogeneration of amines, using a slurry of platinum on activated carbon as a catalyst.
4. Gas–solid: where the solid may take part in the reaction or act as a catalyst. The reduction of iron ores in blast furnaces and the combustion of solid fuels are examples where the solid is a reactant.

Reactor geometry (type)

The reactors used for established processes are usually complex designs which have been developed (have evolved) over a period of years to suit the requirements of the process, and are unique designs. However, it is convenient to classify reactor designs into the following broad categories.

Stirred tank reactors

Stirred tank (agitated) reactors consist of a tank fitted with a mechanical agitator and a cooling jacket or coils. They are operated as batch reactors or continuously. Several reactors may be used in series.

The stirred tank reactor can be considered the basic chemical reactor; modelling on a large scale the conventional laboratory flask. Tank sizes range from a few litres to several thousand litres. They are used for homogeneous and heterogeneous liquid–liquid and liquid–gas reactions; and for reactions that involve finely suspended solids, which are held in suspension by the agitation. As the degree of agitation is under the designer's control, stirred tank reactors are particularly suitable for reactions where good mass transfer or heat transfer is required.

When operated as a continuous process the composition in the reactor is constant and the same as the product stream, and, except for very rapid reactions, this will limit the conversion that can be obtained in one stage.

The power requirements for agitation will depend on the degree of agitation required and will range from about $0.2 \, kW/m^3$ for moderate mixing to $2 \, kW/m^3$ for intense mixing.

Tubular reactor

Tubular reactors are generally used for gaseous reactions, but are also suitable for some liquid-phase reactions.

If high heat-transfer rates are required, small-diameter tubes are used to increase the surface area to volume ratio. Several tubes may be arranged in parallel, connected to a manifold or fitted into a tube sheet in a similar arrangement to a shell and tube heat exchanger. For high-temperature reactions the tubes may be arranged in a furnace.

The pressure-drop and heat-transfer coefficients in empty tube reactors can be calculated using the methods for flow in pipes given in Volume 1.

Packed bed reactors

There are two basic types of packed-bed reactor: those in which the solid is a reactant, and those in which the solid is a catalyst. Many examples of the first type can be found in the extractive metallurgical industries.

In the chemical process industries the designer will normally be concerned with the second type: catalytic reactors. Industrial packed-bed catalytic reactors range in size from small tubes, a few centimetres diameter, to large diameter packed beds. Packed-bed reactors are used for gas and gas–liquid reactions. Heat-transfer rates in large diameter packed beds are poor and where high heat-transfer rates are required fluidised beds should be considered.

Fluidised bed reactors

The essential features of a fluidised bed reactor is that the solids are held in suspension by the upward flow of the reacting fluid; this promotes high mass and heat-transfer rates and good mixing. Heat-transfer coefficients in the order of $200 \, W/m^2 \, °C$ to jackets and internal coils are typically obtained. The solids may be a catalyst; a reactant in fluidised combustion processes; or an inert powder, added to promote heat transfer.

Though the principal advantage of a fluidised bed over a fixed bed is the higher heat-transfer rate, fluidised beds are also useful where it is necessary to transport large quantities of solids as part of the reaction processes, such as where catalysts are transferred to another vessel for regeneration.

Fluidisation can only be used with relatively small sized particles, $< 300 \, \mu$m with gases.

A great deal of research and development work has been done on fluidised bed reactors in recent years, but the design and scale up of large diameter reactors is still an uncertain process and design methods are largely empirical.

The principles of fluidisation processes are covered in Volume 2, Chapter 6. The design of fluidised bed reactors is discussed by Rase (1977).

10.13.2. Design procedure

A general procedure for reactor design is outlined below:

1. Collect together all the kinetic and thermodynamic data on the desired reaction and the side reactions. It is unlikely that much useful information will be gleaned from a literature search, as little is published in the open literature on commercially attractive processes. The kinetic data required for reactor design will normally be obtained from laboratory and pilot plant studies. Values will be needed for the rate of reaction over a range of operating conditions: pressure, temperature, flow-rate and catalyst concentration. The design of experimental reactors and scale-up is discussed by Rase (1977) and Jordan (1968).
2. Collect the physical property data required for the design; either from the literature, by estimation or, if necessary, by laboratory measurements.
3. Identify the predominant rate-controlling mechanism: kinetic, mass or heat transfer. Choose a suitable reactor type, based on experience with similar reactions, or from the laboratory and pilot plant work.
4. Make an initial selection of the reactor conditions to give the desired conversion and yield.
5. Size the reactor and estimate its performance. Use as a guide the design methods given in Volume 3, and other texts on reactor design.

   Exact analytical solutions of the design relationships are rarely possible; semi-empirical methods based on the analysis of idealised reactors will normally have to be used.
6. Select suitable materials of construction.
7. Make a preliminary mechanical design for the reactor: the vessel design, heat-transfer surfaces, internals and general arrangement.
8. Cost the proposed design, capital and operating, and repeat steps 4 to 8, as necessary, to optimise the design.

In choosing the reactor conditions, particularly the conversion, and optimising the design, the interaction of the reactor design with the other process operations must not be overlooked. The degree of conversion of raw materials in the reactor will determine the size, and cost, of any equipment needed to separate and recycle unreacted materials. In these circumstances the reactor and associated equipment must be optimised as a unit.

10.14. References

ALDERS, L. (1955) *Liquid–liquid Extraction* (Elsevier).
AMBLER, C. M. (1952) *Chem. Eng. Prog.* **48** (March) 150. Evaluating the performance of centrifuges.
AMBLER, C. M. (1971) *Chem. Eng., Albany* **78** (Feb. 15th) 55. Centrifuge selection.
BAMFORTH, A. W. (1965) *Industrial Crystallisation* (Leonard Hill).
BEGG, G. A. J. (1966) *Chem. & Process Eng.* **47**, 153. Gas compression in the chemical industry.
BENNETT, J. G. (1936) *J. Inst. Fuel* **10**, 22. Broken coal.
BRADLEY, D. (1960) Institute of Minerals and Metals, International Congress, London, April, Paper 7, Group 2. Design and performance of cyclone thickeners.
BRADLEY, D. (1965a) *Chem. & Process Eng.* 595. Medium-speed centrifuges.
BRADLEY, D. (1965b) *The Hydrocyclone* (Pergamon).
CARBERRY, J. J. (1976) *Chemical and Catalytic Reactor Engineering* (McGraw-Hill).
CHLUMSKY, V. (1965) *Reciprocating and Rotary Compressors* (E. & F. N. Spon).
CLAYTON, C. G., BALL, A. M. and SPACKMAN, R. (1968) *Dispersion and Mixing during Turbulent Flow in a Circular Pipe*. UK Atomic Energy Authority Res. Group Report AERE-R 5569.
CREMER, H. W. and DAVIES, T. (Eds.) (1956) *Chemical Engineering Practice* (Butterworths).
DAHLSTROM, D. A. and CORNELL, C. F. (1971) *Chem. Eng., Albany* **78** (Feb. 15th) 63. Thickening and clarification.
DAY, R. W. and GRICHAR, C. N. (1979) Hydrocyclone separation, in *Handbook of Separation Processes for Chemical Engineers*, Schweitzer, P. A. (Ed.) (McGraw-Hill).
DIMOPLON, W. (1978) *Hyd. Proc.* **57** (May) 221. What process engineers need to know about compressors.
DORMAN, R. G. (1974) *Dust Control and Air Cleaning* (Pergamon).
EVANS, F. L. (1980) *Equipment Design Handbook for Refineries and Chemical Plants*, 2nd ed., Vol 2 (Gulf).
FISCHER, R. (1965) *Chem. Eng., Albany* **72** (Sept. 13th) 179. Agitated evaporators, Part 2, equipment and economics.
FOSSETT, H. and PROSSER, L. E. (1949) *Proc. Inst. Mech. Eng.* **160**, 224. The application of free jets to the mixing of fluids in tanks.
HANSON, C. (1968) *Chem. Eng., Albany* **75** (Aug. 26th) 76. Solvent extraction.
HANSON, C. (Ed.) (1971) *Recent Advances in Liquid–liquid Extraction* (Pergamon).
HENGLEIN, F. A. (1969) *Chemical Technology* (Pergamon).
HICKS, R. W. (1976) *Chem. Eng., Albany* **83** (July 19th) 141. How to select turbine agitators for dispersing gas into liquids.
HIORNS, F. J. (1970) *Brit. Chem. Eng.* **15**, 1565. Advances in comminution.
HOLLAND, F. A. and CHAPMAN, F. S. (1966a) *Liquid Processing and Mixing in Stirred Tanks* (Reinhold).
HOLLAND, F. A. and CHAPMAN, F. S. (1966b) *Pumping of Liquids* (Reinhold).
HOOPER, W. B. (1975) *Chem. Eng., Albany* **82** (Aug. 4th) 103. Predicting flow patterns in plant equipment.
HOOPER, W. B. and JACOBS, L. T. (1979) Decantation, in *Handbook of Separation Techniques for Chemical Engineers*, Schweitzer, P. A. (Ed.) (McGraw-Hill).
JENIKE, A. W. (1967) *Powder Technology* **1**, 237. Quantitive design of mass flow in bins.
JENIKE, A. W. and JOHNSON, J. R. (1970) *Chem. Eng. Prog.* **66** (June) 31. Solids flow in bins and moving beds.
JORDAN, D. J. (1968) *Chemical Process Development*, Vol. 1 (Wiley).
KEEY, R. B. (1972) *Drying—Principles and Practice* (Pergamon).
KEEY, R. B. (1978) *Introduction to Industrial Drying* (Pergamon).
KOCH, W. H. and LIGHT, W. (1977) *Chem. Eng., Albany* **84** (Nov. 7th) 80. New design approach boosts cyclone efficiency.
KOVAT, A. (1964) *Design and Performance of Centrifugal and Axial Flow Pumps and Compressors* (Pergamon).
LACEY, R. E. and LOEB, S. (Eds.) (1972) *Industrial Processing with Membranes* (Wiley).
LARSON, M. A. (1978) *Chem. Eng., Albany* **85** (Feb. 13th) 90. Guidelines for selecting crystallizers.
LAVANCHY, A. C., KEITH, F. W. and BEAMS, J. W. (1964) Centrifugal separation, in *Kirk-Othmer Encyclopedia of Chemical Technology*, 2nd ed. (Interscience).
LEE, J. and BRODSKEY, R. S. (1964) *AIChEJI* **10**, 187. Turbulent motion and mixing in a pipe.
LEVENSPIEL, O. (1972) *Chemical Reaction Engineering*, 2nd ed. (Wiley).
LOWRISON, G. C. (1974) *Crushing and Grinding* (Butterworths).
MAAS, J. H. (1979) Gas–solid separations, in *Handbook of Separation Techniques for Chemical Engineers*, Schweitzer, P. A. (Ed.) (McGraw-Hill).
MAIS, L. G. (1971) *Chem. Eng., Albany* **78** (Feb. 15th) 49. Filter media.
MARSHALL, V. C. (1974) *Comminution* (IChemE, London).
MASTERS, K. (1972) *Spray Drying* (Leonard Hill).
MATTHEWS, C. W. (1971) *Chem. Eng., Albany* **78** (Feb. 15th) 99. Screening.
MEAD, W. J. (1964) *The Encyclopedia of Chemical Process Equipment* (Reinhold).

MEADE, A. (1921) *Modern Gasworks Practice*, 2nd ed. (Benn Bros.).

MEARES, P. (Ed.) (1976) *Membrane Separation Processes* (Elsevier).

MIZRAHI, J. and BARNEA, E. (1973) *Process Engineering* (Jan.) 60. Compact settler gives efficient separation of liquid–liquid dispersions.

MORRIS, B. G. (1966) *Brit. Chem. Eng.* **11**, 347, 846 (in two parts) Application and selection of centrifuges.

MULLIN, J. W. (1961) *Crystallisation*, 2nd ed. (Butterworths).

MUTZENBURG, A. B. (1965) *Chem. Eng., Albany* **72** (Sept. 13th) 175. Agitated evaporators, Part 1, thin-film technology.

NAGATA, S. (1975) *Mixing Principles and Applications* (Halstead Press/Wiley).

NONHEBEL, G. (Ed.) (1972) *Gas Purification Processes for Air Pollution Control*, 2nd ed. (Newnes–Butterworths).

NONHEBEL, G. and MOSS, A. A. H. (1971) *Drying of Solids in the Chemical Industry* (Butterworths).

NÝVLT, J. (1971) *Industrial Crystallisation from Solutions* (Butterworths).

OLDSHUE, J. Y., HIRSHLAND, H. E. and GRETTON, A. T. (1956) *Chem. Eng. Prog.* **52** (Nov.) 481. Side-entering mixers.

ORR, C. (Ed.) (1977) *Filtration: Principles and Practice*, 2 volumes (Dekker).

PARKER, N. H. (1963a) *Chem. Eng., Albany* **70** (June 24th) 115. Aids to dryer selection.

PARKER, N. H. (1963b) *Chem. Eng., Albany* **70** (July 22nd) 135. How to specify evaporators.

PARKER, N. (1965) *Chem. Eng., Albany* **72** (Sept. 13th) 179. Agitated evaporators, Part 2, equipment and economics.

PENNEY, N. R. (1970) *Chem. Eng., Albany* **77** (June 1st) 171. Guide to trouble free mixing.

PERRY, R. H. and CHILTON, C. H. (Eds.) (1973) *Chemical Engineers Handbook*, 5th ed. (McGraw-Hill).

PICKETT, D. J. (1979) *Electrochemical Reactor Design* (Elsevier).

POLLAK, F. (Ed.) (1980) *Pump Users Handbook*, 2nd ed. (Trade & Technical Press).

PORTER, H. F., FLOOD, J. E. and RENNIE, F. W. (1971) *Chem. Eng., Albany* **78** (Feb. 15th) 39. Filter selection.

PORTER, M. C. (1979) Membrane filtration, in *Handbook of Separation Processes for Chemical Engineers*, Schweitzer, P. A. (Ed.) (McGraw-Hill).

POWER, R. B. (1964) *Hyd. Proc.* **43** (March) 138. Steam jet air ejectors.

PRYCE BAYLEY, D. and DAVIES, G. A. (1973) *Chemical Processing* **19** (May) 33. Process applications of knitted mesh mist eliminators.

PURCHAS, D. B. (1967) *Industrial Filtration of Liquids* (Leonard Hill).

PURCHAS, D. B. (1971) *Chemical Processing* **17** (Jan.) 31, (Feb.) 55 (in two parts). Choosing the cheapest filter medium.

RASE, H. F. (1977) *Chemical Reactor Design for Process Plants*, 2 volumes (Wiley).

REDMON, O. C. (1963) *Chem. Eng. Prog.* **59** (Sept.) 87. Cartridge type coalescers.

REID, R. W. (1979) Mixing and kneading equipment, in *Solids Separation and Mixing*, Bhatia, M. V. and Cheremisinoff, P. E. (Eds.) (Technomic).

REISNER, W. and ROTHE, M. E. (1971) *Bins and Bunkers for Handling Bulk Materials* (Trans. Tech. Publications).

ROBERTS, E. J., STAVENGER, P., BOWERSOX, J. P., WALTON, A. K. and MEHTA, M. (1971) *Chem. Eng., Albany* **78** (Feb. 15th) 89. Solid/solid separation.

ROSE, H. E. and WOOD, A. J. (1966) *An Introduction to Electrostatic Precipitation in Theory and Practice*, 2nd ed. (Constable).

ROSENZWEIG, M. D. (1977) *Chem. Eng., Albany* **84** (May 9th) 95. Motionless mixers move into new processing roles.

RUSHTON, J. H., COSTICH, E. W. and EVERETT, H. J. (1950) *Chem. Eng. Prog.* **46**, 467. Power characteristics of mixing impellers.

RYON, A. D., DALEY, F. L. and LOWRIE, R. S. (1959) *Chem. Eng. Prog.* **55** (Oct.) 70. Scale-up of mixer-settlers.

SARGENT, G. D. (1971) *Chem. Eng., Albany* **78** (Feb. 15) 11. Gas/solid separations.

SCHNEIDER, G. G., HORZELLA, T. I., STRIEGEL, P. J. and COOPER, P. J. (1975) *Chem. Eng., Albany* **82** (May 26th) 94. Selecting and specifying electrostatic precipitators.

SCHWEITZER, P. A. (Ed.) (1979) *Handbook of Separation Techniques for Chemical Engineers* (McGraw-Hill).

SUTTLE, H. K. (Ed.) (1969) *Process Engineering Technique Evaluation: Filtration* (Morgan-Grampian).

SIGNALES, B. (1975) *Chem. Eng., Albany* **82** (June 23rd) 141. How to design settling drums.

SMITH, J. M. (1970) *Chemical Engineering Kinetics* (McGraw-Hill).

SMITH, N. (1945) *Gas Manufacture and Utilisation* (British Gas Council).

STAIRMAND, C. J. (1949) *Engineering* **168**, 409. Pressure drop in cyclone separators.

STAIRMAND, C. J. (1951) *Trans. Inst. Chem. Eng.* **29**, 356. Design and performance of cyclone separators.

STEPANOFF, A. J. (1955) *Turboblowers: Theory, design and applications of centrifugal and axial flow compressors and fans* (Wiley).

STEPANOFF, A. J. (1969) *Gravity Flow of Bulk Solids and Transport of Solids in Suspension* (Wiley).

STERBACEK, Z. and TAUSK, P. (1965) *Mixing in the Chemical Industry* (Pergamon).

STRAUSS, N. (1966) *Industrial Gas Cleaning* (Pergamon).

SUTHERLAND, K. S. (1970) *Chemical Processing* **16** (May) 10. How to specify a centrifuge.

TREYBAL, R. E. (1963) *Liquid Extraction*, 2nd ed. (McGraw-Hill).
TROWBRIDGE, M. E. O'K. (1962) *Chem. Engr., London* No. 162 (Aug.) 73. Problems in scaling-up of centrifugal separation equipment.
UHL, W. W. and GRAY, J. B. (Eds.) (1967) *Mixing, Theory and Practice*, 2 volumes (Academic Press).
VAN HOOK, A. (1961) *Crystallisation: Theory and Practice* (Reinhold).
WAKEMAN, R. J. (1975) *Filters and Filtration* (Elsevier).
WARRING, R. H. (1979) *Pumps, Selection Systems and Application* (Trade & Technical Press).
WATERMAN, L. L. (1965) *Chem. Eng. Prog.* **61** (Oct.) 51. Electrical coalescers.
WILKINSON, W. L. and EDWARDS, M. F. (1972) *Chem. Engr., London* No. 264 (Aug.) 310; No. 265 (Sept.) 328 (in two parts). Heat transfer in agitated vessels.
WILLIAMS-GARDNER, A. (1965) *Chem & Process Eng.* **46**, 609. Selection of industrial dryers.
WILLIAMS-GARDNER, A. (1971) *Industrial Drying* (Leonard Hill).
YORK, O. H. (1954) *Chem. Eng. Prog.* **50** (Aug.) 421. Performance of wire-mesh demisters.
ZANKER, A. (1977) *Chem. Eng., Albany* **84** (May 9th) 122. Hydrocyclones: dimensions and performance.

British Standards
BS 410: 1976   Specification for test sieves.
BS 490: ---- Conveyor and elevator belting.
          Part 1: 1972   Rubber and plastic belting of textile construction for general use.
          Part 2: 1975   Rubber and plastics belting of textile construction for use on bucket elevators.
BS 1796: 1976   Method for test sieving.

10.15. Nomenclature

		Dimensions in **MLT**
A_i	Area of interface	L^2
A_s	Surface area of cyclone	L^2
A_1	Area of cyclone inlet duct	L^2
b	Index in equation 10.11	—
c	Index in equation 10.11	—
D	Agitator diameter	L
D_c	Cyclone diameter	L
D_{c_1}	Diameter of standard cyclone	L
D_{c_2}	Diameter of proposed cyclone design	L
D_T	Tank diameter	L
d	Particle diameter	L
d_s	Diameter of solid particle removed in a centrifuge	L
d_1	Mean diameter of particles separated in cyclone under standard conditions	L
d_2	Mean diameter of particles separated in proposed cyclone design	L
d_{50}	Particle diameter for which cyclone is 50 per cent efficient	L
f_c	Friction factor for cyclones	—
K	Constant in equation 10.11	—
L	Cyclone feed volumetric flow-rate	L^3T^{-1}
L_c	Continuous phase volumetric flow-rate	L^3T^{-1}
l	Length of decanter vessel	L
N	Agitator speed	T^{-1}
P	Agitator shaft power	ML^2T^{-3}
ΔP	Press differential (pressure drop)	$ML^{-1}T^{-2}$
p	Agitator blade pitch	L
Q	Volumetric flow-rate of liquid through a centrifuge	L^3T^{-1}
Q_p	Volumetric liquid flow through a pump	L^3T^{-1}
Q_1	Standard flow-rate in cyclone	L^3T^{-1}
Q_2	Proposed flow-rate in cyclone	L^3T^{-1}
r	Radius of decanter vessel	L
r_e	Radius of cyclone exit pipe	L
r_t	Radius of circle to which centre line of cyclone inlet duct is tangential	L
u_c	Velocity of continuous phase in a decanter	LT^{-1}
u_d	Settling (terminal) velocity of dispersed phase in a decanter	LT^{-1}
u_g	Terminal velocity of solid particles settling under gravity	LT^{-1}
$\hat{u}_v$	Maximum allowable vapour velocity in a separating vessel	LT^{-1}

u_1	Velocity in cyclone inlet duct	LT^{-1}
u_2	Velocity in cyclone exit duct	LT^{-1}
w	Width of interface in a decanter	L
z_1	Height to light liquid overflow from a decanter	L
z_2	Height to heavy liquid overflow from a decanter	L
z_3	Height to the interface in a decanter	L
η	Separating efficiency of a centrifuge	—
η_p	Pump efficiency	—
μ	Liquid viscosity	$ML^{-1}T^{-1}$
μ_c	Viscosity of continuous phase	$ML^{-1}T^{-1}$
μ_1	Cyclone test fluid viscosity	$ML^{-1}T^{-1}$
μ_2	Viscosity of fluid in proposed cyclone design	$ML^{-1}T^{-1}$
ρ	Liquid density	ML^{-3}
ρ_f	Gas density	ML^{-3}
ρ_L	Liquid density	ML^{-3}
ρ_s	Density of solid	ML^{-3}
ρ_1	Light liquid density in a decanter	ML^{-3}
ρ_2	Heavy liquid density in a decanter	ML^{-3}
$\Delta\rho$	Difference in density between solid and liquid	ML^{-3}
$\Delta\rho_1$	Density difference under standard conditions in standard cyclone	ML^{-3}
$\Delta\rho_2$	Density difference in proposed cyclone design	ML^{-3}
Σ	Sigma value for centrifuges, defined by equation 10.1	L^2
ϕ	Factor in Fig. 10.48	—
ψ	Parameter in Fig. 10.48	—

CHAPTER 11

Separation Columns
(Distillation and Absorption)

11.1 Introduction

This chapter covers the design of separating columns. Though the emphasis is on distillation processes, the basic construction features, and many of the design methods, also apply to other multistage processes; such as stripping, absorption and extraction.

Distillation is probably the most widely used separation process in the chemical and allied industries; its applications ranging from the rectification of alcohol, which has been practiced since antiquity, to the fractionation of crude oil.

Only a brief review of the fundamental principles that underlie the design procedures will be given; a fuller discussion can be found in Volume 2, and in other text books; Robinson and Gilliland (1950), Norman (1961), Oliver (1966), Smith (1963), King (1971), Hengstebeck (1961).

A good understanding of methods used for correlating vapour–liquid equilibrium data is essential to the understanding of distillation and other equilibrium-staged processes; this subject was covered in Chapter 8.

In recent years, most of the work done to develop reliable design methods for distillation equipment has been carried out by a commercial organisation, Fractionation Research Inc., an organisation set up with the resources to carry out experimental work on full-size columns. Since their work is of a proprietary nature, it is not published in the open literature and it has not been possible to refer to their methods in this book. Fractionation Research's design manuals will, however, be available to design engineers whose companies are subscribing members of the organisation.

Distillation column design

The design of a distillation column can be divided into the following steps:

1. Specify the degree of separation required: set product specifications.
2. Select the operating conditions: batch or continuous; operating pressure.
3. Select the type of contacting device: plates or packing.
4. Determine the stage and reflux requirements: the number of equilibrium stages.
5. Size the column: diameter, number of real stages.
6. Design the column internals: plates, distributors, packing supports.
7. Mechanical design: vessel and internal fittings.

The principal step will be to determine the stage and reflux requirements. This is a relatively simple procedure when the feed is a binary mixture, but a complex and difficult task when the feed contains more than two components (multicomponent systems).

11.2. Continuous distillation: process description

The separation of liquid mixtures by distillation depends on differences in volatility between the components. The greater the relative volatilities, the easier the separation. The basic equipment required for continuous distillation is shown in Fig. 11.1. Vapour flows up the column and liquid counter-currently down the column. The vapour and liquid are brought into contact on plates, or packing. Part of the condensate from the condenser is returned to the top of the column to provide liquid flow above the feed point (reflux), and part of the liquid from the base of the column is vaporised in the reboiler and returned to provide the vapour flow.

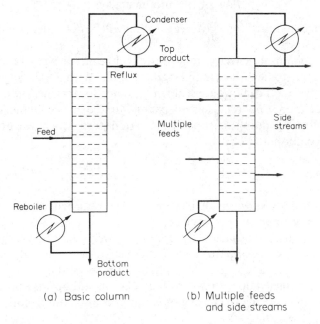

(a) Basic column

(b) Multiple feeds and side streams

FIG. 11.1. Distillation column

In the section below the feed, the more volatile components are stripped from the liquid and this is known as the *stripping section*. Above the feed, the concentration of the more volatile components is increased and this is called the enrichment, or more commonly, the *rectifying section*. Figure 11.1a shows a column producing two product streams, referred to as tops and bottoms, from a single feed. Columns are occasionally used with more than one feed, and with side streams withdrawn at points up the column, Fig. 11.1b. This does not alter the basic operation, but complicates the analysis of the process, to some extent.

If the process requirement is to strip a volatile component from a relatively non-volatile solvent, the rectifying section may be omitted, and the column would then be called a *stripping column*.

In some operations, where the top product is required as a vapour, only sufficient liquid is condensed to provide the reflux flow to the column, and the condenser is referred to as a partial condenser. When the liquid is totally condensed, the liquid returned to the column will have the same composition as the top product. In a partial condenser the reflux will be in equilibrium with the vapour leaving the condenser. Virtually pure top and bottom products can be obtained in a single column from a binary feed, but where the feed contains more than two components, only a single "pure" product can be produced, either from the top or bottom of the column. Several columns will be needed to separate a multicomponent feed into its constituent parts.

11.2.1. Reflux considerations

The reflux ratio, R, is normally defined as:

$$R = \frac{\text{flow returned as reflux}}{\text{flow of top product taken off}}$$

The number of stages required for a given separation will be dependent on the reflux ratio used.

In an operating column the effective reflux ratio will be increased by vapour condensed within the column due to heat leakage through the walls. With a well-lagged column the heat loss will be small and no allowance is normally made for this increased flow in design calculations. If a column is poorly insulated, changes in the internal reflux due to sudden changes in the external conditions, such as a sudden rain storm, can have a noticeable effect on the column operation and control.

Total reflux

Total reflux is the condition when all the condensate is returned to the column as reflux: no product is taken off and there is no feed.

At total reflux the number of stages required for a given separation is the minimum at which it is theoretically possible to achieve the separation. Though not a practical operating condition, it is a useful guide to the likely number of stages that will be needed.

Columns are often started up with no product take-off and operated at total reflux till steady conditions are attained. The testing of columns is also conveniently carried out at total reflux.

Minimum reflux

As the reflux ratio is reduced a *pinch point* will occur at which the separation can only be achieved with an infinite number of stages. This sets the minimum possible reflux ratio for the specified separation (see Volume 2, Chapter 11).

Optimum reflux ratio

Practical reflux ratios will lie somewhere between the minimum for the specified separation and total reflux. The designer must select a value at which the specified separation is achieved at minimum cost. Increasing the reflux reduces the number of stages required, and hence the capital cost, but increases the service requirements (steam and water) and the operating costs. The optimum reflux ratio will be that which gives the lowest annual operating cost. No hard and fast rules can be given for the selection of the design reflux ratio, but for many systems the optimum will lie between 1·2 to 1·5 times the minimum reflux ratio.

For new designs, where the ratio cannot be decided on from past experience, the effect of reflux ratio on the number of stages can be investigated using the short-cut design methods given in this chapter. This will usually indicate the best of value to use in more rigorous design methods.

At low reflux ratios the calculated number of stages will be very dependent on the accuracy of the vapour–liquid equilibrium data available. If the data are suspect a higher than normal ratio should be selected to give more confidence in the design.

11.2.2. Feed-point location

The precise location of the feed point will affect the number of stages required for a specified separation and the subsequent operation of the column. As a general rule, the feed should enter the column at the point that gives the best match between the feed composition (vapour and liquid if two phases) and the vapour and liquid streams in the column. In practice, it is wise to provide two or three feed-point nozzles located round the predicted feed point to allow for uncertainties in the design calculations and data, and possible changes in the feed composition after start-up.

11.2.3. Selection of column pressure

Except when distilling heat-sensitive materials, the main consideration when selecting the column operating-pressure will be to ensure that the dew point of the distillate is above that which can be easily obtained with the plant cooling water. The maximum, summer, temperature of cooling water is usually taken as 30°C. If this means that high pressures will be needed the provision of refrigerated brine cooling should be considered. Vacuum operation is used to reduce the column temperatures for the distillation of heat-sensitive materials and where very high temperatures would otherwise be needed to distill relatively non-volatile materials.

When calculating the stage and reflux requirements it is usual to take the operating pressure as constant throughout the column. In vacuum columns, the column pressure drop will be a significant fraction of the total pressure and the change in pressure up the column should be allowed for when calculating the stage temperatures. This may require a trial and error calculation, as clearly the pressure drop cannot be estimated before an estimate of the number of stages is made.

11.3. Continuous distillation: basic principles

11.3.1. Stage equations

Material and energy balance equations can be written for any stage in a multistage process.

Figure 11.2 shows the material flows into and out of a typical stage n in a distillation column. The equations for this stage are set out below, for any component i.

material balance

$$V_{n+1}y_{n+1} + L_{n-1}x_{n-1} + F_n z_n = V_n y_n + L_n x_n + S_n x_n \qquad (11.1)$$

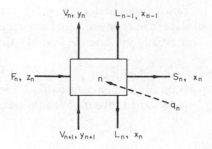

FIG. 11.2. Stage flows

energy balance

$$V_{n+1}H_{n+1} + L_{n-1}h_{n-1} + Fh_f + q_n = V_nH_n + L_nh_n + S_nh_n \qquad (11.2)$$

where V_n = vapour flow from the stage,

V_{n+1} = vapour flow into the stage from the stage below,

L_n = liquid flow from the stage,

L_{n-1} = liquid flow into the stage from the stage above,

F_n = any feed flow into the stage,

S_n = any side stream from the stage,

q_n = heat flow into, or removal from, the stage,

n = any stage, numbered from the top of the column,

z = mol fraction of component i in the feed steam (note, feed may be two-phase),

x = mol fraction of component i in the liquid streams,

y = mol fraction component i in the vapour streams,

H = specific enthalpy vapour phase,

h = specific enthalpy liquid phase,

h_f = specific enthalpy feed (vapour + liquid).

All flows are the total stream flows (mols/unit time) and the specific enthalpies are also for the total stream (J/mol).

It is convenient to carry out the analysis in terms of "equilibrium stages". In an equilibrium stage (theoretical plate) the liquid and vapour streams leaving the stage are taken to be in equilibrium, and their compositions are determined by the vapour–liquid equilibrium relationship for the system (see Chapter 8). In terms of equilibrium constants:

$$y_i = K_i x_i \qquad (11.3)$$

The performance of real stages is related to an equilibrium stage by the concept of plate efficiencies for plate contactors, and "height of an equivalent theoretical plate" for packed columns. Material and energy balance, and equilibrium relationship, equations can be written for each stage in the column, and for the condenser and reboiler. These stage equations form the basis of all the rigorous methods that have been developed for the analysis of staged processes.

11.3.2. Dew points and bubble points

To estimate the stage, and the condenser and reboiler temperatures, procedures are required for calculating dew and bubble points. By definition, a saturated liquid is at its bubble point (any rise in temperature will cause a bubble of vapour to form), and a saturated vapour is at its dew point (any drop in temperature will cause a drop of liquid to form).

Dew points and bubble points can be calculated from a knowledge of the vapour-liquid equilibrium for the system. In terms of equilibrium constants, the bubble point is defined by the equation:

$$\text{bubble point:} \quad \sum y_i = \sum K_i x_i = 1\cdot 0 \tag{11.4}$$

$$\text{and dew point:} \quad \sum x_i = \sum \frac{y_i}{K_i} = 1\cdot 0 \tag{11.5}$$

For multicomponent mixtures the temperature that satisfies these equations, at a given system pressure, must be found by trial and error.

For binary systems the equations can be solved more readily because the component compositions are not independent; fixing one fixes the other.

$$y_a = 1 - y_b \tag{11.6a}$$

$$x_a = 1 - x_b \tag{11.6b}$$

Bubble- and dew-point calculations are illustrated in Example 11.9.

11.3.3. Equilibrium flash calculations

In an equilibrium flash process a feed stream is separated into liquid and vapour streams at equilibrium. The composition of the streams will depend on the quantity of the feed vaporised (flashed). The equations used for equilibrium flash calculations are developed below and a typical calculation is shown in Example 11.1.

Flash calculations are often needed to determine the condition of the feed to a distillation column and, occasionally, to determine the flow of vapour from the reboiler, or condenser if a partial condenser is used.

Single-stage flash distillation processes are used to make a coarse separation of the light components in a feed; often as a preliminary step before a multicomponent distillation column, as in the distillation of crude oil.

Figure 11.3 shows a typical equilibrium flash process. The equations describing this process are:

Material balance, for any component, i

$$Fz_i = Vy_i + Lx_i \tag{11.7}$$

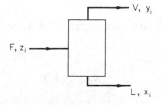

FIG. 11.3. Flash distillation

Energy balance, total stream enthalpies:

$$Fh_f = VH + Lh \tag{11.8}$$

If the vapour–liquid equilibrium relationship is expressed in terms of equilibrium constants, equation 11.7 can be written in a more useful form:

$$Fz_i = VK_ix_i + Lx_i$$

$$= Lx_i\left[\frac{V}{L}K_i + 1\right]$$

from which

$$L = \sum_i \frac{Fz_i}{\left[\dfrac{VK_i}{L} + 1\right]} \tag{11.9}$$

and, similarly,

$$V = \sum_i \frac{Fz_i}{\left[\dfrac{L}{VK_i} + 1\right]} \tag{11.10}$$

The groups incorporating the liquid and vapour flow-rates and the equilibrium constants have a general significance in separation process calculations.

The group L/VK_i is known as the absorption factor A_i, and is the ratio of the mols of any component in the liquid stream to the mols in the vapour stream.

The group VK_i/L is called the stripping factor S_i, and is the reciprocal of the absorption factor.

Efficient techniques for the solution of the trial and error calculations necessary in multicomponent flash calculations are given by several authors; Smith (1963), Oliver (1966), Hengstebeck (1961) and King (1971).

Example 11.1

A feed to a column has the composition given in the table below, and is at a pressure of 14 bar and a temperature of 60°C. Calculate the flow and composition of the liquid and vapour phases. Take the equilibrium data from the Depriester charts given in Chapter 8.

		kmol/h	z_i
Feed	ethane (C_2)	20	0·25
	propane (C_3)	20	0·25
	isobutane (iC_4)	20	0·25
	n-pentane (nC_5)	20	0·25

Solution

For two phases to exist the flash temperature must lie between the bubble point and dew point of the mixture.

From equations 11.4 and 11.5:

$$\sum K_i z_i > 1·0$$

$$\sum \frac{z_i}{K_i} > 1·0$$

Check feed condition

	K_i	$K_i z_i$	z_i / K_i
C_2	3·8	0·95	0·07
C_3	1·3	0·33	0·19
iC_4	0·43	0·11	0·58
nC_5	0·16	0·04	1·56
		$\sum 1·43$	$\sum 2·40$

therefore the feed is a two phase mixture.
Flash calculation

		Try $L/V = 1·5$		Try $L/V = 3·0$	
	K_i	$A_i = L/VK_i$	$V_i = Fz_i/(1 + A_i)$	A_i	V_i
C_2	3·8	0·395	14·34	0·789	11·17
C_3	1·3	1·154	9·29	2·308	6·04
iC_4	0·43	3·488	4·46	6·977	2·51
nC_5	0·16	9·375	1·93	18·750	1·01
		$V_{calc} = 30·02$		$V_{calc} = 20·73$	
		$L/V = \dfrac{80 - 30·02}{30·02} = 1·67$		$L/V = 2·80$	

Hengstebeck's method is used to find the third trial value for L/V. The calculated values are plotted against the assumed values and the intercept with a line at 45° (calculated = assumed) gives the new trial value, 2·4.

	Try $L/V = 2·4$			
	A_i	V_i	$y_i = V_i/V$	$x_i = (Fz_i - V_i)/L$
C_2	0·632	12·26	0·52	0·14
C_3	1·846	7·03	0·30	0·23
iC_4	5·581	3·04	0·13	0·30
nC_5	15·00	1·25	0·05	0·33
	$V_{cal} = 23·58$		1·00	1·00

$L = 80 - 23·58 = 56·42 \, \text{kmol/h}$,
L/V calculated $= 56·42/23·58 = 2·39$ close enough to the assumed value of 2·4.

Adiabatic flash

In many flash processes the feed stream is at a higher pressure than the flash pressure and the heat for vaporisation is provided by the enthalpy of the feed. In this situation the flash temperature will not be known and must be found by trial and error. A temperature must be found at which both the material and energy balances are satisfied.

11.4. Design variables in distillation

It was shown in Chapter 1 that to carry out a design calculation the designer must specify values for a certain number of independent variables to define the problem completely, and that the ease of calculation will often depend on the judicious choice of these design variables.

In manual calculations the designer can use intuition in selecting the design variables and, as he proceeds with the calculation, can define other variables if it becomes clear that the problem is not sufficiently defined. He can also start again with a new set of design variables if the calculations become tortuous. When specifying a problem for a computer method it is essential that the problem is completely and sufficiently defined.

In Chapter 1 it was shown that the number of independent variables for any problem is equal to the difference between the total number of variables and the number of linking equations and other relationships. Examples of the application of this formal procedure for determining the number of independent variables in separation process calculations are given by Gilland and Reed (1942), Kwauk (1956) and Hanson and Somerville (1963). For a multistage, multicomponent, column, there will be a set of material and enthalpy balance equations and equilibrium relationships for each stage, and for the reboiler and condenser; for each component. If there are more than a few stages the task of counting the variables and equations becomes burdensome and mistakes are very likely to be made. A simpler, more practical, way to determine the number of independent variables is the "description rule" procedure given by Hanson et al. (1962). Their description rule states that to determine a separation process completely the number of independent variables which must be set (by the designer) will equal the number that are set in the construction of the column or that can be controlled by external means in its operation. The application of this rule requires the designer to visualise the column in operation and list the number of variables fixed by the column construction; those fixed by the process; and those that have to be controlled for the column to operate steadily and produce product within specification. The method is best illustrated by considering the operation of the simplest type of column: with one feed, no side streams, a total condenser, and a reboiler. The construction will fix the number of stages above and below the feed point (two variables). The feed composition and total enthalpy will be fixed by the processes upstream $(1 + (n - 1)$ variables, where n is the number of components). The feed rate, column pressure and condenser and reboiler duties (cooling water and steam flows) will be controlled (four variables).

$$\text{Total number of variables fixed} = 2 + 1 + (n - 1) + 4 = \underline{\underline{n + 6}}$$

To design the column this number of variables must be specified completely to define the problem, but the same variables need not be selected.

Typically, in a design situation, the problem will be to determine the number of stages required at a specified reflux rate and column pressure, for a given feed, and with the product compositions specified in terms of two key components and one product flow-rate. Counting up the number of variables specified it will be seen that the problem is completely defined:

Feed flow, composition, enthalpy	$= 2 + (n - 1)$
Reflux (sets q_c)	$= 1$
Key component compositions, top and bottom	$= 2$
Product flow	$= 1$
Column pressure	$= 1$
	$\overline{n + 6}$

Note: specifying $(n - 1)$ component compositions completely defines the feed composition as the fractions add up to 1.

In theory any $(n + 6)$ independent variables could have been specified to define the problem, but it is clear that the use of the above variables will lead to a straightforward solution of the problem.

When replacing variables identified by the application of the description rule it is important to ensure that those selected are truly independent, and that the values assigned to them lie within the range of possible, practical, values.

The number of independent variables that have to be specified to define a problem will depend on the type of separation process being considered. Some examples of the application of the description rule to more complex columns are given by Hanson *et al.* (1962).

11.5. Design methods for binary systems

A good understanding of the basic equations developed for binary systems is essential to the understanding of distillation processes.

The distillation of binary mixtures is covered thoroughly in Volume 2, Chapter 11, and the discussion in this section is limited to a brief review of the most useful design methods. Though binary systems are usually considered separately, the design methods developed for multicomponent systems (Section 11.6) can obviously also be used for binary systems. With binary mixtures fixing the composition of one component fixes the composition of the other, and iterative procedures are not usually needed to determine the stage and reflux requirements; simple graphical methods are normally used.

11.5.1. Basic equations

Sorel (1899) first derived and applied the basic stage equations to the analysis of binary systems. Figure 11.4a shows the flows and compositions in the top part of a column. Taking the system boundary to include the stage n and the condenser, gives the following equations:

Material balance
total flows
$$V_{n+1} = L_n + D \tag{11.11}$$

for either component
$$V_{n+1} y_{n+1} = L_n x_n + D x_d \tag{11.12}$$

Energy balance
total stream enthalpies
$$V_{n+1} H_{n+1} = L_n h_n + D h_d + q_c \tag{11.13}$$

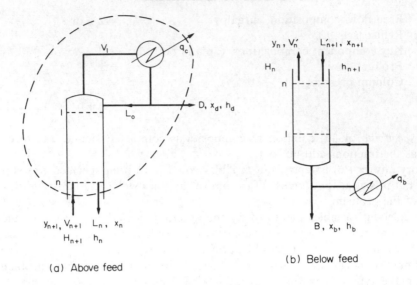

(a) Above feed

(b) Below feed

FIG. 11.4. Column flows and compositions

where q_c is the heat removed in the condenser.

Combining equations 11.11 and 11.12 gives

$$y_{n+1} = \frac{L_n}{L_n + D} x_n + \frac{D}{L_n + D} x_d \tag{11.14}$$

Combining equations 11.11 and 11.13 gives

$$V_{n+1} H_{n+1} = (L_n + D)H_{n+1} = L_n h_n + Dh_d + q_c \tag{11.15}$$

Analogous equations can be written for the stripping section, Fig. 11.6b.

$$x_{n+1} = \frac{V'_n}{V'_n + B} y_n + \frac{B}{V'_n + B} x_b \tag{11.16}$$

and

$$L'_{n+1} h_{n+1} = (V'_n + B)h_{n+1} = V'_n H_n + Bh_b - q_b \tag{11.17}$$

At constant pressure, the stage temperatures will be functions of the vapour and liquid compositions only (dew and bubble points) and the specific enthalpies will therefore also be functions of composition

$$H = f(y) \tag{11.18a}$$

$$h = f(x) \tag{11.18b}$$

Lewis–Sorel method (equimolar overflow)

For most distillation problems a simplifying assumption, first proposed by Lewis (1909), can be made that eliminates the need to solve the stage energy-balance equations. The molar liquid and vapour flow rates are taken as constant in the stripping and rectifying sections. This condition is referred to as equimolar overflow: the molar vapour and liquid flows from each stage are constant. This will only be true where the component molar latent heats of vaporisation are the same and, together with the specific heats, are constant over the range of temperature in the column; there is no significant heat of mixing; and the heat losses are negligible. These conditions are substantially true for practical systems when the components form near-ideal liquid mixtures.

Even when the latent heats are substantially different the error introduced by assuming equimolar overflow to calculate the number of stages is usually small, and acceptable.

With equimolar overflow equations 11.14 and 11.16 can be written without the subscripts to denote the stage number:

$$y_{n+1} = \frac{L}{L+D}x_n + \frac{D}{L+D}x_d \tag{11.19}$$

$$x_{n+1} = \frac{V'}{V'+B}y_n + \frac{B}{V'+B}x_b \tag{11.20}$$

where L = the constant liquid flow in the rectifying section = the reflux flow, L_0, and V' is the constant vapour flow in the stripping section.

Equations 11.19 and 11.20 can be written in an alternative form:

$$y_{n+1} = \frac{L}{V}x_n + \frac{D}{V}x_d \tag{11.21}$$

$$y_n = \frac{L'}{V'}x_{n+1} - \frac{B}{V'}x_b \tag{11.22}$$

where V is the constant vapour flow in the rectifying section = $(L + D)$; and L' is the constant liquid flow in the stripping section = $V' + B$.

These equations are linear, with slopes L/V and L'/V'. They are referred to as *operating lines*, and give the relationship between the liquid and vapour compositions between stages. For an equilibrium stage, the compositions of the liquid and vapour streams leaving the stage are given by the equilibrium relationship.

11.5.2. McCabe–Thiele method

Equations 11.21 and 11.22 and the equilibrium relationship are conveniently solved by the graphical method developed by McCabe and Thiele (1925). The method is discussed fully in Volume 2. A simple procedure for the construction of the diagram is given below and illustrated in Example 11.2.

Procedure

Refer to Fig. 11.5, all compositions are those of the more volatile component.

1. Plot the vapour–liquid equilibrium curve from data available at the column operating pressure. In terms of relative volatility:

$$y = \alpha x/(1 + (\alpha - 1)x) \tag{11.23}$$

where α is the geometric average relative volatility of the lighter (more volatile) component with respect to the heavier component (less volatile).

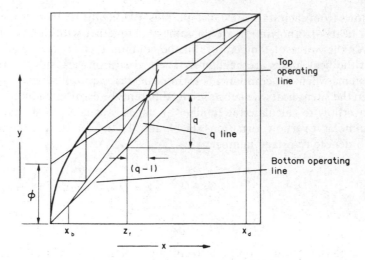

FIG. 11.5. McCabe–Thiele diagram

It is usually more convenient, and less confusing, to use equal scales for the x and y axes.

2. Make a material balance over the column to determine the top and bottom compositions, x_d and x_b, from the data given.

3. The top and bottom operating lines intersect the diagonal at x_d and x_b respectively; mark these points on the diagram.

4. The point of intersection of the two operating lines is dependent on the phase condition of the feed. The line on which the intersection occurs is called the q *line* (see Volume 2). The q line is found as follows:

(i) calculate the value of the ratio q given by

$$q = \frac{\text{heat to vaporise 1 mol of feed}}{\text{molar latent heat of feed}}$$

(ii) plot the q line, slope $= q/(q-1)$, intersecting the diagonal at z_f (the feed composition).

5. Select the reflux ratio and determine the point where the top operating line extended cuts the y axis:

$$\phi = \frac{x_d}{1+R} \tag{11.24}$$

6. Draw in the top operating line, from x_d on the diagonal to ϕ.

7. Draw in the bottom operating line; from x_b on the diagonal to the point of intersection of the top operating line and the q line.

8. Starting at x_d or x_b, step off the number of stages.

Note: The feed point should be located on the stage closest to the intersection of the operating lines.

The reboiler, and a partial condenser if used, act as equilibrium stages. However, when designing a column there is little point in reducing the estimated number of stages to account for this; they can be considered additional factors of safety.

The efficiency of real contacting stages can be accounted for by reducing the height of the steps on the McCabe–Thiele diagram, see diagram Fig. 11.6. Stage efficiencies are discussed in Section 11.10.

The McCabe–Thiele method can be used for the design of columns with side streams and multiple feeds. The liquid and vapour flows in the sections between the feed and take-off points are calculated and operating lines drawn for each section.

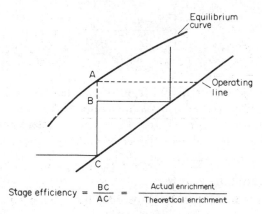

$$\text{Stage efficiency} = \frac{BC}{AC} = \frac{\text{Actual enrichment}}{\text{Theoretical enrichment}}$$

FIG. 11.6. Stage efficiency

Stage vapour and liquid flows not constant

The McCabe–Thiele method can be used when the condition of equimolar overflow cannot be assumed, but the operating lines will not then be straight. They can be drawn by making energy balances at a sufficient number of points to determine the approximate slope of the lines; see Hengstebeck (1961). Alternatively the more rigorous graphical method of Ponchon and Savarit derived in Volume 2 can be used. Nowadays, it should rarely be necessary to resort to complex graphical methods when the simple McCabe–Thiele diagram is not sufficiently accurate, as computer programs will normally be available for the rigorous solution of such problems.

11.5.3. Low product concentrations

When concentrations of the more volatile component of either product is very low the steps on the McCabe–Thiele diagram become very small and difficult to plot. This problem can be overcome by replotting the top or bottom sections to a larger scale, or on log-log paper. In a log plot the operating line will not be straight and must be drawn by plotting points calculated using equations 11.21 and 11.22. This technique is described by Alleva (1962) and is illustrated in Example 11.2.

If the operating and equilibrium lines are straight, and they usually can be taken as such when the concentrations are small, the number of stages required can be calculated using the equations given by Robinson and Gilliland (1950).

For the stripping section:

$$N_s^* = \frac{\log[(K'/s' - 1)(x_r'/x_b - 1)]}{\log[1/s'(K' - 1)]} \tag{11.25}$$

where N_s^* = number of ideal stages required from x_b to some reference point x_r',
$\quad\quad x_b$ = mol fraction of the more volatile component in the bottom product,
$\quad\quad x_r'$ = mol fraction of more volatile component at the reference point,
$\quad\quad s'$ = slope of the bottom operating line,
$\quad\quad K'$ = equilibrium constant for the more volatile component.

For the rectifying section:

$$N_r^* = \frac{\log\left[\dfrac{(1 - s) + x_r/x_d(s - K)}{1 - K}\right]}{\log\,(s/K)} \tag{11.26}$$

where N_r^* = number of stages required from some reference point x_r to the x_d,
$\quad\quad x_d$ = mol fraction of the *least volatile* component in the top product,
$\quad\quad x_r$ = mol fraction of *least volatile* component at reference point,
$\quad\quad K$ = equilibrium constant for the *least volatile* component,
$\quad\quad s$ = slope of top operating line.

Note: at low concentrations $K = \alpha$.
The use of these equations is illustrated in Example 11.3.

Example 11.2

Acetone is to be recovered from an aqueous waste stream by continuous distillation. The feed will contain 10 per cent w/w acetone. Acetone of at least 98 per cent purity is wanted, and the aqueous effluent must not contain more than 50 ppm acetone. The feed will be at 20°C. Estimate the number of ideal stages required.

Solution

There is no point in operating this column at other than atmospheric pressure. The equilibrium data available for the acetone–water system were discussed in Chapter 8, Section 8.4.

The data of Kojima *et al.* will be used.

Mol fraction x, liquid	0·00	0·05	0·10	0·15	0·20	0·25	0·30
Acetone y, vapour	0·00	0·6381	0·7301	0·7716	0·7916	0·8034	0·8124
bubble point °C	100·0	74·80	68·53	65·26	63·59	62·60	61·87

x	0·35	0·40	0·45	0·50	0·55	0·60	0·65
y	0·8201	0·8269	0·8376	0·8387	0·8455	0·8532	0·8615
°C	61·26	60·75	60·35	59·95	59·54	59·12	58·71

x	0·70	0·75	0·80	0·85	0·90	0·95
y	0·8712	0·8817	0·8950	0·9118	0·9335	0·9627
°C	58·29	57·90	57·49	57·08	56·68	56·30

The equilibrium curve can be drawn with sufficient accuracy to determine the stages above the feed by plotting the concentrations at increments of 0·1. The diagram would normally be plotted at about twice the size of Fig. 11.7.

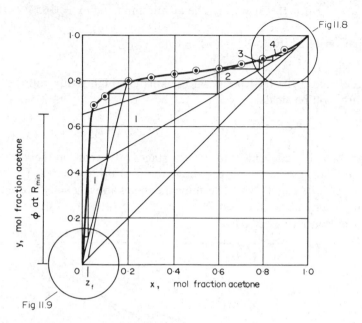

FIG. 11.7. McCabe–Thiele plot, Example 11.2

Molecular weights, acetone 58, water 18

$$\text{Mol fractions acetone feed} = \frac{10/58}{10/58 + 90/18} = 0.033$$

$$\text{top product} = \frac{98/58}{98/58 + 2/18} = 0.94$$

$$\text{bottom product} = 50 \times 10^{-6} \times 18/58 = 15.5 \times 10^{-6}$$

Feed condition (q-line)

Bubble point of feed (interpolated) = 83°C
Latent heats, water 41,360, acetone 28,410 J/mol
Mean specific heats, water 75·3, acetone 128 J/mol °C
Latent heat of feed = $28,410 \times 0.033 + (1 - 0.033)41,360 = 40,933$ J/mol
Specific heat of feed = $(0.033 \times 128) + (1 - 0.033)75.3 = 77.0$ J/mol °C
Heat to vaporise 1 mol of feed = $(83 - 20)77.0 + 40,933 = 45,784$ J

$$q = \frac{45,784}{40,933} = 1.12$$

$$\text{Slope of } q \text{ line} = \frac{1.12}{1.12 - 1} = 9.32$$

For this problem the condition of minimum reflux occurs where the top operating line just touches the equilibrium curve at the point where the q line cuts the curve.

From the Fig. 11.7,

$$\phi \text{ for the operating line at minimum reflux} = 0.65$$

From equation 11.24, $R_{min} = \dfrac{0.94}{0.65} - 1 = 0.45$

Take $R = R_{min} \times 3$

As the flows above the feed point will be small, a high reflux ratio is justified; the condenser duty will be small.

$$\text{At } R = 3 \times 0.45 = 1.35, \quad \phi = \frac{0.94}{1 + 1.35} = 0.4$$

For this problem it is convenient to step the stages off starting at the intersection of the operating lines. This gives three stages above the feed up to $y = 0.8$. The top section is drawn to a larger scale, Fig. 11.8, to determine the stages above $y = 0.8$: three to four stages required; total stages above the feed 7.

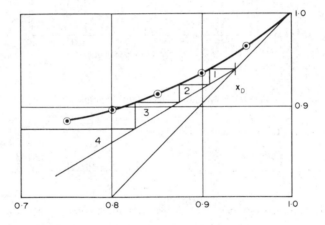

FIG. 11.8. Top section enlarged

Below the feed, one stage is required down to $x = 0.04$. A log-log plot is used to determine the stages below this concentration. Data for log-log plot:

Stripping section

$$y = 3.92(x - x_b) + x_b$$
$$= 3.92x - 45.3 \times 10^{-6}$$

	x	4×10^{-2}	10^{-3}	10^{-4}	4×10^{-5}	2×10^{-5}
Operating line,	y	0.16	3.9×10^{-3}	3.5×10^{-4}	1.2×10^{-4}	3.3×10^{-5}
Equilibrium line,	y	0.64	10.6×10^{-2}	1.6×10^{-3}	6.4×10^{-4}	3.2×10^{-4}

From Fig. 11.9, number of stages required for this section = 6
Total number of stages below feed = 7
Total stages = $7 + 7 = \underline{\underline{14}}$

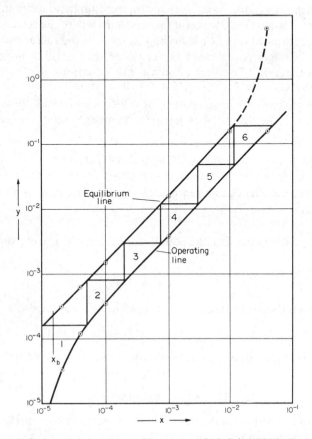

FIG. 11.9. Log–log expansion of lower part of McCabe–Thiele diagram

Example 11.3

For the problem specified in Example 11.2, estimate the number of ideal stages required below an acetone concentration of 0·01 (more volatile component), using the Robinson–Gilliland equation.

Solution

From the McCabe–Thiele diagram in Example 11.2:

$\qquad$ slope of the bottom operating line, $s' = 3{\cdot}92$
$\qquad$ slope of the equilibrium curve, $K'_L = 16$
$\qquad$ $x_b = 15 \times 10^{-6}$

(11.25) $\qquad N_s^* = \dfrac{\log[\,(16/3{\cdot}92 - 1)(0{\cdot}01/15 \times 10^{-6} - 1)\,]}{\log[\,(1/3{\cdot}92(16 - 1))\,]} = \underline{\underline{5{\cdot}7}}$

11.5.4. The Smoker equations

Smoker (1938) derived an analytical equation that can be used to determine the number of stages when the relative volatility is constant. Though his method can be used for any problem for which the relative volatilities in the rectifying and stripping sections can be taken as constant, it is particularly useful for problems where the relative volatility is low; for example, in the separation of close boiling isomers. If the relative volatility is close to one, the number of stages required will be very large, and it will be impractical to draw a McCabe–Thiele diagram. The derivation of the equations are outlined below and illustrated in Example 11.4.

The equations can be easily programmed for solution with small "desk-top" computers. A program written in BASIC for a Hewlett Packard 9830 model is given in Table 10.1.

Derivation of the equations:

A straight operating line can be represented by the equation:

$$y = sx + c \tag{11.27}$$

and in terms of relative volatility the equilibrium values of y are given by:

$$y = \frac{\alpha x}{1 + (\alpha - 1)x} \qquad \text{(equation 11.23)}$$

Eliminating y from these equations gives a quadratic in x:

$$s(\alpha - 1)x^2 + [s + b(\alpha - 1) - \alpha]x + b = 0 \tag{11.28}$$

For any particular distillation problem equation 11.28 will have only one real root k between 0 and 1

$$s(\alpha - 1)k^2 + [s + b(\alpha - 1) - \alpha]k + b = 0 \tag{11.29}$$

k is the value of the x ordinate at the point where the extended operating lines intersect the vapour–liquid equilibrium curve. Smoker shows that the number of stages required is given by the equation:

$$N = \log\left[\frac{x_0^*(1 - \beta x_n^*)}{x_n^*(1 - \beta x_0^*)} \right] \Big/ \log\left(\frac{\alpha}{sc^2} \right) \tag{11.30}$$

where

$$\beta = \frac{sc(\alpha - 1)}{\alpha - sc^2} \tag{11.31}$$

N = number of stages required to effect the separation represented by the concentration change from

$$x_n^* \text{ to } x_0^*; \ x^* = (x - k) \text{ and } x_0^* > x_n^*$$
$$c = 1 + (\alpha - 1)k \tag{11.32}$$

s = slope of the operating line between x_n^* and x_0^*,
α = relative volatility, assumed constant over x_n^* to x_0^*.

For a column with a single feed and no side streams:

Rectifying section

$$x_0^* = x_d - k \tag{11.33}$$
$$x_n^* = z_f - k \tag{11.34}$$

$$s = \frac{R}{R+1} \tag{11.35}$$

$$b = \frac{x_d}{R+1} \tag{11.36}$$

Stripping section

$$x_0^* = z_f - k \tag{11.37}$$

$$x_n^* = x_b - k \tag{11.38}$$

$$s = \frac{Rz_f + x_d - (R+1)x_b}{(R+1)(z_f - x_b)} \tag{11.39}$$

$$b = \frac{(z_f - x_d)x_b}{(R+1)(z_f - x_b)} \tag{11.40}$$

If the feed stream is not at its bubble point, z_f is replaced by the value of x at the intersection of operating lines, given by

$$z_f^* = \frac{b + z_f/(q-1)}{q/(q-1) - s} \tag{11.41}$$

All compositions for the more volatile component.

Example 11.4

A column is to be designed to separate a mixture of ethylbenzene and styrene. The feed will contain 0·5 mol fraction styrene, and a styrene purity of 99·5 per cent is required, with a recovery of 85 per cent. Estimate the number of equilibrium stages required at a reflux ratio of 8. Maximum column bottom pressure 0·20 bar.

Solution

Ethylbenzene is the more volatile component.

$$\text{Antoine equations, ethylbenzene,} \quad \ln P^\circ = 9 \cdot 386 - \frac{3279 \cdot 47}{T - 59 \cdot 95}$$

$$\text{styrene} \quad \ln P^\circ = 9 \cdot 386 - \frac{3328 \cdot 57}{T - 63 \cdot 72}$$

$$P \text{ bar}, \, T \text{ Kelvin}$$

Material balance, basis 100 kmol feed:

at 85 per cent recovery, styrene in bottoms $= 50 \times 0 \cdot 85 = 42 \cdot 5$ kmol

at 99·5 per cent purity, ethylbenzene in bottoms $= \dfrac{42 \cdot 5}{99 \cdot 5} \times 0 \cdot 5 = 0 \cdot 21$ kmol ethylbenzene in the tops $= 50 - 0.21 = 49.79$ kmol

styrene in tops $= 50 - 42.5 = 7.5$ kmol

mol fraction ethylbenzene in tops $= \dfrac{49.79}{49.79 + 7.5} = 0.87$

$z_f = 0.5$, $x_b = 0.005$, $x_d = 0.87$

Column bottom temperature, from Antoine equation for styrene

$$\ln 0.2 = 9.386 - \frac{3328.57}{T - 63.72}$$

$$T = 366\,\text{K},\ 93.3°\text{C}$$

At 93.3°C, vapour pressure of ethylbenzene

$$\ln P° = 9.386 - \frac{3279.47}{366.4 - 59.95} = 0.27\ \text{bar}$$

$$\text{Relative volatility} = \frac{P° \text{ ethylbenzene}}{P° \text{ styrene}} = \frac{0.27}{0.20} = 1.35$$

The relative volatility will change as the compositions and (particularly for a vacuum column) the pressure changes up the column. The column pressures cannot be estimated until the number of stages is known; so as a first trial the relative volatility will be taken as constant, at the value determined by the bottom pressure.

Rectifying section

(11.35) $$s = \frac{8}{8 + 1} = 0.89$$

(11.36) $$b = \frac{0.87}{8 + 1} = 0.097$$

(11.29) $$0.89(1.35 - 1)k^2 + [0.89 + 0.097(1.35 - 1) - 1.35]k + 0.097 = 0$$

$$k = 0.290$$

(11.33) $$x_0^* = 0.87 - 0.29 = 0.58$$

(11.34) $$x_n^* = 0.50 - 0.29 = 0.21$$

(11.32) $$c = 1 + (1.35 - 1)\,0.29 = 1.10$$

(11.31) $$\beta = \frac{0.89 \times 1.10(1.35 - 1)}{1.35 - 0.89 \times 1.1^2} = 1.255$$

(11.30) $$N = \log\left[\frac{0.58(1 - 1.255 \times 0.21)}{0.21(1 - 1.255 \times 0.58)}\right] \Big/ \log(1.35/0.89 \times 1.1^2)$$

$$= \frac{\log 7.473}{\log 1.254} = 8.87,\ \underline{\underline{\text{say } 9}}$$

Stripping section, feed taken as at its bubble point

$$(11.39) \qquad s = \frac{8 \times 0.5 + 0.87 - (8+1)\,0.005}{(8+1)(0.5-0.005)} = 1.084$$

$$(11.40) \qquad b = \frac{(0.5-0.87)\,0.005}{(8+1)(0.5-0.005)} = -4.15 \times 10^{-4} \text{ (essentially zero)}$$

$$(11.29) \quad 1.084(1.35-1)k^2 + [1.084 - 4.15 \times 10^{-4}(1.35-1) - 1.35]k - 4.15 \times 10^{-4}$$
$$k = 0.702$$

TABLE 11.1. *Smoker equation*

```
10 PRINT "NUMBER OF STAGES BY SMOKER EQN."
20 REM ALLOWS DIFFERENT REL. VOLS. IN STRIPPING AND RECTIFYING SECTIONS
30 REM IF FEED NOT AT BOILING POINT, REPLACE FEED MOL FRACTION WITH
40 REM X COORDINATE  OF INTERSECTION Q-LINE AND OPERATING LINES
50 PRINT "MOL.FRACTIONS: FEED, TOPS, BOTTOMS ?"
60 INPUT X1,X2,X3
70 PRINT "XF=";X1;"XD=";X2;"XB=";X3
80 PRINT " AV. REL. VOL.: RECTIFYING; STRIPPING SECTIONS ?"
90 INPUT A1,A2
100 PRINT "REL. VOLS. RECT.   ";A1;"STRIP.=";A2
110 PRINT "REFLUX RATIO ?"
120 INPUT R
130 PRINT "REFLUX RATIO=";R
140 REM RECTIFYING SECTION
150 M1=R/(R+1)
160 B1=X2/(R+1)
170 REM CALC. K
180 A=M1*(A1-1)
190 B=M1+B1*(A1-1)-A1
200 C=B1
210 GOSUB 430
220 REM CALC STAGES
230 C1=1+(A1-1)*K
240 X4=X2-K
250 X5=X1-K
260 G1=(M1*C1*(A1-1))/(A1-M1*C1↑2)
270 N1=LGT(X4*(1-G1*X5)/(X5*(1-G1*X4)))/LGT(A1/(M1*C1↑2))
280 PRINT "NUMBER OF STAGES IN RECTIFYING SECTION";N1
290 REM STRIPPING SECTION
300 M2=(R*X1+X2-(R+1)*X3)/((R+1)*(X1-X3))
310 B2=((X1-X2)*X3)/(R+1)*(X1-X3))
320 A=M2*(A2-1)
330 B=(M2+B2*(A2-1)-A2)
340 C=B2
350 GOSUB 430
360 X6=X1-K
370 X7=X3-K
380 C2=1+(A2-1)*K
390 G2=M2*C2*(A2-1)/(A2-M2*C2↑2)
400 N2=LGT(X6*(1-G2*X7)/(X7*(1-G2*X6)))/LGT(A2/(M2*C2↑2))
410 PRINT " NUMBER OF STAGES STRIPPING SECTION";N2
420 END
430 REM SUB PROG. TO SOLVE QUADRATIC
440 D=(B↑2-4*A*C)
450 IF D<0 THEN 580
460 K1=(-B+SQR(D))/(2*A)
470 K2=(-B-SQR(D))/(2*A)
480 IF K1<0 THEN 520
490 IF K1>1 THEN 520
500 K=K1
510 GOTO 600
520 IF K2<0 THEN 560
530 IF K2>1 THEN 560
540 K=K2
550 GOTO 600
560 PRINT "     ROOTS OUT OF LIMITS, CHECK PROBLEM SPECIFICATION"
570 STOP
580 PRINT " IMAGINARY ROOTS, CHECK PROBLEM SPECIFICATION"
590 STOP
600 RETURN
```

(11.37) $\qquad x_0^* = 0.5 - 0.702 = -0.202$

(11.38) $\qquad x_n^* = 0.005 - 0.702 = -0.697$

(11.32) $\qquad c = 1 + (1.35 - 1)0.702 = 1.246$

(11.31) $$\beta = \frac{1.084 \times 1.246(1.35 - 1)}{1.35 - 1.084 \times 1.246^2} = -1.42$$

(11.30) $$N = \log\left[\frac{-0.202(1 - 0.697 \times 1.42)}{-0.697(1 - 0.202 \times 1.42)}\right] \bigg/ \log\left(\frac{1.35}{1.084 \times 1.246^2}\right)$$

$$= \frac{\log\left[4.17 \times 10^{-3}\right]}{\log 0.8} = 24.6, \text{ say } \underline{\underline{25}}$$

11.6. Multicomponent distillation: general considerations

The problem of determining the stage and reflux requirements for multicomponent distillations is much more complex than for binary mixtures. With a multicomponent mixture, fixing one component composition does not uniquely determine the other component compositions and the stage temperature. Also when the feed contains more than two components it is not possible to specify the complete composition of the top and bottom products independently. The separation between the top and bottom products is specified by setting limits on two "key" components, between which it is desired to make the separation.

The complexity of multicomponent distillation calculations can be appreciated by considering a typical problem. The normal procedure is to solve the stage equations (Section 11.3.1) stage-by-stage, from the top and bottom of the column toward the feed point. For such a calculation to be exact, the compositions obtained from both the bottom-up and top-down calculations must mesh at the feed point and match the feed composition. But the calculated compositions will depend on the compositions assumed for the top and bottom products at the commencement of the calculations. Though it is possible to match the key components, the other components will not match unless the designer was particularly fortunate in choosing the trial top and bottom compositions. For a completely rigorous solution the compositions must be adjusted and the calculations repeated until a satisfactory mesh at the feed point is obtained. Clearly, the greater the number of components, the more difficult the problem. As was shown in Section 11.3.2, trial-and-error calculations will be needed to determine the stage temperatures. For other than ideal mixtures, the calculations will be further complicated by the fact that the component volatilities will be functions of the unknown stage compositions. If more than a few stages are required, stage-by-stage calculations are complex and tedious; as illustrated in Example 11.9.

Before the advent of the modern digital computer, various "short-cut" methods were developed to simplify the task of designing multicomponent columns. A comprehensive summary of the methods used for hydrocarbon systems is given by Edmister (1947 to 1949) in a series of articles in the journal *The Petroleum Engineer*. Though computer programs will normally be available for the rigorous solution of the stage equations, short-cut methods are still useful in the preliminary design work, and as an aid in defining problems for computer solution. Intelligent use of the short-cut methods can reduce the computer time and costs.

The short-cut methods available can be divided into two classes:

1. Simplifications of the rigorous stage-by-stage procedures to enable the calculations to be done using hand calculators, or graphically. Typical examples of this approach are the methods given by Hengstebeck (1961), and the Smith–Brinkley method (1960); which are described in Section 11.7.
2. Empirical methods, which are based on the performance of operating columns, or the results of rigorous designs. Typical examples of these methods are Gilliland's correlation, which is given in Volume 2, Chapter 11, and the Erbar–Maddox correlation given in Section 11.7.3.

11.6.1. Key components

Before commencing the column design, the designer must select the two "key" components between which it is desired to make the separation. The light key will be the component that it is desired to keep out of the bottom product, and the heavy key the component to be kept out of the top product. Specifications will be set on the maximum concentrations of the keys in the top and bottom products. The keys are known as "adjacent keys" if they are "adjacent" in a listing of the components in order of volatility, and "split keys" if some other component lies between them in the order; they will usually be adjacent.

Which components are the key components will normally be clear, but sometimes, particularly if close boiling isomers are present, judgement must be used in their selection. If any uncertainty exists, trial calculations should be made using different components as the keys to determine the pair that requires the largest number of stages for separation (the worst case). The Fenske equation can be used for these calculations; see Section 11.7.3.

The "non-key" components that appear in both top and bottom products are known as "distributed" components; and those that are not present, to any significant extent, in one or other product, are known as "non-distributed" components.

11.6.2. Number of columns

As was mentioned in Section 11.2, in multicomponent distillations it is not possible to obtain more than one pure component, one sharp separation, in a single column. If a multicomponent feed is to be split into two or more virtually pure products several columns will be needed. Impure products can be taken off as side streams; and the removal of a side stream from a stage where a minor component is concentrated will reduce the concentration of that component in the main product.

Where a large number of stages is required it may be necessary to split a column into two separated columns to reduce the height of the column, even though the required separation could theoretically have been obtained in a single column. This may also be done in vacuum distillations to reduce the column pressure drop and limit the bottom temperatures.

11.7. Multicomponent distillation: short-cut methods for stage and reflux requirements

Some of the more useful short-cut procedures which can be used to estimate stage and reflux requirements without the aid of computers are given in this section. Most of the short-cut methods were developed for the design of separation columns for hydrocarbon systems in the petroleum and petrochemical systems industries, and caution must be exercised when applying them to other systems. They usually depend on the assumption of constant relative volatility, and should not be used for severely non-ideal systems.

11.7.1. Pseudo-binary systems

If the presence of the other components does not significantly affect the volatility of the key components, the keys can be treated as a pseudo-binary pair. The number of stages can then be calculated using a McCabe–Thiele diagram, or the other methods developed for binary systems. This simplification can often be made when the amount of the non-key components is small, or where the components form near-ideal mixtures.

Where the concentration of the non-keys is small, say less than 10 per cent, they can be lumped in with the key components. For higher concentrations the method proposed by Hengstebeck (1946) can be used to reduce the system to an equivalent binary system. Hengstebeck's method is outlined below and illustrated in Example 11.5. Hengstebeck's book (1961) should be consulted for the derivation of the method and further examples of its application.

Hengstebeck's method

For any component i the Lewis–Sorel material balance equations (Section 11.5) and equilibrium relationship can be written in terms of the individual component molar flow rates; in place of the component composition:

$$v_{n+1,i} = l_{n,i} + d_i \tag{11.42}$$

$$v_{n,i} = K_{n,i} \frac{V}{L} l_{n,i} \tag{11.43}$$

for the stripping section:

$$l'_{n+1,i} = v'_{n,i} + b_i \tag{11.44}$$

$$v'_{n,i} = K_{n,i} \frac{V'}{L'} l'_{n,i} \tag{11.45}$$

where $l_{n,i}$ = the liquid flow rate of any component i from stage n,
$\quad v_{n,i}$ = the vapour flow rate of any component i from stage n,
$\quad d_i$ = the flow rate of component i in the tops,
$\quad b_i$ = the flow rate of component i in the bottoms,
$\quad K_{n,i}$ = the equilibrium constant for component i at stage n.

The superscript ′ denotes the stripping section.
V and L are the total flow-rates, assumed constant.
To reduce a multicomponent system to an equivalent binary it is necessary to estimate the flow-rate of the key components throughout the column. Hengstebeck makes use of

the fact that in a typical distillation the flow-rates of each of the light non-key components approaches a constant, limiting, rate in the rectifying section; and the flows of each of the heavy non-key components approach limiting flow-rates in the stripping section. Putting the flow-rates of the non-keys equal to these limiting rates in each section enables the combined flows of the key components to be estimated.

Rectifying section

$$L_e = L - \sum l_i \tag{11.46}$$

$$V_e = V - \sum v_i \tag{11.47}$$

Stripping section

$$L'_e = L' - \sum l'_i \tag{11.48}$$

$$V'_e = V' - \sum v'_i \tag{11.49}$$

where V_e and L_e are the estimated flow rates of the combined keys,

l_i and v_i are the limiting liquid and vapour rates of components *lighter* than the keys in the rectifying section,

l'_i and v'_i are the limiting liquid and vapour rates of components *heavier* than the keys in the stripping section.

The method used to estimate the limiting flow-rates is that proposed by Jenny (1939). The equations are:

$$l_i = \frac{d_i}{\alpha_i - 1} \tag{11.50}$$

$$v_i = l_i + d_i \tag{11.51}$$

$$v'_i = \frac{\alpha_i b_i}{\alpha_{LK} - \alpha_i} \tag{11.52}$$

$$l'_i = v'_i + b_i \tag{11.53}$$

where α_i = relative volatility of component i, relative to the heavy key (HK),

α_{LK} = relative volatility of the light key (LK), relative to the heavy key.

Estimates of the flows of the combined keys enable operating lines to be drawn for the equivalent binary system. The equilibrium line is drawn by assuming a constant relative volatility for the light key:

$$y = \frac{\alpha_{LK} x}{1 + (\alpha_{LK} - 1)x} \qquad \text{(equation 11.23)}$$

where y and x refer to the vapour and liquid concentrations of the light key.

Hengstebeck shows how the method can be extended to deal with situations where the relative volatility cannot be taken as constant, and how to allow for variations in the liquid and vapour molar flow rates. He also gives a more rigorous graphical procedure based on the Lewis–Matheson method (see Section 11.8).

Example 11.5

Estimate the number of ideal stages needed in the butane–pentane splitter defined by the compositions given in the table below. The column will operate at a pressure of 8·3 bar, with a reflux ratio of 2·5. The feed is at its boiling point.

Note: a similar problem has been solved by Lyster *et al.* (1959) using a rigorous computer method and it was found that ten stages were needed.

	Feed (f)	Tops (d)	Bottoms (b)
Propane, C_3	5	5	0
i-Butane, iC_4	15	15	0
n-Butane, nC_4	25	24	1
i-Pentane, iC_5	20	1	19
n-Pentane, nC_5	35	0	35
	100	45	55 kmol

Solution

The top and bottom temperatures (dew points and bubble points) were calculated by the methods illustrated in Example 11.9. Relative volatilities are given by equation 8.30:

$$\alpha_i = \frac{K_i}{K_{HK}}$$

Equilibrium constants were taken from the Depriester charts (Chapter 8).

Relative volatilities

	Top	Bottom	Average
Temp. °C	65	120	
C_3	5·5	4·5	5·0
iC_4	2·7	2·5	2·6
(LK) nC_4	2·1	2·0	2·0
(HK) iC_5	1·0	1·0	1·0
nC_5	0·84	0·85	0·85

Calculations of non-key flows
Equations 11.50, 11.51, 11.52, 11.53

	α_i	d_i	$l_i = d_i/(\alpha_i - 1)$	$v_i = l_i + d_i$
C_3	5	5	1·3	6·3
iC_4	2·6	15	9·4	24·4

$$\sum l_i = 10\cdot7 \quad \sum v_i = 30\cdot7$$

	α_i	b_i	$v'_i = \alpha_i b_i/(\alpha_{LK} - \alpha_i)$	$l'_i = v'_i + b_i$
nC_5	0·85	35	25·9	60·9

$$\sum v'_i = 25\cdot9 \quad \sum l'_i = 60\cdot9$$

Flows of combined keys

(11.46) $\qquad L_e = 2\cdot5 \times 45 - 10\cdot7 = 101\cdot8$

(11.47) $\qquad V_e = (2\cdot5 + 1)45 - 30\cdot7 = 126\cdot8$

(11.49) $\qquad V'_e = (2.5 + 1)45 - 25.9 = 131.6$
(11.48) $\qquad L'_e = (2.5 + 1)45 + 55 - 60.9 = 151.6$

Slope of top operating line

$$L_e/V_e = \frac{101.8}{126.8} = 0.8$$

Slope of bottom operating line

$$L'_e/V'_e = \frac{151.6}{131.6} = 1.15$$

$$x_b = \frac{\text{flow LK}}{\text{flow(LK + HK)}} = \frac{1}{19 + 1} = 0.05$$

$$x_d = \frac{24}{24 + 1} = 0.96$$

$$x_f = \frac{25}{25 + 20} = 0.56$$

(11.23) $$y = \frac{2x}{1 + (2 - 1)x} = \frac{2x}{1 + x}$$

x	0	0.20	0.40	0.60	0.80	1.0
y	0	0.33	0.57	0.75	0.89	1.0

The McCabe–Thiele diagram is shown in Fig. 11.10.
Twelve stages required; feed on seventh from base.

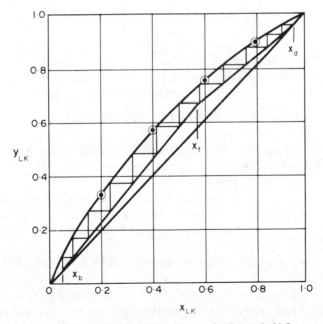

FIG. 11.10. McCabe–Thiele diagram for Example 11.5

11.7.2. Smith–Brinkley method

Smith and Brinkley developed a method for determining the distribution of components in multicomponent separation processes. Their method is based on the solution of the finite-difference equations that can be written for multistage separation processes, and can be used for extraction and absorption processes, as well as distillation. Only the equations for distillation will be given here. The derivation of the equations is given by Smith and Brinkley (1960) and Smith (1963). For any component i (suffix i omitted in the equation for clarity)

$$\frac{b}{f} = \frac{(1 - S_r^{N_r - N_s}) + R(1 - S_r)}{(1 - S_r^{N_r - N_s}) + R(1 - S_r) + GS_r^{N_r - N_s}(1 - S_s^{N_s + 1})} \tag{11.54}$$

where b/f is the fractional split of the component between the feed and the bottoms, and:

N_r = number of equilibrium stages above the feed,
N_s = number of equilibrium stages below the feed,
S_r = stripping factor, rectifying section = $K_i V/L$,
S_s = stripping factor, stripping section = $K_i' V'/L'$,
V and L are the total molar vapour and liquid flow rates, and the superscript ' denotes
     the stripping section.

G depends on the condition of the feed.
If the feed is mainly liquid:

$$G_i = \frac{K_i' L}{K_i L'} \left[\frac{1 - S_r}{1 - S_s} \right]_i \tag{11.55}$$

and the feed stage is added to the stripping section.
If the feed is mainly vapour:

$$G_i = \frac{L}{L'} \left[\frac{1 - S_r}{1 - S_s} \right]_i \tag{11.56}$$

Equation 11.54 is for a column with a total condenser. If a partial condenser is used the number of stages in the rectifying section should be increased by one.

The procedure for using the Smith–Brinkley method is as follows:

1. Estimate the flow-rates L, V and L', V' from the specified component separations and reflux ratio.
2. Estimate the top and bottom temperatures by calculating the dew and bubble points for assumed top and bottom compositions.
3. Estimate the feed point temperature.
4. Estimate the average component K values in the stripping and rectifying sections.
5. Calculate the values of $S_{r,i}$ for the rectifying section and $S_{s,i}$ for the stripping section.
6. Calculate the fractional split of each component, and hence the top and bottom compositions.
7. Compare the calculated with the assumed values and check the overall column material balance.
8. Repeat the calculation until a satisfactory material balance is obtained. The usual procedure is to adjust the feed temperature up and down till a satisfactory balance is obtained.

Examples of the application of the Smith–Brinkley method are given by Smith (1963).

This method is basically a rating method, suitable for determining the performance of an existing column, rather than a design method, as the number of stages must be known.

It can be used for design by estimating the number of stages by some other method and using equation 11.54 to determine the top and bottom compositions. The estimated stages can then be adjusted and the calculations repeated until the required specifications are achieved. However, the Geddes–Hengstebeck method for estimating the component splits, described in Section 11.7.4, is easier to use and satisfactory for preliminary design.

11.7.3. Empirical correlations

The two most frequently used empirical methods for estimating the stage requirements for multicomponent distillations are the correlations published by Gilliland (1940) and by

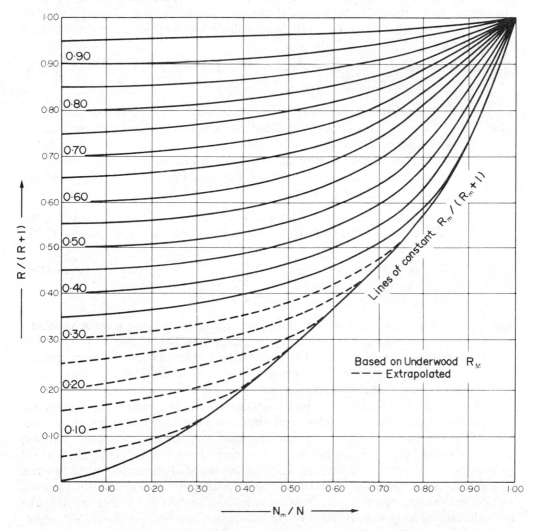

FIG. 11.11. Erbar–Maddox correlation (Erbar and Maddox, 1961)

Erbar and Maddox (1961). These relate the number of ideal stages required for a given separation, at a given reflux ratio, to the number at total reflux (minimum possible) and the minimum reflux ratio (infinite number of stages).

Gilliland's correlation is given in Volume 2, Chapter 11.

The Erbar–Maddox correlation is given in this section, as it is now generally considered to give more reliable predictions. Their correlation is shown in Fig. 11.11; which gives the ratio of number of stages required to the number at total reflux, as a function of the reflux ratio, with the minimum reflux ratio as a parameter. To use Fig. 11.11, estimates of the number of stages at total reflux and the minimum reflux ratio are needed.

Minimum number of stages (Fenske Equation)

The Fenske equation (Fenske, 1932) can be used to estimate the minimum stages required at total reflux. The derivation of this equation for a binary system is given in Volume 2, Chapter 11. The equation applies equally to multicomponent systems and can be written as:

$$\left[\frac{x_i}{x_r}\right]_d = \alpha_i^{N_m}\left[\frac{x_i}{x_r}\right]_b \qquad (11.57)$$

where $\left[\dfrac{x_i}{x_r}\right]$ = the ratio of the concentration of any component i to the concentration of a reference component r, and the suffixes d and b denote the distillate (tops) (d) and the bottoms (b),

N_m = minimum number of stages at total reflux, including the reboiler,

α_i = average relative volatility of the component i with respect to the reference component.

Normally the separation required will be specified in terms of the key components, and equation 11.57 can be rearranged to give an estimate of the number of stages.

$$N_m = \frac{\log\left[\dfrac{x_{LK}}{x_{HK}}\right]_d\left[\dfrac{x_{HK}}{x_{LK}}\right]_b}{\log \alpha_{LK}} \qquad (11.58)$$

where α_{LK} is the average relative volatility of the light key with respect to the heavy key, and x_{LK} and x_{HK} are the light and heavy key concentrations. The relative volatility is taken as the geometric mean of the values at the column top and bottom temperatures. To calculate these temperatures initial estimates of the compositions must be made, so the calculation of the minimum number of stages by the Fenske equation is a trial-and-error procedure. The procedure is illustrated in Example 11.7. If there is a wide difference between the relative volatilities at the top and bottom of the column the use of the average value in the Fenske equation will underestimate the number of stages. In these circumstances, a better estimate can be made by calculating the number of stages in the rectifying and stripping sections separately; taking the feed concentration as the base concentration for the rectifying section and as the top concentration for the stripping section, and estimating the average relative volatilities separately for each section. This procedure will also give an estimate of the feed point location.

Winn (1958) has derived an equation for estimating the number of stages at total reflux, which is similar to the Fenske equation, but which can be used when the relative volatility cannot be taken as constant.

If the number of stages is known, equation 11.57 can be used to estimate the split of components between the top and bottom of the column at total reflux. It can be written in a more convenient form for calculating the split of components:

$$\frac{d_i}{b_i} = \alpha_i^{N_m}\left[\frac{d_r}{b_r}\right] \tag{11.59}$$

where d_i and b_i are the flow-rates of the component i in the tops and bottoms, d_r and b_r are the flow-rates of the reference component in the tops and bottoms.

Note: from the column material balance:

$$d_i + b_i = f_i$$

where f_i is the flow rate of component i in the feed.

Minimum reflux ratio

Colburn (1941) and Underwood (1948) have derived equations for estimating the minimum reflux ratio for multicomponent distillations. These equations are discussed in Volume 2, Chapter 11. As the Underwood equation is more widely used it is presented in this section. The equation can be stated in the form:

$$\sum\frac{\alpha_i x_{i,d}}{\alpha_i - \theta} = R_m + 1 \tag{11.60}$$

where α_i = the relative volatility of component i with respect to some reference
          component, usually the heavy key,
          R_m = the minimum reflux ratio,
          $x_{i,d}$ = concentration of component i in the tops at minimum reflux
and θ is the root of the equation:

$$\sum\frac{\alpha_i x_{i,f}}{\alpha_i - \theta} = 1 - q \tag{11.61}$$

where $x_{i,f}$ = the concentration of component i in the feed, and q depends on the
          condition of the feed and was defined in Section 11.5.2.
The value of θ must lie between the values of the relative volatility of the light and heavy keys, and is found by trial and error.

In the derivation of equations 11.60 and 11.61 the relative volatilities are taken as constant. The geometric average of values estimated at the top and bottom temperatures should be used. This requires an estimate of the top and bottom compositions. Though the compositions should strictly be those at minimum reflux, the values determined at total reflux, from the Fenske equation, can be used. A better estimate can be obtained by replacing the number of stages at total reflux in equation 11.59 by an estimate of the actual number; a value equal to $N_m/0.6$ is often used. The Erbar–Maddox method of estimating the stage and reflux requirements, using the Fenske and Underwood equations, is illustrated in Example 11.7.

Feed-point location

A limitation of the Erbar–Maddox, and similar empirical methods, is that they do not give the feed-point location. An estimate can be made by using the Fenske equation to calculate the number of stages in the rectifying and stripping sections separately, but this requires an estimate of the feed-point temperature. An alternative approach is to use the empirical equation given by Kirkbride (1944):

$$\log\left[\frac{N_r}{N_s}\right] = 0{\cdot}206 \log\left[\left(\frac{B}{D}\right)\left(\frac{x_{f,\,HK}}{x_{f,\,LK}}\right)\left(\frac{x_{b,\,LK}}{x_{d,\,HK}}\right)^2\right] \tag{11.62}$$

where
  N_r = number of stages above the feed, including any partial condenser,
  N_s = number of stages below the feed, including the reboiler,
  B = molar flow bottom product,
  D = molar flow top product,
  $x_{f,\,HK}$ = concentration of the heavy key in the feed,
  $x_{f,\,LK}$ = concentration of the light key in the feed,
  $x_{d,\,HK}$ = concentration of the heavy key in the top product,
  $x_{b,\,LK}$ = concentration of the light key if in the bottom product.

The use of this equation is illustrated in Example 11.8.

11.7.4. Distribution of non-key components (graphical method)

The graphical procedure proposed by Hengstebeck (1946), which is based on the Fenske equation, is a convenient method for estimating the distribution of components between the top and bottom products.

Hengstebeck and Geddes (1958) have shown that the Fenske equation can be written in the form:

$$\log\left(d_i/b_i\right) = A + C \log \alpha_i \tag{11.63}$$

Specifying the split of the key components determines the constants A and C in the equation.

The distribution of the other components can be readily determined by plotting the distribution of the keys against their relative volatility on log-log paper, and drawing a straight line through these two points. The method is illustrated in Example 11.6.

Yaws *et al.* (1979) have shown that the components distributions calculated by equation 11.63 compare well with those obtained by rigorous plate by plate calculations.

Chang (1980) gives a computer program, based on the Geddes–Hengstebeck equation, for the estimation of component distributions.

Example 11.6

Use the Geddes–Hengstebeck method to check the component distributions for the separation specified in Example 11.5

Summary of problem, flow per 100 *kmol feed*

Component	α_i	Feed (f_i)	Distillate (d_i)	Bottoms (b_i)
C_3	5	5		
iC_4	2·6	15		
nC_4 (LK)	2·0	25	24	1
iC_5 (HK)	1·0	20	1	19
nC_5	0·85	35		

Solution

The average volatilities will be taken as those estimated in Example 11.5. Normally, the volatilities are estimated at the feed bubble point, which gives a rough indication of the

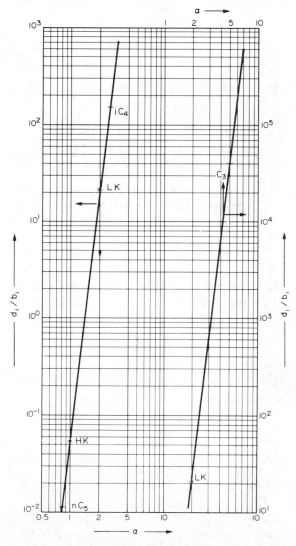

FIG. 11.12. Component Distribution (Example 11.6)

average column temperatures. The dew point of the tops and bubble point of the bottoms can be calculated once the component distributions have been estimated, and the calculations repeated with a new estimate of the average relative volatilities, as necessary.

$$\text{For the light key, } d_i/b_i = 24/1 = 24$$
$$\text{For the heavy key, } d_i/b_i = 1/19 = 0.053$$

These values are plotted on Fig. 11.12.

The distribution of the non-keys are read from Fig. 11.12 at the appropriate relative volatility and the component flows calculated from the following equations:

Overall column material balance

$$f_i = d_i + b_i$$

from which

$$d_i = f_i/(b_i/d_i + 1)$$
$$b_i = f_i/(d_i/b_i + 1)$$

	α_i	f_i	d_i/b_i	d_i	b_i
C_3	5	5	40,000	5	0
iC_4	2·6	15	150	14·9	·0·1
nC_4	2·0	25	21	24	1
iC_5	1·0	20	0·053	1	19
nC_5	0·85	35	0·011	0·4	34·6

As these values are close to those assumed for the calculation of the dew points and bubble points in Example 11.5, there is no need to repeat with new estimates of the relative volatilities.

Example 11.7

For the separation specified in Example 11.5, evaluate the effect of changes in reflux ratio on the number of stages required. This is an example of the application of the Erbar–Maddox method.

Solution

The relative volatilities estimated in Example 11.5, and the component distributions calculated in Example 11.6 will be used for this example.

Summary of data

	α_i	f_i	d_i	b_i
C_3	5	5	5	0
iC_4	2·6	15	14·9	0·1
nC_4 (LK)	2·0	25	24	1
iC_5 (HK)	1	20	1	19
nC_5	0·85	35	0·4	34·6
		100	$D = 45·3$	$B = 54·7$

Solution

Minimum number of stages; Fenske equation, equation 11.58:

$$N_m = \frac{\log[(24/1)(19/1)]}{\log 2} = \underline{\underline{8 \cdot 8}}$$

Minimum reflux ratio; Underwood equations 11.60 and 11.61.
This calculation is best tabulated.
As the feed is at its boiling point $q = 1$

(11.61)
$$\sum \frac{\alpha_i x_{i,f}}{\alpha_i - \theta} = 0$$

$x_{i,f}$	α_i	$\alpha_i x_{i,f}$	$\theta = 1 \cdot 5$	$\theta = 1 \cdot 3$	$\theta = 1 \cdot 35$
				Try	
0·05	5	0·25	0·071	0·068	0·068
0·15	2·6	0·39	0·355	0·300	0·312
0·25	2·0	0·50	1·000	0·714	0·769
0·20	1	0·20	−0·400	−0·667	−0·571
0·35	0·85	0·30	−0·462	−0·667	−0·600
			$\sum = 0 \cdot 564$	−0·252	0·022
					close enough

$$\theta = 1 \cdot 35$$

Equation 11.60

$x_{i,d}$	α_i	$\alpha_i x_{i,d}$	$\dfrac{\alpha_i x_{i,d}}{\alpha_i - \theta}$
0·11	5	0·55	0·15
0·33	2·6	0·86	0·69
0·53	2·0	1·08	1·66
0·02	1	0·02	−0·06
0·01	0·85	0·01	−0·02
			$\sum = 2 \cdot 42$

$$R_m + 1 = 2 \cdot 42$$

$$R_m = \underline{\underline{1 \cdot 42}}$$

$$R_m/(R_m + 1) = 1 \cdot 42/2 \cdot 42 = 0 \cdot 59$$

Specimen calculation, for $R = 2 \cdot 0$

$$R/(R + 1) = 2/3 = 0 \cdot 66$$

from Fig. 11.11

$$N_m/N = 0 \cdot 56$$
$$N = 8 \cdot 8/0 \cdot 56 = \underline{\underline{15 \cdot 7}}$$

for other reflux ratios

R	2	3	4	5	6
N	15·7	11·9	10·7	10·4	10·1

Note: Above a reflux ratio of 4 there is little change in the number of stages required, and the optimum reflux ratio will be near this value.

Example 11.8

Estimate the position of the feed point for the separation considered in Example 11.7, for a reflux ratio of 3.

Solution

Use the Kirkbride equation, equation 11.62. Product distributions taken from Example 11.6,

$$x_{b,\ LK} = 1/54.7 = 0.018$$
$$x_{d,\ HK} = 1/45.3 = 0.022$$

$$\log (N_r/N_s) = 0.206 \log \left[\frac{54.7}{45.3} \left(\frac{0.20}{0.25} \right) \left(\frac{0.018}{0.022} \right)^2 \right]$$

$$\log (N_r/N_s) = 0.206 \log (0.65)$$

$$N_r/N_s = \underline{0.91}$$

for $R = 3$, $N = 12$

number of stages, excluding the reboiler $= \underline{\underline{11}}$

$$N_r + N_s = 11$$
$$N_s = 11 - N_r = 11 - 0.91 N_s$$
$$N_s = 11/1.91 = 5.76, \text{ say } \underline{\underline{6}}$$

Checks with the method used in Example 11.5, where the reflux ratio was 2.5.

Example 11.9

This example illustrates the complexity and trial and error nature of stage-by-stage calculation.

The same problem specification has been used in earlier examples to illustrate the short-cut design methods.

A butane–pentane splitter is to operate at 8.3 bar with the following feed composition:

		x_f	f mol/100 mol feed
Propane,	C_3	0.05	5
Isobutane,	$i\,C_4$	0.15	15
Normal butane,	nC_4	0.25	25
Isopentane,	$i\,C_5$	0.20	20
Normal pentane,	nC_5	0.35	35
Light key	nC_4		
Heavy key	$i\,C_5$		

For a specification of not more than 1 mol of the light key in the bottom product and not more than 1 mol of the heavy key in the top product, and a reflux ratio of 2·5, make a stage-by-stage calculation to determine the product composition and number of stages required.

Solution

Only sufficient trial calculations will be made to illustrate the method used. Basis 100 mol feed.

Estimation of dew and bubble points

$$\text{Bubble point (11.4)} \quad \sum y_i = \sum K_i x_i = 1\cdot0$$

$$\text{Dew point (11.5)} \quad \sum x_i = \sum \frac{y_i}{K_i} = 1\cdot0$$

The K values, taken from the De Priester charts (Chapter 8), are plotted in Fig. (a) for easy interpolation.

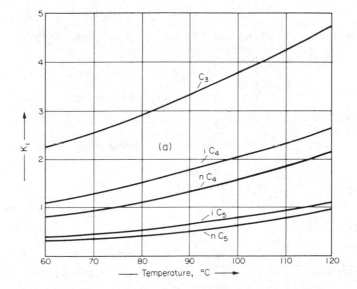

FIG. (a) K-values at 8·3 bar

To estimate the dew and bubble points, assume that nothing heavier than the heavy key appears in the tops, and nothing lighter than the light key in the bottoms.

	d	x_d	b	x_b
C_3	5	0·11	0	—
C_4	15	0·33	0	—
nC_4	24	0·54	1	0·02
iC_5	1	0·02	19	0·34
nC_5	0	—	35	0·64
	45		55	

Bubble-point calculation, bottoms

		Try 100°C		Try 120°C	
	x_b	K_i	$K_i x_i$	K_i	$K_i x_i$
C_3	—	—	—	—	—
i C_4	—	—	—	—	—
n C_4	0·02	1·85	0·04	2·1	0·04
i C_5	0·34	0·94	0·32	1·1	0·37
n C_5	0·64	0·82	0·52	0·96	0·61
		$\sum K_i x_i = 0.88$		1·02	
		temp. too low		close enough	

Dew-point calculation, tops

		Try 70°C		Try 60°C	
	x_d	K_i	y_i/K_i	K_i	y_i/K_i
C_3	0·11	2·6	0·04	2·20	0·24
i C_4	0·33	1·3	0·25	1·06	0·35
n C_4	0·54	0·9	0·60	0·77	0·42
i C_5	0·02	0·46	0·04	0·36	0·01
n C_5	—	—	—	—	—
		$\sum y_i/K_i = 0.94$		1·02	
		temp. too high		close enough	

Bubble point calculation, feed (liquid feed)

		Try 80° C		Try 90° C		Try 85° C	
	x_f	K_i	$x_i K_i$	K_i	$x_i K_i$	K_i	$x_i K_i$
C_3	0·05	2·9	0·15	3·4	0·17	3·15	0·16
i C_4	0·15	1·5	0·23	1·8	0·27	1·66	0·25
n C_4	0·25	1·1	0·28	1·3	0·33	1·21	0·30
i C_5	0·20	0·5	0·11	0·66	0·13	0·60	0·12
n C_5	0·35	0·47	0·16	0·56	0·20	0·48	0·17
			0·93		1·10		1·00
			temp. too low		temp. too high		satisfactory

Stage-by-stage calculations

Top down calculations, assume total condensation with no subcooling

$$y_1 = x_d = x_0$$

It is necessary to estimate the composition of the "non-keys" so that they can be included in the stage calculations. As a first trial the following values will be assumed:

	x_d	d
C_3	0·10	5
$i\,C_4$	0·33	15
nC_4	0·54	24
$i\,C_5$	0·02	1
nC_5	0·001	0·1
		45·1

In each stage calculation it will be necessary to estimate the stage temperatures to determine the K values and liquid and vapour enthalpies. The temperature range from top to bottom of the column will be approximately $120 - 60 = 60°C$. An approximate calculation (Example 11.7) has shown that around fourteen ideal stages will be needed; so the temperature change from stage to stage can be expected to be around 4 to 5°C.

Stage 1

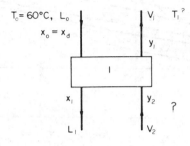

$$L_0 = R \times D \quad = 2·5 \times 45·1 = 112·8$$
$$V_1 = (R+1)D = 3·5 \times 45·1 = 157·9$$

Estimation of stage temperature and outlet liquid composition (x_1)

	y_1	Try $T_1 = 66°C$		Try $T_1 = 65°C$		
		K_i	y_i/K_i	K_i	y_i/K_i	$x_1 = y_i/K_i$ Normalised
C_3	0·10	2·40	0·042	2·36	0·042	0·042
$i\,C_4$	0·33	1·20	0·275	1·19	0·277	0·278
nC_4	0·54	0·88	0·614	0·86	0·628	0·629
$i\,C_5$	0·02	0·42	0·048	0·42	0·048	0·048
nC_5	0·001	0·32	0·003	0·32	0·003	0·003
		$\sum y_i/K_i = 0·982$ too low		0·998 close enough		

Summary of stage equations

$$L_0 + V_2 = L_1 + V_1 \tag{i}$$
$$L_0 x_0 + V_2 y_2 = L_1 x_1 + V_1 y_1 \tag{ii}$$
$$h_0 L_0 + H_2 V_2 = h_1 L_1 + H_1 V_1 \tag{iii}$$
$$h = f(x, T) \tag{iv}$$
$$H = f(x, T) \tag{v}$$

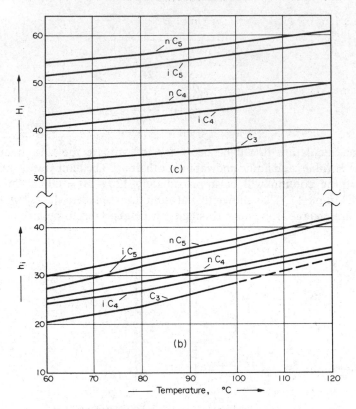

FIGS. (b) and (c). Enthalpy kJ/mol (adapted from J. B. Maxwell, *Data Book of Hydrocarbons* (Van Nostrand, 1962)

The enthalpy relationship is plotted in Figs. (b) and (c).

$$y_i = K_i x_i \qquad (vi)$$

Before a heat balance can be made to estimate L_1 and V_2, an estimate of y_2 and T_2 is needed. y_2 is dependent on the liquid and vapour flows, so as a first trial assume that these are constant and equal to L_0 and V_1; then, from equations (i) and (ii),

$$y_2 = (L_0/V_1)(x_1 - x_0) + y_1$$
$$L_0/V_1 = 112.8/157.9 = 0.71$$

	x_1	x_0	$y_2 = 0.71(x_1 - x_0) + y_1$	y_2 Normalised
C_3	0.042	0.10	0.057	0.057
$i\,C_4$	0.278	0.33	0.294	0.292
nC_4	0.629	0.54	0.604	0.600
$i\,C_5$	0.048	0.02	0.041	0.041
nC_5	0.003	0.001	0.013	0.013

1.009
close enough

Enthalpy data from Figs. (b) and (c) J/mol

	$h_0(T_0 = 60°C)$			$h_1(T_1 = 65°C)$		
	x_0	h_i	$h_i x_i$	x_1	h_i	$h_i x_i$
C_3	0·10	20,400	2040	0·042	21,000	882
iC_4	0·33	23,400	7722	0·278	24,900	6897
nC_4	0·54	25,200	13,608	0·629	26,000	16,328
iC_5	0·02	27,500	550	0·048	28,400	1363
nC_5	0·001	30,000	30	0·003	30,700	92
		$h_0 =$	23,950		$h_1 =$	25,562

	$H_1(T_1 = 65°C)$			$H_2(T_2 = 70°C$ assumed$)$		
	v_1	H_i	$H_i y_i$	y_2	H_i	$H_i y_i$
C_3	0·10	34,000	3400	0·057	34,800	1984
iC_4	0·33	41,000	13,530	0·292	41,300	12,142
nC_4	0·54	43,700	23,498	0·600	44,200	26,697
iC_5	0·02	52,000	1040	0·041	52,500	2153
nC_5	0·001	54,800	55	0·013	55,000	715
		$H_1 =$	41,623		$H_2 =$	43,691

Energy balance (equation iii)

$$23,950 \times 112·8 + 43,691 V_2 = 25,562 L_1 + 41,623 \times 157·9$$
$$43,691 V_2 = 255,626 L_1 + 3,870,712$$

Material balance (equation i)

$$112·8 + V_2 = L_1 + 157·9$$

substituting

$$43,691(L_1 + 45·1) = 25,562 L_1 + 3,870,712$$
$$L_1 = 104·8$$
$$V_2 = 104·8 + 45·1 = 149·9$$
$$L_1/V_2 = 0·70$$

Could revise calculated values for y_2 but L_1/V_2 is close enough to assumed value of 0·71, so there would be no significant difference from first estimate.

Stage 2

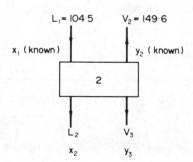

Estimation of stage temperature and outlet liquid composition (x_2).

	y_2	K_i	$x_2 = y_2/K_i$	x_2 Normalised
			$T_2 = 70°C$ (use assumed value as first trial)	
C_3	0·057	2·55	0·022	0·022
i C_4	0·292	1·30	0·226	0·222
nC_4	0·600	0·94	0·643	0·630
i C_5	0·041	0·43	0·095	0·093
nC_5	0·013	0·38	0·034	0·033
			1·020	
			close enough to 1·0	

$$y_3 = L/V(x_2 - x_1) + y_2$$

As a first trial take L/V as $L_1/V_1 = 0.70$

	x_2	x_1	$y_3 = 0.70(x_2 - x_1) + y_2$	y_3 Normalised
C_3	0·022	0·042	0·044	0·043
i C_4	0·222	0·277	0·256	0·251
nC_4	0·630	0·628	0·613	0·601
i C_5	0·093	0·048	0·072	0·072
nC_5	0·033	0·003	0·035	0·034
			1·020	

Enthalpy data from Figs. (*b*) and (*c*)

	x_2	h_i	$h_i x_2$	y_3	H_i	$H_i y_3$
	$h_2(T_2 = 70°C)$			$H_3(T_3 = 75°C$ assumed)		
C_3	0·022	21,900	482	0·043	34,600	1488
i C_4	0·222	25,300	5617	0·251	41,800	10,492
nC_4	0·630	27,000	17,010	0·601	44,700	26,865
i C_5	0·093	29,500	2744	0·072	53,000	3816
nC_5	0·033	31,600	1043	0·035	55,400	1939
	$h_2 = 26,896$			$H_3 = 44,600$		

Energy balance

$$25,562 \times 104·8 + 44,600V_3 = 4369 \times 149·9 + 26,896L_2$$

Material balance

$$104 \cdot 8 + V_3 = 149 \cdot 9 + L_2$$
$$L_2 = 105 \cdot 0$$
$$V_3 = 150 \cdot 1$$
$$L_2 / V_3 = 0 \cdot 70 \text{ checks with assumed value.}$$

Stage 3

As the calculated liquid and vapour flows are not changing much from stage to stage the calculation will be continued with the value of L/V taken as constant at $0 \cdot 7$.

		Try $T_3 = 75°C$ (assumed value)		
	K_i	$x_3 = y_3/K_i$	Normalised	$y_4 = 0 \cdot 7(x_3 - x_2) + y_3$
C_3	2·71	0·016	0·015	0·38
i C_4	1·40	0·183	0·177	0·217
nC_4	1·02	0·601	0·580	0·570
i C_5	0·50	0·144	0·139	0·104
nC_5	0·38	0·092	0·089	0·074
		1·036		1·003
		Close enough		

Stage 4

		Try $T_4 = 81°C$		
	K_i	$x_4 = y_4/K_i$	Normalised	$y_5 = 0 \cdot 7(x_4 - x_3) + y_4$
C_3	2·95	0·013	0·013	0·039
i C_4	1·55	0·140	0·139	0·199
nC_4	1·13	0·504	0·501	0·515
i C_5	0·55	0·189	0·188	0·137
nC_5	0·46	0·161	0·166	0·118
		1·007		1·008
		Close enough		

Stage 5

		Try $T_5 = 85°C$		
	K_i	x_5	Normalised	$y_6 = 0 \cdot 7(x_5 - x_4) + y_5$
C_3	3·12	0·013	0·012	0·038
i C_4	1·66	0·120	0·115	0·179
nC_4	1·20	0·430	0·410	0·450
i C_5	0·60	0·228	0·218	0·159
nC_5	0·46	0·257	0·245	0·192
		1·048		1·018
		Close enough		

Stage 6

	Try $T_6 = 90°C$		Try $T_6 = 92°C$			
	K_i	x_6	K_i	x_6	Normalised	y_7
C_3	3·35	0·011	3·45	0·011	0·011	0·037
i C_4	1·80	0·099	1·85	0·097	0·095	0·166
nC_4	1·32	0·341	1·38	0·376	0·318	0·386
i C_5	0·65	0·245	0·69	0·230	0·224	0·163
nC_5	0·51	0·376	0·53	0·362	0·350	0·268
		1·072		1·026		1·020
		too low		close enough		

Note: ratio of LK to HK in liquid from this stage

$$= \frac{0·386}{0·163} = 2·37$$

Stage 7

	Try $T_6 = 97°C$		
	K_i	x_7	Normalised
C_3	3·65	0·010	0·010
i C_4	1·98	0·084	0·083
nC_4	1·52	0·254	0·251
i C_5	0·75	0·217	0·214
nC_5	0·60	0·447	0·442
		1·012	

$$\text{ratio LK/HK} = \frac{0·251}{0·214} = 1·17$$

This is just below the ratio in the feed

$$= \frac{25}{20} = 1·25$$

So, the feed would be introduced at this stage.

   But the composition of the non-key components on the plate does not match the feed composition.

	x_f	x_7
C_3	0·05	0·10
i C_4	0·15	0·084
nC_4	0·25	0·254
i C_5	0·20	0·217
nC_5	0·35	0·447

So it would be necessary to adjust the assumed top composition and repeat the calculation.

Bottom-up calculation

To illustrate the procedure the calculation will be shown for the reboiler and bottom stage, assuming constant molar overflow.

With the feed at its boiling point and constant molar overflow the base flows can be calculated as follows:

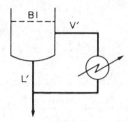

$$V' = V_0 = 157.9$$

$$L' = L_0 + FEED = 112.8 + 100 = 212.8$$

$$V'/L' = \frac{157.9}{212.8} = 0.74$$

It will be necessary to estimate the concentration of the non-key components in the bottom product; as a first trial take:

C_3	iC_4	nC_4	iC_5	nC_5
0.001	0.001	0.02	0.34	0.64

Reboiler

Check bubble-point estimate of 120°C

		Try 120°C		Try 118°C	
	x_B	K_i	$y_B = K_i x_B$	K_i	y_R
C_3	0.001	4.73	0.005	4.60	0.005
iC_4	0.001	2.65	0.003	2.58	0.003
nC_4	0.02	2.10	0.042	2.03	0.041
iC_5	0.34	1.10	0.374	1.06	0.360
nC_5	0.64	0.96	0.614	0.92	0.589
			1.038		0.998
			too high		close enough

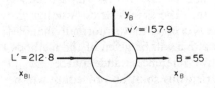

Material balance:

$$x_{B1} L' = y_B V' + x_B B$$

$$x_{B1} = \frac{V'}{L'} y_B + \frac{B}{L'} x_B$$

$$x_{B1} = \frac{157 \cdot 9}{212 \cdot 8} y_B + \frac{55}{212 \cdot 8} x_B$$

$$= 0 \cdot 74 y_B + 0 \cdot 26 x_B$$

Stage 1 from base (B1)

	x_B	y_B	x_{B1}	$x_{B2} = 0 \cdot 74 (y_{1B} - y_B) + x_{1B}$
C_3	0·001	0·005	0·004	0·014
i C_4	0·001	0·003	0·002	0·036
n C_4	0·02	0·041	0·020	0·019
i C_5	0·34	0·361	0·356	0·357
n C_5	0·64	0·590	0·603	0·559
				0·985

The calculation is continued stage-by-stage up the column to the feed point (stage 7 from the top). If the vapour composition at the feed point does not mesh with the top-down calculation, the assumed concentration of the non-keys in the bottom product is adjusted and the calculations repeated.

11.8. Multicomponent systems: rigorous solution procedures (computer methods)

The application of digital computers has made the rigorous solution of the stage equations (Section 11.3.1) a practical proposition, and computer methods for the design of multicomponent separation columns will be available in most design organisations. Programs, and computer time, can also be rented from commercial computing bureaux. A considerable amount of work has been done over the past twenty or so years to develop efficient and reliable computer-aided design procedures for distillation and other staged processes. A detailed discussion of this work is beyond the scope of this book and the reader is referred to the specialist books that have been published on the subject, Smith (1963), Holland (1963, 1975) and Hanson and Sommerville (1963), and to the numerous papers that have appeared in the chemical engineering literature. A good summary of the present state of the art is given by McCormick and Roche (1979).

Several different approaches have been taken to develop programs that are efficient in the use of computer time, and suitable for the full range of multicomponent separation processes that are used in the process industries. A design group will use those methods that are best suited to the processes that it normally handles.

In this section a brief outline will be given of the methods that have been developed; together with authoritative references, and sources of published program listings.

The basic steps in any rigorous solution procedure will be:

1. Specification of the problem; complete specification is essential for computer methods.
2. Selection of values for the iteration variables; for example, estimated stage temperatures, and liquid and vapour flows (the column temperature and flow profiles).
3. A calculation procedure for the solution of the stage equations.
4. A procedure for the selection of new values for the iteration variables for each set of trial calculations.
5. A procedure to test for convergence; to check if a satisfactory solution has been achieved.

It is convenient to consider the methods available under the following four headings:

1. Lewis–Matheson method.
2. Thiele–Geddes method.
3. Relaxation methods.
4. Linear algebra methods.

Rating and design methods

With the exception of the Lewis–Matheson method, all the methods listed above require the specification of the number of stages below and above the feed point. They are therefore not directly applicable to design: where the designer wants to determine the number of stages required for a specified separation. They are strictly what are referred to as "rating methods"; used to determine the performance of existing, or specified, columns. Given the number of stages they can be used to determine product compositions. Iterative procedures are necessary to apply rating methods to the design of new columns. An initial estimate of the number of stages can be made using short-cut methods and the programs used to calculate the product compositions; repeating the calculations with revised estimates till a satisfactory design is obtained.

11.8.1. Lewis–Matheson method

The method proposed by Lewis and Matheson (1932) is essentially the application of the Lewis–Sorel method (Section 11.5.1) to the solution of multicomponent problems. Constant molar overflow is assumed and the material balance and equilibrium relationship equations are solved stage by stage starting at the top or bottom of the column, in the manner illustrated in Example 11.9. To define a problem for the Lewis–Matheson method the following variables must be specified, or determined from other specified variables:

Feed composition, flow rate and condition.
Distribution of the key components.
One product flow.
Reflux ratio.
Column pressure.
Assumed values for the distribution of the non-key components.

The usual procedure is to start the calculation at the top and bottom of the column and proceed toward the feed point. The initial estimates of the component distributions in the products are then revised and the calculations repeated until the compositions calculated from the top and bottom starts mesh, and match the feed at the feed point.

Efficient procedures for adjusting the compositions to achieve a satisfactory mesh at the feed point are given by Hengstebeck (1961). Good descriptions of the Lewis–Matheson method, with examples of manual calculations, are also given in the books by Oliver (1966) and Smith (1963); a simple example is given in Volume 2, Chapter 11.

In some computer applications of the method, where the assumption of constant molar overflow is not made, it is convenient to start the calculations by assuming flow and temperature profiles. The stage component compositions can then be readily determined and used to revise the profiles for the next iteration. With this modification the procedure is similar to the Thiele–Geddes method discussed in the next section.

In general, the Lewis–Matheson method has not been found to be an efficient procedure for computer solutions, other than for relatively straightforward problems. It is not suitable for problems involving multiple feeds, and side-streams, or where more than one column is needed.

The method is suitable for interactive programs run on small "desk-top" computers. Such programs can be "semi-manual" in operation: the computer solving the stage equations, while control of the iteration variables, and convergence is kept by the designer. As the calculations are carried out one stage at a time, only a relatively small computer memory is needed.

11.8.2. Thiele–Geddes method

Like the Lewis–Matheson method, the original method of Thiele and Geddes (1933) was developed for manual calculation. It has subsequently been adapted by many workers for computer applications. The variables specified in the basic method, or that must be derived from other specified variables, are:

> Reflux temperature.
> Reflux flow rate.
> Distillate rate.
> Feed flows and condition.
> Column pressure.
> Number of equilibrium stages above and below the feed point.

The basic procedure used in the Thiele–Geddes method is described in the books by Oliver (1966) and Smith (1963), and its application to computers is described in detail in a series of articles by Lyster et al. (1959) and by Holland (1963).

The method starts with an assumption of the column temperature and flow profiles. The stage equations are then solved to determine the stage component compositions and the results used to revise the temperature profiles for subsequent trial calculations. Efficient convergence procedures have been developed for the Thiele–Geddes method. The so-called "theta method", described by Lyster et al. (1959) and Holland (1963), is recommended. The Thiele–Geddes method can be used for the solution of complex distillation problems, and for other multi-component separation processes. A series of programs for the solution of problems in distillation, extraction, stripping and absorption, which use an iterative procedure similar to the Thiele–Geddes method, are given by Hanson et al. (1962).

11.8.3. Relaxation methods

With the exception of this method, all the methods described solve the stage equations for the steady-state design conditions. In an operating column other conditions will exist at start-up, and the column will approach the "design" steady-state conditions after a period of time. The stage material balance equations can be written in a finite difference form, and procedures for the solution of these equations will model the unsteady-state behaviour of the column.

Rose *et al.* (1958) and Hanson and Sommerville (1963) have applied "relaxation methods" to the solution of the unsteady-state equations to obtain the steady-state values. The application of this method to the design of multistage columns is described by Hanson and Sommerville (1963). They give a program listing and worked examples for a distillation column with side-streams, and for a reboiled absorber.

Relaxation methods are not competitive with the "steady-state" methods in the use of computer time, because of slow convergence. However, because they model the actual operation of the column, convergence should be achieved for all practical problems. The method has the potential of development for the study of the transient behaviour of column designs, and for the analysis and design of batch distillation columns.

11.8.4. Linear algebra methods

The Lewis–Matheson and Thiele–Geddes methods use a stage-by-stage procedure to solve the equations relating the component compositions to the column temperature and flow profiles. However, the development of high-speed digital computers with large memories makes possible the simultaneous solution of the complete set of equations that describe the stage compositions throughout the column.

If the equilibrium relationships and flow-rates are known (or assumed) the set of material balance equations for each component is linear in the component compositions. Amundson and Pontinen (1958) developed a method in which these equations are solved simultaneously and the results used to provide improved estimates of the temperature and flow profiles. The set of equations can be expressed in matrix form and solved using the standard inversion routines available in modern computer systems. Convergence can usually be achieved after a few iterations.

This approach has been further developed by other workers; notably Wang and Henke (1966) and Naphtali and Sandholm (1971).

The linearisation method of Naphtali and Sandholm has been used by Fredenslund *et al.* (1977) for the multicomponent distillation program given in their book. Included in their book, and coupled to the distillation program, are methods for estimation of the liquid–vapour relationships (activity coefficients) using the UNIFAC method (see Chapter 8, Section 16.3). This makes the program particularly useful for the design of columns for new processes, where experimental data for the equilibrium relationships are unlikely to be available. The program is recommended to those who do not have access to their own "in house" programs.

11.9. Batch distillation

Batch operation should be considered when the quantity to be distilled is small; when it is produced at irregular intervals; when a range of products has to be produced; or when the feed composition is likely to vary considerably.

Batch distillation is an unsteady-state process: the composition in the still and the overheads changing as the batch is distilled. The basic theory of batch distillation is covered in Volume 2, Chapter 11, and in several other books, Robinson and Gilliland (1950), Van Winkle (1967), Treybal (1979), Sherwood *et al.* (1975). In the simple theoretical analysis of batch distillation the liquid hold-up in the column is usually neglected. This hold-up can have a significant effect on the separating efficiency and should be taken into account when designing batch columns. The practical design of batch distillation processes is discussed by Hengstebeck (1961), Ellerbe (1979) and Billet (1979).

11.10. Plate efficiency

The designer is concerned with real contacting stages; not the theoretical equilibrium stage assumed for convenience in the mathematical analysis of multistage processes. Equilibrium will rarely be attained in a real stage. The concept of a stage efficiency is used to link the performance of practical contacting stages to the theoretical equilibrium stage.

Three definitions of efficiency are used:

1. Murphree plate efficiency (Murphree, 1925), defined in terms of the vapour compositions by:

$$E_{mV} = \frac{y_n - y_{n-1}}{y_e - y_{n-1}} \qquad (11.64)$$

where y_e is the composition of the vapour that would be in equilibrium with the liquid leaving the plate. The Murphree plate efficiency is the ratio of the actual separation achieved to that which would be achieved in an equilibrium stage (see Fig. 11.6). In this definition of efficiency the liquid and the vapour stream are taken to be perfectly mixed; the compositions in equation 11.64 are the average composition values for the streams.

2. Point efficiency (Murphree point efficiency). If the vapour and liquid compositions are taken at a point on the plate, equation 11.64 gives the local or point efficiency, E_{mv}.

3. Overall column efficiency. This is sometimes confusingly referred to as the overall plate efficiency.

$$E_o = \frac{\text{number of ideal stages}}{\text{number of real stages}} \qquad (11.65)$$

An estimate of the overall column efficiency will be needed when the design method used gives an estimate of the number of ideal stages required for the separation.

In some methods, the Murphree plate efficiencies can be incorporated into the procedure for calculating the number of stages and the number of real stages determined directly.

For the idealised situation where the operating and equilibrium lines are straight, the overall column efficiency and the Murphree plate efficiency are related by an equation derived by Lewis (1936):

$$E_o = \frac{\log\left[1 + E_{mV}\left(\frac{mV}{L} - 1\right)\right]}{\log\left(\frac{mV}{L}\right)} \qquad (11.66)$$

where m = slope of the equilibrium line,
V = molar flow rate of the vapour,
L = molar flow rate of the liquid.

Equation 11.66 is not of much practical use in distillation, as the slopes of the operating and equilibrium lines will vary throughout the column. It can be used by dividing the column into sections and calculating the slopes over each section. For most practical purposes, providing the plate efficiency does not vary too much, a simple average of the plate efficiency calculated at the column top, bottom and feed points will be sufficiently accurate.

11.10.1. Prediction of plate efficiency

Whenever possible the plate efficiencies used in design should be based on experimental values for similar systems, obtained on full-sized columns. There is no entirely satisfactory method for predicting plate efficiencies from the system physical properties and plate design parameters. However, the methods given in this section can be used to make a rough estimate where no reliable experimental values are available. They can also be used to extrapolate data obtained from small-scale experimental columns. If the system properties are at all unusual, experimental confirmation of the predicted values should always be obtained. The small, laboratory scale, glass sieve plate column developed by Oldershaw (1941) has been shown to give reliable values for scale-up. The use of Oldershaw columns is described in papers by Swanson and Gester (1962) and Veatch et al. (1960).

Plate, and overall column, efficiencies will normally be between 30 per cent and 70 per cent, and as a rough guide a figure of 50 per cent can be assumed for preliminary designs.

Efficiencies will be lower for vacuum distillations, as low weir heights are used to keep the pressure drop small (see Section 11.10.4).

Multicomponent systems

The prediction methods given in the following sections, and those available in the open literature, are invariably restricted to binary systems. It is clear that in a binary system the efficiency obtained for each component must be the same. This is not so for a multicomponent system; the heavier components will usually exhibit lower efficiencies than the lighter components.

The following guide rules, adapted from a paper by Toor and Burchard (1960), can be used to estimate the efficiencies for a multicomponent system from binary data:

1. If the components are similar, the multicomponent efficiencies will be similar to the binary efficiency.
2. If the predicted efficiencies for the binary pairs are high, the multicomponent efficiency will be high.
3. If the resistance to mass transfer is mainly in the liquid phase, the difference between the binary and multicomponent efficiencies will be small.
4. If the resistance is mainly in the vapour phase, as it normally will be, the difference between the binary and multicomponent efficiencies can be substantial.

For mixtures of dissimilar compounds the efficiency can be very different from that predicted for each binary pair, and laboratory or pilot–plant studies should be made to confirm any predictions.

11.10.2. O'Connell's correlation

A quick estimate of the overall column efficiency can be obtained from the correlation given by O'Connell (1946), which is shown in Fig. 11.13. The overall column efficiency is correlated with the product of the relative volatility of the light key component (relative to the heavy key) and the molar average viscosity of the feed, estimated at the average column temperature. The correlation was based mainly on data obtained with hydrocarbon systems, but includes some values for chlorinated solvents and water–alcohol mixtures. It has been found to give reliable estimates of the overall column efficiency for hydrocarbon systems; and can be used to make an approximate estimate of the efficiency for other systems. The method takes no account of the plate design parameters; and includes only two physical property variables.

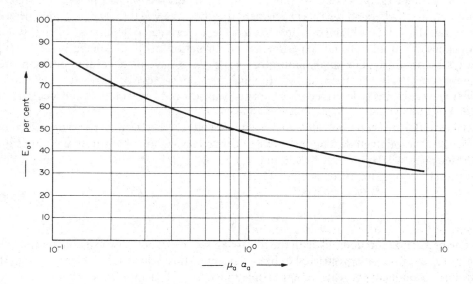

FIG. 11.13. Distillation column efficiencies (bubble-caps) (after O'Connell, 1946)

Eduljee (1958) has expressed the O'Connell correlation in the form of an equation:

$$E_o = 51 - 32 \cdot 5 \log (\mu_a \alpha_a) \qquad (11.67)$$

where μ_a = the molar average liquid viscosity, $mN\,s/m^2$,
α_a = average relative volatility of the light key.

Absorbers

O'Connell gave a similar correlation for the *plate efficiency* of absorbers; Fig. 11.14. Appreciably lower plate efficiencies are obtained in absorption than in distillation.

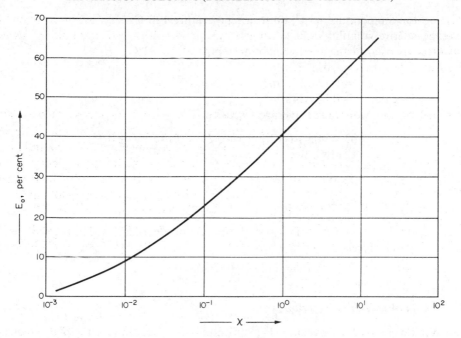

FIG. 11.14. Absorber column efficiencies (bubble-caps) (after O'Connell, 1946)

In O'Connell's paper, the plate efficiency is correlated with a function involving Henry's constant, the total pressure, and the solvent viscosity at the operating temperature.

To convert the original data to SI units, it is convenient to express this function in the following form:

$$x = 0.062 \left[\frac{\rho_s P}{\mu_s \mathscr{H} M_s} \right] = 0.062 \left[\frac{\rho_s}{\mu_s K M_s} \right] \tag{11.68}$$

where $\mathscr{H}$ = the Henry's law constant, $N\,m^{-2}$/mol fraction,

P = total pressure, N/m^2,

μ_s = solvent viscosity, $mN\,s/m^2$,

M_s = molecular weight of the solvent,

ρ_s = solvent density, kg/m^3,

K = equilibrium constant for the solute.

Example 11.10

Using O'Connell's correlation, estimate the overall column efficiency and the number of real stages required for the separation given in Example 11.5.

Solution

From Example 11.5, feed composition, mol fractions:

propane 0·05, i-butane 0·15, n-butane 0·25, i-pentane 0·20, n-pentane 0·35.

Column-top temperature 65°C, bottom temperature 120°C.
Average relative volatility light key = 2·0
Take the viscosity at the average column temperature, 93°C,
viscosities, propane = 0·03 mN s/m²
    butane = 0·12 mN s/m²
    pentane = 0·14 mN s/m²
For feed composition, molar average viscosity

$$= 0·03 \times 0·05 + 0·12 \,(0·15 + 0·25) + 0·14 \,(0·20 + 0·35)$$
$$= 0·13 \,\text{mN s/m}^2$$

$$\alpha_a \mu_a = 2·0 \times 0·13 = 0·26$$

From Fig. 11.13, $E_o = \underline{70}$ per cent

From Example 11.4, number of ideal stages = 12, one ideal stage will be the reboiler, so number of actual stages

$$= (12 - 1)/0·7 = \underline{\underline{16}}$$

11.10.3. Van Winkle's correlation

Van Winkle *et al.* (1972) have published an empirical correlation for the plate efficiency which can be used to predict plate efficiencies for binary systems. Their correlation uses dimensionless groups that include those system variables and plate parameters that are known to effect plate efficiency. They give two equations, the simplest, and that which they consider the most accurate, is given below. The data used to derive the correlation covered both bubble-cap and sieve plates.

$$E_{mV} = 0·07 \, Dg^{0·14} \, Sc^{0·25} \, Re^{0·08} \tag{11.69}$$

where Dg = surface tension number = $(\sigma_L/\mu_L u_v)$,
  u_v = superficial vapour velocity,
  σ_L = liquid surface tension,
  μ_L = liquid viscosity,
  Sc = liquid Schmidt number = $(\mu_L/\rho_L D_{LK})$,
  ρ_L = liquid density,
  D_{LK} = liquid diffusivity, light key component,
  Re = Reynolds number = $(h_w u_v \rho_v/\mu_L \,(\text{FA}))$,
  h_w = weir height,
  ρ_v = vapour density,

$$(\text{FA}) = \text{fractional area} = \frac{(\text{area of holes or risers})}{(\text{total column cross-sectional area})}$$

The use of this method is illustrated in Example 11.13.

11.10.4. AIChE method

This method of predicting plate efficiency, published in 1958, was the result of a five-year study of bubble-cap plate efficiency directed by the Research Committee of the American Institute of Chemical Engineers.

The AIChE method is the most detailed method for predicting plate efficiencies that is available in the open literature. It takes into account all the major factors that are known to affect plate efficiency; this includes:

The mass transfer characteristics of the liquid and vapour phases.

The design parameters of the plate.

The vapour and liquid flow-rates.

The degree of mixing on the plate.

The method is well established, and in the absence of experimental values, or proprietary prediction methods, should be used when more than a rough estimate of efficiency is needed.

The approach taken is semi-empirical. Point efficiencies are estimated making use of the "two-film theory", and the Murphree efficiency estimated allowing for the degree of mixing likely to be obtained on real plates.

The procedure and equations are given in this section without discussion of the theoretical basis of the method. The reader should refer to the AIChE manual, AIChE (1958); or to Smith (1963) who gives a comprehensive account of the method, and extends its use to sieve plates. A brief discussion of the method is given in Volume 2; to which reference can be made for the definition of any unfamiliar terms used in the equations.

AIChE method

The mass transfer resistances in the vapour and liquid phases are expressed in terms of the number of transfer units, $\mathbf{N}_G$ and $\mathbf{N}_L$. The point efficiency is related to the number of transfer units by the equation:

$$\frac{1}{\ln\left(1 - E_{mv}\right)} = -\left[\frac{1}{\mathbf{N}_G} + \frac{mV}{L} \times \frac{1}{\mathbf{N}_L}\right] \tag{11.70}$$

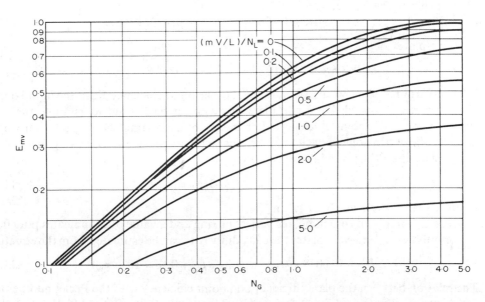

FIG. 11.15. Relationship between point efficiency and number of liquid and vapour transfer units (Equation 11.70)

where m is the slope of the operating line and V and L the vapour and liquid molar flow rates.

Equation 11.70 is plotted in Fig. 11.15.

The number of gas phase transfer units is given by:

$$N_G = (0{\cdot}776 + 4{\cdot}57 \times 10^{-3} h_w - 0{\cdot}24 F_v + 105 L_p)/(\mu_v/\rho_v D_v)^{0{\cdot}5} \tag{11.71}$$

where h_w = weir height, mm,

F_v = the column vapour "F" factor = $u_a \sqrt{\rho_v}$,

u_a = vapour velocity based on the active tray area (bubbling area), see Section 11.13.2, m/s,

L_p = the volumetric liquid flow rate across the plate, divided by the average width of the plate, m^3/sm. The average width can be calculated by dividing the active area by the length of the liquid path Z_L,

μ_v = vapour viscosity, N s/m^2,

ρ_v = vapour density; kg/m^3,

D_v = vapour diffusivity, m^2/s.

The number of liquid phase transfer units is given by:

$$N_L = (4{\cdot}13 \times 10^8 D_L)^{0{\cdot}5} (0{\cdot}21 F_v + 0{\cdot}15) t_L \tag{11.72}$$

where D_L = liquid phase diffusivity, m^2/s,

t_L = liquid contact time, s,

given by:

$$t_L = Z_c Z_L/L_p \tag{11.73}$$

where Z_L = length of the liquid path, from inlet downcomer to outlet weir, m,

Z_c = liquid hold-up on the plate, m^3 per m^2 active area,

given by:

for bubble-cap plates

$$Z_c = 0{\cdot}042 + 0{\cdot}19 \times 10^{-3} h_w + 0{\cdot}014 F_v + 2{\cdot}5 L_p \tag{11.74}$$

for sieve plates

$$Z_c = 0{\cdot}006 + 0{\cdot}73 \times 10^{-3} h_w - 0{\cdot}24 \times 10^{-3} F_v h_w + 1{\cdot}22 L_p \tag{11.75}$$

The Murphree efficiency E_{mV} is only equal to the point efficiency E_{mv} if the liquid on the plate is perfectly mixed. On a real plate this will not be so, and to estimate the plate efficiency from the point efficiency some means of estimating the degree of mixing is needed. The dimensionless Peclet number characterises the degree of mixing in a system. For a plate the Peclet number is given by:

$$Pe = Z_L^2/D_e t_L \tag{11.76}$$

where D_e is the "eddy diffusivity", m^2/s.

A Peclet number of zero indicates perfect mixing and a value of ∞ indicates plug flow.

For bubble-cap and sieve plates the eddy diffusivity can be estimated from the equation:

$$D_e = (0{\cdot}0038 + 0{\cdot}017 u_a + 3{\cdot}86 L_p + 0{\cdot}18 \times 10^{-3} h_w)^2 \tag{11.77}$$

The relation between the plate efficiency and point efficiency with the Peclet number as a parameter is shown in Fig. 11.16a and b. The application of the AIChE method is illustrated in Example 11.12.

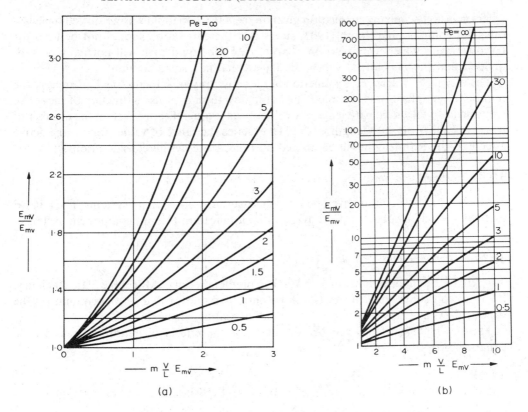

FIG. 11.16. Relationship between plate and point efficiency

Estimation of physical properties

To use the AIChE method, and Van Winkle's correlation, estimates of the physical properties are required. It is unlikely that experimental values will be found in the literature for all systems that are of practical interest. The prediction methods given in Chapter 8, and in the references given in that chapter, can be used to estimate values.

The AIChE design manual recommends the Wilke and Chang (1955) equation for liquid diffusivities and the Wilke and Lee (1955) modification to the Hirschfelder, Bird and Spotz equation for gas diffusivities.

Plate design parameters

The significance of the weir height in the AIChE equations should be noted. The weir height was the plate parameter found to have the most significant effect on plate efficiency. Increasing weir height will increase the plate efficiency, but at the expense of an increase in pressure drop and entrainment. Weir heights will normally be in the range 40 to 100 mm for columns operating at and above atmospheric pressure, but will be as low as 6 mm for vacuum columns. This, in part, accounts for the lower plate efficiencies obtained in vacuum columns.

The length of the liquid path Z_L is taken into account when assessing the plate-mixing performance. The mixing correlation given in the AIChE method was not tested on large-diameter columns, and Smith (1963) states that the correlation should not be used for large-diameter plates. However, on a large plate the liquid path will normally be sub-divided, and the value of Z_L will be similar to that in a small column.

The vapour "F" factor F_v is a function of the active tray area. Increasing F_v decreases the number of gas-phase transfer units. The liquid flow term L_p is also a function of the active tray area, and the liquid path length. It will only have a significant effect on the number of transfer units if the path length is long. In practice the range of values for F_v, the active area, and the path length will be limited by other plate design considerations.

Multicomponent systems

The AIChE method was developed from measurements on binary systems. The AIChE manual should be consulted for advice on its application to multicomponent systems.

11.10.5. Entrainment

The AIChE method, and that of Van Winkle, predict the "dry" Murphree plate efficiency. In operation some liquid droplets will be entrained and carried up the column by the vapour flow, and this will reduce the actual, operating, efficiency.

The dry-plate efficiency can be corrected for the effects of entrainment using the equation proposed by Colburn (1936):

$$E_a = \frac{E_{mV}}{1 + E_{mV}[\psi/(1 - \psi)]} \qquad (11.78)$$

where E_a = actual plate efficiency, allowing for entrainment,

$$\psi = \text{the fractional entrainment} = \frac{\text{entrained liquid}}{\text{gross liquid flow}}.$$

Methods for predicting the entrainment from sieve plates are given in Section 11.13.5, Fig. 11.27; a similar method for bubble-cap plates is given by Bolles (1963)

11.11. Approximate column sizing

An approximate estimate of the overall column size can be made once the number of real stages required for the separation is known. This is often needed to make a rough estimate of the capital cost for project evaluation.

Plate spacing

The overall height of the column will depend on the plate spacing. Plate spacings from 0·15 m(6 in.) to 1 m (36 in.) are normally used. The spacing chosen will depend on the column diameter and operating conditions. Close spacing is used with small-diameter columns, and where head room is restricted; as it will be when a column is installed in a building. For columns above 1 m diameter, plate spacings of 0·3 to 0·6 m will normally be used, and 0·5 m (18 in.) can be taken as an initial estimate. This would be revised, as necessary, when the detailed plate design is made.

A larger spacing will be needed between certain plates to accommodate feed and side-streams arrangements, and for manways.

Column diameter

The principal factor that determines the column diameter is the vapour flow-rate. The vapour velocity must be below that which would cause excessive liquid entrainment or a high-pressure drop. The equation given below, which is based on the well-known Souders and Brown equation, Lowenstein (1961), can be used to estimate the maximum allowable superficial vapour velocity, and hence the column area and diameter,

$$\hat{u}_v = (-0.171 l_t^2 + 0.27 l_t - 0.047)\left[(\rho_L - \rho_v)/\rho_v\right]^{\frac{1}{2}} \qquad (11.79)$$

where $\hat{u}_v$ = maximum allowable vapour velocity, based on the gross (total) column cross-sectional area, m/s,

l_t = plate spacing, m.

The column diameter, D_c, can then be calculated:

$$D_c = \sqrt{\frac{4 \hat{V}_w}{\pi \rho_v \hat{u}_v}} \qquad (11.80)$$

where $\hat{V}_w$ is the maximum vapour rate, kg/s.

This approximate estimate of the diameter would be revised when the detailed plate design is undertaken.

11.12. Plate contactors

Cross-flow plates are the most common type of plate contactor used in distillation and absorption columns. In a cross-flow plate the liquid flows across the plate and the vapour

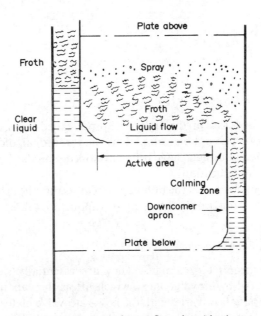

FIG. 11.17. Typical cross-flow plate (sieve)

up through the plate. A typical layout is shown in Fig. 11.17. The flowing liquid is transferred from plate to plate through vertical channels called "downcomers". A pool of liquid is retained on the plate by an outlet weir.

Other types of plate are used which have no downcomers (non-cross-flow plates), the liquid showering down the column through large openings in the plates (sometimes called shower plates). These, and, other proprietary non-cross-flow plates, are used for special purposes, particularly when a low-pressure drop is required.

Three principal types of cross-flow tray are used, classified according to the method used to contact the vapour and liquid.

1. Sieve plate (perforated plate) (Fig. 11.18)

This is the simplest type of cross-flow plate. The vapour passes up through perforations in the plate; and the liquid is retained on the plate by the vapour flow. There is no positive vapour liquid seal, and at low flow-rates liquid will "weep" through the holes, reducing the plate efficiency. The perforations are usually small holes, but larger holes and slots are used.

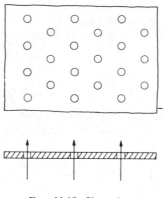

Fig. 11.18. Sieve plate

2. Bubble-cap plates (Fig. 11.19)

In which the vapour passes up through short pipes, called risers, covered by a cap with a serrated edge, or slots. The bubble-cap plate is the traditional, oldest, type of cross-flow plate, and many different designs have been developed. Standard cap designs would now be specified for most applications.

The most significant feature of the bubble-cap plate is that the use of risers ensures that a level of liquid is maintained on the tray at all vapour flow-rates.

3. Valve plates (floating cap plates) (Fig. 11.20)

Valve plates are proprietary designs. They are essentially sieve plates with large-diameter holes covered by movable flaps, which lift as the vapour flow increases.

As the area for vapour flow varies with the flow-rate, valve plates can operate efficiently at lower flow-rates than sieve plates: the valves closing at low vapour rates.

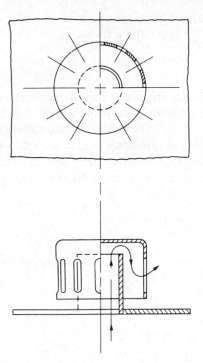

FIG. 11.19.   Bubble-cap

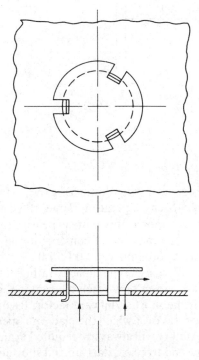

FIG. 11.20.   Simple valve

Some very elaborate valve designs have been developed, but the simple type shown in Fig. 11.20 is satisfactory for most applications.

Liquid flow pattern

Cross-flow trays are also classified according to the number of liquid passes on the plate. The design shown in Fig. 11.21*a* is a single-pass plate. For low liquid flow rates reverse flow plates are used; Fig. 11.21*b*. In this type the plate is divided by a low central partition, and inlet and outlet downcomers are on the same side of the plate. Multiple-pass plates, in which the liquid stream is sub-divided by using several downcomers, are used for high liquid flow-rates and large diameter columns. A double-pass plate is shown in Fig. 11.21*c*.

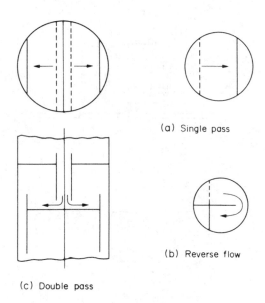

(a) Single pass

(b) Reverse flow

(c) Double pass

FIG. 11.21. Liquid flow patterns on cross-flow trays

11.12.1. Selection of plate type

The principal factors to consider when comparing the performance of bubble-cap, sieve and valve plates are: cost, capacity, operating range, efficiency and pressure drop.

Cost. Bubble-cap plates are appreciably more expensive than sieve or valve plates. The relative cost will depend on the material of construction used; for mild steel the ratios, bubble-cap:valve:sieve, are approximately 3·0:1·5:1·0

Capacity. There is little difference in the capacity rating of the three types (the diameter of the column required for a given flow-rate); the ranking is sieve, valve, bubble-cap.

Operating range. This is the most significant factor. By operating range is meant the range of vapour and liquid rates over which the plate will operate satisfactorily (the stable operating range). Some flexibility will always be required in an operating plant to allow for changes in production rate, and to cover start-up and shut-down conditions. The ratio of the highest to the lowest flow rates is often referred to as the "turn-down" ratio. Bubble-cap plates have a positive liquid seal and can therefore operate efficiently at very low vapour rates.

Sieve plates rely on the flow of vapour through the holes to hold the liquid on the plate, and cannot operate at very low vapour rates. But, with good design, sieve plates can be designed to give a satisfactory operating range; typically, from 50 per cent to 120 per cent of design capacity.

Valve plates are intended to give greater flexibility than sieve plates at a lower cost than bubble-caps.

Efficiency. The Murphree efficiency of the three types of plate will be virtually the same when operating over their design flow range, and no real distinction can be made between them; see Zuiderweg *et al.* (1960).

Pressure drop. The pressure drop over the plates can be an important design consideration, particularly for vacuum columns. The plate pressure drop will depend on the detailed design of the plate but, in general, sieve plates give the lowest pressure drop, followed by valves, with bubble-caps giving the highest.

Summary. Sieve plates are the cheapest and are satisfactory for most applications. Valve plates should be considered if the specified turn-down ratio cannot be met with sieve plates. Bubble-caps should only be used where very low vapour (gas) rates have to be handled and a positive liquid seal is essential at all flow-rates.

11.12.2. Plate construction

The mechanical design features of sieve plates are described in this section. The same general construction is also used for bubble-cap and valve plates. Details of the various types of bubble-cap used, and the preferred dimensions of standard cap designs, can be found in the books by Smith (1963) and Ludwig (1979). The manufacturers' design manuals should be consulted for details of valve plate design; Glitsch (1970) and Koch (1960).

Two basically different types of plate construction are used. Large-diameter plates are normally constructed in sections, supported on beams. Small plates are installed in the column as a stack of pre-assembled plates.

Sectional construction

A typical plate is shown in Fig. 11.22. The plate sections are supported on a ring welded round the vessel wall, and on beams. The beams and ring are about 50 mm wide, with the beams set at around 0·6 m spacing. The beams are usually angle or channel sections, constructed from folded sheet. Special fasteners are used so the sections can be assembled from one side only. One section is designed to be removable to act as a manway. This reduces the number of manways needed on the vessel, which reduces the vessel cost.

Diagrams and photographs, of sectional plates, are given in Volume 2, Chapter 11.

Stacked plates (cartridge plates)

The stacked type of construction is used where the column diameter is too small for a man to enter to assemble the plates, say less than 1·2 m (4 ft). Each plate is fabricated complete with the downcomer, and joined to the plate above and below using screwed rods

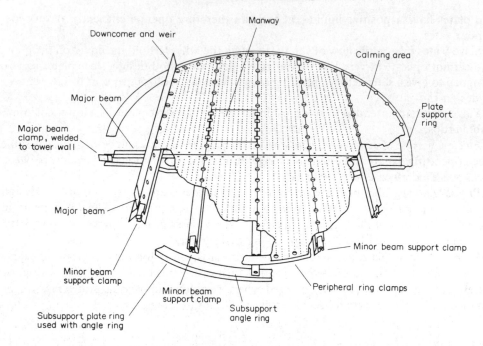

FIG. 11.22. Typical sectional plate construction

(spacers); see Fig. 11.23. The plates are installed in the column shell as an assembly (stack) of ten, or so, plates. Tall columns have to be divided into flanged sections so that plate assemblies can be easily installed and removed. The weir, and downcomer supports, are usually formed by turning up the edge of the plate.

The plates are not fixed to the vessel wall, as they are with sectional plates, so there is no positive liquid seal at the edge of the plate, and a small amount of leakage will occur. In some designs the plate edges are turned up round the circumference to make better contact at the wall. This can make it difficult to remove the plates for cleaning and maintenance, without damage.

Downcomers

The segmental, or chord downcomer, shown in Fig. 11.24a is the simplest and cheapest form of construction and is satisfactory for most purposes. The downcomer channel is formed by a flat plate, called an apron, which extends down from the outlet weir. The apron is usually vertical, but may be sloped (Fig. 11.24b) to increase the plate area available for perforation. If a more positive seal is required at the downcomer at the outlet, an inlet weir can be fitted (Fig. 11.24c) or a recessed seal pan used (Fig. 11.24d). Circular downcomers (pipes) are sometimes used for small liquid flow-rates.

Side-stream and feed points

Where a side-stream is withdrawn from the column the plate design must be modified to provide a liquid seal at the take-off pipe. A typical design is shown in Fig. 11.25a. When the

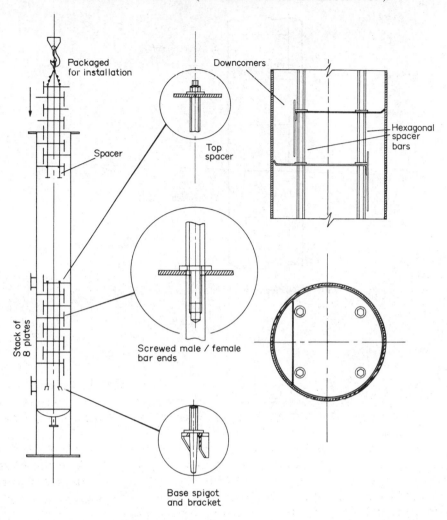

FIG. 11.23. Typical stacked-plate construction

feed stream is liquid it will be normally introduced into the downcomer leading to the feed plate, and the plate spacing increased at this point; Fig. 11.25b.

Structural design

The plate structure must be designed to support the hydraulic loads on the plate during operation, and the loads imposed during construction and maintenance. Typical design values used for these loads are:

Hydraulic load: 600 N/m² live load on the plate, plus 3000 N/m² over the downcomer seal area.

Erection and maintenance: 1500 N concentrated load on any structural member.

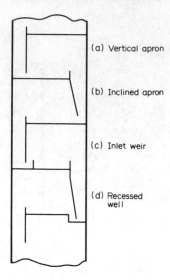

FIG. 11.24. Segment (chord) downcomer designs

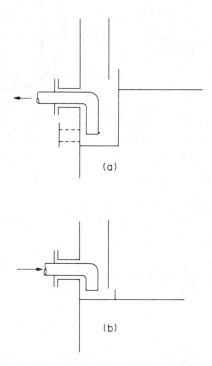

FIG. 11.25. Feed and take-off nozzles

It is important to set close tolerances on the weir height, downcomer clearance, and plate flatness, to ensure an even flow of liquid across the plate. The tolerances specified will depend on the dimensions of the plate but will typically be about 3 mm.

The plate deflection under load is also important, and will normally be specified as not greater than 3 mm under the operating conditions for plates greater than 2·5 m, and proportionally less for smaller diameters.

The mechanical specification of bubble-cap, sieve and valve plates is covered in a series of articles by Glitsch (1960), McClain (1960), Thrift (1960a,b) and Patton and Pritchard (1960).

11.13. Plate hydraulic design

The basic requirements of a plate contacting stage are that it should:

Provide good vapour–liquid contact.
Provide sufficient liquid hold-up for good mass transfer (high efficiency).
Have sufficient area and spacing to keep the entrainment and pressure drop within acceptable limits.
Have sufficient downcomer area for the liquid to flow freely from plate to plate.

Plate design, like most engineering design, is a combination theory and practice. The design methods use semi-empirical correlations derived from fundamental research work combined with practical experience obtained from the operation of commercial columns. Proven layouts are used, and the plate dimensions are kept within the range of values known to give satisfactory performance.

A short procedure for the hydraulic design of sieve plates is given in this section. Design methods for bubble-cap plates are given by Bolles (1963) and Ludwig (1979). Valve plates are proprietary designs and will be designed in consultation with the vendors. Design manuals are available from some vendors; Glistsch (1970) and Koch (1960). Some information on valve plates is given in Volume 2, Chapter 11.

A detailed discussion of the extensive literature on plate design and performance will not be given in this volume. A short review of the literature is given in Volume 2, Chapter 11. Chase (1967) gives a critical review of the literature on sieve plates.

Several design methods have been published for sieve plates; Koch and Kuzniar (1966), Huang and Hodson (1958), Fair (1963). Economopoulos (1978) discusses the use of computers in the design of sieve plates.

Operating range

Satisfactory operation will only be achieved over a limited range of vapour and liquid flow rates. A typical performance diagram for a sieve plate is shown in Fig. 11.26.

The upper limit to vapour flow is set by the condition of flooding. At flooding there is a sharp drop in plate efficiency and increase in pressure drop. Flooding is caused by either the excessive carry over of liquid to the next plate by entrainment, or by liquid backing-up in the downcomers.

The lower limit of the vapour flow is set by the condition of weeping. Weeping occurs when the vapour flow is insufficient to maintain a level of liquid on the plate. "Coning" occurs at low liquid rates, and is the term given to the condition where the vapour pushes the liquid back from the holes and jets upward, with poor liquid contact.

In the following sections gas can be taken as synonymous with vapour when applying the method to the design of plates for absorption columns.

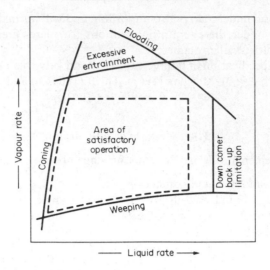

FIG. 11.26. Sieve plate performance diagram

11.13.1. Plate-design procedure

A trial-and-error approach is necessary in plate design: starting with a rough plate layout, checking key performance factors and revising the design, as necessary, until a satisfactory design is achieved. A typical design procedure is set out below and discussed in the following sections. The normal range of each design variable is given in the discussion, together with recommended values which can be used to start the design.

Procedure

1. Calculate the maximum and minimum vapour and liquid flow-rates, for the turn down ratio required.
2. Collect, or estimate, the system physical properties.
3. Select a trial plate spacing (Section 11.11).
4. Estimate the column diameter, based on flooding considerations (Section 11.13.3).
5. Decide the liquid flow arrangement (Section 11.13.4).
6. Make a trial plate layout: downcomer area, active area, hole area, hole size, weir height (Sections 11.13.8 to 11.13.10).
7. Check the weeping rate (Section 11.13.6), if unsatisfactory return to step 6.
8. Check the plate pressure drop (Section 11.13.14), if too high return to step 6.
9. Check downcomer back-up, if too high return to step 6 or 3 (Section 11.13.15).
10. Decide plate layout details: calming zones, unperforated areas. Check hole pitch, if unsatisfactory return to step 6 (Section 11.13.11).
11. Recalculate the percentage flooding based on chosen column diameter.
12. Check entrainment, if too high return to step 4 (section 11.13.5).
13. Optimise design: repeat steps 3 to 12 to find smallest diameter and plate spacing acceptable (lowest cost).
14. Finalise design: draw up the plate specification and sketch the layout.

This procedure is illustrated in Example 11.11.

11.13.2. Plate areas

The following areas terms are used in the plate design procedure:

A_c = total column cross-sectional area,

A_d = cross-sectional area of downcomer,

A_n = net area available for vapour–liquid disengagement, normally equal to $A_c - A_d$, for a single pass plate,

A_a = active, or bubbling, area, equal to $A_c - 2A_d$ for single-pass plates,

A_h = hole area, the total area of all the active holes,

A_p = perforated area (including blanked areas),

A_{ap} = the clearance area under the downcomer apron.

11.13.3. Diameter

The flooding condition fixes the upper limit of vapour velocity. A high vapour velocity is needed for high plate efficiencies, and the velocity will normally be between 70 to 90 per cent of that which would cause flooding. For design, a value of 80 to 85 per cent of the flooding velocity should be used.

The flooding velocity can be estimated from the correlation given by Fair (1961):

$$u_f = K_1 \sqrt{\frac{\rho_L - \rho_v}{\rho_v}} \qquad (11.81)$$

where　　u_f = flooding vapour velocity, m/s, based on the net column cross-sectional area A_n (see Section 11.13.2),

　　　　　K_1 = a constant obtained from Fig. 11.27.

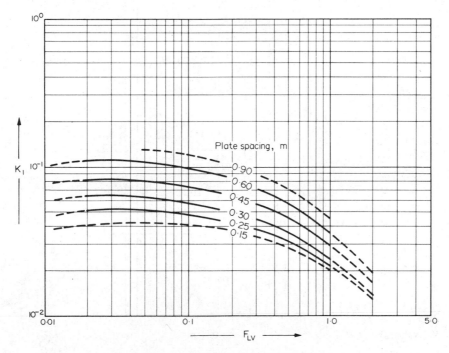

FIG. 11.27. Flooding velocity, sieve plates

The liquid–vapour flow factor F_{LV} in Fig. 11.27 is given by:

$$F_{LV} = \frac{L_w}{V_w} \sqrt{\frac{\rho_v}{\rho_L}}$$ (11.82)

where L_w = liquid mass flow-rate, kg/s,
V_w = vapour mass flow-rate, kg/s.

The following restrictions apply to the use of Fig. 11.27:

1. Hole size less than 6·5 mm. Entrainment may be greater with larger hole sizes.
2. Weir height less than 15 per cent of the plate spacing.
3. Non-foaming systems.
4. Hole: active area ratio greater than 0·10; for other ratios apply the following corrections:

hole: active area	multiply K_1 by
0·10	1·0
0·08	0·9
0·06	0·8

5. Liquid surface tension 0·02 N/m, for other surface tensions σ multiply the value of K_1 by $\left[\dfrac{\sigma}{0·02} \right]^{0·2}$

To calculate the column diameter an estimate of the net area A_n is required. As a first trial take the downcomer area as 12 per cent of the total, and assume that the hole : active area is 10 per cent.

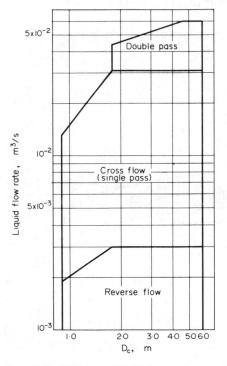

FIG. 11.28. Selection of liquid-flow arrangement

Where the vapour and liquid flow-rates, or physical properties, vary significantly throughout the column a plate design should be made for several points up the column. For distillation it will usually be sufficient to design for the conditions above and below the feed points. Changes in the vapour flow-rate will normally be accommodated by adjusting the hole area; often by blanking off some rows of holes. Different column diameters would only be used where there is a considerable change in flow-rate. Changes in liquid rate can be allowed for by adjusting the liquid downcomer areas.

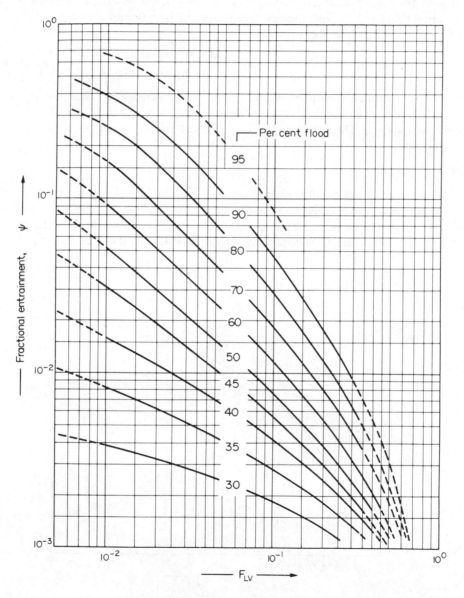

Fig. 11.29. Entrainment correlation for sieve plates (Fair, 1961)

11.13.4. Liquid-flow arrangement

The choice of plate type (reverse, single pass or multiple pass) will depend on the liquid flow-rate and column diameter. An initial selection can be made using Fig. 11.28, which has been adapted from a similar figure given by Huang and Hodson (1958).

11.13.5. Entrainment

Entrainment can be estimated from the correlation given by Fair (1961), Fig. 11.29, which gives the fractional entrainment ψ (kg/kg gross liquid flow) as a function of the liquid–vapour factor F_{LV}, with the percentage approach to flooding as a parameter.

The percentage flooding is given by:

$$\text{percentage flooding} = \frac{u_n \text{ actual velocity (based on net area)}}{u_f \text{ (from equation 11.81)}} \tag{11.83}$$

The effect of entrainment on plate efficiency can be estimated using equation 11.78.

As a rough guide the upper limit of ψ can be taken as 0·1; below this figure the effect on efficiency will be small. The optimum design value may be above this figure, see Fair (1963).

11.13.6. Weep point

The lower limit of the operating range occurs when liquid leakage through the plate holes becomes excessive. This is known as the weep point. The vapour velocity at the weep point is the minimum value for stable operation. The hole area must be chosen so that at the lowest operating rate the vapour flow velocity is still well above the weep point.

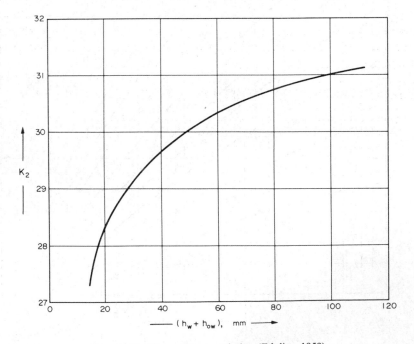

FIG. 11.30. Weep-point correlation (Eduljee, 1959)

Several correlations have been proposed for predicting the vapour velocity at the weep point; see Chase (1967). That given by Eduljee (1959) is one of the simplest to use, and has been shown to be reliable.

The minimum design vapour velocity is given by:

$$\hat{u}_h = \frac{[K_2 - 0.90(25.4 - d_h)]}{(\rho_v)^{\frac{1}{2}}}$$ (11.84)

where $\hat{u}_h$ = minimum vapour velocity through the holes
(based on the hole area), m/s,

d_h = hole diameter, mm,

K_2 = a constant, dependent on the depth of clear liquid on the plate, obtained from Fig. 11.30.

The clear liquid depth is equal to the height of the weir h_w plus the depth of the crest of liquid over the weir h_{ow}; this is discussed in the next section.

11.13.7. Weir liquid crest

The height of the liquid crest over the weir can be estimated using the Francis weir formula (see Volume 1, Chapter 5). For a segmental downcomer this can be written as:

$$h_{ow} = 750\left[\frac{L_w}{\rho_L l_w}\right]^{2/3}$$ (11.85)

where l_w = weir length, m,

h_{ow} = weir crest, mm liquid,

L_w = liquid flow-rate, kg/s.

With segmental downcomers the column wall constricts the liquid flow, and the weir crest will be higher than that predicted by the Francis formula for flow over an open weir. The constant in equation 11.85 has been increased to allow for this effect.

To ensure an even flow of liquid along the weir, the crest should be at least 10 mm at the lowest liquid rate. Serrated weirs are sometimes used for very low liquid rates.

11.13.8. Weir dimensions

Weir height

The height of the weir determines the volume of liquid on the plate and is an important factor in determining the plate efficiency (see Section 11.10.4). A high weir will increase the plate efficiency but at the expense of a higher plate pressure drop. For columns operating above atmospheric pressure the weir heights will normally be between 40 mm to 90 mm (1.5 to 3.5 in.); 40 to 50 mm is recommended. For vacuum operation lower weir heights are used to reduce the pressure drop; 6 to 12 mm ($\frac{1}{4}$ to $\frac{1}{2}$ in.) is recommended.

Inlet weirs

Inlet weirs, or recessed pans, are sometimes used to improve the distribution of liquid across the plate; but are seldom needed with segmental downcomers.

Weir length

With segmental downcomers the length of the weir fixes the area of the downcomer. The chord length will normally be between 0·6 to 0·85 of the column diameter. A good initial value to use is 0·77, equivalent to a downcomer area of 12 per cent.

The relationship between weir length and downcomer area is given in Fig. 11.31.

For double-pass plates the width of the central downcomer is normally 200–250 mm (8–10 in.).

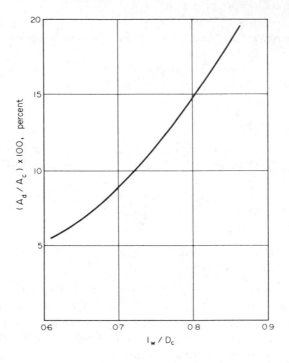

FIG. 11.31. Relation between downcomer area and weir length

11.13.9. *Perforated area*

The area available for perforation will be reduced by the obstruction caused by structural members (the support rings and beams), and by the use of calming zones.

Calming zones are unperforated strips of plate at the inlet and outlet sides of the plate. The width of each zone is usually made the same; recommended values are: below 1·5 m diameter, 75 mm; above, 100 mm.

The width of the support ring for sectional plates will normally be 50 to 75 mm: the support ring should not extend into the downcomer area. A strip of unperforated plate will be left round the edge of cartridge-type trays to stiffen the plate.

The unperforated area can be calculated from the plate geometry. The relationship between the weir chord length, chord height and the angle subtended by the chord is given in Fig. 11.32.

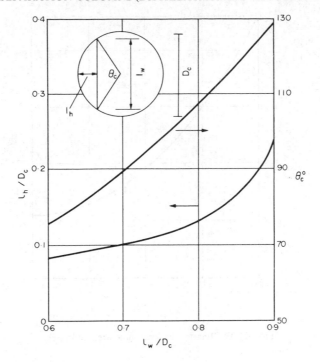

Fig. 11.32. Relation between angle subtended by chord, chord height and chord length

11.13.10. Hole size

The hole sizes used vary from 2·5 to 12 mm; 5 mm is the preferred size. Larger holes are occasionally used for fouling systems. The holes are drilled or punched. Punching is cheaper, but the minimum size of hole that can be punched will depend on the plate thickness. For carbon steel, hole sizes approximately equal to the plate thickness can be punched, but for stainless steel the minimum hole size that can be punched is about twice the plate thickness. Typical plate thicknesses used are: 5 mm ($\frac{3}{16}$ in.) for carbon steel, and 3 mm (12 gauge) for stainless steel.

When punched plates are used they should be installed with the direction of punching upward. Punching forms a slight nozzle, and reversing the plate will increase the pressure drop.

11.13.11. Hole pitch

The hole pitch (distance between the hole centres) l_p should not be less than 2·0 hole diameters, and the normal range will be 2·5 to 4·0 diameters. Within this range the pitch can be selected to give the number of active holes required for the total hole area specified.

Square and equilateral triangular patterns are used; triangular is preferred. The total hole area as a fraction of the perforated area A_p is given by the following expression, for an equilateral triangular pitch:

$$\frac{A_h}{A_p} = 0.9 \left[\frac{d_h}{l_p} \right]^2 \tag{11.86}$$

This equation is plotted in Fig. 11.33.

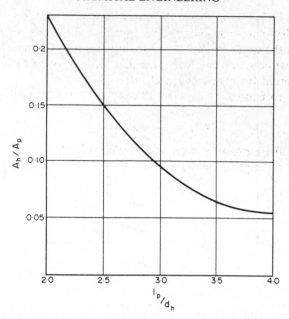

FIG. 11.33. Relation between hole area and pitch

11.13.12. Hydraulic gradient

The hydraulic gradient is the difference in liquid level needed to drive the liquid flow across the plate. On sieve plates, unlike bubble-cap plates, the resistance to liquid flow will be small, and the hydraulic gradient is usually ignored in sieve-plate design. It can be significant in vacuum operation, as with the low weir heights used the hydraulic gradient can be a significant fraction of the total liquid depth. Methods for estimating the hydraulic gradient are given by Fair (1963).

11.13.13. Liquid throw

The liquid throw is the horizontal distance travelled by the liquid stream flowing over the downcomer weir. It is only an important consideration in the design of multiple-pass plates. Bolles (1963) gives a method for estimating the liquid throw.

11.13.14 Plate pressure drop

The pressure drop over the plates is an important design consideration. There are two main sources of pressure loss: that due to vapour flow through the holes (an orifice loss), and that due to the static head of liquid on the plate.

A simple additive model is normally used to predict the total pressure drop. The total is taken as the sum of the pressure drop calculated for the flow of vapour through the dry plate (the dry plate drop h_d); the head of clear liquid on the plate ($h_w + h_{ow}$); and a term to account for other, minor, sources of pressure loss, the so-called residual loss h_r. The residual loss is the difference between the observed experimental pressure drop and the simple sum of the dry-plate drop and the clear-liquid height. It accounts for the two effects: the energy to form the vapour bubbles and the fact that on an operating plate the liquid head will not be clear liquid but a head of "aerated" liquid froth, and the froth density and height will be different from that of the clear liquid.

It is convenient to express the pressure drops in terms of millimetres of liquid. In pressure units:

$$\Delta P_t = 9 \cdot 81 \times 10^{-3} \, h_t \rho_L \tag{11.87}$$

where ΔP_t = total plate pressure drop, Pa(N/m²),

h_t = total plate pressure drop, mm liquid.

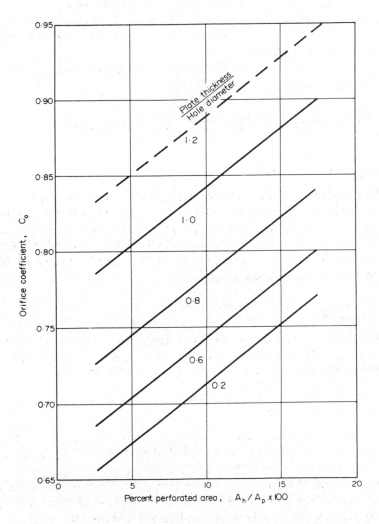

FIG. 11.34. Discharge coefficient, sieve plates (Liebson et al., 1957)

Dry plate drop

The pressure drop through the dry plate can be estimated using expressions derived for flow through orifices (see Volume 2, Chapter 5).

$$h_d = 51 \left[\frac{u_h}{C_0}\right]^2 \frac{\rho_v}{\rho_L} \qquad (11.88)$$

where the orifice coefficient C_0 is a function of the plate thickness, hole diameter, and the hole to perforated area ratio. C_0 can be obtained from Fig. 11.34; which has been adapted from a similar figure by Liebson *et al.* (1957). u_h is the velocity through the holes, m/s.

Residual head

Methods have been proposed for estimating the residual head as a function of liquid surface tension, froth density and froth height. However, as this correction term is small the use of an elaborate method for its estimation is not justified, and the simple equation proposed by Hunt *et al.* (1955) can be used:

$$h_r = \frac{12 \cdot 5 \times 10^3}{\rho_L} \qquad (11.89)$$

Equation 11.89 is equivalent to taking the residual drop as a fixed value of 12·5 mm of water ($\frac{1}{2}$ in.).

Total drop

The total plate drop is given by:

$$h_t = h_d + (h_w + h_{ow}) + h_r \qquad (11.90)$$

If the hydraulic gradient is significant, half its value is added to the clear liquid height.

11.13.15. Downcomer design [back-up]

The downcomer area and plate spacing must be such that the level of the liquid and froth in the downcomer is well below the top of the outlet weir on the plate above. If the level rises above the outlet weir the column will flood.

The back-up of liquid in the downcomer is caused by the pressure drop over the plate (the downcomer in effect forms one leg of a U-tube) and the resistance to flow in the downcomer itself; see Fig. 11.35.

In terms of clear liquid the downcomer back-up is given by:

$$h_b = (h_w + h_{ow}) + h_t + h_{dc} \qquad (11.91)$$

where h_b = downcomer back-up, measured from plate surface, mm,

$\quad h_{dc}$ = head loss in the downcomer, mm.

The main resistance to flow will be caused by the constriction at the downcomer outlet, and the head loss in the downcomer can be estimated using the equation given by Cicalese *et al.* (1947)

$$h_{dc} = 166 \left[\frac{L_{wd}}{\rho_L A_m}\right]^2 \qquad (11.92)$$

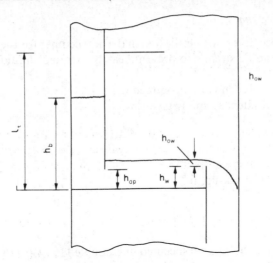

Fig. 11.35. Down-comer back-up

where L_{wd} = liquid flow rate in downcomer, kg/s,
   A_m = either the downcomer area A_d or the clearance area
        under the downcomer A_{ap}; whichever is the smaller, m².

The clearance area under the downcomer is given by:

$$A_{ap} = h_{ap} l_w \qquad (11.93)$$

where h_{ap} is height of the bottom edge of the apron above the plate. This height is normally set at 5 to 10 mm ($\frac{1}{4}$ to $\frac{1}{2}$ in.) below the outlet weir height:

$$h_{ap} = h_w - (5 \text{ to } 10 \text{ mm})$$

Froth height

To predict the height of "aerated" liquid on the plate, and the height of froth in the downcomer, some means of estimating the froth density is required. The density of the "aerated" liquid will normally be between 0·4 to 0·7 times that of the clear liquid. A number of correlations have been proposed for estimating froth density as a function of the vapour flow-rate and the liquid physical properties; see Chase (1967); however, none is particularly reliable, and for design purposes it is usually satisfactory to assume an average value of 0·5 of the liquid density.

This value is also taken as the mean density of the fluid in the downcomer; which means that for safe design the clear liquid back-up, calculated from equation 11.91, should not exceed half the plate spacing l_t, to avoid flooding.

Allowing for the weir height:

$$h_b \not> \tfrac{1}{2}(l_t + h_w) \qquad (11.94)$$

This criterion is, if anything, oversafe, and where close plate spacing is desired a better estimate of the froth density in the downcomer should be made. The method proposed by Thomas and Shah (1964) is recommended.

Downcomer residence time

Sufficient residence time must be allowed in the downcomer for the entrained vapour to disengage from the liquid stream; to prevent heavily "aerated" liquid being carried under the downcomer.

A time of at least 3 seconds is recommended.

The downcomer residence time is given by:

$$t_r = \frac{A_d h_{bc} \rho_L}{L_{wd}}$$

(11.95)

where t_r = residence time, s,

$\quad h_{bc}$ = clear liquid back-up, m.

Example 11.11

Design the plates for the column specified in Example 11.2. Take the minimum feed rate as 70 per cent of the maximum (maximum feed 10,000 kg/h). Use sieve plates.

Solution

As the liquid and vapour flow-rates and compositions will vary up the column, plate designs should be made above and below the feed point. Only the bottom plate will be designed in detail in this example.

From McCabe–Thiele diagram, Example 11.2:

Number of stages = 14
Slope of the bottom operating line = 3·9
Slope of top operating line = 0·75
Top composition 94 per cent mol. 98 per cent w/w.
Bottom composition—essentially water.

Flow-rates

$$\text{Top product} = \frac{0 \cdot 1 \times 10,000}{0 \cdot 98} = 1020 \text{ kg/h}$$

$$\text{Vapour rate} = (1 \cdot 35 + 1)1020 = 2397 \text{ kg/h}$$

$$\text{Bottom product} = 10,000 - 1020 = 8980 \text{ kg/h}$$

Material balance gives:

$$V_m' + 8980 = L_m'$$

and

$$L_m'/V_m' = 3 \cdot 9$$

from which

vapour rate V_m' = 3097 kg/h

liquid rate L_m' = 12,078 kg/h

Physical properties

Estimate base pressure, assume column efficiency of 60 per cent, take reboiler as equivalent to one stage.

$$\text{Number of real stages} = \frac{14 - 1}{0 \cdot 6} = 22$$

Assume 100 mm water, pressure drop per plate.

$$\text{Column pressure drop} = 100 \times 10^{-3} \times 1000 \times 9 \cdot 81 \times 22 = 21{,}580 \text{ Pa}$$
$$\text{Top pressure, 1 atm } (14 \cdot 7 \text{ lb/in}^2) = 101 \cdot 4 \times 10^3 \text{ Pa}$$
$$\text{Estimated bottom pressure} \quad = 101 \cdot 4 \times 10^3 + 21{,}580$$
$$= 122{,}980 \text{ Pa} = \underline{\underline{1 \cdot 23 \text{ bar}}}$$

From steam tables, base temperature 106°C.

$$\rho_v = 0 \cdot 77 \text{ kg/m}^3$$
$$\rho_L = 950 \text{ kg/m}^3$$

Surface tension $57 \times 10^{-3} \text{ N/m}$
Top temperature 57°C (acetone)

$$\rho_L = 780 \text{ kg/m}^3$$

Mol. weight 58

$$\rho_v = \frac{58}{22 \cdot 4} \times \frac{273}{(57 + 273)} = 2 \cdot 14 \text{ kg/m}^3$$

Surface tension $19 \times 10^{-3} \text{ N/m}$

Column diameter

(11.82)
$$F_{LV} \text{ bottom} = 3 \cdot 9 \sqrt{\frac{0 \cdot 77}{950}} = 0 \cdot 11$$

$$F_{LV} \text{ top} \quad = 0 \cdot 75 \sqrt{\frac{2 \cdot 14}{780}} = 0 \cdot 04$$

Take tray spacing as 0·5 m
From Fig. 11.27
$$\text{base } K_1 = 7 \times 10^{-2}$$
$$\text{top } K_1 \quad = 8 \times 10^{-2}$$

Correction for surface tensions

$$\text{base } K_1 = \left(\frac{57}{20}\right)^{0 \cdot 2} \times 7 \times 10^{-2} = 8 \cdot 6 \times 10^{-2}$$

$$\text{top } K_1 \quad = \left(\frac{19}{20}\right)^{0 \cdot 2} \times 8 \times 10^{-2} = 7 \cdot 9 \times 10^{-2}$$

(11.81)   base u_f $= 8.6 \times 10^{-2} \sqrt{\dfrac{950 - 0.77}{0.77}} = 3.03$ m/s

                 top u_f $= 7.9 \times 10^{-2} \sqrt{\dfrac{780 - 2.14}{2.14}} = 1.51$ m/s

Design for 85 per cent flooding at maximum flow rate

         base $\hat{u}_v$ $= 3.03 \times 0.85 = 2.58$ m/s
         top $\hat{u}_v$ $= 1.5 \times 0.85 = 1.28$ m/s

Maximum volumetric flow-rate

$$\text{base} = \frac{3097}{0.77 \times 3600} = 1.12 \text{ m}^3/\text{s}$$

$$\text{top} = \frac{2397}{2.14 \times 3600} = 0.31 \text{ m}^3/\text{s}$$

Net area required

$$\text{bottom} = \frac{1.12}{2.58} = 0.43 \text{ m}^2$$

$$\text{top} = \frac{0.31}{1.28} = 0.24 \text{ m}^2$$

As first trial take downcomer area as 12 per cent of total.
Column cross-sectioned area

$$\text{base} = \frac{0.43}{0.88} = 0.49 \text{ m}^2$$

$$\text{top} = \frac{0.24}{0.88} = 0.27 \text{ m}^2$$

Column diameter

$$\text{base} = \sqrt{\frac{0.49 \times 4}{\pi}} = 0.79 \text{ m}$$

$$\text{top} = \sqrt{\frac{0.27 \times 4}{\pi}} = 0.59 \text{ m}$$

Use same diameter above and below feed, reducing the perforated area for plates above the feed.

Nearest standard pipe size (BS 1600, Pt. 2); outside diameter 812·8 mm (32 in); standard wall thickness 9·52 mm; inside diameter 794 mm.

Liquid flow pattern

$$\text{Maximum volumetric liquid rate} = \frac{12078}{3600 \times 950} = 3.5 \times 10^{-3} \text{ m}^3/\text{s}$$

The plate diameter is outside the range of Fig. 11.28, but it is clear that a single pass plate can be used.

Provisional plate design

Column diameter D_c = 0·79 m
Column area A_c = 0·50 m^2
Downcomer area A_d = 0·12 × 0·50 = 0·06 m^2, at 12 per cent
Net area A_n = $A_c - A_d$ = 0·50 − 0·06 = 0·44 m^2
Active area A_a = $A_c - 2A_d$ = 0·50 − 0·12 = 0·38 m^2
Hole area A_h take 10 per cent A_a as first trial = 0·038 m^2

Weir length (from Fig. 11.31) = 0·76 × 0·79 = 0·60 m

Take weir height 50 mm
Hole diameter 5 mm
Plate thickness 5 mm

Check weeping

$$\text{Maximum liquid rate} = \left(\frac{12{,}078}{3600}\right) = 3·36 \text{ kg/s}$$

Minimum liquid rate, at 70 per cent turn-down = 0·7 × 3·36 = 2·35 kg/s

(11.85) $\quad \text{maximum } h_{ow} = 750\left(\dfrac{3·36}{950 \times 0·60}\right)^{2/3} = 25 \text{ mm liquid}$

$\quad \text{minimum } h_{ow} = 750\left(\dfrac{2·35}{950 \times 0·60}\right)^{2/3} = 19 \text{ mm liquid}$

at minimum rate $h_o + h_{ow} = 50 + 19 = 69$ mm

From Fig. 11.30, $\quad K_2 = 30·5$

(11.84) $\quad \breve{u}_h(\min) = \dfrac{30·5 - 0·90(25·4 - 5)}{(0·77)^{1/2}} = 14 \text{ m/s}$

actual minimum vapour velocity $= \dfrac{\text{minimum vapour rate}}{A_h}$

$$= \frac{0·7 \times 1·12}{0·038} = 20·6 \text{ m/s}$$

So minimum operating rate will be well above weep point.

Plate pressure drop

Dry plate drop
Maximum vapour velocity through holes

$$\hat{u}_h = \frac{1·12}{0·038} = 29·5 \text{ m/s}$$

From Fig. 11.34, for plate thickness/hole dia. $= 1$, and $A_h/A_p \simeq A_h/A_a = 0\cdot1$, $C_0 = 0\cdot84$

(11.88)
$$h_d = 51\left(\frac{29\cdot5}{0\cdot84}\right)^2 \frac{0\cdot77}{950} = 51 \text{ mm liquid}$$

residual head

(11.89)
$$h_r = \frac{12\cdot5 \times 10^3}{950} = 13\cdot2 \text{ mm liquid}$$

total plate pressure drop

$$h_t = 51 + (50 + 25) + 13 = 139 \text{ mm liquid}$$

Note: 100 mm was assumed to calculate the base pressure. The calculation could be repeated with a revised estimate but the small change in physical properties will have little effect on the plate design. 139 mm per plate is considered acceptable.

Downcomer liquid back-up

Downcomer pressure loss
Take $h_{ap} = h_w - 10 = 40$ mm.
Area under apron, $A_{ap} = 0\cdot60 \times 40 \times 10^{-3} = 0\cdot024 \text{ m}^2$.
As this is less than $A_d = 0\cdot06 \text{ m}^2$ use A_{ap} in equation 11.92

$$h_{dc} = 166\left(\frac{3\cdot36}{950 \times 0\cdot024}\right)^2 = 3\cdot6 \text{ mm}$$

say 4 mm.

Back-up in downcomer

(11.91)
$$h_b = (50 + 25) + 139 + 4 = 218 \text{ mm}$$
$$\underline{\underline{0\cdot22 \text{ m}}}$$

$0\cdot22 < \frac{1}{2}$(plate spacing + weir height)
so tray spacing is acceptable

Check residence time

(11.95)
$$t_r = \frac{0\cdot06 \times 0\cdot22 \times 950}{3\cdot36} = 3\cdot7 \text{ s}$$
> 3 s, satisfactory.

Check entrainment

Actual percentage flooding for design area

$$u_v = \frac{1\cdot12}{0\cdot44} = 2\cdot55 \text{ m/s}$$

$$\text{per cent flooding} = \frac{2\cdot55}{3\cdot02} \times 100 = 84, F_{LV} = 0\cdot11$$

From Fig. 11.28, $\psi = 0.02$

well below 0.1, satisfactory.

Trial layout

Use cartridge-type construction. Allow 50 mm unperforated strip round plate edge; 50 mm wide calming zones.

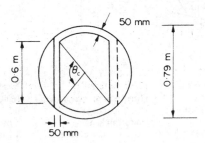

Perforated area

From Fig. 11.32, at $l_w/D_c = 0.76$

$$\theta_c = 99°$$

angle subtended at plate edge by unperforated strip $= 180 - 99 = 81°$

mean length, unperforated edge strips $= (0.79 - 50 \times 10^{-3})\pi \times \dfrac{81}{180} = 1.05$ m

area of unperforated edge strips $= 50 \times 10^{-3} \times 1.05 = 0.053$ m^2

area of calming zones $\quad = 2(50 \times 10^{-3}) \times (0.6 - 2 \times 50 \times 10^{-3})$

$$= \underline{\underline{0.050 \text{ m}^2}}$$

Total area available for perforations A_p

$$= 0.38 - (0.053 + 0.050)$$
$$= 0.277 \text{ m}^2$$

$$\frac{A_h}{A_p} = \frac{0.038}{0.277} = 0.137$$

From Fig. 11.33 $\quad l_p/d_h = 2.6$, satisfactory, within 2.5 to 4.0.

Number of holes

Area of one hole $= 1.964 \times 10^{-5}$ m^2

$$\text{Number of holes} = \frac{0.038}{1.964 \times 10^{-5}} = 1935$$

Plate specification

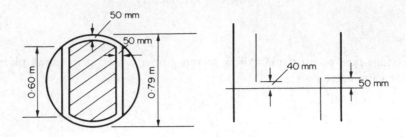

Plate No.	1	Turn-down	70 per cent max rate
Plate I.D.	0·79 m	Plate material	Mild steel
Hole size	5 mm	Downcomer material	Mild steel
Hole pitch	12·5 mm Δ	Plate spacing	0·5 m
Total no. holes	—	Plate thickness	5 mm
Active holes	1935	Plate pressure drop	140 mm liquid = 1·3 kPa
Blanking area	—		

Example 11.12

For the plate design in Example 11.11, estimate the plate efficiency for the plate on which the concentration of acetone is 5 mol per cent. Use the AIChE method.

Solution

Plate will be in the stripping section (see Fig. 11.7).
Plate dimensions:
    active area = $0·38\,\text{m}^2$,
    length between downcomers (Fig. 11.32) (liquid path) = $0·79 - 2 \times 0·095 = 0·60\,\text{m}$,
    weir height = 50 mm.
Flow rates, check efficiency at minimum rates, at column base:

$$\text{vapour} = 0·7 \times 3097 = 2168\,\text{kg/h} = \frac{2168}{18 \times 3600} = 0·0335\,\text{kmol/s},$$

$$\text{liquid} = 0·7 \times 12078 = 8455\,\text{kg/h} = \frac{8455}{18 \times 3600} = 0·1305\,\text{kmol/s}.$$

Assuming constant molar overflow, the flows on the plate will be the same as at the column base.

From the McCabe–Thiele diagram (Fig. 11.7) at $x = 0·05$, assuming 65 per cent plate efficiency, $y_n = 0·49$.

At $x = 0·05$, plate temperature = 75°C, at atmospheric pressure.

$$\text{Densities, acetone} = 760\,\text{kg/m}^3$$

$$\text{water} = 975\,\text{kg/m}^3$$

$$5\text{ mol per cent} = \frac{0·05 \times 58}{0·95 \times 18 + 0·05 \times 58} = 0·145\text{ wt fraction}$$

Taking liquid volumes as additive

$$\rho_L = \frac{760 \times 975}{0.145 \times 975 + (1 - 0.145)760} = 937 \text{ kg/m}^3$$

$$\rho_v = \frac{18(1 - 0.49) + 58(0.49)}{22.4} \times \frac{273}{348} = 1.32 \text{ kg/m}^3$$

$$\mu_L = 0.28 \times 10^{-3} \text{ N s/m}^2 \text{ (taken as water)}$$

Viscosities, water vapour $= 12.1 \times 10^{-6} \text{ N s/m}^2$
acetone vapour $= 9.5 \times 10^{-6} \text{ N s/m}^2$

Use molar average

$$\mu_V = [12.1(1 - 0.49) + 9.5 \times 0.49] \times 10^{-6}$$
$$= 10.8 \times 10^{-6} \text{ N s/m}^2$$

$D_L = 4.64 \times 10^{-9} \text{ m}^2/\text{s}$ (estimated using the Wilke–Chang equation, Chapter 8)

$D_V = 18.6 \times 10^{-6} \text{ m}^2/\text{s}$ (estimated using the Fuller equation, Chapter 8)

Average molecular weights:
  liquid $= 58 \times 0.05 + 18 \times 0.95 = 20.0$,
  vapour $= 58 \times 0.49 + 18(1 - 0.49) = 37.6$.

Volumetric flow-rate, vapour $= \dfrac{0.0335 \times 37.6}{1.32}$

$$= 0.95 \text{ m}^3/\text{s}.$$

Volumetric flow-rate, liquid $= \dfrac{0.1305 \times 20.0}{937}$

$$= 2.79 \times 10^{-3} \text{ m}^3/\text{s}$$

$$u_a = \frac{0.95}{0.38} = 2.5 \text{ m/s}$$

$$F_v = u_a \sqrt{\rho_v} = 2.5 \sqrt{1.32} = 2.87$$

Average width over active area $= \dfrac{0.38}{0.60} = 0.63 \text{ m}$

$$L_p = \frac{2.79 \times 10^{-3}}{0.63} = 4.43 \times 10^{-3} \text{ m}^2/\text{s}$$

(11.71) $N_G = (0.776 + 4.57 \times 10^{-3} \times 50 - 0.24 \times 2.87 + 105 \times 4.43 \times 10^{-3})$

$$\left/ \left(\frac{10.8 \times 10^{-6}}{1.32 \times 18.6 \times 10^{-6}} \right)^{1/2} \right. = 1.17$$

(11.75) $Z_C = 0.006 + 0.73 \times 10^{-3} \times 50 - 0.24 \times 10^{-3} \times 2.87 \times 50 + 1.22$
$\times 4.43 \times 10^{-3} = 0.0135 \text{ m}^3/\text{m}^2$

(11.73) $t_L = \dfrac{0.0135 \times 0.6}{4.43 \times 10^{-3}} = 1.83$

(11.72) $N_L = (4 \cdot 13 \times 10^8 \times 4 \cdot 64 \times 10^{-9})^{0 \cdot 5}(0 \cdot 21 \times 2 \cdot 87 + 0 \cdot 15)1 \cdot 83 = 1 \cdot 9$

(11.77) $D_e = (0 \cdot 0038 + 0 \cdot 017 \times 2 \cdot 5 + 3 \cdot 86 \times 4 \cdot 43 \times 10^{-3} + 0 \cdot 18 \times 10^{-3} \times 50)^2$

$$= 0 \cdot 0052 \text{ m}^2/\text{s}$$

(11.76) $$Pe = \frac{0 \cdot 60^2}{0 \cdot 0052 \times 1 \cdot 83} = 38$$

From the McCabe–Thiele diagram, at $x = 0 \cdot 05$, slope of operating line $= 3 \cdot 9$ and slope of equilibrium line $= 1 \cdot 0$,

$$\text{so, } mV/L \quad = 1/3 \cdot 9 = 0 \cdot 26$$

$$(mV/L)/N_L = \frac{0 \cdot 26}{1 \cdot 9} = 0 \cdot 14$$

From Fig. 11.15 $E_{mv} = 0 \cdot 63$

From Fig. 11.16 $\dfrac{E_{mV}}{E_{mv}} = 1 \cdot 0$.

So $E_{mV} = 63$ per cent (say 60 per cent).

Note: The slope of the equilibrium line is difficult to determine at $x = 0 \cdot 05$, but any error will not greatly affect the value of E_{mV}.

Example 11.13

Calculate the plate efficiency from the plate design considered in Examples 11.11 and 11.12, using Van Winkle's correlation.

Solution

From Examples 11.12 and 11.11:

$\rho_L = 937 \text{ kg/m}^3$,

$\rho_v = 1 \cdot 32 \text{ kg/m}^3$,

$\mu_L = 0 \cdot 28 \times 10^{-3} \text{ N s/m}^2$,

$\mu_v = 10 \cdot 8 \times 10^{-6} \text{ N s/m}^2$,

$D_{LK} = D_L = 4 \cdot 64 \times 10^{-9} \text{ m}^2/\text{s}$,

$h_w = 50 \text{ mm}$,

FA (fractional area) $= A_h/A_c = \dfrac{0 \cdot 038}{0 \cdot 50} = 0 \cdot 076$,

u_v = superficial vapour velocity $= \dfrac{0 \cdot 95}{0 \cdot 50} = 1 \cdot 9 \text{ m/s}$,

$\sigma_L = 57 \times 10^{-3} \text{ N/m}$ (taken as for water),

$Dg = \left(\dfrac{57 \times 10^{-3}}{0 \cdot 28 \times 10^{-3} \times 1 \cdot 9} \right) = 107$,

$Sc = \left(\dfrac{0 \cdot 28 \times 10^{-3}}{937 \times 4 \cdot 67 \times 10^{-9}} \right) = 64$,

$Re = \left(\dfrac{50 \times 10^{-3} \times 1 \cdot 9 \times 1 \cdot 32}{0 \cdot 28 \times 10^{-3} \times 0 \cdot 076} \right) = 5893$,

(11.69) $$E_{mV} = 0.07 \, (107)^{0.14} \, (64)^{0.25} \, (5893)^{0.08}$$
$$= \underline{\underline{0.76}} \; (76 \text{ per cent})$$

11.14. Packed columns

Packed columns are used for distillation, gas absorption, and liquid–liquid extraction; only distillation and absorption will be considered in this section. Stripping (desorption) is the reverse of absorption and the same design methods will apply.

The gas liquid contact in a packed bed column is continuous, not stage-wise, as in a plate column. The liquid flows down the column over the packing surface and the gas or vapour, counter-currently, up the column. In some gas-absorption columns co-current flow is used. The performance of a packed column is very dependent on the maintenance of good liquid and gas distribution throughout the packed bed, and this is an important consideration in packed-column design.

A schematic diagram, showing the main features of a packed absorption column, is given in Fig. 11.36. A packed distillation column will be similar to the plate columns shown in Figs. 11.1, with the plates replaced by packed sections.

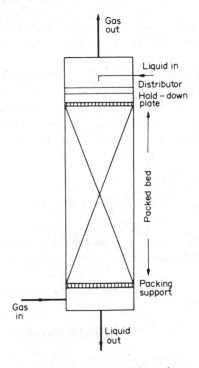

FIG. 11.36. Packed absorption column

Choice of plates or packing

The choice between a plate or packed column for a particular application can only be made with complete assurance by costing each design. However, this will not always be worthwhile, or necessary, and the choice can usually be made, on the basis of experience by considering main advantages and disadvantages of each type; which are listed below:

1. Plate columns can be designed to handle a wider range of liquid and gas flow-rates than packed columns.
2. Packed columns are not suitable for very low liquid rates.
3. The efficiency of a plate can be predicted with more certainty than the equivalent term for packing (HETP or HTU).
4. Plate columns can be designed with more assurance than packed columns. There is always some doubt that good liquid distribution can be maintained throughout a packed column under all operating conditions, particularly in large columns.
5. It is easier to make provision for cooling in a plate column; coils can be installed on the plates.
6. It is easier to make provision for the withdrawal of side-streams from plate columns.
7. If the liquid causes fouling, or contains solids, it is easier to make provision for cleaning in a plate column; manways can be installed on the plates. With small-diameter columns it may be cheaper to use packing and replace the packing when it becomes fouled.
8. For corrosive liquids a packed column will usually be cheaper than the equivalent plate column.
9. The liquid hold-up is appreciably lower in a packed column than a plate column. This can be important when the inventory of toxic or flammable liquids needs to be kept as small as possible for safety reasons.
10. Packed columns are more suitable for handling foaming systems.
11. The pressure drop per equilibrium stage (HETP) can be lower for packing than plates; and packing should be considered for vacuum columns.
12. Packing should always be considered for small diameter columns, say less than 0·6 m, where plates would be difficult to install, and expensive.

Packed-column design procedures

The design of a packed column will involve the following steps:

1. Select the type and size of packing.
2. Determine the column height required for the specified separation.
3. Determine the column diameter (capacity), to handle the liquid and vapour flow rates.
4. Select and design the column internal features: packing support, liquid distributor, redistributors.

These steps are discussed in the following sections, and a packed-column design illustrated in Example 11.14.

11.14.1. Types of packing

The principal requirements of a packing are that it should:

Provide a large surface area: a high interfacial area between the gas and liquid.
Have an open structure: low resistance to gas flow.

Promote uniform liquid distribution on the packing surface.

Promote uniform vapour gas flow across the column cross-section.

Many diverse types and shapes of packing have been developed to satisfy these requirements. They can be divided into two broad classes: stacked packings, which are regular arrangements of the packing elements, and random packings.

Stacked packings, such as grids, have an open structure, and are used for high gas rates, where a low-pressure drop is essential; for example, in cooling towers.

Random packings are more commonly used in the process industries. The principal types of random packings are shown in Fig. 11.37. Design data for these packings are given in Table 11.2. Data on a wider range of packing sizes are given in Volume 2, Chapter 4. The design methods and data given in this section can be used for the preliminary design of packed columns, but for detailed design it is advisable to consult the packing manufacturer's technical literature to obtain data for the particular packing that will be used. The packing manufacturers should be consulted for details of the many special types of packing that are available for special applications.

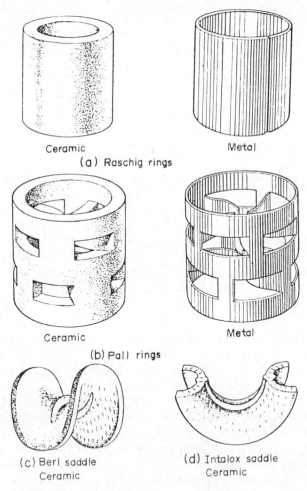

Ceramic   Metal

(a) Raschig rings

Ceramic   Metal

(b) Pall rings

(c) Berl saddle   (d) Intalox saddle
Ceramic   Ceramic

FIG. 11.37. Types of packing (Norton Co.)

TABLE 11.2. *Design data for various packings*

	Size		Bulk density (kg/m^3)	Surface area a (m^2/m^3)	Packing factor F_p
	in.	mm			
Raschig rings	0·50	13	881	368	640
Ceramic	1·0	25	673	190	160
	1·5	38	689	128	95
	2·0	51	651	95	65
	3·0	76	561	69	36
Metal	0·5	13	1201	417	300
(density for carbon steel)	1·0	25	625	207	115
	1·5	38	785	141	83
	2·0	51	593	102	57
	3·0	76	400	72	32
Pall rings	0·625	16	593	341	70
Metal	1·0	25	481	210	48
(density for carbon steel)	1·25	32	385	128	28
	2·0	51	353	102	20
	3·5	76	273	66	16
Plastics	0·625	16	112	341	97
(density for polypropylene)	1·0	25	88	207	52
	1·5	38	76	128	40
	2·0	51	68	102	25
	3·5	89	64	85	16
Intalox saddles	0·5	13	737	480	200
Ceramic	1·0	25	673	253	92
	1·5	38	625	194	52
	2·0	51	609	108	40
	3·0	76	577		22

    Raschig rings, Fig. 11.37*a*, are one of the oldest specially manufactured types of random packing, and are still in general use. Pall rings, Fig. 11.37*b*, are essentially Raschig rings in which openings have been made by folding strips of the surface into the ring. This increases the free area and improves the liquid distribution characteristics. Berl saddles, Fig. 11.37*c*, were developed to give improved liquid distribution compared to Raschig rings, Intalox saddles, Fig. 11.37*d*, can be considered to be an improved type of Berl saddle; their shape makes them easier to manufacture than Berl saddles. The Hypac and Super Intalox packings shown in Fig. 11.38 can be considered improved types of Pall ring and Intalox saddle, respectively.

    Intalox saddles, Super Intalox and Hypac packings are proprietary design, and registered trade marks of the Norton Chemical Process Products Ltd.

    Ring and saddle packings are available in a variety of materials: ceramics, metals, plastics and carbon. Metal and plastics (polypropylene) rings are more efficient than ceramic rings, as it is possible to make the walls thinner.

    Raschig rings are cheaper per unit volume than Pall rings or saddles but are less efficient, and the total cost of the column will usually be higher if Raschig rings are specified. For new columns, the choice will normally be between Pall rings and Berl or Intalox saddles.

    The choice of material will depend on the nature of the fluids and the operating temperature. Ceramic packing will be the first choice for corrosive liquids; but ceramics are unsuitable for use with strong alkalies. Plastic packings are attacked by some organic solvents, and can only be used up to moderate temperatures; so are unsuitable for

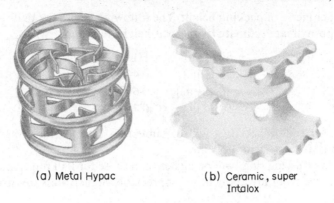

(a) Metal Hypac   (b) Ceramic, super
Intalox

FIG. 11.38. Improved packing types (Norton Co.)

distillation columns. Where the column operation is likely to be unstable metal rings should be specified, as ceramic packing is easily broken. The choice of packings for distillation and absorption is discussed in detail by Eckert (1963).

Packing size

In general, the largest size of packing that is suitable for the size of column should be used, up to 50 mm. Small sizes are appreciably more expensive than the larger sizes. Above 50 mm the lower cost per cubic metre does not normally compensate for the lower mass transfer efficiency. Use of too large a size in a small column can cause poor liquid distribution.

Recommended size ranges are:

Column diameter	Use packing size
< 0·3 m (1 ft)	< 25 mm (1 in.)
0·3 to 0·9 m (1 to 3 ft)	25 to 38 mm (1 to 1·5 in.)
> 0·9 m	50 to 75 mm (2 to 3 in.)

11.14.2. Packed-bed height

Distillation

For the design of packed distillation columns it is simpler to treat the separation as a staged process, and use the concept of the height of an equivalent equilibrium stage to convert the number of ideal stages required to a height of packing. The methods for estimating the number of ideal stages given in Sections 11.5 to 11.8 can then be applied to packed columns.

The height of an equivalent equilibrium stage, usually called the height of a theoretical plate (HETP), is the height of packing that will give the same separation as an equilibrium stage. It has been shown by Eckert (1975) that in distillation the HETP for a given type and size of packing is essentially constant, and independent of the system physical properties; providing good liquid distribution is maintained and the pressure drop is at least above

17 mm water per metre of packing height. The following values for Pall rings can be used to make an approximate estimate of the bed height required.

Size, mm	HETP, m
25 (1 in.)	0·4–0·5
38 (1½ in.)	0·6–0·75
50 (2 in.)	0·75–1·0

The HETP for saddle packings will be similar to that for Pall rings providing the pressure drop is at least 29 mm per m.

The HETP for Raschig rings will be higher than those for Pall rings or saddles, and the values given above will only apply at an appreciably higher pressure drop, greater than 42 mm per m.

Absorption

Though packed absorption and stripping columns can also be designed as staged process, it is usually more convenient to use the integrated form of the differential equations set up by considering the rates of mass transfer at a point in the column. The derivation of these equations given in Volume 2, Chapter 12.

Where the concentration of the solute is small, say less than 10 per cent, the flow of gas and liquid will be essentially constant throughout the column, and the height of packing required, Z, is given by:

$$Z = \frac{G_m}{K_G a P} \int_{y_2}^{y_1} \frac{dy}{y - y_e} \tag{11.96}$$

in terms of the overall gas phase mass transfer coefficient K_G and the gas composition.

Or,

$$Z = \frac{L_m}{K_L a C_t} \int_{x_2}^{x_1} \frac{dx}{x_e - x} \tag{11.97}$$

in terms of the overall liquid-phase mass-transfer coefficient K_L and the liquid composition,

where G_m = molar gas flow-rate per unit cross-sectional area,
L_m = molar liquid flow-rate per unit cross-sectional area,
a = interfacial surface area per unit volume,
P = total pressure,
C_t = total molar concentration,
y_1 and y_2 = the mol fractions of the solute in the gas at the bottom and top of the column, respectively,
x_1 and x_2 = the mol fractions of the solute in the liquid at the bottom and top of the column, respectively,
x_e = the concentration in the liquid that would be in equilibrium with the gas concentration at any point,
y_e = the concentration in the gas that would be in equilibrium with the liquid concentration at any point.

The relation between the equilibrium concentrations and actual concentrations is shown in Fig. 11.39.

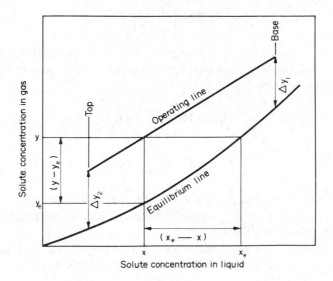

FIG. 11.39. Gas absorption concentration relationships

For design purposes it is convenient to write equations 11.96 and 11.97 in terms of "Transfer Units" (HTU); where the value of integral is the number of transfer units, and the group in front of the integral sign, which has units of length, is the height of a transfer unit.

$$Z = \mathbf{H}_{OG}\mathbf{N}_{OG} \qquad (11.98)$$

or
$$Z = \mathbf{H}_{OL}\mathbf{N}_{OL} \qquad (11.99)$$

where $\mathbf{H}_{OG}$ is the height of an overall gas-phase transfer unit

$$= \frac{G_m}{K_G a P} \qquad (11.100)$$

$\mathbf{N}_{OG}$ is the number of overall gas-phase transfer units

$$= \int_{y_2}^{y_1} \frac{dy}{y - y_e} \qquad (11.101)$$

$\mathbf{H}_{OL}$ is the height of an overall liquid-phase transfer unit

$$= \frac{L_m}{K_L a C_t} \qquad (11.102)$$

$\mathbf{N}_{OL}$ is the number of overall liquid-phase transfer units

$$= \int_{x_2}^{x_1} \frac{dx}{x_e - x} \qquad (11.103)$$

The number of overall gas-phase transfer units is often more conveniently expressed in terms of the partial pressure of the solute gas.

$$N_{OG} = \int_{P_1}^{P_2} \frac{dp}{p - p_e}$$

(11.104)

The relationship between the overall height of a transfer unit and the individual film transfer units H_L and H_G, which are based on the concentration driving force across the liquid and gas films, is given by:

$$H_{OG} = H_G + m \frac{G_m}{L_m} H_L$$

(11.105)

$$H_{OL} = H_L + \frac{L_m}{mG_m} H_G$$

(11.106)

where m is the slope of the equilibrium line and G_m/L_m the slope of the operating line.

The number of transfer units is obtained by graphical or numerical integration of equations 11.101, 11.103 or 11.104.

Where the operating and equilibrium lines are straight, and they can usually be considered to be so for dilute systems, the number of transfer units is given by:

$$N_{OG} = \frac{y_1 - y_2}{\Delta y_{lm}}$$

(11.107)

where Δy_{lm} is the log mean driving force, given by:

$$y_{lm} = \frac{\Delta y_1 - \Delta y_2}{\ln (\Delta y_1/\Delta y_2)}$$

(11.108)

where $\Delta y_1 = y_1 - y_e$,
$\Delta y_2 = y_2 - y_e$.

If the equilibrium curve and operating lines can be taken as straight and the solvent feed essentially solute free, the number of transfer units is given by:

$$N_{OG} = \frac{1}{1 - (mG_m/L_m)} \ln \left[\left(1 - \frac{mG_m}{L_m} \right) \frac{y_1}{y_2} + \frac{mG_m}{L_m} \right]$$

(11.109)

This equation is plotted in Fig. 11.40, which can be used to make a quick estimate of the number of transfer units required for a given separation.

It can be seen from Fig. 11.40 that the number of stages required for a given separation is very dependent on the flow rate L_m. If the solvent rate is not set by other process considerations, Fig. 11.40 can be used to make quick estimates of the column height at different flow rates to find the most economic value. Colburn (1939) has suggested that the optimum value for the term mG_m/L_m will lie between 0·7 to 0·8.

Only physical absorption from dilute gases has been considered in this section. For a discussion of absorption from concentrated gases and absorption with chemical reaction, the reader should refer to Volume 2, or to the books by Treybal (1979) and Sherwood et al. (1975). If the inlet gas concentration is not too high, the equations for dilute systems can be used by dividing the operating line up into two or three straight sections.

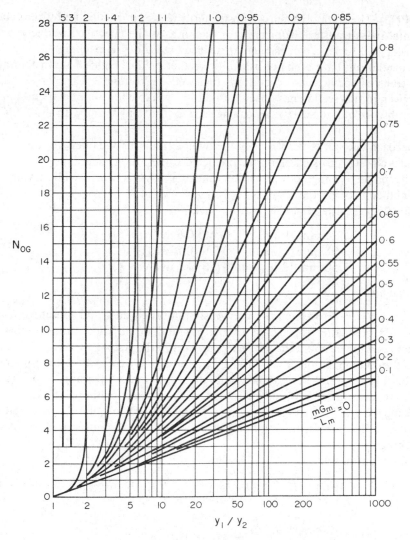

FIG. 11.40. Number of transfer units N_{OG} as a function of y_1/y_2 with mG_m/L_m as parameter

11.14.3. Prediction of the height of a transfer unit (HTU)

There is no entirely satisfactory method for predicting the height of a transfer unit. In practice the value for a particular packing will depend not only on the physical properties and flow-rates of the gas and liquid, but also on the uniformity of the liquid distribution throughout the column, which is dependent on the column height and diameter. This makes it difficult to extrapolate data obtained from small size laboratory and pilot plant columns to industrial size columns. Whenever possible estimates should be based on actual values obtained from operating columns of similar size to that being designed. Experimental values for several systems are given by Eckert (1963) and Cornell et al. (1960). Values for the term K_G, from which the height of the overall gas phase H_{OG} can be calculated, can be found in several publications; Eckert (1975), Norman (1961), Perry and

Chilton (1973). The composite mass transfer term $K_G a$ is normally used when reporting experimental mass-transfer coefficients for packing, as the effective interfacial area for mass transfer will be less than the actual surface area a of the packing.

Many correlations have been published for predicting the height of a transfer unit, and the mass-transfer coefficients; several are reviewed in Volume 2, Chapter 12. The two methods given in this section have been found to be reliable for preliminary design work, and, in the absence of practical values, can be used for the final design with a suitable factor of safety.

The approach taken by the authors of the two methods is fundamentally different, and this provides a useful cross-check on the predicted values. Judgement must always be used when using predictive methods in design, and it is always worth while trying several methods and comparing the results.

Typical values for the HTU of random packings are:

25 mm (1 in.)	0·3 to 0·6 m (1 to 2 ft)	
38 mm ($1\frac{1}{2}$ in.)	0·5 to 0·75 m ($1\frac{1}{2}$ to $2\frac{1}{2}$ ft)	
50 mm (2 in.)	0·6 to 1·0 m (2 to 3 ft)	

Cornell's method

Cornell *et al.* (1960) reviewed the previously published data and presented empirical equations for predicting the height of the gas and liquid film transfer units. Their correlation takes into account the physical properties of the system, the gas and liquid flow-rates; and the column diameter and height. Equations and figures are given for a range of sizes of Raschig rings and Berl saddles. Only those for Berl saddles are given here, as it is unlikely that Raschig rings would be considered for a new column. Though the mass-transfer efficiency of Pall rings and Interlox saddles will be higher than that of the equivalent size Berl saddle, the method can be used to make conservative estimates for these packings.

Cornell's equations are:

$$\mathbf{H}_G = 0·011 \, \psi_h \, (Sc)_v^{0·5} \left(\frac{D_c}{0·305} \right)^{1·11} \left(\frac{Z}{3·05} \right)^{0·33} \bigg/ (L_w^* f_1 f_2 f_3)^{0·5} \tag{11.110}$$

$$\mathbf{H}_L = 0·305 \, \phi_h \, (Sc)_L^{0·5} \, K_3 \left(\frac{Z}{3·05} \right)^{0·15} \tag{11.111}$$

where $\mathbf{H}_G$ = height of a gas-phase transfer unit, m,

$\quad \mathbf{H}_L$ = height of a liquid-phase transfer unit, m,

$\quad (Sc)_v$ = gas Schmidt number = $(\mu_v/\rho_v D_v)$,

$\quad (Sc)_L$ = liquid Schmidt number = $(\mu_L/\rho_L D_L)$,

$\quad D_c$ = column diameter, m,

$\quad Z$ = column height, m,

$\quad K_3$ = percentage flooding correction factor, from Fig. 11.41,

$\quad \psi_h$ = $\mathbf{H}_G$ factor from Fig. 11.42,

$\quad \phi_h$ = $\mathbf{H}_L$ factor from Fig. 11.43,

$\quad L_w^*$ = liquid mass flow-rate per unit area column cross-sectional area, kg/m^2 s,

$\quad f_1$ = liquid viscosity correction factor = $(\mu_L/\mu_w)^{0·16}$,

$\quad f_2$ = liquid density correction factor = $(\rho_w/\rho_L)^{1·25}$,

$\quad f_3$ = surface tension correction factor = $(\sigma_w/\sigma_L)^{0·8}$,

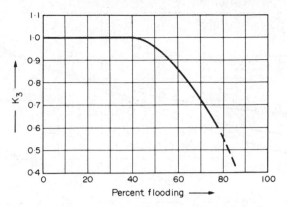

FIG. 11.41. Percentage flooding correction factor

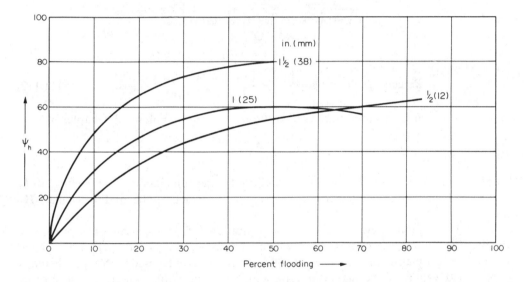

FIG. 11.42. Factor for H_G for Berl saddles

where the suffix w refers to the physical properties of water at 20°C; all other physical properties are evaluated at the column conditions.

The terms $(D_c/0.305)$ and $(Z/3.05)$ are included in the equations to allow for the effects of column diameter and packed-bed height. The "standard" values used by Cornell were 1 ft (0.305 m) for diameter, and 10 ft (3.05 m) for height. These correction terms will clearly give silly results if applied over too wide a range of values. For design purposes the diameter correction term should be taken as a fixed value of 2.3 for columns above 0.6 m (2 ft) diameter, and the height correction should only be included when the distance between liquid redistributors is greater than 3 m. To use Figs. 11.41 and 11.42 an estimate of the column percentage flooding is needed. This can be obtained from Fig. 11.44, where a flooding line has been included with the lines of constant pressure drop.

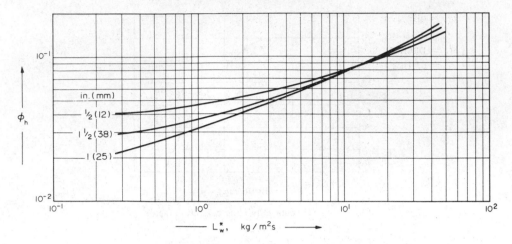

FIG. 11.43. Factor for $\mathbf{H}_L$ for Berl saddles

$$\text{Percentage flooding} = \left[\frac{K_4 \text{ at design pressure drop}}{K_4 \text{ at flooding}} \right]^{1/2} \qquad (11.112)$$

A full discussion of flooding in packed columns is given in Volume 2, Chapter 4.

Onda's method

Onda *et al.* (1968) published useful correlations for the film mass-transfer coefficients k_G and k_L and the effective wetted area of the packing a_w, which can be used to calculate $\mathbf{H}_G$ and $\mathbf{H}_L$.

Their correlations were based on a large amount of data on gas absorption and distillation; with a variety of packings, which included Pall rings and Berl saddles. Their method for estimating the effective area of packing can also be used with experimentally determined values of the mass-transfer coefficients, and values predicted using other correlations.

The equation for the effective area is:

$$\frac{a_w}{a} = 1 - \exp\left[-1\cdot45 \left(\frac{\sigma_c}{\sigma_L} \right)^{0\cdot75} \left(\frac{L_w^*}{a\mu_L} \right)^{0\cdot1} \left(\frac{L_w^{*2} a}{\rho_L^2 g} \right)^{-0\cdot05} \left(\frac{L_w^{*2}}{\rho_L \sigma_L a} \right)^{0\cdot2} \right] \qquad (11.113)$$

and for the mass coefficients:

$$k_L \left(\frac{\rho_L}{\mu_L g} \right)^{1/3} = 0\cdot0051 \left(\frac{L_w^*}{a_w \mu_L} \right)^{2/3} \left(\frac{\mu_L}{\rho_L D_L} \right)^{-1/2} (a d_p)^{0\cdot4} \qquad (11.114)$$

$$\frac{k_G}{a} \frac{RT}{D_v} = K_5 \left(\frac{V_w^*}{a\mu_v} \right)^{0\cdot7} \left(\frac{\mu_v}{\rho_v D_v} \right)^{1/3} (a d_p)^{-2\cdot0} \qquad (11.115)$$

where $K_5 = 5\cdot23$ for packing sizes above 15 mm, and $2\cdot00$ for sizes below 15 mm,

L_w^* = liquid mass flow rate per unit cross-sectional area, kg/mg^{2}s,

V_w^* = gas mass flow rate per unit column cross-sectional area, kg/m^{2}s,

a_w = effective interfacial area of packing per unit volume, m^2/m^3,
a = actual area of packing per unit volume (see Table 11.2), m^2/m^3,
d_p = packing size, m,
σ_c = critical surface tension for the particular packing material given below:

Material	σ_c mN/m
Ceramic	61
Metal (steel)	75
Plastic (polyethylene)	33
Carbon	56

σ_L = liquid surface tension, N/m,
k_G = gas film mass transfer coefficient, $kmol/m^2s$ atm or $kmol/m^2s$ bar,
k_L = liquid film mass transfer coefficient, $kmol/m^2s$ $(kmol/m^3)$ = m/s.

Note: all the groups in the equations are dimensionless.
The units for k_G will depend on the units used for the gas constant:

$$\mathbf{R} = 0{\cdot}08206 \text{ atm } m^3/kmol \text{ K or}$$
$$0{\cdot}08314 \text{ bar } m^3/kmol \text{ K}$$

The film transfer unit heights are given by:

$$\mathbf{H}_G = \frac{G_m}{k_G a_w P} \qquad (11.116)$$

$$\mathbf{H}_L = \frac{L_m}{k_L a_w C_t} \qquad (11.117)$$

where P = column operating pressure, atm or bar,
C_t = total concentration, $kmol/m^3 = \rho_L/$molecular weight solvent,
G_m = molar gas flow-rate per unit cross-sectional area, $kmol/m^2s$,
L_m = molar liquid flow-rate per unit cross-sectional area, $kmol/m^2s$.

Nomographs

A set of nomographs are given in Volume 2, Chapter 12 for the estimation of $\mathbf{H}_G$ and $\mathbf{H}_L$, and the wetting rate. These were taken from a proprietary publication, but are based on a set of similar nomographs given by Czermann *et al.* (1958), who developed the nomographs from correlations put forward by Morris and Jackson (1953) and other workers.

The nomographs can be used to make a quick, rough, estimate of the column height, but are an oversimplification, as they do not take into account all the physical properties and other factors that affect mass transfer in packed columns.

11.14.4. Column diameter (capacity)

The capacity of a packed column is determined by its cross-sectional area. Normally, the column will be designed to operate at the highest economical pressure drop, to ensure good liquid and gas distribution. For random packings the pressure drop will not normally exceed 80 mm of water per metre of packing height. At this value the gas velocity will be about 80 per cent of the flooding velocity. Recommended design values, mm water per m packing, are:

Absorbers and strippers   15 to 50
Distillation, atmospheric and moderate pressure   40 to 80

Where the liquid is likely to foam, these values should be halved.

For vacuum distillations the maximum allowable pressure drop will be determined by the process requirements, but for satisfactory liquid distribution the pressure drop should not be less than 8 mm water per m. If very low bottom pressures are required special low pressure-drop gauze packings should be considered; such as Hyperfil, Multifil or Dixon rings; see Volume 2, Chapter 4.

The column cross-sectional area and diameter for the selected pressure drop can be determined from the generalised pressure-drop correlation given in Fig. 11.44.

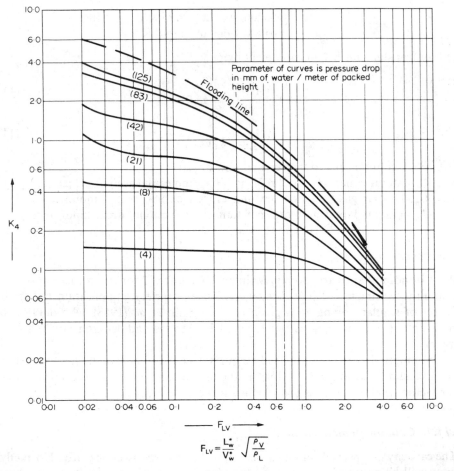

$$F_{LV} = \frac{L_w^*}{V_w^*} \sqrt{\frac{\rho_V}{\rho_L}}$$

FIG. 11.44. Generalised pressure drop correlation, adapted from a figure by the Norton Co. with permission

Figure 11.44 correlates the liquid and vapour flow rates, system physical properties and packing characteristics, with the gas mass flow-rate per unit cross-sectional area; with lines of constant pressure drop as a parameter.

The term K_4 on Fig. 11.44 is the function:

$$K_4 = \frac{42 \cdot 9 \, (V_w^*)^2 \, F_p \, (\mu_L / \rho_L)^{0 \cdot 1}}{\rho_v (\rho_L - \rho_v)} \tag{11.118}$$

where V_w^* = gas mass flow-rate per unit column cross-sectional area, kg/m²s
$\quad F_p$ = packing factor, characteristic of the size and type of packing, see Table 11.2.
$\quad \mu_L$ = liquid viscosity, Ns/m²
$\quad \rho_L, \rho_v$ = liquid and vapour densities, kg/m³

The values of the flow factor F_{LV} given in Fig. 11.44 covers the range that will generally give satisfactory column performance.

The ratio of liquid to gas flow will be fixed by the reflux ratio in distillation; and in gas absorption will be selected to give the required separation with the most economic use of solvent.

Example 11.14

Sulphur dioxide produced by the combustion of sulphur in air is absorbed in water. Pure SO_2 is then recovered from the solution by steam stripping. Make a preliminary design for the absorption column. The feed will be 5000 kg/h of gas containing 8 per cent v/v SO_2. The gas will be cooled to 20°C. A 95 per cent recovery of the sulphur dioxide is required.

Solution

As the solubility of SO_2 in water is high, operation at atmospheric pressure should be satisfactory. The feed-water temperature will be taken as 20°C, a reasonable design value.

Solubility data

From *Chemical Engineers Handbook*, 5th ed., McGraw-Hill, 1973.

	per cent w/w solution	0·05	0·1	0·15	0·2	0·3	0·5	0·7	1·0	1·5
SO_2	Partial press. gas mm Hg	1·2	3·2	5·8	8·5	14·1	26	39	59	92

These figures are plotted in Fig. d.

Partial pressure of SO_2 in the feed $= \dfrac{8}{100} \times 760 = 60 \cdot 8$ mm Hg

Number of stages

Partial pressure in the exit gas at 95 per cent recovery

$$= 60 \cdot 8 \times 0 \cdot 05 = 3 \cdot 06 \text{ mm Hg}$$

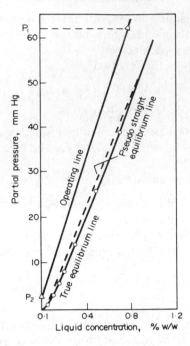

FIG. (d). SO$_2$ absorber design (Example 11.14)

Over this range of partial pressure the equilibrium line is essentially straight so Fig. 11. 40 can be used to estimate the number of stages needed.

The use of Fig. 11.40 will slightly overestimate the number of stages and a more accurate estimate would be made by graphical integration of equation 11.104; but this is not justified in view of the uncertainty in the prediction of the transfer unit height.

Molecular weights: SO$_2$ = 64, H$_2$O = 18, air = 29

Slope of equilibrium line, from Fig. $d = \dfrac{62/760}{1 \cdot 0 \times 10^{-2} \times 18/64} = 29 \cdot 0$

To decide the most economic water flow-rate, the stripper design should be considered together with the absorption design, but for the purpose of this example the absorption design will be considered alone. Using Fig. 11.40 the number of stages required at different water rates will be determined and the "optimum" rate chosen:

$$y_1/y_2 = p_1/p_2 = \frac{60 \cdot 8}{3 \cdot 04} = 20$$

$m\dfrac{G_m}{L_m}$	0·5	0·6	0·7	0·8	0·9	1·0
N_{OG}	3·7	4·1	6·3	8	10·8	1·9

It can be seen that the "optimum" will be between $m\,G_m/L_m = 0 \cdot 6$ to $0 \cdot 8$, as would be expected. Below 0·6 there is only a small decrease in the number of stages required with increasing liquid rate; and above 0·8 the number of stages increases rapidly with decreasing liquid rate.

Check the liquid outlet composition at 0·6 and 0·8:

$$\text{Material balance } L_m x_1 = G_m(y_1 - y_2)$$

$$\text{so } x_1 = \frac{G_m}{L_m}(0\cdot08 \times 0\cdot95) = \frac{m}{29\cdot0}\frac{G_m}{L_m}(0\cdot076)$$

at $m\, G_m/L_m = 0\cdot6$, $x_1 = 1\cdot57 \times 10^{-3}$ mol fraction
at $m\, G_m/L_m = 0\cdot8$, $x_1 = 2\cdot10 \times 10^{-3}$ ''   ''

Use 0·8, as the higher concentration will favour the stripper design and operation, without significantly increasing the number of stages needed in the absorber.

$$N_{OG} = \underline{\underline{8}}$$

Column diameter

The physical properties of the gas can be taken as those for air, as the concentration of SO_2 is low.

$$\text{Gas flow-rate} = \frac{5000}{3600} = 1\cdot39 \text{ kg/s}, \ = \frac{1\cdot39}{29} = 0\cdot048 \text{ kmol/s}$$

$$\text{Liquid flow-rate} = \frac{29\cdot0}{0\cdot8} \times 0\cdot048 = 1\cdot74 \text{ kmol/s}$$

$$= 31\cdot3 \text{ kg/s}.$$

Select 38 mm ($1\frac{1}{2}$ in.) ceramic Intalox saddles.
From Table 11.2, $F_p = 52$,

$$\text{Gas density at } 20°C = \frac{29}{22\cdot4} \times \frac{273}{293} = 1\cdot21 \text{ kg/m}^3$$

Liquid density $\simeq 1000 \text{ kg/m}^3$
Liquid viscosity $= 10^{-3} \text{ N s/m}^2$

$$\frac{L_W^*}{V_W^*}\sqrt{\frac{\rho_v}{\rho_L}} = \frac{31\cdot3}{1\cdot39}\sqrt{\frac{1\cdot21}{10^3}} = 0\cdot78$$

Design for a pressure drop of 20 mm H_2O/m packing

From Fig. 11.44,

$$K_4 = 0\cdot35$$
$$\text{At flooding } K_4 = 0\cdot8$$

$$\text{Percentage flooding} = \sqrt{\frac{0\cdot35}{0\cdot8}} \times 100 = 66 \text{ per cent, satisfactory.}$$

From equation 11.118

$$V_W^* = \left[\frac{K_4 \rho_V(\rho_L - \rho_v)}{42\cdot9 F_p(\mu_L/\rho_L)^{0\cdot1}}\right]^{1/2}$$

$$= \left[\frac{0\cdot35 \times 1\cdot21(1000 - 1\cdot21)}{42\cdot9 \times 52(10^{-3}/10^3)^{0\cdot1}}\right]^{1/2} = 0\cdot87 \text{ kg/m}^2\text{s}$$

$$\text{Column area required} = \frac{1 \cdot 39}{0 \cdot 87} = 1 \cdot 6 \text{ m}^2$$

$$\text{Diameter} = \sqrt{\frac{4}{\pi} \times 1 \cdot 6} = 1 \cdot 43 \text{ m}$$

Round off to 1·50 m

$$\text{Column area} = \frac{\pi}{4} \times 1 \cdot 5^2 = 1 \cdot 77 \text{ m}^2$$

$$\text{Packing size to column diameter ratio} = \frac{1 \cdot 5}{38 \times 10^{-3}} = 39,$$

A larger packing size could be considered.
   Percentage flooding at selected diameter

$$= 66 \times \frac{1 \cdot 6}{1 \cdot 77} = 60 \text{ per cent,}$$

Could consider reducing column diameter.

Estimation of $\mathbf{H}_{OG}$

   Cornell's method

$$D_L = 1 \cdot 7 \times 10^{-9} \text{ m}^2/\text{s}$$
$$D_v = 1 \cdot 45 \times 10^{-5} \text{ m}^2/\text{s}$$
$$\mu_v = 0 \cdot 018 \times 10^{-3} \text{ N s/m}^2$$

$$(Sc)_v = \frac{0 \cdot 018 \times 10^{-3}}{1 \cdot 21 \times 1 \cdot 45 \times 10^{-5}} = 1 \cdot 04$$

$$(Sc)_L = \frac{10^{-3}}{1000 \times 1 \cdot 7 \times 10^{-9}} = 588$$

$$L_W^* = \frac{31 \cdot 3}{1 \cdot 77} = 17 \cdot 6 \text{ kg/s m}^2$$

From Fig. 11.41, at 60 per cent flooding, $K_3 = 0 \cdot 85$.
From Fig. 11.42, at 60 per cent flooding, $\psi_h = 80$.
From Fig. 11.43, at $L_W^* = 17.6$, $\phi_h = 0 \cdot 1$.

$\mathbf{H}_{OG}$ can be expected to be around 1 m, so as a first estimate Z can be taken as 8 m. The column diameter is greater than 0·6 m so the diameter correction term will be taken as 2·3.

$$(11.111) \qquad \mathbf{H}_L = 0 \cdot 305 \times 0 \cdot 1 (588)^{0 \cdot 5} \times 0 \cdot 85 \left(\frac{8}{3 \cdot 05} \right)^{0 \cdot 15} = 0 \cdot 7 \text{ m}$$

As the liquid temperature has been taken as 20°C, and the liquid is water,

$$f_1 = f_2 = f_3 = 1$$

$$(11.110) \qquad \mathbf{H}_G = 0 \cdot 011 \times 80 (1 \cdot 04)^{0 \cdot 5} (2 \cdot 3) \left(\frac{8}{3 \cdot 05} \right)^{0 \cdot 33} \Big/ (17 \cdot 6)^{0 \cdot 5} = 0 \cdot 7 \text{ m}$$

(11.105) $\mathbf{H}_{OG} = 0 \cdot 7 + 0 \cdot 8 \times 0 \cdot 7 = 1 \cdot 3 \text{ m}$

$Z = 8 \times 1 \cdot 3 = 10 \cdot 4 \text{ m}$, close enough to the estimated value.

Onda's method

$\mathbf{R} = 0 \cdot 08314 \text{ bar m}^3/\text{kmol K}.$
Surface tension of liquid, taken as water at $20°C = 70 \times 10^{-3} \text{ N/m}$
$g = 9 \cdot 81 \text{ m/s}^2$
$d_p = 38 \times 10^{-3} \text{ m}$

From Table 11.2, for 38 mm Intalox saddles

$$a = 194 \text{ m}^2/\text{m}^3$$
$$\sigma_c \text{ for ceramics} = 61 \times 10^{-3} \text{ N/m}$$

(11.113)
$$a_W/a = 1 - \exp\left[-1 \cdot 45 \left(\frac{61 \times 10^{-3}}{70 \times 10^{-3}} \right)^{0 \cdot 75} \left(\frac{17 \cdot 6}{194 \times 10^{-3}} \right)^{0 \cdot 1} \left(\frac{17 \cdot 6^2 \times 194}{1000^2 \times 9 \cdot 81} \right)^{-0 \cdot 05} \right.$$
$$\left. \times \left(\frac{17 \cdot 6^2}{1000 \times 70 \times 10^{-3} \times 194} \right)^{0 \cdot 2} \right] = 0 \cdot 71$$

$a_W = 0 \cdot 71 \times 194 = 138 \text{ m}^2/\text{m}^3$

(11.114)
$$k_L \left(\frac{10^3}{10^{-3} \times 9 \cdot 81} \right)^{1/3} = 0 \cdot 0051 \left(\frac{17 \cdot 6}{138 \times 10^{-3}} \right)^{2/3} \left(\frac{10^{-3}}{10^3 \times 1 \cdot 7 \times 10^{-9}} \right)^{-1/2}$$
$$\times (194 \times 38 \times 10^{-3})^{0 \cdot 4}$$

$$k_L = 2 \cdot 5 \times 10^{-4} \text{ m/s}$$

V_W^* on actual column diameter $= \dfrac{1 \cdot 39}{1 \cdot 77} = 0 \cdot 79 \text{ kg/m}^2 \text{ s}$

(11.115) $$k_G \frac{0 \cdot 08314 \times 293}{194 \times 1 \cdot 45 \times 10^{-5}} = 5 \cdot 23 \left(\frac{0 \cdot 79}{194 \times 0 \cdot 018 \times 10^{-3}} \right)^{0 \cdot 7}$$
$$\times \left(\frac{0 \cdot 018 \times 10^{-3}}{1 \cdot 21 \times 1 \cdot 45 \times 10^{-5}} \right)^{1/3} (194 \times 38 \times 10^{-3})^{-2 \cdot 0}$$

$$k_G = 5 \cdot 0 \times 10^{-4} \text{ kmol/sm}^2 \text{ bar}$$

$$G_m = \frac{0 \cdot 79}{29} = 0 \cdot 027 \text{ kmol/m}^2 \text{ s}$$

$$L_m = \frac{17 \cdot 6}{18} = 0 \cdot 98 \text{ kmol/m}^2 \text{ s}$$

(11.116) $$\mathbf{H}_G = \frac{0 \cdot 027}{5 \cdot 0 \times 10^{-4} \times 138 \times 1 \cdot 013} = 0 \cdot 39 \text{ m}$$

$C_T =$ total concentration, as water,

$$= \frac{1000}{18} = 55 \cdot 6 \text{ kmol/m}^3$$

(11.117)
$$\mathbf{H}_L = \frac{0.98}{2.5 \times 10^{-4} \times 138 \times 55.6} = 0.51 \text{ m}$$

(11.105)
$$\mathbf{H}_{OG} = 0.39 + 0.8 \times 0.51 = \underline{0.80 \text{ m}}$$

Use higher value, estimated using Cornell's method, and round up packed bed height to 11 m.

11.14.5. Column internals

The internal fittings in a packed column are simpler than those in a plate column but must be carefully designed to ensure good performance. As a general rule, the standard fittings developed by the packing manufacturers should be specified. Some typical designs are shown in Figs. 11.45 to 11.54; and their use is discussed in the following paragraphs.

Packing support

The function of the support plate is to carry the weight of the wet packing, whilst allowing free passage of the gas and liquid. These requirements conflict; a poorly designed support will give a high pressure drop and can cause local flooding. Simple grid and perforated plate supports are used, but in these designs the liquid and gas have to vie for the same openings. Wide-spaced grids are used to increase the flow area; with layers of larger size packing stacked on the grid to support the small size random packing, Fig. 11.45.

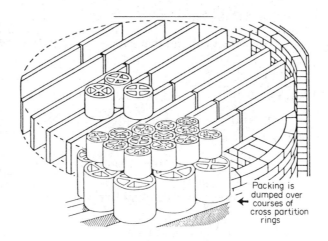

Packing is dumped over courses of cross partition rings

FIG. 11.45. Stacked packing used to support random packing

The best design of packing support is one in which gas inlets are provided above the level where the liquid flows from the bed; such as the gas-injection type shown in Fig. 11.46 and Fig. 11.47. These designs have a low pressure drop and no tendency to flooding. They are available in a wide range of sizes and materials: metals, ceramics and plastics.

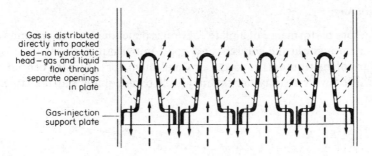

Gas is distributed
directly into packed
bed—no hydrostatic
head—gas and liquid
flow through
separate openings
in plate

Gas-injection
support plate

FIG. 11.46. The principle of the gas-injection packing support

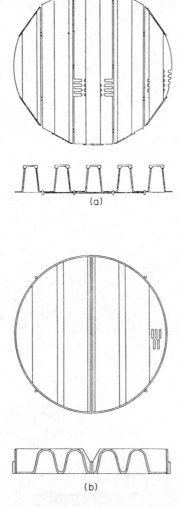

(a)

(b)

Fig. 11.47. Typical designs of gas-injection supports (Norton Co.). (a) Small diameter columns. (b) Large
diameter columns

Liquid distributors

The satisfactory performance of a plate column is dependent on maintaining a uniform flow of liquid throughout the column, and good initial liquid distribution is essential. Various designs of distributors are used. For small-diameter columns a central open feed-pipe, or one fitted with a spray nozzle, may well be adequate; but for larger columns

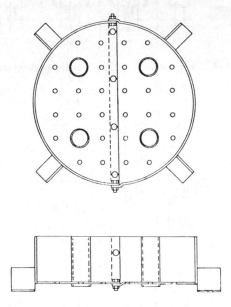

FIG. 11.48. Orifice-type distributor (Norton Co.)

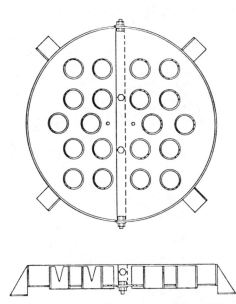

FIG. 11.49. Weir-type distributor (Norton Co.)

more elaborate designs are needed to ensure good distribution at all liquid flow-rates. The two most commonly used designs are the orifice type, shown in Fig. 11.48, and the weir type, shown in Fig. 11.49. In the orifice type the liquid flows through holes in the plate and the gas through short stand pipes. The gas pipes should be sized to give sufficient area for gas flow without creating a significant pressure drop; the holes should be small enough to ensure that there is a level of liquid on the plate at the lowest liquid rate, but large enough to prevent the distributor overflowing at the highest rate. In the weir type the liquid flows

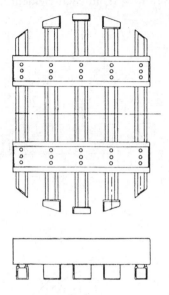

FIG. 11.50. Weir-trough distributors (Norton Co.)

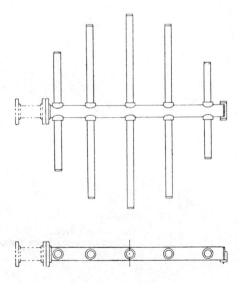

FIG. 11.51. Pipe distributor (Norton Co.)

over notched weirs in the gas stand-pipes. This type can be designed to cope with a wider range of liquid flow rates than the simpler orifice type.

For large-diameter columns, the trough-type distributor shown in Fig. 11.50 can be used, and will give good liquid distribution with a large free area for gas flow.

All distributors which rely on the gravity flow of liquid must be installed in the column level, or maldistribution of liquid will occur.

A pipe manifold distributor, Fig. 11.51, can be used when the liquid is fed to the column under pressure and the flow-rate is reasonably constant. The distribution pipes and orifices should be sized to give an even flow from each element.

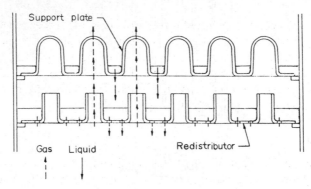

FIG. 11.52. Full redistributor

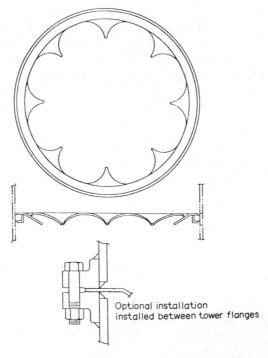

FIG. 11.53. "Wall wiper" redistributor (Norton Co.)

Liquid redistributors

Redistributors are used to collect liquid that has migrated to the column walls and redistribute it evenly over the packing. They will also even out any maldistribution that has occurred within the packing.

A full redistributor combines the functions of a packing support and a liquid distributor; a typical design is shown in Fig. 11.52.

The "wall-wiper" type of redistributor, in which a ring collects liquid from the column wall and redirects it into the centre packing, is occasionally used in small-diameter columns, less than 0·6 m. Care should be taken when specifying this type to select a design that does not unduly restrict the gas flow and cause local flooding. A good design is that shown in Fig. 11.53.

The maximum bed height that should be used without liquid redistribution depends on the type of packing and the process. Distillation is less susceptible to maldistribution than absorption and stripping. As a general guide, the maximum bed height should not exceed 3 column diameters for Raschig rings, and 8 to 10 for Pall rings and saddles. In a large-diameter column the bed height will also be limited by the maximum weight of packing that can be supported by the packing support and column walls; this will be around 8 m.

Hold-down plates

At high gas rates, or if surging occurs through mis-operation, the top layers of packing can be fluidised. Under these conditions ceramic packing can break up and the pieces filter down the column and plug the packing; metal and plastic packing can be blown out of the column. Hold-down plates are used with ceramic packing to weigh down the top layers and prevent fluidisation; a typical design is shown in Fig. 11.54. Bed-limiters are sometimes used with plastics and metal packings to prevent expansion of the bed when operating at a high-pressure drop. They are similar to hold-down plates but are of lighter construction and are fixed to the column walls. The openings in hold-down plates and bed-limiters should be small enough to retain the packing, but should not restrict the gas and liquid flow.

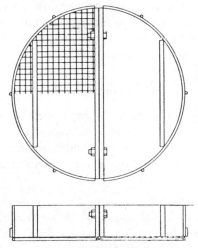

Fig. 11.54. Hold-down plate design (Norton Co.)

Installing packing

Ceramic and metal packings are normally dumped into the column "wet", to ensure a truly random distribution and prevent damage to the packing. The column is partially filled with water and the packing dumped into the water. A height of water must be kept above the packing at all times.

If the columns must be packed dry, for instance to avoid contamination of process fluids with water, the packing can be lowered into the column in buckets or other containers. Ceramic packings should not be dropped from a height of more than half a metre.

Liquid hold-up

An estimate of the amount of liquid held up in the packing under operating conditions is needed to calculate the total load carried by the packing support. The liquid hold-up will depend on the liquid rate and, to some extent, on the gas flow-rate. The packing manufacturers' design literature should be consulted to obtain accurate estimates. As a rough guide, a value of about 25 per cent of the packing weight can be taken for ceramic packings.

11.14.6. Wetting rates

If very low liquid rates have to be used, outside the range of F_{LV} given in Fig. 11.44, the packing wetting rate should be checked to make sure it is above the minimum recommended by the packing manufacturer.

Wetting rate is defined as:

$$\text{wetting rate} = \frac{\text{volumetric liquid rate per unit cross-sectional area}}{\text{packing surface area per unit volume}}$$

A nomograph for the calculation of wetting rates is given in Volume 2, Chapter 4.

Norman (1961) recommends that the liquid rate in absorbers should be kept above $2.7 \text{ kg/m}^2\text{s}$.

If the design liquor rate is too low, the diameter of the column should be reduced. For some processes liquid can be recycled to increase the flow over the packing.

A substantial factor of safety should be applied to the calculated bed height for process where the wetting rate is likely to be low.

11.15. Column auxiliaries

Intermediate storage tanks will normally be needed to smooth out fluctuations in column operation and process upsets. These tanks should be sized to give sufficient hold-up time for smooth operation and control. The hold-up time required will depend on the nature of the process and on how critical the operation is; some typical values for distillation processes are given below:

Operation	Time, minutes		
Feed to a train of columns	10	to	20
Between columns	5	to	10
Feed to a column from storage	2	to	5
Reflux drum	5	to	15

The time given is that for the level in the tank to fall from the normal operating level to the minimum operating level if the feed ceases.

Horizontal or vertical tanks are used, depending on the size and duty. Where only a small hold-up volume is required this can be provided by extending the column base, or, for reflux accumulators, by extending the bottom header of the condenser.

The specification and sizing of surge tanks and accumulators is discussed in more detail by Mehra (1979) and Evans (1980).

11.16. References

AIChE (1958) *Bubble-tray Design Manual* (American Institute of Chemical Engineers).

ALLEVA, R. Q. (1962) *Chem. Eng., Albany* **69** (Aug. 6th) 111. Improving McCabe–Thiele diagrams.

AMUNDSON, N. R. and PONTINEN, A. J. (1958) *Ind. Eng. Chem.* **50,** 730. Multicomponent distillation calculations on a large digital computer.

BILLET, R. (1979) *Distillation Engineering* (Heydon).

BOLLES, W. L. (1963) Tray hydraulics: bubble-cap trays, in *Design of Equilibrium Stage Processes*, Smith, B. D. (McGraw-Hill).

CHANG, H-Y. (1980) *Hyd. Proc.* **59** (Aug.) 79. Computer aids short-cut distillation design.

CHASE, J. D. (1967) *Chem. Eng., Albany* **74** (July 31st) 105 (Aug. 28th) 139 (in two parts). Sieve-tray design.

CICALESE, J. J., DAVIS, J. A., HARRINGTON, P. J., HOUGHLAND, G. S., HUTCHINSON, A. J. L. and WALSH, T. J. (1947) *Pet. Ref.* **26** (May) 495. Study of alkylation-plant isobutane tower performance.

COLBURN, A. P. (1936) *Ind. Eng. Chem.* **28,** 520. Effect of entrainment on plate efficiency in distillation.

COLBURN, A. P. (1939) *Trans. Am. Inst. Chem. Eng.* **35,** 211. The simplified calculation of diffusional processes.

COLBURN, A. P. (1941) *Trans. Am. Inst. Chem. Eng.* **37,** 805. The calculation of minimum reflux ratio in the distillation of multicomponent mixtures.

CORNELL, D., KNAPP, W. G. and FAIR, J. R. (1960) *Chem. Eng. Prog.* **56** (July) 68 (Aug.) 48 (in two parts). Mass transfer efficiency in packed columns.

CZERMANN, J. J., GYOKHEGYI, S. L. and HAY, J. J. (1958) *Pet. Ref.* **37** (April) 165. Designed packed columns graphically.

ECKERT, J. S. (1963) *Chem. Eng. Prog.* **59** (May) 76. A new look at distillation—4 tower packings—comparative performance.

ECKERT, J. S. (1975) *Chem. Eng., Albany* **82** (April 14th) 70. How tower packings behave.

ECONOMOPOULOS, A. P. (1978) *Chem. Eng., Albany* **85** (Dec. 4th) 109. Computer design of sieve tray columns.

EDMISTER, W. C. (1947) Hydrocarbon absorption and fractionation process design methods, a series of articles published in the *Petroleum Engineer* from May 1947 to March 1949 (19 parts). Reproduced in *A Sourcebook of Technical Literature on Distillation* (Gulf).

EDULJEE, H. E. (1958) *Brit. Chem. Eng.* **53,** 14. Design of sieve-type distillation plates.

EDULJEE, H. E. (1959) *Brit. Chem. Eng.* **54,** 320 Design of sieve-type distillation plates.

ELLERBE, R. W. (1979) Batch distillation, in *Handbook of Separation Processes for Chemical Engineers*, Schweitzer, P. A. (Ed.) (McGraw-Hill).

ERBAR, J. H. and MADDOX, R. N. (1961) *Pet. Ref.* **40** (May) 183. Latest score: reflux vs. trays.

EVANS, F. L. (1980) *Equipment Design Handbook for Refineries and Chemical Plants*, vol. 2, 2nd ed. (Gulf).

FAIR, J. R. (1961) *Petro/Chem. Eng.* **33** (Oct.) 45. How to predict sieve tray entrainment and flooding.

FAIR, J. R. (1963) Tray hydraulics: perforated trays, in *Design of Equilibrium Stage Processes*, Smith, B. D. (McGraw-Hill).

FAIR, J. R. and MATTHEWS, R. L. (1958) *Pet. Ref.* **37** (April) 153. Better estimate of entrainment from bubble-cap trays.

FENSKE, M. R. (1932) *Ind. Eng. Chem.* **24,** 482. Fractionation of straight-run gasoline.

FREDENSLUND, A., GMEHLING, J. and RASMUSSEN, P. (1977) *Vapour–liquid Equilibria using UNIFAC* (Elsevier).

GEDDES, R. L. (1958) *AIChE Jl* **4,** 389. General index of fractional distillation power for hydrocarbon mixtures.

GILLILAND, E. R. (1940) *Ind. Eng. Chem.* **32,** 1220. Multicomponent rectification, estimation of the number of theoretical plates as a function of the reflux ratio.

GILLILAND, E. R. and REED, C. E. (1942) *Ind. Eng. Chem.* **34,** 551. Degrees of freedom in multicomponent absorption and rectification.

GLITSCH, H. C. (1960) *Pet. Ref.* **39** (Aug) 91. Mechanical specification of trays.

GLITSCH, H. C. (1970) *Ballast Tray Design Manual*, Bulletin No. 4900 (W. Glistch & Son, Dallas, Texas).

HANSON, D. N., DUFFIN, J. H. and SOMERVILLE, G. E. (1962) *Computation of Multistage Separation Processes* (Reinhold).

HANSON, D. N. and SOMERVILLE, G. F. (1963) *Advances in Chemical Engineering* **4**, 279. Computing multistage vapor–liquid processes.

HENGSTEBECK, R. J. (1946) *Trans. Am. Inst. Chem. Eng.* **42**, 309. Simplified method for solving multicomponent distillation problems.

HENGSTEBECK, R. J. (1961) *Distillation: Principles and design procedures* (Reinhold).

HOLLAND, C. D. (1963) *Multicomponent Distillation* (Prentice-Hall).

HOLLAND, C. D. (1975) *Fundamentals and Modeling of Separation Processes* (Prentice-Hall).

HUANG, C-J. and HODSON, J. R. (1958) *Pet. Ref.* **37** (Feb.) 103. Perforated trays—designed this way.

HUNT, C. D'A., HANSON, D. N. and WILKE, C. R. (1955) *AIChE Jl* **1**, 441. Capacity factors in the performance of perforated-plate columns.

JENNY, F. T. (1939) *Trans. Am. Inst. Chem. Eng.* **35**, 635. Graphical solution of problems in multicomponent fractionation.

KING, C. J. (1971) *Separation Processes* (McGraw-Hill).

KIRKBRIDE, C. G. (1944) *Pet. Ref.* **23** (Sept.) 87(321). Process design procedure for multicomponent fractionators.

KOCH (1960) *Flexitray Design Manual*, Bulletin 960 (Koch Engineering Co., Wichita, Kansas).

KOCH, R. and KUZNIAR, J. (1966) *International Chem. Eng.* **6** (Oct.) 618. Hydraulic calculations of a weir sieve tray.

KWAUK, M. (1956) *AIChE Jl* **2**, 240. A system for counting variables in separation processes.

LEWIS, W. K. (1909) *Ind. Eng. Chem.* **1**, 522. The theory of fractional distillation.

LEWIS, W. K. (1936) *Ind. Eng. Chem.* **28**, 399. Rectification of binary mixtures.

LEWIS, W. K. and MATHESON, G. L. (1932) *Ind. Eng. Chem.* **24**, 494. Studies in distillation.

LIEBSON, I., KELLEY, R. E. and BULLINGTON, L. A. (1957) *Pet. Ref.* **36** (Feb.) 127. How to design perforated trays.

LOWENSTEIN, J. G. (1961) *Ind. Eng. Chem.* **53** (Oct.) 44A. Sizing distillation columns.

LUDWIG, E. E. (1979) *Applied Process Design for Chemical and Petrochemical Plant*, Vol. 2, 2nd ed. (Gulf).

LYSTER, W. N., SULLIVAN, S. L. BILLINGSLEY, D. S. and HOLLAND, C. D. (1959) *Pet. Ref.* **38** (June) 221 (July) 151 (Oct.) 139 and **39** (Aug.) 121 (in four parts). Figure distillation this way.

McCABE, W. L. and THIELE, E. W. (1925) *Ind. Eng. Chem.* **17**, 605. Graphical design of distillation columns.

McCLAIN, R. W. (1960) *Pet. Ref.* **39** (Aug.) 92. How to specify bubble-cap trays.

McCORMICK, J. E. and ROCHE, E. C. (1979) Continuous distillation: separation of multicomponent mixtures, in *Handbook of Separation Processes for Chemical Engineers*, Schweitzer, P. A. (Ed.) (McGraw-Hill).

MEHRA, Y. R. (1979) *Chem. Eng., Albany* **86** (July 2nd) 87. Liquid surge capacity in horizontal and vertical vessels.

MORRIS, G. A. and JACKSON, J. (1953) *Absorption Towers* (Butterworths).

MURPHREE, E. V. (1925) *Ind. Eng. Chem.* **17**, 747. Rectifying column calculations.

NAPHTALI, L. M. and SANDHOLM, D. P. (1971) *AIChE Jl* **17**, 148. Multicomponent separation calculations by linearisation.

NORMAN, W. S. (1961) *Absorption, Distillation and Cooling Towers* (Longmans).

O'CONNELL, H. E. (1946) *Trans. Am. Inst. Chem. Eng.* **42**, 741. Plate efficiency of fractionating columns and absorbers.

OLDERSHAW, C. F. (1941) *Ind. Eng. Chem. (Anal. Ed.)* **13**, 265. Perforated plate columns for analytical batch distillations.

OLIVER, E. D. (1966) *Diffusional Separation Processes* (Wiley).

ONDA, K., TAKEUCHI, H. and OKUMOTO, Y. (1968) *J. Chem. Eng. Japan* **1**, 56. Mass transfer coefficients between gas and liquid phases in packed columns.

PATTON, B. A. and PRITCHARD, B. L. (1960) *Pet. Ref.* **39** (Aug.) 95. How to specify sieve trays.

PERRY, R. H. and CHILTON, C. H. (Eds.) (1973) *Chemical Engineers Handbook*, 5th ed. (McGraw-Hill).

ROBINSON, C. S. and GILLILAND, E. R. (1950) *Elements of Fractional Distillation* (McGraw-Hill).

ROSE, A., SWEENEY, R. F. and SCHRODT, V. N. (1958) *Ind. Eng. Chem.* **50**, 737. Continuous distillation calculations by relaxation method.

SHERWOOD, T. K., PIGFORD, R. L. and WILKE, C. R. (1975) *Mass Transfer* (McGraw-Hill).

SMITH, B. D. (1963) *Design of Equilibrium Stage Processes* (McGraw-Hill).

SMITH, B. D. and BRINKLEY, W. K. (1960) *AIChE Jl* **6**, 446. General short-cut equation for equilibrium stage processes.

SMOKER, E. H. (1938) *Trans. Am. Inst. Chem. Eng.* **34**, 165. Analytical determination of plates in fractionating columns.

SOREL, E. (1899) *Distillation et Rectification Industrielle* (G. Carré et C. Naud).

SOUDERS, M. and BROWN, G. G. (1934) *Ind. Eng. Chem.* **26**, 98. Design of fractionating columns.

SWANSON, R. W. and GESTER, J. A. (1962) *J. Chem. Eng. Data* **7**, 132. Purification of isoprene by extractive distillation.

THIELE, E. W. and GEDDES, R. L. (1933) *Ind. Eng. Chem.* **25**, 289. The computation of distillation apparatus for hydrocarbon mixtures.

THOMAS, W. J. and SHAH, A. N. (1964) *Trans. Inst. Chem. Eng.* **42**, T71. Downcomer studies in a frothing system.

THRIFT, C. (1960a) *Pet. Ref.* **39** (Aug.) 93. How to specify valve trays.

THRIFT, C. (1960b) *Pet. Ref.* **39** (Aug.) 95. How to specify sieve trays.

TOOR, H. L. and BURCHARD, J. K. (1960) *AIChE Jl* **6**, 202. Plate efficiencies in multicomponent systems.

TREYBAL, R. E. (1979) *Mass Transfer Operations*, 3rd ed. (McGraw-Hill).

UNDERWOOD, A. J. V. (1948) *Chem. Eng. Prog.* **44** (Aug.) 603. Fractional distillation of multicomponent mixtures.

VAN WINKLE, M. (1967) *Distillation* (McGraw-Hill).

VAN WINKLE, M., MACFARLAND, A. and SIGMUND, P. M. (1972) *Hyd. Proc.* **51** (July) 111. Predict distillation efficiency.

VEATCH, F., CALLAHAN, J. L., DOL, J. D. and MILBERGER, E. C. (1960) *Chem. Eng. Prog.* **56** (Oct.) 65. New route to acrylonitrile.

WANG, J. C. and HENKE, G. E. (1966) *Hyd. Proc.* **48** (Aug) 155. Tridiagonal matrix for distillation.

WILKE, C. R. and CHANG, P. (1955) *AIChE Jl* **1**, 264. Correlation for diffusion coefficients in dilute solutions.

WILKE, C. R. and LEE, C. Y. (1955) *Ind. Eng. Chem.* **47**, 1253. Estimation of diffusion coefficients for gases and vapours.

WINN, F. W. (1958) *Pet. Ref.* **37** (May) 216. New relative volatility method for distillation calculations.

YAWS, C. L., PATEL, P. M., PITTS, F. H. and FANG, C. S. (1979) *Hyd. Proc.* **58** (Feb.) 99. Estimate multicomponent recovery.

ZUIDERWEG, F. J., VERBURG, H. and GILISSEN, F. A. H. (1960) *First International Symposium on Distillation*, Inst. Chem. Eng. London, 201. Comparison of fractionating devices.

11.17. Nomenclature

		Dimensions in **MLT** θ
A	Constant in equation 11.63	—
A_a	Active area of plate	L^2
A_{ap}	Clearance area under apron	L^2
A_c	Total column cross-sectional area	L^2
A_d	Downcomer cross-sectional area	L^2
A_h	Total hole area	L^2
A_i	Absorption factor	—
A_m	Area term in equation 11.92	L^2
A_n	Net area available for vapour–liquid disengagement	L^2
A_p	Perforated area	L^2
a	Packing surface area per unit volume	L^{-1}
a_w	Effective interfacial area of packing per unit volume	L^{-1}
B	Mols of bottom product per unit time	MT^{-1}
b	Parameter in equation 11.28	—;
b_i	Mols of component i in bottom product	M
C_o	Orifice coefficient in equation 11.88	—
C_T	Total molar concentration	ML^{-3}
c	Parameter defined by equation 11.32	—
D	Mols of distillate per unit time	MT^{-1}
D_c	Column diameter	L
D_e	Eddy diffusivity	L^2T^{-1}
D_L	Liquid diffusivity	L^2T^{-1}
D_{LK}	Diffusivity of light key component	L^2T^{-1}
D_v	Diffusivity of vapour	L^2T^{-1}
d_h	Hole diameter	L
d_i	Mols of component i in distillate per unit time	MT^{-1}
d_p	Size of packing	L
E_a	Actual plate efficiency, allowing for entrainment	—
E_{mV}	Murphree plate efficiency	—
E_{mv}	Murphree point efficiency	—
E_o	Overall column efficiency	—
FA	Fractional area, equation 11.69	—
F_n	Feed rate to stage n	MT^{-1}
F_p	Packing factor	—
F_v	Column 'F' factor $= u_a \sqrt{\rho_v}$	$M^{1/2}L^{-1/2}T^{-1}$
F_{LV}	Column liquid–vapour factor in Fig. 11.27	—

f_i	Mols of component i in feed per unit time	$\mathbf{MT}^{-1}$
f_1	Viscosity correction factor in equation 11.110	—
f_2	Liquid density correction factor in equation 11.110	—
f_3	Surface tension correction factor in equation 11.110	—
G	Feed condition factor defined by equations 11.55 & 11.56	—
G_m	Molar flow-rate of gas per unit area	$\mathbf{ML}^{-2}\mathbf{T}^{-1}$
g	Gravitational acceleration	$\mathbf{LT}^{-2}$
H	Specific enthalpy of vapour phase	$\mathbf{L}^2\mathbf{T}^{-2}$
$\mathbf{H}_G$	Height of gas film transfer unit	$\mathbf{L}$
$\mathbf{H}_L$	Height of liquid film transfer unit	$\mathbf{L}$
$\mathbf{H}_{OG}$	Height of overall gas phase transfer unit	$\mathbf{L}$
$\mathbf{H}_{OL}$	Height of overall liquid phase transfer unit	$\mathbf{L}$
$\mathscr{H}$	Henry's constant	$\mathbf{ML}^{-1}\mathbf{T}^{-2}$
h	Specific enthalpy of liquid phase	$\mathbf{L}^2\mathbf{T}^{-2}$
h_{ap}	Apron clearance	$\mathbf{L}$
h_b	Height of liquid backed-up in downcomer	$\mathbf{L}$
h_{bc}	Downcomer back-up in terms of clear liquid head	$\mathbf{L}$
h_d	Dry plate pressure drop, head of liquid	$\mathbf{L}$
h_{dc}	Head loss in downcomer	$\mathbf{L}$
h_f	Specific enthalpy of feed stream	$\mathbf{L}^2\mathbf{T}^{-2}$
h_{ow}	Height of liquid crest over downcomer weir	$\mathbf{L}$
h_r	Plate residual pressure drop, head of liquid	$\mathbf{L}$
h_t	Total plate pressure drop, head of liquid	$\mathbf{L}$
h_w	Weir height	$\mathbf{L}$
K	Equilibrium constant for least volatile component	—
K'	Equilibrium constant for more volatile component	—
K_G	Overall gas phase mass transfer coefficient	$\mathbf{L}^{-1}\mathbf{T}$
K_i	Equilibrium constant for component i	—
K_L	Overall liquid phase mass transfer coefficient	$\mathbf{LT}^{-1}$
K_n	Equilibrium constant on stage n	—
K_1	Constant in equation 11.81	$\mathbf{LT}^{-1}$
K_2	Constant in equation 11.84	—
K_3	Percentage flooding factor in equation 11.111	—
K_4	Parameter in Fig. 11.44, defined by equation 11.118	—
K_5	Constant in equation 11.115	—
k	Root of equation 11.28	—
k_G	Gas film mass transfer coefficient	$\mathbf{L}^{-1}\mathbf{T}$
k_L	Liquid film mass transfer coefficient	$\mathbf{LT}^{-1}$
L	Liquid flow-rate, mols per unit time	$\mathbf{MT}^{-1}$
L_e	Estimated flow-rate of combined keys, liquid	$\mathbf{MT}^{-1}$
L_m	Molar flow-rate of liquid per unit area	$\mathbf{ML}^{-2}\mathbf{T}^{-1}$
L_p	Volumetric flow-rate across plate divided by average plate width	$\mathbf{L}^2\mathbf{T}^{-1}$
L_w	Liquid mass flow-rate	$\mathbf{MT}^{-1}$
L_w^*	Liquid mass flow-rate per unit area	$\mathbf{ML}^{-2}\mathbf{T}^{-1}$
L_{wd}	Liquid mass flow-rate through downcomer	$\mathbf{MT}^{-1}$
$\underline{l}_i$	Limiting liquid flow-rate of components lighter than the keys in the rectifying section	$\mathbf{MT}^{-1}$
$\underline{l}_i'$	Limiting liquid flow-rates of components heavier than the keys in the stripping section	$\mathbf{MT}^{-1}$
l_h	Weir chord height	$\mathbf{L}$
l_n	Molar liquid flow rate of component from stage n	$\mathbf{MT}^{-1}$
l_p	Pitch of holes (distance between centres)	$\mathbf{L}$
l_t	Plate spacing in column	$\mathbf{L}$
l_w	Weir length	$\mathbf{L}$
M_s	Molecular weight of solvent	—
m	Slope of equilibrium line	—
N	Number of stages	—
$\mathbf{N}_G$	Number of gas-film transfer units	—
$\mathbf{N}_L$	Number of liquid-film transfer units	—
N_m	Number of stages at total reflux	—
$\mathbf{N}_{OG}$	Number of overall gas-phase transfer units	—
$\mathbf{N}_{OL}$	Number of overall liquid-phase transfer units	—

N_r	Number of equilibrium stages above feed	—
N_r^*	Number of stages in rectifying section (equation 11.26)	—
N_s	Number of equilibrium stages below feed	—
N_s^*	Number of stages in stripping section (equation 11.25)	—
n	Stage number	—
P	Total pressure	$\mathbf{ML^{-1}T^{-2}}$
P^0	Vapour pressure	$\mathbf{ML^{-1}T^{-2}}$
ΔP_t	Total plate pressure drop	$\mathbf{ML^{-1}T^{-2}}$
p	Partial pressure	$\mathbf{ML^{-1}T^{-2}}$
q	Heat to vaporise one mol of feed divided by molar latent heat	—
q_b	Heat supplied to reboiler	$\mathbf{ML^2T^{-3}}$
q_c	Heat removed in condenser	$\mathbf{ML^2T^{-3}}$
q_n	Heat supplied to or removed from stage n	$\mathbf{ML^2T^{-3}}$
$\mathbf{R}$	Universal gas constant	$\mathbf{L^2T^{-2}\theta^{-1}}$
R	Reflux ratio	—
R_m	Minimum reflux ratio	—
S_i	Stripping factor	—
S_n	Side stream flow from stage n	$\mathbf{MT^{-1}}$
S_r	Stripping factor for rectifying section (equation 11.54)	—
S_s	Stripping factor for stripping section (equation 11.54)	—
s	Slope of operating line	—
t_L	Liquid contact time	$\mathbf{T}$
t_r	Residence time in downcomer	$\mathbf{T}$
u_a	Vapour velocity based on active area	$\mathbf{LT^{-1}}$
u_f	Vapour velocity at flooding point	$\mathbf{LT^{-1}}$
u_h	Vapour velocity through holes	$\mathbf{LT^{-1}}$
u_n	Vapour velocity based on net cross-sectional area	$\mathbf{LT^{-1}}$
u_v	Superficial vapour velocity (based on total cross-sectional area)	$\mathbf{LT^{-1}}$
V	Vapour flow-rate mols per unit time	$\mathbf{MT^{-1}}$
V_e	Estimated flow-rate of combined keys, vapour	$\mathbf{MT^{-1}}$
V_w	Vapour mass flow-rate	$\mathbf{MT^{-1}}$
V_w^*	Vapour mass flow-rate per unit area	$\mathbf{ML^{-2}T^{-1}}$
$\underline{v_i}$	Limiting vapour flow-rates of components lighter than the keys in the rectifying section	$\mathbf{MT^{-1}}$
$\underline{v_i'}$	Limiting vapour flow-rates of components heavier than the keys in the stripping section	$\mathbf{MT^{-1}}$
v_n	Molar vapour flow-rate of component from stage n	$\mathbf{MT^{-1}}$
x	Mol fraction of component in liquid phase	—
x_A	Mol fraction of component A in binary mixture	—
x_B	Mol fraction of component B in binary mixture	—
x_b	Mol fraction of component in bottom product	—
x_d	Mol fraction of component in distillate	—
x_e	Equilibrium concentration	—
x_i	Mol fraction of component i	—
x_r	Concentration of reference component (equation 11.57)	—
x_n^*	Reference concentration in equation 11.30	—
x_0^*	Reference concentration in equation 11.30	—
x_1	Concentration of solute in solution at column base	—
x_2	Concentration of solute in solution at column top	—
x_r	Reference concentration equations 11.25 and 11.26	—
y	Mol fraction of component in vapour phase	—
y_A	Mol fraction of component A in a binary mixture	—
y_B	Mol fraction of component B in a binary mixture	—
y_e	Equilibrium concentration	—
y_i	Mol fraction of component i	—
Δy	Concentration driving force in the gas phase	—
Δy_{lm}	Log mean concentration driving force	—
y_1	Concentration of solute in gas phase at column base	—
y_2	Concentration of solute in gas phase at column top	—
Z	Height of packing	$\mathbf{L}$
Z_c	Liquid hold-up on plate	$\mathbf{L}$
Z_L	Length of liquid path	$\mathbf{L}$

z_i	Mol fraction of component i in feed stream	—
z_f	Mol fraction of component in feed stream	—
z_f^*	Pseudo feed concentration defined by equation 11.41	—
α	Relative volatility	—
α_i	Relative volatility of component i	—
α_a	Average relative volatility of light key	—
β	Parameter defined by equation 11.31	—
θ	Root of equation 11.61	—
μ	Dynamic viscosity	$\mathbf{ML^{-1}T^{-1}}$
μ_a	Molar average liquid viscosity	$\mathbf{ML^{-1}T^{-1}}$
μ_s	Viscosity of solvent	$\mathbf{ML^{-1}T^{-1}}$
μ_w	Viscosity of water at 20°C	$\mathbf{ML^{-1}T^{-1}}$
ρ	Density	$\mathbf{ML^{-3}}$
ρ_w	Density of water at 20°C	$\mathbf{ML^{-3}}$
σ	Surface tension	$\mathbf{MT^{-2}}$
σ_c	Critical surface tension for packing material	$\mathbf{MT^{-2}}$
σ_w	Surface tension of water at 20°C	$\mathbf{MT^{-2}}$
Φ	Intercept of operating line on Y axis	—
Φ_n	Factor in equation 11.43	—
ψ	Fractional entrainment	—
ψ_h	Factor in equation 11.42	—

Dg	Surface tension number
Pe	Peclet number
Re	Reynolds number
Sc	Schmidt number

Suffixes

L	Liquid
v	Vapour
HK	Heavy key
LK	Light key
b	Bottoms
d	Distillate (Tops)
f	Feed
i	Component number
n	Stage number
1	Base of packed column
2	Top of packed column

Superscripts

$'$	Stripping section of column

CHAPTER 12

Heat-transfer Equipment

12.1. Introduction

The transfer of heat to and from process fluids is an essential part of most chemical processes. The most commonly used type of heat-transfer equipment is the ubiquitous shell and tube heat exchanger; the design of which is the main subject of this chapter.

The fundamentals of heat-transfer theory are covered in Volume 1, Chapter 7; and in many other textbooks; McAdams (1954), Jacob (1957), Kay (1963), Kreith (1976), Fishenden and Saunders (1950). Several useful books have been published on the design of heat-exchanger equipment, and may be consulted for fuller details of the construction of equipment and design methods than can be given in this short chapter; Kern (1950), Ludwig (1965), Fraas and Ozisik (1965), Rohsenow and Hartnett (1973).

As with distillation, work on the development of reliable design methods for heat exchangers has been dominated in recent years by commercial research organisations: Heat Transfer Research Inc. (HTRI) in the United States, and Heat Transfer and Fluid Flow Service (HTFS) operated by the United Kingdom Atomic Energy Authority and National Engineering Laboratory in the United Kingdom. Their methods are of a proprietary nature and are not therefore available in the open literature. They will, however, be available to design engineers in the major operating and contracting companies, whose companies subscribe to these organisations.

The principal types of heat exchanger used in the chemical process and allied industries, which will be discussed in this chapter, are listed below:

1. Double-pipe exchanger: the simplest type, used for cooling and heating.
2. Shell and tube exchangers: used for all applications.
3. Plate and frame exchangers (plate heat exchangers): used for heating and cooling.
4. Air cooled: coolers and condensers.
5. Direct contact: cooling and quenching.

The word "exchanger" really applies to all types of equipment in which heat is exchanged but is often used specifically to denote equipment in which heat is exchanged between two process streams. Exchangers in which a process fluid is heated or cooled by a plant service stream are referred to as heaters and coolers. If the process stream is vaporised the exchanger is called a vaporiser if the stream is essentially completely vaporised; a reboiler if associated with a distillation column; and an evaporator if used to concentrate a solution (see Chapter 10). The term fired exchanger is used for exchangers heated by combustion gases, such as boilers; other exchangers are referred to as "unfired exchangers".

12.2. Basic design procedure and theory

The general equation for heat transfer across a surface is:

$$Q = UA\Delta T_m \tag{12.1}$$

where Q = heat transferred per unit time, W,
         U = the overall heat transfer coefficient, $W/m^2\,°C$,
         A = heat-transfer area, m^2,
         ΔT_m = the mean temperature difference, the temperature driving force, $°C$.

The prime objective in the design of an exchanger is to determine the surface area required for the specified duty (rate of heat transfer) using the temperature differences available.

The overall coefficient is the reciprocal of the overall resistance to heat transfer, which is the sum of several individual resistances. For heat exchange across a typical heat-exchanger tube the relationship between the overall coefficient and the individual coefficients, which are the reciprocals of the individual resistances, is given by:

$$\frac{1}{U_o} = \frac{1}{h_o} + \frac{1}{h_{od}} + \frac{d_o \ln(d_o/d_i)}{2k_w} + \frac{d_o}{d_i} \times \frac{1}{h_{id}} + \frac{d_o}{d_i} \times \frac{1}{h_i} \tag{12.2}$$

where U_o = the overall coefficient based on the outside area of the tube, $W/m^2\,°C$,
         h_o = outside fluid film coefficient, $W/m^2\,°C$,
         h_i = inside fluid film coefficient, $W/m^2\,°C$,
         h_{od} = outside dirt coefficient (fouling factor), $W/m^2\,°C$,
         h_{id} = inside dirt coefficient, $W/m^2\,°C$,
         k_w = thermal conductivity of the tube wall material, $W/m\,°C$,
         d_i = tube inside diameter, m,
         d_o = tube outside diameter, m.

The magnitude of the individual coefficients will depend on the nature of the heat-transfer process (conduction, convection, condensation, boiling or radiation), on the physical properties of the fluids, on the fluid flow-rates, and on the physical arrangement of the heat-transfer surface. As the physical layout of the exchanger cannot be determined until the area is known the design of an exchanger is of necessity a trial and error procedure. The steps in a typical design procedure are given below:

1. Define the duty: heat-transfer rate, fluid flow-rates, temperatures.
2. Collect together the fluid physical properties required: density, viscosity, thermal conductivity.
3. Decide on the type of exchanger to be used.
4. Select a trial value for the overall coefficient, U.
5. Calculate the mean temperature difference, ΔT_m.
6. Calculate the area required from equation 12.1.
7. Decide the exchanger layout.
8. Calculate the individual coefficients.
9. Calculate the overall coefficient and compare with the trial value. If the calculated value differs significantly from the estimated value, substitute the calculated for the estimated value and return to step 6.
10. Calculate the exchanger pressure drop; if unsatisfactory return to steps 7 or 4 or 3, in that order of preference.

11. Optimise the design: repeat steps 4 to 10, as necessary, to determine the cheapest exchanger that will satisfy the duty. Usually this will be the one with the smallest area.

Procedures for estimating the individual heat-transfer coefficients and the exchanger pressure drops are given in this chapter.

12.3. Overall heat-transfer coefficient

Typical values of the overall heat-transfer coefficient for various types of heat exchanger are given in Table 12.1. More extensive data can be found in the books by Perry and Chilton (1973), TEMA (1978), and Ludwig (1965).

Figure 12.1, which is adapted from a similar nomograph given by Frank (1974), can be used to estimate the overall coefficient for tubular exchangers (shell and tube). The film coefficients given in Fig. 12.1 include an allowance for fouling.

TABLE 12.1. *Typical overall coefficients*

Shell and tube exchangers		
Hot fluid	Cold fluid	U (W/m^2 °C)
Heat exchangers		
Water	Water	800–1500
Organic solvents	Organic solvents	100–300
Light oils	Light oils	100–400
Heavy oils	Heavy oils	50–300
Gases	Gases	10–50
Coolers		
Organic solvents	Water	250–750
Light oils	Water	350–900
Heavy oils	Water	60–300
Gases	Water	20–300
Organic solvents	Brine	150–500
Water	Brine	600–1200
Gases	Brine	15–250
Heaters		
Steam	Water	1500–4000
Steam	Organic solvents	500–1000
Steam	Light oils	300–900
Steam	Heavy oils	60–450
Steam	Gases	30–300
Dowtherm	Heavy oils	50–300
Dowtherm	Gases	20–200
Flue gases	Steam	30–100
Flue	Hydrocarbon vapours	30–100
Condensers		
Aqueous vapours	Water	1000–1500
Organic vapours	Water	700–1000
Organics (some non-condensibles)	Water	500–700
Vacuum condensers	Water	200–500
Vaporisers		
Steam	Aqueous solutions	1000–1500
Steam	Light organics	900–1200
Steam	Heavy organics	600–900

TABLE 12.1 (*Cont.*)

Air-cooled exchangers

Process fluid	U (W/m^2 °C)
Water	300–450
Light organics	300–700
Heavy organics	50–150
Gases, 5–10 bar	50–100
10–30 bar	100–300
Condensing hydrocarbons	300–600

Immersed coils

Coil	Pool	
Natural circulation		
Steam	Dilute aqueous solutions	500–1000
Steam	Light oils	200–300
Steam	Heavy oils	70–150
Aqueous solutions	Water	200–500
Light oils	Water	100–150
Agitated		
Steam	Dilute aqueous solutions	800–1500
Steam	Light oils	300–500
Steam	Heavy oils	200–400
Aqueous solutions	Water	400–700
Light oils	Water	200–300

Jacketed vessels

Jacket	Vessel	
Steam	Dil. aqueous soln.	500–700
Steam	Light organics	250–500
Water	Dil. aqueous soln.	200–500
Water	Light organics	200–300

The values given in Table 12.1 and Fig 12.1 can be used for the preliminary sizing of equipment for process evaluation, and as trial values for starting a detailed thermal design.

12.4. Fouling factors (dirt factors)

Most process and service fluids will foul the heat-transfer surfaces in an exchanger to a greater or lesser extent. The deposited material will normally have a relatively low thermal conductivity and will reduce the overall coefficient. It is therefore necessary to oversize an exchanger to allow for the reduction in performance during operation. The effect of fouling is allowed for in design by including the inside and outside fouling coefficients in equation 12.2. Fouling factors are usually quoted as heat-transfer resistances, rather than coefficients. They are difficult to predict and are usually based on past experience. Estimating fouling factors introduces a considerable uncertainty into exchanger design; the value assumed for the fouling factor can overwhelm the accuracy of the predicted values of the other coefficients. Fouling factors are often wrongly used as factors of safety

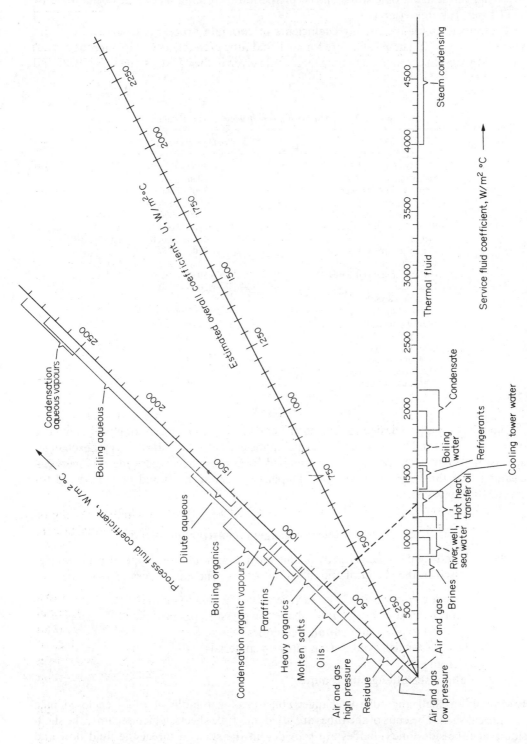

FIG. 12.1. Overall coefficients (join process side duty to service side and read U from centre scale)

in exchanger design. Some work on the prediction of fouling factors has been done by HTRI; see Taborek *et al.* (1972).

Typical values for the fouling coefficients of common process and service fluids are given in Table 12.2. These values are for shell and tube exchangers with plain (not finned) tubes. More extensive data on fouling factors are given in the TEMA standards (1978), and by Ludwig (1965).

TABLE 12.2. *Fouling factors (coefficients), typical values*

Fluid	Coefficient (W/m² °C)
River water	3000–12,000
Sea water	1000–3000
Cooling water (towers)	3000–6000
Towns water (soft)	3000–5000
Towns water (hard)	1000–2000
Steam condensate	1500–5000
Steam (oil free)	4000–10,000
Steam (oil traces)	2000–5000
Refrigerated brine	3000–5000
Air and industrial gases	5000–10,000
Flue gases	2000–5000
Organic vapours	5000
Organic liquids	5000
Light hydrocarbons	5000
Heavy hydrocarbons	2000
Boiling organics	2500
Condensing organics	5000
Heat transfer fluids	5000
Aqueous salt solutions	3000–5000

The selection of the design fouling coefficient will often be an economic decision. The optimum design will be obtained by balancing the extra capital cost of a larger exchanger against the savings in operating cost obtained from the longer operating time between cleaning that the larger area will give. Duplicate exchangers should be considered for severely fouling systems.

12.5. Shell and tube exchangers: construction details

The shell and tube exchanger is by far the most commonly used type of heat-transfer equipment used in the chemical and allied industries. The advantages of this type are:

1. The configuration gives a large surface area in a small volume.
2. Good mechanical layout: a good shape for pressure operation.
3. Uses well-established fabrication techniques.
4. Can be constructed from a wide range of materials.
5. Easily cleaned.
6. Well-established design procedures.

Essentially, a shell and tube exchanger consists of a bundle of tubes enclosed in a cylindrical shell. The ends of the tubes are fitted into tube sheets, which separate the shell-side and tube-side fluids. Baffles are provided in the shell to direct the fluid flow and support the tubes. The assembly of baffles and tubes is held together by support rods and spacers, Fig 12.2.

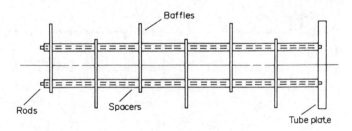

FIG. 12.2. Baffle spacers and tie rods

Exchanger types

The principal types of shell and tube exchanger are shown in Figs. 12.3 to 12.8. Diagrams of other types and full details of their construction can be found in the heat-exchanger standards (see Section 12.5.1.). The standard nomenclature used for shell and tube exchangers is given below; the numbers refer to the features shown in Figs. 12.3 to 12.8.

Nomenclature

Part number
1. Shell
2. Shell cover
3. Floating-head cover
4. Floating-tube plate
5. Clamp ring
6. Fixed-tube sheet (tube plate)
7. Channel (end-box or header)
8. Channel cover
9. Branch (nozzle)
10. Tie rod and spacer
11. Cross baffle or tube-support plate
12. Impingement baffle
13. Longitudinal baffle
14. Support bracket

15. Floating-head support
16. Weir
17. Split ring
18. Tube
19. Tube bundle
20. Pass partition
21. Floating-head gland (packed gland)
22. Floating-head gland ring
23. Vent connection
24. Drain connection
25. Test connection
26. Expansion bellows
27. Lifting ring

The simplest and cheapest type of shell and tube exchanger is the fixed tube sheet design shown in Fig. 12.3. The main disadvantages of this type are that the tube bundle cannot be removed for cleaning and there is no provision for differential expansion of the shell and tubes. As the shell and tubes will be at different temperatures, and may be of different materials, the differential expansion can be considerable and the use of this type is limited to temperature differences up to about 80°C. Some provision for expansion can be made by including an expansion loop in the shell (shown dotted on Fig. 12.3) but their use is limited to low shell pressure; up to about 8 bar. In the other types, only one end of the tubes is fixed and the bundle can expand freely.

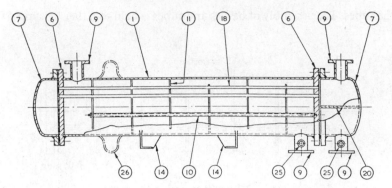

Fig. 12.3. Fixed-tube plate (based on figures from BS 3274: 1960)

The U-tube (U-bundle) type shown in Fig 12.4 requires only one tube sheet and is cheaper than the floating-head types; but is limited in use to relatively clean fluids as the tubes and bundle are difficult to clean. It is also more difficult to replace a tube in this type.

Exchangers with an internal floating head, Figs. 12.5 and 12.6, are more versatile than fixed head and U-tube exchangers. They are suitable for high-temperature differentials and, as the tubes can be rodded from end to end and the bundle removed, are easier to

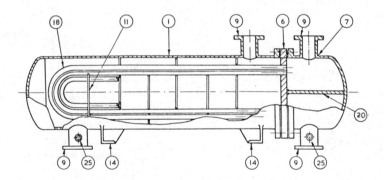

Fig. 12.4. U-tube (based on figures from BS 3274: 1960)

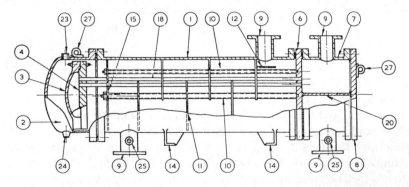

Fig. 12.5. Internal floating head without clamp ring (based on figures from BS 3274: 1960)

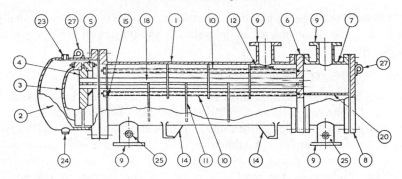

FIG. 12.6. Internal floating head with clamp ring (based on figures from BS 3274: 1960)

clean and can be used for fouling liquids. A disadvantage of the pull-through design, Fig. 12.5, is that the clearance between the outermost tubes in the bundle and the shell must be made greater than in the fixed and U-tube designs to accommodate the floating-head flange, allowing fluid to bypass the tubes. The clamp ring (split flange design), Fig. 12.6, is used to reduce the clearance needed. There will always be a danger of leakage occurring from the internal flanges in these floating head designs.

In the external floating head designs, Fig 12.7, the floating-head joint is located outside the shell, and the shell sealed with a sliding gland joint employing a stuffing box. Because of

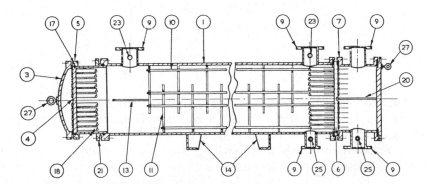

FIG. 12.7. External floating head, packed gland (based on figures from BS 3274: 1960)

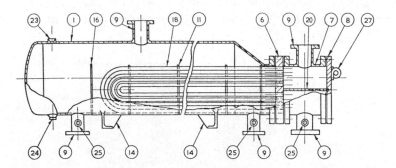

FIG. 12.8. Kettle reboiler with U-tube bundle (based on figures from BS 3274: 1960)

the danger of leaks through the gland, the shell-side pressure in this type is usually limited to about 20 bar, and flammable or toxic materials should not be used on the shell side.

12.5.1. Heat-exchanger standards and codes

The mechanical design features, fabrication, materials of construction, and testing of shell and tube exchangers is covered by British Standard, BS 3274. The standards of the American Tubular Heat Exchanger Manufacturers Association, the TEMA standards, are also universally used. The TEMA standards cover three classes of exchanger: class R covers exchangers for the generally severe duties of the petroleum and related industries; class C covers exchangers for moderate duties in commercial and general process applications; and class B covers exchangers for use in the chemical process industries. The British and American standards should be consulted for full details of the mechanical design features of shell and tube exchangers; only brief details will be given in this chapter.

The standards give the preferred shell and tube dimensions; the design and manufacturing tolerances; corrosion allowances; and the recommended design stresses for materials of construction. The shell of an exchanger is a pressure vessel and will be designed in accordance with the appropriate national pressure vessel code. In the United Kingdom this will be BS 5500, which is discussed in Chapter 13. The dimensions of standard flanges for use with heat exchangers are given in BS 3274, and in the TEMA standards.

In both the American and British standards dimensions are given in feet and inches, so these units have been used in this chapter with the equivalent values in SI units given in brackets.

12.5.2. Tubes

Dimensions

Tube diameters in the range $\frac{5}{8}$ in. (16 mm) to 2 in. (50 mm) are used. The smaller diameters $\frac{5}{8}$ to 1 in. (16 to 25 mm) are preferred for most duties, as they will give more compact, and therefore cheaper, exchangers. Larger tubes are easier to clean by mechanical methods and would be selected for heavily fouling fluids.

The tube thickness (gauge) is selected to withstand the internal pressure and give an adequate corrosion allowance. Steel tubes for heat exchangers are covered by BS 3606 (metric sizes); the standards applicable to other materials are given in BS 3274. Standard diameters and wall thicknesses for steel tubes are given in Table 12.3.

The preferred lengths of tubes for heat exchangers are: 6 ft. (1·83 m), 8 ft (2·44 m), 12 ft

TABLE 12.3. *Standard dimensions for steel tubes*

Outside diameter (mm)	Wall thickness (mm)				
16	1·2	1·6	2·0	—	—
20	—	1·6	2·0	2·6	—
25	—	1·6	2·0	2·6	3·2
30	—	1·6	2·0	2·6	3·2
38	—	—	2·0	2·6	3·2
50	—	—	2·0	2·6	3·2

(3·66 m), and 16 ft (4·88 m). For a given surface area, the use of long tubes will reduce the shell diameter; which will generally result in a lower cost exchanger, particularly for high shell pressures.

The tube size is often determined by the plant maintenance department standards, as clearly it is an advantage to reduce the number of sizes that have to be held in stores for tube replacement.

As a guide, $\frac{3}{4}$ in. (19 mm) is a good trial diameter with which to start design calculations.

Tube arrangements

The tubes in an exchanger are usually arranged in an equilateral triangular, square, or rotated square pattern; see Fig. 12.9.

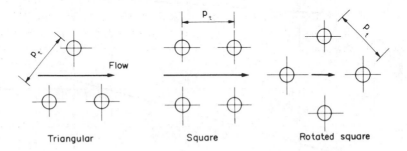

Fig. 12.9. Tube patterns

The triangular and rotated square patterns give higher heat-transfer rates, but at the expense of a higher pressure drop than the square pattern. A square pattern is used where mechanical cleaning of the outside of the tubes is necessary. The recommended tube pitch (distance between tube centres) is 1·25 times the tube outside diameter; and this will normally be used unless process requirements dictate otherwise. Where a square pattern is used for ease of cleaning, the recommended minimum clearance between the tubes is 0·25 in. (6·4 mm).

Tube-side passes

The fluid in the tube is usually directed to flow back and forth in a number of "passes" through groups of tubes arranged in parallel, to increase the length of the flow path. The number of passes is selected to give the required tube-side design velocity. Exchangers are built with from one to up to about sixteen tube passes. The tubes are arranged into the number of passes required by dividing up the exchanger headers (channels) with partition plates (pass partitions).

12.5.3. Shells

The British standard BS 3274 covers exchangers from 6 in. (150 mm) to 42 in. (1067 mm) diameter; and the TEMA standards, exchangers up to 60 in. (1520 mm).

Up to about 24 in. (610 mm) shells are normally constructed from standard, close tolerance, pipe; above 24 in. (610 mm) they are rolled from plate. The minimum shell thickness for various materials is given in BS 3274.

The shell diameter must be selected to give as close a fit to the tube bundle as is practical; to reduce bypassing round the outside of the bundle; see Section 12.9. The clearance required between the outermost tubes in the bundle and the shell inside diameter will depend on the type of exchanger and the manufacturing tolerances; typical values are given in Fig. 12.10.

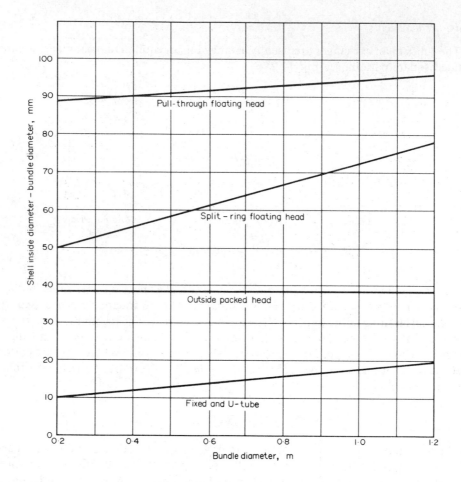

FIG. 12.10. Shell-bundle clearance

12.5.4. Tube-sheet layout (tube count)

The bundle diameter will depend not only on the number of tubes but also on the number of tube passes, as spaces must be left in the pattern of tubes on the tube sheet to accommodate the pass partition plates.

An estimate of the bundle diameter D_b can be obtained from equation 12.3b, which is an empirical equation based on standard tube layouts. The constants for use in this equation, for triangular and square patterns, are given in Table 12.4.

$$N_t = K_1(D_b/d_o)^{n_1},$$ (12.3a)

$$D_b = d_o(N_t/K_1)^{1/n_1},$$ (12.3b)

where N_t = number of tubes,
D_b = bundle diameter, mm,
d_o = tube outside diameter, mm.

TABLE 12.4. *Constants for use in equation 12.3*

Triangular pitch, $p_t = 1.25d_o$

No. passes	1	2	4	6	8
K_1	0·319	0·249	0·175	0·0743	0·0365
n_1	2·142	2·207	2·285	2·499	2·675

Square pitch, $p_t = 1.25d_o$

No. passes	1	2	4	6	8
K_1	0·215	0·156	0·158	0·0402	0·0331
n_1	2·207	2·291	2·263	2·617	2·643

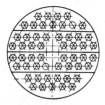 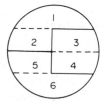

Six tube passes

Four passes

Two passes

FIG. 12.11. Tube arrangements, showing pass-partitions in headers

If U-tubes are used the number of tubes will be slightly less than that given by equation 12.3a, as the spacing between the two centre rows will be determined by the minimum allowable radius for the U-bend. Use of too small a radius will cause too great a thinning of the outer tube wall at the bend. The minimum radius of the bend is normally taken as 2·5 times the tube outside diameter. An estimate of the number of tubes in a U-tube exchanger (twice the actual number of U-tubes), can be made by reducing the number given by equation 12.3a by one centre row of tubes. The number of tubes in the centre row, the row at the shell equator, is given by:

$$\text{Tubes in centre row} = D_b/p_t$$

where p_t = tube pitch.

A tube layout drawing would normally be made before fabrication. A few tubes will have to be omitted from the regular pattern given by the pitch arrangement to accommodate the tie rods; and one or two rows may also be omitted at the top and bottom of the bundle to increase the clearance and flow area at the inlet and outlet nozzles. Typical tube arrangements are shown in Fig. 12.11.

Tube count tables, which give the number of tubes that can be accommodated in standard shell sizes for the commonly used tube sizes, pitches and number of passes, for the various types of exchanger, are available in several books; Perry and Chilton (1973), Kern (1950) and Ludwig (1965).

12.5.5. Shell types (passes)

The principal shell arrangements used are shown in Fig. 12.12. The single shell pass type is by far the most commonly used. Two shell passes are occasionally used where the shell and tube side temperature differences (temperature driving force) will be unfavourable in a single pass; though the same flow arrangement can be obtained by using two, or more, exchanger shells in series.

The divided flow and split-flow types are used to reduce the shell-side pressure drop; where pressure drop, rather than heat transfer, is the controlling factor in the design.

12.5.6. Baffles

Baffles are used in the shell to direct the fluid stream across the tubes, to increase the fluid velocity and so improve the rate of transfer. The most commonly used type of baffle is the single segmental baffle shown in Fig. 12.13a, other types are shown in Figs 12.13b, c and d.

Only the design of exchangers using single segmental baffles will be considered in this chapter.

If the arrangement shown in Fig. 12.13a were used with a horizontal condenser the baffles would restrict the condensate flow. This problem can be overcome either by rotating the baffle arrangement through 90°, or by trimming the base of the baffle, Fig. 12.14.

The term "baffle cut" is used to specify the dimensions of a segmental baffle. The baffle cut is the height of the segment removed to form the baffle, expressed as a percentage of the baffle disc diameter. Baffle cuts from 15 to 45 per cent are used. Generally, a baffle cut

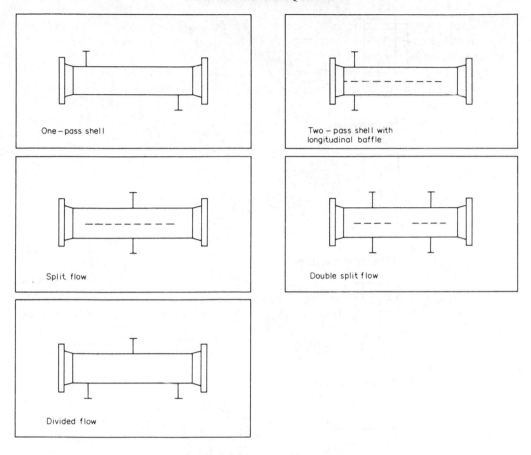

FIG. 12.12. Shell types (pass arrangements)

of 20 to 25 per cent will be the optimum, giving good heat-transfer rates, without excessive drop. There will be some leakage of fluid round the baffle as a clearance must be allowed for assembly. The clearance needed will depend on the shell diameter; typical values, and tolerances, are given in Table 12.5.

Another leakage path occurs through the clearance between the tube holes in the baffle and the tubes. The maximum design clearance will normally be $\frac{1}{32}$ in. (0·8 mm).

The minimum thickness to be used for baffles and support plates are given in the standards. The baffle spacings used range from 0·2 to 1·0 shell diameters. A close baffle spacing will give higher heat transfer coefficients but at the expense of higher pressure drop. The optimum spacing will usually be between 0·3 to 0·5 times the shell diameter.

12.5.7. Support plates and tie rods

Where segmental baffles are used some will be fabricated with closer tolerances, $\frac{1}{64}$ in. (0·4 mm), to act as support plates. For condensers and vaporisers, where baffles are not needed for heat-transfer purposes, a few will be installed to support the tubes.

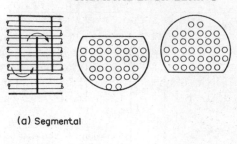

(a) Segmental

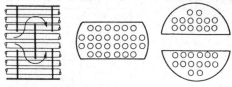

(b) Segmental and strip

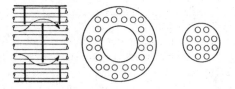

(c) Disc and doughnut

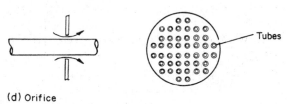

Tubes

(d) Orifice

FIG. 12.13. Types of baffle used in shell and tube heat exchangers

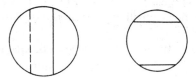

FIG. 12.14.   Baffles for condensers

TABLE 12.5. *Typical baffle clearances and tolerances*

Shell diameter, D_s	Baffle diameter	Tolerance
Pipe shells		
6 to 25 in. (152 to 635 mm)	$D_s - \frac{1}{16}$ in. (1·6 mm)	$+\frac{1}{32}$ in. (0·8 mm)
Plate shells		
6 to 25 in. (152 to 635 mm)	$D_s - \frac{1}{8}$ in. (3·2 mm)	$+0, -\frac{1}{32}$ in. (0·8 mm)
27 to 42 in. (686 to 1067 mm)	$D_s - \frac{3}{16}$ in. (4·8 mm)	$+0, -\frac{1}{16}$ in. (1·6 mm)

The minimum spacings to be used for support plates are given in the standards. The spacing ranges from around 1 m for 16 mm tubes to 2 m for 25 mm tubes.

The baffles and support plate are held together with tie rods and spacers. The number of rods required will depend on the shell diameter, and will range from 4, 16 mm diameter rods, for exchangers under 380 mm diameter; to 8, 12·5 mm rods, for exchangers of 1 m diameter. The recommended number for a particular diameter can be found in the standards.

12.5.8. *Tube sheets* (*plates*)

In operation the tube sheets are subjected to the differential pressure between shell and tube sides. The design of tube sheets as pressure-vessel components is covered by BS 5500 and is discussed in Chapter 13. Design formulae for calculating tube sheet thicknesses are also given in the TEMA standards.

The joint between the tubes and tube sheet is normally made by expanding the tube by rolling with special tools. Tube rolling is a skilled task; the tube must be expanded sufficiently to ensure a sound leak-proof joint, but not overthinned, weakening the tube. The tube holes are normally grooved, Fig. 12.15a, to lock the tubes more firmly in position and to prevent the joint from being loosened by the differential expansion of the shell and tubes. When it is essential to guarantee a leak-proof joint the tubes can be welded to the sheet, Fig. 12.15b. This will add to the cost of the exchanger; not only due to the cost of welding, but also because a wider tube spacing will be needed.

The tube sheet forms the barrier between the shell and tube fluids, and where it is essential for safety or process reasons to prevent any possibility of intermixing due to leakage at the tube sheet joint, double tube-sheets can be used, with the space between the sheets vented; Fig. 12.16.

To allow sufficient thickness to seal the tubes the tube sheet thickness should not be less than the tube outside diameter, up to about 25 mm diameter. Recommended minimum plate thicknesses are given in the standards.

The thickness of the tube sheet will reduce the effective length of the tube slightly, and this should be allowed for when calculating the area available for heat transfer. As a first approximation the length of the tubes can be reduced by 25 mm for each tube sheet.

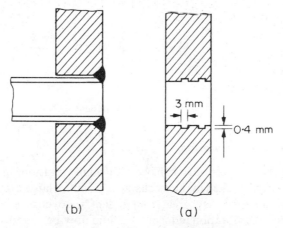

(b) (a)

FIG. 12.15. Tube/tube sheet joints

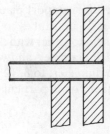

FIG. 12.16. Double-tube sheet

12.5.9. Shell and header nozzles (branches)

Standard pipe sizes will be used for the inlet and outlet nozzles. It is important to avoid flow restrictions at the inlet and outlet nozzles to prevent excessive pressure drop and flow-induced vibration of the tubes. As well as omitting some tube rows (see Section 12.5.4), the baffle spacing is usually increased in the nozzle zone, to increase the flow area. For vapours and gases, where the inlet velocities will be high, the nozzle may be flared, or special designs used, to reduce the inlet velocities; Fig. 12.17a and b. The extended shell design shown in Fig. 12.17b also serves as an impingement plate. Impingement plates are used where the shell-side fluid contains liquid drops, or for high-velocity fluids containing abrasive particles.

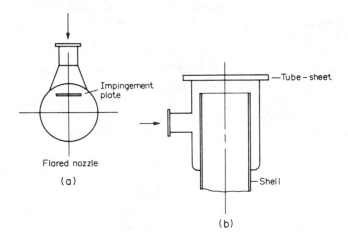

FIG. 12.17. Inlet nozzle designs

12.5.10. Flow-induced tube vibrations

Premature failure of exchanger tubes can occur through vibrations induced by the shell-side fluid flow. Care must be taken in the mechanical design of large exchangers and where high shell velocities are used, say greater than 3 m/s, to ensure that the tubes are adequately supported. The vibration induced by a fluid flowing round a tube is associated with vortex shedding from the down-stream side of the tube. As the vortex is shed the flow pattern and pressure distribution round the tube are changed causing oscillations in the magnitude and direction of the pressure forces acting on the tube. If the frequency of the vortex shedding approaches the natural frequency of the tube a resonance effect can occur and vibrations of considerable magnitude develop, leading to tube failure. Much work has been done on this topic in recent years, but detailed discussion of this is beyond the scope of this book. A critical review of the literature and research is given by Moretti (1973).

Thorngren (1970) gives a method for calculating a dimensionless "damage" number, which is a function of the tube dimensions and material properties, and can be used to predict if damage is likely to occur at a particular shell-side flow velocity. Lord et al. (1970) give a method for predicting the critical shell-side velocity at which flow-induced vibration is likely to occur with a particular support spacing, tube arrangement and dimensions, and tube material.

12.6. Mean temperature difference (temperature driving force)

Before equation 12.1 can be used to determine the heat transfer area required for a given duty, an estimate of the mean temperature difference ΔT_m must be made. This will normally be calculated from the terminal temperature differences: the difference in the fluid temperatures at the inlet and outlet of the exchanger. The well-known "logarithmic mean" temperature difference (see Volume 1, Chapter 7) is only applicable to sensible heat transfer in true co-current or counter-current flow (linear temperature–enthalpy curves). For counter-current flow, Fig. 12.18a, the logarithmic mean temperature is given by:

$$\Delta T_{lm} = \frac{(T_1 - t_2) - (T_2 - t_1)}{\ln \dfrac{(T_1 - t_2)}{(T_2 - t_1)}} \tag{12.4}$$

where ΔT_{lm} = log mean temperature difference,
T_1 = inlet shell-side fluid temperature,
T_2 = outlet shell-side fluid temperature,
t_1 = inlet tube-side temperature,
t_2 = outlet tube-side temperature.

The equation is the same for co-current flow, but the terminal temperature differences will be $(T_1 - t_1)$ and $(T_2 - t_2)$. Strictly, equation 12.4 will only apply when there is no change in the specific heats, the overall heat-transfer coefficient is constant, and there are no heat losses. In design, these conditions can be assumed to be satisfied providing the temperature change in each fluid stream is not large.

In most shell and tube exchangers the flow will be a mixture of co-current, counter-current and cross flow. Figures 12.18b and c show typical temperature profiles for an exchanger with one shell pass and two tube passes (a 1:2 exchanger). Figure 12.18c shows a temperature cross, where the outlet temperature of the cold stream is above that of the hot stream.

The usual practice in the design of shell and tube exchangers is to estimate the "true temperature difference" from the logarithmic mean temperature by applying a correction factor to allow for the departure from true counter-current flow:

$$\Delta T_m = F_t \Delta T_{lm} \tag{12.5}$$

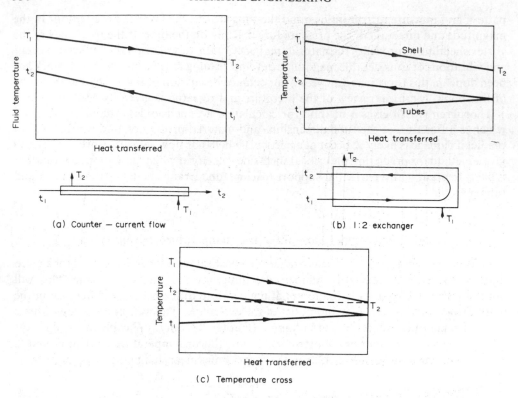

(a) Counter — current flow   (b) 1:2 exchanger

(c) Temperature cross

FIG. 12.18. Temperature profiles

where ΔT_m = true temperature difference, the mean temperature difference for use in the design equation 12.1,

F_t = the temperature correction factor.

The correction factor is a function of the shell and tube fluid temperatures, and the number of tube and shell passes. It is normally correlated as a function of two dimensionless temperature ratios:

$$R = (T_1 - T_2)/(t_2 - t_1) \qquad (12.6)$$

and

$$S = (t_2 - t_1)/(T_1 - t_1) \qquad (12.7)$$

R is equal to the shell-side fluid flow-rate times the fluid mean specific heat; divided by the tube-side fluid flow-rate times the tube-side fluid specific heat.

S is a measure of the temperature efficiency of the exchanger.

For a 1 shell:2 tube pass exchanger, the correction factor is given by:

$$F_t = \frac{\sqrt{(R^2+1)}\ln\left[(1-S)/(1-RS)\right]}{(R-1)\ln\left[\dfrac{2-S[R+1-\sqrt{(R^2+1)}]}{2-S[R+1+\sqrt{(R^2+1)}]}\right]} \qquad (12.8)$$

The derivation of equation 12.8 is given by Kern (1950). The equation for a 1 shell:2 tube

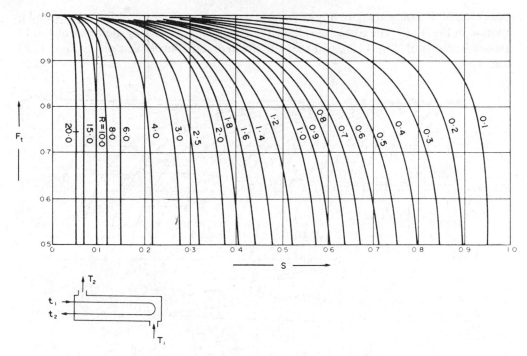

FIG. 12.19. Temperature correction factor: one shell pass; two or more even tube passes

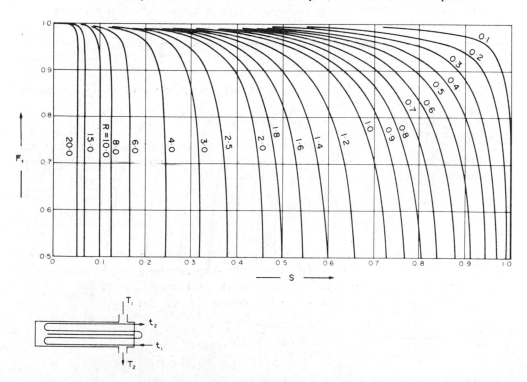

FIG. 12.20. Temperature correction factor: two shell passes; four or multiples of four tube passes

pass exchanger can be used for any exchanger with an even number of tube passes, and is plotted in Fig. 12.19. The correction factor for 2 shell passes and 4, or multiples of 4, tube passes is shown in Fig. 12.20, and that for divided and split flow shells in Figs. 12.21 and 12.22.

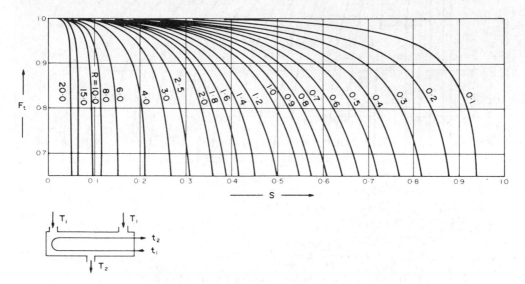

FIG. 12.21. Temperature correction factor: divided-flow shell; two or more even-tube passes

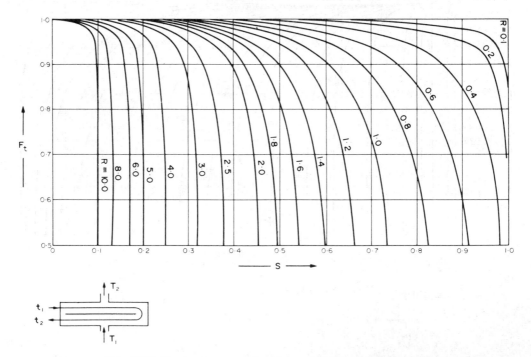

FIG. 12.22. Temperature correction factor, split flow shell, 2 tube pass

Temperature correction factor plots for other arrangements can be found in the TEMA standards and the books by Kern (1950) and Ludwig (1965). Mueller (1973) gives a comprehensive set of figures for calculating the log mean temperature correction factor, which includes figures for cross-flow exchangers.

The following assumptions are made in the derivation of the temperature correction factor F_t in addition to those made for the calculation of the log mean temperature difference:

1. Equal heat transfer areas in each pass.
2. A constant overall heat-transfer coefficient in each pass.
3. The temperature of the shell-side fluid in any pass is constant across any cross-section.
4. There is no leakage of fluid between shell passes.

Though these conditions will not be strictly satisfied in practical heat exchangers, the F_t values obtained from the curves will give an estimate of the "true mean temperature difference" that is sufficiently accurate for most designs. Mueller (1973) discusses these assumptions, and gives F_t curves for conditions when all the assumptions are not met; see also Butterworth (1973) and Emerson (1973).

The shell-side leakage and bypass streams (see Section 12.9) will affect the mean temperature difference, but are not normally taken into account when estimating the correction factor F_t. Fisher and Parker (1969) give curves which show the effect of leakage on the correction factor for a 1 shell pass:2 tube pass exchanger.

The value of F_t will be close to one when the terminal temperature differences are large, but will appreciably reduce the logarithmic mean temperature difference when the temperatures of shell and tube fluids approach each other; it will fall drastically when there is a temperature cross.

Where the F_t curve is near vertical values cannot be read accurately, and this will introduce a considerable uncertainty into the design.

An economic exchanger design cannot normally be achieved if the correction factor F_t falls below about 0·75. In these circumstances an alternative type of exchanger should be considered which give a closer approach to true counter-current flow. The use of two or more shells in series, or multiple shell-side passes, will give a closer approach to true counter-current flow, and should be considered where a temperature cross is likely to occur.

Where both sensible and latent heat is transferred, it will be necessary to divide the temperature profile into sections and calculate the mean temperature difference for each section.

12.7. Shell and tube exchangers: general design considerations

12.7.1. Fluid allocation: shell or tubes

Where no phase change occurs, the following factors will determine the allocation of the fluid streams to the shell or tubes.

Corrosion. The more corrosive fluid should be allocated to the tube-side. This will reduce the cost of expensive alloy or clad components.

Fouling. The fluid that has the greatest tendency to foul the heat-transfer surfaces should be placed in the tubes. This will give better control over the design fluid velocity, and the higher allowable velocity in the tubes will reduce fouling. Also, the tubes will be easier to clean.

Fluid temperatures. If the temperatures are high enough to require the use of special alloys placing the higher temperature fluid in the tubes will reduce the overall cost. At moderate temperatures, placing the hotter fluid in the tubes will reduce the shell surface temperatures, and hence the need for lagging to reduce heat loss, or for safety reasons.

Operating pressures. The higher pressure stream should be allocated to the tube-side. High-pressure tubes will be cheaper than a high-pressure shell.

Pressure drop. For the same pressure drop, higher heat-transfer coefficients will be obtained on the tube-side than the shell-side, and fluid with the lowest allowable pressure drop should be allocated to the tube-side.

Viscosity. Generally, a higher heat-transfer coefficient will be obtained by allocating the more viscous material to the shell-side, providing the flow is turbulent. The critical Reynolds number for turbulent flow in the shell is in the region of 200. If turbulent flow cannot be achieved in the shell it is better to place the fluid in the tubes, as the tube-side heat-transfer coefficient can be predicted with more certainty.

Stream flow-rates. Allocating the fluids with the lowest flow-rate to the shell-side will normally give the most economical design.

12.7.2. Shell and tube fluid velocities

High velocities will give high heat-transfer coefficients but also a high-pressure drop. The velocity must be high enough to prevent any suspended solids settling, but not so high as to cause erosion. High velocities will reduce fouling. Plastics inserts are sometimes used to reduce erosion at the tube inlet. Typical design velocities are given below:

Liquids

Tube-side, process fluids: 1 to 2 m/s, maximum 4 m/s if required to reduce fouling; water: 1·5 to 2·5 m/s.
Shell-side: 0·3 to 1 m/s.

Vapours

For vapours, the velocity used will depend on the operating pressure and fluid density; the lower values in the ranges given below will apply to high molecular weight materials.

Vacuum	50 to 70 m/s
Atmospheric pressure	10 to 30 m/s
High pressure	5 to 10 m/s

12.7.3. Stream temperatures

The closer the temperature approach used (the difference between the outlet temperature of one stream and the inlet temperature of the other stream) the larger will be the heat-transfer area required for a given duty. The optimum value will depend on the application, and can only be determined by making an economic analysis of alternative designs. As a general guide the greater temperature difference should be at least 20°C and the least temperature difference 5 to 7°C for coolers using cooling water, and 3 to 5°C using refrigerated brines. The maximum temperature rise in recirculated cooling water is limited to around 30°C. Care should be taken to ensure that cooling media temperatures are kept well above the freezing point of the process materials. When the heat exchange is between process fluids for heat recovery the optimum approach temperatures will normally not be lower than 20°C.

12.7.4. Pressure drop

In many applications the pressure drop available to drive the fluids through the exchanger will be set by the process conditions, and the available pressure drop will vary from a few millibars in vacuum service to several bars in pressure systems.

When the designer is free to select the pressure drop an economic analysis can be made to determine the exchanger design which gives the lowest operating costs, taking into consideration both capital and pumping costs. However, a full economic analysis will only be justified for very large, expensive, exchangers. The values suggested below can be used as a general guide, and will normally give designs that are near the optimum.

Liquids

$$\text{Viscosity} < 1 \text{ mN s/m}^2 \quad 35 \text{ kN/m}^2$$
$$1 \text{ to } 10 \text{ mN s/m}^2 \quad 50\text{–}70 \text{ kN/m}^2$$

Gas and vapours

High vacuum	0·4–0·8 kN/m^2
Medium vacuum	0·1 × absolute pressure
1 to 2 bar	0·5 × system gauge pressure
Above 10 bar	0·1 × system gauge pressure

When a high-pressure drop is utilised, care must be taken to ensure that the resulting high fluid velocity does not cause erosion or flow-induced tube vibration.

12.7.5. Fluid physical properties

The fluid physical properties required for heat-exchanger design are: density, viscosity, thermal conductivity and temperature-enthalpy correlations (specific and latent heats). Sources of physical property data are given in Chapter 8. The thermal conductivities of commonly used tube materials are given in Table 12.6.

In the correlations used to predict heat-transfer coefficients, the physical properties are usually evaluated at the mean stream temperature. This is satisfactory when the temperature change is small, but can cause a significant error when the change in temperature is large. In these circumstances, a simple, and safe, procedure is to evaluate the heat-transfer coefficients at the stream inlet and outlet temperatures and use the lowest of the two values. Alternatively, the method suggested by Frank (1978) can be used; in which equations 12.1 and 12.3 are combined:

$$Q = \frac{A[U_2(T_1 - t_2) - U_1(T_2 - t_1)]}{\ln\left[\dfrac{U_2(T_1 - t_2)}{U_1(T_2 - t_1)}\right]} \qquad (12.9)$$

where U_1 and U_2 are evaluated the ends of the exchanger. Equation 12.9 is derived by assuming that the heat-transfer coefficient varies linearly with temperature.

If the variation in the physical properties is too large for these simple methods to be used it will be necessary to divide the temperature–enthalpy profile into sections and evaluate the heat-transfer coefficients and area required for each section.

TABLE 12.6. *Conductivity of metals*

Metal	Temperature (°C)	k_w (W/m °C)
Aluminium	0	202
	100	206
Brass	0	97
(70 Cu, 30 Zn)	100	104
	400	116
Copper	0	388
	100	378.
Nickel	0	62
	212	59
Cupro-nickel (10 per cent Ni)	0–100	45
Monel	0–100	30
Stainless steel (18/8)	0–100	16
Steel	0	45
	100	45
	600	36
Titanium	0–100	16

12.8. Tube-side heat-transfer coefficient and pressure drop (single phase)

12.8.1. Heat transfer

Turbulent flow

Heat-transfer data for turbulent flow inside conduits of uniform cross-section are usually correlated by an equation of the form:

$$Nu = C\, Re^a\, Pr^b\, (\mu/\mu_w)^c \qquad (12.10)$$

where Nu = Nusselt number = $\dfrac{h_i d_e}{k_f}$,

Re = Reynolds number = $\dfrac{\rho u_t d_e}{\mu} = \dfrac{G_t d_e}{\mu}$,

Pr = Prandtl number = $\dfrac{C_p \mu}{k_f}$

and: h_i = inside coefficient, $W/m^2 \, °C$,

$\quad d_e$ = equivalent (or hydraulic) diameter, m

$$d_e = \frac{4 \times \text{cross-sectional area for flow}}{\text{wetted perimeter}}$$

$\quad\quad = d_i$ for tubes,

$\quad u_t$ = fluid velocity, m/s,

$\quad k_f$ = fluid thermal conductivity, $W/m°C$,

$\quad G_t$ = mass velocity, mass flow per unit area, $kg/m^2 s$,

$\quad \mu$ = fluid viscosity at the bulk fluid temperature, $N \, s/m^2$,

$\quad \mu_w$ = fluid viscosity at the wall,

$\quad C_p$ = fluid specific heat, heat capacity, $J/kg°C$.

The index for the Reynolds number is generally taken as 0·8. That for the Prandtl number can range from 0·3 for cooling to 0·4 for heating. The index for the viscosity factor is normally taken as 0·14 for flow in tubes, from the work of Sieder and Tate (1936), but some workers report higher values. A general equation that can be used for exchanger design is:

$$Nu = C \, Re^{0.8} \, Pr^{0.33} (\mu/\mu_w)^{0.14} \qquad (12.11)$$

where C = 0·021 for gases,

$\quad\quad$ = 0·023 for non-viscous liquids,

$\quad\quad$ = 0·027 for viscous liquids.

It is not really possible to find values for the constant and indexes to cover the complete range of process fluids, from gases to viscous liquids, but the values predicted using equation 12.11 should be sufficiently accurate for design purposes. The uncertainty in the prediction of the shell-side coefficient and fouling factors will usually far outweigh any error in the tube-side value. Where a more accurate prediction than that given by equation 12.11 is required, and justified, the data and correlations given in the Engineering Sciences Data Unit report are recommended: Nos. 67016 (1967), 68006 (1968) and 68007 (1968).

$\quad$ Butterworth (1977) gives the following equation, which is based on the ESDU work:

$$St = E \, Re^{-0.205} \, Pr^{-0.505} \qquad (12.12)$$

where St = Stanton number $= \dfrac{Nu}{Re \, Pr} = \dfrac{h_i}{\rho u_t C_p}$

and $\quad E = 0.0225 \exp(-0.0225 (\ln Pr)^2)$.

Equation 12.12 is applicable at Reynolds numbers greater than 10,000.

Laminar flow

$\quad$ Below a Reynolds number of about 2000 the flow in pipes will be laminar. Providing the natural convection effects are small, which will normally be so in forced convection, the following equation can be used to estimate the film heat-transfer coefficient:

$$Nu = 1.86 (Re \, Pr)^{0.33} (d_e/L)^{0.33} (\mu/\mu_w)^{0.14} \qquad (12.13)$$

Where L is the length of the tube, metres.

If the Nusselt number given by equation 12.13 is less than 3·5, it should be taken as 3·5.

In laminar flow the length of the tube can have a marked effect on the heat-transfer rate for length to diameter ratios less than 500.

Transition region

In the flow region between laminar and fully developed turbulent flow heat-transfer coefficients cannot be predicted with certainty, as the flow in this region is unstable, and the transition region should be avoided in exchanger design. If this is not practicable the coefficient should be evaluated using both equations 12.11 and 12.13 and the least value taken.

Heat-transfer factor, j_h

It is often convenient to correlate heat-transfer data in terms of a heat transfer "j" factor, which is similar to the friction factor used for pressure drop (see Volume 1, Chapters 3 and 7). The heat-transfer factor is defined by:

$$j_h = St\, Pr^{0·67}(\mu/\mu_w)^{-0·14} \tag{12.14}$$

The use of the j_h factor enables data for laminar and turbulent flow to be represented on the same graph; Fig. 12.23. The j_h values obtained from Fig. 12.23 can be used with equation 12.14 to estimate the heat-transfer coefficients for heat-exchanger tubes and commercial pipes. The coefficient estimated for pipes will normally be conservative (on the high side) as pipes are rougher than the tubes used for heat exchangers, which are finished to closer tolerances. Equation 12.14 can be rearranged to a more convenient form:

$$\frac{h_i d_i}{k_f} = j_h\, Re\, Pr^{0·33}\left(\frac{\mu}{\mu_w}\right)^{0·14} \tag{12.15}$$

Note. Kern (1950), and other workers, define the heat transfer factor as:

$$j_H = Nu\, Pr^{-1/3}(\mu/\mu_w)^{-0·14}$$

The relationship between j_h and j_H is given by:

$$j_H = j_h\, Re$$

Viscosity correction factor

The viscosity correction factor will normally only be significant for viscous liquids.

To apply the correction an estimate of the wall temperature is needed. This can be made by first calculating the coefficient without the correction and using the following relationship to estimate the wall temperature:

$$h_i(t_w - t) = U(T - t) \tag{12.16}$$

where t = tube-side bulk temperature (mean),

t_w = estimated wall temperature,

T = shell-side bulk temperature (mean).

Usually an approximate estimate of the wall temperature is sufficient, but trial-and-error calculations can be made to obtain a better estimate if the correction is large.

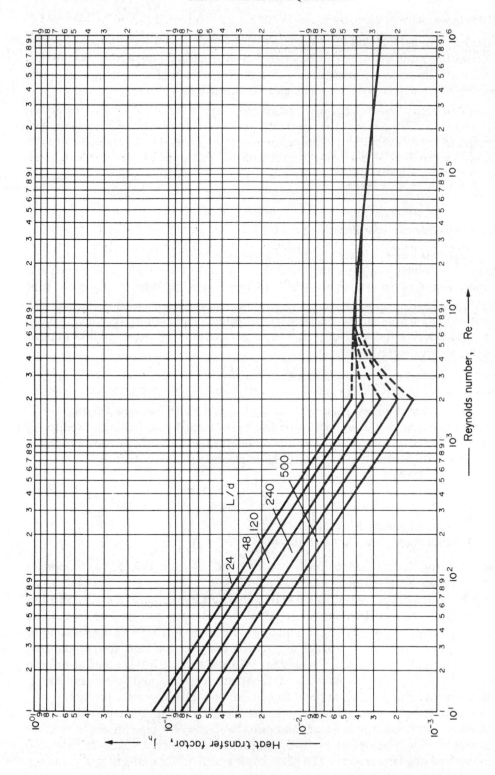

FIG. 12.23. Tube-side heat-transfer factor

Coefficients for water

Though equations 12.11 and 12.13 and Fig. 12.23 may be used for water, a more accurate estimate can be made by using equations developed specifically for water. The physical properties are conveniently incorporated into the correlation. The equation below has been adapted from data given by Eagle and Ferguson (1930):

$$h_i = 4200(1 \cdot 35 + 0 \cdot 02t)u_t^{0 \cdot 8}/d_i^{0 \cdot 2} \qquad (12.17)$$

where h_i = inside coefficient, for water, $W/m^2 \, °C$,
    t = water temperature, $°C$,
    u_t = water velocity, m/s,
    d_i = tube inside diameter, mm.

12.8.2. Tube-side pressure drop

There are two major sources of pressure loss on the tube-side of a shell and tube exchanger: the friction loss in the tubes and the losses due to the sudden contraction and expansion and flow reversals that the fluid experiences in flow through the tube arrangement.

The tube friction loss can be calculated using the familiar equations for pressure-drop loss in pipes (see Volume 1, Chapter 3). The basic equation for isothermal flow in pipes (constant temperature) is:

$$\Delta P = 8j_f (L'/d_i) \frac{\rho u_t^2}{2} \qquad (12.18)$$

where j_f is the dimensionless friction factor and L' is the effective pipe length.

The flow in a heat exchanger will clearly not be isothermal, and this is allowed for by including an empirical correction factor to account for the change in physical properties with temperature. Normally only the change in viscosity is considered:

$$\Delta P = 8j_f (L'd_i) \rho \frac{u_t^2}{2} (\mu/\mu_w)^{-m} \qquad (12.19)$$

$m = 0 \cdot 25$ for laminar flow, $Re < 2100$,
  $= 0 \cdot 14$ for turbulent flow, $Re > 2100$.

Values of j_f for heat exchanger tubes can be obtained from Fig. 12.24. Values for commercial pipes are given in Volume 2, Chapter 3.

The pressure losses due to contraction at the tube inlets, expansion at the exits, and flow reversal in the headers, can be a significant part of the total tube-side pressure drop. There is no entirely satisfactory method for estimating these losses. Kern (1950) suggests adding four velocity heads per pass. Frank (1978) considers this to be too high, and recommends 2·5 velocity heads. Butterworth (1978) suggests 1·8. Lord et al. (1970) take the loss per pass as equivalent to a length of tube equal to 300 tube diameters for straight tubes, and 200 for U-tubes; whereas Evans (1980) appears to add only 67 tube diameters per pass.

The loss in terms of velocity heads can be estimated by counting the number of flow contractions, expansions and reversals, and using the factors for pipe fittings to estimate the number of velocity heads lost. For two tube passes, there will be two contractions, two expansions and one flow reversal. The head loss for each of these effects (see Volume 1,

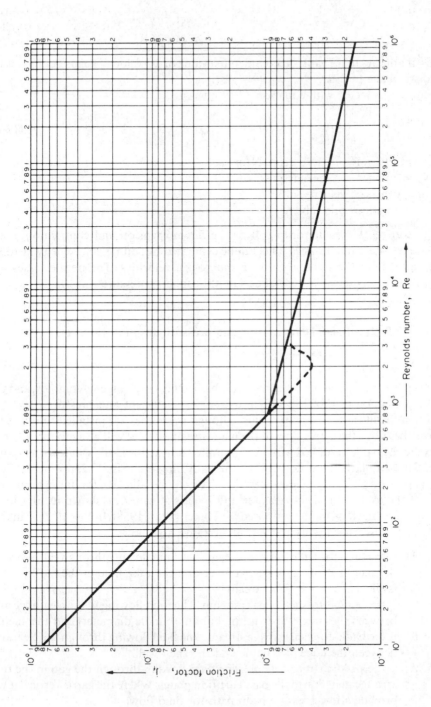

Fig. 12.24. Tube-side friction factors

Note: The friction factor j_f is the same as the friction factor for pipes $\phi \left(= \dfrac{R}{\rho u^2} \right)$, defined in Volume 1 Chapter 3.

Chapter 3) is: contraction 0·5, expansion 1·0, 180° bend 1·5; so for two passes the maximum loss will be

$$2 \times 0·5 + 2 \times 1·0 + 1·5 = 4·5 \text{ velocity heads}$$

$$= \underline{2·25 \text{ per pass}}$$

From this, it appears that Frank's recommended value of 2·5 velocity heads per pass is the most realistic value to use.

Combining this factor with equation 12.19 gives

$$\Delta P_t = N_p [8 j_f (L/d_i)(\mu/\mu_w)^{-m} + 2·5] \frac{\rho u_t^2}{2} \qquad (12.20)$$

where ΔP_t = tube-side pressure drop, N/m^2 (Pa),

N_p = number of tube-side passes,

u_t = tube-side velocity, m/s,

L = length of one tube.

Another source of pressure drop will be the flow expansion and contraction at the exchanger inlet and outlet nozzles. This can be estimated by adding one velocity head for the inlet and 0·5 for the outlet, based on the nozzle velocities. The nozzle losses will normally only be significant for gases at sub-atmospheric pressure.

12.9. Shell-side heat-transfer and pressure drop (single phase)

12.9.1. Flow pattern

The flow pattern in the shell of a segmentally baffled heat exchanger is complex, and this makes the prediction of the shell-side heat-transfer coefficient and pressure drop very much more difficult than for the tube-side. Though the baffles are installed to direct the flow across the tubes, the actual flow of the main stream of fluid will be a mixture of cross flow between the baffles, coupled with axial (parallel) flow in the baffle windows; as shown in Fig. 12.25. Not all the fluid flow follow the path shown in Fig. 12.25; some will leak through gaps formed by the clearances that have to be allowed for fabrication and assembly of the exchanger. These leakage and bypass streams are shown in Fig. 12.26, which is based on the flow model proposed by Tinker (1951, 1958). In Fig. 12.26, Tinker's nomenclature is used to identify the various streams, as follows:

Stream A   is the tube-to-baffle leakage stream. The fluid flowing through the clearance between the tube outside diameter and the tube hole in the baffle.

Stream B   is the actual cross-flow stream.

Stream C   is the bundle-to-shell bypass stream. The fluid flowing in the clearance area between the outer tubes in the bundle (bundle diameter) and the shell.

Stream E   is the baffle-to-shell leakage stream. The fluid flowing through the clearance between the edge of a baffle and the shell wall.

Stream F   is the pass-partition stream. The fluid flowing through the gap in the tube arrangement due to the pass partition plates. Where the gap is vertical it will provide a low-pressure drop path for fluid flow.

Note. There is no stream D.

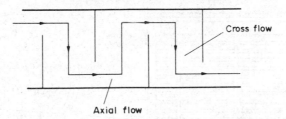

FIG. 12.25. Idealised main stream flow

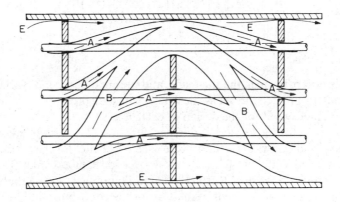

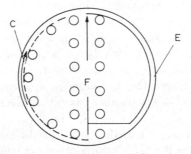

FIG. 12.26. Shell-side leakage and by-pass paths

The fluid in streams C, E and F bypasses the tubes, which reduces the effective heat-transfer area.

Stream C is the main bypass stream and will be particularly significant in pull-through bundle exchangers, where the clearance between the shell and bundle is of necessity large. Stream C can be considerably reduced by using sealing strips; horizontal strips that block the gap between the bundle and the shell, Fig. 12.27. Dummy tubes are also sometimes used to block the pass-partition leakage stream F.

The tube-to-baffle leakage stream A does not bypass the tubes, and its main effect is on pressure drop rather than heat transfer.

The clearances will tend to plug as the exchanger becomes fouled and this will increase the pressure drop; see Section 12.9.6.

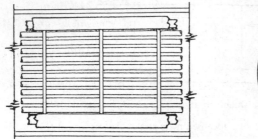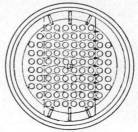

FIG. 12.27. Sealing strips

12.9.2. Design methods

The complex flow pattern on the shell-side, and the great number of variables involved, make it difficult to predict the shell-side coefficient and pressure drop with complete assurance. In methods used for the design of exchangers prior to about 1960 no attempt was made to account for the leakage and bypass streams. Correlations were based on the total stream flow, and empirical methods were used to account for the performance of real exchangers compared with that for cross flow over ideal tube banks. Typical of these "bulk-flow" methods are those of Kern (1950) and Donohue (1955). Reliable predictions can only be achieved by comprehensive analysis of the contribution to heat transfer and pressure drop made by the individual streams shown in Fig. 12.26. Tinker (1951, 1958) published the first detailed stream-analysis method for predicting shell-side heat-transfer coefficients and pressure drop, and the methods subsequently developed have been based on his model. Tinker's presentation is difficult to follow, and his method difficult and tedious to apply in manual calculations. It has been simplified by Devore (1961, 1962); using standard tolerance for commercial exchangers and only a limited number of baffle cuts. Devore gives nomographs that facilitate the application of the method in manual calculations. Mueller (1973) has further simplified Devore's method and gives an illustrative example.

Tinker's model has been used as the basis for the proprietary computer methods developed by Heat Transfer Research Incorporated; see Palen and Taborek (1969), and by Heat Transfer and Fluid Flow Services; see Grant (1973).

Bell (1960, 1963) developed a semi-analytical method based on work done in the co-operative research programme on shell and tube exchangers at the University of

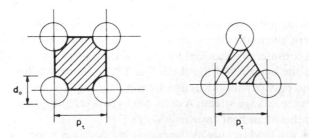

FIG. 12.28. Equivalent diameter, cross-sectional areas and wetted perimeters

Delaware. His method accounts for the major bypass and leakage streams and is suitable for a manual calculation. Bell's method is outlined in Section 12.9.4 and illustrated in Example 12.2.

Though Kern's method does not take account of the bypass and leakage streams, it is simple to apply and is accurate enough for preliminary design calculations, and for designs where uncertainty in other design parameters is such that the use of more elaborate methods is not justified. Kern's method is given in Section 12.9.3 and is illustrated in Example 12.1.

12.9.3. Kern's method

This method was based on experimental work on commercial exchangers with standard tolerances and will give a reasonably satisfactory prediction of the heat-transfer coefficient for standard designs. The prediction of pressure drop is less satisfactory, as pressure drop is more affected by leakage and bypassing than heat transfer. The shell-side heat transfer and friction factors are correlated in a similar manner to those for tube-side flow by using a hypothetical shell velocity and shell diameter. As the cross-sectional area for flow will vary across the shell diameter, the linear and mass velocities are based on the maximum area for cross-flow: that at the shell equator. The shell equivalent diameter is calculated using the flow area between the tubes taken in the axial direction (parallel to the tubes) and the wetted perimeter of the tubes; see Fig. 12.28.

Shell-side j_h and j_f factors for use in this method are given in Figs. 12.29 and 12.30, for various baffle cuts and tube arrangements. These figures are based on data given by Kern (1950) and by Ludwig (1965).

The procedure for calculating the shell-side heat-transfer coefficient and pressure drop for a single shell pass exchanger is given below:

Procedure.

1. Calculate the area for cross-flow A_s for the hypothetical row of tubes at the shell equator, given by:

$$A_s = \frac{(p_t - d_o)D_s l_B}{p_t} \tag{12.21}$$

where p_t = tube pitch,
$\quad d_o$ = tube outside diameter,
$\quad D_s$ = shell inside diameter, m,
$\quad l_B$ = baffle spacing, m.
The term $(p_t - d_o)/p_t$ is the ratio of the clearance between tubes and the total distance between tube centres.

2. Calculate the shell-side mass velocity G_s and the linear velocity u_s:

$$G_s = W_s/A_s$$
$$u_s = G_s/\rho$$

where W_s = fluid flow-rate on the shell-side, kg/s,
$\quad \rho$ = shell-side fluid density, kg/m^3.

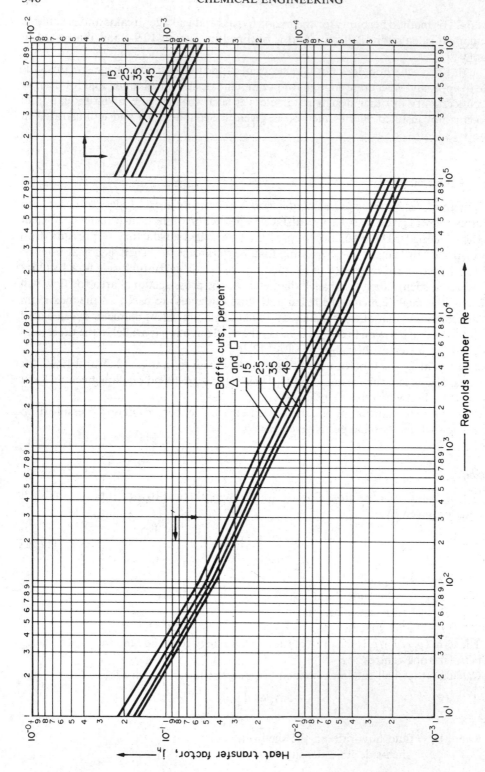

FIG. 12.29. Shell-side heat-transfer factors, segmental baffles

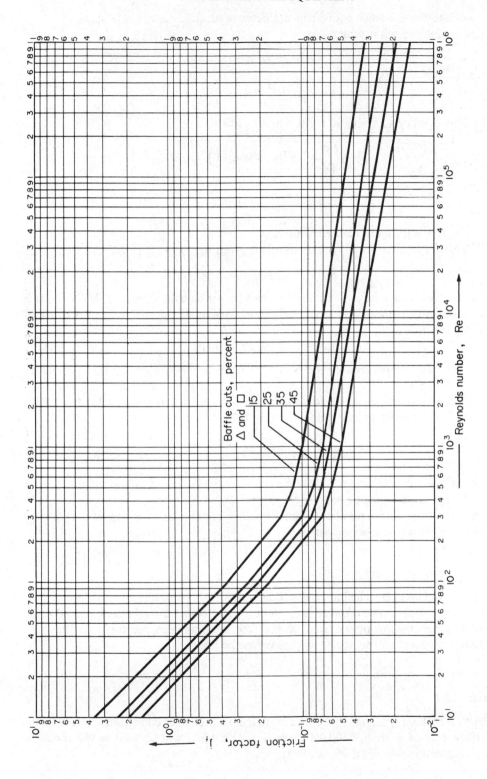

FIG. 12.30. Shell-side friction factors, segmental baffles

3. Calculate the shell-side equivalent diameter (hydraulic diameter), Fig. 12.28.
   For a square pitch arrangement:

$$d_e = 4(p_t^2 - \pi d_o^2/4)/\pi d_o$$

$$= \frac{1 \cdot 27}{d_o}(p_t^2 - 0 \cdot 785 d_o^2) \tag{12.22}$$

For an equilateral triangular pitch arrangement:

$$d_e = 4\left(\frac{p_t}{2} \times 0 \cdot 87 p_t - \tfrac{1}{2}\pi d_o^2/4\right)/\pi d_o/2$$

$$= \frac{1 \cdot 10}{d_o}(p_t^2 - 0 \cdot 917 d_o^2) \tag{12.23}$$

where d_e = equivalent diameter, m.
4. Calculate the shell-side Reynolds number, given by:

$$Re = G_s d_e/\mu = u_s d_e \rho/\mu \tag{12.24}$$

5. For the calculated Reynolds number, read the value of j_h from Fig. 12.29 for the
   selected baffle cut and tube arrangement, and calculate the shell-side heat transfer
   coefficient h_s from:

$$Nu = h_s d_e/k_f = j_h Re Pr^{1/3}(\mu/\mu_w)^{0 \cdot 14} \tag{12.25}$$

The tube wall temperature can be estimated using the method given for the tube-side,
Section 12.8.1.
6. For the calculated shell-side Reynolds number, read the friction factor from Fig.
   12.30 and calculate the shell-side pressure drop from:

$$\Delta P_s = 8 j_f (D_s/d_e)(L/l_B)\frac{\rho u_s^2}{2}\left(\frac{\mu}{\mu_w}\right)^{-0 \cdot 14} \tag{12.26}$$

where L = tube length,
      l_B = baffle spacing.
The term (L/l_B) is the number of times the flow crosses the tube bundle = $(N_b + 1)$,
where N_b is the number of baffles.

Shell nozzle-pressure drop

The pressure loss in the shell nozzles will normally only be significant with gases. The
nozzle pressure drop can be taken as equivalent to $1\tfrac{1}{2}$ velocity heads for the inlet and $\tfrac{1}{2}$ for
the outlet, based on the nozzle area or the free area between the tubes in the row
immediately adjacent to the nozzle, whichever is the least.

Example 12.1

Design an exchanger to sub-cool condensate from a methanol condenser from 95°C to
40°C. Flow-rate of methanol 100,000 kg/h. Brackish water will be used as the coolant,
with a temperature rise from 25° to 40°C.

Solution

Only the thermal design will be considered.
This example illustrates Kern's method.
Coolant is corrosive, so assign to tube-side.

$$\text{Heat capacity methanol} = 2\cdot84 \text{ kJ/kg}^\circ\text{C}$$

$$\text{Heat load} = \frac{100{,}000}{3600} \times 2\cdot84\,(95-40) = 4340 \text{ kW}$$

$$\text{Heat capacity water} = 4\cdot2 \text{ kJ/kg}^\circ\text{C}$$

$$\text{Cooling water flow} = \frac{4340}{4\cdot2\,(40-25)} = 68\cdot9 \text{ kg/s}$$

$$\Delta T_{lm} = \frac{(95-40)-(40-25)}{\ln\dfrac{(95-40)}{(40-25)}} = 31^\circ\text{C}$$

Use one shell pass and two tube passes

(12.6)
$$R = \frac{95-40}{40-25} = 3\cdot67$$

(12.7)
$$S = \frac{40-25}{95-25} = 0\cdot21$$

From Fig. 12.19

$$F_t = 0\cdot85$$

$$\Delta T_m = 0\cdot85 \times 31 = 26^\circ\text{C}$$

From Fig. 12.1

$$U = 600 \text{ W/m}^2\,^\circ\text{C}$$

Provisional area

(12.1)
$$A = \frac{4340 \times 10^3}{26 \times 600} = 278 \text{ m}^2$$

Choose 20 mm o.d., 16 mm i.d., 4·88-m-long tubes ($\frac{3}{4}$ in. × 16 ft), cupro-nickel.
Allowing for tube-sheet thickness, take

$$L = 4\cdot83 \text{ m}$$
$$\text{Area of one tube} = 4\cdot83 \times 20 \times 10^{-3}\,\pi = 0\cdot303 \text{ m}^2$$
$$\text{Number of tubes} = \frac{278}{0\cdot303} = \underline{\underline{918}}$$

As the shell-side fluid is relatively clean use 1·25 triangular pitch.

(12.3b)
$$\text{Bundle diameter } D_b = 20\left(\frac{918}{0\cdot249}\right)^{\frac{1}{2\cdot207}} = 826 \text{ mm}$$

Use a split-ring floating head type.

From Fig. 12.10, bundle diametrical clearance = 68 mm,

$$\text{shell diameter, } D_s = 826 + 68 = 894 \text{ mm}.$$

(*Note.* nearest standard pipe sizes are 863·6 or 914·4 mm).
Shell size could be read from standard tube count tables.

Tube-side coefficient

$$\text{Mean water temperature} = \frac{40 + 25}{2} = 33°C$$

$$\text{Tube cross-sectional area} \quad = \frac{\pi}{4} \times 16^2 = 201 \text{ mm}^2$$

$$\text{Tubes per pass} \quad = \frac{918}{2} = 459$$

$$\text{Total flow area} \quad = 459 \times 201 \times 10^{-6} = 0.092 \text{ m}^2$$

$$\text{Water mass velocity} \quad = \frac{68.9}{0.092} = 749 \text{ kg/s m}^2$$

$$\text{Density water} \quad = 995 \text{ kg/m}^3$$

$$\text{Water linear velocity} \quad = 749/995 = 0.75 \text{ m/s}$$

(12.17) $h_i = 4200(1.35 + 0.02 \times 33)0.75^{0.8}/16^{0.2} = 3852 \text{ W/m}^2 \, °C$

The coefficient can also be calculated using equation 12.15; this is done to illustrate use
of this method.

$$\frac{h_i d_i}{k_f} = j_h Re Pr^{0.33} \left(\frac{\mu}{\mu_w} \right)^{0.14}$$

Viscosity of water = 0·8 mNs/m²
Thermal conductivity = 0·59 W/m°C

$$Re = \frac{\rho u d_i}{\mu} = \frac{995 \times 0.75 \times 16 \times 10^{-3}}{0.8 \times 10^{-3}} = 14,925$$

$$Pr = \frac{C_p \mu}{k_f} = \frac{4.2 \times 10^3 \times 0.8 \times 10^{-3}}{0.59} = 5.7$$

Neglect (μ/μ_w)

$$L/d_i = \frac{4.83 \times 10^3}{16} = 302$$

From Fig. 12.23, $j_h = 3.7 \times 10^{-3}$

$$h_i = \frac{0.59}{16 \times 10^{-3}} \times 3.7 \times 10^{-3} \times 14,925 \times 5.7^{0.33} = 3616 \text{ W/m}^2 \, °C$$

Checks reasonably well with value calculated from equation 12.17; use lower figure.

Shell-side coefficient

Choose baffle spacing $= D_s/5 = \dfrac{894}{5} = 178$ mm.

Tube pitch $= 1.25 \times 20 = 25$ mm

Cross-flow area

(12.21)
$$A_s = \frac{(25-20)}{25} 894 \times 178 \times 10^{-6} = 0.032 \text{ m}^2$$

Mass velocity, $G_S = \dfrac{100{,}000}{3600} \times \dfrac{1}{0.032} = 868 \text{ kg/s m}^2$

Equivalent diameter

(12.23)
$$d_e = \frac{1.1}{20}(25^2 - 0.917 \times 20^2) = 14.4 \text{ mm}$$

Mean shell side temperature $= \dfrac{95+40}{2} = 68°C$

Methanol density $= 750 \text{ kg/m}^3$
Viscosity $= 0.34 \text{ mNs/m}^2$
Heat capacity $= 2.84 \text{ kJ/kg °C}$
Thermal conductivity $= 0.19 \text{ W/m °C}$

(12.24)
$$Re = \frac{G_s d_e}{\mu} = \frac{868 \times 14.4 \times 10^{-3}}{0.34 \times 10^{-3}} = 36{,}762$$

$$Pr = \frac{C_p \mu}{k_f} = \frac{2.84 \times 10^3 \times 0.34 \times 10^{-3}}{0.19} = 5.1$$

Choose 25 per cent baffle cut, from Fig. 12.29

$$j_h = 3.3 \times 10^{-3}$$

Without the viscosity correction term

(12.25) $\quad h_s = \dfrac{0.19}{14.4 \times 10^{-3}} \times 3.3 \times 10^{-3} \times 36{,}762 \times 5.1^{1/3} = 2740 \text{ W/m}^2 \text{ °C}$

Estimate wall temperature

Mean temperature difference $= 68 - 33 = 35°C$
across all resistances

across methanol film $= \dfrac{U}{h_o} \times \Delta T = \dfrac{600}{2740} \times 35 = 8°C$

Mean wall temperature $= 68 - 8 = 60°C$

$$\mu_w = 0.37 \text{ mNs/m}^2$$

$$\left(\frac{\mu}{\mu_w}\right)^{0.14} = 0.99$$

which shows that the correction for a low-viscosity fluid is not significant.

Overall coefficient

Thermal conductivity of cupro-nickel alloys = 50 W/m °C.
Take the fouling coefficients as 6000 W/m² °C

(12.2)
$$\frac{1}{U_o} = \frac{1}{2740} + \frac{1}{6000} + \frac{20 \times 10^{-3} \ln(20/16)}{2 \times 50}$$

$$+ \frac{20}{16} \times \frac{1}{6000} + \frac{20}{16} \times \frac{1}{3616}$$

$$U_o = \underline{\underline{885 \text{ W/m}^2 \text{ °C}}}$$

well above assumed value of 600 W/m² °C.

Pressure drop

Tube-side
From Fig. 12.24, for $Re = 14{,}925$

$$j_f = 4.3 \times 10^{-3}$$

Neglecting the viscosity correction term

(12.20)
$$\Delta P_t = 2\left(8 \times 4.3 \times 10^{-3}\left(\frac{4.83 \times 10^3}{16}\right) + 2.5\right)\frac{995 \times 0.75^2}{2}$$

$$= 7211 \text{ N/m}^2 = 7.2 \text{ kPa} \quad (1.1 \text{ psi})$$

low, could consider increasing the number of tube passes.
   Shell-side

$$\text{Linear velocity} = \frac{G_s}{\rho} = \frac{868}{750} = 1.16 \text{ m/s}$$

From Fig. 12.30, at Re = 36,762

$$j_f = 4 \times 10^{-2}$$

Neglect viscosity correction

(12.26)
$$\Delta P_s = 8 \times 4 \times 10^{-2}\left(\frac{894}{14.4}\right)\left(\frac{4.83 \times 10^3}{178}\right)\frac{750 \times 1.16^2}{2}$$

$$= 272{,}019 \text{ N/m}^2$$
$$= 272 \text{ kPa } (39 \text{ psi}) \text{ too high,}$$

could be reduced by increasing the baffle pitch. Doubling the pitch halves the shell-side velocity, which reduces the pressure drop by a factor of approximately $(1/2)^2$

$$\Delta P_s = \frac{272}{4} = 68 \text{ kPa } (10 \text{ psi}), \qquad\qquad \text{acceptable}$$

This will reduce the shell-side heat-transfer coefficient by a factor of $(1/2)^{0.8}$ ($h_o \propto Re^{0.8} \propto u_s^{0.8}$)

$$h_o = 2740 \times (1/2)^{0.8} = 1573 \text{ W/m}^2 \text{ °C}$$

This gives an overall coefficient of $714\,W/m^2\,°C$—still well above assumed value of $600\,W/m^2\,°C$.

12.9.4. Bell's method

In Bell's method the heat-transfer coefficient and pressure drop are estimated from correlations for flow over ideal tube-banks, and the effects of leakage, bypassing and flow in the window zone are allowed for by applying correction factors.

This approach will give more satisfactory predictions of the heat-transfer coefficient and pressure drop than Kern's method; and, as it takes into account the effects of leakage and bypassing, can be used to investigate the effects of constructional tolerances and the use of sealing strips. The procedure in a simplified and modified form to that given by Bell (1963), is outlined below.

Heat-transfer coefficient

The shell-side heat transfer coefficient is given by:

$$h_s = h_{oc} F_n F_w F_b F_L \tag{12.27}$$

where h_{oc} = heat transfer coefficient calculated for cross-flow over an ideal tube bank, no leakage or bypassing,

F_n = correction factor to allow for the effect of the number of vertical tube rows,

F_w = window effect correction factor,

F_b = bypass stream correction factor,

F_L = leakage correction factor.

The total correction will vary from 0·6 for a poorly designed exchanger with large clearances to 0·9 for a well-designed exchanger.

h_{oc}, ideal cross-flow coefficient

The heat-transfer coefficient for an ideal cross-flow tube bank can be calculated using the heat transfer factors j_h given in Fig. 12.31. Figure 12.31 has been adapted from a similar figure given by Mueller (1973). Mueller includes values for more tube arrangements than are shown in Fig. 12.31. As an alternative to Fig. 12.31, the comprehensive data given in the Engineering Sciences Data Unit report on heat transfer during cross-flow of fluids over tube banks, ESDU No. 73031 (1973), can be used; see Butterworth (1977).

The Reynolds number for cross-flow through a tube bank is given by:

$$Re = \frac{G_s d_o}{\mu} = \frac{u_s \rho d_o}{\mu}$$

where G_s = mass flow rate per unit area, based on the total flow and free area at the bundle equator. This is the same as G_s calculated for Kern's method,

d_o = tube outside diameter.

The heat-transfer coefficient is given by:

$$\frac{h_{oc} d_o}{k_f} = j_h \, Re \, Pr^{1/3} (\mu/\mu_w)^{0.14} \tag{12.28}$$

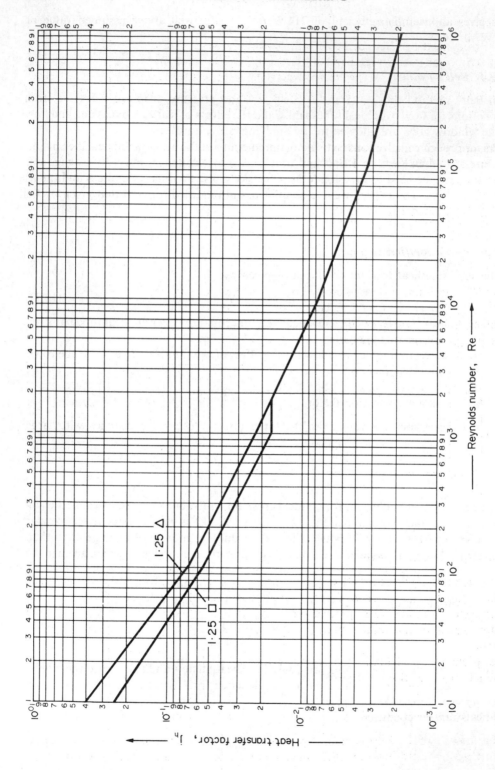

FIG. 12.31. Heat-transfer factor for cross-flow tube banks

F_n, tube row correction factor

The mean heat-transfer coefficient will depend on the number of tubes crossed. Figure 12.31 is based on data for ten rows of tubes. For turbulent flow the correction factor F_n is close to 1·0. In laminar flow the heat-transfer coefficient may decrease with increasing rows of tubes crossed, due to the build up of the temperature boundary layer. The factors given below can be used for the various flow regimes; the factors for turbulent flow are based on those given by Bell (1963).

N_{cv} is number of constrictions crossed = number of tube rows between the baffle tips; see Fig. 12.39, and Section 12.9.5.

1. $Re > 2000$, turbulent;
   take F_n from Fig. 12.32.
2. $Re > 100$ to 2000, transition region,
   take $F_n = 1·0$;
3. $Re < 100$, laminar region,
   $F_n \propto (N_c')^{-0.18}$, $\qquad\qquad\qquad\qquad\qquad\qquad\qquad\qquad$ (12.29)

where N_c' is the number of rows crossed in series from end to end of the shell, and depends on the number of baffles. The correction factor in the laminar region is not well established, and Bell's paper, or the summary given by Mueller (1973), should be consulted if the design falls in this region.

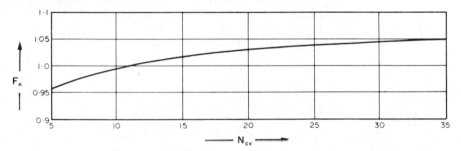

FIG. 12.32. Tube row correction factor, F_n

F_w, window correction factor

This factor corrects for the effect of flow through the baffle window, and is a function of the heat-transfer area in the window zones and the total heat-transfer area. The correction factor is shown in Fig. 12.33 plotted versus R_w, the ratio of the number of tubes in the window zones to the total number in the bundle, determined from the tube layout diagram.

For preliminary calculations R_w can be estimated from the bundle and window cross-sectional areas, see Section 12.9.5.

F_b, by pass correction factor

This factor corrects for the main bypass stream, the flow between the tube bundle and the shell wall, and is a function of the shell to bundle clearance, and whether sealing strips

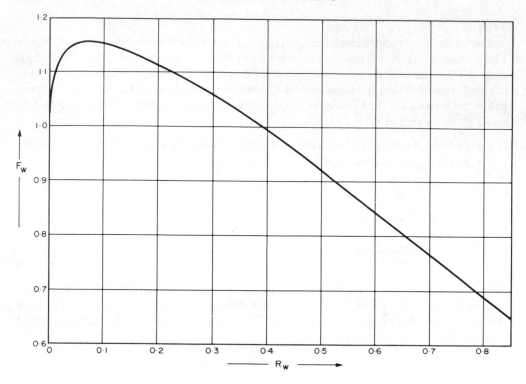

FIG. 12.33. Window correction factor

are used:

$$F_b = \exp\left[-\alpha \frac{A_b}{A_s}(1 - (2N_s/N_{cv})^{1/3})\right] \qquad (12.30)$$

where $\alpha = 1.5$ for laminar flow, $Re < 100$,

$\alpha = 1.35$ for transitional and turbulent flow $Re > 100$,

A_b = clearance area between the bundle and the shell, see Fig. 12.39 and Section 12.9.5,

A_s = maximum area for cross-flow, equation 12.21,

N_s = number of sealing strips encountered by the bypass stream in the cross-flow zone,

N_{cv} = the number of constrictions, tube rows, encountered in the cross-flow section.

Equation 12.30 applies for $N_s \leqslant N_{cv}/2$.
Where no sealing strips are used, F_b can be obtained from Fig. 12.34.

F_L, Leakage correction factor

This factor corrects for the leakage through the tube-to-baffle clearance and the baffle-to-shell clearance.

$$F_L = 1 - \beta_L[(A_{tb} + 2A_{sb})/A_L] \qquad (12.31)$$

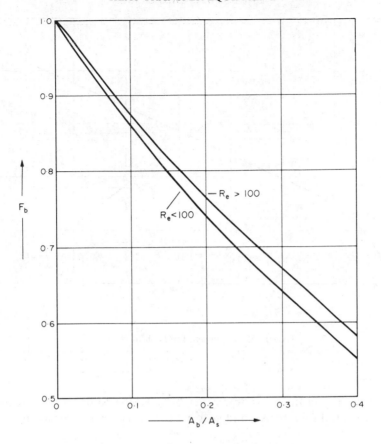

FIG. 12.34. By-pass correction factor

where β_L = a factor obtained from Fig. 12.35,
     A_{tb} = the tube to baffle clearance area, per baffle, see Fig. 12.39 and Section 12.9.5,
     A_{sb} = shell-to-baffle clearance area, per baffle, see Fig. 12.39 and Section 12.9.5,
     A_L = total leakage area = $(A_{tb} + A_{sb})$.

Typical values for the clearances are given in the standards, and are discussed in Section 12.5.6. The clearances and tolerances required in practical exchangers are discussed by Rubin (1968).

Pressure drop

The pressure drops in the cross-flow and window zones are determined separately, and summed to give the total shell-side pressure drop.

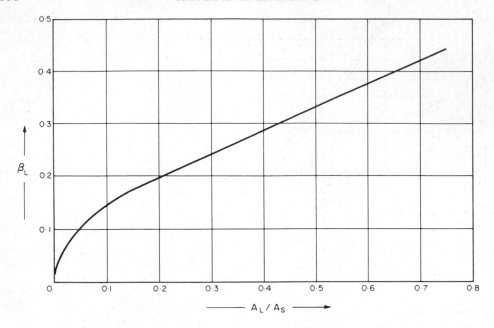

FIG. 12.35. Coefficient for F_L, heat transfer

Cross-flow zones

The pressure drop in the cross-flow zones between the baffle tips is calculated from correlations for ideal tube banks, and corrected for leakage and bypassing.

$$\Delta P_c = \Delta P_i F_b' F_L' \tag{12.32}$$

where ΔP_c = the pressure drop in a cross-flow zone between the baffle tips, corrected for by-passing and leakage,

ΔP_i = the pressure drop calculated for an equivalent ideal tube bank,

F_b' = by-pass correction factor,

F_L' = leakage correction factor.

ΔP_i ideal tube bank pressure drop

The number of tube rows has little effect on the friction factor and is ignored.

Any suitable correlation for the cross-flow friction factor can be used; for that given in Fig. 12.36, the pressure drop across the ideal tube bank is given by:

$$\Delta P_i = 8 j_f N_{cv} \frac{\rho u_s^2}{2} (\mu/\mu_w)^{-0.14} \tag{12.33}$$

where N_{cv} = number of tube rows crossed (in the cross-flow region),

u_s = shell side velocity, based on the clearance area at the bundle equator, equation 12.21,

j_f = friction factor obtained from Fig. 12.36, at the appropriate Reynolds number, $Re = \dfrac{\rho u_s d_o}{\mu}$.

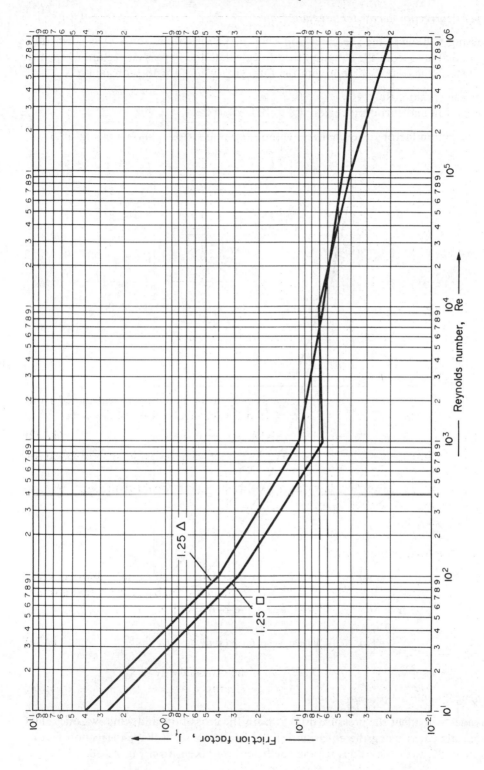

FIG. 12.36. Friction factor for cross-flow tube banks

F'_b, bypass correction factor for pressure drop

Bypassing will affect the pressure drop only in the cross-flow zones. The correction factor is calculated from the equation used to calculate the bypass correction factor for heat transfer, equation 12.30, but with the following values for the constant α.

Laminar region, Re < 100, $\alpha = 5.0$
Transition and turbulent region, Re > 100, $\alpha = 4.0$

The correction factor for exchangers without sealing strips is shown in Fig. 12.37.

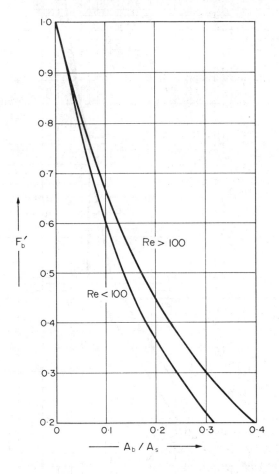

FIG. 12.37. Bypass factor for pressure drop F'_b

F'_L, leakage factor for pressure drop

Leakages will affect the pressure drop in both the cross-flow and window zones. The factor is calculated using the equation for the heat-transfer leakage-correction factor, equation 12.31, with the values for the coefficient β'_L taken from Fig. 12.38.

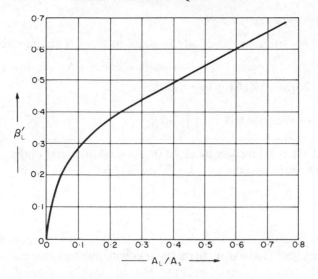

F$_{IG}$. 12.38. Coefficient for F'_L, pressure drop

Window-zone pressure drop

Any suitable method can be used to determine the pressure drop in the window area; see Butterworth (1977). Bell used a method proposed by Colburn. Corrected for leakage, the window drop for turbulent flow is given by:

$$\Delta P_w = F'_L (2 + 0.6 N_{wv}) \frac{\rho u_z^2}{2} \tag{12.34}$$

where u_z = the geometric mean velocity,

$u_z = \sqrt{u_w u_s}$,

u_w = the velocity in the window zone, based on the window area less the area occupied by the tubes A_w, see Section 12.9.5,

$$u_w = \frac{W_s}{A_w \rho} \tag{12.35}$$

W_s = shell-side fluid mass flow, kg/s,

N_{wv} = number of restrictions for cross-flow in window zone, approximately equal to the number of tube rows.

End zone pressure drop

There will be no leakage paths in an end zone (the zone between tube sheet and baffle). Also, there will only be one baffle window in these zones; so the total number of restrictions in the cross-flow zone will be $N_{cv} + N_{wv}$. The end zone pressure drop ΔP_e will therefore be given by:

$$\Delta P_e = \Delta P_i [(N_{wv} + N_{cv})/N_{cv}] F'_b \tag{12.36}$$

Total shell-side pressure drop

Summing the pressure drops over all the zones in series from inlet to outlet gives:

$$\Delta P_s = 2 \text{ end zones} + (N_b - 1) \text{ cross-flow zones} + N_b \text{ window zones}$$

$$\Delta P_s = 2\Delta P_e + \Delta P_c(N_b - 1) + N_b \Delta P_w \qquad (12.37)$$

where N_b is the number of baffles $= \left[\dfrac{L}{l_B} - 1 \right]$.

An estimate of the pressure loss incurred in the shell inlet and outlet nozzles must be added to that calculated by equation 12.37; see Section 12.9.3.

End zone lengths

The spacing in the end zones will often be increased to provide more flow area at the inlet and outlet nozzles. The velocity in these zones will then be lower and the heat transfer and pressure drop will be reduced slightly. The effect on pressure drop will be more marked than on heat transfer, and can be estimated by using the actual spacing in the end zone when calculating the cross-flow velocity in those zones.

12.9.5. Shell and bundle geometry

The bypass and leakage areas, window area, and the number of tubes and tube rows in the window and cross-flow zones can be determined precisely from the tube layout diagram. For preliminary calculations they can be estimated with sufficient accuracy by considering the tube bundle and shell geometry.

With reference to Figs. 12.39 and 12.40:

H_c = baffle cut height $= D_s \times B_c$, where B_c is the baffle cut as a fraction,

H_b = height from the baffle chord to the top of the tube bundle,

B_b = "bundle cut" $= H_b/D_b$,

θ_b = angle subtended by the baffle chord, rads,

D_b = bundle diameter.

Then:

$$H_b = D_b/2 - D_s(0\cdot5 - B_c) \qquad (12.38)$$

$$N_{cv} = (D_b - 2H_b)/p_t' \qquad (12.39)$$

$$N_{wv} = H_b/p_t' \qquad (12.40)$$

where p_t' is the vertical tube pitch

$p_t' = p_t$ for square pitch,

$p_t' = 0.87 p_t$ for equilateral triangular pitch.

The number of tubes in a window zone N_w is given by:

$$N_w = N_t \times R_a' \qquad (12.41)$$

where R_a' is the ratio of the bundle cross-sectional area in the window zone to the total bundle cross-sectional area, R_a' can be obtained from Fig. 12.41, for the appropriate "bundle cut", B_b.

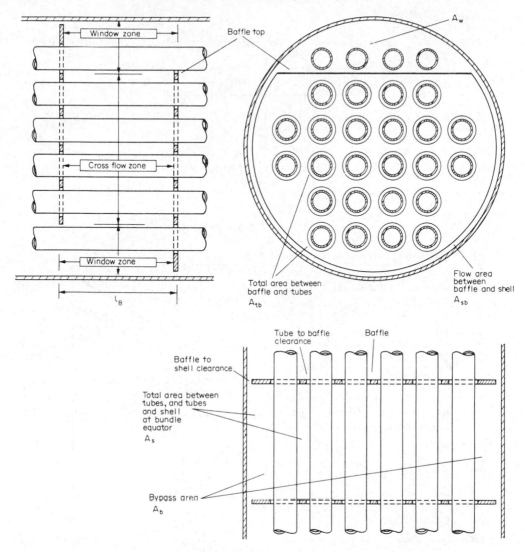

FIG. 12.39. Clearance and flow areas in the shell-side of a shell and tube exchanger

The number of tubes in a cross-flow zone N_c is given by

$$N_c = N_t - 2N_w \qquad (12.42)$$

and

$$R_w = 2N_w/N_t \qquad (12.43)$$

$$A_w = \left(\frac{\pi D_s^2}{4} \times R_a\right) - \left(N_w \frac{\pi d_o^2}{4}\right) \qquad (12.44)$$

R_a is obtained from Fig. 12.41, for the appropriate baffle cut B_c

$$A_{tb} = \frac{c_t \pi d_o}{2} (N_t - N_w) \qquad (12.45)$$

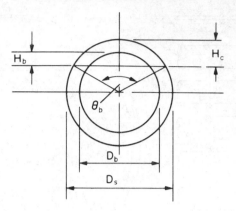

FIG. 12.40. Baffle and tube geometry

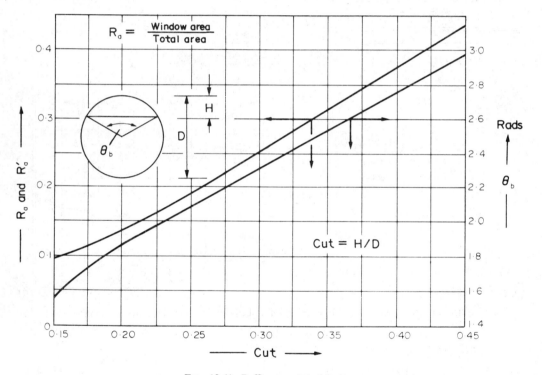

FIG. 12.41. Baffle geometrical factors

where c_t is the diametrical tube-to-baffle clearance; the difference between the hole and tube diameter, typically 0·8 mm.

$$A_{sb} = \frac{c_s D_s}{2} (2\pi - \theta_b) \qquad (12.46)$$

where c_s is the baffle-to-shell clearance, see Table 12.5.

θ_b can be obtained from Fig. 12.41, for the appropriate baffle cut, B_c

$$A_b = l_B(D_s - D_b) \qquad (12.47)$$

where l_B is the baffle spacing.

12.9.6. Effect of fouling on pressure drop

Bell's method gives an estimate of the shell-side pressure drop for the exchanger in the clean condition. In service, the clearances will tend to plug up, particularly the small clearance between the tubes and baffle, and this will increase the pressure drop. Devore (1961) has estimated the affect of fouling on pressure drop by calculating the pressure drop in an exchange in the clean condition and with the clearance reduced by fouling, using Tinker's method. He presented his results as ratios of the fouled to clean pressure drop for various fouling factors and baffle spacings.

The ratios given in Table 12.7, which are adapted from Devore's figures, can be used to make a rough estimate of the effect of fouling on pressure drop.

TABLE 12.7. *Ratio of fouled to clean pressure drop*

Fouling coefficient	Shell diameter/baffle spacing		
(W/m² °C)	1·0	2·0	5·0
Laminar flow			
6000	1·06	1·20	1·28
2000	1·19	1·44	1·55
< 1000	1·32	1·99	2·38
Turbulent flow			
6000	1·12	1·38	1·55
2000	1·37	2·31	2·96
< 1000	1·64	3·44	4·77

12.9.7. Pressure-drop limitations

Though Bell's method will give a better estimate of the shell-side pressure drop than Kern's, it is not sufficiently accurate for the design of exchangers where the allowable pressure drop is the overriding consideration. For such designs the method developed by Devore is recommended, in the absence of access to a reliable computer method.

Example 12.2

Using Bell's method, calculate the shell-side heat transfer coefficient and pressure drop for the exchanger designed in Example 12.1.

Summary of proposed design

Number of tubes	= 918
Shell i.d.	894 mm
Bundle diameter	826 mm
Tube o.d.	20 mm

Pitch 1.25 Δ	25 mm
Tube length	4830 mm
Baffle pitch	356 mm

Physical properties from Example 12.1

Solution

Heat-transfer coefficient

Ideal bank coefficient, h_{oc}

$$A_s = \frac{25-20}{25} \times 894 \times 356 \times 10^{-6} = 0.062 \text{ m}^2$$

$$G_s = \frac{100,000}{3600} \times \frac{1}{0.062} = 448 \text{ kg/s m}^2$$

$$Re = \frac{G_s d_o}{\mu} = \frac{448 \times 20 \times 10^{-3}}{0.34 \times 10^{-3}} = 26,353$$

From Fig. 12.31 $j_h = 5.3 \times 10^{-3}$.
Prandtl number, from Example 12.1 = 5.1.
Neglect viscosity correction factor (μ/μ_w).

(12.28) $h_{oc} = \dfrac{0.19}{20 \times 10^{-3}} \times 5.3 \times 10^{-3} \times 26,353 \times 5.1^{1/3} = 2272 \text{ W/m}^2 \, ^\circ\text{C}$

Tube row correction factor, F_n

Tube vertical pitch $p_t' = 0.87 \times 25 = 21.8$ mm
Baffle cut height $H_c = 0.25 \times 894 = 224$ mm
Height between baffle tips $= 894 - 2 \times 224 = 446$ mm

$$N_{cv} = \frac{446}{21.8} = 20$$

From Fig. 12.32 $F_n = 1.03$.
Window correction factor, F_w

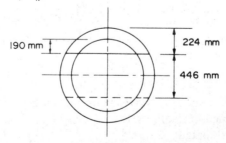

(12.38) $H_b = \dfrac{826}{2} - 894(0.5 - 0.25) = 190$ mm

"Bundle cut" $= \dfrac{190}{826} = 0.23$ (23 per cent)

From Fig. 12.41 at cut of 0·23

$$R'_a = 0·18$$

(12.41)   Tubes in one window area, $N_w = 918 \times 0·18 = 165$

(12.42)   Tubes in cross-flow area $N_c = 918 - 2 \times 165 = 588$

(12.43) $$R_w = \frac{2 \times 165}{918} = 0·36$$

From Fig. 12.33 $F_w = 1·02$.

Bypass correction, F_b

(12.47) $$A_b = (894 - 826)356 \times 10^{-6} = 0·024 \text{ m}^2$$

$$A_b/A_s = \frac{0·024}{0·062} = 0·39$$

(12.30) $$F_b = \exp[-1·35 \times 0·39] = 0·59$$

Very low, sealing strips needed; try one strip for each five vertical rows.

$$N_s/N_{cv} = 1/5$$

(12.30) $$F_b = \exp[-1·35 \times 0·39(1 - (2/5)^{1/3}] = 0·87$$

Leakage correction, F_L

Using clearances as specified in the Standards,

$$\text{tube-to-baffle} \quad \tfrac{1}{32} \text{ in.} = 0·8 \text{ mm}$$
$$\text{baffle-to-shell} \quad \tfrac{3}{16} \text{ in.} = 4·8 \text{ mm}$$

(12.45) $$A_{tb} = \frac{0·8}{2} \times 20\pi(918 - 165) = 18·9 \times 10^3 \text{ mm}^2 = 0·019 \text{ m}^2$$

From Fig. 12.41, 25 per cent cut (0·25), $\theta_b = 2·1$ rads.

(12.46) $$A_{sb} = \frac{4·8}{2} \times 894 \, (2\pi - 2·1) = 8·98 \times 10^3 \text{ mm}^2 = 0·009 \text{ m}^2$$

$$A_L = (0·019 + 0·009) = 0·028 \text{ m}^2$$

$$A_L/A_s = \frac{0·028}{0·062} = 0·45$$

From Fig. 12.35 $\beta_L = 0·3$.

(12.31) $$F_L = 1 - 0·3[(0·019 + 2 \times 0·009)/0·028] = 0·60$$

Shell-Side Coefficient

(12.27) $$h_s = 2272 \times 1·03 \times 1·02 \times 0·87 \times 0·60 = \underline{\underline{1246 \text{ W/m}^2 \, °\text{C}}}$$

Appreciably lower than that predicted by Kern's method.

Pressure drop

Cross-flow zone
From Fig. 12.36 at $Re = 26,353$, for 1.25Δ pitch, $j_f = 5.6 \times 10^{-2}$

$$u_s = \frac{G_s}{\rho} = \frac{448}{750} = 0.60 \text{ m/s}$$

Neglecting viscosity term (μ/μ_w).

(12.33) $\Delta P_i = 8 \times 5.6 \times 10^{-2} \times 20 \times \dfrac{750 \times 0.6^2}{2} = 1209.6 \text{ N/m}^2$

(12.30) $(\alpha = 4.0)$

$$F'_b = \exp\left[-4.0 \times 0.39\left(1 - (2/5)^{1/3}\right)\right]$$
$$= 0.66$$

From Fig. 12.38 $\beta'_L = 0.52$.

(12.31) $F'_L = 1 - 0.52\left[(0.019 + 2 \times 0.009)/0.028\right] = 0.31$

$$\Delta P_c = 1209.6 \times 0.66 \times 0.31 = 248 \text{ N/m}^2$$

Window zone
From Fig. 12.41, for baffle cut 25 per cent (0.25) $R_a = 0.19$.

(12.44) $A_w = \left(\dfrac{\pi}{4} \times 894^2 \times 0.19\right) - \left(165 \times \dfrac{\pi}{4} \times 20^2\right) = 67.4 \times 10^3 \text{ mm}^2$

$$= 0.067 \text{ m}^2$$

$$u_w = \frac{100,000}{3600} \times \frac{1}{750} \times \frac{1}{0.067} = 0.55 \text{ m/s}$$

$$u_z = \sqrt{u_w u_s} = \sqrt{0.55 \times 0.60} = 0.57 \text{ m/s}$$

(12.40) $N_{wv} = \dfrac{190}{21.8} = 8$

(12.34) $\Delta P_w = 0.31(2 + 0.6 \times 8)\dfrac{750 \times 0.57^2}{2} = 257 \text{ N/m}^2$

End zone

(12.36) $\Delta P_e = 1209.6\left[(8 + 20)/20\right]0.66 = 1118 \text{ N/m}^2$

Total pressure drop

Number of baffles $N_b = \dfrac{4830}{356} - 1 = 12$

(12.37) $\Delta P_s = 2 \times 1118 + 248(12 - 1) + 12 \times 257 = 8048 \text{ N/m}^2$

$$= \underline{\underline{8.05 \text{ kPa}}} \quad (1.2 \text{ psi})$$

This for the exchanger in the clean condition. Using the factors given in Table 12.7 to estimate the pressure drop in the fouled condition

$$\Delta P_s = 1.4 \times 8.05 = \underline{\underline{11.3 \text{ kPa}}}$$

Appreciably lower than that predicted by Kern's method. This shows the unsatisfactory nature of the methods available for predicting the shell-side pressure drop.

12.10. Condensers

This section covers the design of shell and tube exchangers used as condensers. Direct contact condensers are discussed in Section 12.13.

The construction of a condenser will be similar to other shell and tube exchangers, but with a wider baffle spacing, typically $l_B = D_s$.

Four condenser configurations are possible:

1. Horizontal, with condensation in the shell, and the cooling medium in the tubes.
2. Horizontal, with condensation in the tubes.
3. Vertical, with condensation in the shell.
4. Vertical, with condensation in the tubes.

Horizontal shell-side and vertical tube-side are the most commonly used types of condenser. A horizontal exchanger with condensation in the tubes is rarely used as a process condenser, but is the usual arrangement for heaters and vaporisers using condensing steam as the heating medium.

12.10.1. Heat-transfer fundamentals

The fundamentals of condensation heat transfer are covered in Volume 1, Chapter 7.

The normal mechanism for heat transfer in commercial condensers is filmwise condensation. Dropwise condensation will give higher heat-transfer coefficients, but is unpredictable; and is not yet considered a practical proposition for the design of condensers for general purposes.

The basic equations for filmwise condensation were derived by Nusselt (1916), and his equations form the basis for practical condenser design. The basic Nusselt equations are derived in Volume 1, Chapter 7. In the Nusselt model of condensation laminar flow is assumed in the film, and heat transfer is assumed to take place entirely by conduction through the film. In practical condensers the Nusselt model will strictly only apply at low liquid and vapour rates, and where the flowing condensate film is undisturbed. Turbulence can be induced in the liquid film at high liquid rates, and by shear at high vapour rates. This will generally increase the rate of heat transfer over that predicted using the Nusselt model. The effect of vapour shear and film turbulence are discussed in Volume 1, Chapter 7; see also Butterworth (1978) and Taborek (1974).

Physical properties

The physical properties of the condensate for use in the following equations, are evaluated at the average condensate film temperature: the mean of the condensing temperature and the tube-wall temperature.

12.10.2. Condensation outside horizontal tubes

$$(h_c)_1 = 0.95 \, k_L \left[\frac{\rho_L(\rho_L - \rho_v)g}{\mu_L \Gamma} \right]^{1/3} \tag{12.48}$$

where $(h_c)_1$ = mean condensation film coefficient, for a single tube, W/m^2°C
  k_L = condensate thermal conductivity, W/m°C,
  ρ_L = condensate density, kg/m^3,
  ρ_v = vapour density, kg/m^3,
  μ_L = condensate viscosity, N s/m^2,
  g = gravitational acceleration, 9·81 m/s^2,
  Γ = the tube loading, the condensate flow per unit length of tube, kg/m s.

In a bank of tubes the condensate from the upper rows of tubes will add to that condensing on the lower tubes. If there are N_r tubes in a vertical row and the condensate is assumed to flow smoothly from row to row, Fig. 12.42a, and if the flow remains laminar, the mean coefficient predicted by the Nusselt model is related to that for the top tube by:

$$(h_c)_{N_r} = (h_c)_1 N_r^{-1/4} \tag{12.49}$$

In practice, the condensate will not flow smoothly from tube to tube, Fig. 12.42b, and the factor of $(N_r)^{-1/4}$ applied to the single tube coefficient in equation 12.49 is considered to be too conservative. Based on results from commercial exchangers, Kern (1950) suggests using an index of 1/6. Frank (1978) suggests multiplying single tube coefficient by a factor of 0·75.

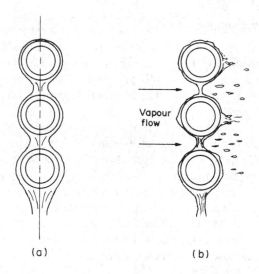

(a)   (b)

FIG. 12.42. Condensate flow over tube banks

Using Kern's method, the mean coefficient for a tube bundle is given by:

$$(h_c)_b = 0.95 k_L \left[\frac{\rho_L(\rho_L - \rho_v)g}{\mu_L \Gamma_h} \right]^{1/3} N_r^{-1/6} \tag{12.50}$$

where $\Gamma_h = \dfrac{W_c}{L N_t}$

and $\quad L$ = tube length,

$\quad\quad W_c$ = total condensate flow,

$\quad\quad N_t$ = total number of tubes in the bundle,

$\quad\quad N_r$ = average number of tubes in a vertical tube row.

N_r can be taken as two-thirds of the number in the central tube row.

For low-viscosity condensates the correction for the number of tube rows is generally ignored.

12.10.3 Condensation inside and outside vertical tubes

For condensation inside and outside vertical tubes the Nusselt model gives:

$$(h_c)_v = 0\cdot926 k_L \left[\frac{\rho_L (\rho_L - \rho_v) g}{\mu_L \Gamma_v} \right]^{1/3} \qquad (12.51)$$

where $(h_c)_v$ = mean condensation coefficient, $W/m^2 {}^\circ C$,

$\quad\quad \Gamma_v$ = vertical tube loading, condensate rate per unit tube perimeter, kg/m s

for a tube bundle

$$\Gamma_v = \frac{W_c}{N_t \pi d_o} \quad \text{or} \quad \frac{W_c}{N_t \pi d_i}$$

Equation 12.51 will apply up to a Reynolds number of 30; above this value waves on the condensate film become important. The Reynolds number for the condensate film is given by:

$$Re_c = \frac{4\Gamma_v}{\mu_L}$$

The presence of waves will increase the heat-transfer coefficient, so the use of equation 12.51 above a Reynolds number of 30 will give conservative (safe) estimates. The effect of waves on condensate film on heat transfer is discussed by Kutateladze (1963).

Above a Reynolds number of around 2000, the condensate film becomes turbulent. The effect of turbulence in the condensate film was investigated by Colburn (1934) and Colburn's results are generally used for condenser design, Fig. 12.43. Equation 12.51 is also shown on Fig. 12.43. The Prandtl number for the condensate film is given by:

$$Pr_c = \frac{C_p \mu_L}{k_L}$$

Figure 12.43 can be used to estimate condensate film coefficients in the absence of appreciable vapour shear. Horizontal and downward vertical vapour flow will increase the rate of heat transfer, and the use of Fig. 12.43 will give conservative values for most practical condenser designs.

Boyko and Kruzhilin (1967) developed a correlation for shear-controlled condensation in tubes which is simple to use. Their correlation gives the mean coefficient between two points at which the vapour quality is known. The vapour quality x is the mass fraction of

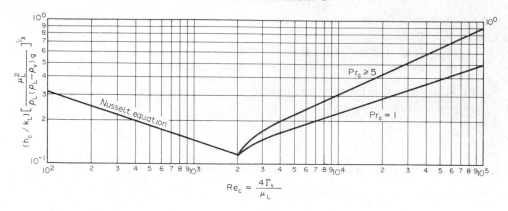

FIG. 12.43. Condensation coefficient for vertical tubes

the vapour present. It is convenient to represent the Boyko–Kruzhilin correlation as:

$$(h_c)_{BK} = h_i' \left[\frac{J_1^{1/2} + J_2^{1/2}}{2} \right] \tag{12.52}$$

$$\text{where } J = 1 + \left[\frac{\rho_L - \rho_v}{\rho_v} \right] x$$

and the suffixes 1 and 2 refer to the inlet and outlet conditions respectively. h_i' is the tube-side coefficient evaluated for single-phase flow of the total condensate (the condensate at point 2). That is, the coefficient that would be obtained if the condensate filled the tube and was flowing alone; this can be evaluated using any suitable correlation for forced convection in tubes; see Section 12.8.

Boyko and Kruzhilin used the correlation:

$$h_i' = 0.021 \, (k_L/d_i) Re^{0.8} \, Pr^{0.43} \tag{12.53}$$

In a condenser the inlet stream will normally be saturated vapour and the vapour will be totally condensed.

For these conditions equation 12.52 becomes:

$$(h_c)_{BK} = h_i' \left[\frac{1 + \sqrt{\rho_L/\rho_v}}{2} \right] \tag{12.54}$$

For the design of condensers with condensation inside the tubes and downward vapour flow, the coefficient should be evaluated using Fig. 12.43 and equation 12.52, and the *higher* value selected.

Flooding in vertical tubes

When the vapour flows up the tube, which will be the usual arrangement for a reflux condenser, care must be taken to ensure that the tubes do not flood. Several correlations have been published for the prediction of flooding in vertical tubes, see Perry and Chilton (1973). One of the simplest to apply, which is suitable for use in the design of condensers handling low-viscosity condensates, is the criterion given by Hewitt and Hall-Taylor

(1970); see also Butterworth (1977). Flooding should not occur if the following condition is satisfied:

$$[u_v^{1/2}\rho_v^{1/4} + u_L^{1/2}\rho_L^{1/4}] < 0.6[gd_i(\rho_L - \rho_v)]^{1/4} \tag{12.55}$$

where u_v and u_L are the velocities of the vapour and liquid, based on each phase flowing in the tube alone; and d_i is in metres. The critical condition will occur at the bottom of the tube, so the vapour and liquid velocities should be evaluated at this point.

Example 12.3

Estimate the heat-transfer coefficient for steam condensing on the outside, and on the inside, of a 25 mm o.d., 21 mm i.d. vertical tube 3·66 m long. The steam condensate rate is 0·015 kg/s per tube and condensation takes place at 3 bar. The steam will flow down the tube.

Solution

Physical properties, from steam tables:

$$\text{Saturation temperature} = 133.5°C$$

$$\rho_L = 931 \text{ kg/m}^3$$
$$\rho_v = 1.65 \text{ kg/m}^3$$
$$k_L = 0.688 \text{ W/m °C}$$
$$\mu_L = 0.21 \text{ mNs/m}^2$$
$$Pr_c = 1.27$$

Condensation outside the tube

$$\Gamma_v = \frac{0.015}{\pi 25 \times 10^{-3}} = 0.191 \text{ kg/s m}$$

$$Re_c = \frac{4 \times 0.191}{0.21 \times 10^{-3}} = 3638$$

From Fig. 12.43

$$\frac{h_c}{k_L}\left[\frac{\mu_L^2}{\rho_L(\rho_L - \rho_v)g}\right]^{1/3} = 1.65 \times 10^{-1}$$

$$h_c = 1.65 \times 10^{-1} \times 0.688\left[\frac{(0.21 \times 10^{-3})^2}{931(931 - 1.65)9.81}\right]^{-1/3}$$

$$= \underline{\underline{6554 \text{ W/m}^2 °C}}$$

Condensation inside the tube

$$\Gamma_v = \frac{0.015}{\pi 21 \times 10^{-3}} = 0.227 \text{ kg/s m}$$

$$Re_c = \frac{4 \times 0.227}{0.21 \times 10^{-3}} = 4324$$

From Fig. 12.43

$$h_c = 1{\cdot}72 \times 10^{-1} \times 0{\cdot}688 \left[\frac{(0{\cdot}21 \times 10^{-3})^2}{931(931 - 1{\cdot}65)9{\cdot}81} \right]^{-1/3}$$

$$= 6832 \text{ W/m}^2{}^\circ\text{C}$$

Boyko–Kruzhilin method

Cross-sectional area of tube $= (21 \times 10^{-3})^2 \dfrac{\pi}{4} = 3{\cdot}46 \times 10^{-4} \text{ m}^2$

Fluid velocity, total condensation

$$u_t = \frac{0{\cdot}015}{931 \times 3{\cdot}46 \times 10^{-4}} = 0{\cdot}047 \text{ m/s}$$

$$Re = \frac{\rho u d_i}{\mu_L} = \frac{931 \times 0{\cdot}047 \times 21 \times 10^{-3}}{0{\cdot}21 \times 10^{-3}}$$

$$= 4376$$

(12.53) $h_i' = 0{\cdot}021 \times \dfrac{0{\cdot}688}{21 \times 10^{-3}} (4376)^{0{\cdot}8}(1{\cdot}27)^{0{\cdot}43} = 624 \text{ W/m}^2\,{}^\circ\text{C}$

(12.54) $h_c = 624 \left[\dfrac{1 + \sqrt{931/1{\cdot}65}}{2} \right] = 7723 \text{ W/m}^2\,{}^\circ\text{C}$

Take higher value, $h_c = \underline{7723 \text{ W/m}^2\,{}^\circ\text{C}}$

Example 12.4

It is proposed to use an existing distillation column, which is fitted with a dephlegmator (reflux condenser) which has 200 vertical, 50 mm i.d., tubes, for separating benzene from a mixture of chlorobenzenes. The top product will be 2500 kg/h benzene and the column will operate with a reflux ratio of 3. Check if the tubes are likely to flood. The condenser pressure will be 1 bar.

Solution

The vapour will flow up and the liquid down the tubes. The maximum flow rates of both will occur at the base of the tube.

Vapour flow $= (3 + 1)2500 = 10{,}000 \text{ kg/h}$
Liquid flow $= 3 \times 2500 = 7500 \text{ kg/h}$

Total area tubes $= \dfrac{\pi}{4} (50 \times 10^{-3})^2 \times 200 = 0{\cdot}39 \text{ m}^2$

Densities at benzene boiling point

$$\rho_L = 840 \text{ kg/m}^3, \ \rho_v = 2{\cdot}7 \text{ kg/m}^3$$

Vapour velocity (vapour flowing alone in tube)

$$u_v = \frac{10{,}000}{3600 \times 0{\cdot}39 \times 2{\cdot}7} = 2{\cdot}64 \text{ m/s}$$

Liquid velocity (liquid alone)

$$u_L = \frac{7500}{3600 \times 0.39 \times 840} = 0.006 \text{ m/s}$$

From equation 12.55 for no flooding

$$[u_v^{1/2} \rho_v^{1/4} + u_L^{1/2} \rho_L^{1/4}] < 0.6 [gd_i(\rho_L - \rho_v)]^{1/4}$$

$$[(2.64)^{1/2}(2.7)^{1/4} + (0.006)^{1/2}(840)^{1/4}] < 0.6[9.81 \times 50 \times 10^{-3}(840 - 2.7)]^{1/4}$$

$$[2.50] < [2.70]$$

Tubes should not flood, but there is little margin of safety.

12.10.4. Condensation inside horizontal tubes

Where condensation occurs in a horizontal tube the heat-transfer coefficient at any point along the tube will depend on the flow pattern at that point. The various patterns that can exist in two-phase flow are shown in Fig. 12.44; and are discussed in Volume 1, Chapter 4. In condensation, the flow will vary from a single-phase vapour at the inlet to a single-phase liquid at the outlet; with all the possible patterns of flow occurring between these points. Bell *et al.* (1970) give a method for following the change in flow pattern as condensation occurs on a Baker flow-regime map. Correlations for estimating the average condensation coefficient have been published by several workers, but there is no generally satisfactory method that will give accurate predictions over a wide flow range. A comparison of the published methods is given by Bell *et al.* (1970).

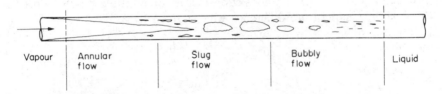

| Vapour | Annular flow | Slug flow | Bubbly flow | Liquid |

FIG. 12.44. Flow patterns, vapour condensing in a horizontal tube

Two flow models are used to estimate the mean condensation coefficient in horizontal tubes: stratified flow, Fig. 12.45*a*, and annular flow, Fig. 12.45*b*. The stratified flow model represents the limiting condition at low condensate and vapour rates, and the annular model the condition at high vapour and low condensate rates. For the stratified flow model, the condensate film coefficient can be estimated from the Nusselt equation, applying a suitable correction for the reduction in the coefficient caused by the accumulation of condensate in the bottom of the tube. The correction factor will typically be around 0.8, so the coefficient for stratified flow can be estimated from:

$$(h_c)_s = 0.76 \, k_L \left[\frac{\rho_L(\rho_L - \rho_v)g}{\mu_L \Gamma_h} \right]^{1/3} \qquad (12.56)$$

The Boyko–Kruzhilin equation, equation 12.52, can be used to estimate the coefficient for annular flow.

For condenser design, the mean coefficient should be evaluated using the correlations for both annular and stratified flow and the *higher* value selected.

(b) Annular flow   (a) Stratified flow

FIG. 12.45. Flow patterns in condensation

12.10.5. Condensation of steam

Steam is frequently used as a heating medium. The film coefficient for condensing steam can be calculated using the methods given in the previous sections; but, as the coefficient will be high and will rarely be the limiting coefficient, it is customary to assume a typical, conservative, value for design purposes. For air-free steam a coefficient of 8000 W/m²°C (1500 Btu/h ft² °F) can be used.

12.10.6. Mean temperature difference

A pure, saturated, vapour will condense at a fixed temperature, at constant pressure. For an isothermal process such as this, the simple logarithmic mean temperature difference can be used in the equation 12.1; no correction factor for multiple passes is needed. The logarithmic mean temperature difference will be given by:

$$\Delta T_{\text{lm}} = \frac{(t_2 - t_1)}{\ln\left[\dfrac{T_{\text{sat}} - t_1}{T_{\text{sat}} - t_2}\right]} \tag{12.57}$$

where T_{sat} = saturation temperature of the vapour,
     t_1 = inlet coolant temperature,
     t_2 = outlet coolant.

When the condensation process is not exactly isothermal but the temperature change is small; such as where there is a significant change in pressure, or where a narrow boiling range multicomponent mixture is being condensed; the logarithmic temperature difference can still be used but the temperature correction factor will be needed for multipass condensers. The appropriate terminal temperatures should be used in the calculation.

12.10.7. Desuperheating and sub-cooling

When the vapour entering the condenser is superheated, and the condensate leaving the condenser is cooled below its boiling point (sub-cooled), the temperature profile will be as shown in Fig. 12.46.

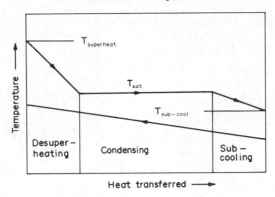

FIG. 12.46. Condensation with desuperheating and sub-cooling

Desuperheating

If the degree of superheat is large, it will be necessary to divide the temperature profile into sections and determine the mean temperature difference and heat-transfer coefficient separately for each section. If the tube wall temperature is below the dew point of the vapour, liquid will condense directly from the vapour on to the tubes. In these circumstances it has been found that the heat-transfer coefficient in the superheating section is close to the value for condensation and can be taken as the same. So, where the amount of superheating is not too excessive, say less than 25 per cent of the latent heat load, and the outlet coolant temperature is well below the vapour dew point, the sensible heat load for desuperheating can be lumped with the latent heat load. The total heat-transfer area required can then be calculated using a mean temperature difference based on the saturation temperature (not the superheat temperature) and the estimated condensate film heat-transfer coefficient.

Sub-cooling of condensate

Some sub-cooling of the condensate will usually be required to control the net positive suction head at the condensate pump (see Chapter 5) or to cool a product for storage. Where the amount of sub-cooling is large, it is more efficient to sub-cool in a separate exchanger. A small amount of sub-cooling can be obtained in a condenser by controlling the liquid level so that some part of the tube bundle is immersed in the condensate.

In a horizontal shell-side condenser a dam baffle can be used, Fig. 12.47a. A vertical condenser can be operated with the liquid level above the bottom tube sheet, Fig. 12.47b.

The temperature difference in the sub-cooled region will depend on the degree of mixing in the pool of condensate. The limiting conditions are plug flow and complete mixing. The temperature profile for plug flow is that shown in Fig. 12.46. If the pool is perfectly mixed, the condensate temperature will be constant over the sub-cooling region and equal to the condensate outlet temperature. Assuming perfect mixing will give a very conservative (safe) estimate of the mean temperature difference. As the liquid velocity will be low in the sub-cooled region the heat-transfer coefficient should be estimated using correlations for natural convection (see Volume 1, Chapter 7); a typical value would be 200 $W/m^2 \,^\circ C$.

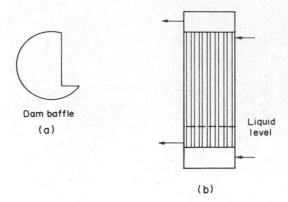

FIG. 12.47. Arrangements for sub-cooling

12.10.8. *Condensation of mixtures*

The correlations given in the previous sections apply to the condensation of a single component; such as an essentially pure overhead product from a distillation column. The design of a condenser for a mixture of vapours is a more difficult task.

The term "mixture of vapours" covers three related situations of practical interest:

1. Total condensation of a multicomponent mixture; such as the overheads from a multicomponent distillation.
2. Condensation of only part of a multicomponent vapour mixture, all components of which are theoretically condensible. This situation will occur where the dew point of some of the lighter components is above the coolant temperature. The uncondensed component may be soluble in the condensed liquid; such as in the condensation of some hydrocarbons mixtures containing light "gaseous" components.
3. Condensation from a non-condensable gas, where the gas is not soluble to any extent in the liquid condensed. These exchangers are often called cooler-condensers.

The following features, common to all these situations, must be considered in the developing design methods for mixed vapour condensers:

1. The condensation will not be isothermal. As the heavy component condenses out the composition of the vapour, and therefore its dew point, change.
2. Because the condensation is not isothermal there will be a transfer of sensible heat from the vapour to cool the gas to the dew point. There will also be a transfer of sensible heat from the condensate, as it must be cooled from the temperature at which it condensed to the outlet temperature. The transfer of sensible heat from the vapour can be particularly significant, as the sensible-heat transfer coefficient will be appreciably lower than the condensation coefficient.
3. As the composition of the vapour and liquid change throughout the condenser their physical properties vary.
4. The heavy component must diffuse through the lighter components to reach the condensing surface. The rate of condensation will be governed by the rate of diffusion, as well as the rate of heat transfer.

Temperature profile

To evaluate the true temperature difference (driving force) in a mixed vapour condenser a condensation curve (temperature vs. enthalpy diagram) must be calculated; showing the change in vapour temperature versus heat transferred throughout the condenser, Fig. 12.48. The temperature profile will depend on the liquid-flow pattern in the condenser. There are two limiting conditions of condensate-vapour flow:

1. Differential condensation: in which the liquid separates from the vapour from which it has condensed. This process is analogous to differential, or Rayleigh, distillation, and the condensation curve can be calculated using methods similar to those for determining the change in composition in differential distillation; see Volume 2, Chapter 11.
2. Integral condensation: in which the liquid remains in equilibrium with the uncondensed vapour. The condensation curve can be determined using procedures similar to those for multicomponent flash distillation given in Chapter 11. This will be a relatively simple calculation for a binary mixture, but complex and tedious for mixtures of more than two components.

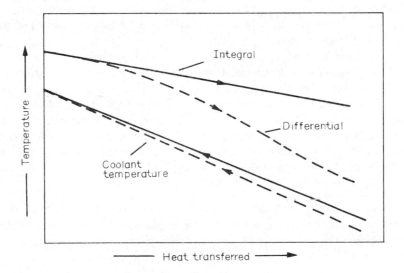

Fig. 12.48. Condensation curves

It is normal practice to assume that integral condensation occurs. The conditions for integral condensation will be approached if condensation is carried out in one pass, so that the liquid and vapour follow the same path; as in a vertical condenser with condensation inside or outside the tubes. In a horizontal shell-side condenser the condensate will tend to separate from the vapour. The mean temperature difference will be lower for differential condensation, and arrangements where liquid separation is likely to occur should generally be avoided for the condensation of mixed vapours.

Where integral condensation can be considered to occur, the use of a corrected logarithmic mean temperature difference based on the terminal temperatures will generally give a conservative (safe) estimate of the mean temperature difference, and can be used in preliminary design calculations.

Estimation of heat-transfer coefficients

Total condensation. For the design of a multicomponent condenser in which the vapour is totally condensed, an estimate of the mean condensing coefficient can be made using the single component correlations with the liquid physical properties evaluated at the average condensate composition. It is the usual practice to apply a factor of safety to allow for the sensible-heat transfer and any resistance to mass transfer. Frank (1978) suggests a factor of 0·65, but this is probably too pessimistic. Kern (1950) suggests increasing the area calculated for condensation alone by the ratio of the total heat (condensing + sensible) to the condensing load. Where a more exact estimate of the coefficient is required, and justified by the data, the rigorous methods developed for partial condensation can be used.

Partial condensation. The methods developed for partial condensation and condensation from a non-condensable gas can be divided into two classes:

1. Empirical methods: approximate methods, in which the resistance to heat transfer is considered to control the rate of condensation, and the mass transfer resistance is neglected. Design methods have been published by Silver (1947), Bell and Ghaly (1973) and Ward (1960).
2. Analytical methods: more exact procedures, which are based on some model of the heat and mass transfer process, and which take into account the diffusional resistance to mass transfer. The classic method is that of Colburn and Hougen (1934); see also Colburn and Drew (1937) and Porter and Jeffreys (1963). The analytical methods are complex, requiring step-by-step, trial and error, calculations, or graphical procedures. They are suited for computer solution using numerical methods; and proprietary design programs are available. Examples of the application of the Colburn and Drew method are given by Kern (1950) and Jeffreys (1961). The method is discussed briefly in Volume 1, Chapter 7.

Approximate methods. The local coefficient for heat transfer can be expressed in terms of the local condensate film coefficient h'_c and the local coefficient for sensible-heat transfer from the vapour (the gas film coefficient) h'_g, by a relationship first proposed by Silver (1947):

$$\frac{1}{h'_{cg}} = \frac{1}{h'_c} + \frac{Z}{h'_g} \qquad (12.58)$$

where h'_{cg} = the local effective cooling-condensing coefficient

and $Z = \dfrac{\Delta H_s}{\Delta H_t} = x C_{p_g} \dfrac{dT}{dH_t}$,

$\dfrac{\Delta H_s}{\Delta H_t}$ = the ratio of the change in sensible heat to the total enthalpy change.

$\dfrac{dT}{dH_t}$ = slope of the temperature–enthalpy curve,

x = vapour quality, mass fraction of vapour,

C_{p_g} = vapour (gas) specific heat.

The term dT/dH_t can be evaluated from the condensation curve; h'_c from the single component correlations; and h'_g from correlations for forced convection.

If this is done at several points along the condensation curve the area required can be determined by graphical or numerical integration of the expression:

$$A = \int_0^{Q_t} \frac{dQ}{U(T_v - t_c)} \qquad (12.59)$$

where Q_t = total heat transferred,
$\quad U$ = overall heat transfer coefficient, from equation 12.1, using h'_{cg},
$\quad T_v$ = local vapour (gas) temperature,
$\quad t_c$ = local cooling medium temperature.

Gilmore (1963) gives an integrated form of equation 12.57, which can be used for the approximate design of partial condensers

$$\frac{1}{h_{cg}} = \frac{1}{h_c} + \frac{Q_g}{Q_t} \frac{1}{h_g} \qquad (12.60)$$

where $\quad h_{cg}$ = mean effective coefficient,
$\quad h_c$ = mean condensate film coefficient, evaluated from the single-component correlations, at the average condensate composition, and total condensate loading,
$\quad h_g$ = mean gas film coefficient, evaluated using the average vapour flow-rate: arithmetic mean of the inlet and outlet vapour (gas) flow-rates,
$\quad Q_g$ = total sensible-heat transfer from vapour (gas),
$\quad Q_t$ = total heat transferred: latent heat of condensation + sensible heat for cooling the vapour (gas) and condensate.

As a rough guide, the following rules of thumb suggested by Frank (1978) can be used to decide the design method to use for a partial condenser (cooler-condenser):

1. Non-condensables < 0·5 per cent: use the methods for total condensation; ignore the presence of the uncondensed portion.
2. Non-condensables >70 per cent: assume the heat transfer is by forced convection only. Use the correlations for forced convection to calculate the heat-transfer coefficient, but include the latent heat of condensation in the total heat load transferred.
3. Between 0·5 to 70 per cent non-condensables: use methods that consider both mechanisms of heat transfer.

In partial condensation it is usually better to put the condensing stream on the shell-side, and to select a baffle spacing that will maintain high vapour velocities, and therefore high sensible-heat-transfer coefficients.

Fog formation. In the condensation of a vapour from a non-condensable gas, if the bulk temperature of the gas falls below the dew point of the vapour, liquid can condense out directly as a mist or fog. This condition is undesirable, as liquid droplets may be carried out of the condenser. Fog formation in cooler-condensers is discussed by Colburn and Edison (1941), and by Steinmeyer (1972), who gives criteria for the prediction of fog formation. Demisting pads can be used to separate entrained liquid droplets.

12.10.9. Pressure drop in condensers

The pressure drop on the condensing side is difficult to predict as two phases are present and the vapour mass velocity is changing throughout the condenser.

A common practice is to calculate the pressure drop using the methods for single-phase flow and apply a factor to allow for the change in vapour velocity. For total condensation, Frank (1978) suggests taking the pressure drop as 40 per cent of the value based on the inlet vapour conditions; Kern (1950) suggests a factor of 50 per cent.

An alternative method, which can also be used to estimate the pressure drop in a partial condenser, is given by Gloyer (1970). The pressure drop is calculated using an average vapour flow-rate in the shell (or tubes) estimated as a function of the ratio of the vapour flow-rate in and out of the shell (or tubes), and the temperature profile.

$$W_s(\text{average}) = W_s(\text{inlet}) \times K_2 \qquad (12.61)$$

K_2 is obtained from Fig. 12.49.

$\Delta T_{\text{in}}/\Delta T_{\text{out}}$ in Fig. 12.49 is the ratio of the terminal temperature differences.

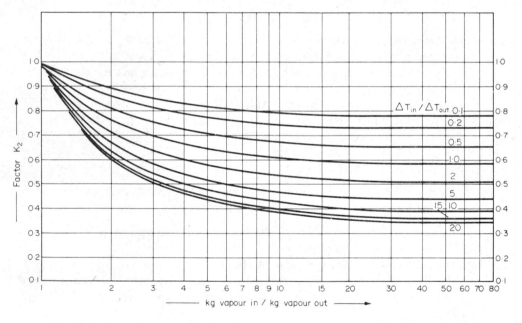

FIG. 12.49. Factor for average vapour flow-rate for pressure-drop calculation (Gloyer, 1970)

These methods can be used to make a crude estimate of the likely pressure drop. A reliable prediction can be obtained by treating the problem as one of two-phase flow. For tube-side condensation the general methods for two-phase flow in pipes can be used; see Collier (1972); and Volume 1, Chapter 4. As the flow pattern will be changing throughout condensation, some form of step-wise procedure will need to be used. Two-phase flow on the shell-side is discussed by Grant (1973), who gives a method for predicting the pressure drop based on Tinker's shell-side flow model.

Pressure drop is only likely to be a major consideration in the design of vacuum condensers; and where reflux is returned to a column by gravity flow from the condenser.

Example 12.5

Design a condenser for the following duty: 45,000 kg/h of mixed light hydrocarbon vapours to be condensed. The condenser to operate at 10 bar. The vapour will enter the condenser saturated at 60°C and the condensation will be complete at 45°C. The average molecular weight of the vapours is 52. The enthalpy of the vapour is 596·5 kJ/kg and the condensate 247·0 kJ/kg. Cooling water is available at 30°C and the temperature rise is to be limited to 10°C. Plant standards require tubes of 20 mm o.d., 16·8 mm i.d., 4·88 m (16 ft) long, of admiralty brass. The vapours are to be totally condensed and no sub-cooling is required.

Solution

Only the thermal design will be done. The physical properties of the mixture will be taken as the mean of those for n-propane (MW = 44) and n-butane (MW = 58), at the average temperature.

$$\text{Heat transferred from vapour} = \frac{45,000}{3600}(596{\cdot}5 - 247{\cdot}0) = 4368{\cdot}8 \text{ kW}$$

$$\text{Cooling water flow} = \frac{4368{\cdot}8}{(40-30)4{\cdot}18} = \underline{\underline{104{\cdot}5 \text{ kg/s}}}$$

Assumed overall coefficient (Table 12.1) = 900 W/m² °C

Mean temperature difference: the condensation range is small and the change in saturation temperature will be linear, so the corrected logarithmic mean temperature difference can be used.

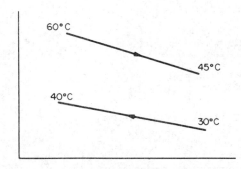

(12.6) $$R = \frac{(60-45)}{(40-30)} = 1{\cdot}5$$

(12.7) $$S = \frac{(40-30)}{(60-30)} = 0{\cdot}33$$

Try a horizontal exchanger, condensation in the shell, four tube passes. For one shell pass, four tube passes, from Fig. 12.19, $F_t = 0.92$

$$\Delta T_{lm} = \frac{(60 - 40) - (45 - 30)}{\ln \dfrac{(60 - 40)}{(45 - 30)}} = 17.4°C$$

$$\Delta T_m = 0.92 \times 17.4 = 16°C$$

$$\text{Trial area} = \frac{4368.8 \times 10^3}{900 \times 16} = 303 \text{ m}^2$$

Surface area of one tube $= 20 \times 10^{-3} \pi \times 4.88 = 0.305 \text{ m}^2$ (ignore tube sheet thickness)

$$\text{Number of tubes} = \frac{303}{0.305} = 992$$

Use square pitch, $P_t = 1.25 \times 20 \text{ mm} = 25 \text{ mm}$.
Tube bundle diameter

(12.3b)
$$D_b = 20 \left(\frac{992}{0.158}\right)^{1/2.263} = 954 \text{ mm}$$

Number of tubes in centre row $N_r = D_b/P_t = \dfrac{954}{25} = 38$

Shell-side coefficient
Estimate tube wall temperature, T_w; assume condensing coefficient of 1500 W/m² °C,

Mean temperature

$$\text{Shell-side} = \frac{60 + 45}{2} = 52.5°C$$

$$\text{Tube-side} = \frac{40 + 30}{2} = 35°C$$

$$(52.5 - T_w)1500 = (52.5 - 35)900$$
$$T_w = 42.0°C$$

$$\text{Mean temperature condensate} = \frac{52.5 + 42.0}{2} = 47°C$$

Physical properties at 47°C

$$\mu_L = 0.16 \text{ mN s/m}^2$$
$$\rho_L = 551 \text{ kg/m}^3$$
$$k_L = 0.13 \text{ W/m°C}$$

vapour density at mean vapour temperature

$$\rho_v = \frac{52}{22 \cdot 4} \times \frac{273}{(273 + 52 \cdot 5)} \times \frac{10}{1} = 19 \cdot 5 \text{ kg/m}^3$$

$$\Gamma_h = \frac{W_c}{LN_t} = \frac{45{,}000}{3600} \times \frac{1}{4 \cdot 88 \times 992} = 2 \cdot 6 \times 10^{-3} \text{ kg/s m}$$

$$N_r = 2/3 \times 38 = 25$$

(12.50)
$$h_c = 0 \cdot 95 \times 0 \cdot 13 \left[\frac{551(551 - 19 \cdot 5)\,9 \cdot 81}{0 \cdot 16 \times 10^{-3} \times 2 \cdot 6 \times 10^{-3}} \right]^{1/3} \times 25^{-1/6}$$

$$= 1375 \text{ W/m}^2 \, {}^\circ\text{C}$$

Close enough to assumed value of 1500 W/m² °C, so no correction to T_w needed.

Tube-side coefficient

Tube cross-sectional area $= \dfrac{\pi}{4}(16 \cdot 8 \times 10^{-3})^2 \times \dfrac{992}{4} = 0 \cdot 055 \text{ m}^2$

Density of water, at 35°C $= 993 \text{ kg/m}^3$

Tube velocity $\qquad = \dfrac{104 \cdot 5}{993} \times \dfrac{1}{0 \cdot 055} = 1 \cdot 91 \text{ m/s}$

(12.17)
$$h_i = 4200(1 \cdot 35 + 0 \cdot 02 \times 35)\,1 \cdot 91^{0 \cdot 8}/16 \cdot 8^{0 \cdot 2}$$
$$= 8218 \text{ W/m}^2 \, {}^\circ\text{C}$$

Fouling factors: as neither fluid is heavily fouling, use 6000 W/m² °C for each side.

$$k_w = 50 \text{ W/m} \, {}^\circ\text{C}$$

Overall coefficient

(12.2)
$$\frac{1}{U} = \frac{1}{1375} + \frac{1}{6000} + \frac{20 \times 10^{-3} \ln\left(\dfrac{20}{16 \cdot 8}\right)}{2 \times 50} + \frac{20}{16 \cdot 8} \times \frac{1}{6000} + \frac{20}{16 \cdot 8} \times \frac{1}{8218}$$

$$U = \underline{\underline{786 \text{ W/m}^2 \, {}^\circ\text{C}}}$$

Significantly lower than the assumed value of 900 W/m² °C.
Repeat calculation using new trial value of 750 W/m² °C.

$$\text{Area} = \frac{4368 \times 10^3}{750 \times 16} = 364 \text{ m}^2$$

$$\text{Number of tubes} = \frac{364}{0 \cdot 305} = 1194$$

$$D_b = 20 \left(\frac{1194}{0 \cdot 158} \right)^{1/2 \cdot 263} = 1035 \text{ mm}$$

Number of tubes in centre row $= \dfrac{1035}{25} = 41$

$\Gamma_h = \dfrac{45,000}{3600} \times \dfrac{1}{4 \cdot 88 \times 1194} = 2 \cdot 15 \times 10^{-3} \, \text{kg/m s}$

$N_r = 2/3 \times 41 = 27$

$h_c = 0 \cdot 95 \times 0 \cdot 13 \left[\dfrac{551(551 - 19 \cdot 5)9 \cdot 81}{0 \cdot 16 \times 10^{-3} \times 2 \cdot 15 \times 10^{-3}} \right]^{1/3} \times 27^{-1/6}$

$\qquad = 1447 \, \text{W/m}^2 \, ^\circ\text{C}$

New tube velocity $= 1 \cdot 91 \times \dfrac{992}{1194} = 1 \cdot 59 \, \text{m/s}$

$h_i = 4200 \, (1 \cdot 35 + 0 \cdot 02 \times 35) \dfrac{1 \cdot 59^{0 \cdot 8}}{16 \cdot 8^{0 \cdot 2}} = 7097 \, \text{W/m}^2 \, ^\circ\text{C}$

$\dfrac{1}{U} = \dfrac{1}{1447} + \dfrac{1}{6000} + \dfrac{20 \times 10^{-3} \ln \left(\dfrac{20}{16 \cdot 8} \right)}{2 \times 50} + \dfrac{20}{16 \cdot 8} \times \dfrac{1}{6000} + \dfrac{20}{16 \cdot 8} \times \dfrac{1}{7097}$

$U = \underline{773 \, \text{W/m}^2 \, ^\circ\text{C}}$

Close enough to estimate, firm up design.

Shell-side pressure drop
Use pull-through floating head, no need for close clearance.
Select baffle spacing = shell diameter, 45 per cent cut.

From Fig. 12.10, clearance = 95 mm.

$$\text{Shell i.d.} = 1035 + 95 = 1130 \, \text{mm}$$

Use Kern's method to make an approximate estimate.

(12.21)   Cross-flow area $A_s = \dfrac{(25 - 20)}{25} 1130 \times 1130 \times 10^{-6}$

$$= 0 \cdot 255 \, \text{m}^2$$

Mass flow-rate, based on inlet conditions

$$G_s = \dfrac{45,000}{3600} \times \dfrac{1}{0 \cdot 255} = 49 \cdot 02 \, \text{kg/s m}^2$$

(12.22)   Equivalent diameter, $d_e = \dfrac{1 \cdot 27}{20} (25^2 - 0 \cdot 785 \times 20^2)$

$$= 19 \cdot 8 \, \text{mm}$$

Vapour viscosity $= 0 \cdot 008 \, \text{mN s/m}^2$

$$Re = \dfrac{49 \cdot 02 \times 19 \cdot 8 \times 10^{-3}}{0 \cdot 008 \times 10^{-3}} = 121,325$$

From Fig. 12.30, $j_f = 2 \cdot 2 \times 10^{-2}$

$$u_s = \frac{G_s}{\rho_v} = \frac{49 \cdot 02}{19 \cdot 5} = 2 \cdot 51 \text{ m/s}$$

Take pressure drop as 50 per cent of that calculated using the inlet flow; neglect viscosity correction.

(12.26)

$$\Delta P_s = \frac{1}{2}\left[8 \times 2 \cdot 2 \times 10^{-2} \left(\frac{1130}{19 \cdot 8}\right)\left(\frac{4 \cdot 88}{1 \cdot 130}\right)\frac{19 \cdot 5 (2 \cdot 51)^2}{2} \right]$$

$$= 1322 \text{ N/m}^2$$

$$= \underline{\underline{1 \cdot 3 \text{ kPa}}}$$

Negligible; more sophisticated method of calculation not justified.

Tube-side pressure drop

$$\text{Viscosity of water} = 0 \cdot 6 \text{ mN s/m}^2$$

$$Re = \frac{u_t \rho d_i}{\mu} = \frac{1 \cdot 59 \times 993 \times 16 \cdot 8 \times 10^{-3}}{0 \cdot 6 \times 10^{-3}} = \underline{\underline{44,208}}$$

From Fig. 12.24, $j_f = 3 \cdot 5 \times 10^{-3}$.

Neglect viscosity correction.

(12.20)

$$\Delta P_t = 4\left[8 \times 3 \cdot 5 \times 10^{-3} \left(\frac{4 \cdot 88}{16 \cdot 8 \times 10^{-3}}\right) + 2 \cdot 5 \right]\frac{993 \times 1 \cdot 59^2}{2}$$

$$= 53,388 \text{ N/m}^2$$

$$= \underline{\underline{53 \text{ kPa}}} \quad (7 \cdot 7 \text{ psi}),$$

acceptable.

12.11. Reboilers and vaporisers

The design methods given in this section can be used for reboilers and vaporisers. Reboilers are used with distillation columns to vaporise a fraction of the bottom product; whereas in a vaporiser essentially all the feed is vaporised.

Three principal types of reboiler are used:

1. Forced circulation, Fig. 12.50: in which the fluid is pumped through the exchanger, and the vapour formed is separated in the base of the column. When used as a vaporiser a disengagement vessel will have to be provided.
2. Thermosyphon, natural circulation, Fig. 12.51: vertical exchangers with vaporisation in the tubes, or horizontal exchangers with vaporisation in the shell. The liquid circulation through the exchanger is maintained by the difference in density between the two-phase mixture of vapour and liquid in the exchanger and the single-phase liquid in the base of the column. As with the forced-circulation type, a disengagement vessel will be needed if this type is used as a vaporiser.
3. Kettle type, Fig. 12.52: in which boiling takes place on tubes immersed in a pool of

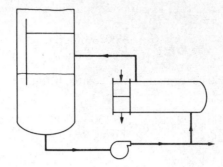

FIG. 12.50. Forced-circulation reboiler

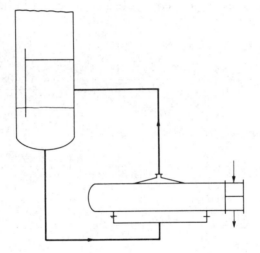

FIG. 12.51. Horizontal thermosyphon reboiler

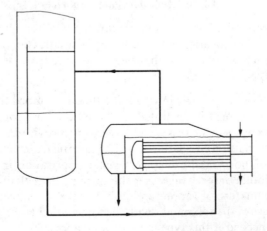

FIG. 12.52. Kettle reboiler

liquid; there is no circulation of liquid through the exchanger. This type is also, more correctly, called a submerged bundle reboiler.

In some applications it is possible to accommodate the bundle in the base of the column, Fig. 12.53; saving the cost of the exchanger shell.

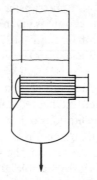

FIG. 12.53. Internal reboiler

Choice of type

The choice of the best type of reboiler or vaporiser for a given duty will depend on the following factors:

1. The nature of the process fluid, particularly its viscosity and propensity to fouling.
2. The operating pressure: vacuum or pressure.
3. The equipment layout, particularly the headroom available.

Forced-circulation reboilers are especially suitable for handling viscous and heavily fouling process fluids; see Chantry and Church (1958). The circulation rate is predictable and high velocities can be used. They are also suitable for low vacuum operations, and for low rates of vaporisation. The major disadvantage of this type is that a pump is required and the pumping cost will be high. There is also the danger that leakage of hot fluid will occur at the pump seal; canned-rotor type pumps can be specified to avoid the possibility of leakage.

Thermosyphon reboilers are the most economical type for most applications, but are not suitable for high viscosity fluids or high vacuum operation. They would not normally be specified for pressures below 0·3 bar. A disadvantage of this type is that the column base must be elevated to provide the hydrostatic head required for the thermosyphon effect. This will increase the cost of the column supporting-structure. Horizontal reboilers require less headroom than vertical, but have more complex pipework. Horizontal exchangers are more easily maintained than vertical, as tube bundle can be more easily withdrawn.

Kettle reboilers have lower heat-transfer coefficients than the other types, as there is no liquid circulation. They are not suitable for fouling materials, and have a high residence time. They will generally be more expensive than an equivalent thermosyphon type as a larger shell is needed, but if the duty is such that the bundle can be installed in the column base, the cost will be competitive with the other types. They are often used as vaporisers, as a separate vapour–liquid disengagement vessel is not needed. They are suitable for vacuum operation, and for high rates of vaporisation, up to 80 per cent of the feed.

12.11.1. Boiling heat-transfer fundamentals

The complex phenomena involved in heat transfer to a boiling liquid are discussed in Volume 1, Chapter 7. A more detailed account is given by Collier (1972), Westwater (1956, 1958), Rohsenow (1973) and Hsu and Graham (1976). Only a brief discussion of the subject will be given in this section: sufficient for the understanding of the design methods given for reboilers and vaporisers.

The mechanism of heat transfer from a submerged surface to a pool of liquid depends on the temperature difference between the heated surface and the liquid; Fig. 12.54. At low-temperature differences, when the liquid is below its boiling point, heat is transferred by natural convection. As the surface temperature is raised incipient boiling occurs, vapour bubbles forming and breaking loose from the surface. The agitation caused by the rising bubbles, and other effects caused by bubble generation at the surface, result in a large increase in the rate of heat transfer. This phenomenon is known as nucleate boiling. As the temperature is raised further the rate of heat transfer increases until the heat flux reaches a critical value. At this point, the rate of vapour generation is such that dry patches occur spontaneously over the surface, and the rate of heat transfer falls off rapidly. At higher temperature differences, the vapour rate is such that the whole surface is blanketed with vapour, and the mechanism of transfer is by conduction through the vapour film. Conduction is augmented at high temperature differences by radiation.

The maximum heat flux achievable with nucleate boiling is known as the critical heat flux. In a system where the surface temperature is not self-limiting, such as a nuclear reactor fuel element, operation above the critical flux will result in a rapid increase in the surface temperature, and in the extreme situation the surface will melt. This phenomenon is known as "burn-out". The heating media used for process plant are normally self-

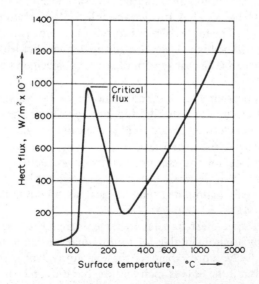

FIG. 12.54. Typical pool boiling curve (water at 1 bar)

limiting; for example, with steam the surface temperature can never exceed the saturation temperature. Care must be taken in the design of electrically heated vaporisers to ensure that the critical flux can never be exceeded.

The critical flux is reached at surprisingly low temperature differences; around 20 to 30°C for water, and 20 to 50°C for light organics.

Estimation of boiling heat-transfer coefficients

In the design of vaporisers and reboilers the designer will be concerned with two types of boiling: pool boiling and convective boiling. Pool boiling is the name given to nucleate boiling in a pool of liquid; such as in kettle-type reboiler or a jacketed vessel. Convective boiling occurs where the vaporising fluid is flowing over the heated surface, and heat transfer takes place both by forced convection and nucleate boiling; as in forced circulation or thermosyphon reboilers.

Boiling is a complex phenomenon, and boiling heat-transfer coefficients are difficult to predict with any certainty. Whenever possible experimental values obtained for the system being considered should be used, or values for a closely related system.

12.11.2. Pool boiling

In the nucleate boiling region the heat-transfer coefficient is dependent on the nature and condition of the heat-transfer surface, and it is not possible to present a universal correlation that will give accurate predictions for all systems. Palen and Taborek (1962) have reviewed the published correlations and compared their suitability for use in reboiler design.

The correlation given by Forster and Zuber (1955) can be used to estimate pool boiling coefficients, in the absence of experimental data. Their equation can be written in the form:

$$h_{nb} = 0.00122 \left[\frac{k_L^{0.79} C_{pL}^{0.45} \rho_L^{0.49}}{\sigma^{0.5} \mu_L^{0.29} \lambda^{0.24} \rho_v^{0.24}} \right] (T_w - T_s)^{0.24} (p_w - p_s)^{0.75} \qquad (12.62)$$

where h_{nb} = nucleate, pool, boiling coefficient, W/m^2°C,

$\quad k_L$ = liquid thermal conductivity, W/m°C,

$\quad C_{pL}$ = liquid heat capacity, J/kg°C,

$\quad \rho_L$ = liquid density, kg/m^3,

$\quad \mu_L$ = liquid viscosity, N s/m^2,

$\quad \lambda$ = latent heat, J/kg,

$\quad \rho_v$ = vapour density, kg/m^3,

$\quad T_w$ = wall, surface temperature, °C,

$\quad T_s$ = saturation temperature of boiling liquid °C,

$\quad p_w$ = saturation pressure corresponding to the wall temperature, T_w, N/m^2,

$\quad p_s$ = saturation pressure corresponding to T_s, N/m^2,

$\quad \sigma$ = surface tension, N/m.

The reduced pressure correlation given by Mostinski (1963) is simple to use and gives values that are as reliable as those given by more complex equations.

$$h_{nb} = 0.104 (P_c)^{0.69} (q)^{0.7} \left[1.8 (P/P_c)^{0.17} + 4(P/P_c)^{1.2} + 10(P/P_c)^{10} \right] \qquad (12.63)$$

where P = operating pressure, bar,
  P_c = liquid critical pressure, bar,
  q = heat flux, W/m^2.

Note. $q = h_{nb}(T_w - T_s)$.

Mostinski's equation is convenient to use when data on the fluid physical properties are not available.

Equations 12.62 and 12.63 are for boiling single component fluids; for mixtures the coefficient will generally be lower than that predicted by these equations. The equations can be used for close boiling range mixtures, say less than $5\,^{\circ}C$; and for wider boiling ranges with a suitable factor of safety (see Section 12.11.6).

Critical heat flux

It is important to check that the design, and operating, heat flux is well below the critical flux. Several correlations are available for predicting the critical flux. That given by Zuber *et al.* (1961) has been found to give satisfactory predictions for use in reboiler and vaporiser design. In SI units, Zuber's equation can be written as:

$$q_c = 0.131 \lambda [\sigma g (\rho_L - \rho_v) \rho_v^2]^{1/4} \tag{12.64}$$

where q_c = maximum, critical, heat flux, W/m^2,
  g = gravitational acceleration, $9.81\ m/s^2$.

Mostinski also gives a reduced pressure equation for predicting the maximum critical heat flux:

$$q_c = 3.67 \times 10^4 P_c (P/P_c)^{0.35} [1 - (P/P_c)]^{0.9} \tag{12.65}$$

Film boiling

The equation given by Bromley (1950) can be used to estimate the heat-transfer coefficient for film boiling on tubes. Heat transfer in the film-boiling region will be controlled by conduction through the film of vapour, and Bromley's equation is similar to the Nusselt equation for condensation, where conduction is occurring through the film of condensate.

$$h_{fb} = 0.62 \left[\frac{k_v^3 (\rho_L - \rho_v) \rho_v g \lambda}{\mu_v d_o (T_w - T_s)} \right]^{1/4} \tag{12.66}$$

where h_{fb} is the film boiling heat-transfer coefficient; the suffix v refers to the vapour phase and d_o is in metres. It must be emphasised that process reboilers and vaporisers will always be designed to operate in the nucleate boiling region. The heating medium would be selected, and its temperature controlled, to ensure that in operation the temperature difference is well below that at which the critical flux is reached. For instance, if direct heating with steam would give too high a temperature difference, the steam would be used to heat water, and hot water used as the heating medium.

Example 12.6

Estimate the heat-transfer coefficient for the pool boiling of water at 2.1 bar, from a surface at $125\,°C$. Check that the critical flux is not exceeded.

Solution

Physical properties, from steam tables:

Saturation temperature, $T_s = 121.8°C$

$$\rho_L = 941.6 \text{ kg/m}^3, \ \rho_v = 1.18 \text{ kg/m}^3$$
$$C_{pL} = 4.25 \times 10^3 \text{ J/kg°C}$$
$$k_L = 687 \times 10^{-3} \text{ W/m °C}$$
$$\mu_L = 230 \times 10^{-6} \text{ N s/m}^2$$
$$\lambda = 2198 \times 10^3 \text{ J/kg}$$
$$\sigma = 55 \times 10^{-3} \text{ N/m}$$
$$p_w \text{ at } 125°C = 2.321 \times 10^5 \text{ N/m}^2$$
$$p_s = 2.1 \times 10^5 \text{ N/m}^2$$

Use the Foster–Zuber correlation, equation 12.62:

$$h_b = 1.22 \times 10^{-3} \left[\frac{(687 \times 10^{-3})^{0.79}(4.25 \times 10^3)^{0.45}(941.6)^{0.49}}{(55 \times 10^{-3})^{0.5}(230 \times 10^{-6})^{0.29}(2198 \times 10^3)^{0.24}1.18^{0.24}} \right]$$
$$\times (125 - 121.8)^{0.24}(2.321 \times 10^5 - 2.10 \times 10^5)^{0.75}$$
$$= \underline{\underline{3736 \text{ W/m}^2 \text{ °C}}}$$

Use the Forster–Zuber correlation, equation 12.62:

$$q_c = 0.131 \times 2198 \times 10^3 \left[55 \times 10^{-3} \times 9.81(941.6 - 1.18)1.18^2 \right]^{1/4}$$
$$= \underline{\underline{1.48 \times 10^6 \text{ W/m}^2}}$$

Actual flux $= (125 - 121.8)3736 = \underline{11,955 \text{ W/m}^2}$,

well below critical flux.

12.11.3. *Convective boiling*

The mechanism of heat transfer in convective boiling, where the boiling fluid is flowing through a tube or over a tube bundle, differs from that in pool boiling. It will depend on the state of the fluid at any point. Consider the situation of a liquid boiling inside a vertical tube; Fig 12.55. The following conditions occur as the fluid flows up the tube.

1. Single-phase flow region: at the inlet the liquid is below its boiling point (sub-cooled) and heat is transferred by forced convection. The equations for forced convection can be used to estimate the heat-transfer coefficient in this region.

2. Sub-cooled boiling: in this region the liquid next to the wall has reached boiling point, but not the bulk of the liquid. Local boiling takes place at the wall, which increases the rate of heat transfer over that given by forced convection alone.

3. Saturated boiling region: in this region bulk boiling of the liquid is occurring in a manner similar to nucleate pool boiling. The volume of vapour is increasing and various flow patterns can form (see Volume 2, Chapter 14). In a long tube, the flow pattern will finally eventually become annular: where the liquid phase is spread over the tube wall and the vapour flows up the central core.

4. Dry wall region: Ultimately, if a large fraction of the feed is vaporised, the wall dries

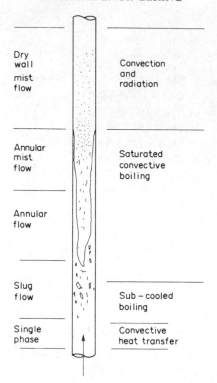

FIG. 12.55. Convective boiling in a vertical tube

out and any remaining liquid is present as a mist. Heat transfer in this region is by convection and radiation to the vapour. This condition is unlikely to occur in commercial reboilers and vaporisers.

Saturated, bulk, boiling is the principal mechanism of interest in the design of reboilers and vaporisers.

A comprehensive review of the methods and correlations available for predicting convective boiling coefficients is given by Rohsenow (1973). The method developed by Chen (1966) is convenient to use, and is outlined below and illustrated in Example 12.7.

In forced-convective boiling the effective heat-transfer coefficient h_{cb} can be considered to be made up of convective and nucleate boiling components; h'_{fc} and h'_{nb}.

$$h_{cb} = h'_{fc} + h'_{nb} \qquad (12.67)$$

The convective boiling coefficient h'_{fc} can be estimated using the equations for single-phase forced-convection heat transfer modified by a factor f_c to account for the effects of two-phase flow:

$$h'_{fc} = h_{fc} \times f_c \qquad (12.68)$$

The forced-convection coefficient h_{fc} is calculated assuming that the liquid phase is flowing in the conduit alone.

The two-phase correction factor f_c is obtained from Fig. 12.56; in which the term $1/X_{tt}$ is

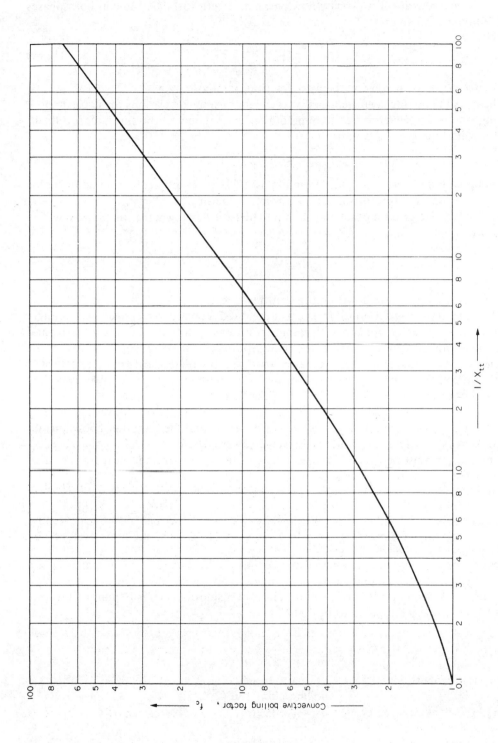

F_IG. 12.56. Convective boiling enhancement factor

the Lockhart–Martinelli two-phase flow parameter with turbulent flow in both phases (see Volume 1, Chapter 4). This parameter is given by:

$$1/X_{tt} = \left[\frac{x}{1-x}\right]^{0.9}\left[\frac{\rho_L}{\rho_v}\right]^{0.5}\left[\frac{\mu_v}{\mu_L}\right]^{0.1} \tag{12.69}$$

where x is the vapour quality, the mass fraction of vapour.

The nucleate boiling coefficient can be calculated using correlations for nucleate pool boiling modified by a factor f_s to account for the fact that nucleate boiling is more difficult in a flowing liquid.

$$h'_{nb} = h_{nb} \times f_s \tag{12.70}$$

The suppression factor f_s is obtained from Fig. 12.57. It is a function of the liquid Reynolds number Re_L and the forced-convection correction factor f_c.

Re_L is evaluated assuming that only the liquid phase is flowing in the conduit, and will be given by:

$$Re_L = \frac{(1-x)Gd_e}{\mu_L} \tag{12.71}$$

where G is the total mass flow rate per unit flow area.

Chen's method was developed from experimental data on forced convective boiling in vertical tubes. It can be applied, with caution, to forced convective boiling in horizontal tubes, and annular conduits (concentric pipes). Butterworth (1977) suggests that, in the absence of more reliable methods, it may be used to estimate the heat-transfer coefficient for forced convective boiling in cross-flow over tube bundles; using a suitable cross-flow correlation to predict the forced-convection coefficient.

A major problem that will be encountered when applying convective boiling correlations to the design of reboilers and vaporisers is that, because the vapour quality changes progressively throughout the exchanger, a step-by-step procedure will be needed. The exchanger must be divided into sections and the coefficient and heat transfer area estimated sequentially for each section.

Example 12.7

A fluid whose properties are essentially those of o-dichlorobenzene is vaporised in the tubes of a forced convection reboiler. Estimate the local heat-transfer coefficient at a point where 5 per cent of the liquid has been vaporised. The liquid velocity at the tube inlet is 2 m/s and the operating pressure is 0·3 bar. The tube inside diameter is 16 mm and the local wall temperature is estimated to be 120°C.

Solution

Physical properties:

$$\text{boiling point } 136°C$$
$$\rho_L = 1170\,\text{kg/m}^3$$
$$\mu_L = 0.45\,\text{mN s/m}^2$$
$$\mu_v = 0.01\,\text{mN s/m}^2$$

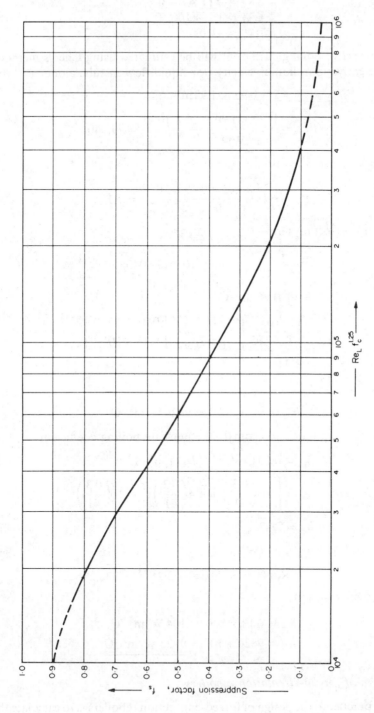

FIG. 12.57. Nucleate boiling suppression factor

$$\rho_v = 1{\cdot}31\,\text{kg/m}^3$$
$$k_L = 0{\cdot}11\,\text{W/m}^\circ\text{C}$$
$$C_{pL} = 1{\cdot}25\,\text{kJ/kg}^\circ\text{C}$$
$$P_c = 41\,\text{bar}$$

The forced-convective boiling coefficient will be estimated using Chen's method. With 5 per cent vapour, liquid velocity (for liquid flow in tube alone)

$$= 2 \times 0{\cdot}95 = 1{\cdot}90\,\text{m/s}$$

$$Re_L = \frac{1170 \times 1{\cdot}90 \times 16 \times 10^{-3}}{0{\cdot}45 \times 10^{-3}} = 79{,}040$$

From Fig. 12.23, $j_h = 3{\cdot}3 \times 10^{-3}$

$$Pr = \frac{1{\cdot}25 \times 10^3 \times 0{\cdot}45 \times 10^{-3}}{0{\cdot}11} = 5{\cdot}1$$

Neglect viscosity correction term.

(12.15) $$h_{fc} = \frac{0{\cdot}11}{16 \times 10^{-3}} \times 3{\cdot}3 \times 10^{-3}(79{,}040)(5{\cdot}1)^{0{\cdot}33}$$

$$= 3070\,\text{W/m}^2\,^\circ\text{C}$$

(12.69) $$\frac{1}{X_{tt}} = \left[\frac{0{\cdot}05}{1-0{\cdot}05}\right]^{0{\cdot}9} \left[\frac{1170}{1{\cdot}31}\right]^{0{\cdot}5} \left[\frac{0{\cdot}01 \times 10^{-3}}{0{\cdot}45 \times 10^{-3}}\right]^{0{\cdot}1}$$

$$= 1{\cdot}44$$

From Fig. 12.56, $f_c = 3{\cdot}2$

$$h'_{fc} = 3{\cdot}2 \times 3070 = 9824\,\text{W/m}^2\,^\circ\text{C}$$

Using Mostinski's correlation to estimate the nucleate boiling coefficient

(12.63) $$h_{nb} = 0{\cdot}104 \times 41^{0{\cdot}69}\left[h_{nb}(136-120)\right]^{0{\cdot}7}$$

$$\times \left[1{\cdot}8\left(\frac{0{\cdot}3}{41}\right)^{0{\cdot}17} + 4\left(\frac{0{\cdot}3}{41}\right)^{1{\cdot}2} + 10\left(\frac{0{\cdot}3}{41}\right)^{10}\right]$$

$$h_{nb} = 7{\cdot}43 h_{nb}^{0{\cdot}7}$$

$$h_{nb} = 800\,\text{W/m}^2\,^\circ\text{C}$$

$$Re_L f_c^{1{\cdot}25} = 79{,}040 \times 3{\cdot}2^{1{\cdot}25} = 338{,}286$$

From Fig. 12.57, $f_s = 0{\cdot}13$,

$$h'_{nb} = 0{\cdot}13 \times 800 = 104\,\text{W/m}^2\,^\circ\text{C}$$
$$h_{cb} = 9824 + 104 = \underline{\underline{9928\,\text{W/m}^2\,^\circ\text{C}}}$$

12.11.4. Design of forced-circulation reboilers

The normal practice in the design of forced-convection reboilers is to calculate the heat-transfer coefficient assuming that the heat is transferred by forced convection only. This will give conservative (safe) values, as any boiling that occurs will invariably increase the rate of heat transfer. In many designs the pressure is controlled to prevent any appreciable vaporisation in the exchanger. A throttle value is installed in the exchanger outlet line, and the liquid flashes as the pressure is let down into the vapour–liquid separation vessel.

If a significant amount of vaporisation does occur, the heat-transfer coefficient can be evaluated using correlations for convective boiling, such as Chen's method.

Conventional shell and tube exchanger designs are used, with one shell pass and two tube passes, when the process fluid is on the shell side; and one shell and one tube pass when it is in the tubes. High tube velocities are used to reduce fouling, 3–9 m/s.

Because the circulation rate is set by the designer, forced-circulation reboilers can be designed with more certainty than natural circulation units.

The critical flux in forced-convection boiling is difficult to predict. Kern (1950) recommends that for commercial reboiler designs the heat flux should not exceed 63,000 W/m² (20,000 Btu/ft² h) for organics and 95,000 W/m² (30,000 Btu/ft² h) for water and dilute aqueous solutions. These values are now generally considered to be too pessimistic.

12.11.5. Design of thermosyphon reboilers

The design of thermosyphon reboilers is complicated by the fact that, unlike a forced-convection reboiler, the fluid circulation rate cannot be determined explicitly. The circulation rate, heat-transfer rate and pressure drop are all interrelated, and iterative design procedures must be used. The fluid will circulate at a rate at which the pressure losses in the system are just balanced by the available hydrostatic head. The exchanger, column base and piping can be considered as the two legs of a U-tube; Fig. 12.58. The driving force for circulation round the system is the difference in density of the liquid in the "cold" leg (the column base and inlet piping) and the two-phase fluid in the "hot" leg (the exchanger tubes and outlet piping).

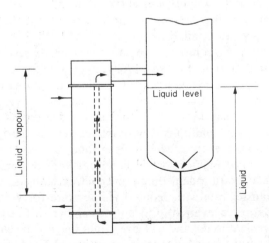

Liquid level

Liquid – vapour

Liquid

FIG. 12.58. Vertical thermosyphon reboiler, liquid and vapour flows

To calculate the circulation rate it is necessary to make a pressure balance round the system.

A typical design procedure will include the following steps:

1. Calculate the vaporisation rate required; from the specified duty.
2. Estimate the exchanger area; from an assumed value for the overall heat-transfer coefficient. Decide the exchanger layout and piping dimensions.
3. Assume a value for the circulation rate through the exchanger.
4. Calculate the pressure drop in the inlet piping (single phase).
5. Divide the exchanger tube into sections and calculate the pressure drop section-by-section up the tube. Use suitable methods for the sections in which the flow is two-phase. Include the pressure loss due to the fluid acceleration as the vapour rate increases. For a horizontal reboiler, calculate the pressure drop in the shell, using a method suitable for two-phase flow.
6. Calculate the pressure drop in the outlet piping (two-phase).
7. Compare the calculated pressure drop with the available differential head; which will depend on the vapour voidage, and hence the assumed circulation rate. If a satisfactory balance has been achieved, proceed. If not, return to step 3 and repeat the calculations with a new assumed circulation rate.
8. Calculate the heat-transfer coefficient and heat-transfer rate section-by-section up the tubes. Use a suitable method for the sections in which the boiling is occurring; such as Chen's method.
9. Calculate the rate of vaporisation from the total heat-transfer rate, and compare with the value assumed in step 1. If the values are sufficiently close, proceed. If not, return to step 2 and repeat the calculations for a new design.
10. Check that the critical heat flux is not exceeded at any point up the tubes.
11. Repeat the complete procedure as necessary to optimise the design.

It can be seen that to design a thermosyphon reboiler using hand calculations would be tedious and time-consuming. The iterative nature of the procedure lends itself to solution by computers. Sarma *et al.* (1973) discuss the development of a computer program for vertical thermosyphon reboiler design, and give algorithms and design equations. Proprietary programs are also available.

In the absence of access to a computer program the rigorous design methods given by Fair (1960, 1963) or Hughmark (1961, 1964, 1969) can be used for thermosyphon vertical reboilers. Collins (1976) gives a design procedure for horizontal thermosyphon reboilers.

Approximate methods can be used for preliminary designs. Fair (1960) and Kern (1950) give methods in which the heat transfer and pressure drop in the tubes are based on the average of the inlet and outlet conditions. This simplifies step 5 in the design procedure but trial-and-error calculations are still needed to determine the circulation rate. Frank and Prickett (1973) programmed Fair's rigorous design method for computer solution and used it, together with operating data on commercial exchangers, to derive a general correlation of heat-transfer rate with reduced temperature for vertical thermosyphon reboilers. Their correlation, converted to SI units, is shown in Fig 12.59.

Figure 12.59 is convenient to use and will give values that are as reliable as other, more complex and time consuming, approximate design methods. The basis and limitations of the correlation are listed below:

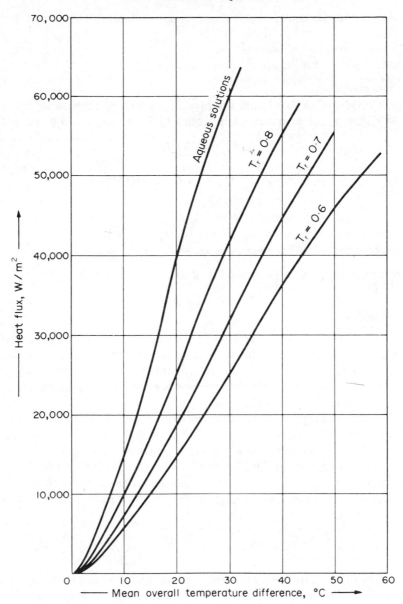

FIG. 12.59. Vertical thermosyphon design correlation

1. Conventional designs: tube lengths 2·5 to 3·7 m (8 to 12 ft) (standard length 2·44 m), preferred diameter 25 mm (1 in.).
2. Liquid level in the sump level with the top tube sheet.
3. Process side fouling coefficient 6000 W/m² °C.
4. Heating medium steam, coefficient including fouling, 6000 W/m² °C.
5. Simple inlet and outlet piping.

6. For reduced temperatures greater than 0·8, use the limiting curve (that for aqueous solutions).
7. Minimum operating pressure 0·3 bar.
8. Inlet fluid should not be appreciably sub-cooled.
9. Extrapolation is not recommended.

For heating media other than steam and process side fouling coefficients different from 6000 W/m²°C, the design heat flux taken from Fig. 12.59 may be adjusted as follows:

$$U' = q'/\Delta T' \tag{12.72}$$

and

$$1/U_c = 1/U' - 1/6000 + 1/h_s - 1/6000 + 1/h_{id}$$

where q' = flux read from Fig. 12.59 at $\Delta T'$,
  h_s = new shell side coefficient W/m²°C,
  h_{id} = fouling coefficient on the process (tube) side W/m²°C,
  U_c = corrected overall coefficient.

The use of Frank and Prickett's method is illustrated in Example 12.8.

Maximum heat flux

Thermosyphon reboilers can suffer from flow instabilities if too high a heat flux is used. The liquid and vapour flow in the tubes is not smooth but tends to pulsate, and at high heat fluxes the pulsations can become large enough to cause vapour locking. A good practice is to install a flow restriction in the inlet line, a valve or orifice plate, so that the flow resistance can be adjusted should vapour locking occur in operation.

Kern recommends that the heat flux in thermosyphon reboilers, based on the total heat-transfer area, should not exceed 37,900 W/m² (12,000 Btu/ft² h). For horizontal thermo-syphon reboilers, Collins recommends a maximum flux ranging from 47,300 W/m² for 20-mm tubes to 56,800 W/m² for 25-mm tubes (15,000 to 18,000 Btu/ft² h). These "rule of thumb" values are now thought to be too conservative; see Skellene *et al.* (1968). A correlation for determining the maximum flux for vertical thermosyphon reboilers is given by Lee *et al.* (1956).

General design considerations

The tube lengths used for vertical thermosyphon reboilers vary from 1·83 m (6 ft) for vacuum service to 3·66 m (12 ft) for pressure operation. A good size for general applications is 2·44 m (8 ft) by 25 mm internal diameter. Larger tube diameters, up to 50 mm, are used for fouling systems.

The top tube sheet is normally aligned with the liquid level in the base of the column; Fig. 12.58. The outlet pipe should be as short as possible, and have a cross-sectional area at least equal to the total cross-sectional area of the tubes.

Example 12.8

Make a preliminary design for a vertical thermosyphon for a column distilling crude aniline. The column will operate at atmospheric pressure and a vaporisation rate of

6000 kg/h is required. Steam is available at 22 bar (300 psig). Take the column bottom pressure as 1·2 bar.

Solution

Physical properties, taken as those of aniline:
Boiling point at 1·2 bar 190°C
Molecular weight 93·13
T_c 699 K
Latent heat 42,000 kJ/kmol
Steam saturation temperature 217°C.

Mean overall $\Delta T = (217 - 190) = 27°C$.

$$\text{Reduced temperature, } T_r = \frac{(190 + 273)}{699} = 0·66$$

From Fig. 12·59, design heat flux = 25,000 W/m^2

$$\text{Heat load} = \frac{6000}{3600} \times \frac{42,000}{93·13} = 751 \text{ kW}$$

$$\text{Area required} = \frac{751 \times 10^3}{25,000} = 30 \text{ m}^2$$

Use 25 mm i.d., 30 mm o.d., 2·44 m long tubes.

$$\text{Area of one tube} = \pi 25 \times 10^{-3} \times 2·44 = 0·192 \text{ m}^2$$

$$\text{Number of tubes} = \frac{30}{0·192} = 157$$

Approximate diameter of bundle, for 1·25 square pitch

(12.3b)
$$D_b = 30 \left[\frac{157}{0·215} \right]^{\frac{1}{2·207}} = 595 \text{ mm}$$

A fixed tube sheet will be used for a vertical thermosyphon reboiler.
From Fig. 12.10, shell diametrical clearance = 14 mm,

$$\text{Shell inside dia.} = 595 + 14 = 609 \text{ mm}$$

Outlet pipe diameter; take area as equal to total tube cross-sectional area

$$= 157(25 \times 10^{-3})^2 \frac{\pi}{4} = 0·077 \text{ m}^2$$

$$\text{Pipe diameter} = \sqrt{\frac{0·077 \times 4}{\pi}} = 0·31 \text{ m}$$

12.11.6. Design of kettle reboilers

Kettle reboilers, and other submerged bundle equipment, are essentially pool boiling devices, and their design is based on data for nucleate boiling.

In a tube bundle the vapour rising from the lower rows of tubes passes over the upper rows. This has two opposing effects: there will be a tendency for the rising vapour to blanket the upper tubes, particularly if the tube spacing is close, which will reduce the heat-transfer rate; but this is offset by the increased turbulence caused by the rising vapour bubbles. Palen and Small (1964) give a detailed procedure for kettle reboiler design in which the heat-transfer coefficient calculated using equations for boiling on a single tube is reduced by an empirically derived tube bundle factor, to account for the effects of vapour blanketing. Later work by Heat Transfer Research Inc., reported by Palen et al. (1972), showed that the coefficient for bundles was usually greater than that estimated for a single tube. On balance, it seems reasonable to use the correlations for single tubes to estimate the coefficient for tube bundles without applying any correction (equations 12.62 or 12.63).

The maximum heat flux for stable nucleate boiling will, however, be less for a tube bundle than for a single tube. Palen and Small (1964) suggest modifying the Zuber equation for single tubes (equation 12.64) with a tube density factor. This approach was supported by Palen et al. (1972).

The modified Zuber equation can be written as:

$$q_{cb} = K_b(p_t/d_o)(\lambda/\sqrt{N_t})[\sigma g(\rho_L - \rho_v)\rho_v^2]^{0.25} \qquad (12.74)$$

where q_{cb} = maximum (critical) heat flux for the tube bundle, W/m^2,
   K_b = 0.44 for square pitch arrangements,
      = 0.41 for equilateral triangular pitch arrangements,
   p_t = tube pitch,
   d_o = tube outside diameter,
   N_t = total number of tubes in the bundle,

Note. For U-tubes N_t will be equal to twice the number of actual U-tubes.

Palen and Small suggest that a factor of safety of 0.7 be applied to the maximum flux estimated from equation 12.74. This will still give values that are well above those which have traditionally been used for the design of commercial kettle reboilers; such as that of 37,900 W/m^2 (12,000 Btu/ft^2 h) recommended by Kern (1950). This has had important implications in the application of submerged bundle reboilers, as the high heat flux allows a smaller bundle to be used, which can then often be installed in the base of the column; saving the cost of shell and piping.

General design considerations

A typical layout is shown in Fig. 12.8. The tube arrangement, triangular or square pitch, will not have a significant effect on the heat-transfer coefficient. A tube pitch of between 1.5 to 2.0 times the tube outside diameter should be used to avoid vapour blanketing. Long thin bundles will be more efficient than short fat bundles.

The shell should be sized to give adequate space for the disengagement of the vapour and liquid. The shell diameter required will depend on the heat flux. The following values can be used as a guide:

Heat flux W/m^2	Shell dia./Bundle dia.
25,000	1.2 to 1.5
25,000 to 40,000	1.4 to 1.8
40,000	1.7 to 2.0

The freeboard between the liquid level and shell should be at least $0.25\,\text{m}$. To avoid excessive entrainment, the maximum vapour velocity $\hat{u}_v$ (m/s) at the liquid surface should be less than that given by the expression:

$$\hat{u}_v < 0.2\left[\frac{\rho_L - \rho_v}{\rho_v}\right]^{1/2} \tag{12.75}$$

When only a low rate of vaporisation is required a vertical cylindrical vessel with a heating jacket or coils should be considered. The boiling coefficients for internal submerged coils can be estimated using the equations for nucleate pool boiling.

Mean temperature differences

When the fluid being vaporised is a single component and the heating medium is steam (or another condensing vapour), both shell and tubes side processes will be isothermal and the mean temperature difference will be simply the difference between the saturation temperatures. If one side is not isothermal the logarithmic mean temperature difference should be used. If the temperature varies on both sides, the logarithmic temperature difference must be corrected for departures from true cross- or counter-current flow (see Section 12.6).

If the feed is sub-cooled, the mean temperature difference should still be based on the boiling point of the liquid, as the feed will rapidly mix with the boiling pool of liquid; the quantity of heat required to bring the feed to its boiling point must be included in the total duty.

Mixtures

The equations for estimating nucleate boiling coefficients given in Section 12.11.1 can be used for close boiling mixtures, say less than $5°C$, but will overestimate the coefficient if used for mixtures with a wide boiling range. Palen and Small (1964) give an empirical correction factor for mixtures which can be used to estimate the heat-transfer coefficient in the absence of experimental data:

$$(h_{nb}) \text{ mixture} = f_m\,(h_{nb}) \text{ single component} \tag{12.76}$$

where $f_m = \exp\left[-0.0083\,(T_{bo} - T_{bi})\right]$
and T_{bo} = temperature of the vapour mixture leaving the reboiler $°C$,
$\quad T_{bi}$ = temperature of the liquid entering the reboiler $°C$.

The inlet temperature will be the saturation temperature of the liquid at the base of the column, and the vapour temperature the saturation temperature of the vapour returned to the column. The composition of these streams will be fixed by the distillation column design specification.

Example 12.9

Design a vaporiser to vaporise $5000\,\text{kg/h}$ n-butane at 5.84 bar. The minimum temperature of the feed (winter conditions) will be $0°C$. Steam is available at 1.70 bar (10 psig).

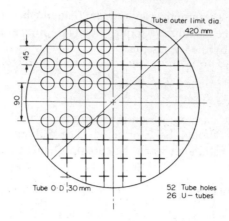

Tube sheet layout, U-tubes, Example 12.9

Solution

Only the thermal design and general layout will be done. Select kettle type.
Physical properties of n-butane at 5·84 bar:
  boiling point = 56·1°C
  latent heat = 326 kJ/kg
  mean specific heat, liquid = 2·51 kJ/kg °C
  critical pressure, P_c = 38 bar
Heat loads:
  sensible heat (maximum) = $(56·1 - 0)2·51 = 140·8$ kJ/kg

$$\text{total heat load} = (140·8 + 326) \times \frac{5000}{3600} = 648·3 \text{ kW,}$$

  add 5 per cent for heat losses
    maximum heat load (duty) = $1·05 \times 648·3$
                             = 681 kW

From Fig. 12.1 assume $U = 1000$ W/m² °C.
Mean temperature difference; both sides isothermal, steam saturation temperature at
1·7 bar = 115·2°C
$$\Delta T_m = 115·2 - 56·1 = 59·1°C$$

$$\text{Area (outside) required} = \frac{681 \times 10^3}{1000 \times 59·1} = 11·5 \text{ m}^2$$

Select 25 mm i.d., 30 mm o.d. plain U-tubes,

Nominal length 4·8 m (one U-tube)

$$\text{Number of U tubes} = \frac{11·5}{(30 \times 10^{-3})\pi 4·8} = 25$$

Use square pitch arrangement, pitch = 1·5 × tube o.d.

$$= 1·5 \times 30 = 45 \text{ mm}$$

Draw a tube layout diagram, take minimum bend radius

$$3.0 \times \text{tube o.d.} = 90 \text{ mm}$$

Proposed layout gives 26 U-tubes, tube outer limit diameter 420 mm.

Boiling coefficient
Use Mostinski's equation:

heat flux, based on estimated area,

$$q = \frac{681}{11.5} = 59.2 \text{ kW/m}^2$$

(12.63)

$$h_{nb} = 0.104 \, (38)^{0.69} \, (59.2 \times 10^3)^{0.7} \left[1.8 \left(\frac{5.84}{38} \right)^{0.17} + 4 \left(\frac{5.84}{38} \right)^{1.2} + 10 \left(\frac{5.84}{38} \right)^{10} \right]$$

$$= 4855 \text{ W/m}^2 \, {}^\circ\text{C}$$

Take steam condensing coefficient as 8000 W/m² °C, fouling coefficient 5000 W/m² °C; butane fouling coefficient, essentially clean, 10,000 W/m² °C.
Tube material will be plain carbon steel, $k_w = 55$ W/m °C

(12.2) $$\frac{1}{U_o} = \frac{1}{4855} + \frac{1}{10,000} + \frac{30 \times 10^{-3} \ln 30/25}{2 \times 55} + \frac{30}{25} \left(\frac{1}{5000} + \frac{1}{8000} \right)$$

$$U_0 = \underline{\underline{1341 \text{ W/m}^2 \, {}^\circ\text{C}}}$$

Close enough to original estimate of 1000 W/m² °C for the design to stand.

Myers and Katz (*Chem. Eng. Prog. Sym. Ser.* **49**(5) 107–114) give some data on the boiling of n-butane on banks of tubes. To compare the value estimate with their values an estimate of the boiling film temperature difference is required:

$$= \frac{1341}{4855} \times 59.1 = 16.3 \, {}^\circ\text{C} \; (29 \, {}^\circ\text{F})$$

Myers data, extrapolated, gives a coefficient of around 3000 Btu/h ft² °F at a 29°F temperature difference = 17,100 W/m² °C, so the estimated value of 4855 is certainly on the safe side.

Check maximum allowable heat flux. Use modified Zuber equation.

Surface tension (estimated) $= 9.7 \times 10^{-3}$ N/m

$\rho_L = 550$ kg/m³

$$\rho_v = \frac{58}{22.4} \times \frac{273}{(273 + 56)} \times 5.84 = 12.6 \text{ kg/m}^3$$

$$N_t = 52$$

For square arrangement $K_b = 0.44$

(12.74) $$q_c = 0.44 \times 1.5 \times \frac{326 \times 10^3}{\sqrt{52}} \, [9.7 \times 10^{-3} \times 9.81(550 - 12.6)12.6^2]^{0.25}$$

$$= 283,224 \text{ W/m}^2$$

$$= 280 \text{ kW/m}^2$$

Applying a factor of 0·7, maximum flux should not exceed $280 \times 0·7 = 196 \text{ kW/m}^2$. Actual flux of $59·2 \text{ kW/m}^2$ is well below maximum allowable.

Layout

From tube sheet layout $D_b = 420$ mm.
Take shell diameter as twice bundle diameter

$$D_s = 2 \times 420 = 840 \text{ mm}.$$

Take liquid level as 500 mm from base,

$$\text{freeboard} = 840 - 500 = 340 \text{ mm,} \quad \text{satisfactory.}$$

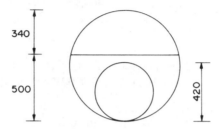

From sketch, width at liquid level $= 0·8$ m.
Surface area of liquid $= 0·8 \times 2·4 = 1·9 \text{ m}^2$.

$$\text{Vapour velocity at surface} = \frac{5000}{3600} \times \frac{1}{12·6} \times \frac{1}{1·9} = \underline{\underline{0·06 \text{ m/s}}}$$

Maximum allowable velocity

(12.75) $$\hat{u}_v = 0·2 \left[\frac{550 - 12·6}{12·6} \right]^{1/2} = \underline{\underline{1·3 \text{ m/s}}}$$

so actual velocity is well below maximum allowable velocity. A smaller shell diameter could be considered.

12.12. Plate heat exchangers

A plate heat exchanger consists of a stack of closely spaced thin metal plates, clamped together in a frame, in a similar arrangement to a plate and frame filter press. A thin gasket seals the plates round their edges. The gap between the plates is normally between about 3 and 6 mm. Corner ports in plates direct the flow from plate to plate. The basic flow arrangements are shown in Fig. 12.60; various combinations of these arrangements are used. The plates are embossed with a pattern of ridges, which increase the rigidity of the plate and improve the heat-transfer performance. Plates are available in a wide range of materials; including stainless steel, aluminium and titanium. Synthetic rubber, plastics, and asbestos-based gasket materials are used. For many metals the cost per unit surface area will be less for plates than tubes. Heat-transfer coefficients are generally higher in plate heat exchangers than shell and tube exchangers; and the units are more compact. Plates are

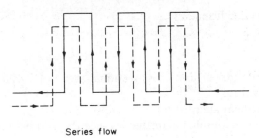

Series flow

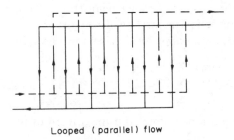

Looped (parallel) flow

FIG. 12.60. Plate heat-exchanger flow arrangements

available with effective areas from 0·03 to 1·3 m², and up to 400 plates can be contained in a large frame.

A plate is not a good shape to resist pressure, and the maximum operating pressure is limited to about 20 bar. The operating temperature is limited by the performance of the available gasket materials to about 250°C.

Plate heat exchangers are used extensively in the food and beverage industries, as they can be readily taken apart for cleaning and inspection. Their use in the chemical industry will depend on the relative cost for the particular application compared with a conventional shell and tube exchanger; see Parker (1964).

Plate heat exchangers are proprietary designs, and will normally be specified in consultation with the manufacturers. Emerson (1967) gives performance data for several proprietary designs.

12.12.1. Thermal design

The equation for forced-convective heat transfer in conduits can be used for plate heat exchangers (equation 12.10).

The values for the constant C and the indices a, b, c will depend on the particular type of plate being used.

Typical values for turbulent flow are given in the equation below, which can be used to make a preliminary estimate of the area required.

$$\frac{h_p d_e}{k_f} = 0 \cdot 26 \, Re^{0 \cdot 65} \, Pr^{0 \cdot 4} \, (\mu/\mu_w)^{0 \cdot 14} \tag{12.77}$$

where h_p = plate film coefficient,

$$Re = \frac{G_p d_e}{\mu},$$

G_p = mass flow rate per unit cross-sectional area = W/A_f,
W = mass flow rate,
A_f = cross-sectional area for flow,
d_e = equivalent (hydraulic) diameter = twice the gap between the plates.

The mean temperature difference will generally be higher in a plate heat exchanger than in a shell and tube exchanger, as the flow arrangement gives a closer approach to true counter-current flow. For a series arrangement (Fig. 12.60) the logarithmic mean temperature difference correction factor F_t will be close to 1.

Buonopane et al. (1963) give graphs for determining the value of F_t for various plate arrangements.

12.12.2. Pressure drop

The plate pressure drop can be estimated using a form of the equation for flow in a conduit (equation 12.18):

$$\Delta P_p = 8 j_f (L_p/d_e) \frac{\rho u_p^2}{2} \qquad (12.78)$$

where L_p = the path length and $u_p = G_p/\rho$.

The value of the friction factor j_f will depend on the design of plate used. For preliminary calculations the following relationship can be used for turbulent flow:

$$j_f = 1 \cdot 25 Re^{-0 \cdot 3} \qquad (12.79)$$

The transition from laminar to turbulent flow will normally occur at a Reynolds number of 100 to 400, depending on the plate design. With some designs, turbulence can be achieved at very low Reynolds numbers, which makes plate heat exchangers very suitable for use with viscous fluids.

12.13. Direct-contact heat exchangers

In direct-contact heat exchange the hot and cold streams are brought into contact without any separating wall, and high rates of heat transfer are achieved.

Applications include: reactor off-gas quenching, vacuum condensers, cooler-condensers, desuperheating and humidification. Water-cooling towers are a particular example of direct-contact heat exchange. In direct-contact cooler-condensers the condensed liquid is frequently used as the coolant, Fig. 12.61.

Direct-contact heat exchangers should be considered whenever the process stream and coolant are compatible. The equipment used is basically simple and cheap, and is suitable for use with heavily fouling fluids and with liquids containing solids; spray chambers, spray columns, and plate and packed columns are used.

There is no general design method for direct contact exchangers. Most applications will involve the transfer of latent heat as well as sensible heat, and the process is one of

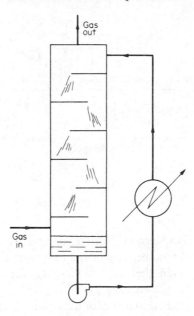

FIG. 12.61. Typical direct-contact cooler (baffle plates)

simultaneous heat and mass transfer. When the approach to thermal equilibrium is rapid, as it will be in many applications, the size of the contacting vessel is not critical and the design can be based on experience with similar processes. For other situations the designer must work from first principles, setting up the differential equations for mass and heat transfer, and using judgement in making the simplifications necessary to achieve a solution. The design procedures used are analogous to those for gas absorption and distillation.

The design and application of direct-contact heat exchangers is discussed by Fair (1961, 1972a, 1972b), who gives practical design methods and data for a range of applications.

The design of water-cooling towers, and humidification, is covered in Volume 1, Chapter 11. The same basic principles will apply to the design of other direct-contact exchangers.

12.14. Finned tubes

Fins are used to increase the effective surface area of heat-exchanger tubing. Many different types of fin have been developed, but the plain transverse fin shown in Fig. 12.62 is the most commonly used type for process heat exchangers. Typical fin dimensions are: pitch 2·0 to 4·0 mm, height 12 to 16 mm; ratio of fin area to bare tube area 15:1 to 20:1.

Finned tubes are used when the heat-transfer coefficient on the outside of the tube is appreciably lower than that on the inside; as in heat transfer from a liquid to a gas, such as in air-cooled heat exchangers.

The fin surface area will not be as effective as the bare tube surface, as the heat has to be conducted along the fin. This is allowed for in design by the use of a fin effectiveness, or fin efficiency, factor. The basic equations describing heat transfer from a fin are derived in

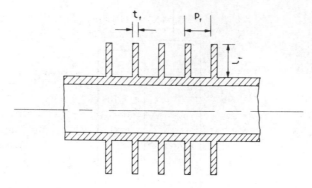

FIG. 12.62. Finned tube

Volume 1, Chapter 7; see also Kern (1950). The fin effectiveness is a function of the fin dimensions and the thermal conductivity of the fin material. Fins are therefore usually made from metals with a high thermal conductivity; for copper and aluminium the effectiveness will typically be between 0·9 to 0·95.

When using finned tubes, the coefficients for the outside of the tube in equation 12.2 are replaced by a term involving fin area and effectiveness:

$$1/h_o + 1/h_{od} = \frac{1}{E_f}(1/h_f + 1/h_{df})\frac{A_o}{A_f} \tag{12.80}$$

where h_f = heat-transfer coefficient based on the fin area,
  h_{df} = fouling coefficient based on the fin area,
  A_o = outside area of the bare tube,
  A_f = fin area,
  E_f = fin effectiveness.

It is not possible to give a general correlation for the coefficient h_f covering all types of fin and fin dimensions. Design data should be obtained from the tube manufacturers for the particular type of tube to be used. Some information is given in Volume 1, Chapter 7. For banks of tubes in cross flow, with plain transverse fins, the correlation given by Briggs and Young (1963) can be used to make an approximate estimate of the fin coefficient.

$$Nu = 0·134Re^{0·681}Pr^{0·33}\left[\frac{p_f - t_f}{l_f}\right]^{0·2}\left[\frac{p_f}{t_f}\right]^{0·1134} \tag{12.81}$$

where p_f = fin pitch,
  l_f = fin height,
  t_f = fin thickness.

The Reynolds number is evaluated for the bare tube (i.e. assuming that no fins exist).

Kern and Kraus (1972) give full details of the use of finned tubes in process heat exchangers design and design methods.

Low fin tubes

Tubes with low transverse fins, about 1 mm high, can be used with advantage as replacements for plain tubes in many applications. The fins are formed by rolling, and the tube outside diameters are the same as those for standard plain tubes. Details are given in the manufacturer's data books, Wolverine (1959); see also Webber (1960).

12.15. Double-pipe heat exchangers

One of the simplest and cheapest types of heat exchanger is the concentric pipe arrangement shown in Fig. 12.63. These can be made up from standard fittings, and are useful where only a small heat-transfer area is required. Several units can be connected in series to extend their capacity.

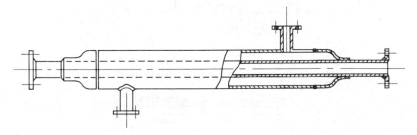

FIG. 12.63. Double-pipe exchanger (constructed for weld fittings)

The correlation for forced convective heat transfer in conduits (equation 12.10) can be used to predict the heat transfer coefficient in the annulus, using the appropriate equivalent diameter:

$$d_e = \frac{4 \times \text{cross-sectional area}}{\text{wetted perimeter}} = \frac{4(d_2^2 - d_1^2)\pi/4}{\pi d_1} = \frac{d_2^2 - d_1^2}{d_1}$$

where d_2 is the inside diameter of the outer pipe and d_1 the outside diameter of the inner pipe.

Some designs of double-pipe exchanger use inner tubes fitted with longitudinal fins.

12.16. Air-cooled exchangers

Air-cooled exchangers should be considered when cooling water is in short supply or expensive. They can also be competitive with water-cooled units even when water is plentiful. Frank (1978) suggests that in moderate climates air cooling will usually be the best choice for minimum process temperatures above 65°C, and water cooling for minimum processes temperatures below 50°C. Between these temperatures a detailed economic analysis would be necessary to decide the best coolant. Air-cooled exchangers are used for cooling and condensing.

Air-cooled exchangers consist of banks of finned tubes over which air is blown or drawn by fans mounted below or above the tubes (forced or induced draft). Typical units are shown in Fig. 12.64. Air-cooled exchangers are packaged units, and would normally be selected and specified in consultation with the manufacturers. Some typical overall coefficients are given in Table 12.1. These can be used to make an approximate estimate of the area required for a given duty. The equation for finned tubes given in Section 12.14 can also be used.

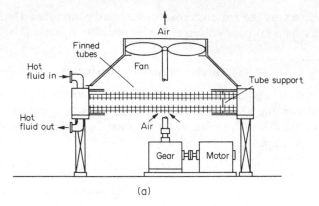

(a)

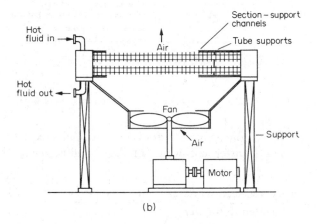

(b)

Fig. 12.64. Air-cooled exchangers

The design and application of air-cooled exchangers is discussed by Rubin (1960) and Lerner (1972). Lerner gives typical values for the overall heat-transfer coefficients as a function of the process fluid viscosity, and sufficient other design data for the preliminary design and costing of air-cooled exchangers.

Details of the constructional features of commercial air-cooled exchangers are given by Ludwig (1965). Design procedures are also given by Kern (1950), and Kern and Kraus (1972).

12.17. Fired heaters (furnaces and boilers)

When high process temperatures are required, directly fired exchangers, heated with fuel oil or gas, are used. There are no national standards covering the design of fired process heaters, equivalent to the standards for shell and tube exchangers. The fired heaters used in the process industries are essentially chambers constructed of refractory brick, containing banks of tubes through which the process fluid to be heated passes. The tubes are generally arranged in rows around the walls of the chamber. A typical design is shown in Fig. 12.65. Heat is transferred mainly by radiation.

The fundamentals of radiation heat transfer are covered in Volume 1, Chapter 7. Details of the general construction features of fired heaters are given by Evans (1980). Design

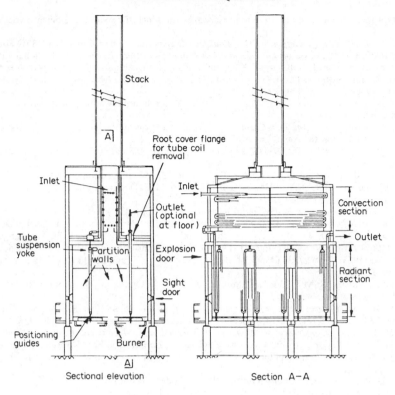

FIG. 12.65. (Foster Wheeler) Multi-zoned pyrolysis furnace

procedures for fired heaters are given by Kern (1950) and Evans (1980). A comprehensive review of the literature on fired heaters is given by Lihou (1975).

Typical applications of fired heaters include the preheating of the feeds to crude petroleum stills; boilers; heaters for high-temperature heat-transfer agents; and high-temperature reactors.

12.18. References

BELL, K. J. (1960) Petro/Chem. **32** (Oct.) C26. Exchanger design: based on the Delaware research report.

BELL, K. J. (1963) Final Report of the Co-operative Research Program on Shell and Tube Heat Exchangers, University of Delaware, Eng. Expt. Sta. Bull. 5 (University of Delaware).

BELL, K. J., TABOREK, J. and FENOGLIO, F. (1970) Chem. Eng. Prog. Symp. Ser. No. 102, **66**, 154. Interpretation of horizontal in-tube condensation heat transfer correlations with a two-phase flow regime map.

BELL, K. J. and GHALY, M. A. (1973) Chem. Eng. Prog. Symp. Ser. No. 131, **69**, 72. An approximate generalized design method for multicomponent/partial condensers.

BOYKO, L. D. and KRUZHILIN, G. N. (1967) Int. J. Heat Mass Transfer **10**, 361. Heat transfer and hydraulic resistance during condensation of steam in a horizontal tube and in a bundle of tubes.

BRIGGS, D. E. and YOUNG, E. H. (1963) Chem. Eng. Prog. Symp. Ser. No. 59, **61**, 1. Convection heat transfer and pressure drop of air flowing across triangular pitch banks of finned tubes.

BROMLEY, L. A. (1950) Chem. Eng. Prog. **46**, 221. Heat transfer in stable film boiling.

BUONOPANE, R. A., TROUPE, R. A. and MORGAN, J. C. (1963) Chem. Eng. Prog. **59** (July) 57. Heat transfer design method for plate heat exchangers.

BUTTERWORTH, D. (1973) Conference on Advances in Thermal and Mechanical Design of Shell and Tube Heat Exchangers, NEL Report No. 590. (National Engineering Laboratory, East Kilbride, Glasgow, UK). A calculation method for shell and tube heat exchangers in which the overall coefficient varies along the length.

BUTTERWORTH, D. (1977) *Introduction to Heat Transfer*, Engineering Design Guide No. 18 (Oxford U.P.).

BUTTERWORTH, D. (1978) *Course on the Design of Shell and Tube Heat Exchangers* (National Engineering Laboratory, East Kilbride, Glasgow, UK). Condensation 1 – Heat transfer across the condensed layer.

CHANTRY, W. A. and CHURCH, D. M. (1958) *Chem. Eng. Prog.* **54** (Oct.) 64. Design of high velocity forced circulation reboilers for fouling service.

CHEN, J. C. (1966) *Ind. Eng. Chem. Proc. Des. Dev.* **5**, 322. A correlation for boiling heat transfer to saturated fluids in convective flow.

COLBURN, A. P. (1934) *Trans. Am. Inst. Chem. Eng.* **30**, 187. Note on the calculation of condensation when a portion of the condensate layer is in turbulent motion.

COLBURN, A. P. and DREW, T. B. (1937) *Trans. Am. Inst. Chem. Eng.* **33**, 197. The condensation of mixed vapours.

COLBURN, A. P. and EDISON, A. G. (1941) *Ind. Eng. Chem.* **33**, 457. Prevention of fog in condensers.

COLBURN, A. P. and HOUGEN, O. A. (1934) *Ind. Eng. Chem.* **26**, 1178. Design of cooler condensers for mixtures of vapors with non-condensing gases.

COLLIER, J. G. (1972) *Convective Boiling and Condensation* (McGraw-Hill).

COLLINS, G. K. (1976) *Chem. Eng., Albany* **83** (July 19th) 149. Horizontal-thermosiphon reboiler design.

DEVORE, A. (1961) *Pet. Ref.* **40** (May) 221. Try this simplified method for rating baffled exchangers.

DEVORE, A. (1962) *Hyd. Proc. & Pet. Ref.* **41** (Dec.) 103. Use nomograms to speed exchanger design.

DONOHUE, D. A. (1955) *Pet. Ref.* **34** (Aug.) 94, (Oct.) 128, (Nov.) 175, and **35** (Jan.) 155, in four parts. Heat exchanger design.

EAGLE, A. and FERGUSON, R. M. (1930) *Proc. Roy. Soc.* A. **127**, 540. On the coefficient of heat transfer from the internal surfaces of tube walls.

EMERSON, W. H. (1967) *Thermal and Hydrodynamic Performance of Plate Heat Exchangers*, NEL. Reports Nos. 283, 284, 285, 286 (National Engineering Laboratories, East Kilbride, Glasgow, UK).

EMERSON, W. H. (1973) *Conference on Advances in Thermal and Mechanical Design of Shell and Tube Exchangers*, NEL Report No. 590. (National Engineering Laboratory, East Kilbride, Glasgow, UK). Effective tube-side temperature in multi-pass heat exchangers with non-uniform heat-transfer coefficients and specific heats.

EVANS, F. L. (1980) *Equipment Design Handbook*, Vol. 2, 2nd ed. (Gulf).

FAIR, J. R. (1961) *Petro./Chem. Eng.* **33** (Aug.) 57. Design of direct contact gas coolers.

FAIR, J. R. (1960) *Pet. Ref.* **39** (Feb.) 105. What you need to design thermosiphon reboilers.

FAIR, J. R. (1963) *Chem. Eng., Albany* **70** (July 8th) 119, (Aug. 5th) 101, in two parts. Vaporiser and reboiler design.

FAIR, J. R. (1972a) *Chem. Eng. Prog. Sym. Ser.* No. 118, **68**, 1. Process heat transfer by direct fluid-phase contact.

FAIR, J. R. (1972b) *Chem. Eng., Albany* **79** (June 12th) 91. Designing direct-contact cooler/condensers.

FISHENDEN, M. and SAUNDERS, O. A. (1950) *An Introduction to Heat Transfer* (Clarendon Press).

FISHER, J. and PARKER, R. O. (1969) *Hyd. Proc.* **48** (July) 147. New ideas on heat exchanger design.

FORSTER, K. and ZUBER, N. (1955) *AIChE Jl* **1**, 531. Dynamics of vapour bubbles and boiling heat transfer.

FRAAS, A. P. and OZISIK, M. N. (1965) *Heat Exchanger Design* (Wiley).

FRANK, O. and PRICKETT, R. D. (1973) *Chem. Eng., Albany* **80** (Sept. 3rd) 103. Designing vertical thermosiphon reboilers.

FRANK, O. (1974) *Chem. Eng., Albany* **81** (May 13th) 126. Estimating overall heat transfer coefficients.

FRANK, O. (1978) Simplified design procedure for tubular exchangers, in *Practical Aspects of Heat Transfer*, Chem. Eng. Prog. Tech. Manual (Am. Inst. Chem. Eng.).

GILMORE, G. H. (1963) Chapter 10 in *Chemical Engineers Handbook*, 4th ed., Perry, R. H., Chilton, C. H. and Kirkpatrick, S. P. (Eds.) (McGraw-Hill).

GLOYER, W. (1970) *Hydro. Proc.* **49** (July) 107. Thermal design of mixed vapor condensers.

GRANT, I. D. R. (1973) *Conference on Advances in Thermal and Mechanical Design of Shell and Tube Exchangers*, NEL Report No. 590 (National Engineering Laboratory, East Kilbride, Glasgow, UK.). Flow and pressure drop with single and two phase flow on the shell-side of segmentally baffled shell-and-tube exchangers.

HEWITT, G. F. and HALL-TAYLOR, N. S. (1970) *Annular Two-phase Flow* (Pergamon).

HSU, Y. and GRAHAM, R. W. (1976) *Transport Processes in Boiling and Two-phase Flow* (McGraw-Hill).

HUGHMARK, G. A. (1961) *Chem. Eng. Prog.* **57** (July) 43. Designing thermosiphon reboilers.

HUGHMARK, G. A. (1964) *Chem. Eng. Prog.* **60** (July) 59. Designing thermosiphon reboilers.

HUGHMARK, G. A. (1969) *Chem. Eng. Prog.* **65** (July) 67. Designing thermosiphon reboilers.

JACOB, M. (1957) *Heat Transfer*, 2 vols. (Wiley).

JACOBS, J. K. (1961) *Hyd. Proc. & Pet. Ref.* **40** (July) 189. Reboiler selection simplified.

JEFFREYS, G. V. (1961) *A Problem in Chemical Engineering Design* (Inst. Chem. Eng., London).

KAY, J. M. (1963) *An Introduction to Fluid Mechanics and Heat Transfer: with applications in chemical and mechanical process engineering,* 2nd ed. (Wiley).

KERN, D. Q. (1950) *Process Heat Transfer* (McGraw-Hill).

KERN, D. Q. and KRAUS, A. D. (1972) *Extended Surface Heat Transfer* (McGraw-Hill).

KREITH, F. (1976) *Principles of Heat Transfer,* 3rd ed. (Harper & Row).

KUTATELADZE, S. S. (1963) *Fundamentals of Heat Transfer* (Academic Press).

LEE, D. C., DORSEY, J. W., MOORE, G. Z. and MAYFIELD, F. D. (1956) *Chem. Eng. Prog.* **52** (April) 160. Design data for thermosiphon reboilers.

LERNER, J. E. (1972) *Hyd. Proc.* **51** (Feb.) 93. Simplified air cooler estimating.

LIHOU, D. (1975) *Heaters for Chemical Reactors* (Inst. Chem. Eng., London).

LORD, R. C., MINTON, P. E. and SLUSSER, R. P. (1970) *Chem. Eng., Albany* **77** (June 1st) 153. Guide to trouble free heat exchangers.

LUDWIG, E. E. (1965) *Applied Process Design for Chemical and Petroleum Plants,* Vol. 3 (Gulf).

MCADAMS, W. H. (1954) *Heat Transmission,* 3rd ed. (McGraw-Hill).

MCKEE, H. R. (1970) *Ind. Eng. Chem.* **62** (Dec.) 76. Thermosiphon reboilers – a review.

MORETTI, P. M. (1973) *Am. Inst. Chem. Eng. 74th National Meeting, New Orleans, March 14th.* A critical review of the literature on flow-induced vibrations in heat exchangers.

MOSTINSKI, I. L. (1963) *Teploenergetika* **4,** 66; English abstract in *Brit. Chem. Eng.* **8,** 580 (1963). Calculation of boiling heat transfer coefficients, based on the law of corresponding states.

MUELLER, A. C. (1973) Heat Exchangers, section 18 in Rosenow, W. M. and Hartnell, H. P. (Eds.) *Handbook of Heat Transfer* (McGraw-Hill).

NUSSELT, W. (1916) *Z. Ver. duet. Ing.* **60,** 541, 569. Die Oberflächenkondensation des Wasserdampfes.

PALEN, J. W. and SMALL, W. M. (1964) *Hyd. Proc.* **43** (Nov.) 199. A new way to design kettle reboilers.

PALEN, J. W. and TABOREK, J. (1962) *Chem. Eng. Prog.* **58** (July) 39. Refinery kettle reboilers.

PALEN, J. W. and TABOREK, J. (1969) *Chem. Eng. Prog. Sym. Ser.* No. 92, **65,** 53. Solution of shell side flow pressure drop and heat transfer by stream analysis method.

PALEN, J. W., YARDEN, A. and TABOREK, J. (1972) *Chem. Eng. Symp. Ser.* No. 118, **68,** 50. Characteristics of boiling outside large-scale horizontal multitube boilers.

PARKER, D. V. (1964) *Brit. Chem. Eng.* **1,** 142. Plate heat exchangers.

PERRY, R. H. and CHILTON, C. H. (Eds.) (1973) *Chemical Engineers Handbook,* 5th ed. (McGraw-Hill).

PORTER, K. E. and JEFFREYS, G. V. (1963) *Trans. Inst. Chem. Eng.* **41,** 126. The design of cooler condensers for the condensation of binary vapours in the presence of a non-condensable gas.

ROHSENOW, W. M. (1973) Boiling, in Rosenow, W. M. and Hartnett, J. P. (Eds.) *Handbook of Heat Transfer* (McGraw-Hill).

ROHSENOW, W. M. and HARTNETT, H. P. (Eds.) (1973) *Handbook of Heat Transfer* (McGraw-Hill).

RUBIN, F. L. (1960) *Chem. Eng., Albany* **67** (Oct. 31st) 91. Design of air cooled heat exchangers.

RUBIN, F. L. (1968) *Chem. Eng. Prog.* **64** (Dec.) 44. Practical heat exchange design.

SARMA, N. V. L. S., REDDY, P. J. and MURTI, P. S. (1973) *Ind. Eng. Chem. Proc. Des. Dev.* **12,** 278. A computer design method for vertical thermosyphon reboilers.

SIEDER, E. N. and TAIE, G. E. (1936) *Ind. Eng. Chem* **28,** 1429. Heat transfer and pressure drop of liquids in tubes.

SILVER, L. (1947) *Trans. Inst. Chem. Eng.* **25,** 30. Gas cooling with aqueous condensation.

STEINMEYER, D. E. (1972) *Chem. Eng. Prog.* **68** (July) 64. Fog formation in partial condensers.

SKELLENE, K. R., STERNLING, C. V., CHURCH, D. M. and SNYDER, N. H. (1968) *Chem. Eng. Prog. Symp. Ser.* No. 82, **64,** 102. An experimental study of vertical thermosiphon reboilers.

TABOREK, J. (1974) Design methods for heat transfer equipment: a critical survey of the state of the art, in Afgan, N. and Schlünder, E. V. (Eds.), *Heat Exchangers: Design and Theory Source Book* (McGraw-Hill).

TABOREK, J., AOKI, T., RITTER, R. B. and PALEN, J. W. (1972) *Chem. Eng. Prog.* **68** (Feb.) 59, (July) 69, in two parts. Fouling: the major unresolved problem in heat transfer.

TEMA (1978) *Standards of the Tubular Heat Exchanger Manufactures Association,* 6th ed. (Tubular Heat Exchanger Manufactures Association, New York).

THORNGREN, J. T. (1970) *Hyd. Proc.* **49** (April) 129. Predict exchanger tube damage.

TINKER, T. (1951) *Proceedings of the General Discussion on Heat Transfer,* p. 89, Inst. Mech. Eng., London. Shell-side characteristics of shell and tube heat exchangers.

TINKER, T. (1958) *Trans. Am. Soc. Mech. Eng.* **80** (Jan.) 36. Shell-side characteristics of shell and tube exchangers.

WARD, D. J. (1960) *Petro./Chem. Eng.* **32,** C-42. How to design a multiple component partial condenser.

WEBBER, W. O. (1960) *Chem Eng., Albany* **53** (Mar. 21st) 149. Under fouling conditions finned tubes can save money.

WESTWATER, J. W. (1956) *Advances in Chemical Engineering* **1,** 1. Boiling liquids.

WESTWATER, J. W. (1958) *Advances in Chemical Engineering* **2,** 1. Boiling liquids.

WOLVERINE (1959) *Engineering Data Book* (Wolverine Tube Division, Calumet and Hecla, Inc., Michigan).

ZUBER, N., TRIBUS, M. and WESTWATER, J. W. (1961) *Second International Heat Transfer Conference*, Paper 27, p. 230, Am. Soc. Mech. Eng. The hydrodynamic crisis in pool boiling of saturated and sub-cooled liquids.

British Standards
BS 3274: 1960 Tubular heat exchangers for general purposes.
BS 3606: 1978 Specification for steel tubes for heat exchangers.
BS 5500: 1976 Unfired fusion welded pressure vessels.

Engineering Sciences Data Unit Reports
No. 67016 (1967) Forced convection heat transfer in circular tubes. Part I, turbulent flow.
No. 68006 (1968) Forced convection heat transfer in circular tubes. Part II, laminar and transitional flow.
No. 68007 (1968) Forced convection in circular tubes. Part III, further data for turbulent flow.
No. 69004 (1969) Convective heat transfer during forced cross flow of fluids over a circular cylinder, including free convection effects.
No. 73031 (1973) Convective heat transfer during cross flow of fluids over plain tube banks.

Engineering Sciences Data Unit, 251–259 Regent Street, London W1R 7AD.

12.19. Nomenclature

Dimensions
in **MLTθ**

A	Heat transfer area	L^2
A_0	Clearance area between bundle and shell	L^2
A_f	Fin area	L^2
A_L	Total leakage area	L^2
A_o	Outside area of bare tube	L^2
A_s	Cross-flow area between tubes	L^2
A_{sb}	Shell-to-baffle clearance area	L^2
A_{tb}	Tube-to-baffle clearance area	L^2
a	Index in equation 12.10	—
B_c	Baffle cut	—
B_b	Bundle cut	—
b	Index in equation 12.10	—
C	Constant in equation 12.10	—
C_p	Heat capacity at constant pressure	$L^2T^{-2}\theta^{-1}$
C_{p_g}	Heat capacity of gas	$L^2T^{-2}\theta^{-1}$
C_{p_L}	Heat capacity of liquid phase	$L^2T^{-2}\theta^{-1}$
c	Index in equation 12.10	—
c_s	Shell-to-baffle diametrical clearance	L
c_t	Tube-to-baffle diametrical clearance	L
D_b	Bundle diameter	L
D_s	Shell diameter	L
d_e	Equivalent diameter	L
d_i	Tube inside diameter	L
d_o	Tube outside diameter	L
d_1	Outside diameter of inner of concentric tubes	L
d_2	Inside diameter of outer of concentric tubes	L
E_f	Fin efficiency	—
F_b	Bypass correction factor, heat transfer	—
F'_b	Bypass correction factor, pressure drop	—
F_L	Leakage correction factor, heat transfer	—
F'_L	Leakage correction factor, pressure drop	—
F_n	Tube row correction factor	—
F_t	Log mean temperature difference correction factor	—
F_w	Window effect correction factor	—
f_c	Two-phase flow factor	—
f_m	Temperature correction factor for mixtures	—
f_s	Nucleate boiling suppression factor	—

G	Total mass flow-rate per unit area	$ML^{-2}T^{-1}$
G_p	Mass flow-rate per unit cross-sectional area between plates	$ML^{-2}T^{-1}$
G_s	Shell-side mass flow-rate per unit area	$ML^{-2}T^{-1}$
G_t	Tube-side mass flow-rate per unit area	$ML^{-2}T^{-1}$
g	Gravitational acceleration	LT^{-2}
H_b	Height from baffle chord to top of tube bundle	L
H_c	Baffle cut height	L
H_s	Sensible heat of stream	ML^2T^{-3}
H_t	Total heat of stream (sensible + latent)	ML^2T^{-3}
h_c	Heat-transfer coefficient in condensation	$MT^{-3}\theta^{-1}$
$(h_c)_1$	Mean condensation heat-transfer coefficient for a single tube	$MT^{-3}\theta^{-1}$
$(h_c)_b$	Heat-transfer coefficient for condensation on a horizontal tube bundle	$MT^{-3}\theta^{-1}$
$(h_c)_{N_r}$	Mean condensation heat-transfer coefficient for a tube in a row of tubes	$MT^{-3}\theta^{-1}$
$(h_c)_v$	Heat-transfer coefficient for condensation on a vertical tube	$MT^{-3}\theta^{-1}$
$(h_c)_{BK}$	Condensation coefficient from Boko–Kruzhilin correlation	$MT^{-3}\theta^{-1}$
$(h_c)_s$	Condensation heat transfer coefficient for stratified flow in tubes	$MT^{-3}\theta^{-1}$
h'_c	Local condensing film coefficient, partial condenser	$MT^{-3}\theta^{-1}$
h_{cb}	Convective boiling-heat transfer coefficient	$MT^{-3}\theta^{-1}$
h_{cg}	Local effective cooling-condensing heat-transfer coefficient, partial condenser	$MT^{-3}\theta^{-1}$
h_{df}	Fouling coefficient based on fin area	$MT^{-3}\theta^{-1}$
h_f	Heat-transfer coefficient based on fin area	$MT^{-3}\theta^{-1}$
h_{fb}	Film boiling heat-transfer coefficient	$MT^{-3}\theta^{-1}$
h'_{fc}	Forced-convection coefficient in equation 12.67	$MT^{-3}\theta^{-1}$
h'_g	Local sensible-heat-transfer coefficient, partial condenser	$MT^{-3}\theta^{-1}$
h_i	Film heat-transfer coefficient inside a tube	$MT^{-3}\theta^{-1}$
h'_i	Inside film coefficient in Boyko–Kruzhilin correlation	$MT^{-3}\theta^{-1}$
h_{id}	Fouling coefficient on inside of tube	$MT^{-3}\theta^{-1}$
h_{nb}	Nucleate boiling-heat-transfer coefficient	$MT^{-3}\theta^{-1}$
h'_{nb}	Nucleate boiling coefficient in equation 12.67	$MT^{-3}\theta^{-1}$
h_o	Heat-transfer coefficient outside a tube	$MT^{-3}\theta^{-1}$
h_{oc}	Heat-transfer coefficient for cross flow over an ideal tube bank	$MT^{-3}\theta^{-1}$
h_{od}	Fouling coefficient on outside of tube	$MT^{-3}\theta^{-1}$
h_p	Heat-transfer coefficient in a plate heat exchanger	$MT^{-3}\theta^{-1}$
h_s	Shell-side heat-transfer coefficient	$MT^{-3}\theta^{-1}$
j_h	Heat transfer factor defined by equation 12.14	—
j_H	Heat-transfer factor defined by equation 12.15	—
j_f	Friction factor	—
K_1	Constant in equation 12.3, from Table 12.4	—
K_2	Constant in equation 12.61	—
K_b	Constant in equation 12.74	—
k_f	Thermal conductivity of fluid	$MLT^{-3}\theta^{-1}$
k_L	Thermal conductivity of liquid	$MLT^{-3}\theta^{-1}$
k_v	Thermal conductivity of vapour	$MLT^{-3}\theta^{-1}$
k_w	Thermal conductivity of tube wall material	$MLT^{-3}\theta^{-1}$
L'	Effective tube length	L
L_P	Path length in a plate heat exchanger	L
l_B	Baffle spacing (pitch)	L
l_f	Fin height	L
N_b	Number of baffles	—
N_c	Number of tubes in cross flow zone	—
N'_c	Number of tube rows crossed from end to end of shell	—
N_{cv}	Number of constrictions crossed	—
N_r	Number of tubes in a vertical row	—
N_s	Number of sealing strips	—
N_t	Number of tubes in a tube bundle	—
N_w	Number of tubes in window zone	—
N_{wv}	Number of restrictions for cross flow in window zone	—
P	Total pressure	$ML^{-1}T^{-2}$
P_c	Critical pressure	$ML^{-1}T^{-2}$
ΔP_c	Pressure drop in cross flow zone[1]	$ML^{-1}T^{-2}$
ΔP_e	Pressure drop in end zone[1]	$ML^{-1}T^{-2}$

ΔP_i	Pressure drop for cross flow over ideal tube bank[1]	$\mathbf{ML^{-1}T^{-2}}$
ΔP_p	Pressure drop in a plate heat exchanger[1]	$\mathbf{ML^{-1}T^{-2}}$
ΔP_s	Shell-side pressure drop[1]	$\mathbf{ML^{-1}T^{-2}}$
ΔP_t	Tube-side pressure drop[1]	$\mathbf{ML^{-1}T^{-2}}$
ΔP_w	Pressure drop in window zone[1]	$\mathbf{ML^{-1}T^{-2}}$
p_i	Fin pitch	$\mathbf{L}$
p_s	Saturation vapour pressure	$\mathbf{ML^{-1}T^{-2}}$
p_t	Tube pitch	$\mathbf{L}$
p_t'	Vertical tube pitch	$\mathbf{L}$
p_w	Saturation vapour pressure corresponding to wall temperature	$\mathbf{ML^{-1}T^{-2}}$
Q	Heat transferred in unit time	$\mathbf{ML^2T^{-3}}$
Q_g	Sensible-heat-transfer rate from gas phase	$\mathbf{ML^2T^{-3}}$
Q_t	Total heat-transfer rate from gas phase	$\mathbf{ML^2T^{-3}}$
q	Heat flux (heat-transfer rate per unit area)	$\mathbf{MT^{-3}}$
q'	Uncorrected value of flux from Fig. 12.59	$\mathbf{MT^{-3}}$
q_c	Maximum (critical) flux for a single tube	$\mathbf{MT^{-3}}$
q_{cb}	Maximum flux for a tube bundle	$\mathbf{MT^{-3}}$
R	Dimensionless temperature ratio defined by equation 12.6	—
R_a	Ratio of window area to total area	—
R_a'	Ratio of bundle cross-sectional area in window zone to total cross-sectional area of bundle	—
R_w	Ratio number of tubes in window zones to total number	—
S	Dimensionless temperature ratio defined by equation 12.7	—
T	Shell-side temperature	θ
T_r	Reduced temperature	—
T_s	Saturation temperature	θ
T_{sat}	Saturation temperature	θ
T_v	Vapour (gas) temperature	θ
T_w	Wall (surface) temperature	θ
T_1	Shell-side inlet temperature	θ
T_2	Shell-side exit temperature	θ
ΔT	Temperature difference	θ
ΔT_{lm}	Logarithmic mean temperature difference	θ
ΔT_m	Mean temperature difference in equation 12.1	θ
ΔT_s	Temperature change in vapour (gas) stream	θ
t	Tube-side temperature	θ
t_c	Local coolant temperature	θ
t_f	Fin thickness	$\mathbf{L}$
t_1	Tube-side inlet temperature	θ
t_2	Tube-side exit temperature	θ
U	Overall heat-transfer coefficient	$\mathbf{MT^{-3}\theta^{-1}}$
U'	Uncorrected overall coefficient, equation 12.72	$\mathbf{MT^{-3}\theta^{-1}}$
U_c	Corrected overall coefficient, equation 12.72	$\mathbf{MT^{-3}\theta^{-1}}$
U_o	Overall heat-transfer coefficient based on tube outside area	$\mathbf{MT^{-3}\theta^{-1}}$
u	Fluid velocity	$\mathbf{LT^{-1}}$
u_L	Liquid velocity, equation 12.55	$\mathbf{LT^{-1}}$
u_p	Fluid velocity in a plate heat exchanger	$\mathbf{LT^{-1}}$
u_s	Shell-side fluid velocity	$\mathbf{LT^{-1}}$
u_t	Tube-side fluid velocity	$\mathbf{LT^{-1}}$
u_v	Vapour velocity, equation 12.55	$\mathbf{LT^{-1}}$
$\hat{u}_v$	Maximum vapour velocity in kettle reboiler	$\mathbf{LT^{-1}}$
u_w	Velocity in window zone	$\mathbf{LT^{-1}}$
u_z	Geometric mean velocity	$\mathbf{LT^{-1}}$
W	Mass flow-rate of fluid	$\mathbf{MT^{-1}}$
W_c	Total condensate mass flow-rate	$\mathbf{MT^{-1}}$
W_s	Shell-side fluid mass flow-rate	$\mathbf{MT^{-1}}$
X_{tt}	Lockhart–Martinelli two-phase flow parameter	—
x	Mass fraction of vapour	—
Z	Ratio of change in sensible heat of gas stream to change in total heat of gas stream (sensible + latent)	—
α	Factor in equation 12.30	—
β_L	Factor in equation 12.31, for heat transfer	—
β_L'	Factor in equation 12.31, for pressure drop	—
θ_b	Angle subtended by baffle chord	—

λ	Latent heat	L^2T^{-2}
μ	Viscosity at bulk fluid temperature	$ML^{-1}T^{-1}$
μ_L	Liquid viscosity	$ML^{-1}T^{-1}$
μ_v	Vapour viscosity	$ML^{-1}T^{-1}$
μ_w	Viscosity at wall temperature	$ML^{-1}T^{-1}$
ρ	Fluid density	ML^{-3}
ρ_L	Liquid density	ML^{-3}
ρ_v	Vapour density	ML^{-3}
σ	Surface tension	MT^{-2}
Γ	Tube loading	$ML^{-1}T^{-1}$
Γ_h	Condensate loading on a horizontal tube	$ML^{-1}T^{-1}$
Γ_v	Condensate loading on a vertical tube	$ML^{-1}T^{-1}$

Dimensionless numbers

Nu	Nusselt number
Pr	Prandtl number
Pr_c	Prandtl number for condensate film
Re	Reynolds number
Re_c	Reynolds number for condensate film
Re_L	Reynolds number for liquid phase
St	Stanton number

(1) *Note*: in Volumes 1, 2 and 3, this symbol is used for pressure difference, and pressure drop (negative pressure gradient) indicated by a minus sign. In this chapter, as the symbol is only used for pressure drop, the minus sign has been omitted for convenience.

CHAPTER 13

Mechanical Design of Process Equipment

13.1. Introduction

This chapter covers those aspects of the mechanical design of chemical plant that are of particular interest to chemical engineers. The main topic considered is the design of pressure vessels. The design of storage tanks, centrifuges and heat-exchanger tube sheets are also discussed briefly.

The chemical engineer will not usually be called on to undertake the detailed mechanical design of a pressure vessel. Vessel design is a specialised subject, and will be carried out by mechanical engineers who are conversant with the current design codes and practices, and methods of stress analysis. However, the chemical engineer will be responsible for developing and specifying the basic design information for a particular vessel, and needs to have a general appreciation of pressure vessel design to work effectively with the specialist designer.

The basic data needed by the specialist designer will be:

1. Vessel function.
2. Process materials and services.
3. Operating and design temperature and pressure.
4. Materials of construction.
5. Vessel dimensions and orientation.
6. Type of vessel heads to be used.
7. Openings and connections required.
8. Specification of heating and cooling jackets or coils.
9. Type of agitator.
10. Specification of internal fittings.

There is no strict definition of what constitutes a pressure vessel, but it is generally accepted that any closed vessel over 150 mm diameter subject to a pressure difference of more than 1 bar should be designed as a pressure vessel.

It is not possible to give a completely comprehensive account of vessel design in one chapter. The design methods and data given should be sufficient for the preliminary design of conventional vessels. Sufficient for the chemical engineer to check the feasibility of a proposed equipment design; to estimate the vessel cost for an economic analysis; and to determine the vessel's general proportions and weight for plant layout purposes. For a more detailed account of pressure vessel design the reader should refer to the books by Brownell and Young (1959), Harvey (1974), Bickell and Ruiz (1967) and Gill (1970).

An elementary understanding of the principles of the "Strength of Materials" will be needed to follow this chapter. Readers who are not familiar with the subject should consult one of the many textbooks available; such as those by Ryder (1969), Case and Chilver

622

(1971), Timoshenko and Young (1968) and Faupel and Fisher (1981). The book by Faupel and Fisher is particularly recommended as a general introduction to mechanical design for chemical engineers.

13.1.1. Classification of pressure vessels

For the purposes of design and analysis, pressure vessels are sub-divided into two classes depending on the ratio of the wall thickness to vessel diameter: thin-walled vessels, with a thickness ratio of less than 1:10; and thick-walled above this ratio.

The principal stresses (see Section 13.3.1) acting at a point in the wall of a vessel, due to a pressure load, are shown in Fig 13.1. If the wall is thin, the radial stress σ_3 will be small and can be neglected in comparison with the other stresses, and the longitudinal and circumferential stresses σ_1 and σ_2 can be taken as constant over the wall thickness. In a thick wall, the magnitude of the radial stress will be significant, and the circumferential stress will vary across the wall. The majority of the vessels used in the chemical and allied industries are classified as thin-walled vessels. Thick-walled vessels are used for high pressures, and are discussed in Section 13.15.

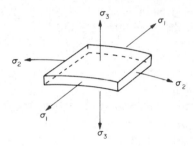

FIG. 13.1. Principal stresses in pressure-vessel wall

13.2. Pressure vessel codes and standards

In all the major industrialised countries the design and fabrication of thin-walled pressure vessels is covered by national standards and codes of practice. A brief summary of the European, American and Japanese codes is given in *Hydrocarbon Processing* (1978). In many countries the codes and standards are legally enforceable.

In the United Kingdom, though not a statutory requirement, all conventional pressure vessels for use in the chemical and allied industries will invariably be designed and fabricated in accordance with the British Standard specification for fusion-welded pressure vessels, BS 5500; or an equivalent code, such as the American Society of Mechanical Engineers code, section VIII (the "ASME" code). The codes and standards cover design, materials of construction, fabrication (manufacture and workmanship), inspection and testing; and form the basis of agreement between the manufacturer and customer, and the customer's insurance company.

In this chapter reference will mainly be made to the current British Standard BS 5500; which has superseded two earlier standards on pressure vessels, BS 1500 and BS 1515. The design of vessels for nuclear reactors is covered by BS 3915. The current (1980) edition of BS 5500 covers vessels fabricated in carbon and alloy steels, and aluminium. The design of vessels in reinforced plastics is covered by BS 4994.

The ASME code is divided into sections, which cover unfired vessels, boilers, nuclear, and fibre-glass-reinforced plastic vessels. A comprehensive review of the ASME code requirements is given by Chuse (1977); a brief review is given by Nolan (1973).

The national codes and standards dictate the minimum requirements, and give general guidance for design and construction; any extension beyond the minimum code requirement will be determined by agreement between the manufacturer and customer.

The codes and standards are drawn up by committees of engineers experienced in vessel design and manufacturing techniques; and are a blend of theory, experiment and experience. They periodically are reviewed, and revisions issued to keep abreast of developments in design, stress analysis, fabrication and testing. The latest version of the appropriate national code or standard should always be consulted before undertaking the design of any pressure vessel.

13.3. Fundamental principles and equations

This section has been included to provide a basic understanding of the fundamental principles that underlie the design equations given in the sections that follow. The derivation of the equations is given in outline only. A full discussion of the topics covered can be found in any text on the "Strength of Materials".

13.3.1. Principal stresses

The state of stress at a point in a structural member under a complex system of loading is described by the magnitude and direction of the principal stresses. The principal stresses are the maximum values of the normal stresses at the point; which act on planes on which the shear stress is zero. In a two-dimensional stress system, Fig. 13.2, the principal stresses at any point are related to the normal stresses in the x and y directions σ_x and σ_y and the shear stress τ_{xy} at the point by the following equation:

$$\text{Principal stresses, } \sigma_1, \sigma_2 = \tfrac{1}{2}(\sigma_y + \sigma_x) \pm \tfrac{1}{2}\sqrt{[(\sigma_y - \sigma_x)^2 + 4\tau_{xy}^2]} \qquad (13.1)$$

The maximum shear stress at the point is equal to half the algebraic difference between the principal stresses:

$$\text{Maximum shear stress} = \tfrac{1}{2}(\sigma_1 - \sigma_2) \qquad (13.2)$$

Compressive stresses are conventionally taken as negative; tensile as positive.

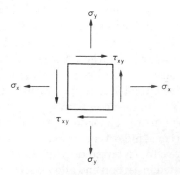

FIG. 13.2. Two-dimensional stress system

13.3.2. Theories of failure

The failure of a simple structural element under unidirectional stress (tensile or compressive) is easy to relate to the tensile strength of the material, as determined in a standard tensile test, but for components subjected to combined stresses (normal and shear stress) the position is not so simple, and several theories of failure have been proposed. The three theories most commonly used are described below:

Maximum principal stress theory: which postulates that a member will fail when one of the principal stresses reaches the failure value in simple tension, σ'_e. The failure point in a simple tension is taken as the yield-point stress, or the tensile strength of the material, divided by a suitable factor of safety.

Maximum shear stress theory: which postulates that failure will occur in a complex stress system when the maximum shear stress reaches the value of the shear stress at failure in simple tension.

For a system of combined stresses there are three shear stresses maxima:

$$\tau_1 = \pm \frac{\sigma_1 - \sigma_2}{2} \tag{13.3a}$$

$$\tau_2 = \pm \frac{\sigma_2 - \sigma_3}{2} \tag{13.3b}$$

$$\tau_3 = \pm \frac{\sigma_3 - \sigma_1}{2} \tag{13.3c}$$

In the tensile test, $$\tau_e = \frac{\sigma'_e}{2} \tag{13.4}$$

The maximum shear stress will depend on the sign of the principal stresses as well as their magnitude, and in a two-dimensional stress system, such as that in the wall of a thin-walled pressure vessel, the maximum value of the shear stress may be given by putting $\sigma_3 = 0$ in equations 13.3b and c.

The maximum shear stress theory is often called Tresca's, or Guest's, theory.

Maximum strain energy theory: which postulates that failure will occur in a complex stress system when the total strain energy per unit volume reaches the value at which failure occurs in simple tension.

The maximum shear-stress theory has been found to be suitable for predicting the failure of ductile materials under complex loading and is the criterion normally used in the pressure-vessel design.

13.3.3. Elastic stability

Under certain loading conditions failure of a structure can occur not through gross yielding or plastic failure, but by buckling, or wrinkling. Buckling results in a gross and sudden change of shape of the structure; unlike failure by plastic yielding, where the structure retains the same basic shape. This mode of failure will occur when the structure is not elastically stable: when it lacks sufficient stiffness, or rigidity, to withstand the load.

The stiffness of a structural member is dependent not on the basic strength of the material but on its elastic properties (E and v) and cross-sectional shape of the member.

The classic example of failure due to elastic instability is the buckling of tall thin columns (struts), which is described in any elementary text on the "Strength of Materials".

For a structure that is likely to fail by buckling there will be a certain critical value of load below which the structure is stable; if this value is exceeded catastrophic failure through buckling can occur.

The walls of pressure vessels are usually relatively thin compared with the other dimensions and can fail by buckling under compressive loads.

Elastic buckling is the decisive criterion in the design of thin-walled vessels under external pressure.

13.3.4. Membrane stresses in shells of revolution

A shell of revolution is the form swept out by a line or curve rotated about an axis. (A solid of revolution is formed by rotating an area about an axis.) Most process vessels are made up from shells of revolution: cylindrical and conical sections; and hemispherical, ellipsoidal and torispherical heads; Fig. 13.3.

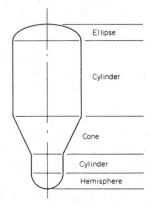

FIG. 13.3. Typical vessel shapes

The walls of thin vessels can be considered to be "membranes"; supporting loads without significant bending or shear stresses; similar to the walls of a balloon.

The analysis of the membrane stresses induced in shells of revolution by internal pressure gives a basis for determining the minimum wall thickness required for vessel shells. The actual thickness required will also depend on the stresses arising from the other loads to which the vessel is subjected.

Consider the shell of revolution of general shape shown in Fig. 13.4, under a loading that is rotationally symmetric; that is, the load per unit area (pressure) on the shell is constant round the circumference, but not necessarily the same from top to bottom.

Let P = pressure,

      t = thickness of shell,

      σ_1 = the meridional (longitudinal) stress, the stress acting along a meridian,

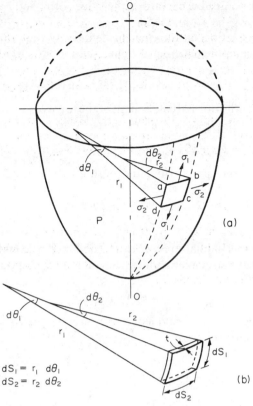

$$dS_1 = r_1 \, d\theta_1$$
$$dS_2 = r_2 \, d\theta_2$$

Element a, b, c, d

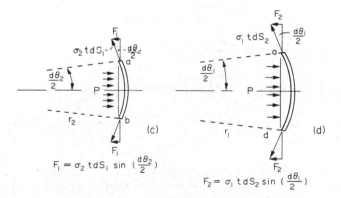

$$F_1 = \sigma_2 \, t \, dS_1 \sin \left(\frac{d\theta_2}{2} \right)$$

$$F_2 = \sigma_1 \, t \, dS_2 \sin \left(\frac{d\theta_1}{2} \right)$$

FIG. 13.4(a)(b). Stress in a shell of revolution (c)(d). Forces acting on sides of element abcd

σ_2 = the circumferential or tangential stress, the stress acting along parallel circles (often called the hoop stress),

r_1 = the meridional radius of curvature,

r_2 = circumferential radius of curvature.

Note: the vessel has a double curvature; the values of r_1 and r_2 are determined by the shape.

Consider the forces acting on the element defined by the points a, b, c, d. Then the normal component (component acting at right angles to the surface) of the pressure force on the element

$$= P[2r_1 \sin(d\theta_1/2)][2r_2 \sin(d\theta_2/2)]$$

This force is resisted by the normal component of the forces associated with the membrane stresses in the walls of the vessel (given by, force = stress × area)

$$= 2\sigma_2 t \, dS_1 \sin(d\theta_2/2) + 2\sigma_1 t \, dS_2 \sin(d\theta_1/2)$$

Equating these forces and simplifying, and noting that in the limit $d\theta/2 \to dS/2r$, and $\sin d\theta \to d\theta$, gives:

$$\frac{\sigma_1}{r_1} + \frac{\sigma_2}{r_2} = \frac{P}{t} \tag{13.5}$$

An expression for the meridional stress σ_1 can be obtained by considering the equilibrium of the forces acting about any circumferential line, Fig. 13.5. The vertical component of the pressure force

$$= P\pi(r_2 \sin\theta)^2$$

This is balanced by the vertical component of the force due to the meridional stress acting in the ring of the wall of the vessel

$$= 2\sigma_1 t\pi(r_2 \sin\theta)\sin\theta$$

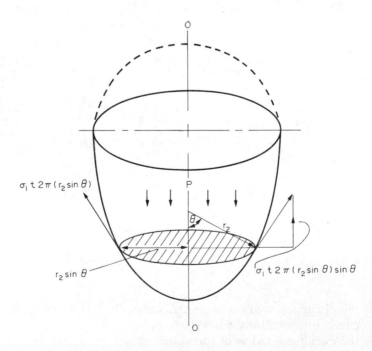

FIG. 13.5. Meridional stress, force acting at a horizontal plane

Equating these forces gives:

$$\sigma_1 = \frac{Pr_2}{2t} \tag{13.6}$$

Equations 13.5 and 13.6 are completely general for any shell of revolution.

Cylinder (Fig. 13.6a)

A cylinder is swept out by the rotation of a line parallel to the axis of revolution, so:

$$r_1 = \infty$$
$$r_2 = D/2$$

where D is the cylinder diameter.

Substitution in equations 13.5 and 13.6 gives:

$$\sigma_2 = \frac{PD}{2t} \tag{13.7}$$

$$\sigma_1 = \frac{PD}{4t} \tag{13.8}$$

Sphere (Fig. 13.6b)

$$r_1 = r_2 = D/2$$

hence:

$$\sigma_1 = \sigma_2 = \frac{PD}{4t} \tag{13.9}$$

Cone (Fig. 13.6c)

A cone is swept out by a straight line inclined at an angle α to the axis.

$$r_1 = \infty$$

$$r_2 = r/\cos\alpha$$

substitution in equations 13.5 and 13.6 gives:

$$\sigma_2 = \frac{Pr}{t\cos\alpha} \tag{13.10}$$

$$\sigma_1 = \frac{Pr}{2t\cos\alpha} \tag{13.11}$$

The maximum values will occur at $r = D_2/2$.

Ellipsoid (Fig. 13.6d)

For an ellipse with major axis $2a$ and minor axis $2b$, it can be shown that (see any standard geometry text):

$$r_1 = \frac{r_2^3 b^2}{a^4}$$

From equations 13.5 and 13.6

$$\sigma_1 = \frac{Pr_2}{2t} \qquad \text{(equation 13.6)}$$

$$\sigma_2 = \frac{P}{t}\left[r_2 - \frac{r_2^2}{2r_1}\right] \qquad (13.12)$$

At the crown (top)

$$r_1 = r_2 = a^2/b$$

$$\sigma_1 = \sigma_2 = \frac{Pa^2}{2tb} \qquad (13.13)$$

At the equator (bottom) $r_2 = a$, so $r_1 = b^2/a$

so

$$\sigma_1 = \frac{Pa}{2t} \qquad (13.13)$$

$$\sigma_2 = \frac{P}{t}\left[a - \frac{a^2}{2b^2/a}\right] = \frac{Pa}{t}\left[1 - \frac{1}{2}\frac{a^2}{b^2}\right] \qquad (13.14)$$

It should be noted that if $\frac{1}{2}(a/b)^2 > 1$, σ_2 will be negative (compressive) and the shell could fail by buckling. This consideration places a limit on the practical proportions of ellipsoidal heads.

Torus (Fig. 13.6e)

A torus is formed by rotating a circle, radius r_2, about an axis.

$$\sigma_1 = \frac{Pr_2}{2t} \qquad \text{(equation 13.6)}$$

$$r_1 = R/\sin\theta = \frac{R_o + r_2\sin\theta}{\sin\theta}$$

and

$$\sigma_2 = \frac{Pr_2}{t}\left[1 - \frac{r_2\sin\theta}{2(R_o + r_2\sin\theta)}\right] \qquad (13.15)$$

On the centre line of the torus, point c, $\theta = 0$ and

$$\sigma_2 = \frac{Pr_2}{t} \qquad (13.16)$$

At the outer edge, point a, $\theta = \pi/2$, $\sin\theta = 1$ and

$$\sigma_2 = \frac{Pr_2}{2t}\left[\frac{2R_o + r_2}{R_o + r_2}\right] \qquad (13.17)$$

the minimum value.

At the inner edge, point b, $\theta = 3\pi/2$, $\sin\theta = -1$ and

$$\sigma_2 = \frac{Pr_2}{2t}\left[\frac{2R_o - r_2}{R_o - r_2}\right] \qquad (13.18)$$

the maximum value.

So σ_2 varies from a maximum at the inner edge to a minimum at the outer edge.

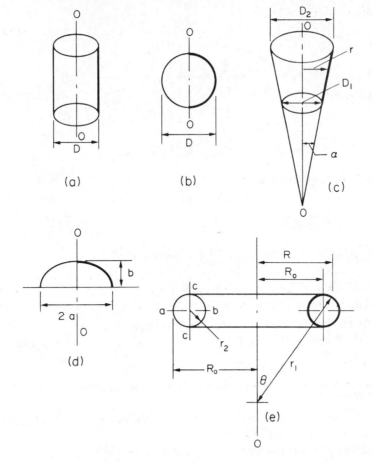

FIG. 13.6. Shells of revolution

Torispherical heads

A torispherical shape, which is often used as the end closure of cylindrical vessels, is formed from part of a torus and part of a sphere, Fig. 13.7. The shape is close to that of an ellipse but is easier and cheaper to fabricate.

In Fig. 13.7 R_k is the knuckle radius (the radius of the torus) and R_c the crown radius (the radius of the sphere). For the spherical portion:

$$\sigma_1 = \sigma_2 = \frac{PR_c}{2t} \qquad (13.19)$$

For the torus:

$$\sigma_1 = \frac{PR_k}{2t} \qquad (13.20)$$

σ_2 depends on the location, and is a function of R_c and R_k; it can be calculated from equations 13.15 and 13.9.

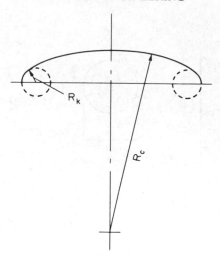

FIG. 13.7. Torisphere

The ratio of the knuckle radius to crown radius should be made not less than 6/100 to avoid buckling. The stress will be higher in the torus section than the spherical section.

13.3.5. Flat plates

Flat plates are used as covers for manholes, as blind flanges, and for the ends of small diameter and low pressure vessels.

For a uniformly loaded circular plate supported at its edges, the slope ϕ at any radius x is given by:

$$\phi = -\frac{dw}{dx} = -\frac{1}{\mathbf{D}}\frac{Px^3}{16} + \frac{C_1 x}{2} + \frac{C_2}{x} \qquad (13.21)$$

(The derivation of this equation can be found in any text on the strength of materials.)
Integration gives the deflection w:

$$w = \frac{Px^4}{64\mathbf{D}} - C_1 \frac{x^2}{4} - C_2 \ln x + C_3 \qquad (13.22)$$

where P = intensity of loading (pressure),
     x = radial distance to point of interest,
    $\mathbf{D}$ = flexual rigidity of plate = $\dfrac{Et^3}{12(1-v^2)}$,
    t = plate thickness,
    v = Poisson's ratio for the material,
    E = modulus of elasticity of the material (Young's modulus).

C_1, C_2, C_3 are constants of integration which can be obtained from the boundary conditions at the edge of the plate.

Two limiting situations are possible:

1. When the edge of the plate is rigidly clamped, not free to rotate; which corresponds to a heavy flange, or a strong joint.
2. When the edge is free to rotate (simply supported); corresponding to a weak joint, or light flange.

1. Clamped edges (Fig. 13.8a)

The edge (boundary) conditions are:

$$\phi = 0 \text{ at } x = 0$$
$$\phi = 0 \text{ at } x = a$$
$$w = 0 \text{ at } x = a$$

where a is the radius of the plate.
Which gives:

$$C_2 = 0, \quad C_1 = \frac{Pa^2}{8D}, \text{ and } C_3 = \frac{Pa^4}{64D}$$

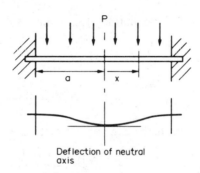

Deflection of neutral axis

(a) Clamped edges

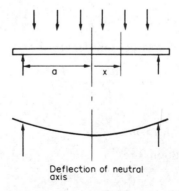

Deflection of neutral axis

(b) Simply supported

FIG. 13.8. Flat circular plates

hence

$$\phi = \frac{Px}{16\mathbf{D}}(a^2 - x^2) \tag{13.23}$$

and

$$w = \frac{P}{64\mathbf{D}}(x^2 - a^2)^2 \tag{13.24}$$

The maximum deflection will occur at the centre of the plate at $x = 0$

$$\hat{w} = \frac{Pa^4}{64\mathbf{D}} \tag{13.25}$$

The bending moments per unit length due to the pressure load are related to the slope and deflection by:

$$M_1 = \mathbf{D}\left[\frac{d\phi}{dx} + v\frac{\phi}{x}\right] \tag{13.26}$$

$$M_2 = \mathbf{D}\left[\frac{\phi}{x} + v\frac{d\phi}{dx}\right] \tag{13.27}$$

Where M_1 is the moment acting along cylindrical sections, and M_2 that acting along diametrical sections.

Substituting for ϕ and $d\phi/dx$ in equations 13.26 and 13.27 gives:

$$M_1 = \frac{P}{16}[a^2(1 + v) - x^2(3 + v)] \tag{13.28}$$

$$M_2 = \frac{P}{16}[a^2(1 + v) - x^2(1 + 3v)] \tag{13.29}$$

The maximum values will occur at the edge of the plate, $x = a$.

$$\hat{M}_1 = -\frac{Pa^2}{8}, \quad \hat{M}_2 = -v\frac{Pa^2}{8}$$

The bending stress is given by:

$$\sigma_b = \frac{M_1}{I'} \times \frac{t}{2}$$

where I' = second moment of area per unit length = $t^3/12$, hence

$$\hat{\sigma}_b = \frac{6\hat{M}_1}{t^2} = \frac{3}{4}\frac{Pa^2}{t^2} \tag{13.30}$$

2. *Simply supported plate (Fig. 13.8b)*

The edge (boundary) conditions are:

$$\phi = 0 \text{ at } x = 0$$
$$w = 0 \text{ at } x = a$$
$$M_1 = 0 \text{ at } x = a \text{ (free to rotate)}$$

which gives C_2 and $C_3 = 0$.

Hence

$$\phi = -\frac{1}{\mathbf{D}}\frac{Px^3}{16} + \frac{C_1 x}{2}$$

and

$$\frac{d\phi}{dx} = -\frac{1}{\mathbf{D}}\left[\frac{3Px^2}{16}\right] + \frac{C_1}{2}$$

Substituting these values in equation 13.26, and equating to zero at $x = a$, gives:

$$C_1 = \frac{Pa^2 (3+v)}{8\mathbf{D} (1+v)}$$

and hence

$$M_1 = \frac{P}{16}(3+v)(a^2 - x^2) \tag{13.31}$$

The maximum bending moment will occur at the centre, where $M_1 = M_2$

so

$$\hat{M}_1 = \hat{M}_2 = \frac{P(3+v)a^2}{16} \tag{13.32}$$

and

$$\hat{\sigma}_b = \frac{6\hat{M}_1}{t^2} = \frac{3}{8}(3+v)\frac{Pa^2}{t^2} \tag{13.33}$$

General equation for flat plates

A general equation for the thickness of a flat plate required to resist a given pressure load can be written in the form:

$$t = CD\sqrt{P/f} \tag{13.34}$$

where $f =$ the maximum allowable stress (the design stress),
    $D =$ the effective plate diameter,
    $C =$ a constant, which depends on the edge support.
The limiting value of C can be obtained from equations 13.30 and 13.33. Taking Poisson's ratio as 0.3, a typical value for steels, then if the edge can be taken as completely rigid $C = 0.43$, and if it is essentially free to rotate $C = 0.56$.

13.3.6. Dilation of vessels

Under internal pressure a vessel will expand slightly. The radial growth can be calculated from the elastic strain in the radial direction. The principal strains in a two-dimensional system are related to the principal stresses by:

$$\varepsilon_1 = \frac{1}{E}(\sigma_1 - v\sigma_2) \tag{13.35}$$

$$\varepsilon_2 = \frac{1}{E}(\sigma_2 - v\sigma_1) \tag{13.36}$$

The radial (diametrical strain) will be the same as the circumferential strain ε_2. For any shell of revolution the dilation can be found by substituting the appropriate expressions for the circumferential and meridional stresses in equation 13.36.

The diametrical dilation $\Delta = D\varepsilon_1$.
For a cylinder

$$\sigma_1 = \frac{PD}{4t}$$

$$\sigma_2 = \frac{PD}{2t}$$

substitution in equation 13.36 gives:

$$\Delta_c = \frac{PD^2}{4tE}(2 - v) \qquad (13.37)$$

For a sphere (or hemisphere)

$$\sigma_1 = \sigma_2 = \frac{PD}{4t}$$

and

$$\Delta_s = \frac{PD^2}{4tE}(1 - v) \qquad (13.38)$$

So for a cylinder closed by a hemispherical head of the same thickness the difference in dilation of the two sections, if they were free to expand separately, would be:

$$\Delta_c - \Delta_s = \frac{PD^2}{4tE}$$

13.3.7. Secondary stresses

In the stress analysis of pressure vessels and pressure vessel components stresses are classified as primary or secondary. Primary stresses can be defined as those stresses that are necessary to satisfy the conditions of static equilibrium. The membrane stresses induced by the applied pressure and the bending stresses due to wind loads are examples of primary stresses. Primary stresses are not self-limiting; if they exceed the yield point of the material, gross distortion, and in the extreme situation, failure of the vessel will occur.

Secondary stresses are those stresses that arise from the constraint of adjacent parts of the vessel. Secondary stresses are self-limiting; local yielding or slight distortion will satisfy the conditions causing the stress, and failure would not be expected to occur in one application of the loading. The "thermal stress" set up by the differential expansion of parts of the vessel, due to different temperatures or the use of different materials, is an example of a secondary stress. The discontinuity that occurs between the head and the cylindrical section of a vessel is a major source of secondary stress. If free, the dilation of the head would be different from that of the cylindrical section (see Section 13.2.6); they are constrained to the same dilation by the welded joint between the two parts. The induced bending moment and shear force due to the constraint give rise to secondary bending and shear stresses at the junction. The magnitude of these discontinuity stresses can be estimated by analogy with the behaviour of beams on elastic foundations; see Hetenyi (1958) and Harvey (1974).

Other sources of secondary stresses are the constraints arising at flanges, supports and the change of section due to reinforcement at a nozzle or opening (see Section 13.6).

Though secondary stresses do not affect the "bursting strength" of the vessel, they are an important consideration when the vessel is subject to repeated pressure loading. If local yielding has occurred, residual stress will remain when the pressure load is removed, and repeated pressure cycling can lead to fatigue failure.

13.4. General design considerations: pressure vessels

13.4.1. Design pressure

A vessel must be designed to withstand the maximum pressure to which it is likely to be subjected in operation.

For vessels under internal pressure, the design pressure is normally taken as the pressure at which the relief device is set. This will normally be 5 to 10 per cent above the normal working pressure, to avoid spurious operation during minor process upsets. When deciding the design pressure, the hydrostatic pressure in the base of the column should be added to the operating pressure, if significant.

Vessels subject to external pressure should be designed to resist the maximum differential pressure that is likely to occur in service. Vessels likely to be subjected to vacuum should be designed for a full negative pressure of 1 bar, unless fitted with an effective, and reliable, vacuum breaker.

13.4.2. Design temperature

The strength of metals decreases with increasing temperature (see Chapter 7) so the maximum allowable design stress will depend on the material temperature. The design temperature at which the design stress is evaluated should be taken as the maximum working temperature of the material, with due allowance for any uncertainty involved in predicting vessel wall temperatures.

13.4.3. Materials

Pressure vessels are constructed from plain carbon steels, low and high alloy steels, other alloys, clad plate, and reinforced plastics.

Selection of a suitable material must take into account the suitability of the material for fabrication (particularly welding) as well as the compatibility of the material with the process environment.

The pressure vessel design codes and standards include lists of acceptable materials; in accordance with the appropriate material standards.

Carbon and alloy steels for pressure vessel construction are covered by the following British standards: BS 1501, plates; BS 1502, sections and bars; BS 1503, forgings; BS 1504 castings.

13.4.4. Design stress (nominal design strength)

For design purposes it is necessary to decide a value for the maximum allowable stress (nominal design strength) that can be accepted in the material of construction.

This is determined by applying a suitable "design stress factor" (factor of safety) to the

maximum stress that the mate   under
standard test conditions. The   design
methods, the loading, the qua

For materials not subject to   e yield
stress (or proof stress), or the t   l at the
design temperature.

Property	Carbon–manganese, low alloy steels	stainless steels	metals
Minimum yield stress or 0·2 per cent proof stress, at the design temperature	1·5	1·5	1·5
Minimum tensile strength, at room temperature	2·35	2·5	4·0
Mean stress to produce rupture at 10^5 h at the design temperature	1·5	1·5	1·0

TABLE 13.2. *Typical design stresses for plate*
(The appropriate material standards should be consulted for particular grades and plate thicknesses)

Material	Tensile strength (N/mm^2)	Design stress at temperature °C (N/mm^2)									
		0 to 50	100	150	200	250	300	350	400	450	500
Carbon steel (semi-killed or silicon killed)	360	135	125	115	105	95	85	80	70		
Carbon–manganese steel (semi-killed or silicon killed)	460	180	170	150	140	130	115	105	100		
Carbon–molybdenum steel, 0·5 per cent Mo	450	180	170	145	140	130	120	110	110		
Low alloy steel (Ni, Cr, Mo, V)	550	240	240	240	240	240	235	230	220	190	170
Stainless steel 18Cr/8Ni unstabilised (304)	510	165	145	130	115	110	105	100	100	95	90
Stainless steel 18Cr/8Ni Ti stabilised (321)	540	165	150	140	135	130	130	125	120	120	115
Stainless steel 18Cr/8Ni Mo 2½ per cent (316)	520	175	150	135	120	115	110	105	105	100	95

For materials subject to conditions at which the creep is likely to be a consideration, the design stress is based on the creep characteristics of the material: the average stress to produce rupture after 10^5 hours, or the average stress to produce a 1 per cent strain after 10^5 hours, at the design temperature. Typical design stress factors for pressure components are shown in Table 13.1.

The nominal design strengths (allowable design stresses), for use with the design methods given in the standard, are listed in BS 5500 for the range of materials covered by the standard. The standard should be consulted for the principles and design stress factors used in determining the nominal design strengths.

Typical design stress values for some common materials are shown in Table 13.2. These may be used for preliminary designs. The standards and codes should be consulted for the values to be used for detailed vessel design.

13.4.5. Welded joint efficiency, and construction categories

The strength of a welded joint will depend on the type of joint and the quality of the welding.

The soundness of welds is checked by visual inspection and by non-destructive testing (radiography).

The possible lower strength of a welded joint compared with the virgin plate is usually allowed for in design by multiplying the allowable design stress for the material by a "welding joint factor" J. The value of the joint factor used in design will depend on the type of joint and amount of radiography required by the design code. Typical values are shown in Table 13.3. Taking the factor as 1·0 implies that the joint is equally as strong as the virgin plate; this is achieved by radiographing the complete weld length, and cutting out and remaking any defects. The use of lower joint factors in design, though saving costs on radiography, will result in a thicker, heavier, vessel, and the designer must balance any cost savings on inspection and fabrication against the increased cost of materials.

TABLE 13.3. *Maximum allowable joint efficiency*

Type of joint	Degree of radiography		
	100 per cent	spot	none
Double-welded butt or equivalent	1·0	0·85	0·7
Single-weld butt joint with bonding strips	0·9	0·80	0·65

The national codes and standards divide vessel construction into different categories, depending on the amount of non-destructive testing required. The higher categories require 100 per cent radiography of the welds, and allow the use of highest values for the weld-joint factors. The lower-quality categories require less radiography, but allow only lower joint-efficiency factors, and place restrictions on the plate thickness and type of materials that can be used. The highest category will invariably be specified for process-plant pressure vessels.

The standards should be consulted to determine the limitations and requirements of the construction categories specified. Welded joint efficiency factors are not used, as such, in the design equations given in BS 5500; instead limitations are placed on the values of the nominal design strength (allowable design stress) for materials in the lower construction category. The standard specifies three construction categories:

Category 1: the highest class, requires 100 per cent non-destructive testing (NDT) of the welds; and allows the use of all materials covered by the standard, with no restriction on the plate thickness.

Category 2: requires less non-destructive testing but places some limitations on the materials which can be used and the maximum plate thickness.

Category 3: the lowest class, requires only visual inspection of welds, but the material nominal design strengths are limited to about half those allowed in categories 1 and 2, and the materials are restricted to carbon and carbon-manganese steels less than 16 mm thick, and austenitic stainless steels less than 25 mm thick.

13.4.6. Corrosion allowance

The "corrosion allowance" is the additional thickness of metal added to allow for material lost by corrosion and erosion, or scaling (see Chapter 7). The allowance to be used should be agreed between the customer and manufacturer. Corrosion is a complex phenomenon, and it is not possible to give specific rules for the estimation of the corrosion allowance required for all circumstances. The allowance should be based on experience with the material of construction under similar service conditions to those for the proposed design. For carbon and low-alloy steels, where severe corrosion is not expected, a minimum allowance of 2·0 mm should be used; where more severe conditions are anticipated this should be increased to 4·0 mm. Most design codes and standards specify a minimum allowance of 1·0 mm.

13.4.7. Design loads

A structure must be designed to resist gross plastic deformation and collapse under all the conditions of loading. The loads to which a process vessel will be subject in service are listed below. They can be classified as major loads, that must always be considered in vessel design, and subsidiary loads. Formal stress analysis to determine the effect of the subsidiary loads is only required in the codes and standards where it is not possible to demonstrate the adequacy of the proposed design by other means; such as by comparison with the known behaviour of existing vessels.

Major loads

1. Design pressure: including any significant static head of liquid.
2. Maximum weight of the vessel and contents, under operating conditions.
3. Maximum weight of the vessel and contents under the hydraulic test conditions.
4. Wind loads.
5. Earthquake (seismic) loads.
6. Loads supported by, or reacting on, the vessel.

Subsidiary loads

1. Local stresses caused by supports, internal structures and connecting pipes.
2. Shock loads caused by water hammer, or by surging of the vessel contents.
3. Bending moments caused by eccentricity of the centre of the working pressure relative to the neutral axis of the vessel.
4. Stresses due to temperature differences and differences in the coefficient expansion of materials.
5. Loads caused by fluctuations in temperature and pressure.

A vessel will not be subject to all these loads simultaneously. The designer must determine what combination of possible loads gives the worst situation, and design for that loading condition.

13.4.8. Minimum practical wall thickness

There will be a minimum wall thickness required to ensure that any vessel is sufficiently rigid to withstand its own weight, and any incidental loads. As a general guide the wall thickness of any vessel should not be less than the values given below; the values include a corrosion allowance of 2 mm:

Vessel diameter (m)	Minimum thickness (mm)
1	5
1 to 2	7
2 to 2·5	9
2·5 to 3·0	10
3·0 to 3·5	12

13.5. The design of thin-walled vessels under internal pressure

13.5.1. Cylinders and spherical shells

For a cylindrical shell the minimum thickness required to resist internal pressure can be determined from equation 13.7; the cylindrical stress will be the greater of the two principal stresses.

If D_i is internal diameter and e the minimum thickness required, the mean diameter will be $(D_i + e)$; substituting this for D in equation 13.7 gives:

$$e = \frac{P_i(D_i + e)}{2f}$$

where f is the design stress and P_i the internal pressure. Rearranging gives:

$$e = \frac{P_i D_i}{2f - P_i} \tag{13.39}$$

This is the form of the equation given in the British Standard, BS 5500.

An equation for the minimum thickness of a sphere can be obtained from equation 13.9:

$$e = \frac{P_i D_i}{4f - P_i} \qquad (13.40)$$

The equation for a sphere given in BS 5500 is:

$$e = \frac{P_i D_i}{4f - 1 \cdot 2 P_i} \qquad (13.41)$$

The equation given in the British Standard BS 5500 differs slightly from equation 13.40, as it is derived from the formula for thick-walled vessels; see Section 13.15.

If a welded joint factor is used equations 13.39 and 13.40 are written:

$$e = \frac{P_i D_i}{2Jf - P_i} \qquad (13.39a)$$

and

$$e = \frac{P_i D_i}{4Jf - 1 \cdot 2 P_i} \qquad (13.40b)$$

where J is the joint factor.

Any consistent set of units can be used for equations 13.39a to 13.40b.

13.5.2. Heads and closures

The ends of a cylindrical vessel are closed by heads of various shapes. The principal types used are:

1. Flat plates and formed flat heads; Fig. 13.9.
2. Hemispherical heads; Fig. 13.10a.
3. Ellipsoidal heads; Fig. 13.10b.
4. Torispherical heads; Fig. 13.10c.

Hemispherical, ellipsoidal and torispherical heads are collectively referred to as domed heads. They are formed by pressing or spinning; large diameters are fabricated from formed sections. Torispherical heads are often referred to as dished ends.

The preferred proportions of domed heads are given in the standards and codes.

Choice of closure

Flat plates are used as covers for manways, and as the channel covers of heat exchangers. Formed flat ends, known as "flange-only" ends, are manufactured by turning over a flange with a small radius on a flat plate, Fig. 13.9a. The corner radius reduces the abrupt change of shape, at the junction with the cylindrical section; which reduces the local stresses to some extent: "Flange-only" heads are the cheapest type of formed head to manufacture, but their use is limited to low-pressure and small-diameter vessels.

Standard torispherical heads (dished ends) are the most commonly used end closure for vessels up to operating pressures of 15 bar. They can be used for higher pressures, but above 10 bar their cost should be compared with that of an equivalent ellipsoidal head. Above 15 bar an ellipsoidal head will usually prove to be the most economical closure to use.

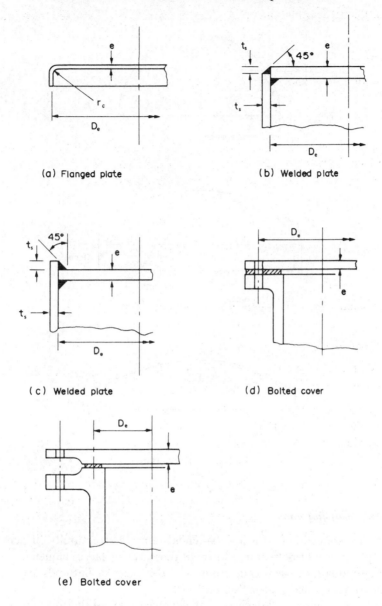

(a) Flanged plate

(b) Welded plate

(c) Welded plate

(d) Bolted cover

(e) Bolted cover

FIG. 13.9. Flat-end closures

A hemispherical head is the strongest shape; capable of resisting about twice the pressure of a torispherical head of the same thickness. The cost of forming a hemispherical head will, however, be higher than that for a shallow torispherical head. Hemispherical heads are used for high pressures.

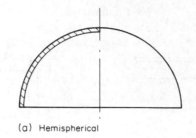

(a) Hemispherical

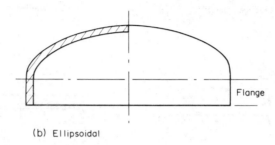

Flange

(b) Ellipsoidal

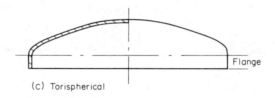

Flange

(c) Torispherical

FIG. 13.10. Domed heads

13.5.3. Design of flat ends

Though the fabrication cost is low, flat ends are not a structurally efficient form, and very thick plates would be required for high pressures or large diameters.

The design equations used to determine the thickness of flat ends are based on the analysis of stresses in flat plates; Section 13.3.5.

The thickness required will depend on the degree of constraint at the plate periphery. The minimum thickness required is given by:

$$e = C_p D_e \sqrt{\frac{P_i}{f}} \qquad (13.42)$$

where C_p = a design constant, dependent on the edge constraint,
  D_e = nominal plate diameter,
  f = design stress.

Any consistent set of units can be used.

Values for the design constant C_p and the nominal plate diameter D_e are given in the design codes and standards for various arrangements of flat end closures (BS 5500, clause 3.5.5).

The values of the design constant and nominal diameter for the typical designs shown in Fig. 13.9 are given below:

(a) Flanged-only end, for diameters less than 0·6 m and corner radii at least equal to 0·25e, C_p can be taken as 0·45; D_e is equal to D_i.

(b, c) Plates welded to the end of the shell with a fillet weld, angle of fillet 45° and depth equal to the plate thickness, take C_p as 0·55 and $D_e = D_i$.

(d) Bolted cover with a full face gasket (see Section 13.10), take $C_p = 0\cdot4$ and D_e equal to the bolt circle diameter.

(e) Bolted end cover with a narrow-face gasket, take $C_p = 0\cdot55$ and D_e equal to the mean diameter of the gasket.

13.5.4. Design of domed ends

Design equations and charts for the various types of domed heads are given in the codes and standards (BS 5500, clause 3.5.2), and should be used for detailed design. The codes and standards cover both unpierced and pierced heads. Pierced heads are those with openings or connections. The head thickness must be increased to compensate for the weakening effect of the holes where the opening or branch is not locally reinforced (see Section 13.6).

For convenience, simplified design equations are given in this section. These are suitable for the preliminary sizing of unpierced heads and for heads with fully compensated openings or branches.

Hemispherical heads

It can be seen by examination of equations 13.7 and 13.9, that for equal stress in the cylindrical section and hemispherical head of a vessel the thickness of the head need only be half that of the cylinder. However, as the dilation of the two parts would then be different, discontinuity stresses would be set up at the head and cylinder junction. For no difference in dilation between the two parts (equal diametrical strain) it can be shown that for steels (Poisson's ratio = 0·3) the ratio of the hemispherical head thickness to cylinder thickness should be 7/17. However, the stress in the head would then be greater than that in the cylindrical section; and the optimum thickness ratio is normally taken as 0·6; see Brownell and Young (1959).

Ellipsoidal heads

Most standard ellipsoidal heads are manufactured with a major and minor axis ratio of 2:1. For this ratio, the following equation can be used to calculate the minimum thickness required:

$$e = \frac{P_i D_i}{2Jf - 0\cdot2P_i} \tag{13.43}$$

Torispherical heads

There are two junctions in a torispherical end closure: that between the cylindrical section and the head, and that at the junction of the crown and the knuckle radii. The bending and shear stresses caused by the differential dilation that will occur at these points must be taken into account in the design of the heads. One approach taken is to use the basic equation for a hemisphere and to introduce a stress concentration, or shape, factor to allow for the increased stress due to the discontinuity. The stress concentration factor is a function of the knuckle and crown radii.

$$e = \frac{P_i R_c C_s}{2fJ + P_i(C_s - 0 \cdot 2)} \tag{13.44}$$

where C_s = stress concentration factor for torispherical heads = $\frac{1}{4}(3 + \sqrt{R_c/R_k})$,
$\quad R_c$ = crown radius,
$\quad R_k$ = knuckle radius.

The ratio of the knuckle to crown radii should not be less than 0·06, to avoid buckling; and the crown radius should not be greater than the diameter of the cylindrical section. Any consistent set of units can be used with equations 13.43 and 13.44. For formed heads (no joints in the head) the joint factor J is taken as 1·0.

Flanges (skirts) on domed heads

Formed domed heads are made with a short straight cylindrical section, called a flange or skirt; Fig. 13.10. This ensures that the weld line is away from the point of discontinuity between the head and the cylindrical section of the vessel.

13.5.5. Conical sections and end closures

Conical sections (reducers) are used to make a gradual reduction in diameter from one cylindrical section to another of smaller diameter.

Conical ends are used to facilitate the smooth flow and removal of solids from process equipment; such as, hoppers, spray-dryers and crystallisers.

From equation 13.10 it can be seen that the thickness required at any point on a cone is related to the diameter by the following expression:

$$e = \frac{P_i D_c}{2fJ - P_i} \cdot \frac{1}{\cos \alpha} \tag{13.45}$$

where D_c is the diameter of the cone at the point,
$\quad \alpha$ = half the cone apex angle.
This equation will only apply at points away from the cone to cylinder junction. Bending and shear stresses will be caused by the different dilation of the conical and cylindrical sections. This can be allowed for by introducing a stress concentration factor, in a similar manner to the method used for torispherical heads,

$$e = \frac{C_c P_i D_c}{2fJ - P_i} \tag{13.46}$$

The design factor C_c is a function of the half apex angle α:

α	20°	30°	45°	60°
C_c	1·00	1·35	2·05	3·20

A formed section would normally be used for the transition between a cylindrical section and conical section; except for vessels operating at low pressures, or under hydrostatic pressure only. The transition section would be made thicker than the conical or cylindrical section and formed with a knuckle radius to reduce the stress concentration at the transition, Fig. 13.11. The thickness at the knuckle can be calculated using equation 13.46, and that for the conical section away from the transition from equation 13.45.

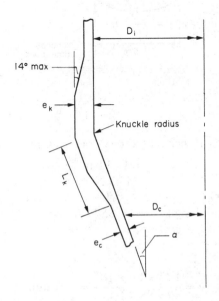

FIG. 13.11. Conical transition section

The length of the thicker section L_k depends on the cone angle and is given by:

$$L_k = \sqrt{\frac{D_i e_k}{4 \cos \alpha}} \qquad (13.47)$$

where e_k is the thickness at the knuckle.

Design procedures for conical sections are given in the codes and standards (BS 5500, clause 3.5.3).

Example 13.1

Estimate the thickness required for the component parts of the vessel shown in the sketch on page 648. The vessel is to operate at a pressure of 14 bar (absolute) and temperature of 300°C. The material of construction will be plain carbon steel. Welds will be fully radiographed. A corrosion allowance of 2 mm should be used.

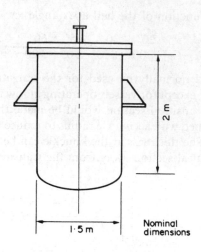

1·5 m   Nominal
               dimensions

Solution

Design pressure, take as 10 per cent above operating pressure, $= (14 - 1) \times 1\cdot 1$

$$= 14\cdot 3 \text{ bar}$$
$$= 1\cdot 43 \text{ N/mm}^2$$

Design temperature 300°C.
From Table 13.2, typical design stress $= 85 \text{ N/mm}^2$.

Cylindrical section

(13.39) $e = \dfrac{1\cdot 43 \times 1\cdot 5 \times 10^3}{2 \times 85 - 1\cdot 43} = 12\cdot 7 \text{ mm}$

add corrosion allowance $12\cdot 7 + 2 = 14\cdot 7$
say <u>15 mm plate</u>

Domed head

(i) Try a standard dished head (torisphere):

crown radius $R_c = D_i = 1\cdot 5$ m
knuckle radius $= 6$ per cent $R_c = 0\cdot 09$ m

A head of this size would be formed by pressing: no joints, so $J = 1$.

(13.44) $C_s = \frac{1}{4}\left(3 + \sqrt{R_c/R_k}\right) = \frac{1}{4}\left(3 + \sqrt{1\cdot 5/0\cdot 09}\right)$

$$= 1\cdot 77$$

(13.44) $e = \dfrac{1\cdot 43 \times 1\cdot 5 \times 10^3 \times 1\cdot 77}{2 \times 85 + 1\cdot 43\,(1\cdot 77 - 0\cdot 2)} = \underline{\underline{22\cdot 0 \text{ mm}}}$

(ii) Try a "standard" ellipsoidal head, ratio major:minor axes = 2:1

(13.43)
$$e = \frac{1\cdot43 \times 1\cdot5 \times 10^3}{2 \times 85 - 0\cdot2 \times 1\cdot43}$$

$$= \underline{12\cdot7 \text{ mm}}$$

So an ellipsoidal head would probably be the most economical. Take as same thickness as wall 15 mm.

Flat head

Use a full face gasket $C_p = 0\cdot4$
D_e = bolt circle diameter, take as approx. 1·7 m.

(13.42)
$$e = 0\cdot4 \times 1\cdot7 \times 10^3 \sqrt{1\cdot43/85} = \underline{88\cdot4 \text{ mm}}$$

Add corrosion allowance and round-off to $\underline{90 \text{ mm.}}$

This shows the inefficiency of a flat cover. It would be better to use a flanged domed head.

13.6. Compensation for openings and branches

All process vessels will have openings for connections, manways, and instrument fittings. The presence of an opening weakens the shell, and gives rise to stress concentrations. The stress at the edge of a hole will be considerably higher than the average stress in the surrounding plate. To compensate for the effect of an opening, the wall thickness is increased in the region adjacent to the opening. Sufficient reinforcement must be provided to compensate for the weakening effect of the opening without significantly altering the general dilation pattern of the vessel at the opening. Over-reinforcement will reduce the flexibility of the wall, causing a "hard spot", and giving rise to secondary stresses; typical arrangements are shown in Fig. 13.12.

The simplest method of providing compensation is to weld a pad or collar around the opening, Fig. 13.12a. The outer diameter of the pad is usually between $1\frac{1}{2}$ to 2 times the diameter of the hole or branch. This method, however, does not give the best disposition of the reinforcing material about the opening, and in some circumstances high thermal stress can arise due to the poor thermal conductivity of the pad to shell junction.

At a branch, the reinforcement required can be provided, with or without a pad, by allowing the branch, to protrude into the vessel, Fig. 13.12b. This arrangement should be used with caution for process vessels, as the protrusion will act as a trap for crud, and local corrosion can occur. Forged reinforcing rings, Fig. 13.12c, provide the most effective method of compensation, but are expensive. They would be used for any large openings and branches in vessels operating under severe conditions.

Calculation of reinforcement required

The "equal area method" is the simplest method used for calculating the amount of reinforcement required, and is allowed in most design codes and standards. The principle

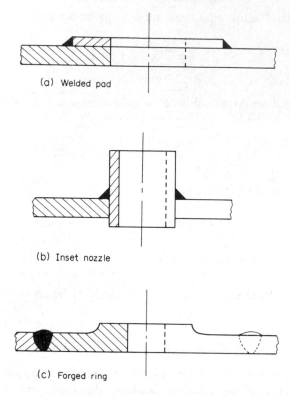

(a) Welded pad

(b) Inset nozzle

(c) Forged ring

FIG. 13.12. Types of compensation for openings

used is to provide reinforcement local to the opening, equal in cross-sectional area to the area removed in forming the opening, Fig. 13.13. If the actual thickness of the vessel wall is greater than the minimum required to resist the loading, the excess thickness can be taken into account when estimating the area of reinforcement required. Similarly with a branch connection, if the wall thickness of the branch or nozzle is greater than the minimum required, the excess material in the branch can be taken into account. Any corrosion allowance must be deducted when determining the excess thickness available as compensation. The standards and codes differ in the areas of the branch and shell considered to be effective for reinforcement, and should be consulted to determine the actual area allowed and the disposition of the various types of reinforcement. Figure 13.14 can be used for preliminary calculations. For branch connections of small diameter the reinforcement area can usually be provided by increasing the wall thickness of the branch pipe. Some design codes and standards do not require compensation for connections below 89 mm (3 in.) diameter.

If anything, the equal area method tends to over-estimate the compensation required and in some instances the additional material can reduce the fatigue life of the vessel. More sophisticated methods for determining the compensation required have been introduced into the latest editions of the codes and standards (BS 5500, clause 3.5.4). BS 5500 also allows the use of the equal area method (BS 5500, appendix F). A critical discussion of the methods that are used in the various national codes and standards for calculating the

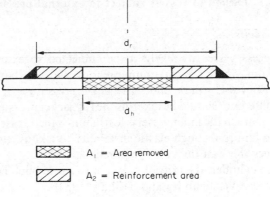

A_1 = Area removed

A_2 = Reinforcement area

$A_2 = A_1$

d_r = 1.5 to 2.0 x d_h

FIG. 13.13. Equal-area method of compensation

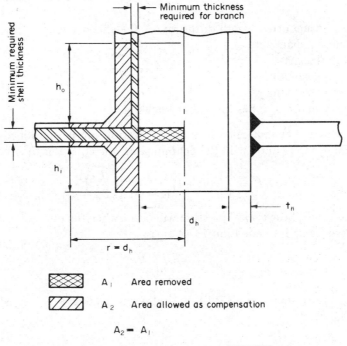

A_1 Area removed

A_2 Area allowed as compensation

$A_2 = A_1$

Max. allowed h_0 and $h_i = 0.64 \sqrt{(d_h + t_n)t_n}$

All dimensions shown are in the fully corroded condition (i.e. less corrosion allowance)

FIG. 13.14. Branch compensation

compensation for openings and branches is given in the British Standards Institute publication PD 6437 (1969).

The equal-area method is generally used for estimating the increase in thickness required to compensate for multiple openings.

13.7. Design of vessels subject to external pressure

13.7.1. Cylindrical shells

Two types of process vessel are likely to be subjected to external pressure: those operated under vacuum, where the maximum pressure will be 1 bar (atm); and jacketed vessels, where the inner vessel will be under the jacket pressure. For jacketed vessels, the maximum pressure difference should be taken as the full jacket pressure, as a situation may arise in which the pressure in the inner vessel is lost. Thin-walled vessels subject to external pressure are liable to failure through elastic instability (buckling) and it is this mode of failure that determines the wall thickness required.

For an open-ended cylinder, the critical pressure to cause buckling P_c is given by the following expression; see Windenburg and Trilling (1934):

$$P_c = \frac{1}{3}\left[n^2 - 1 + \frac{2n^2 - 1 - v}{n^2\left(\frac{2L}{\pi D_o}\right)^2 - 1}\right]\frac{2E}{(1-v^2)}\left(\frac{t}{D_o}\right)^3 + \frac{2Et/D_o}{(n^2-1)\left[n^2\left(\frac{2L}{\pi D_o}\right)^2 + 1\right]^2}$$

(13.48)

where L = the unsupported length of the vessel, the effective length,
$\quad D_o$ = external diameter,
$\quad t$ = wall thickness,
$\quad E$ = Young's modulus,
$\quad v$ = Poisson's ratio,
$\quad n$ = the number of lobes formed at buckling.

For long tubes and cylindrical vessels this expression can be simplified by neglecting terms with the group $(2L/\pi D_o)^2$ in the denominator; the equation then becomes:

$$P_c = \frac{1}{3}\left[(n^2 - 1)\frac{2E}{(1-v^2)}\right]\left(\frac{t}{D_o}\right)^3$$

(13.49)

The minimum value of the critical pressure will occur when the number of lobes is 2, and substituting this value into equation 13.49 gives:

$$P_c = \frac{2E}{1-v^2}\left(\frac{t}{D_o}\right)^3$$

(13.50)

For most pressure-vessel materials Poisson's ratio can be taken as 0·3; substituting this in equation 13.50 gives:

$$P_c = 2·2E(t/D_o)^3$$

(13.51)

For short closed vessels, and long vessels with stiffening rings, the critical buckling pressure will be higher than that predicted by equation 13.51. The effect of stiffening can be taken into account by introducing a "collapse coefficient", K_c, into equation 13.51.

$$P_c = K_c E(t/D_o)^3$$

(13.52)

where K_c is a function of the diameter and thickness of the vessel, and the effective length L' between the ends or stiffening rings; and is obtained from Fig. 13.16. The effective length for some typical arrangements is shown in Fig. 13.15.

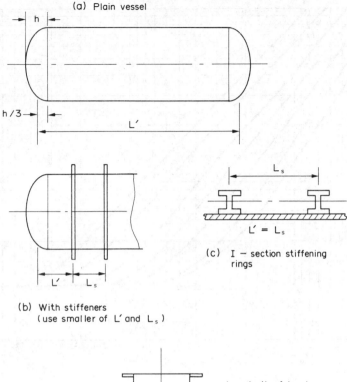

(a) Plain vessel

(b) With stiffeners
(use smaller of L' and L$_s$)

(c) I − section stiffening
rings

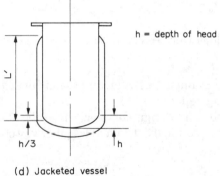

h = depth of head

(d) Jacketed vessel

FIG. 13.15. Effective length, vessel under external pressure

It can be shown (see Southwell, 1913) that the critical distance between stiffeners, L_c, beyond which stiffening will not be effective is given by:

$$L_c = \frac{4\pi \sqrt{6} D_o}{27} \left[(1 - v^2)^{1/4} \right] (D_o/t)^{1/2} \tag{13.53}$$

Substituting $v = 0\cdot3$ gives:

$$L_c = 1\cdot11 D_o (D_o/t)^{1/2} \tag{13.54}$$

Any stiffening rings used must be spaced closer than L_c. Equation 13.52 can be used to determine the critical buckling pressure and hence the thickness required to resist a given

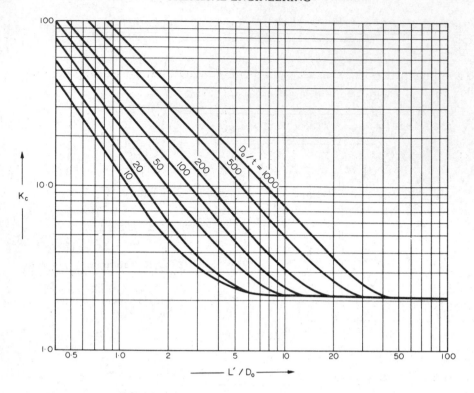

FIG. 13.16. Collapse coefficients for cylinderical shells (after Brownell and Young, 1959)

external pressure; see Example 13.2. A factor of safety of at least 3 should be applied to the values predicted using equation 13.52.

The design methods and design curves given in the standards and codes (BS 5500, clause 3.6) should be used for the detailed design of vessels subject to external pressure.

Out of roundness

Any out-of-roundness in a shell after fabrication will significantly reduce the ability of the vessel to resist external pressure. A deviation from a true circular cross-section equal to the shell thickness will reduce the critical buckling pressure by about 50 per cent. The ovality (out-of-roundness) of a cylinder is measured by:

$$\text{Ovality} = \frac{2(D_{max} - D_{min})}{(D_{max} + D_{min})} \times 100, \text{ per cent}$$

For vessels under external pressure this should not normally exceed 1·5 per cent.

13.7.2. Design of stiffening rings

The usual procedure is to design stiffening rings to carry the pressure load for a distance of $\frac{1}{2}L_s$ on each side of the ring, where L_s is the spacing between the rings. So, the load per unit length on a ring F_r will be given by:

$$F_r = P_e L_s \qquad (13.55)$$

where P_e is the external pressure.

The critical load to cause buckling in a ring under a uniform radial load F_c is given by the following expression; see Faupel and Fisher (1981):

$$F_c = \frac{24EI_r}{D_r^3} \qquad (13.56)$$

where I_r = second moment of area of the ring cross-section,
        D_r = diameter of the ring (approximately equal to the shell outside diameter).

Combining equations 13.55 and 13.56 will give an equation from which the required dimensions of the ring can be determined:

$$P_e L_s \not> \frac{24EI_r}{D_r^3} \div \text{(factor of safety)} \qquad (13.57)$$

In calculating the second moment of area of the ring some allowance is normally made for the vessel wall; the use of I_r calculated for the ring alone will give an added factor of safety.

In vacuum distillation columns, the plate-support rings will act as stiffening rings and strengthen the vessel; see Example 13.2.

13.7.3. Vessel heads

The critical buckling pressure for a sphere subject to external pressure is given by (see Timoshenko, 1936):

$$P_c = \frac{2Et^2}{R_s^2 \sqrt{3(1 - v^2)}} \qquad (13.58)$$

where R_s is the outside radius of the sphere. Taking Poisson's ratio as 0·3 gives:

$$P_c = 1 \cdot 21E \, (t/R_s)^2 \qquad (13.59)$$

This equation gives the critical pressure required to cause general buckling; local buckling can occur at a lower pressure. Karman and Tsien (1939) have shown that the pressure to cause a "dimple" to form is about one-quarter of that given by equation 13.59, and is given by:

$$P_c' = 0 \cdot 365E \, (t/R_s)^2 \qquad (13.60)$$

A generous factor of safety is needed when applying equation 13.60 to the design of heads under external pressure. A value of 6 is typically used, which gives the following equation for the minimum thickness:

$$e = 4R_s \sqrt{(P_e/E)} \qquad (13.61)$$

Any consistent system of units can be used with equation 13.61.

Torispherical and ellipsoidal heads can be designed as equivalent hemispheres. For a torispherical head the radius R_s is taken as equivalent to the crown radius R_c. For an ellipsoidal head the radius can be taken as the maximum radius of curvature; that at the top, given by:

$$R_s = \frac{a^2}{b} \qquad (13.62)$$

where $2a$ = major axis = D_o (shell o.d.),
    $2b$ = minor axis = $2h$,
    h = height of the head from the tangent line.
Because the radius of curvature of an ellipse is not constant the use of the maximum radius will over-size the thickness required.

Design methods for heads under external pressure are given in the standards and codes (BS 5500, clause 3.6).

Example 13.2

A vacuum distillation column is to operate under a top pressure of 50 mm Hg. The plates are supported on rings 75 mm wide, 10 mm deep. The column diameter is 1 m and the plate spacing 0·5 m. Check if the support rings will act as effective stiffening rings. The material of construction is carbon steel and the maximum operating temperature 50°C. If the vessel thickness is 10 mm, check if this is sufficient.

Solution

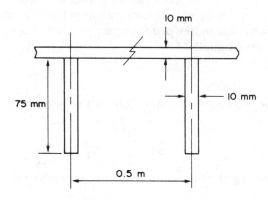

Take the design pressure as 1 bar external.

From equation 13.55 the load on each ring = 0.5×10^5 N/m.

Taking E for steel at 50°C as 200,000 N/mm², = 2×10^{11} N/m², and using a factor of safety of 6, the second moment of area of the ring to avoid buckling is given by: equation 13.57

$$0.5 \times 10^5 = \frac{24 \times 2 \times 10^{11} \times I_r}{1^3 \times 6}$$

$$I_r = 6.25 \times 10^{-8} \text{ m}^4$$

For a rectangular section, the second moment of area is given by

$$I = \frac{\text{breadth} \times \text{depth}^3}{12}$$

so I_r for the support rings = $\dfrac{10 \times (75)^3 \times 10^{-12}}{12}$

$$= 3.5 \times 10^{-7} \text{ m}^4$$

and the support ring is of an adequate size to be considered as a stiffening ring.

$$L'/D_o = 0.5/1 = 0.5$$
$$D_o/t = 1000/10 = 100$$

From Fig. 13.16 $K_c = 75$
From equation 13.52

$$P_c = 75 \times 2 \times 10^{11}(1/100)^3 = \underline{\underline{15 \times 10^6 \, \text{N/m}^2}}$$

which is well above the maximum design pressure of $10^5 \, \text{N/m}^2$.

13.8. Design of vessels subject to combined loading

Pressure vessels are subjected to other loads in addition to pressure (see Section 13.4.7) and must be designed to withstand the worst combination of loading without failure. It is not practical to give an explicit relationship for the vessel thickness to resist combined loads. A trial thickness must be assumed (based on that calculated for pressure alone) and the resultant stress from all loads determined to ensure that the maximum allowable stress intensity is not exceeded at any point.

The main sources of load to consider are:

1. Pressure.
2. Dead weight of vessel and contents.
3. Wind.
4. Earthquake (seismic).
5. External loads imposed by piping and attached equipment.

The primary stresses arising from these loads are considered in the following paragraphs, for cylindrical vessels; Fig. 13.17.

Primary stresses

1. The longitudinal and circumferential stresses due to pressure (internal or external), given by:

$$\sigma_h = \frac{PD_i}{2t} \tag{13.63}$$

$$\sigma_L = \frac{PD_i}{4t} \tag{13.64}$$

2. The direct stress σ_w due to the weight of the vessel, its contents, and any attachments. The stress will be tensile (positive) for points below the plane of the vessel supports, and compressive (negative) for points above the supports, see Fig. 13.18. The dead-weight stress will normally only be significant, compared to the magnitude of the other stresses, in tall vessels.

$$\sigma_w = \frac{W}{\pi(D_i + t)t} \tag{13.65}$$

where W is the total weight which is supported by the vessel wall at the plane considered, see Section 13.8.1.

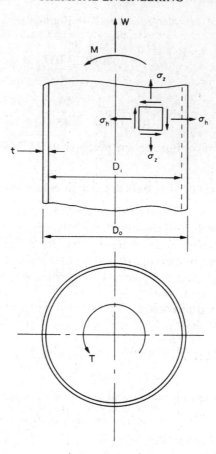

FIG. 13.17. Stresses in a cylindrical shell under combined loading

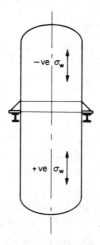

FIG. 13.18. Stresses due to dead-weight loads

3. Bending stresses resulting from the bending moments to which the vessel is subjected. Bending moments will be caused by the following loading conditions:
   (a) The wind loads on tall self-supported vessels (Section 13.8.2).
   (b) Seismic (earthquake) loads on tall vessels (Section 13.8.3).
   (c) The dead weight and wind loads on piping and equipment which is attached to the vessel, but offset from the vessel centre line (Section 13.8.4).
   (d) For horizontal vessels with saddle supports, from the disposition of dead-weight load (see Section 13.9.1).
   The bending stresses will be compressive or tensile, depending on location, and are given by:

$$\sigma_b = \pm \frac{M}{I_v}\left(\frac{D_i}{2} + t\right) \tag{13.66}$$

where M_v is the total bending moment at the plane being considered and I_v the second moment of area of the vessel about the plane of bending.

$$I_v = \frac{\pi}{64}(D_o^4 - D_i^4) \tag{13.67}$$

4. Torsional shear stresses τ resulting from torque caused by loads offset from the vessel axis. These loads will normally be small, and need not be considered in preliminary vessel designs.
   The torsional shear stress is given by:

$$\tau = \frac{T}{I_p}\left(\frac{D_i}{2} + t\right) \tag{13.68}$$

where T = the applied torque,

$$I_p = \text{polar second moment of area} = \frac{\pi}{32}(D_o^4 - D_i^4)$$

Principal stresses

The principal stresses will be given by:

$$\sigma_1 = \tfrac{1}{2}[\sigma_h + \sigma_z + \sqrt{(\sigma_h - \sigma_z)^2 + 4\tau^2}] \tag{13.69}$$

$$\sigma_2 = \tfrac{1}{2}[\sigma_h + \sigma_z - \sqrt{(\sigma_h - \sigma_z)^2 + 4\tau^2}] \tag{13.70}$$

where σ_z = total longitudinal stress
$\quad = \sigma_L + \sigma_w \pm \sigma_b$

σ_w should be counted as positive if tension and negative if compressive.
τ is not usually significant.
   The third principal stress, that in the radial direction σ_3, will usually be negligible for thin-walled vessels (see Section 13.1.1). As an approximation it can be taken as equal to one-half the pressure loading

$$\sigma_3 = 0.5P \tag{13.71}$$

σ_3 will be compressive (negative).

Allowable stress intensity

The maximum intensity of stress allowed will depend on the particular theory of failure adopted in the design method (see Section 13.3.2). The maximum shear-stress theory is normally used for pressure vessel design, and is the criterion used in BS 5500.

Using this criterion the maximum stress intensity at any point is taken for design purposes as the numerically greatest value of the following:

$$(\sigma_1 - \sigma_2)$$
$$(\sigma_1 - \sigma_3)$$
$$(\sigma_2 - \sigma_3)$$

The vessel wall thickness must be sufficient to ensure the maximum stress intensity does not exceed the design stress (nominal design strength) for the material of construction, at any point.

BS 5500 allows a stress intensity of 1·25 times the nominal design strength when the combined loading includes wind and earthquake loads.

Compressive stresses and elastic stability

Under conditions where the resultant axial stress σ_z due to the combined loading is compressive, the vessel may fail by elastic instability (buckling) (see Section 13.3.3). Failure can occur in a thin-walled process column under an axial compressive load by buckling of the complete vessel, as with a strut (Euler buckling); or by local buckling, or wrinkling, of the shell plates. Local buckling will normally occur at a stress lower than that required to buckle the complete vessel. A column design must be checked to ensure that the maximum value of the resultant axial stress does not exceed the critical value at which buckling will occur.

For a curved plate subjected to an axial compressive load the critical buckling stress σ_c is given by (see Timoshenko, 1936):

$$\sigma_c = \frac{E}{\sqrt{3(1 - v^2)}} \left(\frac{t}{R_p} \right) \tag{13.72}$$

where R_p is the radius of curvature.

Taking Poisson's ratio as 0·3 gives:

$$\sigma_c = 0 \cdot 60 E (t/R_p) \tag{13.73}$$

By applying a suitable factor of safety, equation 13.72 can be used to predict the maximum allowable compressive stress to avoid failure by buckling. A large factor of safety is required, as experimental work has shown that cylindrical vessels will buckle at values well below that given by equation 13.72. For steels at ambient temperature $E = 200,000$ N/mm^2, and equation 13.72 with a factor of safety of 12 gives:

$$\sigma_c = 2 \times 10^4 (t/D_o) \ N/mm^2 \tag{13.74}$$

The maximum compressive stress in a vessel wall should not exceed that given by equation 13.74; or the maximum allowable design stress for the material, whichever is the least.

Stiffening

As with vessels under external pressure, the resistance to failure buckling can be increased significantly by the use of stiffening rings, or longitudinal strips. Methods for estimating the critical buckling stress for stiffened vessels are given in the standards and codes (BS 5500, appendix B and clause 3.6.3).

Loading

The loads to which a vessel may be subjected will not all occur at the same time. For example, it is the usual practice to assume that the maximum wind load will not occur simultaneously with a major earthquake.

The vessel must be designed to withstand the worst combination of the loads likely to occur in the following situations:

1. During erection (or dismantling) of the vessel.
2. With the vessel erected but not operating.
3. During testing (the hydraulic pressure test).
4. During normal operation.

13.8.1. Weight loads

The major sources of dead weight loads are:

1. The vessel shell.
2. The vessel fittings: manways, nozzles.
3. Internal fittings: plates (plus the fluid on the plates); heating and cooling coils.
4. External fittings: ladders, platforms, piping.
5. Auxiliary equipment which is not self-supported; condensers, agitators.
6. Insulation.
7. The weight of liquid to fill the vessel. The vessel will be filled with water for the hydraulic pressure test; and may fill with process liquid due to misoperation.

Note: for vessels on a skirt support (see Section 13.9.2), the weight of the liquid to fill the vessel will be transferred directly to the skirt.

The weight of the vessel and fittings can be calculated from the preliminary design sketches. The weights of standard vessel components: heads, shell plates, manways, branches and nozzles, are given in various handbooks; Megyesy (1977) and Brownell and Young (1959).

For preliminary calculations the approximate weight of a cylindrical vessel with domed ends, and uniform wall thickness, can be estimated from the following equation:

$$W_v = C_v \pi \rho_m D_m g (H_v + 0.8 D_m) t \times 10^{-3} \qquad (13.75)$$

where W_v = total weight of the shell, excluding internal fittings, such as plates, N,
   C_v = a factor to account for the weight of nozzles, manways, internal supports, etc; which can be taken as
   = 1·08 for vessels with only a few internal fittings,
   = 1·15 for distillation columns, or similar vessels, with several manways, and with plate support rings, or equivalent fittings,

H_v = height, or length, between tangent lines (the length of the cylindrical
       section), m,
 g = gravitational acceleration, $9\cdot81$ m/s^2,
 t = wall thickness, mm
ρ_m = density of vessel material, kg/m^3,
D_m = mean diameter of vessel = $(D_i + t \times 10^{-3})$, m.

For a steel vessel, equation 13.75 reduces to:

$$W_v = 240C_v D_m (H_v + 0\cdot8 D_m)t \qquad (13.76)$$

The following values can be used as a rough guide to the weight of fittings; see Nelson
(1963):

(a) caged ladders, steel, 360 N/m length,
(b) plain ladders, steel, 150 N/m length,
(c) platforms, steel, for vertical columns, $1\cdot7$ kN/m^2 area,
(d) contacting plates, steel, including typical liquid loading, $1\cdot2$ kN/m^2 plate area.

Typical values for the density of insulating materials are (all kg/m^3):

Foam glass	150
Mineral wool	130
Fibreglass	100
Calcium silicate	200

These densities should be doubled to allow for attachment fittings, sealing, and moisture
absorption.

13.8.2. Wind loads (tall vessels)

Wind loading will only be important on tall columns installed in the open. Columns and
chimney-stacks are usually free standing, mounted on skirt supports, and not attached to
structural steel work. Under these conditions the vessel under wind loading acts as a
cantilever beam, Fig. 13.19. For a uniformly loaded cantilever the bending moment at any
plane is given by:

$$M_x = \frac{wx^2}{2} \qquad (13.77)$$

where x is the distance measured from the free end and w the load per unit length
(Newtons per metre run).

So the bending moment, and hence the bending stress, will vary parabolically from zero
at the top of the column to a maximum value at the base. For tall columns the bending
stress due to wind loading will often be greater than direct stress due to pressure, and will
determine the plate thickness required. The most economical design will be one in which
the plate thickness is progressively increased from the top to the base of the column. The
thickness at the top being sufficient for the pressure load, and that at the base sufficient for
the pressure plus the maximum bending moment.

Any local increase in the column area presented to the wind will give rise to a local,
concentrated, load, Fig. 13.20. The bending moment at the column base caused by a

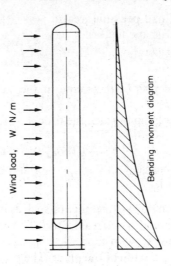

FIG. 13.19. Wind loading on a tall column

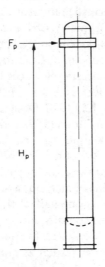

FIG. 13.20. Local wind loading

concentrated load is given by:

$$M_p = F_p H_p \tag{13.78}$$

where F_p = local, concentrated, load,
  H_p = the height of the concentrated load above the column base.

Dynamic wind pressure

The load imposed on any structure by the action of the wind will depend on the shape of the structure and the wind velocity.

$$P_w = \tfrac{1}{2} C_d \rho_a u_w^2 \tag{13.79}$$

where P_w = wind pressure (load per unit area),
   C_d = drag coefficient (shape factor),
   ρ_a = density of air,
   u_w = wind velocity.

The drag coefficient is a function of the shape of the structure and the wind velocity (Reynolds number).

For a smooth cylindrical column or stack the following semi-empirical equation can be used to estimate the wind pressure:

$$P_w = 0.05 u_w^2 \qquad (13.79a)$$

where P_w = wind pressure, N/m^2,
   u_w = wind speed, km/h.

If the column outline is broken up by attachments, such as ladders or pipe work, the factor of 0.05 in equation 13.79a should be increased to 0.07, to allow for the increased drag.

A column must be designed to withstand the highest wind speed that is likely to be encountered at the site during the life of the plant. The probability of a given wind speed occurring can be predicted by studying meteorological records for the site location.

Maps showing the wind speeds to be used in the design of structures at locations in the United Kingdom are given in the British Standards Code of Practice BS CP 3: 1972 "Basic Data for the Design of Buildings, Chapter V Loading: Part 2 Wind Loads". Typical values are around 50 m/s (112 miles per hour). The code of practice also gives methods estimating the dynamic wind pressure on buildings and structures of various shapes. Data and design methods are also given in the Engineering Sciences Data Unit (ESDU) reports on wind engineering, and in the Building Research Establishment Digest No. 119 (1970) "The assessment of wind loads", HMSO, July 1970. Design loadings for locations in the United States are given by Brownell and Young (1959) and Megyesy (1977).

A wind speed of 160 km/h (100 mph) can be used for preliminary design studies; equivalent to a wind pressure of 1280 N/m^2 (25 lb/ft^2).

At any site, the wind velocity near the ground will be lower than that higher up (due to the boundary layer), and in some design methods a lower wind pressure is used at heights below about 20 m; typically taken as one-half of the pressure above this height.

The loading per unit length of the column can be obtained from the wind pressure by multiplying by the effective column diameter: the outside diameter plus an allowance for the thermal insulation and attachments, such as pipes and ladders.

$$F_w = P_w D_{\text{eff}} \qquad (13.80)$$

An allowance of 0.4 m should be added for a caged ladder. The calculation of the wind load on a tall column, and the induced bending stresses, is illustrated in Example 13.3. Further examples of the design of tall columns are given by Brownell (1963) and Henry (1973).

Deflection of tall columns

Tall columns sway in the wind. The allowable deflection will normally be specified as less than 150 mm per 30 metres of height (6 in. per 100 ft).

For a column with a uniform cross-section, the deflection can be calculated using the formula for the deflection of a uniformly loaded cantilever. A method for calculating the deflection of a column where the wall thickness is not constant is given by Tang (1968).

Wind-induced vibrations

Vortex shedding from tall thin columns and stacks can induce vibrations which, if the frequency of shedding of eddies matches the natural frequency of the column, can be severe enough to cause premature failure of the vessel by fatigue. The effect of vortex shedding should be investigated for free standing columns with height to diameter ratios greater than 10. Methods for estimating the natural frequency of columns are given by Freese (1959) and DeGhetto and Long (1966); see also the BRE digest No. 119 (1970) and the ESDU manuals on wind engineering.

Helical strakes (strips) are fitted to the tops of tall smooth chimneys to change the pattern of vortex shedding and so prevent resonant oscillation. The same effect will be achieved on a tall column by distributing any attachments (ladders, pipes and platforms) around the column.

13.8.3. Earthquake loading

The movement of the earth's surface during an earthquake produces horizontal shear forces on tall self-supported vessels, the magnitude of which increases from the base upward. The total shear force on the vessel will be given by:

$$F_s = a_e(W/g) \tag{13.81}$$

where a_e = the acceleration of the vessel due to the earthquake,
    g = the acceleration due to gravity,
    W = total weight of the vessel.

The term (a_e/g) is called the seismic constant C_e, and is a function of the natural period of vibration of the vessel and the severity of the earthquake. Values of the seismic constant have been determined empirically from studies of the damage caused by earthquakes, and are available for those geographical locations which are subject to earthquake activity. Values for sites in the United States, and procedures for determining the bending stresses induced in tall columns, are given by Brownell and Young (1959) and Megyesy (1977).

Earthquake loads are not normally considered in the design of vessels for sites in the United Kingdom, as the probability of an earthquake occurring of sufficient severity to cause significant damage is negligible.

13.8.4. Eccentric loads (tall vessels)

Ancillary equipment attached to a tall vessel will subject the vessel to a bending moment if the centre of gravity of the equipment does not coincide with the centre line of the vessel (Fig. 13.21). The moment produced by small fittings, such as ladders, pipes and manways, will be small and can be neglected. That produced by heavy equipment, such as reflux condensers and side platforms, can be significant and should be considered. The moment is given by:

$$M_e = W_e L_o \tag{13.82}$$

where W_e = dead weight of the equipment,
    L_o = distance between the centre of gravity of the equipment and the column centre line.

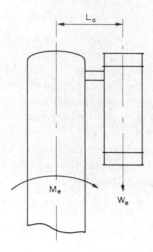

FIG. 13.21. Bending moment due to offset equipment

13.8.5. Torque

Any horizontal force imposed on the vessel by ancillary equipment, the line of thrust of which does not pass through the centre line of the vessel, will produce a torque on the vessel. Such loads can arise through wind pressure on piping and other attachments. However, the torque will normally be small and usually can be disregarded. The pipe work and the connections for any ancillary equipment will be designed so as not to impose a significant load on the vessel.

Example 13.3

Make a preliminary estimate of the plate thickness required for the distillation column specified below:

Height, between tangent lines   50 m
Diameter   2 m
Skirt support, height   3 m
100 sieve plates, equally spaced
Insulation, mineral wool   75 mm thick
Material of construction, stainless steel, design stress 135 N/mm^2 at design temperature 200°C
Operating pressure 10 bar (absolute)
Vessel to be fully radiographed (joint factor 1).

Solution

Design pressure; take as 10 per cent above operating pressure

$$= (10 - 1) \times 1 \cdot 1 = 9 \cdot 9 \text{ bar, say 10 bar}$$
$$= 1 \cdot 0 \text{ N/mm}^2$$

Minimum thickness required for pressure loading

$$(13.39) \qquad = \frac{1 \times 2 \times 10^3}{2 \times 135 - 1} = 7 \cdot 4 \text{ mm}$$

A much thicker wall will be needed at the column base to withstand the wind and dead weight loads.

As a first trial, divide the column into five sections (courses), with the thickness increasing by 2 mm per section. Try 10, 12, 14, 16, 18 mm.

Dead weight of vessel

Though equation 13.76 only applies strictly to vessels with uniform thickness, it can be used to get a rough estimate of the weight of this vessel by using the average thickness in the equation, 14 mm.

Take $C_v = 1 \cdot 15$, vessel with plates,
$D_m = 2 + 14 \times 10^{-3} = 2 \cdot 014$ m,
$H_v = 50$ m,
$t = 14$ mm

$$(13.76) \qquad W_v = 240 \times 1 \cdot 15 \times 2 \cdot 014 \, (50 + 0 \cdot 8 \times 2 \cdot 014) 14$$
$$= 401643 \text{ N}$$
$$= 402 \text{ kN}$$

Weight of plates:
plate area $= \pi/4 \times 2^2 = 3 \cdot 14 \text{ m}^2$
weight of a plate (see page 662) $= 1 \cdot 2 \times 3 \cdot 14 = 3 \cdot 8 \text{ kN}$
100 plates $= 100 \times 3 \cdot 8 = 380 \text{ kN}$
Weight of insulation:
mineral wool density $= 130 \text{ kg/m}^3$
approximate volume of insulation $= \pi \times 2 \times 50 \times 75 \times 10^{-3}$
$= 23 \cdot 6 \text{ m}^3$
weight $= 23 \cdot 6 \times 130 \times 9 \cdot 81 = 30,049 \text{ N}$
double this to allow for fittings, etc. $= 60 \text{ kN}$
Total weight:

shell	402
plates	380
insulation	60
	842 kN

Wind loading

Take dynamic wind pressure as 1280 N/m^2.
Mean diameter, including insulation $= 2 + 2(14 + 75) \times 10^{-3}$
$= 2 \cdot 18$ m

$$(13.81) \qquad \text{Loading (per linear metre)} \; F_w = 1280 \times 2 \cdot 18 = 2790 \text{ N/m}$$

Bending moment at bottom tangent line:

$$(13.77) \qquad M_x = \frac{2790}{2} \times 50^2 = 3,487,500 \text{ Nm}$$

Analysis of stresses

At bottom tangent line
Pressure stresses:

(13.64)
$$\sigma_L = \frac{1 \cdot 0 \times 2 \times 10^3}{4 \times 18} = 27 \cdot 8 \text{ N/mm}^2$$

(13.63)
$$\sigma_h = \frac{1 \times 2 \times 10^3}{2 \times 18} = 55 \cdot 6 \text{ N/mm}^2$$

Dead weight stress:

(13.65)
$$\sigma_w = \frac{W_v}{\pi (D_i + t)t} = \frac{842 \times 10^3}{\pi (2000 + 18)18}$$

$$= 7 \cdot 4 \text{ N/mm}^2 \text{ (compressive)}$$

Bending stresses:

$$D_o = 2000 + 2 \times 18 = 2036 \text{ mm}$$

(13.67)
$$I_v = \frac{\pi}{64}(2036^4 - 2000^4) = 5 \cdot 81 \times 10^{10} \text{ mm}^4$$

(13.66)
$$\sigma_b = \pm \frac{3{,}487{,}500 \times 10^3}{5 \cdot 81 \times 10^{10}} \left(\frac{2000}{2} + 18 \right)$$

$$= \pm 61 \cdot 1 \text{ N/mm}^2$$

The resultant longitudinal stress is:

$$\sigma_z = \sigma_L + \sigma_w \pm \sigma_b$$

σ_w is compressive and therefore negative.

σ_z (upwind) $= 27 \cdot 8 - 7 \cdot 4 + 61 \cdot 1 = +81 \cdot 5 \text{ N/mm}^2$.

σ_z (downwind) $= 27 \cdot 8 - 7 \cdot 4 - 61 \cdot 1 = -40 \cdot 7 \text{ N/mm}^2$.

As there is no torsional shear stress, the principal stresses will be σ_z and σ_h.

The radial stress is negligible, $\simeq \dfrac{P_i}{2} = 0 \cdot 5 \text{ N/mm}^2$.

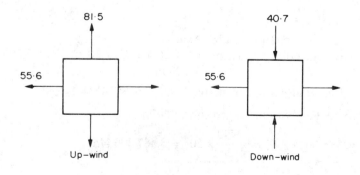

The greatest difference between the principal stress will be on the down-wind side

$$(55 \cdot 6 - (-40 \cdot 7)) = \underline{\underline{96 \cdot 5 \text{ N/mm}^2}},$$

well below the maximum allowable design stress

Check elastic stability (buckling)

Critical buckling stress:

(13.74)
$$\sigma_c = 2 \times 10^4 \left(\frac{18}{2036} \right) = \underline{\underline{176 \cdot 8 \text{ N/mm}^2}}$$

The maximum compressive stress will occur when the vessel is not under pressure $= 7 \cdot 4 + 61 \cdot 1 = 68 \cdot 5$, well below the critical buckling stress.

So design is satisfactory. Could reduce the plate thickness and recalculate.

13.9. Vessel supports

The method used to support a vessel will depend on the size, shape, and weight of the vessel; the design temperature and pressure; the vessel location and arrangement; and the internal and external fittings and attachments. Horizontal vessels are usually mounted on two saddle supports; Fig. 13.22. Skirt supports are used for tall, vertical columns; Fig. 13.23. Brackets, or lugs, are used for all types of vessel; Fig. 13.24. The supports must be designed to carry the weight of the vessel and contents, and any superimposed loads, such as wind loads. Supports will impose localised loads on the vessel wall, and the design must be checked to ensure that the resulting stress concentrations are below the maximum allowable design stress. Supports should be designed to allow easy access to the vessel and fittings for inspection and maintenance.

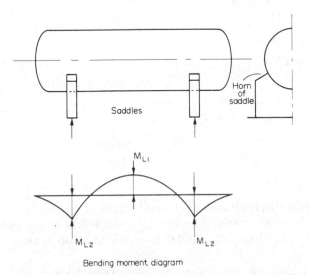

FIG. 13.22. Horizontal cylindrical vessel on saddle supports

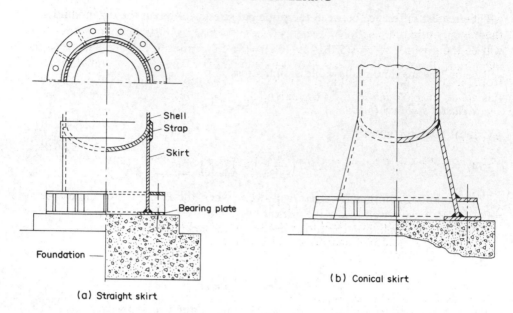

Shell
Strap
Skirt

Bearing plate

Foundation

(a) Straight skirt

(b) Conical skirt

Fig. 13.23. Typical skirt-support designs

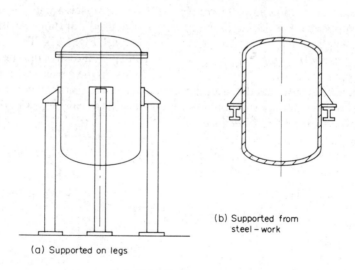

(a) Supported on legs

(b) Supported from
steel-work

Fig. 13.24 Bracket supports

13.9.1. Saddle supports

Though saddles are the most commonly used support for horizontal cylindrical vessels, legs can be used for small vessels. A horizontal vessel will normally be supported at two cross-sections; if more than two saddles are used the distribution of the loading is uncertain.

A vessel supported on two saddles can be considered as a simply supported beam, with an essentially uniform load, and the distribution of longitudinal axial bending moment will be as shown in Fig. 13.22. Maxima occur at the supports and at mid-span. The theoretical optimum position of the supports to give the least maximum bending moment will be the position at which the maxima at the supports and at mid-span are equal in magnitude. For a uniformly loaded beam the position will be at 21 per cent of the span, in from each end. The saddle supports for a vessel will usually be located nearer the ends than this value, to make use of the stiffening effect of the ends.

Stress in the vessel wall

The longitudinal bending stress at the mid-span of the vessel is given by:

$$\sigma_{b1} = \frac{M_{L1}}{I_h} \times \frac{D}{2} \simeq \frac{4M_{L1}}{\pi D^2 t} \qquad (13.83)$$

where M_{L1} = longitudinal bending stress at the mid-span,
$\quad I_h$ = second moment of area of the shell,
$\quad D$ = shell diameter,
$\quad t$ = shell thickness.

The resultant axial stress due to bending and pressure will be given by:

$$\sigma_z = \frac{PD}{4t} \pm \frac{4M_{L1}}{\pi D^2 t} \qquad (13.84)$$

The magnitude of the longitudinal bending stress at the supports will depend on the local stiffness of the shell; if the shell does not remain circular under load a portion of the upper part of the cross-section is ineffective against longitudinal bending; see Fig. 13.25. The stress is given by:

$$\sigma_{b2} = \frac{4M_{L2}}{C_h \pi D^2 t} \qquad (13.85)$$

where M_{L2} = longitudinal bending moment at the supports,
$\quad C_h$ = an empirical constant; varying from 1·0 for a completely stiff shell to about 0·1 for a thin, unstiffened, shell.

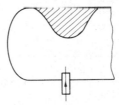

FIG. 13.25. Saddle supports: shaded area is ineffective against longitudinal bending in an unstiffened shell

The ends of the vessels will stiffen the shell if the position of the saddles is less than $D/4$ from the ends. Ring stiffeners, located at the supports, are used to stiffen the shells of long thin vessels. The rings may be fitted inside or outside the vessel.

In addition to the longitudinal bending stress, a vessel supported on saddles will be subjected to tangential shear stresses, which transfer the load from the unsupported sections of the vessel to the supports; and to circumferential bending stresses. All these stresses need to be considered in the design of large, thin-walled, vessels, to ensure that the resultant stress does not exceed the maximum allowable design stress or the critical buckling stress for the material. A detailed stress analysis is beyond the scope of this book. A complete analysis of the stress induced in the shell by the supports is given by Zick (1951). Zick's method forms the basis of the design methods given in the national codes and standards (BS 5500 appendix G3). The method is also given by Brownell and Young (1959) and Megyesy (1977).

Design of saddles

The saddles must be designed to withstand the load imposed by the weight of the vessel and contents. They are constructed of bricks or concrete, or are fabricated from steel plate. The contact angle should not be less than 120°, and will not normally be greater than 150°. Wear plates are often welded to the shell wall to reinforce the wall over the area of contact with the saddle.

The dimensions of a typical "standard" saddle designs are given in Fig. 13.26. To take up any thermal expansion of the vessel, such as that in heat exchangers, the anchor bolt holes in one saddle can be slotted.

Procedures for the design of saddle supports are given by Brownell and Young (1959) and Megyesy (1977).

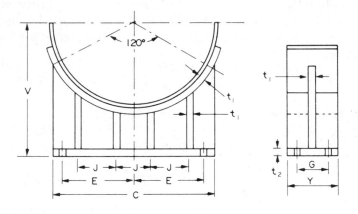

Vessel diam. (m)	Maximum weight (kN)	Dimensions (m)						mm			
		V	Y	C	E	J	G	t_2	t_1	Bolt diam.	Bolt holes
0·6	35	0·48	0·15	0·55	0·24	0·190	0·095	6	5	20	25
0·8	50	0·58	0·15	0·70	0·29	0·225	0·095	8	5	20	25
0·9	65	0·63	0·15	0·81	0·34	0·275	0·095	10	6	20	25
1·0	90	0·68	0·15	0·91	0·39	0·310	0·095	11	8	20	25
1·2	180	0·78	0·20	1·09	0·45	0·360	0·140	12	10	24	30

All contacting edges fillet welded

FIG. 13.26a. Standard steel saddles, for vessels up to 1·2 m (adapted from Bhattacharyya, 1976)

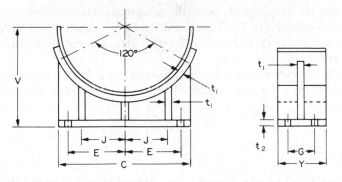

Vessel diam. (m)	Maximum weight (kN)	Dimensions (m)						mm			
		V	Y	C	E	J	G	t_2	t_1	Bolt diam.	Bolt holes
1·4	230	0·88	0·20	1·24	0·53	0·305	0·140	12	10	24	30
1·6	330	0·98	0·20	1·41	0·62	0·350	0·140	12	10	24	30
1·8	380	1·08	0·20	1·59	0·71	0·405	0·140	12	10	24	30
2·0	460	1·18	0·20	1·77	0·80	0·450	0·140	12	10	24	30
2·2	750	1·28	0·225	1·95	0·89	0·520	0·150	16	12	24	30
2·4	900	1·38	0·225	2·13	0·98	0·565	0·150	16	12	27	33
2·6	1000	1·48	0·225	2·30	1·03	0·590	0·150	16	12	27	33
2·8	1350	1·58	0·25	2·50	1·10	0·625	0·150	16	12	27	33
3·0	1750	1·68	0·25	2·64	1·18	0·665	0·150	16	12	27	33
3·2	2000	1·78	0·25	2·82	1·26	0·730	0·150	16	12	27	33
3·6	2500	1·98	0·25	3·20	1·40	0·815	0·150	16	12	27	33

All contacting edges fillet welded

FIG. 13.26b. Standard steel saddles, for vessels greater than 1·2 m (adapted from Bhattacharyya, 1976)

13.9.2. Skirt supports

A skirt support consists of a cylindrical or conical shell welded to the base of the vessel. A flange at the bottom of the skirt transmits the load to the foundations. Typical designs are shown in Fig. 13.23. Openings must be provided in the skirt for access and for any

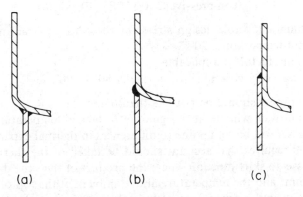

FIG. 13.27. Skirt-support welds

connecting pipes; the openings are normally reinforced. The skirt may be welded to the bottom head of the vessel. Fig. 13.27a; or welded flush with shell, Fig. 13.27b; or welded to the outside of the vessel shell, Fig. 13.27c. The arrangement shown in Fig. 13.27b is usually preferred.

Skirt supports are recommended for vertical vessels as they do not impose concentrated loads on the vessel shell; they are particularly suitable for use with tall columns subject to wind loading.

Skirt thickness

The skirt thickness must be sufficient to withstand the dead-weight loads and bending moments imposed on it by the vessel; it will not be under the vessel pressure.

The resultant stresses in the skirt will be:

$$\sigma_s \text{ (tensile)} = \sigma_{bs} - \sigma_{ws} \tag{13.86}$$

and

$$\sigma_s \text{ (compressive)} = \sigma_{bs} + \sigma_{ws} \tag{13.87}$$

where σ_{bs} = bending stress in the skirt = $\dfrac{4M_s}{\pi(D_s + t_s)t_s D_s}$, $\tag{13.88}$

σ_{ws} = the dead weight stress in the skirt,

$$= \frac{W}{\pi(D_s + t_s)t_s} \tag{13.89}$$

where M_s = maximum bending moment, evaluated at the base of the skirt (due to wind, seismic and eccentric loads, see Section 13.8),

W = total weight of the vessel and contents (see Section 13.8),

D_s = inside diameter of the skirt, at the base,

t_s = skirt thickness.

The skirt thickness should be such that under the worst combination of wind and dead-weight loading the following design criteria are not exceeded:

$$\sigma_s \text{ (tensile)} \not> f_s J \sin \theta_s \tag{13.90}$$

$$\sigma_s \text{ (compressive)} \not> 0.125 E (t_s/D_s) \sin \theta_s \tag{13.91}$$

where f_s = maximum allowable design stress for the skirt material, normally taken at ambient temperature, 20°C,

J = weld joint factor, if applicable,

θ_s = base angle of a conical skirt, normally 80° to 90°.

The minimum thickness should be not less than 6 mm.

Where the vessel wall will be at a significantly higher temperature than the skirt, discontinuity stresses will be set up due to differences in thermal expansion. The British Standard BS 5500 requires that account should be taken of the thermal discontinuity stresses at the vessel to skirt junction where the product of the skirt diameter (mm), the skirt thickness (mm), and the temperature above ambient at the top of the skirt exceeds 1.6×10^7 (mm² °C). Similar criteria are given in the other national codes and standards.

Methods for calculating the thermal stresses in skirt supports are given by Weil and Murphy (1960) and Bergman (1963).

Base ring and anchor bolt design

The loads carried by the skirt are transmitted to the foundation slab by the skirt base ring (bearing plate). The moment produced by wind and other lateral loads will tend to overturn the vessel; this will be opposed by the couple set up by the weight of the vessel and the tensile load in the anchor bolts. A variety of base ring designs is used with skirt supports. The simplest types, suitable for small vessels, are the plain flange ring and rolled-angle rings shown in Fig. 13.28a and b. For larger columns a double ring stiffened by gussets, Fig. 13.28c, or chair supports, Fig. 13.30, are used. Design methods for base rings, and methods for sizing the anchor bolts, are given by Brownell and Young (1959). For preliminary design, the short-cut method and nomographs given by Scheiman (1963) can be used. Scheiman's method is based on a more detailed procedure for the design of base rings and foundations for columns and stacks given by Marshall (1958). Scheiman's method is outlined below and illustrated in Example 13.4.

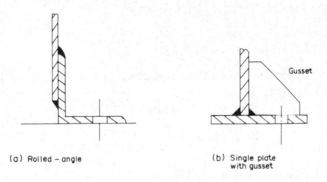

(a) Rolled – angle   (b) Single plate with gusset

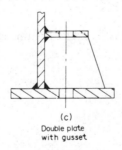

(c) Double plate with gusset

FIG. 13.28. Flange ring designs

The anchor bolts are assumed to share the overturning load equally, and the bolt area required is given by:

$$A_b = \frac{1}{N_b f_b}\left[\frac{4M_s}{D_b} - W\right]$$

(13.92)

where A_b = area of one bolt at the root of the thread, mm^2,
  N_b = number of bolts,
  f_b = maximum allowable bolt stress, N/mm^2;
    typical design value 125 N/mm^2 (18,000 psi),
  M_s = bending (overturning) moment at the base, Nm,
  W = weight of the vessel, N,
  D_b = bolt circle diameter, m.

Scheiman gives the following guide rules which can be used for the selection of the anchor bolts:

1. Bolts smaller than 25 mm (1 in.) diameter should not be used.
2. Minimum number of bolts 8.
3. Use multiples of 4 bolts.
4. Bolt pitch should not be less than 600 mm (2 ft).

If the minimum bolt pitch cannot be accommodated with a cylindrical skirt, a conical skirt should be used.

The base ring must be sufficiently wide to distribute the load to the foundation. The total compressive load on the base ring is given by:

$$F_b = \left[\frac{4M_s}{\pi D_s^2} + \frac{W}{\pi D_s} \right] \tag{13.93}$$

where F_b = the compressive load on the base ring, Newtons per linear metre,
  D_s = skirt diameter, m.

The minimum width of the base ring is given by:

$$L_b = \frac{F_b}{f_c} \times \frac{1}{10^3} \tag{13.94}$$

where L_b = base ring width, mm (Fig. 13.29),
  f_c = the maximum allowable bearing pressure on the concrete foundation pad, which will depend on the mix used, and will typically range from 3·5 to 7 N/mm^2 (500 to 1000 psi).

The required thickness for the base ring is found by treating the ring as a cantilever beam.

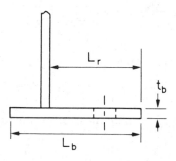

FIG. 13.29. Flange ring dimensions

The minimum thickness is given by:

$$t_b = L_r \sqrt{\frac{3f'_c}{f_r}}$$

(13.95)

where L_r = the distance from the edge of the skirt to the outer edge of the ring, mm; Fig. 13.29,

t_b = base ring thickness, mm,

f'_c = actual bearing pressure on base, N/mm^2,

f_r = allowable design stress in the ring material, typically 140 N/mm^2.

Standard designs will normally be used for the bolting chairs. The design shown in Fig. 13.30 has been adapted from that given by Scheiman.

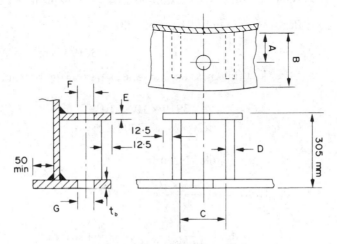

All contacting edges fillet welded

Bolt size	Root area	A	B	C	D	E	F	G
				Dimensions mm				
M24	353	45	76	64	13	19	30	36
M30	561	50	76	64	13	25	36	42
M36	817	57	102	76	16	32	42	48
M42	1120	60	102	76	16	32	48	54
M48	1470	67	127	89	19	38	54	60
M56	2030	75	150	102	25	45	60	66
M64	2680	83	152	102	25	50	70	76
70	—	89	178	127	32	64	76	83
76	—	95	178	127	32	64	83	89

Bolt size = Nominal dia. (BS 4190: 1967)

FIG. 13.30 Anchor bolt chair design

Example 13.4

Design a skirt support for the column specified in Example 13.3.

Solution

Try a straight cylindrical skirt ($\theta_s = 90°$) of plain carbon steel, design stress 135 N/mm^2 and Young's modulus 200,000 N/mm^2 at ambient temperature.

The maximum dead weight load on the skirt will occur when the vessel is full of water.

$$\text{Approximate weight} = \left(\frac{\pi}{4} \times 2^2 \times 50 \right) 1000 \times 9\cdot81$$

$$= 1,540,951 \text{ N}$$

$$= 1541 \text{ kN}$$

Weight of vessel, from Example 13.3 = 842 kN

Total weight = 1541 + 842 = 2383 kN

Wind loading, from Example 13·4 = 2·79 kN/m

(13.77)   Bending moment at base of skirt = $2\cdot79 \times \dfrac{53^2}{2}$

$$= 3919 \text{ kNm}$$

As a first trial, take the skirt thickness as the same as that of the bottom section of the vessel, 18 mm.

(13.88) $\sigma_{bs} = \dfrac{4 \times 3919 \times 10^3 \times 10^3}{\pi(2000 + 18)2000 \times 18}$

$$= 68\cdot7 \text{ N/mm}^2$$

(13.89) $\sigma_{ws} \text{ (test)} = \dfrac{1543 \times 10^3}{\pi(2000 + 18)18} = 13\cdot5 \text{ N/mm}^2$

(13.89) $\sigma_{ws} \text{ (operating)} = \dfrac{842 \times 10^3}{\pi(2000 + 18)18} = 7\cdot4 \text{ N/mm}^2$

Note: the "test" condition is with the vessel full of water for the hydraulic test. In estimating total weight, the weight of liquid on the plates has been counted twice. The weight has not been adjusted to allow for this as the error is small, and on the "safe side".

(13.87)   Maximum $\hat\sigma_s$ (compressive) = 68·7 + 13·5 = 82·2 N/mm^2

(13.86)   Maximum $\hat\sigma_s$ (tensile)   = 68·7 − 7·4 = 61·3 N/mm^2

Take the joint factor J as 0·85.
Criteria for design:

(13.90) $\hat\sigma_s \text{ (tensile)} \not> f_s J \sin\theta$

$$61\cdot3 \not> 0\cdot85 \times 135 \sin 90$$

$$61\cdot3 \not> 115$$

(13.91) $\hat\sigma_s \text{ (compressive)} \not> 0\cdot125 E (t_s/D_s) \sin\theta$

$$82\cdot2 \not> 0\cdot125 \times 200,000 (18/2000) \sin 90$$

$$82\cdot2 \not> 225$$

Both criteria are satisfied, add 2 mm for corrosion, gives a design thickness of 20 mm.

Base ring and anchor bolts

Approximate pitch circle dia., say, 2·2 m
Circumference of bolt circle = 2200π
Number of bolts required, at minimum
recommended bolt spacing $= \dfrac{2200\pi}{600} = 11\cdot5$

Closest multiple of 4 = 12 bolts
Take bolt design stress = 125 N/mm^2
$M_s = 3919$ kN m

Take $W =$ operating value = 842 kN.

(13.92) $$A_b = \frac{1}{12 \times 125}\left[\frac{4 \times 3919 \times 10^3}{2\cdot2} - 842 \times 10^3\right]$$

$$= 4190 \text{ mm}^2$$

$$\text{Bolt root dia.} = \sqrt{\frac{4190 \times 4}{\pi}} = 73 \text{ mm, looks too large.}$$

Total compressive load on the base ring per unit length

(13.93) $$F_b = \left[\frac{4 \times 3919 \times 10^3}{\pi \times 2\cdot0^2} + \frac{842 \times 10^3}{\pi \times 2\cdot0}\right]$$

$$= 1381 \times 10^3 \text{ N/m}$$

Taking the bearing pressure as 5 N/mm^2

(13.94) $$L_b = \frac{1381 \times 10^3}{5 \times 10^3} = 276 \text{ mm}$$

Rather large—consider a flared skirt.
Take the skirt bottom dia. as 3 m

$$\text{Skirt base angle } \theta_s = \tan^{-1}\frac{3}{\frac{1}{2}(3-2)} = 80\cdot5°$$

Keep the skirt thickness the same as that calculated for the cylindrical skirt. Highest stresses will occur at the top of the skirt; where the values will be close to those calculated for the cylindrical skirt. Sin 80·5° = 0·99, so this term has little effect on the design criteria.

Assume bolt circle dia. = 3·2 m.
Take number of bolts as 16.

$$\text{Bolt spacing} = \frac{\pi \times 3\cdot2 \times 10^3}{16} = 628 \text{ mm} \quad \text{satisfactory.}$$

$$A_b = \frac{1}{16 \times 125}\left[\frac{4 \times 3919 \times 10^3}{3\cdot2} - 842 \times 10^3\right]$$

$$= \underline{\underline{2029 \text{ mm}^2}}$$

Use M56 bolts (BS 4190:1967) root area = 2030 mm^2,

$$F_b = \left[\frac{4 \times 3919 \times 10^3}{\pi \times 3^2} + \frac{842 \times 10^3}{\pi \times 3}\right]$$

$$= 644 \text{ kN/m}.$$

$$L_b = \frac{644 \times 10^3}{5 \times 10^3} = 129 \text{ mm}$$

This is the minimum width required; actual width will depend on the chair design.
    Actual width required (Fig. 13.30):

$$= L_r + t_s + 50 \text{ mm}$$
$$= 150 + 20 + 50 = \underline{\underline{220 \text{ mm}}}$$

Actual bearing pressure on concrete foundation:

$$f'_c = \frac{644 \times 10^3}{220 \times 10^3} = 2.93 \text{ N/mm}^2$$

(13.95)
$$t_b = 150 \sqrt{\frac{3 \times 2.93}{140}} = 37.6 \text{ mm}$$

round off to $\underline{40 \text{ mm}}$

Chair dimensions from Fig. 13.30 for bolt size M56.
Skirt to be welded flush with outer diameter of column shell.

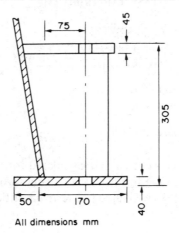

All dimensions mm

13.9.3. Bracket supports

Brackets, or lugs, can be used to support vertical vessels. The bracket may rest on the building structural steel work, or the vessel may be supported on legs; Fig. 13.24.
    The main load carried by the brackets will be the weight of the vessel and contents; in addition the bracket must be designed to resist the load due to any bending moment due to wind, or other loads. If the bending moment is likely to be significant skirt supports should be considered in preference to bracket supports.

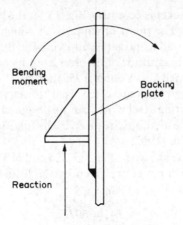

FIG. 13.31. Loads on a bracket support

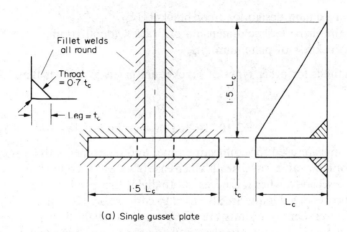

(a) Single gusset plate

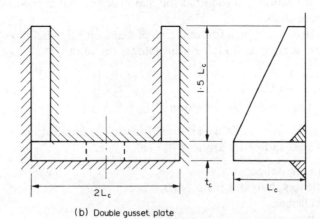

(b) Double gusset plate

FIG. 13.32. Bracket designs

As the reaction on the bracket is eccentric, Fig. 13.31, the bracket will impose a bending moment on the vessel wall. The point of support, at which the reaction acts, should be made as close to the vessel wall as possible; allowing for the thickness of any insulation. Methods for estimating the magnitude of the stresses induced in the vessel wall by bracket supports are given by Brownell and Young (1959) and by Wolosewick (1951); see also BS 5500, appendix G2. Backing plates are often used to carry the bending loads.

The brackets, and supporting steel work, can be designed using the usual methods for structural steelwork. Suitable methods are given in the Institute of Welding Handbook for structural steelwork, I. Weld. (1952).

Typical bracket designs are shown in Figs. 13.32a and b. The loads which steel brackets with these proportions will support are given by the following formula:

Single-gusset plate design, Fig. 13.32a:

$$F_{bs} = 60 L_c t_c \qquad (13.96)$$

Double-gusset plate design, Fig. 13.32b:

$$F_{bs} = 120 L_c t_c \qquad (13.97)$$

where F_{bs} = maximum design load per bracket, N,
L_c = the characteristic dimension of bracket (depth), mm,
t_c = thickness of plate, mm.

Design methods for other types of bracket are given by Brownell and Young (1959).

13.10. Bolted flanged joints

Flanged joints are used for connecting pipes and instruments to vessels, for manhole covers, and for removable vessel heads when ease of access is required. Flanges may also be used on the vessel body, when it is necessary to divide the vessel into sections for transport or maintenance. Flanged joints are also used to connect pipes to other equipment, such as pumps and valves. Screwed joints are often used for small-diameter pipe connections, below 40 mm. Flanged joints are also used for connecting pipe sections where ease of assembly and dismantling is required for maintenance, but pipework will normally be welded to reduce costs.

Flanges range in size from a few millimetres diameter for small pipes, to several metres diameter for those used as body or head flanges on vessels.

13.10.1. Types of flange, and selection

Several different types of flange are used for various applications. The principal types used in the process industries are:

1. Welding-neck flanges.
2. Slip-on flanges, hub and plate types.
3. Lap-joint flanges.
4. Screwed flanges.
5. Blank, or blind, flanges.

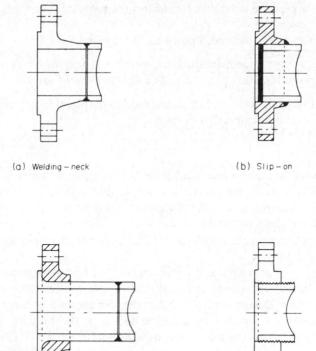

(a) Welding – neck

(b) Slip – on

(c) Lap – joint

(d) Screwed

FIG. 13.33. Flange types

Welding-neck flanges, Fig. 13.33*a*: have a long tapered hub between the flange ring and the welded joint. This gradual transition of the section reduces the discontinuity stresses between the flange and branch, and increases the strength of the flange assembly. Welding-neck flanges are suitable for extreme service conditions; where the flange is likely to be subjected to temperature, shear and vibration loads. They will normally be specified for the connections and nozzles on process vessels and process equipment.

Slip-on flanges, Fig. 13.33*b*: slip over the pipe or nozzle and are welded externally, and usually also internally. The end of the pipe is set back from 0 to 2·0 mm. The strength of a slip-on flange is from one-third to two-thirds that of the corresponding standard welding-neck flange. Slip-on flanges are cheaper than welding-neck flanges and are easier to align, but have poor resistance to shock and vibration loads. Slip-on flanges are generally used for pipe work. Figure 13.33*b* shows a forged flange with a hub; for light duties slip-on flanges can be cut from plate.

Lap-joint flanges, Fig. 13.33*c*: are used for piped work. They are economical when used with expensive alloy pipe, such as stainless steel, as the flange can be made from inexpensive carbon steel. Usually a short lapped nozzle is welded to the pipe, but with some schedules of pipe the lap can be formed on the pipe itself, and this will give a cheap method of pipe assembly.

Lap-joint flanges are sometimes known as "Van-stone flanges".

Screwed flanges, Fig. 13.33*d*: are used to connect screwed fittings to flanges. They are also sometimes used for alloy pipe which is difficult to weld satisfactorily.

Blind flanges (blank flanges): are flat plates, used to blank off flange connections, and as covers for manholes and inspection ports.

13.10.2. Gaskets

Gaskets are used to make a leak-tight joint between two surfaces. It is impractical to machine flanges to the degree of surface finish that would be required to make a satisfactory seal under pressure without a gasket. Gaskets are made from "semi-plastic" materials; which will deform and flow under load to fill the surface irregularities between the flange faces, yet retain sufficient elasticity to take up the changes in the flange alignment that occur under load.

A great variety of proprietary gasket materials is used, and reference should be made to the manufacturers' catalogues and technical manuals when selecting gaskets for a particular application. Design data for some of the more commonly used gasket materials are given in Table 13.4. Further data can be found in the pressure vessel codes and standards (BS 5500, clause 3.8) and in various handbooks, Perry and Chilton (1973). The minimum seating stress y is the force per unit area (pressure) on the gasket that is required to cause the material to flow and fill the surface irregularities in the gasket face.

TABLE 13.4. *Gasket materials*
(Based on a similar table in BS 5500: 1976)

Gasket material		Gasket factor m	Min. design seating stress $y(N/mm^2)$	Sketches	Minimum gasket width (mm)
Rubber without fabric or a high percentage of asbestos fibre; hardness:					
below 75° IRH		0·50	0		10
75° IRH or higher		1·00	1·4		
Asbestos with a suitable binder	3·2 mm thick	2·00	11·0		
for the operating conditions	1·6 mm thick	2·75	25·5		10
	0·8 mm thick	3·50	44·8		
Rubber with cotton fabric insertion		1·25	2·8		10
	3-ply	2·25	15·2		
Rubber with asbestos fabric insertion, with or without wire reinforcement	2-ply	2·50	20·0		10
	1-ply	2·75	25·5		
Vegetable fibre		1·75	7·6		10

TABLE 13.4. (*Cont.*)

Gasket material		Gasket factor m	Min. design seating stress $y(N/mm^2)$	Sketches	Minimum gasket width (mm)
Spiral-wound metal, asbestos filled	Carbon	2·50	20·0		
	Stainless or monel	3·00	31·0		10
Corrugated metal, asbestos inserted or Corrugated metal, jacketed asbestos filled	Soft aluminium	2·50	20·0		
	Soft copper or brass	2·75	25·5		
	Iron or soft steel	3·00	31·0		10
	Monel or 4 to 6 per cent chrome	3·25	37·9		
	Stainless steels	3·50	44·8		
Corrugated metal	Soft aluminium	2·75	25·5		
	Soft copper or brass	3·00	31·0		
	Iron or soft steel	3·25	37·9		10
	Monel or 4 to 6 per cent chrome	3·50	44·8		
	Stainless steels	3·75	52·4		
Flat metal jacketed asbestos filled	Soft aluminium	3·25	37·9		
	Soft copper or brass	3·50	44·8		
	Iron or soft steel	3·75	52·4		
	Monel	3·50	55·1		10
	4 to 6 per cent chrome	3·75	62·0		
	Stainless steels	3·75	62·0		
Grooved metal	Soft aluminium	3·25	37·9		
	Soft copper or brass	3·50	44·8		
	Iron or soft steel	3·75	52·4		10
	Monel or 4 to 6 per cent chrome	3·75	62·0		
	Stainless steels	4·25	69·5		
Solid flat metal	Soft aluminium	4·00	60·6		
	Soft copper or brass	4·75	89·5		
	Iron or soft steel	5·50	124		6
	Monel or 4 to 6 per cent chrome	6·00	150		
	Stainless steels	6·50	179		
Ring joint	Iron or soft steel	5·50	124		
	Monel or 4 to 6 per cent chrome	6·00	150		6
	Stainless steels	6·50	179		

The gasket factor m is the ratio of the gasket stress (pressure) under the operating conditions to the internal pressure in the vessel or pipe. The internal pressure will force the flanges' faces apart, so the pressure on the gasket under operating conditions will be lower than the initial tightening-up pressure. The gasket factor gives the minimum pressure that must be maintained on the gasket to ensure a satisfactory seal.

The following factors must be considered when selecting a gasket material:

1. The process conditions: pressure, temperature, corrosive nature of the process fluid.

2. Whether repeated assembly and disassembly of the joint is required.
3. The type of flange and flange face (see Section 13.10.3).

Up to pressures of 20 bar, the operating temperature and corrosiveness of the process fluid will be the controlling factor in gasket selection. Vegetable fibre and synthetic rubber gaskets can be used at temperatures of up to 100°C. Solid polyfluorocarbon (Teflon) and compressed asbestos gaskets can be used to a maximum temperature of about 260°C. Metal-reinforced gaskets can be used up to around 450°C. Plain soft metal gaskets are normally used for higher temperatures.

13.10.3. Flange faces

Flanges are also classified according to the type of flange face used. There are two basic types:

1. Full-faced flanges, Fig. 13.34a: where the face contact area extends outside the circle of bolts; over the full face of the flange.
2. Narrow-faced flanges, Fig. 13.34b, c, d: where the face contact area is located within the circle of bolts.

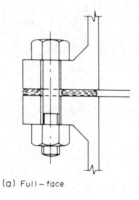

(a) Full-face

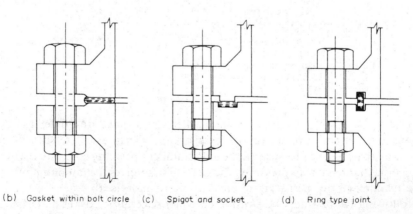

(b) Gasket within bolt circle   (c) Spigot and socket   (d) Ring type joint

Fig. 13.34. Flange types and faces

Full face, wide-faced, flanges are simple and inexpensive, but are only suitable for low pressures. The gasket area is large, and an excessively high bolt tension would be needed to achieve sufficient gasket pressure to maintain a good seal at high operating pressures.

The raised face, narrow-faced, flange shown in Fig. 13.34b is probably the most commonly used type of flange for process equipment.

Where the flange has a plain face, as in Fig. 13.34b, the gasket is held in place by friction between the gasket and flange surface. In the spigot and socket, and tongue and grooved faces, Fig. 13.34c, the gasket is confined in a groove, which prevents failure by "blow-out". Matched pairs of flanges are required, which increases the cost, but this type is suitable for high pressure and high vacuum service. Ring joint flanges, Fig. 13.34d, are used for high temperatures and high pressure services.

13.10.4. Flange design

Standard flanges will be specified for most applications (see Section 13.10.5). Special designs would be used only if no suitable standard flange were available; or for large flanges, such as the body flanges of vessels, where it may be cheaper to size a flange specifically for the duty required rather than to accept the nearest standard flange, which of necessity would be over-sized.

Figure 13.35 shows the forces acting on a flanged joint. The bolts hold the faces together, resisting the forces due to the internal pressure and the gasket sealing pressure. As these forces are offset the flange is subject to a bending moment. It can be considered as a cantilever beam with a concentrated load. A flange assembly must be sized so as to have sufficient strength and rigidity to resist this bending moment. A flange that lacks sufficient rigidity will rotate slightly, and the joint will leak; Fig. 13.36.

Design procedures and work sheets for non-standard flanges are given in the national codes and standards (BS 5500, clause 3.8). The design methods given in the current British Standard BS 5500 and the American code, ASME section VIII, are based on a theoretical analysis of the stresses in flanges published by Waters et al. (1934, 1937), who modified an earlier analysis by Waters and Taylor (1927). The design methods given in the codes and standards are discussed and compared by Rose (1970).

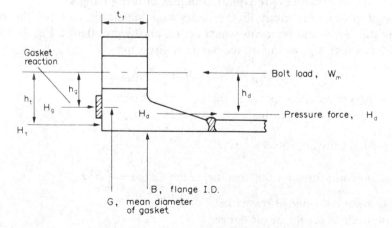

FIG. 13.35. Forces acting on an integral flange

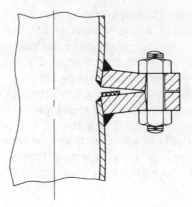

FIG. 13.36. Deflection of a weak flange (exaggerated)

In the analysis of Waters *et al.*, the flange is considered as a flat plate, and the hub and nozzle (or pipe section) as a beam on an elastic foundation. They assume that no plastic deformation occurs and that the bolt load remains constant. Lake and Boyd (1957) presented a method of analysis that took account of the plastic strain that will occur in most practical flanges. Their design method was used as the basis of the flange-design procedure given in the earlier British Standard, BS 1500, (now withdrawn). The method allowed lighter flange designs than the current British Standard and American code methods. It was generally considered satisfactory for flange sizes up to 1·5 m (5 ft) diameter; see Rose (1970).

For design purposes, the flanges are classified as integral or loose flanges.

Integral flanges are those in which the construction is such that the flange obtains support from its hub and the connecting nozzle (or pipe). The flange assembly and nozzle neck form an "integral" structure. A welding-neck flange would be classified as an integral flange.

Loose flanges are attached to the nozzle (or pipe) in such a way that they obtain no significant support from the nozzle neck and cannot be classified as an integral attachment. Screwed and lap-joint flanges are typical examples of loose flanges.

The design procedures given in the codes and standards can be illustrated by considering the forces and moments which act on an integral flange, Fig. 13.35.

The total moment M_{op} acting on the flange is given by:

$$M_{op} = H_d h_d + H_t h_t + H_g h_g \qquad (13.98)$$

Where H_g = gasket reaction (pressure force), $= \pi G(2b)mP_i$

H_t = pressure force on the flange face $= H - H_d$,

H = total pressure force $= \dfrac{\pi}{4}G^2 P_i$,

H_d = pressure force on the area inside the flange $= \dfrac{\pi}{4}B^2 P_i$,

$\quad G$ = mean diameter of the gasket,

$\quad B$ = inside diameter of the flange,

$\quad 2b$ = effective gasket pressure width,

b = effective gasket sealing width,
h_d, h_g and h_t are defined in Fig. 13.35.

The minimum required bolt load under the operating conditions is given by:

$$W_{m1} = H + H_g \tag{13.99}$$

The forces and moments on the flange must also be checked under the bolting-up conditions.

The moment M_{atm} is given by:

$$M_{atm} = W_{m2} h_g \tag{13.100}$$

where W_{m2} is the bolt load required to seat the gasket, given by:

$$W_{m2} = y \pi G b \tag{13.101}$$

where y is the gasket seating pressure (stress).
The flange stresses are given by:

longitudinal hub stress, $\qquad \sigma_{hb} = F_1 M \qquad$ (13.102)
radial flange stress, $\qquad \sigma_{rd} = F_2 M \qquad$ (13.103)
tangential flange stress, $\qquad \sigma_{tg} = F_3 M - F_4 \sigma_{rd} \qquad$ (13.104)

where M is taken as M_{op} or M_{atm}, whichever is the greater; and the factors F_1 to F_4 are functions of the flange type and dimensions, and are obtained from equations and graphs given in the codes and standards (BS 5500, clause 3.8).

The flange must be sized so that the stresses given by equations 13.102 to 13.104 satisfy the following criteria:

$$\sigma_{hb} \not> 1 \cdot 5 f_{fo} \tag{13.105}$$
$$\sigma_{rd} \not> f_{fo} \tag{13.106}$$
$$\tfrac{1}{2}(\sigma_{hb} + \sigma_{rd}) \not> f_{fo} \tag{13.107}$$
$$\tfrac{1}{2}(\sigma_{hb} + \sigma_{tg}) \not> f_{fo} \tag{13.108}$$

where f_{fo} is the maximum allowable design stress for the flange material at the operating conditions.

The minimum bolt area required A_{bf} will be given by:

$$A_{bf} = W_m / f_b \tag{13.109}$$

where W_m is the greater value of W_{m1} or W_{m2}, and f_b the maximum allowable bolt stress. Standard size bolts should be chosen, sufficient to give the required area. The bolt size will not normally be less than 12 mm, as smaller sizes can be sheared off by over-tightening.

The bolt spacing must be selected to give a uniform compression of the gasket. It will not normally be less than 2·5 times the bolt diameter, to give sufficient clearance for tightening with a wrench or spanner. The following formula can be used to determine the maximum bolt spacing:

$$p_b = 2d_b + \frac{6t_f}{(m + 0 \cdot 5)} \tag{13.110}$$

where p_b = bolt pitch (spacing), mm,
$\quad d_b$ = bolt diameter, mm,
$\quad t_f$ = flange thickness, mm,
$\quad m$ = gasket factor.

13.10.5. *Standard flanges*

Standard flanges are available in a range of types, sizes and materials; and are used extensively for pipes, nozzles and other attachments to pressure vessels.

The proportions of standard flanges are set out in the various codes and standards. A typical example of a standard flange design is shown in Fig. 13.37. The relevant British Standards are: BS 1560; Part 2: 1970, which covers cast and forged steel pipe flanges, and flanges for fittings, for the petroleum industry (and allied industries); and BS 4504: 1969, which covers flanges for pipes and fittings for general use.

In the United States, flanges are covered by standards issued by the American National Standards Institute (ANSI). An abstract of the American standards is given in Perry and Chilton (1973), and in the TEMA standards (see Chapter 12).

STEEL SLIP-ON BOSS FLANGE FOR WELDING
Nominal pressure 6 bar

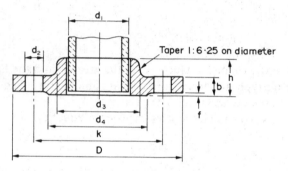

Nom. size	Pipe o.d. d_1 ≈	Flange			Raised face		Bolting	Drilling			Boss
		D	b	h	d_4	f		No.	d_2	k	d_3
6	10·2	65	10	18	25	2	M10	4	11	40	18
8	13·5	70	10	18	30	2	M10	4	11	45	22
10	17·2	75	12	20	35	2	M10	4	11	50	25
15	21·3	80	12	20	40	2	M10	4	11	55	30
20	26·9	90	14	24	50	2	M10	4	11	65	40
25	33·7	100	14	24	60	2	M10	4	11	75	50
32	42·4	120	14	26	70	2	M12	4	14	90	60
40	48·3	130	14	26	80	3	M12	4	14	100	70
50	60·3	140	14	28	90	3	M12	4	14	110	80
65	76·1	160	14	32	110	3	M12	4	14	130	100
80	88·9	190	16	34	128	3	M16	4	18	150	110
100	114·3	210	16	40	148	3	M16	4	18	170	130
125	139·7	240	18	44	178	3	M16	8	18	200	160
150	168·3	265	18	44	202	3	M16	8	18	225	185
200	219·1	320	20	44	258	3	M16	8	18	280	240
250	273	375	22	44	312	3	M16	12	18	335	295
300	323·9	440	22	44	365	4	M20	12	22	395	355

Fɪɢ. 13.37. Typical standard flange design, (Reproduced from BS 4504: Part 1: 1967) (All dimensions mm)

In BS 4504 and 1560: Part 2, the flange dimensions are given in metric units (SI). BS 4504 covers flange sizes from 6 mm to 4000 mm; and BS 1560 sizes from 12 mm to 610 mm (24 in.). Flanges above 24 in. are covered by BS 3293: 1960; which complements BS 1560, but gives the dimensions in inches. All these standards cover carbon steel and low alloy steels; BS 1560 also includes stainless steels; and BS 4504: Part 2, copper alloys.

Standard flanges are designated by a class, or rating, number; which corresponds to the primary service rating for the flange. BS 4504 provides for twelve classes: 2·5, 6, 10, 16, 25, 40, 64, 100, 160, 250, 320, 400 bar. BS 1560 provides for seven classes: 150, 300, 400, 600, 900, 1500 and 2500; where these class numbers represent the nominal ratings in pound per square inch gauge. Standard flanges are often referred to as 150 lb, 300 lb, etc., flanges.

The flange class number required for a particular duty will depend on the design pressure and temperature, and the flange material. The reduction in strength at elevated temperatures is allowed for by selecting a flange with a higher nominal rating than the design pressure. For example, for a design pressure of 10 bar (150 psi) a BS 1560, carbon steel class 150 flange would be specified for temperatures below 275°C; whereas for a temperature of, say, 400°C a class 300 flange would be specified. The pressure–temperature rating for forged carbon steel flanges to BS 4504 are given in Table 13.5. Pressure–temperature ratings for the full range of materials covered can be obtained from the standards. The dimensions of steel welding-neck flanges to BS 4504, for classes 6, 10, 25, 40, 100 and 250 bar are given in Appendix F. The appropriate standard should be consulted for the dimensions of other flange types and materials, and for the full specification of standard flanges.

TABLE 13.5. *Typical pressure–temperature ratings for forged steel flanges, BS 4504: Part 1: 1969*

Nominal pressure (bar)	Design pressure at temperature °C (bar)						
	− 10 to 120	200	250	300	350	400	450
25	25	25	24	21	17	14	9
40	40	40	38	33	28	23	15
64	64	64	61	53	44	36	24
100	100	100	95	82	70	5/	37
160	160	160	152	132	112	92	60
250	250	250	238	206	174	142	93
320	320	320	305	264	222	180	120
400	400	400	382	278	226	200	150

Material BS 1503-161, Grade 28B

13.11. Heat-exchanger tube-plates

The tube-plates (tube-sheets) in shell and tube heat exchangers support the tubes, and separate the shell and tube side fluids (see Chapter 12). One side is subject to the shell-side pressure and the other the tube-side pressure. The plates must be designed to support the maximum differential pressure that is likely to occur. Radial and tangential bending stresses will be induced in the plate by the pressure load and, for fixed-head exchangers, by the load due to the differential expansion of the shell and tubes.

A tube-plate is essentially a perforated plate with an unperforated rim, supported at its periphery. The tube holes weaken the plate and reduce its flexual rigidity. The equations developed for the stress analysis of unperforated plates (Section 13.3.5) can be used for perforated plates by substituting "virtual" (effective) values for the elastic constants E and v, in place of the normal values for the plate material. The virtual elastic constants E' and v' are functions of the plate ligament efficiency, Fig. 13.38; see O'Donnell and Langer (1962). The ligament efficiency of a perforated plate is defined as:

$$\lambda = \frac{p_h - d_h}{p_h} \tag{13.111}$$

where p_h = hole pitch,
     d_h = hole diameter.

The "ligament" is the material between the holes (that which holds the holes together). In a tube-plate the presence of the tubes strengthens the plate, and this is taken into account when calculating the ligament efficiency by using the inside diameter of the tubes in place of the hole diameter in equation 13.111.

Design procedures for tube-plates are given in BS 5500, and in the TEMA heat exchanger standards (see Chapter 12). BS 5500 distinguishes between exchangers with confined heads and those with unconfined heads; and gives separate design procedures for each class. Confined heads are those which are immersed in the shell-side fluid; for example, a floating head. Unconfined heads are those in exchangers in which the tube-plates are fixed at both ends of the shell. The tube-plate must be thick enough to resist the bending and shear stresses caused by the pressure load and any differential expansion of the shell and tubes. The minimum plate thickness to resist bending can be estimated using an equation of similar form to that for flat plate end closures (Section 13.5.3).

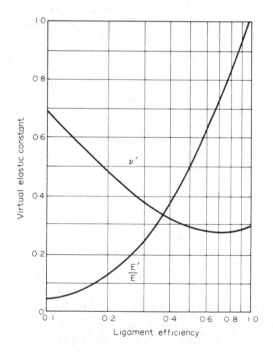

FIG. 13.38. Virtual elastic constants

$$t_p = C_{ph} D_p \sqrt{\frac{\Delta P'}{\lambda f_p}} \qquad (13.112)$$

where t_p = the minimum plate thickness,
$\Delta P'$ = the effective tube plate design pressure,
λ = ligament efficiency,
f_p = maximum allowable design stress for the plate,
C_{ph} = a design factor,
D_p = plate diameter.

The value of the design factor C_{ph} will depend on the type of head, the edge support (clamped or simply supported), the plate dimensions, and the elastic constants for the plate and tube material, and can be obtained from the design charts and equations given in BS 5500, clause 3.9.

The tube-sheet design pressure $\Delta P'$ depends on the type of exchanger. For an exchanger with confined heads or U-tubes it is taken as the maximum difference between the shell-side and tube-side operating pressures; with due consideration being given to the possible loss of pressure on either side. For exchangers with unconfined heads (plates fixed to the shell) the load on the tube-sheets due to differential expansion of the shell and tubes must be added to that due to the differential pressure.

The shear stress in the tube-plate can be calculated by equating the pressure force on the plate to the shear force in the material at the plate periphery. The minimum plate thickness to resist shear is given by:

$$t_p = \frac{0.155 \, D_p \Delta P'}{\lambda \tau_p} \qquad (13.113)$$

where τ_p = the maximum allowable shear stress, taken as half the maximum allowable design stress for the material (see Section 13.3.2).

The design plate thickness is taken as the greater of the values obtained from equations 13.112 and 13.113 and must be greater than the minimum thickness given below (BS 5500, clause 3.9):

Tube o.d. (mm)	Minimum plate thickness (mm)
25	0.75 × tube o.d.
25–30	22
30–40	25
40–50	30

For exchangers with unconfined heads the longitudinal stresses in the tubes and shell must be checked to ensure that the maximum allowable design stress for the materials are not exceeded. Methods for calculating these stresses are given in the standards (BS 5500, clause 3.9.5).

A detailed account of the methods used for the stresses analysis of tube sheets is given by Bickell and Ruiz (1967).

13.12. Welded joint design

Process vessels are built up from preformed parts: cylinders, heads, and fittings, joined by fusion welding. Riveted construction was used extensively in the past (prior to the 1940s) but is now rarely seen.

Cylindrical sections are usually made up from plate sections rolled to the required curvature. The sections (strakes) are made as large as is practicable to reduce the number of welds required. The longitudinal welded seams are offset to avoid a conjunction of welds at the corners of the plates.

Many different forms of welded joint are needed in the construction of a pressure vessel. Some typical forms are shown in Figs. 13.39 to 13.41.

The design of a welded joint should satisfy the following basic requirements:

1. Give good accessibility for welding and inspection.
2. Require the minimum amount of weld metal.
3. Give good penetration of the weld metal; from both sides of the joint, if practicable.
4. Incorporate sufficient flexibility to avoid cracking due to differential thermal expansion.

The preferred types of joint, and recommended designs and profiles, are given in the codes and standards (BS 5500, Appendix E).

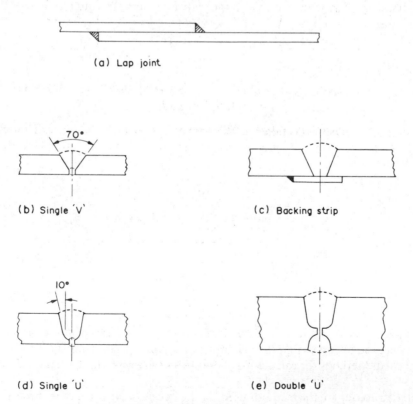

(a) Lap joint

70°

(b) Single 'V'

(c) Backing strip

10°

(d) Single 'U'

(e) Double 'U'

FIG. 13.39. Weld profiles; (b to e) butt welds

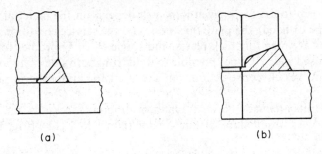

(a)   (b)

Set – on branches

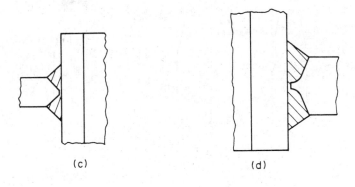

(c)   (d)

Set – in branches

FIG. 13.40. Typical weld profiles—Branches

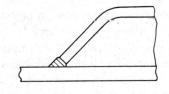

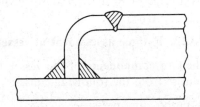

FIG. 13.41. Typical construction methods for welded jackets

The correct form to use for a given joint will depend on the material, the method of welding (machine or hand), the plate thickness, and the service conditions. Double-sided V- or U-sections are used for thick plates, and single V- or U-profiles for thin plates. A backing strip is used where it is not possible to weld from both sides. Lap joints are seldom used for pressure vessels construction, but are used for atmospheric pressure storage tanks.

Where butt joints are made between plates of different thickness, the thicker plate is reduced in thickness with a slope of not greater than 1 in 4 (14°) (Fig. 13.42).

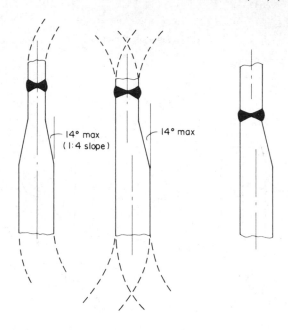

FIG. 13.42. Transition between plates of unequal thickness

The local heating, and consequent expansion, that occurs during welding can leave the joint in a state of stress. These stresses are relieved by post-welding heat treatment. Not all vessels will be stress relieved. Guidance on the need for post-welding heat treatment is given in the codes and standards (BS 5500, clause 4.4), and will depend on the service and conditions, materials of construction, and plate thickness.

To ensure that a satisfactory quality of welding is maintained, welding-machine operators and welders working on the pressure parts of vessels are required to pass welder approval tests; which are designed to test their competence to make sound welds. In the UK the testing of welders is covered by British Standards BS 4870, BS 4871, and BS 4872.

13.13. Fatigue assessment of vessels

During operation the shell, or components of the vessel, may be subjected to cyclic stresses. Stress cycling can arise from the following causes:

1. Periodic fluctuations in operating pressure.
2. Temperature cycling.

3. Vibration.
4. "Water hammer".
5. Periodic fluctuation of external loads.

A detailed fatigue analysis is required if any of these conditions is likely to occur to any significant extent. Fatigue failure will occur during the service life of the vessel if the endurance limit (number of cycles for failure) at the particular value of the cyclic stress is exceeded. The codes and standards should be consulted to determine when a detailed fatigue analysis must be undertaken (BS 5500, Appendix C); see also Langer (1971).

13.14. Pressure tests

The national pressure vessel codes and standards require that all pressure vessels be subjected to a pressure test to prove the integrity of the finished vessel. A hydraulic test is normally carried out, but a pneumatic test can be substituted under circumstances where the use of a liquid for testing is not practical. Hydraulic tests are safer because only a small amount of energy is stored in the compressed liquid. A standard pressure test is used when the required thickness of the vessel parts can be calculated in accordance with the particular code or standard. The vessel is tested at a pressure above the design pressure, typically 25 to 30 per cent. The test pressure is adjusted to allow for the difference in strength of the vessel material at the test temperature compared with the design temperature, and for any corrosion allowance.

Formulae for determining the appropriate test pressure are given in the codes and standards; such as that in BS 5500:

$$\text{Test pressure} = 1{\cdot}25\left[P_d\frac{f_a}{f_n} \times \frac{t}{(t-c)} \right] \tag{13.114}$$

where P_d = design pressure, N/mm^2,
  f_a = nominal design strength (design stress) at the test temperature, N/mm^2,
  f_n = nominal design strength at the design temperature, N/mm^2,
  c = corrosion allowance, mm,
  t = actual plate thickness, mm.

When the required thickness of the vessel component parts cannot be determined by calculation in accordance with the methods given, the codes and standards require that a hydraulic proof test be carried out. In a proof test the stresses induced in the vessel during the test are monitored using strain gauges, or similar techniques. The requirements for the proof testing of vessels are set out in the codes and standards.

13.15. High-pressure vessels

High pressures are required for many commercial chemical processes. For example, the synthesis of ammonia is carried out at reactor pressures of up to 1000 bar, and high-density polyethylene processes operate up to 1500 bar.

Only a brief discussion of the design of vessels for operation at high pressures will be given in this section; sufficient to show the fundamental limitations of single-wall (monobloc) vessels, and the construction techniques that are used to overcome this limitation. A full discussion of the design and construction of high-pressure vessels and ancillary equipment (pumps, compressors, valves and fittings) is given in the books by Tongue (1959) and Manning and Labrow (1974); see also the safety code published by the High Pressure Technology Association (1975).

13.15.1. *Fundamental equations*

Thick walls are required to contain high pressures, and the assumptions made in the earlier sections of this chapter to develop the design equations for "thin-walled" vessels will not be valid. The radial stress will not be negligible and the tangential (hoop) stress will vary across the wall.

Consider the forces acting on the elemental section of the wall of the cylinder shown in Fig. 13.43. The cylinder is under an internal pressure P_i and an external pressure P_e. The conditions for static equilibrium, with the forces resolved radially, give:

$$\sigma_r r \delta\phi - 2\sigma_t \, \delta r \sin \frac{\delta\phi}{2} - (\sigma_r + \delta\sigma_r)(r + \delta r)\delta\phi = 0$$

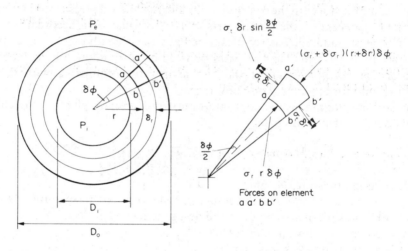

FIG. 13.43. Thick cylinder

multiplying out taking the limit gives:

$$\sigma_t + r \frac{\mathrm{d}\sigma_r}{\mathrm{d}r} + \sigma_r = 0 \tag{13.115}$$

A second equation relating the radial and tangential stresses can be written if the longitudinal strain ε_L and stress σ_L are taken to be constant across the wall; that is, that there is no distortion of plane sections, which will be true for sections away from the ends. The longitudinal strain is given by (see Section 13.3.6):

$$\varepsilon_L = \frac{1}{E}\left[\sigma_L - (\sigma_t - \sigma_r)v\right] \tag{13.116}$$

If ε_L and σ_L are constant, then the term $(\sigma_t - \sigma_r)$ must also be constant, and can be written as:

$$(\sigma_t - \sigma_r) = 2A \tag{13.117}$$

where A is an arbitrary constant.

Substituting for σ_t in equation 13.115 gives:

$$2\sigma_r + r\frac{d\sigma_r}{dr} = -2A$$

and integrating

$$\sigma_r = -A + \frac{B'}{r^2} \tag{13.118}$$

where B' is the constant of integration.

In terms of the cylinder diameter, the equations can be written as:

$$\sigma_r = -A + \frac{B}{d^2} \tag{13.119}$$

$$\sigma_t = A + \frac{B}{d^2} \tag{13.120}$$

These are the fundamental equations for the design of thick cylinders and are often referred to as Lamé's equations, as they were first derived by Lamé and Clapeyron (1833). The constants A and B are determined from the boundary conditions for the particular loading condition.

Most high-pressure process vessels will be under internal pressure only, the atmospheric pressure outside a vessel will be negligible compared with the internal pressure. The boundary conditions for this loading condition will be:

$$\sigma_r = P_i \text{ at } d = D_i$$
$$\sigma_r = 0 \text{ at } d = D_o$$

Substituting these values in equation 13.119 gives

$$P_i = -A + \frac{B}{D_i^2}$$

and

$$0 = -A + \frac{B}{D_o^2}$$

subtracting gives

$$P_i = B\left[\frac{1}{D_i^2} - \frac{1}{D_o^2}\right]$$

hence

$$B = P_i\frac{(D_i^2 D_o^2)}{(D_o^2 - D_i^2)}$$

and

$$A = P_i\frac{D_i^2}{(D_o^2 - D_i^2)}$$

Substituting in equations 13.119 and 13.120 gives:

$$\sigma_r = P_i\left[\frac{D_i^2 (D_o^2 - d^2)}{d^2 (D_o^2 - D_i^2)}\right] \tag{13.121}$$

$$\sigma_t = P_i\left[\frac{D_i^2 (D_o^2 + d^2)}{d^2 (D_o^2 - D_i^2)}\right] \tag{13.122}$$

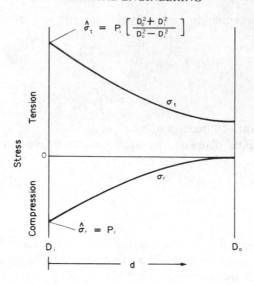

FIG. 13.44. Stress distribution in wall of a monobloc cylinder

The stress distribution across the vessel wall is shown plotted in Fig. 13.44. The maximum values will occur at the inside surface, at $d = D_i$.

Putting $K = D_o/D_i$, the maximum values are given by:

$$\hat{\sigma}_r = P_i \text{ (compressive)} \tag{13.123}$$

$$\hat{\sigma}_t = P_i\left[\frac{K^2+1}{K^2-1}\right] \tag{13.124}$$

An expression for the longitudinal stress can be obtained by equating forces in the axial direction:

$$\sigma_L \frac{\pi}{4}(D_o^2 - D_i^2) = P_i \frac{\pi D_i^2}{4}$$

hence

$$\sigma_L = \frac{P_i D_i^2}{(D_o^2 - D_i^2)} = \frac{P_i}{(K^2 - 1)} \tag{13.125}$$

The maximum shear stress will be given by (see Section 13.3.1):

$$\hat{\tau} = \tfrac{1}{2}(\hat{\sigma}_t + \hat{\sigma}_r) = \frac{P_i K^2}{(K^2 - 1)} \tag{13.126}$$

Theoretical maximum pressure

If the maximum shear stress theory is taken as the criterion of failure (Section 13.3.2), then the maximum pressure that a monobloc vessel can be designed to withstand without failure is given by:

$$\hat{\tau} = \frac{\sigma_e'}{2} = \frac{P_i K^2}{(K^2 - 1)}$$

hence

$$\hat{P}_i = \frac{\sigma_e'}{2}\left[\frac{K^2-1}{K^2}\right] \tag{13.127}$$

where σ'_e is the elastic limit stress for the material of construction divided by a suitable factor of safety. As the wall thickness is increased the term $(K^2 - 1)/K^2$ tends to 1,

and
$$\hat{P}_i = \sigma'_e/2 \tag{13.128}$$

which sets an upper limit on the pressure that can be contained in a monobloc cylinder.

Manning (1947) has shown that the maximum shear strain energy theory of failure (due to Mises (1913)) gives a closer fit to experimentally determined failure pressures for monobloc cylinders than the maximum shear stress theory. This criterion of failure gives:

$$\hat{P}_i = \sigma'_e/\sqrt{3} \tag{13.129}$$

From Fig. 13.44 it can be seen that the stress falls off rapidly across the wall and that the material in the outer part of the wall is not being used effectively. The material can be used more efficiently by prestressing the wall. This will give a more uniform stress distribution under pressure. Several different "prestressing" techniques are used; the principal methods are described briefly in the following sections.

13.15.2. Compound vessels

Shrink-fitted cylinders

Compound vessels are made by shrinking one cylinder over another. The inside diameter of the outer cylinder is made slightly smaller than the outer diameter of the inner cylinder, and is expanded by heating to fit over the inner. On cooling the outer cylinder contracts and places the inner under compression. The stress distribution in a two-cylinder compound vessel is shown in Fig. 13.45; more than two cylinders may be used.

Shrink-fitted compound cylinders are used for small-diameter vessels, such as compressor cylinder barrels. The design of shrink-fitted compound cylinders is discussed by Manning (1947).

Multilayer vessels

Multilayer vessels are made by wrapping several layers of relatively thin plate round a central tube. The plates are heated, tightened and welded, and this gives the desired stress distribution in the compound wall. The vessel is closed with forged heads. A typical design is shown in Fig. 13.46. This construction technique is discussed by Jasper and Scudder (1941).

Wound vessels

Cylindrical vessels can be reinforced by winding on wire or thin ribbons. Winding on the wire under tension places the cylinder under compression. For high-pressure vessels special interlocking strips are used, such as those shown in Fig. 13.47. The interlocking gives strength in the longitudinal direction and a more uniform stress distribution. The strips may be wound on hot to increase the prestressing. This type of construction is described by Birchall and Lake (1947). Wire winding was used extensively for the barrels of large guns.

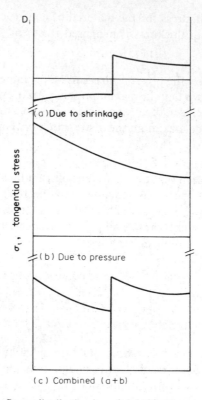

FIG. 13.45. Stress distribution in a shrink-fitted compound cylinder

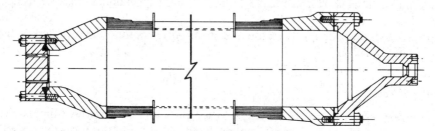

FIG. 13.46. Multilayer construction

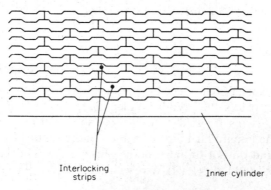

FIG. 13.47. Strip wound vessel

13.15.3. Autofrettage

Autofrettage is a technique used to prestress the inner part of the wall of a monobloc vessel, to give a similar stress distribution to that obtained in a shrink-fitted compound cylinder. The finished vessel is deliberately over-pressurised by hydraulic pressure. During this process the inner part of the wall will be more highly stressed than the outer part and will undergo plastic strain. On release of the "autofrettage" pressure the inner part, which is now over-size, will be placed under compression by the elastic contraction of the outer part, which gives a residual stress distribution similar to that obtained in a two-layer shrink-fitted compound cylinder. After straining the vessel is annealed at a relative low temperature, approximately 300°C. The straining also work-hardens the inner part of the wall. The vessel can be used at pressures up to the "autofrettage" pressure without further permanent distortion.

The autofrettage technique is discussed by Manning (1950).

13.16. Liquid storage tanks

Vertical cylindrical tanks, with flat bases and conical roofs, are universally used for the bulk storage of liquids at atmospheric pressure. Tank sizes vary from a few hundred gallons (tens of cubic metres) to several thousand gallons (several hundred cubic metres).

The main load to be considered in the design of these tanks is the hydrostatic pressure of the liquid, but the tanks must also be designed to withstand wind loading and, for some locations, the weight of snow on the tank roof.

The minimum wall thickness required to resist the hydrostatic pressure can be calculated from the equations for the membrane stresses in thin cylinders (Section 13.3.4):

$$e_s = \frac{\rho_L H_L g}{2 f_t J} \frac{D_t}{10^3} \qquad (13.130)$$

where e_s = tank thickness required at depth H_L, mm,

$\quad H_L$ = liquid depth, m,

$\quad \rho_L$ = liquid density, kg/m³,

$\quad J$ = joint factor (if applicable),

$\quad g$ = gravitational acceleration, 9·81 m/s²,

$\quad f_t$ = design stress for tank material, N/mm²,

$\quad D_t$ = tank diameter, m.

The liquid density should be taken as that of water (1000 kg/m³), unless the process liquid has a greater density.

For small tanks a constant wall thickness would normally be used, calculated at the maximum liquid depth.

With large tanks, it is economical to take account of the variation in hydrostatic pressure with depth, by increasing the plate thickness progressively from the top to bottom of the tank. Plate widths of 2 m (6 ft) are typically used in tank construction.

The roofs of large tanks need to be supported by a steel framework; supported on columns in very large-diameter tanks.

The design and construction of atmospheric storage tanks for the petroleum industry are covered by British Standard BS 2654, and the American Petroleum Industry standard API 650. The design of storage tanks is discussed in detail by Brownell and Young (1959); see also the papers by Debham et al. (1968) and Zick and McGrath (1968).

13.17. Mechanical design of centrifuges

13.17.1. Centrifugal pressure

The fluid in a rotating centrifuge exerts pressure on the walls of the bowl or basket. The minimum wall thickness required to contain this pressure load can be determined in a similar manner to that used for determining the wall thickness of a pressure vessel under internal pressure. If the bowl contains a single homogeneous liquid, Fig. 13.48a, the fluid pressure is given by:

$$P_f = \tfrac{1}{2}\rho_L\omega^2(R_1^2 - R_2^2) \tag{13.131}$$

where P_f = centrifugal pressure, N/m^2,
  ρ_L = liquid density, kg/m^3,
  ω = rotational speed of the centrifuge, radians/s,
  R_1 = inside radius of the bowl, m,
  R_2 = radius of the liquid surface, m.

For design, the maximum fluid pressure will occur when the bowl is full, $R_2 = 0$.

If the centrifuge is separating two immiscible liquids, Fig. 13.48b, the pressure will be given by:

$$P_f = \tfrac{1}{2}\omega^2\left[\rho_{L1}(R_1^2 - R_i^2) + \rho_{L2}(R_i^2 - R_2^2)\right] \tag{13.132}$$

where ρ_{L1} = density of the heavier liquid, kg/m^3,
  ρ_{L2} = density of the lighter liquid, kg/m^3,
  R_i = radius of the interface between the two liquids, m.

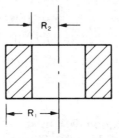

(a) Single fluid

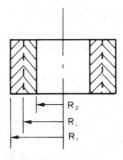

(b) Two fluids

FIG. 13.48. Centrifugal fluid pressure

If the machine is separating a solid–liquid mixture, the mean density of the slurry in the bowl should be used in equation 13.131.

The shell of an empty centrifuge bowl will be under stress due to the rotation of the bowl's own mass; this "self-pressure" P_m is given by:

$$P_m = \tfrac{1}{2}\omega^2 \rho_m \left[(R_1 + t)^2 - R_1^2\right] \qquad (13.133)$$

where ρ_m = density of the bowl material, kg/m³,
t = bowl wall thickness, m.

The minimum wall thickness required can be estimated using the equations for membrane stress derived in Section 13.3.4. For a solid bowl

$$e_c = \frac{P_t R_1}{f_m \times 10^3} \qquad (13.134)$$

where P_t = the total (maximum) pressure (fluid + self-pressure), N/m²,
f_m = maximum allowable design stress for the bowl material, N/mm²,
e_c = wall thickness, mm.

With a perforated basket the presence of the holes will weaken the wall. This can be allowed for by introducing a "ligament efficiency" into equation 13.134 (see Section 13.11)

$$e_c = \frac{P_t R_1}{f_m \times 10^3 \lambda} \qquad (13.135)$$

where λ = ligament efficiency = $\dfrac{p_h - d_h}{p_h}$,

p_h = hole pitch,
d_h = hole diameter.

Equations 13.134 and 13.135 can also be used to estimate the maximum safe load (or speed) for an existing centrifuge, if the service is to be changed.

In deriving these equations no account was taken of the strengthening effect of the bottom and top rings of the bowl or basket; so the equations will give estimates that are on the safe side. Strengthening hoops or bands are used on some basket designs.

The mechanical design of centrifugal separators is covered by British Standard BS 767.

13.17.2. Bowl and spindle motion: critical speed

Centrifuges are classified according to the form of mounting used: fixed or free spindle. With fixed-spindle machines, the bearings are rigidly mounted; whereas, in a free spindle, or self-balancing, machine a degree of "free-play" is allowed in the spindle mounting. The amount of movement of the spindle is restrained by some device, such as a rubber buffer. This arrangement allows the centrifuge to operate with a certain amount of out-of-balance loading without imposing an undue load on the bearings. Self-balancing centrifuges can be under or over-driven; that is, with the drive mounted below or above the bowl. Severe vibration can occur in the operation of fixed-spindle centrifuges and these are often suspended on rods, supported from columns mounted on an independent base, to prevent the vibration being transmitted to the building structure.

Critical speed

If the centre of gravity of the rotating load does not coincide with the axis of rotation of the bowl an uneven force will be exerted on the machine spindle. In a self-balancing machine (or a suspended fixed-spindle machine) this will cause the spindle to deflect from the vertical position and the bowl will develop a whirling vibration. The phenomenon is analogous with the whirling of the shafts in other rotating machinery; such as compressors, pumps, and agitators; which is considered under the general heading of the "whirling of shafts" in standard texts on the "Theory of Machines".

The simple analysis given below is based on that used to determine the whirling speed of a shaft with a single concentrated mass. Figure 13.49 shows the position of the centre of gravity of a rotating mass m_c with an initially displacement h_c. Let x_c be the additional displacement caused by the action of centrifugal force, and s the retoring force, assumed to be proportional to the displacement. The radial outward centrifugal force due to the displacement of the centre of the gravity from the axis of rotation will be $= m_c\omega^2(x + h_c)$. This is balanced by the inward action of the restoring force $= sx_c$.

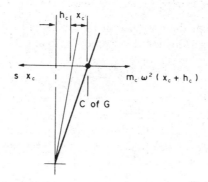

FIG. 13.49. Displacement of centre of gravity of a centrifuge bowl

Equating the two forces:

$$m_c\omega^2(x_c + h_c) = sx_c$$

from which

$$\frac{x_c}{h_c} = \frac{1}{(s/m_c\omega^2) - 1} \tag{13.136}$$

It can be seen by inspection of equation 13.136 that the deflection (the ratio x_c/h_c) will become indefinitely large when the term $s/m_c\omega^2 = 1$; the corresponding value of ω is known as the critical, or whirling, speed. Above the critical speed the term $s/m_c\omega^2$ becomes negative, and x_c/h_c tends to a limiting value of -1 at high speeds. This shows that if the centrifuge is run at speeds in excess of the critical speed the tendency is for the spindle to deflect so that the axis of rotation passes through the centre of gravity of the system. The sequence of events as a self-balancing centrifuge is run up to speed is shown in Fig 13.50. In practice, a centrifuge is accelerated rapidly to get through the critical speed range quickly, and the observed deflections are not great.

It can be seen from equation 13.136 that the critical speed of a centrifuge will depend on the mass of the bowl and the magnitude of the restoring force; it will also depend on the

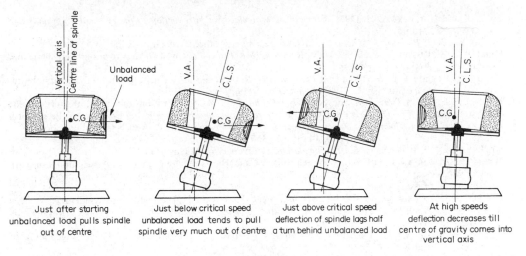

Just after starting	Just below critical speed	Just above critical speed	At high speeds
unbalanced load pulls spindle	unbalanced load tends to pull	deflection of spindle lags half	deflection decreases till
out of centre	spindle very much out of centre	a turn behind unbalanced load	centre of gravity comes into
			vertical axis

FIG. 13.50. Diagram of action of self-balancing centrifugal, showing motion of centre of gravity and unbalanced load with increasing speed

dimensions of the machine and the length of the spindle. The critical speed of a simple system can be calculated, but for a complex system, such as loaded centrifuges, the critical speed must be determined by experiment. It can be shown that the critical speed of a rotating system corresponds with the natural frequency of vibration of the system.

A low critical speed is desired, as less time is then spent accelerating the bowl through the critical range. Suspended fixed-spindle centrifuges generally have a low critical speed.

Precession

In addition to the whirling vibration due to an out-of-balance force, another type of motion can occur in a free-spindle machine. When the bowl or basket is tilted the spindle may move in a circle. This slow gyratory motion is known as "precession", and is similar to the "precession" of a gyroscope. It is usually most pronounced at high speeds, above the critical speed.

A complete analysis of the motion of centrifuges is given by Alliott (1924, 1926).

13.18. References

ALLIOT, E. A. (1924) *Trans. Inst. Chem. Eng.* **2**, 39. Self-balancing centrifugals.

ALLIOT, E. A. (1926) *Centrifugal dryers and separators* (Benn).

BHATTACHARYYA, B. C. (1976) *Introduction to Chemical Equipment Design, Mechanical Aspects* (Indian Institute of Technology).

BERGMAN, D. J. (1963) *Trans. Am. Soc. Mech. Eng.* (*J. Eng. for Ind.*) **85**, 219. Temperature gradients for skirt supports of hot vessels.

BICKELL, M. B. and RUIZ, C. (1967) *Pressure Vessel Design and Analysis* (Macmillan).

BIRCHALL, H. and LAKE, G. F. (1947) *Proc. Inst. Mech. Eng.* **56**, 349. An alternative form of pressure vessel of novel construction.

BROWNELL, L. E. (1963) *Hyd. Proc. & Pet. Ref.* **42** (June) 109. Mechanical design of tall towers.

BROWNELL, L. E. and YOUNG, E. H. (1959) *Process Equipment Design: Vessel design* (Wiley).

CASE, J. and CHILVER, A. H. (1971) *Strength of Materials: an introduction to the mechanics of solids and structures* (Arnold).

CHUSE, R. (1977) *Pressure Vessels: the ASME code simplified*, 5th ed. (McGraw-Hill).

DEBHAM, J. B., RUSSEL, J. and WIILS, C. M. R. (1968) *Hyd. Proc.* **47** (May) 137. How to design a 600,000 b.b.l. tank.

DEGHETTO, K. and LONG, W. (1966) *Hyd. Proc. and Pet. Ref.* **45** (Feb.) 143. Check towers for dynamic stability.

FAUPEL, J. H. and FISHER, F. E. (1981) *Engineering Design*, 2nd ed. (Wiley).

FREESE, C. E. (1959) *Trans. Am. Soc. Mech. E.* (*J. Eng. Ind.*) **81**, 77. Vibrations of vertical pressure vessels.

GILL, S. S. (Ed.) (1970) *The Stress Analysis of Pressure Vessels and Pressure Vessel Components* (Pergamon).

HARVEY, J. F. (1974) *Theory and Design of Modern Pressure Vessels*, 2nd ed. (Van Nostrand–Reinhold).

HENRY, B. D. (1973) *Aust. Chem. Eng.* **14** (Mar.) 13. The design of vertical, free standing process vessels.

HETENYI, M. (1958) *Beams on Elastic Foundations* (University of Michigan Press).

HIGH PRESS. TECH. ASSOC. (1975) *High Pressure Safety Code* (High Pressure Technology Association, London).

JASPER, McL. T. and SCUDDER, C. M. (1941) *Trans. Am. Inst. Chem. Eng.* **37**, 885. Multi-layer construction of thick wall pressure vessels.

I. WELD. (1952) *Handbook for Welded Structural Steel Work*, 4th ed. (The Institute of Welding).

KARMAN, VON T. and, TSIEN, H-S. (1939) *J. Aeronautical Sciences* **7** (Dec.) 43. The buckling of spherical shells by external pressure.

LAKE, G. F. and BOYD, G. (1957) *Proc. Inst. Mech. Eng.* **171**, No. 31, 843. Design of bolted flanged joints of pressure vessels.

LAMÉ, G. and CLAPEYRON, B. P. E. (1833) *Mém presintes par Divers Savart* **4**, Paris.

LANGER, B. T. (1971) Design of vessels involving fatigue, in *Pressure Vessel Engineering Technology*, Nichols, R. W. (Ed.) (Elsevier).

MANNING, W. R. D. (1947) *Engineering* **163** (May 2nd) 349. The design of compound cylinders for high pressure service.

MANNING, W. R. D. (1950) *Engineering* **169** (April 28th) 479, (May 5th) 509, (May 15th) 562, in three parts. The design of cylinders by autofrettage.

MANNING, W. R. D. and LABROW, S. (1974) *High Pressure Engineering* (Leonard Hill).

MARSHALL, V. O. (1958) *Pet. Ref.* **37** (May) (supplement). Foundation design handbook for stacks and towers.

MEGYESY, E. F. (1977) *Pressure Vessel Handbook*, 4th ed. (Pressure Vessel Handbook Publishing Inc., Tulsa, USA).

MISES VON R. (1913) *Math. Phys. Kl.*, 582. Göttinger nachrichten.

NELSON, J. G. (1963) *Hyd. Proc. & Pet. Ref.* **42** (June) 119. Use calculation form for tower design.

NOLAN, R. W. (1973) Storage and process vessels, in *Chemical Engineers Handbook*, 5th ed. Perry, R. H. and Chilton, C. H. (Eds.) (McGraw-Hill).

O'DONNELL, W. J. and LANGER, B. F. (1962) *Trans. Am. Soc. Mech. Eng.* (*J. Eng. Ind.*) **84**, 307. Design of perforated plates.

PERRY, R. H. and CHILTON, C. H. (Eds.) (1973) *Chemical Engineers Handbook*, 5th ed. (McGraw-Hill).

RYDER, G. H. (1969) *Strength of Materials*, 3rd ed. (Macmillan).

ROSE, R. T. (1970) Flanges, in *The Stress Analysis of Pressure Vessels and Pressure Vessel Components*, Gill, S. S. (Ed.) (Pergamon).

SCHEIMAN, A. D. (1963) *Hyd. Proc. & Pet. Ref.* **42** (June) 130. Short cuts to anchor bolting and base ring sizing.

SOUTHWELL, R. V. (1913) *Phil. Trans.* **213A**, 187. On the general theory of elastic stability.

TANG, S. S. (1968) *Hyd. Proc.* **47** (Nov.) 230. Shortcut methods for calculating tower deflections.

TIMOSHENKO, S. (1936) *Theory of Elastic Stability* (McGraw-Hill).

TIMOSHENKO, S. and YOUNG, D. H. (1968) *Elements of Strength of Materials*, 5th ed. (Van Nostrand).

TONGUE, H. (1959) *The Design and Construction of High Pressure Chemical Plant*, 2nd ed. (Chapman & Hall).

WATERS, E. O., WESSTROM, D. B. and WILLIAMS, F. S. G. (1934) *Mechanical Engineering* **56**, 736. Design of bolted flanged connections.

WATERS, E. O., WESSTROM, D. B. ROSSHEIM, D. B. and WILLIAMS, F. S. G. (1937) *Trans. Am. Soc. Mech. Eng.* **59**, 161. Formulas for stresses in bolted flange connections.

WATERS, E. O. and TAYLOR, J. H. (1927) *Mechanical Engineering* **49** (May) 531. The strength of pipe flanges.

WEIL, N. A. and MURPHY, J. J. (1960) *Trans. Am. Soc. Mech. Eng.* (*J. Eng. Ind.*) **82** (Jan.) 1. Design and analysis of welded pressure vessel skirt supports.

WINDENBURG, D. F. and TRILLING, D. C. (1934) *Trans. Am. Soc. Mech. Eng.* **56**, 819. Collapse by instability of thin cylindrical shells under external pressure.

WOLOSEWICK, F. E. (1951) *Pet. Ref.* **30** (July) 137, (Aug.) 101, (Oct.) 143, (Dec.) 151, in four parts. Supports for vertical pressure vessels.

ZICK, L. P. (1951) *Welding J. Research Supplement* **30**, 435. Stresses in large horizontal cylindrical pressure vessels on two saddle supports.

ZICK, L. P. and McGRATH, R. V. (1968) *Hyd. Proc.* **47** (May) 143. New design approach for large storage tanks.

British Standards
BS 767: 1961 Hydro-extractors and centrifugal machines.
BS 1501: ----- Steels for fired and unfired pressure vessels. Plates.
       Part 1: 1964 Carbon and carbon manganese steels. Imperial units.
       Part 2: 1970 Alloy steels. Imperial units.
       Part 3: 1973 Corrosion and heat resisting steel: Imperial units.
BS 1502: 1968 Steels for fired and unfired pressure vessels. Sections and bars.
BS 1503: 1980 Steels for fired and unfired pressure vessels. Forgings.
BS 1504: 1976 Specification for steel castings for pressure purposes.
BS 1560: Part 2: 1970 Steel pipe flanges and flanged fittings for the petroleum industry.
BS 2654: 1973 Vertical steel welded storage tanks with butt-welded shells for the petroleum industry.
BS 3293: 1960 Carbon steel pipe flanges (over 24 in nominal size) for the petroleum industry.
BS 3915: 1965 Carbon and low alloy steel pressure vessels for primary circuits of nuclear reactors.
BS 4504: ----- Flanges and bolting for pipes, valves and fittings. Metric series
       Part 1: 1969 Ferrous.
       Part 2: 1974 Copper alloy and composite flanges.
BS 4741: 1971 Vertical cylindrical welded steel storage tanks for low-temperature service. Single-wall tanks for temperatures down to −50 °C.
BS 4870: ----- Approval testing of welding procedures
       Part 1: 1974 Fusion welding of steel.
BS 4871: ----- Approval testing of welders working to approval welding procedures
       Part 1: 1974 Fusion welding of steel.
BS 4872: ----- Approval testing of welders when welding procedure approval is not required
       Part 1: 1972 Fusion welding of steel.
       Part 2: 1976 TIG or MIG welding of aluminium and its alloys.
BS 4882: 1973 Bolting for flanges and pressure containing purposes.
BS 4994: 1973 Vessels and tanks in reinforced plastics.
BS 5500: 1976 Unfired fusion welded pressure vessels.

Codes of practice
CP 3: ----- Code of basic data for the design of buildings
       Chapter V: Part 2: 1972 Wind loads.

Special issues
PD 6437: 1969 A review of design methods given in present and codes and design proposals for nozzles and openings in pressure vessels.
PD 6438: 1969 A review of present methods for design of bolted flanges for pressure vessels.
PD 6439: 1969 A review of methods for calculating stresses due to local loads and local attachments of pressure vessels.

ASME Boiler and Pressure Vessel Code
Section   I. Power boilers
       II. Material specifications
     III. Nuclear power plant components
     IV. Heating boilers
      V. Nondestructive examination
     VI. Recommended rules for care and operation of heating boilers
    VII. Recommended rules for the care of power boilers
   VIII. Pressure vessels: Division 1
             Division 2, alternative rules
     IX. Welding qualifications
      X. Fiberglass-reinforced plastic pressure vessels
    XI. Rules for inservice inspection of nuclear coolant systems.

American Society of Mechanical Engineers, New York, USA.

13.19. Nomenclature

Dimensions
in **MLT**

A	Arbitrary constant in equation 13.117	$ML^{-1}T^{-2}$
A_{bf}	Total bolt area required for a flange	L^2

A_1	Area removed in forming hole	L^2
A_2	Area of compensation	L^2
a	Diameter of flat plate	L
$2a$	Major axis of ellipse	L
a_e	Acceleration due an earthquake	LT^{-2}
B	Inside diameter of flange	L
B	Arbitrary constant in equation 13.120	MLT^{-2}
B'	Constant of integration in equation 13.118	MLT^{-2}
b	Effective sealing width of gasket	L
$2b$	Minor axis of ellipse	L
C	Constant in equation 13.34	—
C_c	Design factor in equation 13.46	—
C_d	Drag coefficient in equation 13.79	—
C_e	Seismic constant	—
C_h	Constant in equation 13.85	—
C_p	Constant in equation 13.34	—
C_{ph}	Design factor in equation 13.112	—
C_s	Design factor in equation 13.44	—
c	Corrosion allowance	L
D	Diameter	L
$\mathbf{D}$	Flexual rigidity	ML^2T^{-2}
D_b	Bolt circle diameter	L
D_c	Diameter of cone at point of interest	L
D_e	Nominal diameter of flat end	L
D_{eff}	Effective diameter of column for wind loading	L
D_i	Internal diameter	L
D_m	Mean diameter	L
D_o	Outside diameter	L
D_p	Plate diameter, tube-sheet	L
D_r	Diameter of stiffening ring	L
D_s	Skirt internal diameter	L
D_t	Tank diameter	L
d	Diameter at point of interest, thick cylinder	L
d_b	Bolt diameter	L
d_h	Hole diameter	L
d_r	Diameter of reinforcement pad	L
E	Young's modulus	$ML^{-1}T^{-2}$
e	Minimum plate thickness	L
e_c	Minimum thickness of conical section	L
e_k	Minimum thickness of conical transition section	L
e_m	Minimum wall thickness, centrifuge	L
e_s	Minimum thickness of tank	L
F_b	Compressive load on base ring, per unit length	MT^{-2}
F_{bs}	Load supported by bracket	MLT^{-2}
F_c	Critical buckling load for a ring, per unit length	MT^{-2}
F_p	Local, concentrated, wind load	MLT^{-2}
F_r	Load on stiffening ring, per unit length	MT^{-2}
F_s	Shear force due an earthquake	MLT^{-2}
F_w	Loading due to wind pressure, per unit length	MT^{-2}
F_1	Factor in equation 13.102	L^{-3}
F_2	Factor in equation 13.103	L^{-3}
F_3	Factor in equation 13.104	L^{-3}
F_4	Factor in equation 13.104	—
f	Maximum allowable stress (design stress)	$ML^{-1}T^{-2}$
f_a	Nominal design strength at test temperature	$ML^{-1}T^{-2}$
f_b	Maximum allowable bolt stress	$ML^{-1}T^{-2}$
f_c	Maximum allowable bearing pressure	$ML^{-1}T^{-2}$
f_c'	Actual bearing pressure	$ML^{-1}T^{-2}$
f_f	Maximum allowable design stress for flange material	$ML^{-1}T^{-2}$
f_m	Maximum allowable stress for centrifuge material	$ML^{-1}T^{-2}$
f_n	Nominal design strength at design temperature	$ML^{-1}T^{-2}$
f_p	Maximum allowable design stress for plate	$ML^{-1}T^{-2}$

f_r	Maximum allowable design stress for ring material	$ML^{-1}T^{-2}$
f_s	Maximum allowable design stress for skirt material	$ML^{-1}T^{-2}$
f_t	Maximum allowable design stress for tank material	$ML^{-1}T^{-2}$
G	Mean diameter of gasket	L
g	Gravitational acceleration	LT^{-2}
H	Total pressure force on flange	MLT^{-2}
H_d	Pressure force on area inside flange	MLT^{-2}
H_g	Gasket reaction	MLT^{-2}
H_L	Liquid depth	L
H_p	Height of local load above base	L
H_t	Pressure force on flange face	MLT^{-2}
H_v	Height (length) of cylindrical section between tangent lines	L
h	Height of domed head from tangent line	L
h_c	Initial displacement of shaft	L
h_d	Moment arm of force H_d	L
h_g	Moment arm of force H_g	L
h_i	Internal height of branch allowed as compensation	L
h_o	External height of branch allowed as compensation	L
h_t	Moment arm of force H_t	L
I	Second moment of area (moment of inertia)	L^4
I'	Second moment of area per unit length	L^3
I_h	Second moment of area of shell, horizontal vessel	L^4
I_p	Polar second moment of area	L^4
I_r	Second moment of area of ring	L^4
I_v	Second moment of area of vessel	L^4
J	Joint factor, welded joint	—
K	Ratio of diameters of thick cylinder $= D_o/D_i$	—
K_c	Collapse coefficient in equation 13.52	—
L	Unsupported length of vessel	L
L'	Effective length between stiffening rings	L
L_c	Critical distance between stiffening rings	L
L_k	Length of conical transition section	L
L_o	Distance between centre line of equipment and column	L
L_r	Distance between edge of skirt to outer edge of flange	L
M	Bending moment	ML^2T^{-2}
M_{atm}	Moment acting on flange during bolting up	ML^2T^{-2}
M_e	Bending moment due to offset equipment	ML^2T^{-2}
M_{L1}	Longitudinal bending moment at mid-span	ML^2T^{-2}
M_{L2}	Longitudinal bending moment at saddle support	ML^2T^{-2}
M_{op}	Total moment acting on flange	ML^2T^{-2}
M_s	Bending moment at base of skirt	ML^2T^{-2}
M_v	Bending moment acting on vessel	ML^2T^{-2}
M_x	Bending moment at point x from free end of column	ML^2T^{-2}
M_1	Bending moment acting along cylindrical sections	ML^2T^{-2}
M_2	Bending moment acting along diametrical sections	ML^2T^{-2}
m_c	Displaced mass, centrifuge	M
N_b	Number of bolts	—
n	Number of lobes	—
P	Pressure	$ML^{-1}T^{-2}$
P_c	Critical buckling pressure	$ML^{-1}T^{-2}$
P'_c	Critical pressure to cause local buckling in a spherical shell	$ML^{-1}T^{-2}$
P_d	Design pressure	$ML^{-1}T^{-2}$
P_e	External pressure	$ML^{-1}T^{-2}$
P_f	Centrifugal pressure	$ML^{-1}T^{-2}$
P_i	Internal pressure	$ML^{-1}T^{-2}$
P_m	Self-pressure, centrifuge	$ML^{-1}T^{-2}$
P_t	Total pressure acting on centrifuge wall	$ML^{-1}T^{-2}$
P_w	Wind pressure loading	$ML^{-1}T^{-2}$
$\Delta P'$	Effective tube-plate design pressure difference	$ML^{-1}T^{-2}$
p_b	Bolt pitch	L
p_h	Hole pitch	L
R_c	Crown radius	L
R_k	Knuckle radius	L

R_i	Radius of interface	$\mathbf{L}$
R_o	Major radius of torus	$\mathbf{L}$
R_p	Radius of curvature of plate	$\mathbf{L}$
R_s	Outside radius of sphere	$\mathbf{L}$
R_1	Inside radius of centrifuge bowl	$\mathbf{L}$
R_2	Radius of liquid surface	$\mathbf{L}$
r	Radius	$\mathbf{L}$
r_1	Meridional radius of curvature	$\mathbf{L}$
r_2	Circumferential radius of curvature	$\mathbf{L}$
s	Resisting force per unit displacement	$\mathbf{MT^{-2}}$
T	Torque	$\mathbf{ML^2T^{-2}}$
t	Thickness of plate (shell)	$\mathbf{L}$
t_b	Thickness of base ring	$\mathbf{L}$
t_c	Thickness of bracket plate	$\mathbf{L}$
t_f	Thickness of flange	$\mathbf{L}$
t_n	Actual thickness of branch	$\mathbf{L}$
t_p	Tube-plate thickness	$\mathbf{L}$
t_s	Skirt thickness	$\mathbf{L}$
u_w	Wind velocity	$\mathbf{LT^{-1}}$
W	Total weight of vessel and contents	$\mathbf{MLT^{-2}}$
W_e	Weight of ancillary equipment	$\mathbf{MLT^{-2}}$
W_m	Greater value of W_{m1} and W_{m2} in equation 13.109	$\mathbf{MLT^{-2}}$
W_{m1}	Minimum bolt load required under operating conditions	$\mathbf{MLT^{-2}}$
W_{m2}	Minimum bolt load required to seal gasket	$\mathbf{MLT^{-2}}$
W_v	Weight of vessel	$\mathbf{MLT^{-2}}$
w	Deflection of flat plate	$\mathbf{L}$
w	Loading per unit length	$\mathbf{MT^{-2}}$
x	Radius from centre of flat plate to point of interest	$\mathbf{L}$
x	Distance from free end of cantilever beam	$\mathbf{L}$
x_c	Displacement caused by centrifugal force	$\mathbf{L}$
y	Minimum seating pressure for gasket	$\mathbf{ML^{-1}T^{-2}}$
α	Cone half cone apex angle	—
Δ	Dilation	$\mathbf{L}$
Δ_c	Dilation of cylinder	$\mathbf{L}$
Δ_s	Dilation of sphere	$\mathbf{L}$
ε	Strain	—
$\varepsilon_1, \varepsilon_2$	Principal strains	—
θ	Angle	—
θ_s	Base angle of conical section	—
λ	Ligament efficiency	—
v	Poisson's ratio	—
ρ_m	Density of vessel material	$\mathbf{ML^{-3}}$
ρ_a	Density of air	$\mathbf{ML^{-3}}$
ρ_L	Liquid density	$\mathbf{ML^{-3}}$
ρ_{L1}	Density of heavier liquid	$\mathbf{ML^{-3}}$
ρ_{L2}	Density of lighter liquid	$\mathbf{ML^{-3}}$
σ	Normal stress	$\mathbf{ML^{-1}T^{-2}}$
σ_b	Bending stress	$\mathbf{ML^{-1}T^{-2}}$
σ_{b1}	Bending stress at mid-span	$\mathbf{ML^{-1}T^{-2}}$
σ_{b2}	Bending stress at saddle supports	$\mathbf{ML^{-1}T^{-2}}$
σ_e	Stress at elastic limit of material	$\mathbf{ML^{-1}T^{-2}}$
σ_e'	Elastic limit stress divided by factor of safety	$\mathbf{ML^{-1}T^{-2}}$
σ_h	Circumferential (hoop) stress	$\mathbf{ML^{-1}T^{-2}}$
σ_{hb}	Longitudinal hub stress	$\mathbf{ML^{-1}T^{-2}}$
σ_L	Longitudinal stress	$\mathbf{ML^{-1}T^{-2}}$
σ_r	Radial stress	$\mathbf{ML^{-1}T^{-2}}$
σ_{rd}	Radial flange stress	$\mathbf{ML^{-1}T^{-2}}$
σ_s	Stress in skirt support	$\mathbf{ML^{-1}T^{-2}}$
σ_t	Tangential (hoop) stress	$\mathbf{ML^{-1}T^{-2}}$
σ_{tg}	Tangential flange stress	$\mathbf{ML^{-1}T^{-2}}$
σ_{ws}	Stress in skirt due to weight of vessel	$\mathbf{ML^{-1}T^{-2}}$
σ_x	Normal stress in x direction	$\mathbf{ML^{-1}T^{-2}}$

σ_y	Normal stress in y direction	$ML^{-1}T^{-2}$
σ_z	Axial stresses in vessel	$ML^{-1}T^{-2}$
$\sigma_1, \sigma_2, \sigma_3$	Principal stresses	$ML^{-1}T^{-2}$
τ	Torsional shear stress	$ML^{-1}T^{-2}$
τ_{xy}	Shear stress	$ML^{-1}T^{-2}$
τ_1, τ_2, τ_3	Shear stress maxima	$ML^{-1}T^{-2}$
ϕ	Slope of flat plate	—
ϕ	Angle	—
ω	Rotational speed	T^{-1}

Superscript

^ Maximum

CHAPTER 14

General Site Considerations

14.1. Introduction

In the discussion of process and equipment design given in the previous chapters, no reference was made to the plant site. A suitable site must be found for a new project; the site and equipment layout planned; and provision made for the ancillary buildings and services needed for plant operation. These subjects are discussed briefly in this chapter, to complete this introduction to chemical engineering design.

14.2. Plant location and site selection

The location of the plant can have a crucial effect on the profitability of a project, and the scope for future expansion. Many factors must be considered when selecting a suitable site, and only a brief review of the principal factors will be given in this section. Site selection for chemical process plants is discussed in more detail by Vilbrandt and Dryden (1959), Rase and Barrow (1964) and Merims (1966). The principal factors to consider are:

1. Location, with respect to the marketing area.
2. Raw material supply.
3. Transport facilities.
4. Availability of labour.
5. Availability of utilities: water, fuel, power.
6. Availability of suitable land.
7. Effluent disposal.
8. Local community considerations.
9. Climate.
10. Political and strategic considerations.

Marketing area

For materials that are produced in bulk quantities; such as cement, mineral acids, and fertilisers, where the cost of the product per tonne is relatively low and the cost of transport a significant fraction of the sales price, the plant should be located close to the primary market. This consideration will be less important for low volume production, high-priced products; such as pharmaceuticals.

In an international market, there may be an advantage to be gained by locating the plant within an area with preferential tariff agreements; such as the European Economic Community (EEC).

Raw materials

The availability and price of suitable raw materials will often determine the site location. Plants producing bulk chemicals are best located close to the source of the major raw material; where this is also close to the marketing area.

Transport

The transport of materials and products to and from the plant will be an overriding consideration in site selection.

If practicable, a site should be selected that is close to at least two major forms of transport: road, rail, waterway (canal or river), or a sea port. Road transport is being increasingly used, and is suitable for local distribution from a central warehouse. Rail transport will be cheaper for the long-distance transport of bulk chemicals.

Air transport is convenient and efficient for the movement of personnel and essential equipment and supplies, and the proximity of the site to a major airport should be considered.

Availability of labour

Labour will be needed for construction of the plant and its operation. Skilled construction workers will usually be brought in from outside the site area, but there should be an adequate pool of unskilled labour available locally; and labour suitable for training to operate plant. Skilled tradesmen will be needed for plant maintenance. Local trade union customs and restrictive practices will have to be considered when assessing the availability and suitability of the local labour for recruitment and training.

Utilities (Services)

Chemical processes invariably require large quantities of water for cooling and general process use, and the plant must be located near a source of water of suitable quality. Process water may be drawn from a river, from wells, or purchased from a local authority.

At some sites, the cooling water required can be taken from a river or lake or from the sea; at other locations cooling towers will be needed.

Electrical power will be needed at all sites. Electrochemical processes that require large quantities of power; for example, aluminium smelters, need to be located close to a cheap source of power.

A competitively priced fuel must be available on site for steam and power generation.

Effluent disposal

All industrial processes produce waste products, and full consideration must be given to the difficulties and cost of their disposal. The disposal of toxic and harmful effluents will be covered by local regulations, and the appropriate authorities must be consulted during the initial site survey to determine the standards that must be met.

Local community considerations

The proposed plant must fit in with and be acceptable to the local community. Full consideration must be given to the safe location of the plant so that it does not impose a significant additional risk to the community.

On a new site, the local community must be able to provide adequate facilities for the plant personnel: schools, banks, housing, and recreational and cultural facilities.

Land (site considerations)

Sufficient suitable land must be available for the proposed plant and for future expansion. The land should ideally be flat, well drained and have suitable load-bearing characteristics. A full site evaluation should be made to determine the need for piling or other special foundations.

Climate

Adverse climatic conditions at a site will increase costs. Abnormally low temperatures will require the provision of additional insulation and special heating for equipment and pipe runs. Stronger structures will be needed at locations subject to high winds (cyclone/hurricane areas) or earthquakes.

Political and strategic considerations

Capital grants, tax concessions, and other inducements are often given by governments to direct new investment to preferred locations; such as areas of high unemployment. The availability of such grants can be the overriding consideration in site selection.

14.3. Site layout

The process units and ancillary buildings should be laid out to give the most economical flow of materials and personnel around the site. Hazardous processes must be located at a safe distance from other buildings. Consideration must also be given to the future expansion of the site. The ancillary buildings and services required on a site, in addition to the main processing units (buildings), will include:

1. Storages for raw materials and products: tank farms and warehouses.
2. Maintenance workshops.
3. Stores, for maintenance and operating supplies.
4. Laboratories for process control.
5. Fire stations and other emergency services.
6. Utilities: steam boilers, compressed air, power generation, refrigeration, transformer stations.
7. Effluent disposal plant.
8. Offices for general administration.
9. Canteens and other amenity buildings, such as medical centres.
10. Car parks.

When roughing out the preliminary site layout, the process units will normally be sited first and arranged to give a smooth flow of materials through the various processing steps, from raw material to final product storage. Process units are normally spaced at least 30 m apart; greater spacing may be needed for hazardous processes.

The location of the principal ancillary buildings should then be decided. They should be arranged so as to minimise the time spent by personnel in travelling between buildings. Administration offices and laboratories, in which a relatively large number of people will be working, should be located well away from potentially hazardous processes. Control rooms will normally be located adjacent to the processing units, but with potentially hazardous processes may have to be sited at a safer distance.

The siting of the main buildings will determine the layout of the plant roads, pipe alleys and drains. Access roads will be needed to each building for construction, and for operation and maintenance.

Utility buildings should be sited to give the most economical run of pipes to and from the process units.

Cooling towers should be sited so that under the prevailing wind the plume of condensate spray drifts away from the plant area and adjacent properties.

The main storage areas should be placed between the loading and unloading facilities and the process units they serve. Storage tanks containing hazardous materials should be sited at least 70 m (200 ft) from the site boundary.

A typical plot plan is shown in Fig. 14.1.

A comprehensive discussion of site layout is given by Mecklenburgh (1973); see also Kaess (1970) and House (1969).

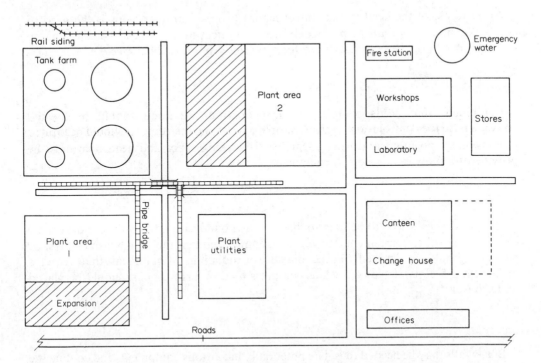

FIG. 14.1. A typical site plan

14.4. Plant layout

The economic construction and efficient operation of a process unit will depend on how well the plant and equipment specified on the process flow-sheet is laid out.

A detailed account of plant layout techniques cannot be given in this short section. A fuller discussion can be found in the book by Mecklenburgh (1973) and in a series of articles by Kern (1977, 1978).

The principal factors to be considered are:

1. Economic considerations: construction and operating costs.
2. The process requirements.
3. Convenience of operation.
4. Convenience of maintenance.
5. Safety.
6. Future expansion.

Costs

The cost of construction can be minimised by adopting a layout that gives the shortest run of connecting pipe between equipment, and the least amount of structural steel work. However, this will not necessarily be the best arrangement for operation and maintenance.

Process requirements

An example of the need to take into account process considerations is the need to elevate the base of columns to provide the necessary net positive suction head to a pump (see Chapter 5) or the operating head for a thermosyphon reboiler (see Chapter 12).

Operation

Equipment that needs to have frequent operator attention should be located convenient to the control room. Valves, sample points, and instruments should be located at convenient positions and heights. Sufficient working space and headroom must be provided to allow easy access to equipment.

Maintenance

Heat exchangers need to be sited so that the tube bundles can be easily withdrawn for cleaning and tube replacement. Vessels that require frequent replacement of catalyst or packing should be located on the outside of buildings. Equipment that requires dismantling for maintenance, such as compressors and large pumps, should be placed under cover.

Safety

Blast walls may be needed to isolate potentially hazardous equipment, and confine the effects of an explosion.

At least two escape routes for operators must be provided from each level in process buildings.

Plant expansion

Equipment should be located so that it can be conveniently tied in with any future expansion of the process.

Space should be left on pipe alleys for future needs, and service pipes over-sized to allow for future requirements.

General considerations

Open, structural steelwork, buildings are normally used for process equipment; closed buildings are only used for process operations that require protection from the weather.

The arrangement of the major items of equipment will usually follow the sequence given on the process flow-sheet: with the columns and vessels arranged in rows and the

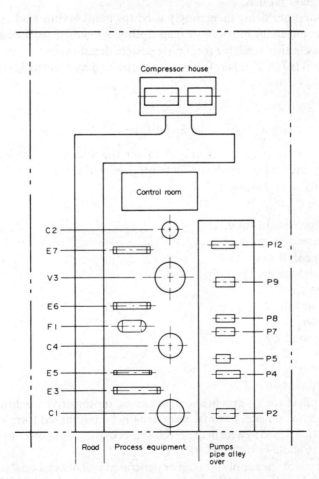

Fig. 14.2. A typical plant layout

ancillary equipment, such as heat exchangers and pumps, positioned along the outside. A typical preliminary layout is shown in Fig. 14.2.

14.4.1 Techniques used in site and plant layout

Cardboard cut-outs of the equipment outlines can be used to make trial plant layouts. Simple models, made up from rectangular and cylindrical blocks, can be used to study alternative layouts in plan and elevation. Cut-outs and simple block models can also be used for site layout studies. Once the layout of the major pieces of equipment has been decided, the plan and elevation drawings can be made and the design of the structural steel-work and foundations undertaken.

Large-scale models, to a scale of at least 1:30, are normally made for major projects. These models are used for piping design and to decide the detailed arrangement of small items of equipment, such as valves, instruments and sample points. Piping isometric diagrams are taken from the finished models. The models are also useful on the construction site, and for operator training. Proprietary kits of parts are available for the construction of plant models.

Digital computers are being increasingly used for plant layout studies. Programs are available from commercial bureaux to help optimise layouts; see Leesley and Newall (1972). Programs are also available for piping design, detailing and quantities; see Daniel (1971) and Spitzer (1971). The use of computer graphics may eventually replace the use of models for plant layout.

14.5. Utilities

The word "Utilities" is now generally used for the ancillary services needed in the operation of any production process. These services will normally be supplied from a central site facility; and will include:

1. Electricity.
2. Steam, for process heating.
3. Cooling water.
4. Water for general use.
5. Demineralised water.
6. Compressed air.
7. Inert-gas supplies.
8. Refrigeration.
9. Effluent disposal facilities.

Electricity

The power required for electrochemical processes; motor drives, lighting, and general use, may be generated on site, but will more usually be purchased from the local supply company (the national grid system in the UK). The economics of power generation on site are discussed by Caudle (1975).

The voltage at which the supply is taken or generated will depend on the demand. For a large site the supply will be taken at a very high voltage, typically 11,000 or 33,000 V.

Transformers will be used to step down the supply voltage to the voltages used on the site. In the United Kingdom a three-phase 415-V system is used for general industrial purposes, and 240-V single-phase for lighting and other low-power requirements. If a number of large motors is used, a supply at an intermediate high voltage will also be provided, typically 6,000 or 11,000 V.

A detailed account of the factors to be considered when designing electrical distribution systems for chemical process plants, and the equipment used (transformers, switch gear and cables), is given by Clay (1960); see also Silverman (1964).

Steam

The steam for process heating is usually generated in water tube boilers; using the most economical fuel available. The process temperatures required can usually be obtained with low-pressure steam, typically 2·5 bar (25 psig), and steam is distributed at a relatively low mains pressure, typically around 8 bar (100 psig). Higher steam pressures, or proprietary heat-transfer fluids, such as Dowtherm (see Conant and Seifert, 1963), will be needed for high process temperatures. The generation, distribution and utilisation of steam for process heating in the manufacturing industries is discussed in detail by Lyle (1963).

Combined heat and power systems

The energy costs on a large site can be reduced if the electrical power required is generated on site and the exhaust steam from the turbines used for process heating. The overall thermal efficiency of such systems can be in the range 70 to 80 per cent; compared with the 30 to 40 per cent obtained from a conventional power station, where the heat in the exhaust steam is wasted in the condenser. Whether a combined heat and power system scheme is worth considering for a particular site will depend on the size of the site, the cost of fuel, the balance between the power and heating demands; and particularly on the availability of, and cost of, standby supplies and the price paid for any surplus power electricity generated. The economics of combined heat and power schemes for chemical process plant sites in the United Kingdom is discussed by Grant (1979).

On any site it is always worth while considering driving large compressors or pumps with steam turbines and using the exhaust steam for local process heating.

Cooling water

Natural and forced-draft cooling towers (see Volume 1, Chapter 11) are generally used to provide the cooling water required on a site; unless water can be drawn from a convenient river or lake in sufficient quantity. Sea water, or brackish water, can be used at coastal sites, but if used directly will necessitate the use of more expensive materials of construction for heat exchangers (see Chapter 7).

Water for general use

The water required for general purposes on a site will usually be taken from the local mains supply, unless a cheaper source of suitable quality water is available from a river, lake or well.

Demineralised water

Demineralised water, from which all the minerals have been removed by ion-exchange, is used where pure water is needed for process use, and as boiler feed-water. Mixed and multiple-bed ion-exchange units are used; one resin converting the cations to hydrogen and the other removing the acid radicals. Water with less than 1 part per million of dissolved solids can be produced (see Volume 3, Chapter 7).

Refrigeration

Refrigeration will be needed for processes that require temperatures below those that can be economically obtained with cooling water. For temperatures down to around 10°C chilled water can be used. For lower temperatures, down to -30°C, salt brines (NaCl and $CaCl_2$) are used to distribute the "refrigeration" round the site from a central refrigeration machine. Vapour compression machines are normally used.

Compressed air

Compressed air will be needed for general use, and for the pneumatic controllers that are usually used for chemical process plant control. Air is normally distributed at a mains pressure of 6 bar (100 psig). Rotary and reciprocating single-stage or two-stage compressors are used. Instrument air must be dry and clean (free from oil).

Inert gases

Where large quantities of inert gas are required for the inert blanketing of tanks and for purging (see Chapter 9) this will usually be supplied from a central facility. Nitrogen is normally used, and is manufactured on site in an air liquefraction plant, or purchased as liquid in tankers.

Effluent disposal

Facilities will be required at all sites for the disposal of waste materials without creating a public nuisance.

Flammable materials can be disposed of by burning. Special incinerators may well be needed to avoid air pollution. A diagram of a typical incinerator for burning liquid or gaseous wastes is given in Chapter 3.

Small quantities of non-inflammable liquid or solid waste can be disposed of by dumping and burial at approved tips. The disposal of toxic wastes is controlled by legislation; in the United Kingdom, the Disposal of Poisonous Wastes Act, 1972.

The disposal of aqueous wastes to public sewers, and surface waters (streams, rivers and estuaries), is also controlled by legislation. Strict controls are placed on the nature of the effluent that can be discharged.

The legislation covering the disposal of industrial waste in the United Kingdom is covered in detail in the books by Tearle (1973), McLoughlin (1976a) and Walker (1979). In a second book McLoughlin (1976b) covers the legislation relating to pollution control in the European Community. The principal legislation in the United Kingdom is the Control of Pollution Act, 1974, which is discussed by Garner (1975).

Gaseous effluents which contain toxic or noxious substances will need treatment before discharge to the atmosphere; or must be discharged from stacks tall enough to dilute and disperse the effluent harmlessly. Gaseous pollutants can be removed by scrubbing or adsorption (see Volume 3, Chapter 7). Finely divided solids can be removed by scrubbing, filtration, or using electrostatic precipitators (see Chapter 10). The treatment and disposal of gaseous effluents is discussed by Nonhebel (1972).

14.5.1. *Aqueous waste treatment*

The principal factors which determine the nature of an aqueous industrial effluent, and on which strict controls will be placed by the responsible authority (local authority or Water Board) are:

1. pH.
2. Suspended solids.
3. Toxicity.
4. Biological oxygen demand.

The pH can be adjusted by the addition of acid or alkali. Lime is frequently used to neutralise acidic effluents.

Suspended solids can be removed by settling, using clarifiers (see Chapter 10, and Volume 2, Chapter 5).

For some effluents it will be possible to reduce the toxicity to acceptable levels by dilution. Other effluents will need chemical treatment. Some data on the toxicity of specific chemicals and trade wastes to fish is given by Klein (1962).

The oxygen concentration in a water course must be maintained at a level sufficient to support aquatic life. For this reason, the biological oxygen demand of an effluent is of utmost importance. It is measured by a standard test: the BOD5 (five-day biological oxygen demand). This test measures the quantity of oxygen which a given volume of the effluent (when diluted with water containing suitable bacteria, essential inorganic salts, and saturated with oxygen) will absorb in 5 days, at a constant temperature of $20°C$. The results are reported as parts of oxygen absorbed per million parts effluent (ppm). The BOD5 test is a rough measure of the strength of the effluent: the organic matter present. It does not measure the total oxygen demand, as any nitrogen compounds present will not be completely oxidised in 5 days. The Ultimate Oxygen Demand (UOD) can be determined by conducting the test over a longer period, up to 90 days. If the chemical composition of the effluent is known, or can be predicted from the process flow-sheet, the UOD can be estimated by assuming complete oxidation of the carbon present to carbon dioxide and the nitrogen present to nitrate:

$$UOD = 2·67C + 4·57N$$

where C and N are the concentrations of carbon and nitrogen in ppm.

A full description of the procedures for carrying out the standard BOD tests, and other tests carried out to monitor and control effluent quality, is given by Klein (1959). Activated sludge processes (see Volume 3, Chapter 5) are frequently used to reduce the biological oxygen demand of an aqueous effluent before discharge. The use of activated sludge processes for the treatment of industrial effluents is discussed by Eckenfelder and O'Connor (1961).

A BOD5 limit of 20 ppm is typically placed on industrial effluent discharged into water courses.

14.6. References

CAUDLE, P. G. (1975) *Chemistry & Industry* (Sept. 6th) 717 The comparative economics of self generated and purchased power.

CLAY, Sir HENRY (1960) Electrical installations, in *Chemical Engineering Practice*, vol. 10, *Ancillary Services*, Cremer, H. W. and Watkins, S. B. (Eds.) (Butterworths).

CONANT, A. R. and SEIFERT, W. F. (1963) *Chem. Eng. Prog.* **59** (May) 46. High temperature heating media: Dowtherm.

DANIEL, P. T. (1971) *Chem. Engr, London* No. 252 (Aug.) 297. An integrated system of pipework estimating, detailing and control (ISOPEDAC): some experiences during development and implementation.

ECKENFELDER, W. N. and O'CONNOR, D. J. (1961) *Biological Waste Treatment* (Pergamon Press).

GARNER, J. F. (1975) *Control of Pollution Act, 1974* (Butterworths).

GRANT, C. D. (1979) *Energy Conservation in the Chemical and Petroleum Industries* (I Chem E/Godwin).

HOUSE, F. F. (1969) *Chem. Eng., Albany* **76** (July 28) 120. Engineers guide to plant layout.

KAESS, D. (1970) *Chem. Eng., Albany* **77** (June 1st) 122. Guide to trouble free plant layouts.

KERN, R. (1977) *Chem. Eng., Albany* **84**:
(May 23rd) 130. How to manage plant design to obtain minimum costs.
(July 4th) 123. Specifications are the key to successful plant design.
(Aug. 15th) 153. Layout arrangements for distillation columns.
(Sept. 12th) 169. How to find optimum layout for heat exchangers.
(Nov. 7th) 93. Arrangement of process and storage vessels.
(Dec. 5th) 131. How to get the best process-plant layouts for pumps and compressors.

KERN, R. (1978) *Chem. Eng., Albany* **85**:
(Jan. 30th) 105. Pipework design for process plants.
(Feb. 27th) 117. Space requirements and layout for process furnaces.
(April 10th) 127. Instrument arrangements for ease of maintenance and convenient operation.
(May 8th) 191. How to arrange plot plans for process plants.
(July 17th) 123. Arranging the housed chemical process plant.
(Aug. 14th) 141. Controlling the cost factor in plant design.

KLEIN, L. (1959) *River Pollution*, Vol. 1. *Chemical Analysis* (Butterworths).

KLEIN, L. (1962) *River Pollution*, Vol. 2. *Causes and Effects* (Butterworths).

KLEIN, L. (1966) *River Pollution*, Vol. 3. *Control* (Butterworths).

LEESLEY, M. E. and NEWALL, R. G. (1972) *Inst. Chem. Eng. Symp. Ser.* No. 35, 2.20. The determination of plant layout by interactive computer methods.

LYLE, O. (1963) *The Efficient Use of Steam* (HMSO).

MCLOUGHLIN, I. (1976a) *Law and Practice Relating to Pollution Control in the U.K.* (Graham Trotman).

MCLOUGHLIN, I. (1976b) *Law and Practice Relating to Pollution Control in the European Community* (Graham Trotman).

MECKLENBURGH, J. C. (1973) *Plant Layout; a guide to the layout of process plant and sites* (Leonard Hill).

MERIMS, R. (1966) Plant location and site considerations, in *The Chemical Plant*, Landau, R. (Ed.) (Reinhold).

NONHEBEL, G. (1972) *Gas Purification Processes for Air Pollution Control* (Butterworths).

RASE, H. F. and BARROW, M. H. (1964) *Project Engineering of Process Plants* (Wiley).

SILVERMAN, D. (1964) *Chem. Eng., Albany* **71** (May 25th) 131, (June 22nd) 133, (July 6th) 121, (July 20th) 161, in four parts. Electrical design.

SPITZER, H. (1971) *Chem. Engr, London* No. 252 (Aug.) 305. The computer approach to pipe detailing.

TEARLE, K. (1973) *Industrial Pollution Control* (Business Books).

VILBRANDT, F. C. and DRYDEN, C. E. (1959) *Chemical Engineering Plant Design*, 4th ed. (McGraw-Hill).

WALKER, A. (1979) *Law of Industrial Pollution Control* (Godwin).

APPENDIX A

Graphical Symbols for Piping Systems and Plant

Based on BS 1553: Part 1: 1976

Scope

This part of BS 1553 specifies graphical symbols for use in flow and piping diagrams for process plant.

Symbols (or elements of symbols) for use in conjunction with other symbols

Description	Symbol	Description	Symbol
Mechanical linkage		Access point	
Weight device		Equipment branch: general symbol. Note. The upper representation does no necessarily imply a flange, merely the termination point. Where a breakable connection is required the branch/pipe would be as shown in the lower symbol	
Electrical device			
Vibratory or loading device (any type)		Equipment penetration (fixed)	
Spray device		Equipment penetration (removable)	
Rotary movement		Boundary line	
Stirring device		Point of change	
Fan		Discharge to atmosphere	

Basic and developed symbols for plant and equipment

Heat transfer equipment

Heat exchanger (basic symbols) Alternative:	
Shell and tube: fixed tube sheet	
Shell and tube : U tube or floating head	
Shell and tube : kettle reboiler	
Air – blown cooler	
Plate type	
Double pipe type	
Heating /cooling coil (basic symbol)	
Fired heater / boiler (basic symbol)	

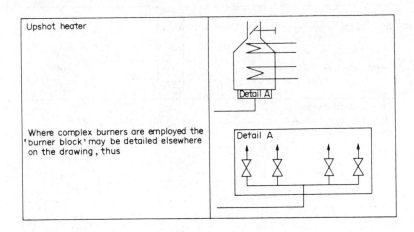

Upshot heater	
Where complex burners are employed the 'burner block' may be detailed elsewhere on the drawing, thus	

Vessels and tanks

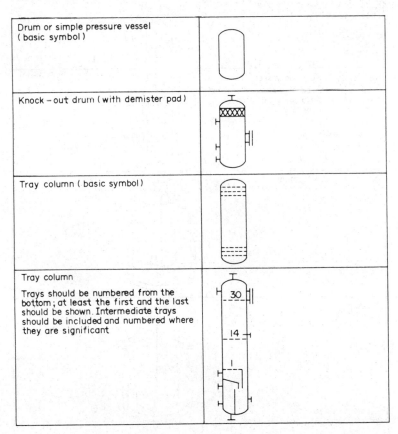

Drum or simple pressure vessel (basic symbol)	
Knock-out drum (with demister pad)	
Tray column (basic symbol)	
Tray column

Trays should be numbered from the bottom; at least the first and the last should be shown. Intermediate trays should be included and numbered where they are significant | |

Fluid contacting vessel (basic symbol)	
Fluid contacting vessel Support grids and distribution details may be shown	
Reaction or absorption vessel (basic symbol)	
Reaction or absorption vessel Where it is necessary to show more than one layer of material alternative hatching should be used	
Autoclave (basic symbol)	
Autoclave	

Open tank (basic symbol)	
Open tank	
Clarifier or settling tank	
Sealed tank	
Covered tank	
Tank with fixed roof (with draw-off sump)	
Tank with floating roof (with roof drain)	
Storage sphere	
Gas holder (basic symbol for all types)	

Pumps and compressors

Rotary pump, fan or simple compressor (basic symbol)	
Centrifugal pump or centrifugal fan	
Centrifugal pump (submerged suction)	
Positive displacement rotary pump or rotary compressor	
Positive displacement pump (reciprocating)	
Axial flow fan	
Compressor : centrifugal /axial flow (basic symbol)	
Compressor : centrifugal / axial flow	
Compressor : reciprocating (basic symbol)	
Ejector / injector (basic symbol)	

Solids handling

Size reduction	
Breaker gyratory	
Roll crusher	
Pulverizer : ball mill	
Mixing (basic symbol)	
Kneader	
Ribbon blender	
Double cone blender	
Filter (basic symbol, simple batch)	
Filter press (basic symbol)	
Rotary filter, film drier or flaker	

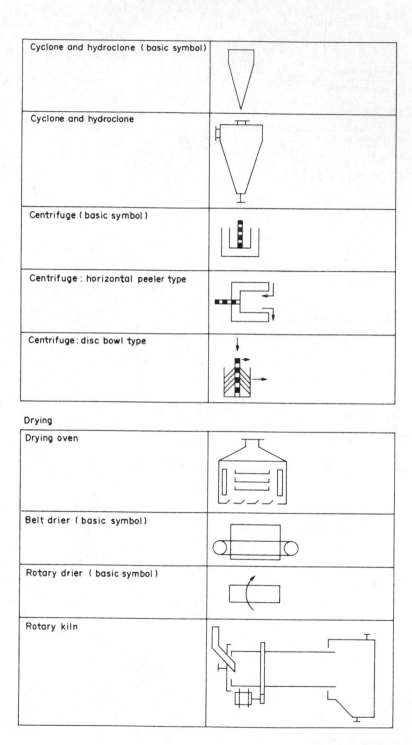

Cyclone and hydroclone (basic symbol)	
Cyclone and hydroclone	
Centrifuge (basic symbol)	
Centrifuge : horizontal peeler type	
Centrifuge : disc bowl type	

Drying

Drying oven	
Belt drier (basic symbol)	
Rotary drier (basic symbol)	
Rotary kiln	

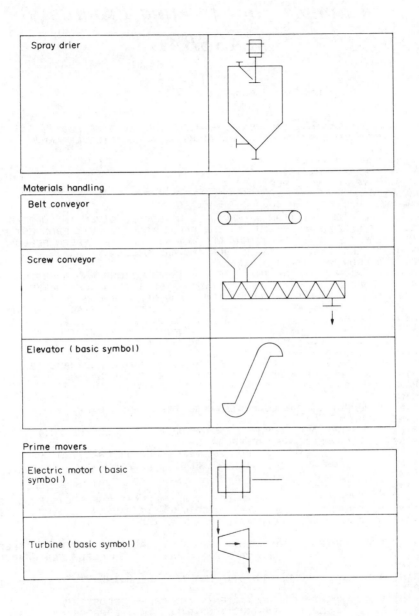

Spray drier	
Materials handling	
Belt conveyor	
Screw conveyor	
Elevator (basic symbol)	
Prime movers	
Electric motor (basic symbol)	
Turbine (basic symbol)	

A Simple Flow-sheeting Program
MASSBAL

PROGRAM MASSBAL

```
PROG MM1
100   PRINT "MASS BALANCE CALCULATIONS USING NAGIEV'S METHOD"
110   PRINT "DO YOU WANT FULL INSTRUCTIONS ?, ANSWER YES OR  NO"
120   INPUT A$
130   IF A$="NO" THEN 1470
140   IF A$="YES" THEN 170
150   PRINT " YES OR NO EXPECTED"
160   GOTO 110
170   REM INSTRUCTIONS FOR RUNNING PROGRAMS
180   REM  A FULL DESCRIPTION OF NAGIEV'S METHOD IS GIVEN IN CHAPTER 4
190   PRINT "THE PROGRAMS CALCULATE THE FEEDS TO EACH UNIT FROM ESTIMATES"
200   PRINT "OF THE SPLIT FRACTION COEFFICIENTS AND THE FRESH FEEDS"
210   PRINT "THREE SEPARATE BASIC PROGRAMS ARE USED"
220   PRINT " PROG MM1 (THIS PROGRAM)"
230   PRINT " WHICH CONTAINS THIS LIST OF INSTRUCTIONS AND A PROCEDURE TO "
240   PRINT " SET UP A DATA FILE CONTAINING THE INITIAL ESTIMATES OF THE"
250   PRINT " SPLIT FRACTION COEFFICIENTS AND FRESH FEEDS"
260   PRINT " PROG MM2, WHICH CONTAINS A PROCEDURE  TO ENABLE THE VALUES IN THE"
270   PRINT " DATA FILE TO BE AMMENDED AS REQUIRED"
280   PRINT " PROG MM3, WHICH CONTAINS A ROUTINE TO SOLVE THE   "
290   PRINT " EQUATIONS, AND PRINT OUT THE RESULTS"
295   PRINT
300   PRINT " IN BRIEF, THE PROCEDURE FOR USING THESE PROGRAMS IS: "
310   PRINT " 1.  DRAW UP A BLOCK DIAGRAM OF THE PROCESS",
320   PRINT "     ONE BLOCK FOR EACH PROCESSING UNIT"
330   PRINT
340   PRINT " 2. PUT IN THE CONNECTIONS BETWEEN THE BLOCKS"
350   PRINT
360   PRINT " 3. FOR EACH COMPONENT, DRAW UP A TABLE OF THE UNIT "
370   PRINT "    CONNECTIONS, AS BELOW: "
380   PRINT "    TO UNIT NO.     FROM UNIT NO.     SPLIT FRACTION COEFF. "
390   PRINT "    (ROW NO. )       (COLN.  NO. )         (ALPHA)"
400   PRINT "       M                N                  A"
410   PRINT
420   PRINT    " THIS GIVES THE ADDRESS AND VALUE OF THE SPLIT FRACTIONS"
430   PRINT    " FOR EACH COMPONENT, AND MAKES IT EASIER TO TYPE THESE"
440   PRINT    " INTO THE PROGRAMS WHEN REQUESTED"
445   PRINT
450   PRINT " 4. FOR EACH COMPONENT, DRAW UP A TABLE OF THE FRESH FEEDS INTO"
460   PRINT "    EACH UNIT. REMEMBER ANY PSUEDO FRESH FEEDS FOR REACTORS"
470   PRINT
480   PRINT " 5. CALL PROGRAM MM1 (IF NOT ALREADY USING IT)"
490   PRINT
500   PRINT " 6. TYPE IN THE COEFICIENTS AND FRESH FEEDS"
510   PRINT "    FOLLOW THE INSTRUCTIONS GIVEN IN THE PROGRAM"
520   PRINT "    THE PROGRAM INCLUDES A ROUTINE FOR CORRECTING MISTAKES"
530   PRINT
540   PRINT " 7. CALL PROGRAM MM3"
550   PRINT "    NOTE: THIS PROGRAM GIVES THE OPTION OF A COMPLETE"
560   PRINT "    PRINT OUT OF THE RESULTS, OR JUST THE RESULTS FOR"
570   PRINT "    SELECTED UNITS. THIS WILL SAVE TIME WHEN ITERATING"
580   PRINT "     ON VALUES THAT EFFECT ONLY A FEW UNITS"
```

```
590   PRINT
600   PRINT " 8.  CHECK THAT THE CALCULATED FLOWS AND COMPOSITIONS"
610   PRINT "     CONFORM TO THE PROCESS PARAMETERS (DESIGN CONSTRAINTS)"
620   PRINT
630   PRINT " 9.  CALL PROGRAM MM2, THE EDITING PROGRAM, MAKE ANY CHANGES"
640   PRINT "     IN THE VALUES OF THE COEFFIIENTS AND FRESH FEEDS TO GET"
650   PRINT "     THE COMPOSTIONS AND FLOWS TO FIT THE DESIGN CONSTRAINTS"
660   PRINT
670   PRINT " 10.  CALL PROGRAM MM3; CHECK OUTPUT AGAINST CONSTRAINTS"
680   PRINT
690   PRINT " 11.  ITERATE ROUND STEPS 8 TO 10 AS NECESARY"
700   PRINT
710   PRINT
720   PRINT " NOTE:  IT IS ONLY NECESSARY TO CALL PROGRAM MM1 TO INPUT THE"
730   PRINT " INITIAL VALUES FOR A NEW PROBLEM"
740   PRINT
750   PRINT " NOTES ON THE WAY THE PROGRAMS HANDLE AND FILE DATA"
755   PRINT
760   PRINT
770   PRINT " (1) PROGRAM MM1: DATA INPUT AND FILES"
780   PRINT " TAKING ONE COMPONENT AT A TIME, THIS PROGRAM SETS UP"
790   PRINT " IN CORE AN IDENTITY MATRIX (ONES ON THE LEADING DIAGONAL)"
800   PRINT " WITH DIMENSIONS EQUAL TO THE NUMBER OF UNITS"
810   PRINT " WHEN THE ROW NUMBER, COLUMN NUMBER, AND VALUE OF THE"
820   PRINT " NON-ZERO COEFFICIENTS ARE TYPED IN,   THE PROGRAM "
830   PRINT " SUBSTITUTES THESE VALUES AT THE CORRECT POSITIONS IN THE"
840   PRINT " MATRIX"
850   PRINT " NOTE: ALL COEFFICIENTS WILL BE NEGATIVE AND MUST BE TYPED"
860   PRINT " IN AS SUCH"
870   PRINT " IF ANY UNIT HAS A SELF-RECYCLE STREAM,   THE VALUE ON THE"
880   PRINT " LEADING DIAGONAL WILL NOT BE ONE AND MUST BE TYPED IN"
890   PRINT " AS (1 - COEFF.) AT THE CORRECT POSITION (ROW NO. = COLN. NO.)"
900   PRINT " WHEN THE MATRIX IS COMPLETE IT IS SCANNED AND ALL THE"
910   PRINT " NON-ZERO VALUES AND THEIR ADDRESSES ARE FILED. "
920   PRINT " THE FRESH FEEDS ARE INPUT AND FILED IN THE SAME WAY. "
930   PRINT " A ZERO VECTOR IS SET UP TO RECEIVE THE VALUES. "
940   PRINT " ONLY THE VALUES FOR THOSE UNITS THAT HAVE A FRESH FEED"
950   PRINT " OF THE PARTICULAR COMPONENT NEED BE TYPED IN, IT IS NOT"
960   PRINT " NECESSARY TO TYPE IN ANY ZERO VALUES"
970   PRINT
980   PRINT " THE 'BASIC' LANGUAGE FILE FACILITY IS USED TO SET UP"
990   PRINT " THE DATA FILES AUTOMATICALLY. "
1000  PRINT " THE PROGRAM INCLUDES A ROUTINE TO ALLOW THE CORRECTION"
1010  PRINT " OF ANY INPUT ERRORS BEFORE THE VALUES ARE FILED"
1015  PRINT
1020  PRINT " UNITS AND NEW FRESH FEEDS CAN BE ADDED WHEN EDITING"
1030  PRINT " BUT NOT NEW UNITS. TO ADD NEW UNITS IT IS NECESSARY TO"
1040  PRINT " REDIMENSION THE MATRIX. TO ADD NEW UNITS START AGAIN "
1050  PRINT " AT PROGRAM MM1"
1055  PRINT
1056  PRINT
1060  PRINT " (2) PROGRAM MM2: EDITING FILES"
```

```
1070    PRINT " THIS PROGRAM CONTAINS A ROUTINE TO GAIN RANDOM ACCESS TO"
1080    PRINT " THE DATA FILE.  THIS ENABLES ANY COMPONENT SUB-FILE TO"
1090    PRINT " BE PICKED OUT FOR EDITING.  IT DOES THIS BY WRITING THE"
1100    PRINT " VALUES INTO A SECOND FILE (DATA FILE 2) UNTIL THE"
1110    PRINT " REQUIRED SUB-FILE IS REACHED , THE DATA IN THE "
1120    PRINT " FILE REQUIRED IS THEN READ INTO THE COEFFICIENT MATRIX"
1130    PRINT " AND FRESH FEED VECTOR, FOR EDITING"
1140    PRINT " AFTER EDITING, THE VALUES FROM THE CORRETED MATRIX"
1150    PRINT " AND VECTOR ARE WRITTEN TO FILE 2, TOGETHERE WITH THE"
1160    PRINT " REMAINING DATA FROM FILE 1.  THE DATA FROM FILE 2 IS"
1170    PRINT " THEN WRITTEN BACK INTO FILE 1"
1180    PRINT
1190    PRINT " TO CHECK THE CONTENTS OF A DATA FILE THE FILE CAN BE"
1200    PRINT " LISTED AS A DATA FILE"
1210    PRINT
1220    PRINT " NEW COEFFICIENTS (REPRESENTING NEW CONNECTIONS BETWEEN"
1230    PRINT " BETWEEN UNITS) AND NEW FRESH FEEDS CAN BE ADDED WHEN"
1240    PRINT " EDITING A FILE, BUT NOT NEW UNITS.  TO ADD NEW UNITS"
1250    PRINT " THE MATRIX MUST BE REDIMENSIONED BY STARTING AGAIN"
1260    PRINT " AT PROGRAM MM1"
1270    PRINT
1280    PRINT
1290    PRINT " (3) PROGRAM MM3: CALCULATION AND PRINT"
1300    PRINT "   THE CALCUATION PROCEDURE USES AN EFFICIENT ALGORITHM"
1310    PRINT "   FOR THE SOLUTION OF SPARSE MATRICES.  TO SAVE CORE SPACE"
1320    PRINT "   THE DATA IS STORED IN CORE IN 4 VECTORS:              "
1330    PRINT
1340    PRINT "     VECTOR A :  CONTAINS THE COEFFICIENT VALUES"
1350    PRINT "     VECTOR B :  THE FRESH FEEDS"
1360    PRINT "     VECTOR Z :  THE COLUMN ADDRESS OF EACH COEFFICIENT"
1370    PRINT "     VECTOR L :  THE POSITION IN THE Z VECTOR OF THE FIRST"
1380    PRINT "               COEFFICIENT IN EACH ROW"
1390    PRINT
1400    PRINT "   THE PROGRAM READS THE DATA FROM THE DATA FILE, SETS UP"
1410    PRINT "   THESE VECTORS, AND SOLVES THE SET OF SIMULTANEOUS"
1420    PRINT "   EQUATIONS TO DETERMINE THE FEEDS TO EACH UNIT , ONE "
1430    PRINT "   COMPONENT AT A TIME.  THE RESULTS ARE STORED IN A MATRIX"
1440    PRINT "   AND WHEN ALL COMPONENT FLOWS HAVE BEEN CALCULATED, THE"
1450    PRINT "   TOTAL FLOW TO EACH UNIT AND THE PERCENTAGE OF EACH "
1460    PRINT "   COMPONENT ARE CALCULATED AND PRINTED OUT"
1470    REM PROG. TO SET UP DATA FILES
1480    DEFINE FILE #1="MDATA1", ASC SEP
1490    DEFINE FILE #2="MDATA2", ASC SEP
1500    PRINT "PROG. TO SET UP DATA FILES; COEFF. AND FRESH FEEDS"
1510    PRINT " NUMBER OF UNITS ? (MAX 50)"
1520    INPUT NO
1530    PRINT " NUMBER OF COMPONENTS ? (MAX 20)"
1540    INPUT C2
1550    REWIND #1
1560    REWIND #2
1570    WRITE #1, NO, C2
1580    FOR E=1 TO C2
```

```
1590    PRINT "FOR COMPONENT"; E
1600    MAT D=IDN(NO, NO)
1610    PRINT " NUMBER OF NON-ZERO COEFFS. ? (EXCL. 1'S ON DIAGONAL)"
1620    INPUT N5
1630    N1=N5+NO
1640    PRINT " NUMBER OF NON-ZERO FRESH FEEDS ? "
1650    INPUT N2
1660    MAT B=ZER(NO)
1670    WRITE #1, N1, N2
1680    PRINT " INPUT COEFFS. : ROW NO., COLN NO., VALUE"
1690    FOR E1=1 TO N5
1700    PRINT " NEXT M, N, A"
1710    INPUT R, S, D
1720    D(R, S)=D
1730    NEXT E1
1740    PRINT "NUMBER OF CORRECTIONS"
1750    INPUT N
1760    FOR E1=1 TO N
1770    PRINT " NEXT M, N, A"
1780    INPUT R, S, D
1790    D(R, S)=D
1800    NEXT E1
1810    PRINT " INPUT FRESH FEEDS, ROW NO., VALUE"
1820    FOR E1=1 TO N2
1830    PRINT " NEXT M, B"
1840    INPUT R, B
1850    B(R)=B
1860    NEXT E1
1870    PRINT "NUMBER OF CORRECTIONS ?"
1880    INPUT N
1890    FOR E1=1 TO N
1900    PRINT " NEXT M, A"
1910    INPUT R, B
1920    B(R)=B
1930    NEXT E1
1940    REM SUB PROG TO FILE COEFFS
1950    FOR R=1 TO NO
1960    FOR S=1 TO NO
1970    IF D(R, S)=0 THEN 1990
1980    WRITE #1, R, S, D(R, S)
1990    NEXT S
2000    NEXT R
2010    REM SUB PROG TO FILE FEEDS
2020    FOR R=1 TO NO
2030    IF B(R)=0 THEN 2050
2040    WRITE #1, R, B(R)
2050    NEXT R
2060    NEXT E
2070    DIM D(50, 50), B(50)
2080    REWIND #1
2090    CLOSE #1
2100    STOP
```

```
PROG MM2
100   REM PROG TO EDIT DATA FILES
105   DEFINE FILE #1="MDATA1", ASC SEP
110   DEFINE FILE #2="MDATA2", ASC SEP
120   REWIND #2
130   REWIND #1
140   PRINT " INPUT THE NUMBER OF THE COMPONENT TO BE ALTERED"
150   INPUT C1
160   REM TRANSFER DATA TO FILE 2 - UP TO COMP. WANTED
170   READ #1, NO, C2
180   ON END #1 GOTO 1090
190   WRITE #2, NO, C2
200   FOR E=1 TO (C1-1)
210   F1=1
220   F2=2
230   GOSUB 970
240   NEXT E
260   REM SETTING UP MATRIX FOR CORRECTION
270   READ #1, N1, N2
280   ON END #1 GOTO 1090
290   MAT D=ZER(NO, NO)
300   FOR I=1 TO N1
310   READ #1, R, S, D
320   ON END #1 GOTO 1090
330   D(R, S)=D
340   NEXT I
350   MAT B=ZER(NO)
360   FOR I=1 TO N2
370   READ #1, R, B
380   ON END #1 GOTO 1090
390   B(R)=B
400   NEXT I
410   REM CORRECTIONS TO COEFFICIENTS, AND ADDING NEW VALUES
420   PRINT " INPUT THE NUMBER OF COEFFS. TO BE ALTERED"
430   INPUT P1
440   PRINT " INPUT NEW VALUE; ROW NO. , COLN. NO.< VALUE"
```

```
450    FOR E1=1 TO P1
460    PRINT "NEXT M, N, A"
470    INPUT R, S, D
480    D(R, S)=D
490    NEXT E1
500    PRINT " INPUT THE NUMBER OF COEFFS. ADDED OR DELETED"
510    PRINT " ADDITIONS ?"
520    INPUT P8
530    PRINT "DELETIONS ?"
540    INPUT P9
550    N1=N1+P8-P9
560    PRINT " NUMBER OF FRESH FEEDS TO BE ALTERED ?"
570    INPUT P2
580    PRINT " INPUT NEW VALUE; ROW NO., VALUE"
590    FOR I=1 TO P2
600    PRINT " NEXT M, B"
610    INPUT R, B
620    B(R)=B
630    NEXT I
640    PRINT " INPUT THE NUMBER ON FEEDS ADDED OR SUBTRACTED"
650    PRINT " ADDITIONS ?"
660    INPUT P6
670    PRINT " DELETIONS ?"
680    INPUT P7
690    N2=N2+P6-P7
700    REM FILE CORRECTED MATICES, FILE 2
710    WRITE #2, N1, N2
720    FOR E3=1 TO NO
730    FOR E4=1 TO NO
740    IF D(E3, E4)=0 THEN 760
750    WRITE #2, E3, E4, D(E3, E4)
760    NEXT E4
770    NEXT E3
780    FOR E3=1 TO NO
790    IF B(E3)=0 THEN 810
```

```
800    WRITE #2,E3,B(E3)
810    NEXT E3
820    REM WRITING REST OF FILE TO FILE 2
822    F1=1
824    F2=2
830    FOR E=1 TO (C2-C1)
840    GOSUB 970
850    NEXT E
860    REM SWITCHING FILES BACK
870    REWIND #2
880    REWIND #1
890    F1=2
900    F2=1
902    READ #F1,NO,C2
904    WRITE #F2,NO,C2
910    FOR E=1 TO C2
940    GOSUB 970
950    NEXT E
960    DIM D(50,50),B(50)
970    READ #F1,N1,N2
980    ON END #F1 GOTO 1090
990    WRITE #F2,N1,N2
1000   FOR E5=1 TO N1
1010   READ #F1,R,S,D
1020   WRITE #F2,R,S,D
1030   NEXT E5
1040   FOR E5=1 TO N2
1050   READ #F1,R,B
1060   WRITE #F2,R,B
1070   NEXT E5
1080   RETURN
1090   PRINT " END OF FILE FOUND "
1095   CLOSE #1
1100   STOP
```

```
PROG MM3
100 REM PROG TO READ FROM FILES TO GUNN'S VECTORS
120 DEFINE FILE #1 = "MDATA1", ASCSEP
125 DEFINE FILE #2 = "MDATA2", ASCSEP
130 READ#1, NO, C2
132 MAT W=ZER(NO, (C2+1))
135 FOR E=1 TO C2
140 R1=1
150 READ#1, N1, N2
160 FOR E1=1 TO N1
170 READ #1, R, S, D
180 D(E1)=D
190 Z(E1)=S
200 IF R1-R <>0 THEN 220
210 L(R)=E1
220 R1=R+1
230 NEXT E1
240 REM SETTING UP FRESH FEED VECTOR
250 MAT B=ZER(NO)
260 FOR E2=1 TO N2
270 READ#1, R, B
280 B(R)=B
290 NEXT E2
580 REM GUNN'S CALC. PROCEDURE
585 L(NO+1)=N1
590 FOR I=1 TO N1
600 D(200-N1+I)=D(I)
610 Z(200-N1+I)=Z(I)
620 NEXT I
630 I1=200-N1
640 FOR I=1 TO NO
650 J0=L(I)
660 J1=L(I+1)
670 IF I<>NO THEN 690
680 J1=J1+1
690 FOR I2=J0 TO J1-1
700 Z(I2)=Z(I2+I1)
710 D(I2)=D(I2+I1)
720 NEXT I2
730 IF I<1.5 THEN 1380
740 IF Z(J0)>I-0.1 THEN 1380
750 J9=0
755 I3=0
760 J8=1
765 K3=1
770 C=D(J0)/D(L(Z(J0)))
780 B1=B(Z(J0))
790 M3=L(Z(J0))+1
800 M4=L(Z(J0)+1)-1
810 IFM4>=M3 THEN 840
820 K8=3
830 GO TO 900
```

```
840 FOR I5=M3 TO M4
850 FOR K=J8 TO J1-J0-1
860 IF Z(J0+K+K3-1)>Z(I5) THEN 1080
870 IF Z(J0+K+K3-1)=Z(I5) THEN 1020
880 IF K3<>1 THEN 1240
890 K8=1
900 I3=I3-1
910 FOR I6=J0 TO J1-2
920 D(I6)=D(I6+1)
930 Z(I6)=Z(I6+1)
940 NEXT I6
950 FOR J6=I+1 TO N0+1
960 L(J6)=L(J6)+I3
970 NEXT J6
980 I1=I1-I3
990 J1=J1+I3
1000 I3=0
1005 K3=0
1010 ON K8 GOTO 1240,1250,1340
1020 J9=J9+1
1030 D(J0+K+K3-1)=D(J0+K+K3-1)-C*D(I5)
1040 J8=K+1-K3
1050 IF K3<>1 THEN 1250
1060 K8=2
1070 GOTO 900
1080 J9=J9+1
1090 IF K3<>1 THEN 1140
1100 K3=0
1110D(J0)=-C*D(I5)
1120 Z(J0)=Z(I5)
1130 GOTO 1250
1140 I3=I3+1
1150 FOR J5=J0+K-1 TO J1-1
1160 J6=J1+J0+K-J5
1170 D(J6)=D(J6-1)
1180 Z(J6)=Z(J6-1)
1190 NEXT J5
1200D(J6-1)=-C*D(I5)
1210 Z(J6-1)=Z(I5)
1220 K8=2
1230 GOTO 950
1240 NEXT K
1250 NEXT I5
1260 M5=M4-M3+1-J9
1270 FOR I6=1 TO M5
1280 D(J1+I6-1)=-C*D(M4+I6-M5)
1290 Z(J1+I6-1)=Z(M4+I6-M5)
1300 I3=I3+1
1310 NEXT I6
1320 K8=3
1330 GOTO 950
1340 B(I)=B(I)-C*B1
```

```
1350 IF I1 > 0 THEN 740
1380 NEXT I
1390 N3=NO-1
1400 FOR I=1 TO N3
1410 B(NO-I+1)=B(NO-I+1)/(D(L(NO-I+1)))
1420 FOR I6=1 TO NO-1
1430 M1=L(I6)+1
1440 M2=L(I6+1)-1
1450 FOR I8=M1 TO M2
1460 IF Z(I8)<>Z(L(NO-I+1)) THEN 1480
1470 B(I6)=B(I6)-D(I8)*B(NO-I+1)
1480 NEXT I8
1490 NEXT I6
1500 NEXT I
1510 B(1)=B(1)/D(1)
1520 REM SUBPROG. TO STORE RESULTS AND CALC. TOTALS
1530 FOR E3=1 TO NO
1540 IF E>1 THEN 1560
1550 W(E3,(C2+1))=0
1560 W(E3,E)=B(E3)
1570 W(E3,(C2+1)) = W(E3,(C2+1)) + W(E3,E)
1580 NEXT E3
1590 NEXT E
1600 DIM B(50),Z(201),D(201),L(50),W(50,1)
1700 REM DECISION ON PRINT OUT
1710 PRINT"RESULTS FOR ALL UNITS WANTED ?"
1720 INPUT B$
1730 IF B$="NO" THEN 1800
1740 FOR C3=1 TO NO
1750 PRINT "UNIT"; C3
1760 GO SUB 2000
1770 NEXT C3
1780 GO TO 1890
1800 PRINT" HOW MANY UNITS WANTED ?"
1810 INPUT C4
1820 FOR E7=1 TO C4
1830 PRINT"NEXT UNIT ?"
1840 INPUT C3
1850 PRINT" UNIT"; C3
1860 PRINT
1870 GO SUB 2000
1880 NEXT E7
1890 STOP
2000 REM SUB PROG TO CALCULATE PERCENTS AND PRINT RESULTS
2010 PRINT"COMPONENT            FLOW              %"
2015 PRINT
2020 FOR E6=1 TO C2
2030 PRINT TAB(5);E6;TAB(20);W(C3,E6);TAB(40);W(C3,E6)*100/W(C3,(C2+1))
2040 NEXT E6
2050 PRINT" TOTAL";TAB(20);W(C3,(C2+1))
2060 RETURN
```

APPENDIX C

Corrosion Chart

An R indicates that the material is resistant to the named chemical up to the temperature shown, subject to the limitations given in the notes. The notes are given at the end of the table.

A *blank* indicates that the material is unsuitable. ND indicates that no data was available for the particular combination of material and chemical.

This chart is reproduced with the permission of IPC Industrial Press Ltd.

METALS

Each metal cell gives the three readings at 20° 60° 100° Centigrade.

	Aluminium (a)	Aluminium Bronze	Brass (b)	Cast Iron (c)	Copper	Gunmetal and Bronze (d)	High Si Iron (14% Si) (c)	Lead	Mild Steel BSS 15	Nickel (cast)
Centigrade	20° 60° 100°	20° 60° 100°	20° 60° 100°	20° 60° 100°	20° 60° 100°	20° 60° 100°	20° 60° 100°	20° 60° 100°	20° 60° 100°	20° 60° 100°
Acetaldehyde	R R R	R R R	R R R	R ND ND	R R R	R R R	R R R	R ND	No data	R R R
Acetic acid (10%)	R R	R R R			R R R	R R R	R R R	R ND		R[20] R R
Acetic acid (glac. & anh.)	R[1] R R	R R R			R R R	R R R	R R R	R ND		R R R
Acetic anhydride	R[1] R R	R R R		R R R	R R R	R R R	R R			R R R
Acetone	R R R	R R R	R R R	R R R	R R R	R R R	R R R	R R R	R[11]	R R R
Other ketones	R R R	R R R	R R R	R R R	R R R	R R R	R R R	R R R	No data	R R R
Acetylene	R R R		R R R[82]	R R R				R R R	R R	R R R
Acid fumes	R[2] R R	R[2] R[2] R[2]							R[2] R	R R R
Alcohols (most fatty)	R[1] R R	R R R	R R R	R[24] R R	R R R	R R R	R R R	R R	R R R	R R R
Aliphatic esters	R R R	R R R	R R R	R R R	R R R	R R R	R R R	R R	No data	R R R
Alkyl chlorides	No data	No data		R[11] R	R R R	R R R	R R R	R	R[11]	R R R
Alum	R R R	R R R			R R R	R R R	R R R	R R[18] R[10]		
Aluminium chloride	R[11] ND ND	R[20] R[20]			R R R	R R R	R R R	R R	R[4.10]	R
Ammonia, anhydrous	R R R	R R R		R	R R R[83]	R R R		R R	R R R[62]	R R R
Ammonia, aqueous	R R R			R R				R R R	R R	
Ammonium chloride	R[84] R R			R				R R R	R R R	R R R
Amyl acetate	R R R	R R R		R[11] R R	R R R	R R R	R R R	R[4] ND ND	No data	R R R
Aniline	R R R			R R R	R R R	R R R		R R R	No data	R R R
Antimony trichloride		No data		R[11] R R	No data	R		R[11] R R	R[4] R	R[11] R R
Aqua regia								R	R[4] R	
Aromatic solvents	R R R	R R R	R R R	R R R	R R R	R R R	R R R	R R	R[11]	R R R
Beer	R R R	R R R	R R R	R R ND	R R R	R R R	R R ND			R R R
Benzoic acid	R R R	R R R	R R R		R R R	R R R	R R R	R[4]		R R R
Boric acid	R R	R R R	R R R		R R R	R R R	R R R	R[4]		R R R
Brines, saturated	R R R	R R R		R[84]	R R R[20]	R R		R R R	R R R[62]	R R R
Bromine	R[11] R R	R[20]		R[11] R				R[24]		
Calcium chloride	R R R	R R R		R R	R R R	R R R	R R R	R[4]		R[20] R R
Carbon disulphide	R R R	R		R R R	R R R	R R R	R R R	R[4]		R
Carbonic acid	R R R	R R R			R R R	R R R	R R R	R R R	ND	R
Carbon tetrachloride	R	R R R	R R R	R[11] R R	R R R	R R R	R R R	R R R	R R	R R R
Caustic soda & potash		R			R R	R R R	R R R	R R[11]	R[11] R	R R R
Chlorates of Na, K, Ba	R[11] R R	R R R			R R R	R R R	R R R			R R R
Chlorine, dry	R R R	R R R	R R R	R R R	R R R	R R R	R R R	R R R[4]	R R R	R R R
Chlorine, wet								R R R	R R R	
Chlorides of Na, K, Mg	R R R	R R R			R R R[20]	R R R	R R R	R R R	R[4] R[4.22]	R R R
Chloroacetic acids		No data			No data	No data			No data	R[11] R R
Chlorobenzene	R ND ND	R R R	No data	R R R	No data	R R R	R R R	R R	R[11] R	R R R
Chloroform	R[1] R R	R R R	R R R	R R	R R R	R R R	No data	R R	R[11] R	R R R
Chlorosulphonic acid		R[20] R[20] R[20]	No data	R[11] R			R R	R[4]	R	
Chromic acid (80%)							R R	R[4]	R	
Citric acid	R R R	R R R			R R R	R R R	R R R	R R R	R R[25]	R R R
Copper salts (most)		R R R			R R R	R R R	R[16] R R	R[16] R		R
Cresylic acids (50%)	R R R	R R R			R R R	R R R	R R R	R R		R R R
Cyclohexane	R R R	R R R	R R R	R R R	R R R	R R R	R R R	R R	No data	R R R
Detergents, synthetic	R R R	No data	R R R	No data	R R R	R R R	No data	R R	No data	R R R
Emulsifiers (all conc.)	R R R	R R R	No data	No data	R R R	No data	No data	R R	No data	No data
Ether	R[1] R R	R R R	R R R	R R R	R R R	R R R	R R R	R R	R R R	R R R
Fatty acids (> C$_6$)	R R R	R R R			R R R	R R R	R R R	R R R	R[4] R R[58]	R R R
Ferric chloride								R	R[4]	
Ferrous sulphate		R[20] R[20] R[20]						R R R	R R R	
Fluorinated refrigerants, aerosols, e.g. *Freon*	R[11] ND ND	R R R	R R R	R R R	R R R	R R R	R R R	R R	R[11] ND ND	No data
Fluorine, dry	R R R	R R R[11]			R R R	R R R		R[4] R	R R R	R R R
Fluorine, wet				No data				R R ND		R R R
Fluosilicic acid								R R R[58]		R[20] R R
Formaldehyde (40%)	R	R R R	R	R	R R R	R R R	R R R	No data	R	R R R
Formic acid	R	R R R	No data		R R R	R R R	R R R	R[30] R[36]		R R R

METALS

Nickel-Copper Alloys (e)			Ni Resist (High Ni Iron) (c)			Platinum			Silver			Stainless Steel 18/8 (f)			Molybdenum Stainless Steel 18/8 (f)			Austenitic Ferric Stainless Steel (x)			Tantalum			Tin (g)			Titanium			Zirconium		
20°	60°	100°	20°	60°	100°	20°	60°	100°	20°	60°	100°	20°	60°	100°	20°	60°	100°	20°	60°	100°	20°	60°	100°	20°	60°	100°	20°	60°	100°	20°	60°	100°
R	R	R	R	ND		R	R	R	R	R	R	R	R	R	R	R	R	R	R	R	R	R	R	R	R		R	R	R	R	R	R
R	R	R	R			R	R	R	R	R	R	R	R	R	R	R	R	R	R	R	R	R	R	R			R	R	R	R	R	R
R	R					R	R	R	R	R	R	R	R		R	R	R^{84}	R	R		R	R	R				R	R	R	R	R	R
R	R		R			R	R	R	R	R	R	R^{80}			R	R		R	R	R	R	R	R				R	R	R	R	R	R
R	R	R	R	R	R	R	R	R	R	R	R	R	R	R	R	R	R	R	R	R	R	R	R	R	R	R	R	R	R	R	R	R
R	R	R	R	R	R	R	R	R	R	R	R	R	R	R	R	R	R	R	R	R	R	R	R	R	R	R	R	R	R	R	R	R
			R	R	R							R	R	R	R	R	R	R	R	R	R	R	R	R	R		R	ND	ND	R	R	R
R^2	R	R				R	R	R				R^2	R	R	R^2	R	R	R^{102}	R^{102}	R^{102}	R^5	R	R	R^{44}	R	R	R^2	R^2	R^2	R^2	R	R
R	R	R	R	R	R	R	R	R	R^{16}	R	R	R	R	R	R	R	R	R	R	R	R	R	R				R^{93}	R^{93}	R^{93}	R	R	R
R	R	R	R	R	R	R	R	R	R	R	R	R	R	R	R	R	R	R	R	R	R	R	R	R			R	R	R	R	R	R
R	R	R	R	R	R	R	R	R	R	R	R	R^{11}	R	R	R^{11}	R	R	R	R	R	R	R	R	R			R	R	R	R	R	R
R	R		R	R		R	R	R	R	R	R	R	R^{13}		R	R		R	R		R	R	R	R			R	R	R	R	R	R
R						R	R	R	R	R	R	R^{84}			R^{84}	R					R	R	R	R^{57}			R	R	R^{10}	R	R	R
R	R		R	R	R	R	R	R	R	R	R	R	R	R	R	R	R	R	R	R	R	R	R	R	R	R	R	R	R	R	R	R
			R	R	R	R	R	R	R^{30}	R	R	R	R	R	R	R	R	R	R	R	R	R	R	R^{13}			R	R	R	R	R	R
R			R	R	R	R	R	R	R^{73}	R	R	R^{84}			R^{84}	R		R			R	R	R'	R			R	R	R	R	R	R
R	R		R	R	R	R	R	R	R	R	R	R	R	R	R	R	R	R	R	R	R	R	R	R			R	R	R	R	R	R
R			R	R	R	R	R	R	R	R	R	R	R	R	R	R	R	R	R	R	R	R	R	R			R	R	ND	R	R	R
R	ND	ND				R	R	R	R	R	R	R^{11}			R^{11}	R^{11}		R	R		R	R	R				R	R	ND	No	data	
																					R	R	R				R					
R	R	R	R	R	R	R	R	R	R	R	R	R	R	R	R	R	R	R	R	R	R	R	R	R	R	R	R	R	R	R	R	R
R	R	R	R	R	ND	R	R	R	R	R	R	R	R	R	R	R	R	R	R	R	R	R	R	R	R		R	R	R	R	R	R
R	R	R				R	R	R	R	R	R	R	R	R	R	R	R	R	R	R	R	R	R	R			R	R	R	R	R	R
R	R		R	R	R	R	R	R	R	R	R	R	R	R	R	R	R	No	data		R	R	R	R	R	R	R	R	R	R	R	R
R	R	R	R	R	R	R	R	R	R	R	R	R^{42}			R^{42}			R	R	R	R	R	R	R			R^{90}			R^{90}	R	R
R	R	R	R	R	R	R	R	R	R	R	R				R^{42}						R	R	R	R			R	R	R	R	R	R
R	R		R	R	R	R	R	R	R	R	R	R	R	ND	R	R	ND	R	R	R	R	R	R	R	R	R	R	R	R	R	R	R
R	R	R	R	R	R	R	R	R	R	R	R	R	R	R	R	R	R	R	R	R	R	R	R	R	R	R	R	R	R	R	R	R
R	R	R	R	R	R	R	R	R	R	R	R	R^{11}	R	R	R^{11}	R	R	R	R	R	R	R	R	R^{11}	R	R	R	R	R	R	R	R
R	R	R	R	R	R	R	R	R	R	R	R	R	R	R^{13}	R	R	R^{13}	R^{103}	R^{103}		R^{10}	R	R				R	R^{19}	R^{15}	R	R	R
R			R	R	R	R	R	R	R^{16}	R	R	R^{16}	R	R	R	R	R	R	R	R	R^{25}	R	R			R^{79}	R^{79}	R^{79}	R^{25}	R^{25}	R^{25}
R	R	R	R	R	R	R	R	R	R	R	R	R	R	R	R	R	R				R	R	R							R^{91}	R	R
						R	R	R	R	R	R										R	R	R				R	R	R			
R	R	R	R	R	R	R	R	R	R^{70}	R	R	R^{84}			R^{84}	R		R^{56}	R^{56}		R	R	R	R^{57}	R	R	R	R	R	R	R	R
R			ND	ND	ND	R	R	R	R	R	R							R	R	R	R	R	R				R	R^2	R^2	R	R	R
R	R		R	R	R	R	R	R	R	R	R	R^{11}	R	ND	R^{11}	R	R	R	R	R	R	R	R	No	data		No	data		R	R	R
R	R	R	R	R	R	R	R	R	R	R	R	R^{11}	R	R	R^{11}	R	R	R	R	R	R	R	R	R^{11}	R	R	R	R	ND	R	R	R
R	R					R	R	R	R	R	R				R^{84}			No	data		R	R	R				R	R	R	R	R	R
						R^{30}	R	R	R^{30}	R	R										R	R	R				R	R	R	R	R	R^{19}
R	R					R	R	R	R	R	R	R^{13}	R	R	R	R	R^{13}	R	R	R	R	R	R	R^{20}	R	R	R	R	R$^{19}_{27}$	R	R	R
R						R	R	R	R^{30}	R	R	R^{16}	R	R	R^{16}	R	R	R^{16}	R^{16}	R^{16}	R	R	R				R	R	R	R^{16}	R	R
R			R	ND	ND	R	R	R	R	R	R	R	R	R	R	R	R	R	R	R	R	R	R	R			R	ND	ND	R	R	R
R	R	R	R	R	R	R	R	R	R	R	R	R	R	R	R	R	R	R	R	R	R	R	R	R	R		R	R	ND	R	R	R
No	data		No	data		R	R	R	R	R	R	R	R	R	R	R	R	R	R	R	R			R			No	data		R	R	R
R	R	R	R	R	R	R	R	R	R	R	R	R	R	R	R	R	R	R	R	R	R	R	R	R	R	R	R	R	ND	R	R	R
R	R	R	R	R	R	R	R	R	R	R	R	R	R	R	R	R	R	R	R	R	R	R	R	R	R	R	R	R	R	R	R	R
						R	R														R	R	R				R	R	R			
R	R		R			R	R	R				R	R	R	R	R	R	R	R	R	R	R	R				R	R	R	R	R	R
																					R	R	R									
R	R	R	R	R	R	R	R	R	R	R	R	R^{11}	R	R	R^{11}	R	R	R	R	R	R	R	R	R	R	R	R	R	R	R	R	R
R	R	R	No	data		R	R	R				R	ND	ND	R	ND	ND	R	R								R^5	R	R			
R	R	R	No	data		R	R	R							R	ND	ND															
R	R	R	R^{32}	R^{32}	ND	R	R	R	R	R	R										R	R	R	R	R		R	R	R	R	R	R
R	R	R	R	R	R	R	R	R	R	R	R	R	R	R	R	R	R	R	R	R	R	R	R	R	R		R	R	R	R	R	R
R	R					R	R	R	R	R	R	R			R	R		R	R		R	R	R				R$^{67}_{69}$	R	R$^{10}_{20}$	R	R	R

METALS

Centigrade	Aluminium (a) 20°	60°	100°	Aluminium Bronze 20°	60°	100°	Brass (b) 20°	60°	100°	Cast Iron (c) 20°	60°	100°	Copper 20°	60°	100°	Gunmetal and Bronze (d) 20°	60°	100°	High Si Iron (14% Si) (c) 20°	60°	100°	Lead 20°	60°	100°	Mild Steel BSS 15 20°	60°	100°	Nickel (cast) 20°	60°	100°
Fruit juices	R	R	R	R	R	R							R	R	R	R	R	R	R	R	R				No data			R	R	R
Gelatine	R	R	R	R	R	R	R	R	R	R	R	R	R	R	R	R	R	R	R	R	R	R	R		No data			R	R	R
Glycerine	R	R	R	R	R	R	R	R	R	R	R	R	R	R	R	R	R	R	R	R	R	R	R		R	R		R	R	R
Glycols	R	R	R	R	R	R	R	R	R	R	R	R	R	R	R	R	R	R	R	R	R	R	R					R	R	R
Hexamine																			R	R	R	R			No data			R	R	R
Hydrazine	R	ND	ND							No data									No data			ND			R	R	R	R	ND	ND
Hydrobromic acid (50%)										ND	ND																			
Hydrochloric acid (10%)				R															R	R								R		
Hydrochloric acid (conc.)				R[62]																		R[4.11]								
Hydrocyanic acid	R	R	R	R[20]	R[20]	R[20]													R	R	R							R	R	R
Hydrofluoric acid (40%)				R[62]																		R						R[20]		
Hydrofluoric acid (75%)				R[62]																								R		
Hydrogen peroxide (30%)	R	R	R																R									R		
Hydrogen peroxide (30–90%)	R	R	R																									R		
Hydrogen sulphide	R	R	R	R[11]	R	R	R[11]	R	R	R			R[11]	R	R	R[11]	R	R	R	R	R	R	R		R[11]	R	R	R[11]	R	R
Hypochlorites				R															R	R	R	R[4.34.76]								
Lactic acid (100%)	R	R	R	R	R								R	R		R[4]	R[4]		R	R	R	ND						R	R	R
Lead acetate	R[11]	R	R	No data						No data																		R	R	R
Lime (CaO)	R[11]			R	R	R	R	R	R	R	R	R	R	R	R	R	R	R	R	R	R	R[4]			R[11]	R	R	R	R	R
Maleic acid	R	R	R	No data			No data						R	R	R	No data			R	R	ND				No data			R	R	R
Meat juices	R	R	R	R	R	R				No data									No data			No data			No data			No data		
Mercuric chloride																			R											
Mercury													R	R	R				R	R	R				R	R	R	R[27]	R	R
Milk & its products	R	R	R	R	R	R				No data						R	R	R	No data									R	R	R
Moist air	R	R	R	R	R	R							R	R	R	R	R	R	R	R	R	R	R	R				R	R	R
Molasses	R	R	R	R[30]	R[30]	R[30]	R[30]	R		R	R	R	R[30]	R	R	R[30]	R	R	R	R	R				No data			R	R	R
Naphtha	R	R	R	R	R	R	R	R	R	R	R	R	R	R	R	R	R	R	No data			R	R		R			R	R	R
Naphthalene	R	R	R	No data			No data			R	R	R	R	R	R	No data			R	R	R	R	R		R			R	R	R
Nickel salts				No data						No data						R	R	R	R	R	R	R	R					R[40]	R	R
Nitrates of Na, K, NH₃	R	R	R	R[73]	R[73]	R[73]				R[11]	R	R							R	R	R							R	R	R
Nitric acid (< 25%)																			R											
Nitric acid (50%)																			R	R	R									
Nitric acid (95%)	R	R	R																R	R	R	R								
Nitric acid, fuming	R	ND	ND																R	R	R	R								
Oils, essential	R	R	R	R	R	R	R	R	R	R	R	R	R	R	R	R	R	R	R	R	R	No data			R	R		R	R	R
Oils, mineral	R	R	R	R	R	R	R	R	R	R	R	R	R	R	R	R	R	R	R	R	R	R	R		R	R		R	R	R
Oils, vegetable & animal	R	R	R	R	R	R	R	R	R	R	R	R	R	R	R	R	R	R	R	R	R	R	R		R	R		R	R	R
Oxalic acid	R[50]			R	R	R	No data						R	R	R	R	R	R	R	R	R	R[4]						R	R	R
Ozone	R	R	R	No data			No data			R	ND	ND	No data			No data			R	R	R	No data			R[11]	R	R	No data		
Paraffin wax	R	R	R	R	R	R	R	R	R	R	R	R	R	R	R	R	R	R	R	R	ND	R	R		R	R		R	R	R
Perchloric acid										No data									R	R	R									
Phenol	R	R	R	R	R	R	R	R	R	R	R	R	R	R	R	R	R	R	R	R	R	R[4]	R	R[19]	No data			R	R	R
Phosphoric acid (25%)	R			R	R	R							R	R	R				R	R	R	R	R	R						
Phosphoric acid (50%)				R	R	R													R	R	R	R	R	R[4]						
Phosphoric acid (95%)				R	R	R													R	R	R	R	R	R						
Phosphorus chlorides				R[11]	R[11]	R[11]				R[11]	R								R	R	R	R	R	R[11]	R	R	R	R	R	R
Phosphorus pentoxide	R[11]	ND	ND	No data															R	R	R				R[11]	R	R	No data		
Phthalic acid	R	R	R	R	R	R				No data			R	R	R	R	R	R	R	R	R	R	R		No data			R	R	R
Picric acid	R	ND	ND																R	R	R	R[4]						R[11]		
Pyridine	R	R	R	No data						R	R	R							R	R	R	R	R		No data			R	R	R
Sea water	R	R	R	R	R	R	R[62]	R	R	R[84]			R	R	R	R	R	R	R	R	R	R	R	R				R	ND	ND
Silicic acid	R	R	R				No data						R	R		No data			R	R	ND	R	R		No data			R	R	ND
Silicone fluids	R	R	R	R	R	R	R	R	R	R	R	R	R	R	R	R	R	R	No data						No data			R	R	R
Silver nitrate										ND	ND								R	R	R									
Sodium carbonate	R[42]	R		R	R	R[4]	R	R	R	R[11]	R	R	R	R	R	R	R	R	R	R	R	R[4]			R	R	R	R	R	R
Sodium peroxide										R[10]	R	R							R[10]	R[10]	R[10]							R	R	R

METALS

Nickel-Copper Alloys (e)			Ni Resist (High Ni Iron) (c)			Platinum			Silver			Stainless Steel 18/8 (f)			Molybdenum Stainless Steel 18/8 (f)			Austernitic Ferric Stainless Steel (x)			Tantalum			Tin (g)			Titanium			Zirconium		
20°	60°	100°	20°	60°	100°	20°	60°	100°	20°	60°	100°	20°	60°	100°	20°	60°	100°	20°	60°	100°	20°	60°	100°	20°	60°	100°	20°	60°	100°	20°	60°	100°
R			R	R		R	R	R	R	R	R	R[94]	R	R	R	R	R	R	R	R	R	R	R	R	R	R	R[8]	ND	ND	R	R	R
R			R	R	R	R	R	R	R	R	R	R[16]	R	R	R[16]	R	R	R	R	R	R	R	R	R	R	R	R[9]	R	ND	R	R	R
R	R	R	R	R	R	R	R	R	R	R	R	R	R	R	R	R	R	R	R	R	R	R	R	R	R	R	R	R	ND	R	R	R
R	R	R	R	R	R	R	R	R	R	R	R	R	R	R	R	R	R	R	R	R	R	R	R	R	R	ND		No data		R	R	R
R				No data		R[19]	R	R				R	R	R	R	R	R	R	R	R		No data		R[13]				No data		R	ND	ND
R						R[70]	R	R	R[45]												R	R	R				R	R	ND	R[30]	R	R
R	R		R			R[70]	R	R	R	R	R										R	R	R				R[49/78]	R		R		
R						R[70]	R	R													R	R	R				R[78]	R[78]	ND	R[92]	R[92]	R[92]
R			R	R	R	R	R	R	R[98]	R	R	R	R		R	R		R	R		R	R	R	R	R			No data			No data	
R	R	R				R	R	R	R	R	R																					
R	R	R				R	R	R	R	R	R																					
R	R					R[87]	R	R				R	R	R	R	R	R	R	R	R	R	R	R	R	R					R	R	R
R[87]						R[87]	R	R				R	R	R	R[87]	R[63]		R	R		R	R	R	R	R	R				R	R	R
R	R		R	R		R	R	R				R	R	R	R	R	R	R[11]	R[11]	R[11]	R	R	R	R[11]	R	R	R	R	R	R	R	R
			R[7]			R	R	R	R[48]												R	R	R				R	R	R	R	R	R
R	R		R			R	R	R	R	R	R	R			R	R		R	R		R	R	R				R	R	R	R	R	R
R	R			No data		R	R	R	R	R	R	R	R	R	R	R	R	R	R	R	R	R	R				R	ND	ND		No data	
R	R	R	R	R	R	R	R	R	R	R	R	R	R	R	R	R	R	R	R	R		No data		R				No data		R	R	R
R	R		R	R	R	R	R	R	R	R	R	R[13]	R[13]	R[13]	R[13]	R	R	R	R	R	R	R	R	R[20]	R	R	R	R	R	R	R	R
	No data		R	R	R	R	R	R	R	R	R	R	R		R	R		R	R	R	R	R	R	R	R	R		No data		R	R	R
				ND	ND	R	R	R										R	R	R	R	R	R				R	R	R[7]	R	R	R
R	R	R	R	R	R	R	R	R				R	R	R	R	R	R	R	R	R	R	R	R				R	R	ND	R	R	R
R			R	R	R	R	R	R	R[86]	R	R	R	R	R	R	R	R	R	R	R	R	R	R	R	R	R		No data		R	R	R
R	R	R	R	R	R	R	R	R	R	R	R	R	R	R	R	R	R	R	R	R	R	R	R	R	R	R	R	R	R	R	R	R
R	R	R	R	R	R	R	R	R	R	R	R	R	R	R	R	R	R	R	R	R	R	R	R	R	R	R		No data		R	R	R
R	R	R	R	R	R	R	R	R	R	R	R	R	R	R	R	R	R	R	R	R	R	R	R	R	R	R	R	R	ND	R	R	R
R	R	R	R	ND	ND	R	R	R	R	R	R	R[16]	R	R	R[16]	R	R	R	R	R	R	R	R	R	R	R	R	R	R[31/32]	R	R	R
R	R		R	R	R	R	R	R	R[30]	R	R	R	R	R	R	R	R	R	R	R	R	R	R	R	R	R	R	R	R	R	R	R
						R	R	R				R	R	R	R	R	R	R	R	R	R	R	R				R	R	R	R	R	R
						R	R	R				R	R	R	R	R	R	R	R	R	R	R	R				R	R	R	R	R	R
						R	R	R				R			R			R			R	R	R				R	R	R	R	R	R
												R	R[80]					R			R	R	R	R	ND					R	R	R
R	R	R	R	R	R	R	R	R	R	R	R	R	R	R	R	R	R	R	R	R	R	R	R	R	R	R	R	R	ND	R	R	R
R	R	R	R	R	R	R	R	R	R	R	R	R	R	R	R	R	R	R	R	R	R	R	R	R	R	R	R	R	ND	R	R	R
R	R	R	R	R	R	R	R	R	R	R	R	R	R	R	R	R	R	R	R	R	R	R	R	R	R	R	R	R	ND	R	R	R
R	R		R	R	R	R	R	R	R	R	R	R[13]			R[10]	R		R	R	R	R	R	R	R[20]	R	R	R[23]			R	R	R
			R	R	R	R	R	R	R[11]	R	R	R	R	R	R	R	R	R	R	R	R[99]	R	R	R	R	ND		No data			No data	
R	R	R	R	R	R	R	R	R	R	R	R	R	R	R	R	R	R	R	R	R	R	R	R	R	R	R	R	R	R	R	R	R
						R	R	R	R	R	R										R	R	R				R	R	ND	R[32]	R	R
R			R	R	R	R	R	R	R	R	R	R	R	R	R	R	R	R	R	R	R	R	R	R	R		R	R	ND	R	R	R
R	R					R	R	R	R	R	R	R	R	R	R	R	R	R	R	R	R	R	R				R	R	R[49/78]	R	R	R
R	R					R	R	R	R	R	R	R	R	R	R	R	R	R	R	R	R	R	R				R[49/78]	R	R	R	R	R
R						R	R	R	R	R	R	R	R		R	R		R[104]	R[104]	R[104]	R[39]	R	R				R[49/78]			R		
R			R[11]	ND	ND	R	R	R	R	R	R										R	R	R				R[11]	R[11]	ND	R		
R			R	R	R	R	R	R	R	R	R	R	R	R[11]	R	R	R[11]	R	R	R	R	ND	ND	R	R			No data			No data	
R	R	R	R	ND	ND	R	R	R	R	R	R		No data			No data			No data		R	R	R	R			R	ND	ND	R	ND	ND
R				ND	ND	R	R	R	R	R	R	R	R	R	R	R	R	R	R	R	R	R	R					No data			No data	
R			R	R	R	R	R	R	R	R	R	R	R	R	R	R	R	R	R	R	R	R	R	R	R	ND	R	R	ND	R	R	R
R[57]	R	R	R	R	R	R	R	R	R[70]	R	R	R[57]			R[57]			R	R	R	R	R	R	R	R	R	R	R	R	R	R	R
R	R	R	R	R	R	R	R	R	R	R	R	R	R	R	R	R	R	R	R	R	R	R	R	R	R	R	R	R	R	R	R	R
R	R	R		No data		R	R	R	R	R	R	R	R	R	R	R	R	R	R	R	R	R	R	R	R	R	R	R	R	R	R	R
				ND	ND	ND	R	R	R	R	R	R	R	R	R	R	R	R	R	R	R	R	R				R[19]	R[19]	ND	R	R	R
R	R	R				R	R	R	R	R	R	R	R	R	R	R	R	R	R	R	R	R	R	R			R	R	R	R	R	R
	No data		R[10]	R[10]	R[10]	R	R	R	R	R	R	R[10]			R[10]	R	R	R	R	R	R[10]	R[10]						No data		R	R	R

METALS

	Aluminium (a)			Aluminium Bronze			Brass (b)			Cast Iron (c)			Copper			Gunmetal and Bronze (d)			High Si Iron (14% Si) (c)			Lead			Mild Steel BSS 15			Nickel (cast)		
Centigrade	20°	60°	100°	20°	60°	100°	20°	60°	100°	20°	60°	100°	20°	60°	100°	20°	60°	100°	20°	60°	100°	20°	60°	100°	20°	60°	100°	20°	60°	100°
Sodium silicate	R	R	R	R	R	R	R	R	R	R	R	R	R	R	R	R	R	R	R	R	R	R	R		R	R	R	R	R	R
Sodium sulphide										R	R	R							R	ND		R[4]	R	R				R	R	R
Stannic chloride				R[11]															R	R										
Starch	R	R	R	R	R	R	No data			R	R	R	R	R	R	R	R	R	R	R	R	No data			No data			R	R	R
Sugar, syrups, jams	R	R	R	R	R	R	R	R	R	R	R	ND	R	R	R	R	R	R	R	R	R	No data			No data			R	R	R
Sulphamic acid	R[50]			No data															R	R					No data			No data		
Sulphates (Na, K, Mg, Ca)	R	R	R	R	R	R	R	R	R	R	R	R	R	R	R	R	R	R	R	R	R	R	R					R	R	R
Sulphites	R	R	R	R	R	R				R[38]	R	R	R	R	R	R	R	R	R[38]	R		R	R					R	R	R
Sulphonic acids	No data			No data			No data			R[11]						No data			R	R	R	R	R		No data			No data		
Sulphur	R	R	R							R	R								R	R	R				R	R	R	R	R	R
Sulphur dioxide, dry	R	R	R	R	R	R	R	R	R	R	R		R	R	R	R	R	R				R	R	R	R	R	R	R	R	R
Sulphur dioxide, wet	R[4]	R	R	R	R	R																R	R	R						
Sulphur trioxide				R[11]	R	R	R[11]	R	R				R[11]	R	R	R[11]	R	R	R	R	R	R	R[4]	R	R[11]	R	R	R	R	R
Sulphuric acid (< 50%)				R	R	R							R	R	R				R	R	R	R	R	R						
Sulphuric acid (70%)				R	R[62]					R									R	R	R	R	R	R	R					
Sulphuric acid (95%)				R[62]									R	R					R	R	R	R	R	R	R					
Sulphuric acid, fuming	R[4]												R	R	R				R	R	R	R	R					R	R	
Sulphur chlorides										R[11]	R[11]								No data			R[4]						No data		
Tallow	R	R	R	R	R	R	No data			R	R	R	R	R	R	No data			R	R	R	R	R		No data					
Tannic acid (10%)	R	R	R	R	R	R	R	R	R				R	R	R	R	R	R	R	R	R							R	ND	ND
Tartaric acid	R	R	R	R	R	R	R	R	R				R	R	R	R	R	R	R	R	R	R[4]	R	R				R[20]	R	R
Trichlorethylene	R	R	R	R	R	R	R	R	R	R	R		R	R	R	R	R	R	R	R	ND	R	R		R[11]	R		R	R	R
Vinegar	R	R	R	R	R	R													R	R	R							R	R	R
Water, distilled	R	R	R	R[53]	R					R	R	R	R[53]	R		R[53]	R	R	R	R	R	R[53]	R	R	R[53]	R	R	R	R	R
Water, soft	R[43]	R	R	R	R	R	R	R	R	R			R	R	R	R	R	R	R	R	R				R[53]	R	R	R	R	R
Water, hard	R[43]	R	R	R	R	R	R	R	R	R	R		R	R	R	R	R	R	R	R	R	R	R	R	R	R	R	R	R	R
Yeast	R	R	R	No data			No data			R	R		R	R	R	R	R	R	R	R	R	No data			No data			R	R	R
Zinc chloride				R	R	R													R			R[4]	R					R[20]	R	R

METALS

Nickel-Copper Alloys (e)			Ni Resist (High Ni Iron) (c)			Platinum			Silver			Stainless Steel 18/8 (f)			Molybdenum Stainless Steel 18/8 (f)			Austernitic Ferric Stainless Steel (x)			Tantalum			Tin (g)			Titanium			Zirconium		
20°	60°	100°	20°	60°	100°	20°	60°	100°	20°	60°	100°	20°	60°	100°	20°	60°	100°	20°	60°	100°	20°	60°	100°	20°	60°	100°	20°	60°	100°	20°	60°	100°
R	R	R	R	R	R	R	R	R	R	R	R	R	R	R	R	R	R	R	R	R	R	R	R	R	R	R	R	R	ND	R	R	R
No data			R	R	R	R	R	R				R	R		R	R		R	R	R		R	R	R				R	ND	R[10]	R	R	R
						R	R	R	R[48]	R	R							No data			R	R	R				R[15]	R[15]	R[15]	R[15]	R[15]	R[15]
R	R	R	R	R	R	R	R	R	R	R	R	R[94]	R	R	R	R	R	R	R	R	R	R	R	R	R	R	R	R	ND	R	R	R
R	R	R	R	R	R	R	R	R	R	R	R	R[94]	R	R	R	R	R	R	R	R	R	R	R	R	R	R	No data			R	R	R
R			R[20]			R	R	R	R	R	R	R[44]			R	R[37]		R	R		R	R	R							No data		
R	R	R	R	R	R	R	R	R	R	R	R	R	R	R	R	R	R	R	R	R	R	R	R	R	R	R	R	R	R	R	R	R
			R[38]	R	R	R	R	R	R	R	R	R	R	R	R	R	R	R	R	R	R	R	R	R	R		R	ND	R	R	R	R
R	R		No data			R	R	R	No data			No data			No data			No data			R	R					No data			R	R	R
R	R	R	R	R	R	R	R	R	R[11]	R	R	R	R	R	R	R	R	R	R	R	R	R	R	R	R	R	R	R	R	R	R	R
R	R		R	R	R	R	R	R	R	R	R	R	R	R	R	R	R	R	R	R	R	R	R	R	R	R	R	R	ND	R	R	R
					ND	R	R	R	R	R	R	R			R	R		R	R		R	R	R				R	R	R	R	R	R
			R	ND	ND	R	R	R	R[11]	R	R				R[11]	R	R	R[11]	R[11]	R[11]	R	R	R									
R			R[20]			R	R	R	R	R	R				R[10]			R			R	R	R				R[49]78	R	R	R	R	R
R			R			R	R	R	R	R								R			R	R	R				R[49]78	R		R	R	R
R						R	R	R				R			R			R	R		R	R	R									
						R	R	R				R	R[80]		R[80]	R		R	R													
R	R		R	R	R	R	R	R										No data			R	R	R				No data					
R	R	R	R	R	R	R	R	R	R	R	R	R	R	R	R	R	R	R	R	R	R	R	R	R	R		R	R	R	R	R	R
R	R	R				R	R	R	R	R	R	R	R	R	R	R	R	R	R	R	R	R	R	R	R	R	R	R	R	R	R	R
R	R	R	R	R	R	R	R	R	R[70]	R		R	R	R	R	R	R	R	R	R	R	R	R	R[20]	R	R	R	R	R[19]	R	R	R
R	R	R	R	R	R	R	R	R	R	R	R	R[11]	R	R	R[11]	R	R	R	R	R	R	R	R	R[11]	R	R	R	R	R	R	R	R
R	R	R	R			R	R	R	R	R	R	R	R	R	R	R	R	R	R	R	R	R	R	R			R	R	R	R	R	R
R	R	R	R	R	R	R	R	R	R	R	R	R	R	R	R	R	R	R	R	R	R	R	R	R	R	R	R	R	R	R	R	R
R	R	R	R	R	R	R	R	R	R	R	R	R	R	R	R[84]	R	R	R	R	R	R	R	R	R	R	R	R	R	R	R	R	R
R	R	R	R	R	R	R	R	R	R	R	R	R	R	R	R[84]	R	R	R	R	R	R	R	R	R[57]	R	R	R	R	R	R	R	R
No data			R	R	R	R	R	R	R	R	R	R	R	R	R	R	R	R	R	R	R	R	R	R	R		No data			R	R	R
R	R	R	R	R	R	R	R	R	R	R	R										R	R	R				R	R	R[52]	R	R	R

THERMOPLASTIC RESINS

Centigrade	Acrylic Sheet (e.g. Perspex)			Acrylonitrile Butadiene Styrene Resins (l)			Nylon 66 Fibre (m)			Nylon 66 Plastics (m)			PCTFE			PTFE (n)			PVDF (y)			Rigid Unplasticised PVC			Plasticised PVC		
	20°	60°	100°	20°	60°	100°	20°	60°	100°	20°	60°	100°	20°	60°	100°	20°	60°	100°	20°	60°	100°	20°	60°	100°	20°	60°	100°
Acetaldehyde							R	ND	ND	R	R[50]	ND	R	R	ND	R	R	R	R	R	R	R[6]					
Acetic acid (10%)	R	R[50]		R						R[50]			R	R	R	R	R	R	R	R	R	R	R		R		
Acetic acid (glac. & anh.)													R	R	R[50]	R	R	R	R	R		R[50]					
Acetic anhydride	R[50]						R	R	R	No data			R	R		R	R	R	R	ND	ND						
Acetone							R	R	R	R	R		R	R[37]		R	R	R	R[106]	ND	ND						
Other ketones							R	R	R	R	ND	ND	R	R[37]		R	R	R									
Acetylene	No data			No data			No data			No data			No data			R	R	R	R	ND	ND	R	R		No data		
Acid fumes	R	R[68]											R	R	R	R	R	R	R	R	R	R	R		No data		
Alcohols (most fatty)							R	R	R	R	R[50]	R[50]	R	R	R	R	R	R	R	R	R	R[33]			No data		
Aliphatic esters							R	R	R	R	ND					R[50]	R		R	R	R				No data		
Alkyl chlorides	No data						R	R	R	R[46]	ND	ND	R	ND	ND	No data			R	R	R	No data			No data		
Alum	R	R		R	R		R	R	R	R	R	R	R	R	R	R	R	R	R	R	R	R	R		R	R	
Aluminium chloride	R	R[68]		R	R		R[43]	R	R	R	ND	ND	R	R	R	R[50]	R	R	R	R	R	R	R		R	R	
Ammonia, anhydrous				R			No data			R	ND	ND	R	R	R	R	R	R	R[107]	R[107]	R[107]	R	R				
Ammonia, aqueous	R	R[4]		R			R	R	ND	R	ND	ND	R	R	R	R	R	R	R[107]	R[107]	R[107]	R	R		No data		
Ammonium chloride	R	R		R	R		R	R	R	R	ND	ND	R	R	R	R[50]	R	R	R	R	R	R	R		R	R	
Amyl acetate							R	R	R	R	ND	ND	R	R		R	R	R	R	R	R						
Aniline										R[50]			R	R	ND	R	R	R	R	R	R						
Antimony trichloride	R[68]	R		R	R					R[50]	ND	ND	No data			No data			R	R	R	R	R		R	R	
Aqua regia													R	R	R	R	R	R	R	R	R	R	R[13]		No data		
Aromatic solvents							R	R	R	R	R[50]	R	R[14]	R		R[50]	R	R	R	R	R				No data		
Beer	R	R		R	R		R	R	R	R	R	R	R	R	R	R	R	R	R	R	R	R	ND		R		
Benzoic acid	R	ND		R	R		No data			R[50]			R	R	ND	R	R	R	R	R	R	R	R[80]		ND		
Boric acid	R	R[68]		R	R		R[43]	R	R	R	R	R	R	R	R	R	R	R	R	R	R	R	R		R		
Brines, saturated	R	R		R	R		R	R	R	R	R	R	R	R	R	R	R	R	R	R	R	R	R		R	R	
Bromine													R	R	R	R[14]	R	R	R	R	R						
Calcium chloride	R	R		R	R		R[43]	R		R[50]	ND	ND	R	R	R	R	R	R	R	R	R	R	R		R	R	
Carbon disulphide							R	R	ND	R[50]	ND	ND	R	R	ND	R	R	R	R	R	ND						
Carbonic acid	R	R		R	R		No data			R	R	ND	R	R	R	R	R	R	R	R	R	R	R		R	R	
Carbon tetrachloride							R	R	R	R	ND	ND	R			R[14]	R	R	R	R	R	R[14]					
Caustic soda & potash	R	R		R	R		R	R	R	R	R	R	R	R	R	R	R	R	R[107]	R[107]	R[107]	R	R		No data		
Chlorates of Na, K, Ba	R	R[68]		R	R					R	R	ND	R	R	R	R	R	R	R	R	R	R			No data		
Chlorine, dry	ND			R	R								R	R	R	R	R	R	R	R	R	R			No data		
Chlorine, wet	R[4]			R	R								R	R	R	R	R	R	R	R	R				No data		
Chlorides of Na, K, Mg	R	R		R	R		R	R	R	R	R	R	R	R	R	R	R	R	R	R	R	R	R		R	R	
Chloroacetic acids	No data												R	R	R	R[2]	R[2]	R[2]	R	ND	ND	R			No data		
Chlorobenzene							R	R	R	R	ND	ND	R			R[14]	R	R	R	R	R						
Chloroform							R	R	R	R	R		R			R[14]	R	R	R	R	R						
Chlorosulphonic acid													R	R	ND	R	R	R	R	R	R	ND					
Chromic acid (80%)				R									R	R	R	R	R	R	No data			R[19]	R[19]				
Citric acid	R	R		R	R		R[43]	R		R[50]	ND	ND	R	R	R	R	R	R	R	R	R	R	R		R		
Copper salts (most)	R[68]	R		R	R		R	R	R[31][48]	R	R	R	R	R	R	R	R	R	R	R	R	R	R		R	R	
Cresylic acids (50%)													R	R	ND	R	R	R	R	R	R						
Cyclohexane							R	R	R	R	ND	ND	R	R	R	R	R	R	R	R	R				No data		
Detergents, synthetic	R	R		R			R	R	R	R	R	R	R	R	R	R	R	R	R	R	R	R	R		R		
Emulsifiers (all conc.)	R	R		No data			R	R	R	R	R	R	R	R	R	R	R	R	R	R	R	R	R		R	R	
Ether							R	R	R	R	ND	ND				R	R	R	R	R	R				No data		
Fatty acids ($>C_6$)	R	ND		R	R		R	R	ND	R	ND	ND	R	R	R	R	R	R	R	R	R	R	R		No data		
Ferric chloride	R	R		R	R		R	R[43]		R[30,50]			R	R	R	R[50]	R	R	R	R	R	R	R		R	R	
Ferrous sulphate	R	R		R	R		R	R	R	R	R	R	R	R	R	R	R	R	R	R	R	R	R		R	R	
Fluorinated refrigerants, aerosols, e.g. Freon	No data						No data			R	ND	ND	R			R[14]	R	R	R	R	R						
Fluorine, dry	No data												R	R		R[48]	R	R	R	R	R						
Fluorine, wet	No data												R	R		No data			R	R	R	R	R				
Fluosilicic acid	No data						R[43]	R					No data			No data			R	R	R	R[15]	R				
Formaldehyde (40%)	R	ND		R	R					R[50]	R[50]		R	R	ND	R	R	R	R	R	R	R	R[30]		R		
Formic acid	R[10]			R[32]	R[10]								R			R	R	R	R	R	R	R			No data		

THERMOPLASTIC RESINS															THERMOSETTING RESINS														
Polyethylene Low Density			Polyethylene High Density			Polycarbonate Resins			Polypropylene			Polystyrene			Melamine Resins (o)			Furane Resin			Epoxy Resins (p)			Phenol Form-aldehyde Resins (r)			Polyester Resins		
20°	60°	100°	20°	60°	100°	20°	60°	100°	20°	60°	100°	20°	60°	100°	20°	60°	100°	20°	60°	100°	20°	60°	100°	20°	60°	100°	20°	60°	100°
R[27]			R	R[80]					R	R	ND	No data						No data			R	R		R	ND	ND	No data		
R	R		R[56]	R		R	R	ND	R	R	ND	R			R[4]	ND	ND				R	R	ND	R	R	R	R[23]		
R[27]			R[56]	R[50,56]		No data			R	R		No data			No data						R[30]	ND		R	R		R[30]		
ND			R	R[50]					R	R		No data									R[68]	ND		R	ND	ND			
			R	R					R	R	ND				R	R	ND	R	R	ND	R	R	R	R					
No data			R[56]	R					R	R	ND				R	R		R	R		R	R	R	R					
No data			R	R		R	ND	ND	No data			No data			R	R	R	No data			No data			No data			No data		
No data			R	R		No data			R[2]	R		R[2]	R		R	R		R	R	R	R[2]	R[30]		R	R	R	No data		
R[27]			R[56]	R		R[46]	ND	ND	R	R	ND	R	R[33]		R	R		R	R		R[50]	R[30,71]		R	R		R	R	
			R	R		ND			R						R	R		R	R		R[50]	R[30,71]		R	R		No data		
No data			R	R		R									R	R		R	R		R[30,71]	R		No data			No data		
			R	R											R	R		R	R		R	R							
R	R		R	R		R	ND	ND	R	R	R	R	R		R	R		R	R	R	R	R		R	R	R	R	R[30]	R[65]
R	R		R	R		R	ND	ND	R	R	R	R	R	R	R	R		ND	ND	ND	R	R		R	R	R	R	R[30]	
R	R		R	R		ND			R	R	R	R	ND	ND	R			ND	ND	ND							R[30]		
R	R		R	R					R	R	R	R	R		R						R	ND		R	R	R	R[30]		
R	R		R	R		R	ND	ND	R	R	R	R	R		R			R	R	R	R	ND		R	R	R	R[30]		
			R[50]	R		ND						R			R	R		R	R	R	R[30]	ND		R	R		R[30]		
			R[50]	R					R	R	R				ND			R	R	ND	R[30]	ND		R	ND	ND			
R	R		R	R		R[7]	ND	ND	R	R	ND	No data			ND	ND		No data			R[68]	ND		R	ND	ND			
									R[56]																				
No data			R[80]						R						R	R		R	R	R	R	R[4,30]		R	R		R	R	
R	R		R	R		R	ND	ND	R	R	ND	R	R		R	ND	ND	No data			R	R		R	R	R	R	R	R
R	R		R[1]	R		R			R	R	ND	R			R			R	R	R	R	R		R	R	R	R[30]	R	
R	R		R	R		R	ND	ND	R	R	R	R			R	ND	ND	R	R	R	R	R		R	R	R	R	R[30]	R[65]
R	R		R	R		R	R	ND	R	R	ND	R			R	ND	ND	R	R	R	R	R		R	R	R	R	R[65]	
R	R		R	R		R	R	R	R	R	R	R	R		R	ND	ND	R	R	R	R	R		R	R		R	R[30]	
			R[50]									R			R			R	R	R				R	R		R		
R	R		R	R		R	ND	ND	R	R	ND	R	R		R	R	R	R	R	R	R[30]	R[30]		R	R	R	R	R	
												R			R			R	R	ND	R	R	ND	R	R	ND	R[30]		
R	R		R[19]	R		R[7]						R	R	R	R[10]			R	R	R	R	R[19]					R[13]		
R	R		R	R		R[7]	ND	ND	R	R	ND	No data			ND	ND		No data			R	R		R	ND	ND	R	R[30]	R[65]
			R[80]			ND									No data			No data			R	R[4,30]					R	R[30]	
			R[50]																								R	R[30]	R[65]
R	R		R	R		R	ND	ND	R	R	R	R	R		R	R		R	R	R	R	R		R	R	R	R	R[30]	R[65]
No data			R	R		ND			R	R	ND	No data						R	ND	ND	R[44,30]	ND		No data			R[30]		
			R[80]												R	R		R	R	R	R[30]	ND		R	ND	ND			
			R[50]												R	R		R	R	ND	R[30]	ND		R	R	R			
			ND			ND						No data			No data						No data								
R	R		R[50]			ND			R	ND	ND				R	ND	ND	R	R	R	R[4,30]						R[10]		
R	R		R[56]	R		R	R		R	R	ND	R			R	ND	ND	R	R	R	R	R		R	R	R	R	R	R[30]
R	R		R	R		R	R		R	R	ND	R			No data			R	R	R	R	R		R	R	R	R	R[30]	
			R[13]	R					No data			No data						No data						R	R				
No data			R[50]	R[50]		R			R						R	R		R	R	ND	R[68]	R[68]		R	R		No data		
R[56]	R		R[56]	R		R[98]	R[98]		R	R	R	R	R		R	R		R	R	R	R	R		R	ND	ND	R[62]	R[62]	
R[56]	R		R[56]	R		ND			R	R	ND	No data			R	R		R	R	R	R	R		R	R	R	No data		
			R[56]	R[50,56]					R[50]	R	ND				R	R		R	ND	ND	R	ND							
No data			R[56]	R		ND			R	R	ND	R			R	R		R	R	R	R	R		R	R	R	R	R[30]	
R	R		R	R		R	R		R	R	R	R	R		No data			R	ND	ND	R	R		R	R	R	R	R[30]	R[65]
R	R		R	R		R	ND		R	R	R	R	R		No data			R	ND	ND	R	R		R	R	R	R	R[30]	R[65]
																		R	ND	ND									
R[50]			R[50]			R			No data						R	R		R	ND	ND	R	R		R	ND	ND	No data		
						ND									No data			No data			R[30]	R					No data		
						ND									No data			No data			R[30]	R[4,30]					No data		
R[3]	R[3]		R[6]	R		ND			R	R	ND	No data			R	R	ND	R	R	ND				R	ND	ND	R[15]		
R	R		R[56]	R		R	R		R	R	ND	R						R	R	ND	R[4,30]			R	R	R	R[30]	R	
R	R		R[56]	R		R[32]	R[32]		R	ND	ND	R			R						R[4,30]			R			R[15]		

Note (printed vertically in the Polyethylene High Density 100° column): H. D. Polyethylene is suitable for a number of applications at 100°C for limited periods, depending on the environment.

THERMOPLASTIC RESINS

Centigrade	Acrylic Sheet (e.g. Perspex) 20°	60°	100°	Acrylonitrile Butadiene Styrene Resins (l) 20°	60°	100°	Nylon 66 Fibre (m) 20°	60°	100°	Nylon 66 Plastics (m) 20°	60°	100°	PCTFE 20°	60°	100°	PTFE (n) 20°	60°	100°	PVDF (y) 20°	60°	100°	Rigid Unplasticised PVC 20°	60°	100°	Plasticised PVC 20°	60°	100°
Fruit juices	R^{68}	R		R	R		R	R	ND	R	R	R	R	R	R	R	R	R	R	R	R	R	R				
Gelatine	R	R		R	R		R	R	ND	R	R	ND	R	R	R	R	R	ND	R	R	R	R	R	R	No data		
Glycerine	R	R		R	R		R	R	R	R	R^{50}	ND	R	R	R	R	R	R	R	R	R	R	R		R		
Glycols	R	ND					R	R	R	R^{50}	R^{50}	ND	R	R	R	R	R	R	R	R	R	R	R		No data		
Hexamine	No data						R^{43}	R	R	No data			R	R	R	R	R	R	R	R	R				No data		
Hydrazine	No data			No data			No data			No data			No data			R	R	R	No data			No data			No data		
Hydrobromic acid (50%)	R	R		R									R	R	R	R	R	R	R	R	R	R^{32}	R		R	R	
Hydrochloric acid (10%)	R	R		R	R								R	R	R	R	R	R	R	R	R	R	R		R	R	
Hydrochloric acid (conc.)	R	R^{50}		R									R	R	R	R	R	R	R	R	R	R^3	R^3		R	ND	
Hydrocyanic acid				R^{10}	R^{10}								R	R	ND	R	R	R	R	R	R	R	R	R	No data		
Hydrofluoric acid (40%)				R									R	R	R	R	R	R	R	R	R	R^{30}			R		
Hydrofluoric acid (75%)													R	R	R	R	R	R	R	R	R	R^{19}					
Hydrogen peroxide (30%)				R									R	R	R	R	R	R	R	R	R	R	R		R		
(30–90%)													R	R	ND	R	R	R	R	R	R	R	R^{30}		R		
Hydrogen sulphide	R	ND		No data			No data			R	ND	ND	R	R	R	R	R	R	R	R	R	R	R		R		
Hypochlorites	R^{34}	ND		R	R								R	R	R	R	R	R	R^{107}	R^{107}	R^{107}	R^{50}	R		No data		
Lactic acid (100%)	R	R^{68}		R	R		R						R	R	ND	R	R	R	R	R	R	R^{15}	R				
Lead acetate	R^{68}	R		R	R		R^{43}	R	R	R^{50}	R^{50}	ND	No data			R	R	R	R	R	R	R	R	R	R	R	
Lime (CaO)	R	R		R	R		R	R	R	R	R	R	R	R	R	R	R	R	R	R	R	R	R	R	No data		
Maleic acid	R^{68}	R		R	R		R^{43}	R	R	R^{50}	ND	ND	R	R	ND	R	R	R	R	R	R	R	R^{37}				
Meat juices	R	R		R	R		No data			R	R	R	R	R	ND	R	R	R	R	R	R	R	R		No data		
Mercuric chloride	R	ND		R			R^{43}	R	R	R^{50}			R	R	R	R	R	R	R	R	R	R	R				
Mercury	R	R		R	R		R	R	ND	R	R	R	R	R	R	R	R	R	R	R	R	R	R	R			
Milk & its products	R	R		R			R	R	R	R	R	R	R	R	R	R	R	R	R	R	R	R					
Moist air	R	R		R	R		R	R	R	R	R	R	R	R	R	R	R	R	R	R	R	R	R		No data		
Molasses	R	R		R			R	R	ND	R	R	R	R	R	R	R	R	R	R	R	R	R	R		R	R	
Naphtha				R			R	R	R	No data						R	R	R	R	R	R	R	R		No data		
Naphthalene	R^4						No data			R	ND	ND				R	R	R	R	R	R						
Nickel salts	R	R		R	R		R^{43}	R^{31}	R	No data			R	R	R	R	R	R	R	R	R	R	R		R	R	
Nitrates of Na, K, NH₃	R	R		R	R		No data			R	R	ND	R	R	R	R	R	R	R	R	R	R	R		R	R	
Nitric acid (<25%)	R	R^{10}											R	R	R	R	R	R	R	R	R	R	R^{37}		R	ND	
Nitric acid (50%)													R	R	R	R	R	R	R	R	R	R	R^{30}		R	ND	
Nitric acid (95%)													R	R	R	R	R	R	R	ND	ND						
Nitric acid, fuming													R	R	ND	R	R	R	R			R^{66}	R				
Oils, essential	R^{62}	R^{62}		R	R		R	R	R	R	R	R	No data			R	R	R	R	R	R	R^{62}	R		No data		
Oils, mineral	R	R		R	R		R	R	R	R	R	R	R	R	R	R	R	R	R	R	R	R	R		No data		
Oils, vegetable & animal	R			R	R		R	R	R	R	R	R	R	R	ND	R	R	R	R	R	R	R	R		No data		
Oxalic acid	R	R		R			R	ND	ND	R^{50}	ND	ND	R	R	R	R	R	R	R	R	R	R	R		R		
Ozone	R	ND					No data			R^{50}	ND	ND	R	R	R	R	R	R	R	R	R	R	R		R		
Paraffin wax	R	R		R	R		R	R	R	R	R	R	R	R	R	R	R	R	R	R	R	R	R		No data		
Perchloric acid	No data			No data									R	R	R	R	R	R	R	R	R	R^{52}	R^{10}		ND		
Phenol													R	R		R	R	R	R	R	R	R	R		ND		
Phosphoric acid (25%)	R	R		R									R	R	R	R	R	R	R	R	R	R	R		R	R	
Phosphoric acid (50%)	R			R									R	R	R	R	R	R	R	R	R	R	R		No data		
Phosphoric acid (95%)				R									R	R	R	R	R	R	R	R	R	R^{55}	R		No data		
Phosphorus chlorides	No data			No data			No data			No data			R	ND	ND	R	R	R	R	R	R				No data		
Phosphorus pentoxide	R^{68}	R		No data			No data			R	R	ND	No data			No data			R	R	R	R	R^{68}		R	ND	
Phthalic acid	No data			No data			No data			R			R	R	ND	R	R	R	R	R	R	R	R				
Picric acid	R^{68}			No data			No data			R^{50}			No data			R	R	R	R	R	R	R^{10}	R		R^{105}	ND	
Pyridine	No data						R	R	R	R	ND	ND	R	R	ND	R	R	R	R	R	R	ND			No data		
Sea water	R	R		R	R		R	R	R	R	R	R	R	R	R	R	R	R	R	R	R	R	R		R	R	
Silicic acid	No data			R	R		R	ND	ND	No data			R	R	R	R	R	R	R	R	R	R	R		No data		
Silicone fluids	R^4_{30}	R		No data			R	ND	ND	R	ND	ND	R	R	R	R	R	R	R	R	R	No data			No data		
Silver nitrate	R	R		R	R^{64}		R	R	R	R	ND	ND	R	R	R	R	R	R	R	R	R	R	R		No data		
Sodium carbonate	R	R		R	R		R	R	R	R	ND	ND	R	R	R	R	R	R	R	R	R	R	R		R	ND	
Sodium peroxide	R^4			R	ND								R	R	R	R	R	R	R	R	R	R	R		R	R	

THERMOPLASTIC RESINS															THERMOSETTING RESINS														
Polyethylene Low Density			Polyethylene High Density			Polycarbonate Resins			Polypropylene			Polystyrene			Melamine Resins (o)			Furane Resin			Epoxy Resins (p)			Phenol Form-aldehyde Resins (r)			Polyester Resins		
20°	60°	100°	20°	60°	100°	20°	60°	100°	20°	60°	100°	20°	60°	100°	20°	60°	100°	20°	60°	100°	20°	60°	100°	20°	60°	100°	20°	60°	100°
R	R		R	R		R			R	R	ND	R[13]			R	R		No data			R	R[4,30]		R	R	R	R	R	R[65]
No data			R	R		ND			R[30]	R	ND	No data			R	R		No data			R	R		No data			No data		
R	R		R[56]	R		R			R	R	R[30]	R	R		R	R		R	R	R	R	R		R	R	R	R	R	R[65]
R	R		R[56]	R		R			R	R	ND	R	R		R	R		R	R	R	R	R	R	R	R	R	R	ND	ND
No data			R[56]	R		ND			No data			No data			R			ND	ND	ND	R	R[4,30]		R	R	R	No data		
No data			R[56]	R		ND			R[13]	ND	ND	No data			ND	ND		R	R		R	R		No data			No data		
R	R		R	R		ND			R[56]	R[27]	R[27]	No data						R	ND	ND	R	R[4,30]		R	ND	ND	No data		
R	R		R	R		R	R		R	R	R[50]	R			R			R	R	R	R	R[4,30]		R	R	R	R[30]	R	R[65]
R	R		R[14]	R[14]					R	R		R									R[4,30]			R	ND	ND			
R	R		R	R		ND			R	R	ND				No data			No data			R[4,30]			No data			No data		
R	R		R	R[80]		R[32]			R	ND	ND										No data			R	ND	ND	R[10]		
R	R		R	R[80]		ND			R[56]	ND											No data								
R			R	R		R						R			R						R[30]	ND	ND	R	ND	ND	R		
R			R			ND						No data									R[4,30]	ND	ND						
R	R		R[13]	R		ND			R	R	ND	R			R			R	R	ND	R	R[30]		R	ND	ND	R	R[65]	
R	R		R	R		R[7,34]			R	R	R	R			R[10]						R[4,30]			R			R	ND	
R	R		R[56]	R[56]		R	R		R	R	ND	R			R			No data			R[4,30]			R			R[30]	R	
R	R		R	R		ND			R[7]	R	ND	R	ND	ND	ND	ND		R	R	R	R	R		R	ND	ND	R	R	R[65]
R	R		R	R					R[7,11]	R	R	R			R	R		R	R	R	R	R		No data			R	R[30]	ND
R	R		R[19]	R		ND			R	R	R	R	R	ND	R			R	R	R	R	R		R			R	R[13]	R
No data			R	R		ND			R	R	R	R	ND	ND	R	R		No data			R[30]	R[30]		No data			No data		
R	R		R	R		R	R		R[6]	R	ND	R	R		No data			No data			R	R		R	R		R	R	R[30]
R	R		R	R		R			R	R	R	R	R		R	R		R	ND	ND	R	R		No data			R	R	R
R	R		R	R		R			R	R	R	R	R		R	R					R	R		R	R	R	R	R	R
R	R		R	R		R			R	R	R	R	R		R	R	R	R	R	R	R	R		R	R	R	R	R	R
R	R		R	R		ND			R	R	ND	No data			R	R		No data			R	R		R	R	R	R	R	R
			R	R[80]		ND			R[30]	ND	ND				R	R		R	R	R	R	R	R	R	R	R	R	R[30]	
			R	ND		ND			R	R	R				R	R		R	R	R	R	R	R	R	R	R	R	R[30]	
R	R		R	R		ND			R	R	R[30]	R	R		No data			No data			R	R		R	R	R	R	R	R[65]
R	R		R	R		R[7,75]			R[7]	R	ND	R[34]	R[34]	ND	R[23]	R		R	R	R	R	R		R[75]			R	R[30]	R[65]
R	R		R[80]	R		R			R	R	R				R[10]						R[10]						R		
R[50]			R[80]																										
R[50]			R[50]			ND			R	R	R[50]				R	R		No data			R	R		R	R	R	No data		
R[50]	R		R[50]	R		ND			R[30]						R	R		R	R	R	R	R		R	R	R	R	R	R
R[50]			R[62]			R	R		R	R		R			R	R		No data			R	R		R	R	R	No data		
R	R		R[56]	R		R			R	R		R	R		R			R	ND	ND	R[30]	ND		R	ND	ND	R	R[30]	
						R			No data						No data			No data			No data			No data			No data		
ND	ND		R	R[50]		ND	R		R	R	ND	R			R	R		R	R	R	R	R		R	R	R	R	R	R
R[10]	R[10]		R	R[66]		R	ND														R[44,30]						R[10]		
			R	R					R	R	ND	R[10]									R[44,30]			R	R				
R	R		R	R		R	R		R	R	R	R	R		R			R	R	R	R	R[30]		R	R		R	R	R[65]
R	R		R	R		R	R		R	R	ND	R	R		R			No data			R[4,30]			R	R		R	R[30]	R[65]
R[50]			R[50]			R	R		R	R	ND	No data			R			ND			R[4,30]			R	R		No data		
R	R		R	R[50]					R	ND	ND	No data						R	R	ND	No data			R	ND	ND	No data		
R	R		R	R		ND			R	R	ND	No data						No data			R[30]			No data			No data		
No data			R	R		ND			R	R	ND	No data			R			R	R	R	R	R[30]		R	ND	ND	R	R	R[65]
R[13]			R	R		ND			R	R	ND	No data			R	ND		No data			R	ND		R	ND	ND	R[10]	ND	ND
No data			R	R[80]					R	ND	ND	No data			R	ND	ND				R	R							
R	R		R	R		R			R	R	R	R	R		R			R	R	R	R	R		R	R	R	R	R	R
R	R		R	R		R	ND		R	R	ND	R	R		R			No data			R	R		No data			No data		
R[50]			R[56]	R[56]		R	R	R	R	R	ND	No data			R	R		R	R	R	R	R		No data			No data		
R	R		R	R		R	ND		R	R	ND	R	R		R			No data			R	R		R	R		R	R[30]	R[65]
R	R		R	R					R	R	ND	R			R[10]	R		R	R	R	R	R		R	R		R	R	
R	R		R[13]	ND		ND			No data			No data			R	R		No data			R	R		R			No data		

Vertical footnote (between Polycarbonate and Polypropylene columns): These results at 20°C refer to low stress moulded parts made from Makrolon. The chemical resistance of this material may be affected by mechanical stresses and high temperatures. The polymer may however be used at relatively high temperatures for a thermoplastic: it has good heat resistance up to 135°C. Makrolon is the polycarbonic acid ester of 4,4'-dihydroxydipheny-2-2'-propane.

THERMOPLASTIC RESINS

	Acrylic Sheet (e.g. Perspex)			Acrylonitrile Butadiene Styrene Resins (l)			Nylon 66 Fibre (m)			Nylon 66 Plastics (m)			PCTFE			PTFE (n)			PVDF (y)			Rigid Unplasticised PVC			Plasticised PVC		
Centigrade	20°	60°	100°	20°	60°	100°	20°	60°	100°	20°	60°	100°	20°	60°	100°	20°	60°	100°	20°	60°	100°	20°	60°	100°	20°	60°	100°
Sodium silicate	R	R		R			R	R	R	R	R	R	R	R	R	R	R	R	R	R	R	R	R		R	R	
Sodium sulphide	R	R^{68}		R			R	R	ND	R	ND	ND	R	R	R	R	R	R	R	R	R	R	R		R	R	
Stannic chloride	R^{68}	R								R^{50}	ND	ND	R	R	R	R	R	R	R	R	R	R	R		R	R	
Starch	R	R		R			R	R	R	R	R	R	R	R	R	R	R	R	R	R	R	R	R		R	R	
Sugar, syrups, jams	R	R		R	R		R	R	R	R	R	R	R	R	R	R	R	R	R	R	R	R	R		No data		
Sulphamic acid	No data			No data			R^{43}	R	R	No data			No data			No data			R	R	R	No data			No data		
Sulphates (Na, K, Mg, Ca)	R	R		R	R		R	R	R	R	R	R	R	R	R	R	R	R	R	R	R	R	R		R	R	
Sulphites	R	R		R			No data			No data			R	R	R	R	R	R	R	R	R	R	R		No data		
Sulphonic acids	No data			No data									No data			R	R	R	R	R	R	R	R^{50}		No data		
Sulphur	R	R^{68}		R	ND		No data			R	ND	ND	R	R	R	R	R	R	R	R	R	R	R		No data		
Sulphur dioxide, dry	R	R^{68}		R			No data			R	ND	ND	R	R	R	R	R	R	R	R	R	R	R		R	R	
Sulphur dioxide, wet	R	R^{68}		R			No data			R^{50}			R	R	R	R	R	R	R	R	R	R	R^{50}		ND		
Sulphur trioxide	No data			R	R								No data			R	R	R	R	R	R	R^{13}	R		No data		
Sulphuric acid (< 50%)	R^{25}	R^{32}											R	R	R	R	R	R	R	R	R	R	R		R	ND	
Sulphuric acid (70%)													R	R	R	R	R	R	R	R	R	R	R		No data		
Sulphuric acid (95%)													R	R	R	R	R	R	R	R	R	R	R^{50}		No data		
Sulphuric acid, fuming													R	R	R	R	R	R	R								
Sulphur chlorides	No data			No data						No data			No data			R^{30}	R	R	No data			ND			No data		
Tallow	R^{68}	R		R	R		R	R	R	R	ND	ND	R	R	R	R	R	R	R	R	R	R	R		R	ND	
Tannic acid (10%)	R	ND		No data			R	ND	ND	R	ND	ND	R	R	ND	R	R	R	R	R	R	R	R		R	ND	
Tartaric acid	R	R		R	R		R	ND	ND	R	R^{50}	ND	R	R	ND	R	R	R	R	R	R	R	R		R	ND	
Trichlorethylene				R	R		R	R	R	R	R	R^{50}				R^{14}	R	R	R	R	R						
Vinegar	R	R^{68}		R	R					R^{50}	ND	ND	R	R	R	R	R	R	R	R	R	R	R		R	ND	
Water, distilled	R	R		R	R		R	R	R	R	R^{50}	R	R	R	R	R	R	R	R	R	R	R	R		No data		
Water, soft	R	R		R	R		R	R	R	R	R^{50}	R	R	R	R	R	R	R	R	R	R	R	R		R	R	
Water, hard	R	R		R	R		R	R	R	R	R^{50}	R	R	R	R	R	R	R	R	R	R	R	R		R	R	
Yeast	R	R^{68}		R			R	R	R	R	ND	ND	R	R	R	R	R	R	R	R	R	R	ND		R	ND	
Zinc chloride	R	R^{68}		R	R		R^{43}	R	R				R	R	R	R	R	R	R	R	R	R	R		R	R	

THERMOPLASTIC RESINS / THERMOSETTING RESINS

Polyethylene Low Density			Polyethylene High Density			Polycarbonate Resins			Polypropylene			Polystyrene			Melamine Resins (o)			Furane Resin			Epoxy Resins (p)			Phenol Formaldehyde Resins (r)			Polyester Resins		
20°	60°	100°	20°	60°	100°	20°	60°	100°	20°	60°	100°	20°	60°	100°	20°	60°	100°	20°	60°	100°	20°	60°	100°	20°	60°	100°	20°	60°	100°
R	R		R	R		ND			R	R	R	No data			R			R	R	R	R	R		R			R	R	R[65]
R	R		R	R		No data			R	R	R	No data			R			R	R	R	R	R		No data			R	R	R[65]
R	R		R	R		No data			R[7]	R	R	R[1]	R		ND	ND		R	R	ND	R[68]			R	ND	ND	No data		
R	R		R	R		R	ND		R	R	R	R	R		R	R		No data			R	R		R	R	R	No data		
R	R		R	R		R			R	R	ND	R	R		R	R		No data			R	R		R	R	R	No data		
ND	ND		No data			No data			R	R	ND	No data			R	ND	ND	R	ND	ND	No data			No data			No data		
R	R		R	R		R	R	ND	R	R	ND	R	R		R	R		R	R		R	R[30]		R	R	R	R	R[30]	R[65]
R[34]			R	R		No data			R	R	ND	R	R		R	R		R	R		R	R		R	R		R	R[30]	R[65]
No data			No data			No data			No data			No data			No data			R	R	ND	R	R[44]		No data			No data		
R	R		R	R		No data			R	R	ND	No data			R			R	R	R	R	R		R	ND	ND	R	ND	ND
R			R	R		ND			R	R	ND				R			R	R	R	R			R	ND	ND	R[30]	ND	ND
						ND			No data												R[11]			R	ND	ND	R[30]	ND	ND
R	R		R	R		R	R		R	R		R			R[10]			R	R	R	R	R[30]		R	R	R	R[30]	R	
R	R[50]		R	R[50]		R	R	ND	R												R[30]						R[30]	R	
R[50]			R[50]	R[80]					R[56]																				
			R[60]			R[60]																							
ND	ND		No data			No data			No data			No data			ND	ND		R	R		No data			R	ND	ND	No data		
R			R[50]	ND	ND	R	ND		R	R	ND	R			R	R		No data			R	R		R	R	R	No data		
R	R		R[56]	R		ND			R	R	ND				R	R		No data			R	R[30]		R	R	R	R	R	R
R	R[10]		R	R		R	R		R	R	ND	R	R		R			No data			R	R[30]		R	R	R	R	R	R
			R[50]												R	R		R	R	R	R	R	R				R[30]		
R	R		R	R		No data			R	R	ND	R			R			No data			R	R[4]		R	ND	ND	R[30]		
R	R		R	R		R	R	R	R	R	R	R	R		R	R		R	R	R	R	R		R	R	R	R	R	R[30]
R	R		R	R		R	R	R	R	R	R	R	R		R	R		R	R	R	R	R		R	R	R	R	R	R[30]
R	R		R	R		R	R	R	R	R	R	R	R		R	R		R	R	R	R	R		R	R	R	R	R	R[30]
R	ND		R	R		ND			R	R	ND	No data			R	R		No data			R	R		No data			No data		
R	R		R	R		R	R		R	R	ND	R	R		R			R	R	R	R	ND		R	ND	ND	R	R[30]	R[65]

RUBBERS

	Butyl Rubber and Halo-Butyl Rubber			Ethylene Propylene Rubber (q)			Hard Rubber (Ebonite) (h)			Soft Natural Rubber (h)			Neoprene (i)			Nitrile Rubber			Chlorosulphonated Polyethylene			Polyurethane Rubber (v)			Silicone Rubbers (k)		
	20°	60°	100°	20°	60°	100°	20°	60°	100°	20°	60°	100°	20°	60°	100°	20°	60°	100°	20°	60°	100°	20°	60°	100°	20°	60°	100°
Acetaldehyde	R	R	ND	R	R	ND	R	R	R	R[80]	R[80]	ND										ND	ND		R	R	R
Acetic acid (10%)	R[14]	R	R	R	R[14]	ND	R	R	R				R	R	R[14]	R	R		R	R	ND	R[80]	R[80]		R	R	R
Acetic acid (glac. & anh.)	R[14]	R	R	R[14]	R[14]	ND	R	R[14]	R				R[95]			R[4]			R[85]			R[80]			R[17]	R	R
Acetic anhydride	R[80]	R	R		No data		R	R[30]					R	R	ND				R	R	ND				R	R	R
Acetone	R	R		R[60]	R[60]		R	R	R	R[60]	R	ND							R[15]	ND	ND				R[17]	R	R
Other ketones	R[13]	R	R	R[60]	R[60]		R[13][80]	R	R	R[30][60]	R														R[17]	R	R
Acetylene	R	R[80]			No data		R[80]	R	R				R[14]	R	R	R	ND	ND	R[14]	R	R	ND	ND			No data	
Acid fumes	R[2]	R	R	R[2]	R[2]	R[2]	R[2]	R	R	R[2]	R	R[2,80]	R[2]	R	R	R[2]			R	R	R[2]	R[2]	R[2]		R[2]	R	R
Alcohols (most fatty)	R	R		R[60]	R[60]		R[30][60]	R	R	R[60]	R		R	R	R[14]	R	R	R	R	R	R	R[4]	R[4]		R	R	R[30]
Aliphatic esters														No data											R[30]	R	R
Alkyl chlorides																									R[21]	R	R
Alum	R	R	R	R	R	R	R	R	R	R	R	R	R	R	R	R	R	R	R	R	R	R	R		R	R	R
Aluminium chloride	R	R	R	R	R	R	R	R	R	R	R	R	R	R	R	R	R	R	R	R	R	R	R		R	R	R[4]
Ammonia, anhydrous	R	R	ND	R	R	ND	R	R	R	R[80]			R	R	R	R	R		R[10]	ND	ND	R[80]			R	R	R
Ammonia, aqueous	R	R	R	R	R	R	R	R	R	R	R	R[80]	R	R	R	R	R		R	R	R	R[30]	R[80]		R	R	R
Ammonium chloride	R	R	R	R	R	R	R	R	R	R	R	R	R	R	R	R	R	R	R	R	R	R	R		R	R	R
Amyl acetate	R[80]																								R[21]	R	R
Aniline	R	R	ND																						R	R	R
Antimony trichloride	R	R	R		No data		R	R	R					No data			No data		R	R	R	ND	ND		R	R	R
Aqua regia																			R[50]								
Aromatic solvents																R[62]	R								R[21]	R	R
Beer	R	R	R	R	R	R	R	R	R	R	R	R	R	R	R	R	R	R	R[86]	R	R	R	R		R	R	R
Benzoic acid	R	R	R	R	R	R	R	R	R	R	R	R	R	R	R	R	R	R		No data		R	R		R	R	R
Boric acid	R	R	R	R	R	R	R	R	R	R	R	R	R	R	R	R	R	R	R	R	R	R	R		R	R	R
Brines, saturated	R	R	R	R	R	R	R	R	R	R	R	R	R	R	R	R	R	R	R	R	R	R	R		R	R	R
Bromine					No data																						
Calcium chloride	R	R	R	R	R	R	R	R	R	R	R	R	R	R	R	R	R	R	R	R	R	R	R		R	R	R
Carbon disulphide																R	ND	ND				R	R		R	R	R
Carbonic acid	R	R	R	R	R	R	R	R	R	R	R	R	R	R	R	R	R	R	R	R	R	R	R		R	R	R
Carbon tetrachloride																									R[21]	R	R
Caustic soda & potash	R	R	R	R	R	R	R	R	R[13]	R	R	R	R	R	R	R	R	R	R	R	R	R[30]	R[30]		R	R	R[30]
Chlorates of Na, K, Ba	R	R	R	R	R	R	R	R	R	R	R	R		No data			No data		R	R	R	R	R		R	R[30]	R
Chlorine, dry	R[50]	R	R	R[50]	R[50]	R[50]	R[30]	R	R											No data					R	R	R
Chlorine, wet	R[80]	R	R	R[50]	R[50]	R[50]	R[13]	R	R										R[3]	ND	ND				R	R	R
Chlorides of Na, K, Mg	R	R	R	R	R	R	R	R	R	R	R	R[80]	R	R	R	R	R	R	R	R	R	R	R		R	R	R
Chloroacetic acids	R[10]						R[2]	R	R				R	R	ND				R			R[80]			R	R	R
Chlorobenzene																									R[21]	R	
Chloroform																											
Chlorosulphonic acid	R[13]	R[13]		R[13]	R[13]		R[13]	R	ND											No data		R[30]	ND			No data	
Chromic acid (80%)																			R[30]	R	ND				R[19]	R	R
Citric acid	R	R	R	R	R	R	R	R	R				R	R	R	R	R	R	R	R	R	R	R		R	R	R
Copper salts (most)	R	R	R	R	R	R	R	R	R	R	R	R	R	R	R	R	R	ND	R	R	R	R	R		R	R	R
Cresylic acids (50%)	R[4]				No data														R			ND	ND		R[21]	R	R
Cyclohexane																R	R	R							R[21]	ND	ND
Detergents, synthetic	R[13]	R	R	R[13]	R[13]	R[13]	R	R	R	R[80]	R[80]	R	R	R	R	R	R	R	R	R	R	R[30]	R[30]		R	R	R
Emulsifiers (all conc.)	R	R	R		No data		R	R[4]	R[4]		No data		R[30]	R	R	R	R	R	R[30]	R	R	ND	ND		R	R	R
Ether																R											
Fatty acids (>C6)	R[1]	R[80]	R	R[80]	R[80]		R[80]	R[13][80]					R	R	R	R	R[4]		R	R	R	R[4][30]	R[80]		R	R	R
Ferric chloride	R	R	R	R	R	R	R	R	R	R[80]	R	R	R	R	R	R	R	R	R	R	R	R	R		R	R	R
Ferrous sulphate	R	R	R	R	R	R	R	R	R	R	R	R	R	R	R	R	R	R	R	R	R	R	R		R	R	R
Fluorinated refrigerants					No data																	ND	ND				
aerosols, e.g. *Freon*	R[4]	ND	ND		No data			No data			No data		R[30]	R	ND	R[30]	R	R	R[30]	ND	ND	ND	ND		R[4,21]		
Fluorine, dry	R[80]	ND	ND	R[80]	ND	ND	R[13]	R[13]	ND					No data						No data							
Fluorine, wet	R[80]	ND	ND	R[80]	ND	ND	R[13]	R[13]	ND					No data						No data							
Fluosilicic acid	R	R	R	R	R	R	R	R	R	R	R	R	R	R	R		No data		R	R	R	R	R		R	R	R
Formaldehyde (40%)	R[80]			R[80]			R	R[30]	R	R[14]			R	ND	ND	R			R			ND	ND		R	R	R
Formic acid	R[13]	R	R	R[14]	R[14]		R	R[80]		R[80]			R	R	R	R			R	R	R	R[80]	R[80]		R	ND	ND

MISCELLANEOUS

Concrete (s)			Glass (t)			Graphite (u)			Porcelain and Stoneware			Vitreous Enamel (w)			Wood (z)		
20°	60°	100°	20°	60°	100°	20°	60°	100°	20°	60°	100°	20°	60°	100°	20°	60°	100°
No data			No data			R	R	R	R	R	R	R	R	R	R	R	R
			R	R	R	R	R	R	R	R	R	R	R	R	R	R	
			R	R	R	R	R	R	R	R	R	R	R	R			
			No data			R	R	R	R	R	R	R	ND	ND			
R	R	R	R	R	R	R	R	R	R	R	R	R	R	R	R	R	R
R	R	R	No data			R	R	R	R	R	R	R	R	R	R	R	R
R	R	R	R	R	R	R	R	R	R	R	R	R	R	R	No data		
			R[5]	R	R	R	R	R	R	R	R	R[5][11]	R	ND			
R	R	R				R	R	R	R	R	R	R	R	R	R	R	
			R	R	R	R	R	R	R	R	R	R	R	R	R	R	R
R	R	R	R	R	R	R	R	R	R	R	R	R	R	ND	R	R	
			R	R	R	R	R	R	R	R	R	R	R	R	R	R	
			R[30]			R	R	R	R	R	R	R	R	ND			
No data			No data			R	R	R	R	R	R	R	R	R			
R	R	R	R	R	R[50]	R	R	R	R	R	R	R	ND				
			R	R	R	R	R	R	R	R	R	R	ND		R	R	
			R	R	R	R	R	R	R	R	R	R	R	R	R	R	R
No data			R			R	R	R	R	R	R	R	R	R	R	R	
No data			R	R	R	R	R	R	R	R	R	R	R	R			
			R	R	R				R	R	R	R	R	ND			
R	R	R	R	R	R	R	R	R	R	R	R	R	R	R	R	R	R
R	R	R[35]	R	R	R	R	R	R	R	R	R	R	R	R	R	R	R
No data			R	R	R	R	R	R	R	R	R	R	R	ND	R	R	R
R	R	R	R	R	R	R	R	R	R	R	R	R	R	ND	R	R	R
R	R	R	R	R	R	R	R	R	R	R	R	R	R	ND	R	R	
			R	R	R				R	R	R	No data					
R[44]	R	R	R	ND	ND	R	R	R	R	R	R	R	ND		R	R	
R	R	R	ND	ND		R	R	R	R	R	R	R	R	R			
R[50]	R	R	R	R	R	R	R	R	R	R	R	R	R	R	R	R	R
R	R	R	R	R	R	R	R	R	R	R	R	R	R	R	R	R	R
R[72]	R	R	ND			R[13]	R	R	R[10]			R[10]					
R	R	R	R	R	R	R	R	R	R	R	R	No data			R	ND	ND
No data			No data			R	R	R	R	R	R	R	R	R			
				ND					R	R	R	No data					
R[12]	R	R	R	R	R	R	R	R	R	R	R	R	R	R	R	R	
			R	R	R	R	R	R	R	R	R	R	ND	ND			
R	R	R	R	R	R	R	R	R	R	R	R	R	R	R	R	R	
R	R	R	R	R	R	R	R	R	R	R	R	R	R	R	R	R	R
			No data			R	R	R	R	R	R	No data					
			R	R	ND				R	R	R	R	ND	ND			
			R	R	ND	R	R	R	R	R	R	R	R	R	R	R	
R[51]	R	R	R	R	R	R	R	R	R	R	R	R	R	R	R	R	R
			No data			R	R	R	R	R	R	R	R	ND			
R	R	R	No data			R	R	R	R	R	R	R	R	R	R	R	R
R[81]	R	R	R[13]	R[30]		R	R	R	R	R	R	R	R	R	R	R	
No data			R	R	R	R	R	R	R	R	R	R	R	R	R	R	
R	R	R	R	R	R	R	R	R	R	R	R	R	R	R	R		
			R	R	R	R	R	R	R	R	R	R	R	R	R	R	R
			R	R	R	R[13]	R	R	R	R	R	R	ND		R[14]		
			R	R	R	R	R	R	R	R	R	R	ND		R[14]		
R	R	R	No data			R	R	R	R[39]	R	R	No data			R	R	R
No data						R	R	R	R	R	R	R	ND		No data		
No data						R	R	R	R	R	R	No data			No data		
R	R	R				R	R	R							No data		
R[80]			R	R	R	R	R	R	R	R	R	R	R	R	R	R	
			R	R	R	R	R	R	R	R	R	R	R	R	R	R	R

NOTES

Explanatory notes at lower temperatures may be taken to apply also at higher temperatures unless otherwise shown.

1 Not anhydrous
2 Depending on the acid
3 35%
4 Fair resistance
5 Not HF fumes
6 Up to 40%
7 Saturated solution
8 Pineapple and grapefruit juices 20°C
9 Photographic emulsions up to 20°C
10 10%
11 Anhydrous
12 Not Mg
13 Depending on concentration
14 Discoloration and/or swelling and softening
15 Up to 25%
16 Not chloride/not if chloride ions present
17 Not fluorinated silicone rubbers
18 Up to 60%
19 Up to 50%
20 Not aerated solutions
21 Fluorinated silicone rubbers only
22 ND for Mg
23 5%
24 Pure only
25 Up to 30%
26 If no iron salts or free chlorine
27 May crack under stressed conditions
28 45%
29 55%
30 Depending upon composition
31 Chloride
32 20%
33 Depending on alcohol
34 Data for sodium
35 Fresh
36 Over 85%
37 Some attack at high temperature
38 Neutral
39 Attacked by fluoride ions
40 Sulphate and nitrate
41 Softening point
42 In strong solutions only when inhibited
43 Depending on water conditions
44 Dilute
45 Up to 15%
46 Not methyl
47 Drawn wire
48 Some attack, but protective coating forms
49 Using anodic passivation techniques
50 Some attack/absorption/slow erosion
51 Not sulphate
52 70%
53 In absence of dissolved O_2 and CO_2
54 75%
55 80%
56 May cause stress cracking
57 Pitting possible in stagnant solutions
58 In presence of H_2SO_4
59 Not ethyl
60 May discolour liquid
61 The material can cause decomposition
62 Depending on type
63 95%
64 Slight plating will occur
65 Not recommended under certain conditions of temperature, etc.
66 65%
67 Aerated solution
68 Estimated effect
69 Up to 90%
70 Not oxidising conditions
71 Not lower members of series
72 Not high alumina cement concrete

RUBBERS

	Butyl Rubber and Halo-Butyl Rubber			Ethylene Propylene Rubber (q)			Hard Rubber (Ebonite) (h)			Soft Natural Rubber (h)			Neoprene (i)			Nitrile Rubber			Chlorosulphonated Polyethylene			Polyurethane Rubber (v)			Silicone Rubbers (k)		
	20°	60°	100°	20°	60°	100°	20°	60°	100°	20°	60°	100°	20°	60°	100°	20°	60°	100°	20°	60°	100°	20°	60°	100°	20°	60°	100°
Fruit juices	R^{80}	R^{80}		R^{60}	R^{60}		R^{65}	R	R	R	R	R	R	R	R	R	R	R	R	R	R	R	R		R	R	R
Gelatine	R	R	R	R	R		R	R	R	R	R	R	R	R	R	R	R	R	R	R	R	R	R		R	R	R
Glycerine	R	R	R	R	R	R	R	R	R	R	R	R	R	R	R	R	R	R	R	R	R	R	R		R	R	R
Glycols	R	R	R	R	R	R	R	R	R	R	R	R	R	R	R	R	R	R^4	R	R	R	R	R		R	R	R
Hexamine	R	R	ND	No data			R	R	ND	R	R	ND	R	R^{37}	R	No data			R	R	R	ND	ND		No data		
Hydrazine	R	R	ND	R	ND	ND	R						R^{44}	ND	ND	R			No data			ND	ND		No data		
Hydrobromic acid (50%)	R	R	R	R	R	ND	R	R^{37}	R^{37}	R^{65}			R	ND	ND	R			R	R	ND	R^{15}	R^{15}_{80}				
Hydrochloric acid (10%)	R	R	R	R	R	R	R	R	R	R	R	R	R	R	R	R	R		R	R		R	R		R	R	R
Hydrochloric acid (conc.)	R^4	R	R	R	R^4	R^{80}	R	R^{37}	R^{37}	R	R^{80}		R	R^{95}	ND				R	R					R^{30}		
Hydrocyanic acid	R	R	R	R	R	R	R	R	R	R	R	R	R	R	R	R			R	R	ND	R	R		No data		
Hydrofluoric acid (40%)	R^{30}	R	R	R^{30}	R^{30}	ND	R^{30}	R	R	R^{30}	R^{30}		R	ND	ND				R	R	ND						
Hydrofluoric acid (75%)				R^{30}	ND	ND							R						R	R	ND						
Hydrogen peroxide (30%)	R^{80}	R	R	R^{80}	ND	ND	R^{80}						R	R		R			R^{87}	R^{87}		R^4	ND		R	R	R
(30–90%)	R^{80}	R^{87}	ND				R^{87}						R	R					R^{87}	R^{87}					R^{30}	R	R
Hydrogen sulphide				R	R	R	R	R	R	R	R		R	R	R	R	R	ND	ND	ND	R	R	R		No data		
Hypochlorites	R^{30}	R^{80}	R	R	R	ND	R^{30}_{13}	R		$R^{80,76}_{13}$			R	R		R			R	R	R	R^{30}	ND				
Lactic acid (100%)	R	R	ND	R	R	ND	R	R	R	$R^{14,80}$			R	R	R	R	R		R	R	R	R^{23}	R^{23}		R	R	R
Lead acetate	R	R	R	R	R	R	R	R	R	R^{80}	R	R	R	R	R	R	ND	ND	No data			R	R		R	R^{30}	R
Lime (CaO)	R	R	R	R	R	R	R	R	R	R	R	R	R	R	R	R	R	R	R	R	R	R	R		R	R	R
Maleic acid	R	R	R	R	R	R	R	R	R	R	R	R	R	R	R	ND			No data			R	R		R	R	R
Meat juices	R	R	R	R	R	R	R^{13}	R	R	R^{13}	R	R	R	R	R	R	R	R	R	R	R	R	R		R	R	R
Mercuric chloride	R	R	R	R	R	R	R	R	R	R	R	R	R	R		R	R	R	R	R	R	R	R		R	R	R
Mercury	R	R	R	R	R	R	R	R	R	R	R	R	R	R		R	R	R	R	R	R	R	R		R	R	R
Milk & its products	R^{80}	R	R	R^{80}	R^{80}		R	R	R	R	R	R	R	R	R	R	R	R	R	R	R	R	R		R	R	R
Moist air	R	R	R	R	R	R	R	R	R	R	R	R	R	R	R	R	R	R	R	R	R	R	R		R	R	R
Molasses	R	R	R	R	R	R	R	R	R	R	R	R	R	R	R	R	ND	ND	R^{86}	R	R	R	R		R	R	R
Naphtha													R	R		R	R		R^{38}						R^{21}	R	R
Naphthalene																									R^{21}	R	R
Nickel salts	R	R	R	R	R	R	R	R	R	R	R	R	R	R	R	R	R	R	R	R	R	R	R		R	R	R
Nitrates of Na, K, NH_3	R	R	R	R	R	R	R	R	R	R	R	R	R	R	R	R	R	R	R	R	R	R	R		R	R	R
Nitric acid (< 25%)	R^{23}	R^{23}	R^{23}				R^{101}	R		R^{101}			R	R		R			R	R					R	R	R
Nitric acid (50%)																			R	R					R^{21}		
Nitric acid (95%)																			R						R		
Nitric acid, fuming																			R^{89}								
Oils, essential	R^{14}	R	ND	R^{60}	ND	ND	R^{14}			No data			R^{14}	ND	ND	R^4	R^4		R^{30}			R	R		R^{30}	R	R
Oils, mineral				R^{80}	R^{80}	R^{80}							R	ND	ND	R	R	R	R^{30}			R	R		R^{30}	R	R
Oils, vegetable & animal	R	R^{14}		R^{14}	R^{14}		R^{80}	R	R	R^{14}	R		R^{14}	ND	ND	R	R	R	R^{30}			R	R		R	R	R
Oxalic acid	R	R	R	No data			R	R	R	R^{80}	R	R	R	R	R	R	R		ND	ND		ND	ND		R	R	R
Ozone	R	R		R	R		R	R	R	R	R	R	R	R	R	R^{30}			R	R	R	R	R		R	R	R
Paraffin wax	R	R	R	R	R	ND	R	R	R	R	R^{14}		R	R	R	R	R	R	R	R	R	R	R		R	R	R
Perchloric acid	R			R															No data			ND	ND		No data		
Phenol	R	R^{80}	ND	R^{80}			R^{13}_{80}	R					R	R	R	R	R		R	R	R				R	R	R
Phosphoric acid (25%)	R	R	R	R	R	R	R	R	R^{60}	R	R	R^{60}	R	R	R	R	R	ND	R	R	R	R	R		R	R	R
Phosphoric acid (50%)	R	R	R	R	R	R	R	R	R^{60}	R	R	R^{60}	R	R	R	R			R	R	R	R	R		R	R	R
Phosphoric acid (95%)	R	R	R	R	R	R	R^{36}	R	R^{60}	R^{36}	R	R^{60}	R	R	R	R			R	R	R				R^{30}	R	R
Phosphorus chlorides	R			No data			No data			No data			No data									ND	ND		No data		
Phosphorus pentoxide	R	R	ND	R	R	ND	R	R		R	R		R	R	R				No data			ND	ND		R	R	R
Phthalic acid	R^{13}	R		R^{13}	R^{13}	ND	R^{80}	R^{80}	R^{80}				R	R	R	R	R	R	No data			R	ND		R	R	R
Picric acid	R^{80}	R	R	No data			R	R^{30}	R	R	R^{30}		R	R	R	R^{13}			R	R	R	ND	ND		No data		
Pyridine	R^4																								No data		
Sea water	R	R	R	R	R	R	R	R	R	R	R	R	R	R	R	R	R	R	R	R	R	R	R		R	R	R
Silicic acid	R	R	R	R	R	R	R	R	R	R	R	R	R	R	R	ND			R	R	R	R	R		R	R	R
Silicone fluids	R	R	ND	No data			R	R	R	R	R	R	R	R	R	R	R	R	R	R	R	R	ND		R^{21}_{30}	R	R
Silver nitrate	R	R	R	R^{60}	R^{60}	R^{60}	R^{61}			R^{80}	R		R	R	R	R	R	ND	R	R	R	R	R		R	R	R
Sodium carbonate	R	R	R	R	R	R	R	R	R	R	R	R	R	R	R	R	R	R	R	R	R	R	R		R	R	R
Sodium peroxide	R	R	R	R	R	ND	R^{13}	R		R^{80}	R^{80}	ND	R	R^{97}	ND	R^{13}			R	R	R	ND	ND		R	R	R

MISCELLANEOUS

Concrete (s)			Glass (t)			Graphite (u)			Porcelain and Stoneware			Vitreous Enamel (w)			Wood (z)		
20°	60°	100°	20°	60°	100°	20°	60°	100°	20°	60°	100°	20°	60°	100°	20°	60°	100°
			R	R	R	R	R	R	R	R	R	R	R	R	R	R	R
R	R	R	R	R	R	R	R	R	R	R	R	R	R	R	R	R	R
			R	R	R	R	R	R	R	R	R	R	R	R	R		
			R	R	R	R	R	R	R	R	R	R	R	R	R		
No data			R	ND	ND	R	R	R	R	R	R	R	R	R	R	R	R
No data			R	R	R	R	R	R	R	R	R	R	R	R	No data		
			R	R	R^{50}	R	R	R	R	R	R	No data					
			R	R	R^{50}	R	R	R	R	R	R	R	R	R	R		
			R	R	R^{50}	R	R	R	R	R	R	R	R	ND			
			R	R		R	R	R	R	R	R	R	ND	ND	R	R	
						R	R	R									
						R	R	R									
No data			R	R	R	R	R	R	R	R	R	R	R	R			
No data			R	R	R	R	R	R	R	R	R	R	R	R			
						R	R	R	R	R	R	R	R	R	R	R	R
R^{72}	R	R	R	R	R	R	R^{37}	R^{37}	R	R	R	R	R	ND			
			No data			R	R	R	R	R	R	R	R	R	R		
No data			R	R	R	R	R	R	R	R	R	R	R	R	R	R	R
R	R	R	R	R	R	R	R	R	R	R	R	R	R	ND	R^{14}		
No data			No data			R	R	R	R	R	R	R	R	R	R	R	R
No data			R	R	R	R	R	R	R	R	R	R	R	R	R	R	R
			R	R	R	R	R	R	R	R	R	No data			R	R	
R	R	R	R	R	R	R	R	R	R	R	R	R	R	ND	R	R	R
R^{35}	R	R	R	R	R	R	R	R	R	R	R	R	R	R	R	R	R
R	R	R	R	R	R	R	R	R	R	R	R	R	R	R	R^{14}	R^{14}	R^{14}
R			R	R	R	R	R	R	R	R	R	R	R	R	R	R	
R	R	R	No data			R	R	R	R	R	R	R	R	R	R	R	R
R	R	R	No data			R	R	R	R	R	R	R	R	R	R	R	R
No data			R	R	R	R	R	R	R	R	R	No data			R	R	R
R^{73}	R	R	R	ND	ND	R	R	R	R	R	R	R	ND	ND	R	R	
			R	R	R^{50}	R	R	R	R	R	R	R	R	R			
			R	R	R^{50}				R	R	R	R	R	ND			
			R	R	R^{50}				R	R	R	R	R	ND			
			R	R	R^{30}				R	R	R	R	ND	ND			
No data			R	R	R	R	R	R	R	R	R	R	R	R	R	R	R
R	R	R	R	R	R	R	R	R	R	R	R	R	R	R	R	R	R
			R	R	R	R	R	R	R	R	R	R	R	R	R	R	R
R	R	R	R	R	R	R	R	R	R	R	R	R	R	R	R		
R	R	R	R	R	R				R	R	R	No data			No data		
R	R	R	R	R	R	R	R	R	R	R	R	R	R	R	R	R	R
			R	R	R	R^{15}	R^{10}		R	R	R	No data					
			R	ND	ND	R	R	R	R	R	R	R	R	R			
			R	R	R	R	R	R	R	R	R	R	R	R			
			R	R	R	R	R	R	R	R	R	R	R	ND			
			R	R	R	R	R	R	R	R	R^{4}	R	R	ND			
			R	ND	ND	R	R	R	R	R	R	R	R	R	No data		
			R	R	R	R	R	R	R	R	R	No data					
			R	R	R	R	R	R	R	R	R	R	R	R	R	R	
			R	R	R	R	R	R	R	R	R	R	R	R	No data		
No data			R	R	R	R	R	R	R	R	R	R	R	R	R^{14}		
R	R	R	R	R	R	R	R	R	R	R	R	R	R	R	R	R	R
R	R	R	R	R	ND	R	R	R	R	R	R	R	R	ND	R	R	R
R^{74}	R	R	R	R	R	R	R	R	R	R	R	R	R	R	R	R	R
No data			R	R	R	R	R	R	R	R	R	R	R	R	R	R	
R^{72}	R	R	R			R	R	R	R	R	R	R	R	R			
R^{72}	R	R	R	ND	ND	No data			R	R	R	No data					

73 Not ammonium
74 Not chlorsilanes
75 Data for ammonium
76 Data for calcium
77 Data for potassium
78 In presence of heavy metal ions
79 ND for Ba
80 Limited service
81 Except those containing sulphate
82 Provided less than 70% copper
83 Water less than 150 ppm
84 May cause some localized pitting
85 60% in one month
86 Low taste and odour
87 Catalyses decomp. of H_2O_2
88 65%
89 1–2 days
90 Wet gas
91 Less than 0·005% water
92 In absence of heavy metal ions oxidising agents
93 Stress corrosion in MeOH and halides (not in other alcohols)
94 When free of SO_2
95 50% swell in 28 days
96 60% swell in 3 days
97 Could be dangerous in black loaded compounds
98 Not alkaline
99 Ozone 2% Oxygen 98%
100 This is the softening point
101 Nitric acid less than 5% concentration
102 Acid fumes dry. Attack might occur if moisture present and concentrated condensate built up
103 Stainless steels not normally recommended for caustic applications
104 In the absence of impurities
105 10% w/w in alcohol
106 Swelling with some ketones
107 Some stress cracking at high pH

(a) **Aluminium:** In many cases where the chart indicates that aluminium is a suitable material there is some attack, but the corrosion is slight enough to allow aluminium to be used economically.

(b) **Brass:** Some types of brass have less corrosion resistance than is shown on the chart, others have more, e.g. Al brass.

(c) **Cast iron:** This is considered to be resistant if the material corrodes at a rate of less than 0·25 mm per annum. When choosing cast iron, Ni-Resist or high Si iron for a particular application the very different physical properties of these materials must be taken into account.

(d) **Gunmetal:** The data refer only to high tin gunmetals.

(e) **Nickel–copper alloys:** The physical properties are for annealed material. Both the tensile strength and hardness can vary with form and heat treatment condition.

(f) **Stainless steels:** Less expensive 13% chromium steels may be used for some applications instead of 18/8 steels. Under certain conditions the addition of titanium increases the corrosion resistance of 18/8 steels. Also, it produces materials which can be welded without the need for subsequent heat treatment. These steels are, however, inferior in corrosion resistance to the more expensive 18/8/Mo steels.

(g) **Tin:** Data refer to pure or lightly alloyed tin; not to discontinuous tin coatings.

(h) **Soft natural rubber and ebonite:** Performance at higher temperatures depends on method of compounding.

(i) **Neoprene:** Brush or spray applied 1·5 mm thick, and properly cured.

(k) **Silicone rubbers:** Withstand temperatures ranging from −90°C to above 250°C and are resistant to many oils and chemicals. In some cases particularly good resistance is shown by the fluorinated type.

RUBBERS

	Butyl Rubber and Halo-Butyl Rubber			Ethylene Propylene Rubber (g)			Hard Rubber (Ebonite) (h)			Soft Natural Rubber (h)			Neoprene (i)			Nitrile Rubber			Chlorosulphonated Polyethylene			Polyurethane Rubber (v)			Silicone Rubbers (k)		
	20°	60°	100°	20°	60°	100°	20°	60°	100°	20°	60°	100°	20°	60°	100°	20°	60°	100°	20°	60°	100°	20°	60°	100°	20°	60°	100°
Sodium silicate	R	R	R	R	R	R	R	R	R	R	R	R	R	R	R	R	R	ND	R	R	R	R	R		R	R	R
Sodium sulphide	R	R	R	R	R	R	R	R	R	R	R	R	R	R	R	R	R	ND	No data			R	ND		R	R	R
Stannic chloride	R	R	R	R	R	R	R	R	R	R	R	R	R	R	R	R	R	R	R	R	R	R	R		R	R	R
Starch	R	R	R	R	R	R	R	R	R	R	R	R	R	R	R	R	R	R	R	R	R	R	R		R	R	R
Sugar, syrups, jams	R[13]	R	R	R[60]	R[60]	R[60]	R[13]	R	R	R[13]	R	R	R	R	R	R	R	R	R	R	R	R	R		R	R	R
Sulphamic acid	No data			R	R	ND	R[13]	R		No data			R	ND	ND	No data			R	R	R	R	R		No data		
Sulphates (Na, K, Mg, Ca)	R	R	R	R	R	R	R	R	R	R	R	R	R	R	R	R	R	R	R	R	R	R	R		R	R	R
Sulphites	R	R	R	R	R	R	R	R	R	R[80]	R[80]		R	R	R	R	R	R	R	R	R	R	R		R	R	R
Sulphonic acids	R[13]	R		R[13]	R[13]	R[13]	R[2]	R[2]	R[2]				R	R	R	ND			R	R	R	R	ND		No data		
Sulphur	R	R	R	R	R	R	R						R	R	R	R[30]	R	R	R	R	R	ND	ND		R	R	R
Sulphur dioxide, dry	R	R	R	R	R	R	R	R	R				R	R	R							ND	ND		R	R	ND
Sulphur dioxide, wet	R	R	R	R	R	R	R	R	R[4]				R	R	R				R[4]	R	R	ND	ND		R	R	ND
Sulphur trioxide																No data									R	R	R
Sulphuric acid (< 50%)	R	R	R	R	R	R	R	R	R	R	R		R	R		R			R	R	R	R[25]	R[25/80]		R	R	R
Sulphuric acid (70%)				R[80]			R[66]						R			R			R	R					No data		
Sulphuric acid (95%)																R											
Sulphuric acid, fuming																											
Sulphur chlorides																No data			No data								
Tallow	R	R	R[4]	R	R[4]	ND	R	R	R	R	R		R	R	R	R	R	R	R	R	R	R	ND		R[30]	R	R
Tannic acid (10%)	R	R	R	R	R	R	R	R	R	R	R	R	R	R	R	R	R	R	R	R	R	R	R		R	R	R
Tartaric acid	R	R	R	R	R	R	R	R	R	R	R	R	R	R	R	R	R		R	R	R	R	R		R	R	R
Trichlorethylene																R[65]									R[21]	R	R
Vinegar	R	R	R	R	R[14]		R	R	R	R[80]	R		R	R	R	R	R[37]	R				R[80]	R[80]		R	R	R
Water, distilled	R	R	R	R	R	R	R[30]	R	R	R[30]	R	R	R	R	R	R	R	R	R	R	R	R	R		R	R	R
Water, soft	R	R	R	R	R	R	R	R	R	R	R	R	R	R	R	R	R	R	R	R	R	R	R		R	R	R
Water, hard	R	R	R	R	R	R	R	R	R	R	R	R	R	R	R	R	R	R	R	R	R	R	R		R	R	R
Yeast	R	R	R	No data			R	R	R	R	R	R	R	R	R	R	ND	ND	R	R	R	ND	ND		R	R	R
Zinc chloride	R	R	R	R	R	R	R	R	R	R	R	R	R	R	R	R	R	ND	R	R	R	R	R		R	R	R

MISCELLANEOUS

	Concrete (s)			Glass (t)			Graphite (u)			Porcelain and Stoneware			Vitreous Enamel (w)			Wood (z)		
	20°	60°	100°	20°	60°	100°	20°	60°	100°	20°	60°	100°	20°	60°	100°	20°	60°	100°
	R	R	R				R	R	R	R	R	R	No data			R		
				R	R	R	R	R	R	R	R	R	No data					
				R	R	R	R	R	R	R	R	R	R	R	R			
	R	R	R	R	R	R	R	R	R	R	R	R	R	R	R	R	R	R
				R	R	R	R	R	R	R	R	R	R	R	R	R	R	R
	No data			R	R	R	R	R	R	R	R	R	No data			No data		
				R	R	R	R	R	R	R	R	R	R	R	ND	R	R	
				R	R	R	R	R	R	R	R	R	R	R	R	R		
				R	R	R	R	R	R	R	R	R	R	R	ND			
	R	R	R	R	R	R	R	R	R	R	R	R	R	R	R	R	R	
	No data			R	R	ND	R	R	R	R	R	R	R	R	R			
				R	R		R	R	R	R	R	R	R	R	R			
				R	R	R				R	R	R	R	R	R			
				R	R	R	R	R	R	R	R	R	R	R	R	R[10]		
				R	R	R	R	R	R	R	R	R	R	R	R			
				R	R	R	R[2]	R		R	R	R	R	R	R			
				R	R	R				R	R	R	R	ND	ND			
	No data			R	R	R	R	R	R	R	R	R	R	R	ND	No data		
	R[80]			No data			R	R	R	R	R	R	R	R	R	R	R	R
				R	ND	ND	R	R	R	R	R	R	R	R	R	R	R	R
	R	R	R	R	R	R	R	R	R	R	R	R	R	R	R	R	R	R
				R	R	R	R	R	R	R	R	R	R	R	R	R	R	
	R[50]	R		R	R	R	R	R	R	R	R	R	R	R	R	R	R	R
	R[50]	R		R	R	R	R	R	R	R	R	R	R	R	R	R	R	R
	R	R	R	R	R	R	R	R	R	R	R	R	R	R	R	R	R	R
	R	R	R	R	R	R	R	R	R	R	R	R	R	R	R	R	R	R
				R	ND	ND	R	R	R	R	R	R	R	R	ND			
				820[100]			3500						> 400[30]					
				10,000			2·4 m			1000–			30					
										9500								
	2·3–			2·20–			1·92[30]			2–2·6			30			0·4–1·0		
	2·5			2·25[30]									30					
				30			35[30]						30					
							(Shore D)											

(l) **Acrylonitrile butadiene styrene resins**: The information refers to a general purpose moulding grade material.

(m) **Nylon**: Prolonged heating may cause oxidation and embrittlement. Data on nylon 66 plastics refer to *Maranyl* products. Other nylons, such as types 6 and 610, can behave differently, e.g. towards aqueous solutions of salts.

(n) **P.T.F.E.**: Is attacked by alkali metals (molten or in solution) and by certain rare fluorinated gases at high temperatures and pressures. Some organic and halogenated solvents can cause swelling and slight dimensional changes but the effects are physical and reversible.

(o) **Melamine resins**: The information refers mainly to laminates surfaced with melamine resins, Melamine coating resins are always used in conjunction with alkyd resins and the specifications will depend on the alkyd resin used.

(p) **Epoxy resins**: Data are for cold curing systems.

(q) The information given is based on compounds made from ethylene propylene terpolymer rubber.

(r) **Phenol formaldehyde resins**: These are of several types and care should be taken that the right type is chosen.

(s) **Concrete**: Usually made from Portland cement, but if made from Ciment Fondu or gypsum slag cement might have superior resistance in particular applications.

(t) **Glass**: The information refers to heat-resistant borosilicate glass.

(u) **Graphite**: Data refer to resin-impregnated graphite. Other specially treated graphites have improved corrosion resistance to many chemicals.

(v) Chemical resistance of polyurethanes is dependent on the particular structure of the material and is not necessarily applicable to all polyurethanes. Specially designed polyurethanes can be used at higher temperatures than 60°C but chemical resistance is temperature dependent.

(w) **Vitreous enamel**: Special enamels may be required to withstand particular reagents.

(x) Data is based on Ferralium alloy 255.

(y) Data is based on Solef.

(z) **Wood**: The behaviour of wood depends both on the species used and on the physical conditions of service. Aqueous solutions of some chemicals may cause more rapid degradation. Organic solvents may dissolve out resins, etc. Hydrogen peroxide (over 50 % w/w) produces a fire risk.

Physical Property Data Bank

Inorganic compounds are listed in alphabetical order of the principal element in the empirical formula.
Organic compounds with the same number of carbon atoms are grouped together, and arranged in order of the number of hydrogen atoms, with other atoms in alphabetical order

```
NO     = Number in list
MOLWT  = Molecular weight
TFP    = Normal freezing point, deg C
TBP    = Normal boiling point, deg C
TC     = Critical temperature, deg K
PC     = Critical pressure, bar
VC     = Critical volume, cubic metre/mol
LDEN   = Liquid density, kg/cubic metre
TDEN   = Reference temperature for liquid density, deg C
HVAP   = Heat of vaporisation at normal boiling point, J/mol
VISA, VISB = Constants in the liquid viscosity equation:
```

$$\text{LOG[viscosity]} = [VISA] * [(1/T) - (1/VISB)], \quad \text{viscosity mNs/sq.m,} \quad T \text{ deg K}$$

```
DELHF = Standard enthalpy of formation at 298 K, kJ/mol
DELGF = Standard Gibbs energy of formation at 298 K, kJ/mol
```

CPVAPA, CPVAPB, CPVAPC, CPVAPD = Constants in the ideal gas heat capacity equation:

$$Cp = CPVAPA + (CPVAPB) * T + (CPVAPC) * T^{**2} + (CPVAPD) * T^{**3},$$
$$Cp \quad \text{J/mol K,} \quad T \text{ deg K}$$

ANTA, ANTB, ANTC = Constants in the Antione equation:

$$\text{Ln(vapour pressure)} = ANTA - ANTB/(T + ANTC), \quad \text{vap. press. mmHg,} \quad T \text{ deg K,}$$
to convert mmHg to N/sq.m multiply by 133.32

```
TMN    = Minimum temperature for Antione constants, deg C
TMX    = Maximum temperature for Antione constants, deg C
```

Most of the values in this data bank were taken, with the permission of the publishers, from: The Properties of Gases and Liquids, by Reid, R.C., Sherwood, T.K. and Prausnitz, J.M., 3rd ed., McGraw-Hill (1977).

NO	FORMULA	COMPOUND NAME	MOLWT	TFP	TBP	TC	PC	VC	LDEN	TDEN	HVAP
1	AR	ARGON	39.948	-189.9	-185.9	150.8	48.7	0.075	1373	-183	6531
2	BCL3	BORON TRICHLORIDE	117.169	-107.3	12.5	452.0	38.7		1350	11	
3	BF3	BORON TRIFLUORIDE	67.805	-126.7	-99.9	260.8	49.9				
4	BR2	BROMINE	159.808	-7.2	58.7	584.0	103.4	0.127	3119	20	30187
5	CLNO	NITROSYL CHLORIDE	65.459	-59.7	-5.5	440.0	91.2	0.139	1420	-12	25707
6	CL2	CHLORINE	70.906	-101.0	-34.5	417.0	77.0	0.124	1563	-34	20432
7	CL3P	PHOSPHORUS TRICHLORIDE	137.333	-112.2	75.8	563.0		0.260	1574	21	
8	CL4SI	SILICON TETRACHLORIDE	169.898	-68.9	57.2	507.0	37.5	0.326	1480	20	27549
9	D2	DEUTERIUM	4.032	-254.5	-249.5	38.4	16.6	0.060	165	-250	1223
10	D2O	DEUTERIUM OXIDE	20.031	3.8	101.4	644.0	216.6	0.056	1105	20	41366
11	F2	FLUORINE	37.997	-219.7	-188.2	144.3	52.2	0.066	1510	-188	6531
12	F3N	NITROGEN TRIFLUORIDE	71.002	-206.8	-129.1	234.0	45.3		1537	-129	
13	F4SI	SILICON TETRAFLUORIDE	104.080	-90.2	-86.2	259.0	37.2		1660	-95	
14	F6S	SULPHUR HEXAFLUORIDE	146.050	-50.7	-63.9	318.7	37.6	0.198	1830	-50	
15	HBR	HYDROGEN BROMIDE	80.912	-86.1	-67.1	363.2	85.5	0.100	2160	-57	17668
16	HCL	HYDROGEN CHLORIDE	36.461	-114.2	-85.1	324.6	83.1	0.081	1193	-85	16161
17	HF	HYDROGEN FLUORIDE	20.006	-83.2	19.5	461.0	64.8	0.069	967	20	6699
18	HI	HYDROGEN IODIDE	127.912	-50.8	-35.6	424.0	83.1	0.131	2803	-36	19778
19	H2	HYDROGEN	2.016	-259.2	-252.8	33.2	13.0	0.065	71	-253	904
20	H2O	WATER	18.015	-85.6	100.0	647.3	220.5	0.056	998	20	40683
21	H2S	HYDROGEN SULPHIDE	34.080	-77.8	-60.4	373.2	89.4	0.098	993	-60	18673
22	H3N	AMMONIA	17.031	1.5	-33.5	405.6	112.8	0.072	639	0	23362
23	H4N2	HYDRAZINE	32.045		113.5	653.0	146.9	0.096	1008	20	44799
24	HE(4)	HELIUM-4	4.003	-269.0	-269.0	5.2	2.3	0.057	123	-269	92
25	I2	IODINE	253.808	113.6	184.3	819.0	116.5	0.155	3740	180	41868
26	KR	KRYPTON	83.800	-157.4	-153.4	209.4	55.0	0.091	2420	-153	9667
27	NO	NITRIC OXIDE	30.006	-163.7	-151.8	180.0	64.8	0.058	1280	-152	13816
28	NO2	NITROGEN DIOXIDE	46.006	-11.3	21.1	431.4	101.3	0.170	1447	20	19071
29	N2	NITROGEN	28.013	-209.9	-195.8	126.2	33.9	0.089	805	-195	5581
30	N2O	NITROUS OXIDE	44.013	-90.9	-88.5	309.6	72.4	0.097	1226	-90	16559
31	NE	NEON	20.183	-248.7	-246.2	44.4	27.6	0.042	1204	-246	1842
32	O2	OXYGEN	31.999	-218.8	-183.0	154.6	50.5	0.073	1149	-183	6824
33	O2S	SULPHUR DIOXIDE	64.063	-75.5	-10.2	430.8	78.8	0.122	1455	-10	24932
34	O3	OZONE	47.998	-192.7	-111.9	261.0	55.7	0.089	1356	-112	11179
35	O3S	SULPHUR TRIOXIDE	80.058	16.8	44.8	491.0	82.1	0.130	1780	45	40679
36	XE	XENON	131.300	-111.9	-108.2	289.7	58.4	0.118	3060	-108	13013
37	CBRF3	TRIFLUOROBROMOMETHANE	148.910	-168.2	-59.2	340.2	39.7	0.200	1750	-115	15516
38	CCLF3	CHLOROTRIFLUOROMETHANE	104.459	-181.2	-81.5	302.0	39.2	0.180			19979
39	CCL2F2	DICHLORODIFLUOROMETHANE	120.914	-157.8	-29.8	385.0	41.2	0.217			24409
40	CCL2O	PHOSGENE	98.916	-128.2	7.6	455.0	56.7	0.190	1361	20	24786
41	CCL3F	TRICHLOROFLUOROMETHANE	137.368	-111.2	23.8	471.2	44.1	0.248			
42	CCL4	CARBON TETRACHLORIDE	153.823	-23.8	76.5	556.4	45.6	0.276	1584	25	30019
43	CF4	CARBON TETRAFLUORIDE	88.005	-186.8	-128.0	227.6	37.4	0.140			11974
44	CO	CARBON MONOXIDE	28.010	-205.1	-191.5	132.9	35.0	0.093	803	-192	6046
45	COS	CARBONYL SULPHIDE	60.070	-138.9	-50.3	375.0	58.8	0.140	1274	-99	
46	CO2	CARBON DIOXIDE	44.010	-56.6	-78.5	304.2	73.8	0.094	777	20	26754
47	CS2	CARBON DISULPHIDE	76.131	-111.9	46.2	552.0	79.0	0.170	1293	0	
48	CHCLF2	CHLORODIFLUOROMETHANE	86.469	-160.2	-40.8	369.2	49.8	0.165	1230	16	20205
49	CHCL2F	DICHLOROFLUOROMETHANE	102.923	-135.2	8.8	451.6	51.7	0.197	1380	9	24953
50	CHCL3	CHLOROFORM	119.378	-63.6	61.1	536.4	54.7	0.239	1489	20	29726

NO	TMX	TMN	ANTC	ANTB	ANTA	CPVAPD	CPVAPC	CPVAPB	CPVAPA	DELGF	DELHF	VISB	VISA
1	-179	-192	-5.84	700.51	15.2330	0.000	5.166E-08	-3.211E-05	20.804	0.00	0.00	58.76	107.57
2													
3													
4	81	-14	-51.56	2582.32	15.8441	4.534E-09	-1.191E-05	1.125E-02	33.859	0.00	0.00	292.79	387.82
5	12	-63	-23.46	2520.70	16.9505	1.014E-08	-3.339E-05	4.471E-02	34.097	66.99	52.63	172.35	191.96
6	-9	-101	-27.01	1978.32	15.9610	1.547E-08	-3.869E-05	3.383E-02	26.929	0.00	0.00		
7													
8	91	-35	-43.15	2634.16	15.8019	-3.684E-09	1.169E-05	-6.405E-03	30.250	0.00	0.00	8.38	19.67
9	-248	-254	0.00	157.89	13.2954					-234.80	-249.41	304.58	757.92
10	-182	-214	-6.00	714.10	15.6700	1.204E-08	-3.462E-05	3.656E-02	23.216	0.00	0.00	52.52	84.20
11	-118	-170	-15.37	1155.69	15.6107					-127.19	-124.68		
12													
13													
14	-53	-114	-11.16	2524.78	19.3785	-6.238E-09	1.722E-05	-9.462E-03	30.647	-1117.88	-1221.71	180.75	251.29
15	-52	-89	-47.86	1242.53	14.4687	-3.897E-09	1.246E-05	-7.201E-03	30.291	-53.30	-36.26	166.32	88.08
16	-73	-136	-14.45	1714.25	16.5040	2.503E-09	-2.032E-06	6.611E-03	29.061	-95.33	-92.36	277.74	372.78
17	40	-67	15.06	3404.49	17.6958	-1.353E-08	2.972E-05	-1.427E-02	31.158	-273.40	-271.30	199.62	438.74
18	-17	-58	-85.06	957.96	12.9149	7.645E-09	1.380E-05	9.273E-03	27.143	1.59	26.38	285.43	155.15
19	-248	-259	3.19	164.90	13.6333	-3.596E-09	1.055E-05	1.923E-03	32.243	0.00	0.00	5.39	13.82
20	168	11	-46.13	3816.44	18.3006	-1.176E-08	2.432E-05	1.436E-03	31.941	-228.77	-242.00	283.16	658.25
21	-43	-83	-26.06	1768.69	16.1040	-1.184E-08	1.707E-05	2.383E-02	27.315	-33.08	-20.18	165.54	342.79
22	-12	-94	-32.98	2132.50	16.9481	6.024E-08	-1.657E-04	1.894E-01	9.768	-16.16	-45.72	169.63	349.02
23	70	15	-45.15	3877.65	17.9899					158.64	95.25	290.88	524.98
24	-269	-269	1.79	33.73	12.2514	2.834E-09	-6.987E-06	6.514E-03	35.592	0.00	0.00		
25	214	110	-68.16	3709.23	16.1597	-4.186E-09	9.746E-06	-9.378E-04	29.345	0.00	0.00	520.55	559.62
26	-144	-160	-8.71	958.75	15.2677	-2.930E-10	-2.080E-05	4.835E-02	24.233	0.00			
27	-133	-178	-4.88	1572.52	16.1314	-1.168E-08	2.679E-05	-1.356E-02	31.150	86.75	90.43		
28	47	-43	3.65	4141.29	20.5324	1.830E-08	-5.777E-05	7.280E-02	21.621	52.00	33.87	230.21	406.20
29	-183	-219	-6.60	588.72	14.9542					0.00	0.00	46.14	90.30
30	-73	-129	-25.99	1506.49	16.1271				20.786	103.71	81.60		
31	-244	-249	-2.61	180.47	14.0099	-1.065E-08	-1.745E-05	-3.680E-06	28.106				
32	-173	-210	-6.45	734.55	15.4075	-1.328E-08	-4.961E-05	6.698E-02	23.852	0.36	0.0	51.50	85.68
33	7	-78	-35.97	2302.35	16.7680	1.697E-08	-6.242E-05	8.009E-02	20.545	-300.01	-297.05	208.42	397.85
34	-99	-164	-22.16	1272.18	15.7427	3.242E-08	-1.120E-04	1.459E-01	16.370	162.91	142.77	120.34	313.79
35	59	17	-36.66	3995.70	20.8403					-370.62	-395.53	315.99	1372.80
36	-95	-115	-14.50	1303.92	12.2958								
37						4.458E-08	-1.576E-04	1.911E-01	22.814	-623.00	-649.37		
38						4.341E-08	-1.508E-04	1.782E-01	31.598	-654.40	-695.01		
39	68	-60	-43.15	2167.31	15.7565	5.070E-08	-1.373E-04	1.360E-01	28.089	-442.54	-481.48	165.55	215.09
40	27	-33	-36.30	2401.61	15.8516	4.146E-08	-1.416E-04	1.630E-01	40.985	-206.91	-221.06		
41	101	-20	-45.99	2808.19	15.8742	8.842E-08	-2.269E-04	2.048E-01	40.717	-245.51	-284.70		
42	-125	-180	-13.06	1244.55	16.0543	4.513E-08	-1.625E-04	2.285E-02	13.980	-58.28	-100.48	290.84	540.15
43	-165	-210	-13.15	530.22	14.3686	-1.271E-08	-2.789E-05	-1.285E-02	30.869	-889.03	-933.66		
44						2.453E-08	-7.017E-05	7.984E-02	23.567	-137.37	-110.62	48.90	94.06
45						1.715E-08	-5.601E-05	7.343E-02	19.795	-165.76	-138.50		
46	-69	-119	-0.16	3103.39	22.5898	2.672E-08	-7.666E-05	8.126E-02	27.444	-394.65	-393.77	185.24	578.08
47	69	-45	-31.62	2690.85	15.9844	3.058E-08	-1.169E-04	1.618E-01	17.300	66.95	117.15	200.22	274.08
48	-33	-48	-41.30	1704.80	15.5602	3.263E-08	-1.058E-04	1.058E-02	23.664	-470.89	-502.00		
49							-1.199E-04	1.581E-01		-268.37	-298.94		
50	97	-13	-46.16	2696.79	15.9732	6.657E-08	-1.840E-04	1.893E-01	24.003	-68.58	-101.32	246.50	394.81

NO	FORMULA	COMPOUND NAME	MOLWT	TFP	TBP	TC	PC	VC	LDEN	TDEN	HVAP	NO
51	CHN	HYDROGEN CYANIDE	27.026	-13.3	25.7	456.8	53.9	0.139	688	20	25234	51
52	CH2BR2	DIBROMOMETHANE	173.835	-52.6	96.8	583.0	71.9		2500	20		52
53	CH2CL2	DICHLOROMETHANE	84.993	-95.1	39.8	510.0	60.8	0.193	1317	25	28010	53
54	CH2O	FORMALDEHYDE	30.026	-117.2	-19.2	408.0	65.9		815	-20	23027	54
55	CH2O2	FORMIC ACID	46.025	8.3	100.6	580.0			1226	15	21939	55
56	CH3BR	METHYL BROMIDE	94.939	-93.7	3.5	464.0	86.1	0.139	1737	-5	23928	56
57	CH3CL	METHYL CHLORIDE	50.488	-97.8	-24.3	416.3	66.8	0.139	915	20	21436	57
58	CH3F	METHYL FLUORIDE	34.033	-141.8	-78.4	317.8	58.8	0.124	843	-60		58
59	CH3I	METHYL IODIDE	141.939	-66.5	42.4	528.0	65.9	0.190	2279	20	27214	59
60	CH3NO2	NITROMETHANE	61.041	-28.6	101.2	588.0	63.1	0.173	1138	20	34436	60
61	CH4	METHANE	16.043	-182.5	-161.5	190.6	46.0	0.099	425	-161	8185	61
62	CH4O	METHANOL	32.042	-97.7	64.6	512.6	81.0	0.118	791	20	35278	62
63	CH4S	METHYL MERCAPTAN	48.107	-123.2	5.9	470.0	72.3	0.145	866	20	24577	63
64	CH5N	METHYL AMINE	31.058	-93.5	-6.4	430.0	74.6	0.140	703	-14	26000	64
65	CH6N2	METHYL HYDRAZINE	46.072		90.8	567.0	80.4	0.271				65
66	C2CLF5	CHLOROPENTAFLUOROETHANE	154.467	-106.2	-39.2	353.2	31.6	0.252	1310	25	19469	66
67	C2CL2F4	1,1-DICHLORO-1,2,2,2-TETRAFLUOROETHANE	170.992	-94.2	3.8	418.6	33.0	0.294	1455	4		67
68	C2CL2F4	1,2-DICHLORO-1,1,2,2-TETRAFLUOROETHANE	170.922	-93.9	3.7	418.9	32.6	0.293	1480	16	23279	68
69	C2CL3F3	1,2-DICHLORO-1,1,2-TETRAFLUOROETHANE	187.380	-35.0	47.5	487.2	34.1	0.304	1580	20	27507	69
70	C2CL4	TETRACHLOROETHYLENE	165.834	-22.2	121.1	620.0	44.6	0.290	1620	20	34750	70
71	C2CL4F2	1,1,2,2-TETRACHLORO-1,2-DIFLUOROETHANE	203.831	24.8	92.5	551.0			1645	25		71
72	C2F4	TETRAFLUOROETHYLENE	100.016	-142.5	-75.7	306.4	39.4	0.175	1519	-76	16161	72
73	C2F6	HEXAFLUOROETHANE	138.012	-100.8	-78.3	292.8		0.224	1590	-78		73
74	C2N2	CYANOGEN	52.035	-27.9	-20.7	400.0	59.8					74
75	C2HCL3	TRICHLOROETHYLENE	131.389	-116.4	87.2	571.0	49.1	0.256	1462	20	31401	75
76	C2HF3O2	TRIFLUOROACETIC ACID	114.024	-15.3	72.4	491.3	32.6		1535	0		76
77	C2H2	ACETYLENE	26.038	-80.8	-84.0	308.3	61.4	0.113	615	-84	16957	77
78	C2H2F2	1,1-DIFLUOROETHYLENE	64.035	-135.8	-85.7	302.8	44.6	0.154				78
79	C2H2O	KETENE	42.038		-41.2	380.0	64.8	0.145			20641	79
80	C2H3CL	VINYL CHLORIDE	62.499	-153.8	-13.4	429.7	56.0	0.169	969	-14	20641	80
81	C2H3CLF2	1-CHLORO-1,1-DIFLUOROETHANE	100.490	-131.2	-9.8	410.2	41.2	0.231	1100	30		81
82	C2H3CLO	ACETYL CHLORIDE	78.498	-113.0	50.7	508.0	58.8	0.204	1104	20	28680	82
83	C2H3CL3	1,1,2-TRICHLOROETHANE	133.400	-36.7	113.7	602.0	41.5	0.294	1441	20	33327	83
84	C2H3F	VINYL FLUORIDE	46.044	-143.2	-72.2	327.8	52.4	0.144				84
85	C2H3F3	1,1,1-TRIFLUOROETHANE	84.041	-111.3	-47.7	346.2	37.6	0.221			19176	85
86	C2H3N	ACETONITRILE	41.053	-43.9	81.6	548.0	48.3	0.173	782	20	31401	86
87	C2H3NO	METHYL ISOCYANATE	57.052		38.8	491.0	55.7		958	20	29601	87
88	C2H4	ETHYLENE	28.054	-169.2	-103.8	282.4	50.4	0.129	577	-110	13553	88
89	C2H4CL2	1,1-DICHLOROETHANE	98.960	-97.0	57.2	523.0	50.7	0.240	1168	25	28721	89
90	C2H4CL2	1,2-DICHLOROETHANE	98.960	-35.7	83.4	561.0	53.7	0.220	1250	16	32029	90
91	C2H4F2	1,1-DIFLUOROETHANE	66.051	-117.0	-24.8	386.6	45.0	0.181			21353	91
92	C2H4O	ACETALDEHYDE	44.054	-123.0	20.3	461.0	55.7	0.154	778	20		92
93	C2H4O	ETHYLENE OXIDE	44.054	-112.2	10.4	469.0	71.9	0.140	899	20	25749	93
94	C2H4O2	ACETIC ACID	60.052	16.6	117.9	594.4	57.9	0.171	1049	20	25623	94
95	C2H4O2	METHYL FORMATE	60.052	-99.0	31.7	487.2	60.0	0.172	974	25	23697	95
96	C2H5BR	ETHYL BROMIDE	108.966	-118.6	38.3	503.8	62.3	0.215	1451	20	28219	96
97	C2H5CL	ETHYL CHLORIDE	64.515	-136.4	12.2	460.4	52.7	0.199	896	20	26502	97
98	C2H5F	ETHYL FLUORIDE	48.060	-143.3	-37.7	375.3	50.3	0.169			24702	98
99	C2H5N	ETHYLENE IMIDE	43.069	-78.2	56.6		55.7		833	25	32071	99
100	C2H6	ETHANE	30.070	-183.3	-88.7	305.4	48.8	0.148	548	-90	14717	100

NO	VISA	VISB	DELHF	DELGF	CPVAPA	CPVAPB	CPVAPC	CPVAPD	ANTA	ANTB	ANTC	TMN	TMX	NO
51	194.70	145.31	130.63	120.20	21.863	6.062E-02	-4.961E-05	1.815E-08	16.5138	2585.80	-37.15	-39	57	51
52	428.91	294.57	-4.19	-5.61	12.954	1.623E-01	-1.302E-04	4.207E-08	16.3029	2622.44	-41.70	-44	59	52
53	359.55	225.13	-95.46	-68.91	23.475	3.156E-02	-2.985E-05	-2.300E-08	16.4775	2204.13	-30.15	-88	-2	53
54	319.83	171.35	-115.97	-109.99	11.715	1.357E-01	-8.411E-05	2.016E-08	16.9882	3599.58	-26.09	-2	136	54
55	729.35	325.72	-378.86	-351.23	14.428	1.091E-01	-5.401E-05	9.583E-09	16.0252	2271.71	-34.83	-58	53	55
56	298.15	211.15	-37.68	-28.18	13.875	1.014E-01	-3.888E-05	2.566E-09	16.1052	2077.97	-29.55	-93	-7	56
57	426.45	193.56	-86.37	-62.93	13.825	8.616E-02	-2.070E-05	-1.984E-09	16.3428	1704.41	-19.27	-132	-64	57
58			-234.04	-210.14	10.806	1.389E-01	-1.041E-04	3.485E-08	16.0905	2629.55	-36.50	-13	52	58
59	336.19	229.95	13.98	15.66	7.423	1.977E-01	-1.081E-04	2.085E-08	16.2193	2972.64	-64.15	5	136	59
60	452.80	261.21	-74.78	-6.95	19.251	5.212E-02	1.197E-05	-1.131E-08	15.2243	597.84	-7.16	-180	-153	60
61	114.14	57.60	-74.86	-50.87	21.152	7.092E-02	2.587E-05	-2.851E-08	18.5875	3626.55	-34.29	-16	91	61
62	555.30	260.64	-201.30	-162.62	13.268	1.456E-01	-8.545E-05	-2.075E-08	16.1909	2338.38	-34.44	-73	27	62
63			-22.99	-9.92	11.476	1.427E-01	-5.334E-05	4.752E-09	17.2622	2484.83	-32.92	-61	38	63
64	311.80	176.30	-23.03	32.28					15.1424	2319.84	-91.70	-3	127	64
65			85.41	177.98										65
66														66
67			-898.49											67
68			-745.67											68
69			-12.14											69
70	392.58	281.82		22.61										70
71			-659.00	-624.13	27.834	3.491E-01	-2.890E-04	8.139E-08						71
72			-1343.96	-1258.22	40.453	3.278E-01	-2.751E-04	7.820E-08						72
73			309.10	297.39	38.778	3.439E-01	-2.950E-04	8.507E-08						73
74			-5.86	19.89	61.140	2.874E-01	-2.420E-04	6.904E-08	15.7343	1848.90	-30.88	-98	-43	74
75	145.67	196.60			45.971	2.255E-01	-2.293E-04	6.382E-08	15.8424	2532.61	-45.67	-23	87	75
76			226.88	209.34	29.010	2.277E-01	-2.036E-04	6.778E-08	16.1642	3259.29	-52.15	34	187	76
77			-345.41	-321.71	26.816	3.457E-01	-2.869E-04	8.135E-08	15.8800	1574.60	-27.22	-133	-63	77
78			-61.13	-60.33	35.935	9.252E-02	-3.222E-05	2.949E-08	15.6422	1512.94	-26.94	-103	-73	78
79			35.17	51.54	30.174	2.286E-01	-2.229E-04	8.243E-08	16.1827	3028.13	-43.15	-13	127	79
80	276.90	167.04			26.821	7.578E-02	-5.007E-05	1.412E-08	16.3481	1637.14	-19.77	-79	-71	80
81			-244.09	-206.37	3.073	2.444E-01	-2.099E-04	7.021E-08	16.0197	1849.21	-35.15	-103	-18	81
82			-138.58	-77.54	6.385	1.638E-01	-1.084E-04	2.698E-08	14.9601	1803.84	-43.15	-88	17	82
83	346.72	304.43			5.949	2.019E-01	-1.536E-04	4.773E-08	15.7514	2447.33	-55.53	-36	82	83
84			-746.09	-679.22	16.818	2.756E-01	-1.992E-04	5.304E-08	16.0381	3110.79	-56.16	29	155	84
85	334.91	210.05	87.92	105.67	25.020	1.710E-01	-9.855E-05	2.219E-08	15.8965	1814.91	-29.92	-3	27	85
86	616.78	227.47	-90.02	68.16	6.322	3.430E-01	-2.957E-04	9.792E-08	16.2874	2945.47	-49.15	-13	117	86
87	168.98	93.94	52.33	-73.14	5.744	3.140E-01	-2.597E-04	8.415E-08	16.3258	2480.37	-56.31	-43	67	87
88	412.27	239.10	-130.00	-73.90	20.482	1.083E-01	-4.492E-05	3.202E-07	15.5368	1347.01	-18.15	-153	-91	88
89	473.95	277.98	-129.79	-436.52	35.764	1.039E-01	-5.819E-06	-1.687E-08	16.0842	2697.29	-45.03	-31	75	89
90	319.27	186.56	-494.04	-133.39	3.806	1.565E-01	-3.488E-05	1.755E-08	16.1764	2927.17	-50.22	-33	100	90
91	368.70	192.82	-166.47	-13.10	12.472	2.695E-01	-2.049E-04	6.301E-08	16.1871	2095.35	-29.15	-35	47	91
92	341.88	194.22	-52.67	-297.39	20.486	2.310E-01	-1.438E-04	3.388E-08	16.2418	2465.15	-37.15	-63	37	92
93	600.94	306.21	-435.13	-26.33	8.675	2.395E-01	-1.457E-04	3.394E-08	16.7400	3405.57	-29.01	-73	157	93
94	363.19	212.70	-350.02	-60.04	7.716	1.822E-01	-1.006E-04	2.380E-08	16.8080	2590.87	-56.34	17	94	94
95	369.80	220.68	-297.37	-209.67	-7.519	2.222E-01	-1.256E-04	2.591E-08	16.5104	2511.68	-42.60	-48	51	95
96	320.94	190.83	-64.06	178.11	1.432	2.548E-01	-1.753E-04	4.948E-08	15.9338	2332.01	-41.44	-47	60	96
97			-111.79		-0.553	2.700E-01	-1.949E-04	5.702E-08	15.9800	1966.89	-36.48	-73	37	97
98			-261.67		4.346	2.348E-01	-1.472E-04	3.804E-08	16.0686	2610.44	-27.00	-103	-21	98
99	156.60	95.57	123.51		-20.771	2.606E-01	-1.839E-04	5.547E-08	16.4227		-63.15	-25	86	99
100			-84.74	-32.95	5.409	1.781E-01	-6.937E-05	8.712E-09	15.6637	1511.42	-17.16	-143	-74	100

NO	FORMULA	COMPOUND NAME	MOLWT	TFP	TBP	TC	PC	VC	LDEN	TDEN	HVAP	NO
101	C2H6O	DIMETHYL ETHER	46.069	-141.5	-24.7	400.0	53.7	0.178	667	20	21520	101
102	C2H6O	ETHANOL	46.069	-114.1	78.3	516.2	63.8	0.167	789	20	38770	102
103	C2H6O2	ETHYLENE GLYCOL	62.069	-13.0	197.2	645.0	77.0	0.186	1114	20	52544	103
104	C2H6S	ETHYL MERCAPTAN	62.134	-147.9	35.0	499.0	54.9	0.207	839	20	26796	104
105	C2H6S	DIMETHYL SULPHIDE	62.130	-98.3	37.3	503.0	55.3	0.201	848	20	26963	105
106	C2H7N	ETHYL AMINE	45.085	-81.2	16.5	456.0	56.2	0.178	683	20	28052	106
107	C2H7N	DIMETHYL AMIDE	45.085	-92.2	6.8	437.6	53.1	0.187	656	20	26502	107
108	C2H7NO	MONOETHANOLAMINE	61.084	10.3	170.3	614.0	44.6	0.196	1016	20	50242	108
109	C2H8N2	ETHYLENEDIAMINE	60.099	10.8	117.2	593.0	62.8	0.206	896	20	41868	109
110	C3H3N	ACRYLONITRILE	53.064	-83.7	77.3	536.0	35.5	0.210	806	20	32657	110
111	C3H4	PROPADIENE	40.065	-136.3	-34.5	393.0	54.7	0.162	658	-35	18631	111
112	C3H4	METHYL ACETYLENE	40.065	-102.7	-23.2	402.4	56.2	0.164	706	-50	22148	112
113	C3H4O	ACROLEIN	56.064	-87.2	52.8	506.0	51.7		839	20	28345	113
114	C3H4O2	ACRYLIC ACID	72.064	11.8	140.8	615.0	56.7	0.210	1051	20	46055	114
115	C3H4O2	VINYL FORMATE	72.064	-57.7	46.4	475.0	57.8	0.210	963	20	32155	115
116	C3H5CL	ALLYL CHLORIDE	76.526	-134.5	45.1	514.0	47.6	0.234	937	20	27110	116
117	C3H5CL3	1,2,3-TRICHLOROPROPANE	147.432	-14.7	155.8	651.0	39.5	0.348	1389	20	38435	117
118	C3H5N	PROPIONITRILE	55.080	-92.7	97.3	564.4	41.8	0.230	782	20	32280	118
119	C3H6	CYCLOPROPANE	42.081	-127.5	-32.8	397.3	54.9	0.170	563	15	20055	119
120	C3H6	PROPYLENE	42.081	-185.3	-47.8	365.0	46.2	0.181	612	-50	18422	120
121	C3H6CL2	1,2-DICHLOROPROPANE	112.987	-100.5	96.3	577.0	44.6	0.226	1150	20	31401	121
122	C3H6O	ACETONE	58.080	-95.0	56.2	508.1	47.0	0.209	790	20	29140	122
123	C3H6O	ALLYL ALCOHOL	58.080	-129.2	96.8	545.0	57.1	0.203	855	15	39984	123
124	C3H6O	PROPIONALDEHYDE	58.080	-80.2	47.8	496.0	47.6	0.223	797	20	28303	124
125	C3H6O	PROPYLENE OXIDE	58.080	-112.2	34.3	482.2	49.2	0.186	829	20	27005	125
126	C3H6O	VINYL METHYL ETHER	58.080	-121.7	4.8	436.0	47.6	0.205	750	20	19050	126
127	C3H6O2	PROPIONIC ACID	74.080	-20.7	140.8	612.0	53.7	0.230	993	20	32238	127
128	C3H6O2	ETHYL FORMATE	74.080	-79.4	54.2	508.4	47.4	0.229	927	16	30145	128
129	C3H6O2	METHYL ACETATE	74.080	-98.2	56.7	506.8	46.9	0.228	934	20	30145	129
130	C3H7CL	PROPYL CHLORIDE	78.542	-122.8	46.4	503.0	45.8	0.254	891	20	27256	130
131	C3H7CL	ISOPROPYL CHLORIDE	78.452	-117.2	35.7	485.0	47.2	0.230	862	20	26293	131
132	C3H8	PROPANE	44.097	-187.7	-42.1	369.8	42.5	0.203	582	-42	18786	132
133	C3H8O	N-PROPYL ALCOHOL	60.096	-126.3	97.2	536.7	51.7	0.218	804	20	41784	133
134	C3H8O	ISOPROPYL ALCOHOL	60.096	-88.5	82.2	508.3	47.6	0.220	786	20	39858	134
135	C3H8O	METHYL ETHYL ETHER	60.096	-139.2	7.3	437.8	44.0	0.221	700	20	24702	135
136	C3H8O2	METHYLAL	76.096	-105.2	41.8	497.0			888	18		136
137	C3H8O2	1,2-PROPANEDIOL	76.096	-60.2	187.3	625.0	60.8	0.237	1036	20	54177	137
138	C3H8O2	1,3-PROPANEDIOL	76.096	-26.8	214.4	658.0	59.8	0.241	1053	20	56522	138
139	C3H8O3	GLYCEROL	92.095	17.8	289.8	726.0	66.9	0.255	1261	20	61127	139
140	C3H8S	METHYL ETHYL SULPHIDE	76.157	-106.2	66.6	533.0	42.6	0.233	837	20	29517	140
141	C3H9N	N-PROPYL AMINE	59.112	-83.2	48.6	497.0	47.4	0.233	717	20	29726	141
142	C3H9N	ISOPROPYL AMINE	59.112	-95.3	32.4	476.0	50.7	0.229	688	20	27214	142
143	C3H9N	TRIMETHYL AMINE	59.112	-117.2	2.9	433.2	40.7	0.254	633	20	24116	143
144	C4H2O3	MALEIC ANHYDRIDE	93.058	52.8	197.6	455.0	47.6	0.202	1310	60	24493	144
145	C4H4	VINYL ACETYLENE	52.076	-45.6	4.9	490.2	55.0	0.218	710	0	27105	145
146	C4H4O	FURAN	68.075	-85.7	31.3	490.2	56.9	0.219	938	20	31485	146
147	C4H4S	THIOPHENE	84.136	-38.3	84.1	579.4	56.9	0.219	1071	16	31485	147
148	C4H5N	ALLYL CYANIDE	67.091	-86.5	118.8	585.0	39.5	0.265	835	20	34332	148
149	C4H5N	PYRROLE	67.091		129.8	640.0			967	21		149
150	C4H6	ETHYLACETYLENE	54.092	-125.8	8.0	463.7	47.1	0.220	650	16	24995	150

NO	VISA	VISB	DELHF	DELGF	CPVAPA	CPVAPB	CPVAPC	CPVAPD	ANTA	ANTB	ANTC	TMN	TMX	NO
101			-184.18	-113.00	17.015	1.790E-01	-5.233E-05	-1.917E-09	16.8467	2361.44	-17.10	-94	-8	101
102	686.64	300.88	-234.96	-168.39	9.014	2.140E-01	-8.390E-05	1.373E-09	18.9119	3803.98	-41.68	-3	96	102
103	1365.00	402.41	-389.58	-304.67	35.697	2.483E-01	-1.497E-04	3.010E-08	20.2501	6022.18	-28.25	91	221	103
104	419.60	206.21	-46.14	-4.69	14.922	2.350E-01	-1.368E-04	3.161E-08	16.0001	2497.23	-41.77	-49	57	104
105	267.34	184.24	-37.56	6.95	24.304	1.874E-01	-6.874E-05	4.098E-09	16.0007	2511.56	-42.35	-47	58	105
106	340.54	192.44	-46.05	37.30	3.693	2.751E-01	-1.583E-04	3.808E-08	17.0073	2616.73	-37.30	-58	43	106
107			-18.84	68.04	-0.172	2.695E-01	-1.329E-04	2.339E-08	16.2653	2358.77	-35.15	-55	37	107
108	1984.10	367.03	-201.72		9.311	3.009E-01	-1.817E-04	4.655E-08	17.8174	3988.33	-86.93	71	204	108
109	839.76	316.41			38.297	2.407E-01	-4.337E-05	-3.948E-08	16.4082	3108.49	-72.15	19	152	109
110	343.31	210.42	185.06	195.44	10.693	2.207E-01	-1.565E-04	4.601E-08	15.9253	2782.21	-51.15	-18	112	110
111			192.26	202.52	9.906	1.977E-01	-1.181E-04	2.782E-08	13.1563	1054.72	-77.08	-99	-16	111
112			185.56	194.56	14.708	1.864E-01	-1.173E-04	3.224E-08	15.6227	1850.66	-44.07	-90	-6	112
113	388.17	217.14	-70.92	-65.19	11.970	2.105E-01	-1.070E-04	1.905E-08	15.9057	2606.53	-45.15	-38	87	113
114	733.02	307.15	-336.45	-286.25	1.742	1.838E-01	-9.975E-05	6.975E-08	15.5617	3319.18	-80.15	-33	177	114
115	428.40	224.83			27.813	3.046E-01	-3.559E-05	-2.335E-07	16.6531	2569.68	-63.15	-43	77	115
116	368.27	210.61	-0.63	43.63	2.529	3.622E-01	-2.278E-04	7.293E-08	15.9772	2531.92	-47.15	42	77	116
117	818.63	342.88	-185.89	-97.85	26.883	2.245E-01	-1.100E-04	8.788E-08	16.1246	3417.27	-69.15	42	197	117
118	366.77	225.86	50.66	96.21	15.403	3.813E-01	-2.881E-04	1.954E-08	15.9571	2940.86	-55.15	-3	-28	118
119			53.34	104.46	-35.240	2.345E-01	-1.160E-04	1.035E-08	15.8599	1971.04	-26.65	-93	135	119
120	273.84	131.63	20.43		3.710	3.654E-01	-2.603E-04	7.741E-08	15.7027	1807.53	-26.15	-113	-33	120
121	514.36	281.03	62.76		10.450	2.605E-01	-2.031E-04	2.046E-08	16.0385	2985.07	-52.16	15	135	121
122	367.25	209.68	-165.80		-6.301	2.614E-01	-2.037E-04	5.321E-08	16.6513	2940.46	-35.93	-32	77	122
123	793.52	307.26	-217.71		-1.105	2.256E-01	-2.031E-04	2.126E-08	16.9066	2928.20	-85.15	13	127	123
124	343.44	219.33	-132.09		11.723	2.341E-01	-1.300E-04	4.823E-08	16.2315	2659.02	-44.15	-38	77	124
125	377.43	213.36	-192.17		-8.457	3.689E-01	-1.988E-04	1.062E-08	15.3227	2107.58	-64.87	-48	67	125
126	318.41	180.98	-92.82		15.629	2.316E-01	-9.696E-05	9.876E-08	14.4602	1980.22	-25.15	-83	42	126
127	535.04	299.32	-455.44	-369.57	5.669	2.245E-01	-2.864E-04	-5.359E-08	17.3789	3723.42	-67.48	42	177	127
128	400.91	226.23	-371.54		24.673	3.625E-01	-2.119E-04	2.914E-08	16.1611	2603.30	-54.15	-33	87	128
129	408.62	224.03	-409.72		16.550	3.487E-01	-4.341E-05	7.448E-08	16.1295	2601.92	-56.15	-28	77	129
130	374.77	215.00	-130.21	-50.70	-3.345	3.062E-01	-2.508E-04	5.861E-08	16.0384	2581.48	-42.95	-43	67	130
131	306.25	212.24	-146.54	-62.55	1.842	3.325E-01	-2.243E-04	3.214E-08	15.7260	2490.48	-43.15	-48	-24	131
132	222.67	133.41	-103.92	-23.49	-4.224	1.886E-01	-1.586E-04	4.295E-08	17.5439	1872.46	-25.16	-109	127	132
133	951.04	327.83	-256.57	-161.90	2.470	2.685E-01	-1.855E-04	-9.261E-08	16.6929	3166.38	-80.15	12	111	133
134	1139.70	323.44	-272.60	-173.50	32.427	3.689E-01	-6.405E-05	8.951E-09	18.5435	3640.20	-112.40	0	37	134
135	303.82	171.66	-216.58	-117.73	18.669	2.508E-01	-1.024E-04	8.951E-08	15.8237	1161.63	-52.58	-68	42	135
136	1404.20	426.74	-424.25		0.632	4.211E-01	-2.981E-04	8.951E-08	20.5324	6091.95	-22.46	84	210	136
137	1813.00	406.96	-409.09		8.269	3.675E-01	-2.161E-04	5.053E-08	15.2917	3888.84	-123.20	107	252	137
138	3337.10	406.00	-585.31		8.424	4.442E-01	-3.198E-04	9.378E-08	17.2392	4487.04	-140.20	167	327	138
139					19.527	2.890E-01	-1.209E-04	3.586E-08	15.9765	2722.95	-48.37	-23	87	139
140			-59.66	11.43	6.691	3.498E-01	-1.822E-04	3.586E-08	15.9957	2551.72	-49.15	-38	77	140
141	433.64	228.46	-72.43	39.82	-7.486	4.175E-01	-2.825E-04	8.348E-08	15.3637	2582.35	-40.15	-34	64	141
142			-83.82		-8.206	3.971E-01	-2.187E-04	4.622E-08	16.0499	2230.51	-39.15	-58	32	142
143			-23.86	98.98										143
144	952.48	365.81	304.80	306.18	-13.075	3.484E-01	-2.184E-04	4.839E-08	16.2747	3765.65	-82.15	79	243	144
145			-34.71	0.88	-6.757	2.840E-01	-2.265E-04	7.460E-08	16.0100	2203.57	-43.15	-73	32	145
146	389.40	222.70	115.81	126.86	-35.529	4.320E-01	-3.454E-04	1.074E-07	16.0612	244.70	-45.41	-35	90	146
147	498.60	264.90			-30.606	4.479E-01	-3.771E-04	1.252E-07	16.0243	2869.07	-51.80	-13	107	147
148	521.30	252.03			21.700	2.571E-01	-1.192E-04	1.229E-08	16.0019	3128.75	-58.15	127	157	148
149			108.35						16.7966	3457.47	-62.73	57	167	149
150			165.29	202.22	12.548	2.743E-01	-1.544E-04	3.449E-08	16.0605	2271.42	-40.30	-73	27	150

NO	FORMULA	COMPOUND NAME	MOL.WT	TFP	TBP	TC	PC	VC	LDEN	TDEN	HVAP	NO
151	C4H6	DIMETHYL ACETYLENE	54.092	-32.3	27.0	488.6	50.9	0.221	691	20	26670	151
152	C4H6	1,2-BUTADIENE	54.092	-136.2	10.8	443.7	45.0	0.219	652	20	24283	152
153	C4H6	1,3-BUTADIENE	54.092	-108.9	-4.5	425.0	43.3	0.221	621	20	22483	153
154	C4H6O2	VINYL ACETATE	86.091	-100.2	72.8	525.0	43.6	0.265	932	20		154
155	C4H6O3	ACETIC ANHYDRIDE	102.089	-74.2	138.8	569.0	46.8	0.290	1087	20	41240	155
156	C4H6O4	DIMETHYL OXALATE	118.090	-53.8	163.4	628.0	39.8		1150	15		156
157	C4H6O4	SUCCINIC ACID	118.090	-182.8	234.8							157
158	C4H7N	BUTYRONITRILE	69.107	-112.2	117.8	582.2	37.9	0.285	792	20	34415	158
159	C4H7O2	METHYL ACRYLATE	86.091	-76.5	80.3	536.0	42.6	0.265	956	20	32029	159
160	C4H8	1-BUTENE	56.108	-185.4	-6.3	419.6	37.2	0.240	595	20	21930	160
161	C4H8	CIS-2-BUTENE	56.108	-138.9	3.7	435.6	42.0	0.234	621	20	23362	161
162	C4H8	TRANS-2-BUTENE	56.108	-105.6	0.8	428.6	41.0	0.238	604	20	22772	162
163	C4H8	CYCLOBUTANE	56.108	-90.8	12.5	459.9	49.9	0.210	694	20	24200	163
164	C4H8	ISOBUTYLENE	56.108	-140.4	-6.9	417.9	40.0	0.239	594	20	22131	164
165	C4H8O	N-BUTYRALDEHYDE	72.107	-96.4	74.8	524.0	40.5	0.278	802	20	31527	165
166	C4H8O	ISOBUTYRALDEHYDE	72.107	-65.0	63.8	513.0	41.5	0.274	789	20	31401	166
167	C4H8O	METHYL ETHYL KETONE	72.107	-86.7	79.6	535.6	41.5	0.267	805	20	31234	167
168	C4H8O	TETRAHYDROFURAN	72.107	-108.5	65.9	540.2	51.9	0.224	889	20	29601	168
169	C4H8O	VINYL ETHYL ETHER	72.107	-115.3	35.6	475.0	40.7	0.260	793	20	26502	169
170	C4H8O2	N-BUTYRIC ACID	88.107	-5.3	163.2	628.0	52.7	0.292	958	20	42035	170
171	C4H8O2	1,4-DIOXANE	88.107	11.8	101.3	587.0	52.1	0.238	1033	20	36383	171
172	C4H8O2	ETHYL ACETATE	88.107	-83.6	77.1	523.2	38.3	0.286	901	20	32238	172
173	C4H8O2	ISOBUTYRIC ACID	88.107	-46.0	154.7	609.0	40.5	0.292	968	20	41156	173
174	C4H8O2	METHYL PROPIONATE	88.107	-87.5	79.3	530.6	40.0	0.282	915	20	32573	174
175	C4H8O2	N-PROPYL FORMATE	88.107	-92.9	80.5	538.0	40.6	0.285	911	16	32490	175
176	C4H9CL	1-CHLOROBUTANE	92.569	-123.1	78.4	542.0	36.9	0.312	886	20	30019	176
177	C4H9CL	2-CHLOROBUTANE	92.569	-131.4	68.2	520.6	39.5	0.305	873	20	29224	177
178	C4H9CL	2-CHLORO-2-METHYL PROPANE	92.569	-25.4	50.8	507.0	39.5	0.295	842	20	27424	178
179	C4H9N	PYRROLIDINE	71.123	-4.8	86.5	568.6	56.1	0.249	852	22		179
180	C4H9NO	MORPHOLINE	87.122	-138.4	128.2	618.0	54.7	0.253	1000	20	37681	180
181	C4H10	N-BUTANE	58.124	-159.6	-0.5	425.2	38.0	0.255	579	20	22408	181
182	C4H10	ISOBUTANE	58.124	-159.6	-11.9	408.1	36.5	0.263	557	20	21311	182
183	C4H10O	N-BUTANOL	74.123	-89.3	117.7	562.7	44.2	0.274	810	20	43124	183
184	C4H10O	2-BUTANOL	74.123	-114.7	99.5	536.0	41.9	0.268	807	20	40821	184
185	C4H10O	ISOBUTANOL	74.123	-108.0	107.8	547.7	43.0	0.273	802	20	42077	185
186	C4H10O	2-METHYL-2-PROPANOL	74.123	25.6	82.4	506.2	39.7	0.275	787	20	39063	186
187	C4H10O	ETHYL ETHER	74.123	-116.3	34.5	466.7	36.4	0.280	713	20	26712	187
188	C4H10O2	1,2-DIMETHOXYETHANE	90.123	-71.2	85.4	536.0	30.7	0.271	867	20	31443	188
189	C4H10O3	DIETHYLENE GLYCOL	106.122	-8.2	245.8	681.0	46.6	0.316	1116	20	57234	189
190	C4H10S	DIMETHYL SULPHIDE	90.184	-104.0	92.1	557.0	39.6	0.318	837	20	31778	190
191	C4H10OS2	DIMETHYL DISULPHIDE	122.244	-101.5	154.0	642.0			998	20	37723	191
192	C4H11N	N-BUTYL AMINE	73.139	-49.1	77.4	524.0	41.5	0.288	739	20	32113	192
193	C4H11N	ISOBUTYL AMINE	73.139	-85.2	67.4	516.0	42.6	0.284			30982	193
194	C4H11N	DIETHYL AMINE	73.139	-49.8	55.4	496.6	37.1	0.301	707	20	27842	194
195	C5H4O2	FURFURAL	96.085	-31.0	161.7	657.1	49.2	0.269	1196	25	35169	195
196	C5H5N	PYRIDINE	79.102	-41.7	115.3	620.0	56.3	0.254	983	20	27005	196
197	C5H8	CYCLOPENTENE	68.119	-135.1	44.2	506.0	40.7		772	20	27591	197
198	C5H8	1,2-PENTADIENE	68.119	-137.3	44.8	496.0	37.9	0.276	693	20	27047	198
199	C5H8	1-TRANS-3-PENTADIEN-	68.119	-87.5	42.0	496.0	37.9	0.275	676	20	25163	199
200	C5H8	1,4-PENTADIENE	68.119	-148.3	25.9	478.0	37.9	0.276	661	20		200

NO	VISA	VISB	DELHF	DELGF	CPVAPA	CPVAPB	CPVAPC	CPVAPD	ANTA	ANTB	ANTC'	TMN	TMX	NO
151			146.41	185.56	15.927	2.381E-01	-1.069E-04	1.753E-08	16.2821	2536.78	-37.34	-33	47	151
152			162.32	198.56	11.200	2.723E-01	-1.468E-04	3.089E-08	16.1039	2397.26	-30.88	-28	32	152
153	300.59	163.12	110.24	150.77	-1.687	3.418E-01	-2.340E-04	6.334E-08	15.7727	2142.66	-34.30	-58	17	153
154	457.89	235.35	-316.10		15.160	2.795E-01	-8.804E-05	-1.660E-08	16.1003	2744.68	-56.15	-18	106	154
155	502.33	286.04	-576.10	-477.00	-23.128	5.087E-01	-3.580E-04	9.834E-08	16.3982	3287.56	-75.11	35	164	155
156														156
157	0.00		34.08	108.73										157
158	438.04	256.84			15.211	1.507E+01	4.689E-02	-3.143E-04	16.2092	3202.21	-56.16	34	160	158
159	451.02	245.30			15.165	3.205E-01	-1.637E-04	-2.982E-08	16.1088	2788.43	-59.15	-13	117	159
160	256.30	151.86	-0.13	71.34	-2.994	3.532E-01	-8.804E-05	-1.660E-08	15.7564	2132.42	-33.15	-83	22	160
161	268.94	155.34	-6.99	65.90	0.440	2.953E-01	-1.017E-04	-4.463E-08	15.8171	2210.71	-36.15	-73	32	161
162	259.01	153.30	-11.18	63.01	18.317	2.563E-01	-7.012E-05	-6.154E-10	15.8177	2212.32	-33.15	-73	27	162
163			-26.67	110.11	-50.254	5.024E-01	-3.557E-04	-8.989E-09	15.9254	2359.09	-31.78	-83	17	163
164			-16.91	58.11	16.052	2.804E-01	-1.091E-04	1.047E-07	15.7528	2125.75	-33.15	-18	17	164
165	472.31	233.42	-205.15	-114.84	14.080	3.457E-01	-1.722E-04	9.097E-09	16.1668	2839.09	-50.15	-26	107	165
166	464.06	253.64	-215.87	-121.42	24.463	3.355E-01	-2.057E-04	2.887E-09	15.9888	2676.98	-51.15	-16	97	166
167	423.84	231.67	-238.52	-146.16	10.944	3.559E-01	-1.900E-04	6.368E-08	15.5986	3150.42	-36.65	-3	103	167
168	419.79	189.02	-184.34		-19.104	5.162E-01	-4.131E-04	3.919E-08	16.1069	2768.38	-46.90	-48	97	168
169	349.95	301.13	-140.26		17.279	3.236E-01	-1.471E-04	1.454E-07	15.8911	2449.26	-44.15	62	67	169
170	660.36		-476.16	-180.91	11.740	4.137E-01	-2.430E-04	2.149E-08	15.9240	4130.93	-70.55	2	197	170
171	427.38	235.98	-315.27	-327.62	-53.574	4.071E-01	-4.085E-04	5.530E-08	16.1327	2966.88	-62.15	-13	137	171
172	588.65	311.24	-443.21		7.235	4.668E-01	-2.091E-04	1.062E-07	16.1516	2790.50	-57.15	57	112	172
173	442.88	238.39	-484.25		9.814	3.139E-01	-3.719E-04	-2.854E-08	16.7792	3385.49	-94.15	-13	192	173
174					18.204	4.496E-01	-9.353E-05	-3.719E-08	16.1693	2804.06	-58.92	7	112	174
175	452.97	246.09	-147.38	-38.81	-2.613	4.559E-01	-2.936E-04	-1.827E-08	15.7671	2593.95	-69.69	-18	87	175
176	783.72	260.03	-161.61	-53.51	-3.433	4.651E-01	-2.980E-04	8.080E-08	15.9750	2826.26	-49.05	-23	112	176
177	480.77	237.30	-183.38	-64.14	-3.931	5.388E-01	-3.886E-04	8.256E-08	15.9907	2753.43	-47.15	-38	102	177
178	543.41	253.35	-3.60	114.76	-51.531	3.313E-01	-3.239E-04	7.871E-08	15.8121	2567.15	-44.15	27	87	178
179					-42.802	3.847E-01	-2.666E-04	7.527E-08	15.9444	2717.03	-67.90	-78	127	179
180	914.14	332.75	-126.23	-17.17	9.487	4.180E-01	-1.108E-04	4.199E-08	16.2364	3171.35	-71.15	-86	167	180
181	265.84	160.20	-134.61	20.85	-1.390	4.245E-01	-1.846E-04	-2.821E-09	15.6782	2154.90	-34.42	15	17	181
182	302.51	170.20	-274.86	-150.85	3.266	4.689E-01	-2.241E-04	2.895E-08	15.5381	2032.73	-33.15	25	7	182
183	984.54	341.12	-292.82	-167.72	5.753	7.172E-01	-2.328E-04	4.685E-08	17.2160	3137.00	-94.43	20	131	183
184	1441.70	331.50	-283.40	-167.43	-7.708	3.358E-01	-2.883E-04	4.773E-08	17.2102	3026.03	-86.65	20	120	184
185	1199.10	343.85	-312.63	-177.77	-48.613	3.567E-01	-7.084E-04	7.230E-08	16.8712	2874.73	-100.30	-48	115	185
186	972.10	363.38	-252.38	-122.42	21.424	3.444E-01	-1.035E-04	2.919E-07	16.8548	2658.29	-95.50	-11	103	186
187	353.14	190.58			32.234	3.959E-01	-1.335E-04	-9.357E-09	16.0828	2511.29	-41.95	129	67	187
188			-571.50	17.75	73.060	4.601E-01	-1.467E-04	8.398E-08	16.0241	2869.79	-53.15	-13	120	188
189	1943.00	385.24	-83.53	22.27	13.595	4.475E-01	-1.779E-04	2.649E-08	17.0326	4122.52	-122.50	39	287	189
190	407.59	233.32	-74.69	49.24	26.896	4.429E-01	-2.709E-04	5.970E-08	15.9531	2896.27	-54.49	-14	117	190
191			-92.11		5.079	4.429E-01	-2.407E-04	7.599E-08	16.0607	3421.57	-64.19	-22	182	191
192	472.06	243.98			9.491		-2.109E-04	2.332E-08	16.6085	3012.70	-48.96	-31	100	192
193					2.039		-2.183E-04	3.653E-08	16.1419	2704.16	-56.15		100	193
194	473.89	229.29	-72.43	72.14					16.0545	2595.01	-53.15		77	194
195	618.50	291.58	140.26	190.35	18.196	2.819E-01	-6.523E-05	-5.476E-08	18.7949	5365.88	5.40	77	277	195
196	396.83	218.66	32.95	110.66	39.791	4.927E-01	-3.557E-04	1.004E-07	16.0910	3095.13	-61.15	12	152	196
197			145.70	210.55	-41.512	4.630E-01	-2.579E-04	5.434E-08	15.9356	2583.07	-39.70	-29	105	197
198			77.87	146.85	8.826	3.879E-01	-2.280E-04	5.246E-08	15.9297	2544.34	-44.30	-23	67	198
199			105.51	170.36	30.689	2.811E-01	-6.711E-05	-2.351E-08	15.9182	2541.69	-41.43	-23	67	199
200					6.996	3.951E-01	-2.373E-04	5.597E-08	15.7392	2344.02	-41.69	-33	47	200

NO	FORMULA	COMPOUND NAME	MOLWT	TFP	TBP	TC	PC	VC	LDEN	TDEN	HVAP	NO
201	C5H8	1-PENTYNE	68.119	-105.7	40.1	493.4	40.5	0.278	690	20		201
202	C5H8	2-METHYL-1,3-BUTADIENE	68.119	-146.0	34.0	484.0	38.5	0.276	681	20	26084	202
203	C5H8	3-METHYL-1,2-BUTADIENE	68.119	-113.7	40.8	496.0	41.1	0.267	686	20	27256	203
204	C5H8O	CYCLOPENTONE	84.118	-50.7	130.7	626.0	53.7	0.268	950	20	36593	204
205	C5H8O2	ETHYL ACRYLATE	100.118	-72.2	99.8	552.0	37.5	0.320	921	20	33285	205
206	C5H10	CYCLOPENTANE	70.135	-93.9	49.2	511.6	45.1	0.260	745	20	27315	206
207	C5H10	1-PENTENE	70.135	-165.3	29.9	464.7	40.5	0.300	640	20	25213	207
208	C5H10	CIS-2-PENTENE	70.135	-151.4	36.9	476.0	36.5	0.300	656	20	26126	208
209	C5H10	TRANS-2-PENTENE	70.135	-140.3	36.3	475.0	36.6	0.300	649	20	26084	209
210	C5H10	2-METHYL-1-BUTENE	70.135	-137.6	31.1	465.0	34.5	0.294	650	20	25514	210
211	C5H10	2-METHYL-2-BUTENE	70.135	-133.8	38.5	470.0	34.5	0.318	662	20	26322	211
212	C5H10	3-METHYL-1-BUTENE	70.135	-168.5	20.1	450.0	35.2	0.300	627	20	24116	212
213	C5H10O	VALERALDEHYDE	86.134	-91.2	102.8	554.0	35.5	0.333	810	20	33662	213
214	C5H10O	METHYL N-PROPYL KETONE	86.134	-77.2	102.3	564.0	38.9	0.301	806	20	33494	214
215	C5H10O	METHYL ISOPROPYL KETONE	86.134	-92.2	94.2	553.4	38.5	0.310	803	20	30647	215
216	C5H10O	DIETHYL KETONE	86.134	-39.0	101.9	561.0	37.4	0.336	814	20	33746	216
217	C5H10O2	N-VALERIC ACID	102.134	-34.2	185.5	651.0	38.5	0.340	939	20	49823	217
218	C5H10O2	ISOBUTYL FORMATE	102.134	-95.2	98.4	551.0	38.8	0.350	885	20	34206	218
219	C5H10O2	N-PROPYL ACETATE	102.134	-95.2	101.6	549.4	33.3	0.345	887	20	34206	219
220	C5H10O2	ETHYL PROPIONATE	102.134	-73.9	98.8	546.0	33.6	0.345	895	16	34248	220
221	C5H10O2	METHYL BUTYRATE	102.134	-84.8	102.6	554.4	34.8	0.340	898	20	34101	221
222	C5H10O2	METHYL ISOBUTYRATE	102.134	-87.8	92.2	540.8	34.3	0.339	891	20	33386	222
223	C5H11N	PIPERIDINE	85.150	-10.5	106.5	594.0	47.6	0.289	862	20	34248	223
224	C5H12	N-PENTANE	72.151	-129.8	36.0	469.6	33.7	0.304	626	20	25791	224
225	C5H12	2-METHYL BUTANE	72.151	-159.3	27.8	460.4	33.8	0.306	620	20	24702	225
226	C5H12	2,2-DIMETHYL PROPANE	72.151	-16.6	9.4	433.8	32.0	0.303	591	20	22768	226
227	C5H12O	1-PENTANOL	88.150	-78.2	137.8	586.0	38.5	0.326	815	20	44380	227
228	C5H12O	2-METHYL-1-BUTANOL	88.150	-70.2	128.7	571.0	38.5	0.322	819	20	45217	228
229	C5H12O	3-METHYL-1-BUTANOL	88.150	-117.2	131.2	579.5	38.5	0.329	810	20	44129	229
230	C5H12O	2-METHYL-2-BUTANOL	88.150	-8.8	102.0	545.0	39.5	0.319	809	20	40612	230
231	C5H12O	2,2-DIMETHYL-1-PROPANOL	88.150	53.8	113.1	549.0	39.5	0.319	783	54	43124	231
232	C5H12O	ETHYL PROPYL ETHER	88.150	-126.8	63.6	500.6	32.5	0.312	733	20	30522	232
233	C6F6	PERFLUOROBENZENE	186.056		80.2	516.7	33.0	0.442				233
234	C6F12	PERFLUOROCYCLOHEXANE	300.047		52.5	457.2	24.3	0.401				234
235	C6F14	PERFLUORO-N-HEXANE	338.044	-87.2	57.1	451.7	19.0					235
236	C6H3CL3	1,2,4-TRICHLOROBENZENE	181.449	16.8	213.0	734.9	39.8	0.401				236
237	C6H4CL2	O-DICHLOROBENZENE	147.004	-17.1	180.4	697.3	41.0	0.360	1306	20	39691	237
238	C6H4CL2	M-DICHLOROBENZENE	147.004	-24.8	172.8	684.0	38.5	0.359	1288	20	38644	238
239	C6H4CL2	P-DICHLOROBENZENE	147.004	53.1	174.1	685.0	39.5	0.372	1248	55	38812	239
240	C6H5BR	BROMOBENZENE	157.010	-30.9	156.0	670.0	45.2	0.324	1495	20		240
241	C6H5CL	CHLOROBENZENE	112.559	-45.6	131.7	632.4	45.2	0.308	1106	20	36572	241
242	C6H5F	FLUOROBENZENE	96.104	-39.2	85.3	560.1	45.5	0.271	1024	20		242
243	C6H5I	IODOBENZENE	204.011	-31.4	188.2	721.0	45.2	0.351	1855	4	39523	243
244	C6H6	BENZENE	78.114	5.5	80.1	562.1	48.9	0.259	885	16	30781	244
245	C6H6O	PHENOL	94.113	40.8	181.8	694.2	61.3	0.229	1059	40	45636	245
246	C6H7N	ANILINE	93.129	-6.2	184.3	699.0	53.1	0.270	1022	20	41868	246
247	C6H7N	4-METHYL PYRIDINE	93.129	3.7	145.3	646.0	44.6	0.311	955	20	37472	247
248	C6H10	1,5-HEXADIENE	82.146	-141.2	59.4	507.0	34.5	0.328	692	20	27470	248
249	C6H10	CYCLOHEXENE	82.146	-103.5	82.9	560.4	43.5	0.292	816	16	30480	249
250	C6H10O	CYCLOHEXANONE	98.145	-31.2	155.6	629.0	38.5	0.312	951	15	39775	250

NO	VISA	VISB	DELHF	DELGF	CPVAPA	CPVAPB	CPVAPC	CPVAPD	ANTA	ANTB	ANTC	TMN	TMX	NO
201			144.44	210.59	18.066	3.503E-01	-1.913E-04	4.097E-08	16.0429	2515.62	-45.97	-43	62	201
202	328.49	182.48	75.78	145.95	-34.122	3.584E-01	-3.337E-04	1.000E-07	15.8548	2467.40	-39.64	-23	57	202
203			129.79	198.75	14.687	3.597E-01	-1.975E-04	4.262E-08	15.9880	2541.83	-42.26	-23	62	203
204	574.71	303.44	-192.76		-40.641	5.225E-01	-3.034E-04	7.130E-08	16.0897	3193.92	-66.15	27	167	204
205	438.08	256.84		38.64	16.810	3.689E+00	-1.381E-04	-5.731E-09	16.0890	2974.94	-58.15		136	205
206	406.69	231.67	-77.29	79.17	-53.625	5.426E-01	-3.030E-04	6.485E-08	15.8574	2588.48	-41.79	-43	72	206
207	305.25	174.70	-20.93	71.89	-0.134	4.329E-01	-2.317E-04	4.680E-08	15.7646	2405.96	-39.63	-53	52	207
208	305.31	175.72	71.89	69.95	-13.151	4.601E-01	-2.540E-04	5.455E-08	15.8251	2459.05	-42.56	-53	57	208
209	349.33	176.62	-28.09	65.65	1.947	4.181E-01	-2.177E-04	4.404E-08	15.9011	2495.97	-40.18	-53	57	209
210	369.27	193.39	-31.78	59.70	10.572	3.997E-01	-1.946E-04	3.313E-08	15.8260	2426.42	-40.36	-47	52	210
211	322.47	180.43	-36.34	74.82	11.803	3.509E-01	-1.116E-04	-5.807E-09	15.9238	2521.53	-40.31	-63	62	211
212			-42.58	-108.35	21.742	3.889E-01	-2.007E-04	4.010E-08	15.7179	2333.61	-36.33	-46	42	212
213	521.30	252.03	-28.97	-137.16	1.147	4.329E-01	-2.107E-04	3.162E-08	16.1623	3030.20	-58.15		139	213
214	437.94	243.03			14.239	4.802E-01	-2.818E-04	6.661E-08	16.0031	2934.87	-62.25	-2	137	214
215			-227.97	135.36	30.011	3.939E-01	-1.906E-04	6.665E-08	14.1779	1993.12	-103.20		133	215
216	409.17	236.65	-258.83	-357.43	13.389	5.032E-01	-2.931E-04	3.397E-08	16.8138	3410.51	-40.15	77	127	216
217	729.09	341.13	-490.69		19.850	5.033E-01	-2.931E-04	6.619E-08	17.6306	4092.15	-86.55	5	222	217
218						4.500E-01	-1.436E-04	-7.402E-09	16.2292	2980.47	-64.15	7	136	218
219	489.53	255.83	-466.03	-323.72	15.420	4.034E-01	-1.686E-04	-1.439E-08	16.2291	2980.47	-64.15	5	137	219
220	463.31	248.72	-470.18		19.854	5.034E-01	-1.436E-04	-7.402E-09	16.1620	2935.11	-64.16		123	220
221	479.35	246.09												221
222	451.21	254.66												222
223	772.79	313.49	-49.03	-8.37	-53.068	6.288E-01	-3.357E-04	6.426E-08	16.1004	3015.46	-61.15	7	143	223
224	313.66	182.48	-146.54	-14.82	-3.626	4.873E-01	-2.580E-04	5.304E-08	15.8333	2477.07	-39.94	-53	57	224
225	367.32	191.58	-154.58	-15.24	-9.525	5.066E-01	-2.729E-04	5.723E-08	15.6338	2348.67	-40.05	-57	49	225
226	355.54	196.35	-166.09	-146.12	-16.592	5.551E-01	-3.306E-04	7.632E-08	16.2069	2034.15	-45.37	-13	32	226
227	1151.10	349.61	-298.94	-165.71	3.869	5.045E-01	-2.639E-04	5.120E-08	16.5270	3026.89	-105.00	37	138	227
228	1259.40	252.89	-302.71		-9.483	5.677E-01	-3.481E-04	8.637E-08	16.2708	2752.19	-116.30	34	129	228
229	1148.80	349.51	-302.29	-165.38	-9.542	5.681E-01	-3.484E-04	8.649E-08	16.7127	3026.43	-104.10	25	153	229
230	1502.00	336.75	-329.92	-125.52	-12.087	6.096E-01	-4.203E-04	1.228E-07	15.0113	1988.08	-137.80	25	102	230
231			-293.08		12.154	5.396E-01	-3.159E-04	7.121E-08	18.1336	3694.96	-65.00	55	133	231
232	399.87	213.39							16.3549	2423.41	-62.28	-27	87	232
233			-957.27	-879.9E	36.283	5.267E-01	-4.546E-04	1.455E-07	16.1940	2827.53	-57.66	-3	117	233
234									13.9087	1374.07	-136.80	7	127	234
235									15.8307	2488.59	-59.73	-3	57	235
236									16.8979	4452.50	-53.00	127	327	236
237	554.35	319.07	29.98	82.73	-14.361	6.087E-01	-5.622E-04	2.072E-07	16.2799	3798.23	-59.84	58	210	237
238	402.20	300.89	26.46	78.63	-14.302	5.505E-01	-4.513E-04	1.429E-07	16.8173	4104.13	-43.15	53	202	238
239	483.82	302.03	23.03	77.20	-13.590	5.493E-01	-4.505E-04	1.426E-07	16.1135	3626.83	-64.64	54	204	239
240	508.18	312.42	105.09	138.62	-14.344	5.534E-01	-4.559E-04	1.447E-07	15.7972	3313.00	-67.71	47	177	240
241	477.76	276.22	51.87	99.23	-28.805	5.350E-01	-4.080E-04	1.211E-07	16.0676	3295.12	-55.60	47	147	241
242	452.06	252.89	-116.64	-69.08	-33.888	5.631E-01	-4.521E-04	1.426E-07	16.5487	3181.78	-37.59	-23	97	242
243	565.72	331.21	162.66	187.90	-38.728	5.668E-01	-4.433E-04	1.355E-07	16.1454	3776.53	-64.38	17	197	243
244	545.64	265.34	82.98	129.75	-29.274	5.564E-01	-4.509E-04	1.443E-07	15.9008	2788.51	-52.36	72	104	244
245	1405.50	370.07	-96.67	-32.91	-33.917	5.982E-01	-3.017E-04	7.130E-07	15.4279	3490.89	-98.59	67	208	245
246	1074.60	357.21	86.92	166.80	-35.843	6.384E-01	-4.827E-04	1.526E-07	16.6748	3857.52	-73.15	27	227	246
247	500.97	285.50	102.28		-40.516	4.881E-01	-5.133E-04	1.633E-07	16.2143	3409.40	-62.65	9	187	247
248			83.74		-17.430		-2.798E-04	5.451E-08	16.1351	2728.54	-45.45	27	77	248
249	506.92	264.54	-5.36	106.93	-68.651	7.251E-01	-5.413E-04	1.644E-07	15.8243	2813.53	-49.98		87	249
250	787.38	336.47	-230.27	-90.81	-37.807	5.539E-01	-1.953E-04	-1.534E-08						250

NO	FORMULA	COMPOUND NAME	MOLWT	TFP	TBP	TC	PC	VC	LDEN	TDEN	HVAP	NO
251	C6H12	CYCLOHEXANE	84.162	6.5	80.7	553.4	40.7	0.308	779	20	29977	251
252	C6H12	METHYLCYCLOPENTANE	84.162	-142.5	71.8	532.7	37.9	0.319	754	16	29098	252
253	C6H12	1-HEXENE	84.162	-139.9	63.4	504.0	31.7	0.350	673	20	28303	253
254	C6H12	CIS-2-HEXENE	84.162	-141.2	68.8	518.0	32.8	0.351	687	20	29140	254
255	C6H12	TRANS-2-HEXENE	84.162	-133.2	67.8	516.0	32.7	0.351	678	20	28931	255
256	C6H12	CIS-3-HEXENE	84.162	-137.9	66.4	517.0	32.8	0.350	680	20	28721	256
257	C6H12	TRANS-3-HEXENE	84.162	-113.5	67.1	519.0	32.5	0.350	677	20	28973	257
258	C6H12	2-METHYL-2-PENTENE	84.162	-135.1	67.3	518.0	32.8	0.351	691	16	29015	258
259	C6H12	3-METHYL-CIS-2-PENTENE	84.162	-134.8	67.7	518.0	32.8	0.351	694	20	28847	259
260	C6H12	3-METHYL-TRANS-2-PENTENE	84.162	-138.5	70.4	521.0	32.9	0.350	698	20	29308	260
261	C6H12	4-METHYL-CIS-2-PENTENE	84.162	-134.2	56.4	490.0	30.4	0.360	669	20	27591	261
262	C6H12	4-METHYL-TRANS-2-PENTENE	84.162	-141.2	58.5	493.0	30.4	0.360	669	20	27968	262
263	C6H12	2,3-DIMETHYL-1-BUTENE	84.162	-157.3	55.6	501.0	32.4	0.343	678	20	27424	263
264	C6H12	2,3-DIMETHYL-2-BUTENE	84.162	-74.3	73.2	524.0	33.6	0.351	708	20	29655	264
265	C6H12	3,3-DIMETHYL-1-BUTENE	84.162	-115.2	41.2	490.0	32.5	0.340	653	20	25665	265
266	C6H12O	CYCLOHEXANOL	100.161	24.8	161.1	625.0	37.5	0.327	942	30	45511	266
267	C6H12O	METHYL ISOBUTYL KETONE	100.161	-84.1	116.4	571.0	32.7	0.371	801	0	35588	267
268	C6H12O2	N-BUTYL ACETATE	116.160	-73.5	126.0	579.0	31.4	0.400	898	0	36006	268
269	C6H12O2	ISOBUTYL ACETATE	116.160	-98.9	116.8	561.0	30.4	0.414	875	20	35873	269
270	C6H12O2	ETHYL BUTYRATE	116.160	-93.2	120.8	566.0	31.4	0.395	879	20	34332	270
271	C6H12O2	ETHYL ISOBUTYRATE	116.160	-88.2	111.0	553.0	30.4	0.410	869	20	35023	271
272	C6H12O2	N-PROPYL PROPIONATE	116.160	-75.9	122.5	578.0			881	20	36383	272
273	C6H14	N-HEXANE	86.178	-95.4	68.7	507.4	29.7	0.370	659	20	28872	273
274	C6H14	2-METHYL PENTANE	86.178	-153.7	60.2	497.5	30.1	0.367	653	20	27800	274
275	C6H14	3-METHYL PENTANE	86.178	-118.2	63.2	504.4	31.2	0.367	664	20	28093	275
276	C6H14	2,2-DIMETHYL BUTANE	86.178	-99.9	49.7	488.7	30.8	0.359	649	20	26322	276
277	C6H14	2,3-DIMETHYL BUTANE	86.178	-128.6	58.0	499.9	31.3	0.358	662	20	27298	277
278	C6H14O	1-HEXANOL	102.177	-44.0	157.0	610.0	40.5	0.381	819	20	48567	278
279	C6H14O	ETHYL BUTYL ETHER	102.177	-103.2	92.3	531.0	30.4	0.390	749	20	31820	279
280	C6H14O	DIISOPROPYL ETHER	102.177	-85.5	68.3	500.0	28.8	0.386	724	20	29349	280
281	C6H15N	DIPROPYLAMINE	101.193	-63.2	109.2	550.0	31.4	0.407	738	20	37011	281
282	C6H15N	TRIETHYLAMINE	101.193	-114.8	89.5	535.0	30.4	0.390	728	20	31401	282
283	C7F14	PERFLUOROMETHYLCYCLOHEXANE	350.055		76.3	486.8	23.3	0.664	1733	20		283
284	C7F16	PERFLUORO-N-HEPTANE	388.051	-78.2	82.5	474.8	16.2					284
285	C7H5N	BENZONITRILE	103.124	-13.2	190.8	699.4	42.2		1010	15		285
286	C7H6O	BENZALDEHYDE	106.124	-57.2	178.8	695.0	46.6		1045	20	42705	286
287	C7H6O2	BENZOIC ACID	122.124	122.4	249.8	752.0	45.6	0.341	1075	130	50660	287
288	C7H8	TOLUENE	92.141	-95.2	110.6	591.7	41.7	0.316	867	20	33201	288
289	C7H8O	METHYL PHENYL ETHER	108.140	-37.5	153.6	641.0	46.6		996	20		289
290	C7H8O	BENZYL ALCOHOL	108.140	-15.4	205.4	677.0	50.1	0.334	1041	25	50535	290
291	C7H8O	O-CRESOL	108.140	30.9	191.0	697.6	45.6	0.282	1028	40	45217	291
292	C7H8O	M-CRESOL	108.140	12.2	202.2	705.8	51.5	0.310	1034	20	47436	292
293	C7H8O	P-CRESOL	108.140	34.7	201.9	704.6			1019	40	47478	293
294	C7H9N	2,3-DIMETHYLPYRIDINE	107.156		160.8	655.4			942	25		294
295	C7H9N	2,5-DIMETHYLPYRIDINE	107.156		157.0	644.2			938	0		295
296	C7H9N	3,4-DIMETHYLPYRIDINE	107.156		179.1	683.8			954	25		296
297	C7H9N	3,5-DIMETHYLPYRIDINE	107.156		171.9	667.0			939	25		297
298	C7H9N	METHYLPHENYLAMINE	107.156	-57.2	195.9	701.0	52.0		989	20		298
299	C7H9N	O-TOLUIDINE	107.156	-14.8	200.1	694.0	37.5	0.343	998	20	45364	299
300	C7H9N	M-TOLUIDINE	107.156	-30.4	203.3	709.0	41.5	0.343	989	20	45636	300

NO	VISA	VISB	DELHF	DELGF	CPVAPA	CPVAPB	CPVAPC	CPVAPD	ANTA	ANTB	ANTC	TMN	TMX	NO
251	653.62	290.84	-123.22	31.78	-54.541	6.112E-01	-2.523E-04	1.321E-08	15.7527	2766.63	-50.50	7	107	251
252	440.52	243.24	-105.93	35.80	-50.108	6.380E-01	-3.642E-04	8.013E-08	15.8023	2731.00	-47.11	-23	102	252
253	357.43	197.74	-41.70	87.50	-1.746	5.300E-01	-2.902E-04	6.054E-08	15.8089	2654.81	-47.30	-33	87	253
254	344.33	197.95	-52.38	76.28	-9.810	5.300E-01	-2.717E-04	6.827E-08	16.2057	2897.97	-39.30	-30	97	254
255	344.33	197.95	-53.93	76.49	-32.925	5.929E-01	-5.618E-04	2.004E-07	15.8727	2701.31	-48.62	-28	92	255
256	344.33	197.95	-47.65	83.07	-21.729	5.811E-01	-3.361E-04	7.456E-08	15.8384	2680.52	-48.40	-28	92	256
257	344.33	197.95	-54.47	77.67	-4.338	5.509E-01	-3.282E-04	8.047E-08	15.9288	2718.68	-47.77	-28	92	257
258	344.33	197.95	-59.79	71.26	-14.750	5.668E-01	-3.340E-04	7.963E-08	15.9423	2725.89	-47.64	-28	97	258
259			-57.78	73.27	-14.750	5.668E-01	-3.340E-04	7.963E-08	15.9124	2731.79	-46.76	-25	91	259
260			-58.70	71.34	-14.750	5.668E-01	-3.340E-04	7.963E-08	15.9484	2750.50	-48.33	-23	93	260
261			-50.37	82.19	-1.675	5.375E-01	-3.044E-04	6.753E-08	15.7527	2580.52	-46.56	-35	79	261
262			-54.39	79.67	12.627	5.154E-01	-3.007E-04	7.326E-08	15.8425	2631.65	-46.00	-33	81	262
263			-55.77	79.09	7.025	5.585E-01	-3.696E-04	1.063E-07	15.8012	2612.69	-43.78	-38	87	263
264			-59.24	75.91	2.294	4.827E-01	-2.198E-04	3.041E-08	16.0043	2798.63	-47.71	-23	102	264
265			-43.17	98.22	-12.556	5.484E-01	-2.915E-04	5.208E-08	15.3755	2326.80	-48.24	-48	67	265
266			-294.75	-117.98	-55.534	7.213E-01	-4.086E-04	8.235E-08	15.7165	2893.66	-70.75	12	152	266
267	473.65	259.03	-284.03		3.894	6.656E-01	-3.318E-04	2.231E-08	16.1836	3151.09	-69.15	22	162	267
268	537.58	272.30	-486.76		13.620	5.488E-01	-2.278E-04	-7.913E-10	16.1714	3092.83	-66.15	16	154	268
269	533.99	270.49	-495.47		7.310	5.740E-01	-2.575E-04	1.101E-08	15.9987	3127.60	-60.15	15	159	269
270	489.95	264.22			21.508	4.927E-01	-1.938E-04	3.558E-09						270
271														271
272	362.79	207.09	-167.30	-0.25	-4.413	5.819E-01	-3.118E-04	6.493E-08	16.8641	3558.18	-47.86	-19	147	272
273	384.13	208.27	-174.42	-5.02	-10.567	6.183E-01	-3.572E-04	8.084E-08	15.8366	2697.55	-48.78	-28	97	273
274	372.11	207.55	-171.74	-2.14	-2.386	5.689E-01	-3.869E-04	5.032E-08	15.7476	2614.38	-46.58	-33	97	274
275	438.44	226.67	-185.68	-9.63	-16.634	6.292E-01	-3.480E-04	6.820E-08	15.5536	2653.43	-46.02	-43	77	275
276	444.19	228.86	-177.90	-4.10	-14.608	6.150E-01	-3.375E-04	6.426E-08	15.6802	2489.50	-43.81	-38	81	276
277	1179.40	354.94	-317.78	-135.65	4.811	5.890E-01	-3.009E-04	4.156E-08	16.0994	2595.44	-44.25	-35	157	277
278	443.32	234.68			23.626	5.367E-01	-2.528E-04	8.844E-08	16.0477	4055.45	-76.49	-8	127	278
279	410.58	219.67	-319.03	-121.96	7.503	5.849E-01	-3.026E-04	7.071E-08	16.3417	2921.52	-55.15	-24	91	279
280	561.11	257.39			6.460	5.292E-01	-3.390E-04		15.5939	2895.73	-43.15	-29	149	280
281	355.52	214.48	-99.65	110.36	-18.430	7.155E-01	-4.392E-04	1.092E-07	15.8853	3259.08	-55.15	-13	127	281
282										2882.38	-51.15			282
283			-2898.10	-3089.31	-26.004	5.731E-01	-4.429E-04	1.349E-07						283
284			-3386.70	261.05	-12.142	4.961E-01	-2.844E-04	5.166E-08						284
285	686.84	314.66	218.97	22.40	-51.292	6.292E-01	-4.237E-04	1.062E-07	15.7130	2610.57	-61.93	17	112	285
286	2617.60	407.88	-36.80	-210.55	-24.355	5.124E-01	-2.765E-04	4.911E-08	15.9747	2719.68	-64.50	-3	117	286
287	467.33	255.24	-290.40	122.09					16.3501	3748.62	-66.12	27	187	287
288	388.84	325.85	50.03		-7.398	5.480E-01	-3.357E-04	7.770E-08	17.1634	4190.70	-125.20	132	287	288
289	1088.00	365.21		-37.10	-32.276	7.004E-01	-5.924E-04	2.124E-07	16.0137	3096.52	-53.67	97	137	289
290	1533.40	365.61	-94.08	-40.57	-45.008	7.264E-01	-6.029E-04	2.077E-07	16.2394	3430.82	-69.58	112	330	290
291	1785.60	370.75	-128.70	-30.90	-40.633	7.054E-01	-5.756E-04	1.967E-07	17.4582	4384.81	-73.15	97	207	291
292	1826.90	372.68	-132.43						15.9148	3305.37	-108.00	97	207	292
293			-125.48						17.2878	4274.42	-74.09	147	167	293
294			68.29						17.1989	4219.74	-111.30	77	162	294
295			66.44						17.1492	3545.14	-33.04	127	187	295
296			70.05						16.3046	4237.04	-63.59	127	187	296
297	915.12	332.74	72.81						16.9517	4106.95	-41.65	47	207	297
298	1085.10	356.46	85.41	199.33					16.8850	3756.28	-44.45	102	227	298
299	928.12	354.07							16.3066	4072.58	-80.71	82	227	299
300					-15.989	5.681E-01	-3.032E-04	4.643E-08	16.7834	4080.32	-73.15			300

NO	FORMULA	COMPOUND NAME	MOLWT	TFP	TBP	TC	PC	VC	LDEN	TDEN	HVAP	NO
301	C7H9N	P-TOLUIDINE	107.156	43.7	200.1	667.0	37.2	0.390	964	50	44799	301
302	C7H14	CYCLOHEPTANE	98.189	-8.2	118.7	589.0			810	20	33076	302
303	C7H14	1,1-DIMETHYLCYCLOPENTANE	98.189	-69.8	87.8	547.0	34.5	0.360	759	16	30312	303
304	C7H14	CIS-1,2-DIMETHYLCYCLOPENTANE	98.189	-53.9	99.5	564.8	34.5	0.368	777	16	31719	304
305	C7H14	TRANS-1,2-DIMETHYLCYCLOPENTANE	98.189	-117.6	91.8	553.2	34.5		756	16	30878	305
306	C7H14	ETHYLCYCLOPENTANE	98.189	-138.5	103.4	569.5	33.9	0.375	771	16	32301	306
307	C7H14	METHYLCYCLOHEXANE	98.189	-126.6	100.9	572.1	34.8	0.368	774	16	31150	307
308	C7H14	1-HEPTENE	98.189	-118.9	93.6	537.2	28.4	0.440	679	20	31108	308
309	C7H14	2,3,3-TRIMETHYL-1-BUTENE	98.189	-109.9	77.8	533.0	28.4		705	20	28889	309
310	C7H16	N-HEPTANE	100.250	-90.6	98.4	540.2	27.4	0.432	684	20	31719	310
311	C7H16	2-METHYLHEXANE	100.205	-118.3	90.0	530.3	27.3	0.421	679	20	30689	311
312	C7H16	3-METHYLHEXANE	100.205	-173.2	91.8	535.2	28.1	0.404	687	20	30815	312
313	C7H16	2,2-DIMETHYLPENTANE	100.205	-123.8	79.2	520.4	27.8	0.416	674	20	29182	313
314	C7H16	2,3-DIMETHYLPENTANE	100.205		89.7	537.3	29.1	0.393	695	20	30409	314
315	C7H16	2,4-DIMETHYLPENTANE	100.205	-119.2	80.5	519.7	27.4	0.418	673	20	29517	315
316	C7H16	3,3-DIMETHYLPENTANE	100.205	-134.5	86.0	536.3	29.5	0.414	693	20	29668	316
317	C7H16	3-ETHYLPENTANE	100.205	-118.6	93.4	540.6	28.9	0.416	698	20	30978	317
318	C7H16	2,2,3-TRIMETHYLBUTANE	100.205	-24.9	80.8	531.1	29.6	0.398	690	20	28968	318
319	C7H16O	1-HEPTANOL	116.204	-34.0	176.3	633.0	30.4	0.435	822	20	48148	319
320	C8H4O3	PHTHALIC ANHYDRIDE	148.118	130.8	286.8	810.0	47.6	0.368			49614	320
321	C8H8	STYRENE	104.152	-30.7	145.1	647.0	39.9		906	20	36844	321
322	C8H8O	METHYL PHENYL KETONE	120.151	19.6	201.7	701.0	38.5	0.376	1032	15	43124	322
323	C8H8O2	METHYL BENZOATE	136.151	-12.4	199.0	692.0	36.5	0.396	1083	20	36844	323
324	C8H10	O-XYLENE	106.168	-25.2	144.4	630.2	37.3	0.369	880	20	36383	324
325	C8H10	M-XYLENE	106.168	-47.9	139.1	617.0	35.5	0.376	864	20	36006	325
326	C8H10	P-XYLENE	106.168	13.2	138.3	616.2	35.2	0.379	861	20	35588	326
327	C8H10	ETHYL BENZENE	106.168	-95.0	136.1	617.1	36.1	0.374	867	20		327
328	C8H10O	O-ETHYLPHENOL	122.167	-3.4	204.5	703.0			1037	0	48106	328
329	C8H10O	M-ETHYLPHENOL	122.167	-4.2	218.4	716.4			1025	0	50828	329
330	C8H10O	P-ETHYLPHENOL	122.167	44.8	217.8	716.4					50660	330
331	C8H10O	ETHYL PHENYL ETHER	122.167	-30.2	169.8	647.0	34.2		979	4	47311	331
332	C8H10O	2,3-XYLENOL	122.167	74.8	216.9	722.8					47143	332
333	C8H10O	2,4-XYLENOL	122.167	24.8	210.8	707.6					46892	333
334	C8H10O	2,5-XYLENOL	122.167	74.8	211.1	723.0					44380	334
335	C8H10O	2,6-XYLENOL	122.167	48.8	200.9	701.0					49823	335
336	C8H10O	3,4-XYLENOL	122.167	64.8	226.8	729.8					49404	336
337	C8H10O	3,5-XYLENOL	122.167	63.8	221.6	715.6						337
338	C8H11N	N,N-DIMETHYLANILINE	121.183	2.4	193.5	687.0	36.3		956	20		338
339	C8H16	1,1-DIMETHYLCYCLOHEXANE	112.216	-33.5	119.5	591.0	29.7		785	16	32615	339
340	C8H16	CIS-1,2-DIMETHYLCYCLOHEXANE	112.216	-50.1	129.7	606.0	29.7	0.416	796	20	33662	340
341	C8H16	TRANS-1,2-DIMETHYLCYCLOHEXANE	112.216	-88.2	123.4	596.0	29.7		776	20	32908	341
342	C8H16	CIS-1,3-DIMETHYLCYCLOHEXANE	112.216	-75.6	120.1	591.0	29.7		766	20	32825	342
343	C8H16	TRANS-1,3-DIMETHYLCYCLOHEXANE	112.216	-90.2	124.4	598.0	29.7		785	20	33871	343
344	C8H16	CIS-1,4-DIMETHYLCYCLOHEXANE	112.216	-87.5	124.3	598.0	29.7		783	20	33787	344
345	C8H16	TRANS-1,4-DIMETHYLCYCLOHEXANE	112.216	-37.0	119.3	590.0	29.7		763	20	32615	345
346	C8H16	ETHYLCYCLOHEXANE	112.216	-111.4	131.7	609.0	30.3	0.450	788	20	34332	346
347	C8H16	1,1,2-TRIMETHYLCYCLOPENTANE	112.216		113.7	579.5	29.4				32615	347
348	C8H16	1,1,3-TRIMETHYLCYCLOPENTANE	112.216		104.8	569.5	28.3				31694	348
349	C8H16	CIS,CIS,TRANS-1,2,4-TRIMETHYLCYCLOPENTANE	112.216		117.8	579.0	28.8				33076	349
350	C8H16	CIS,TRANS,CIS-1,2,4-TRIMETHYLCYCLOPENTANE	112.216		109.2	571.0	28.1				33076	350

NO	TMX	TMN	ANTC	ANTB	ANTA	CPVAPD	CPVAPC	CPVAPB	CPVAPA	DELGF	DELHF	VISB	VISA	NO
301	227	77	-72.15	4041.04	16.6968	7.561E-08	-4.203E-04	7.867E-01	-76.187	63.05	-119.41	356.02	738.90	301
302	162	57	-56.80	3066.05	15.7818	1.010E-07	-4.500E-04	7.670E-01	-57.891	39.06	-138.37			302
303	117	-13	-51.20	2807.94	15.6973	1.014E-07	-4.484E-04	7.615E-01	-55.643	45.76	-129.62			303
304	127	-3	-52.94	2922.30	15.7729	1.017E-07	-4.479E-04	7.590E-01	-54.521	38.39	-136.78			304
305	117	-13	-51.46	2861.53	15.7594	1.004E-07	-4.396E-04	7.511E-01	-55.312	44.59	-127.15	249.72	433.81	305
306	129	-3	-52.47	2990.13	15.8581	9.365E-08	-4.438E-04	7.841E-01	-61.919	27.30	-154.87	271.58	528.41	306
307	127	-3	-51.75	2926.04	15.7105	7.607E-08	-3.511E-04	6.296E-01	-3.303	95.88	-62.34	214.32	368.69	307
308	127	-8	-53.97	2895.51	15.8894						-86.54			308
309	102	-20	-49.56	2719.47	15.6536									309
310	127	-3	-56.51	2911.32	15.8737	7.657E-08	-3.650E-04	6.761E-01	-5.146	8.00	-187.90	232.53	436.73	310
311	117	-9	-53.60	2845.06	15.8261	1.836E-07	-6.288E-04	8.641E-01	-39.389	3.22	-195.06	225.13	417.46	311
312	117	-8	-53.93	2855.66	15.8133	7.833E-08	-3.734E-04	8.837E-01	-7.046	4.61	-192.43			312
313	105	-19	-49.85	2740.15	15.6917	1.735E-07	-6.359E-04	8.955E-01	-50.099	0.08	-206.28	226.19	417.37	313
314	115	-11	-51.33	2850.64	15.7815	7.833E-08	-3.734E-04	7.047E-01	-7.046	0.67	-199.38			314
315	105	-17	-51.52	2744.78	15.7179	7.833E-08	-3.734E-04	6.837E-01	-7.046	3.10	-202.14			315
316	112	-13	-47.83	2829.10	15.7190	7.833E-08	-3.734E-04	6.837E-01	-7.046	2.64	-201.68			316
317	119	-7	-53.26	2882.44	15.8317	1.004E-07	-4.421E-04	7.519E-01	-22.944	11.01	-189.79			317
318	106	-19	-47.10	2764.40	15.6398					4.27	-204.94			318
319	176	60	-146.60	2626.42	15.3068	6.045E-08	-3.446E-04	6.778E-01	4.907	-121.00	-332.01	361.83	1287.00	319
320	342	136	-83.15	4467.01	15.9984	1.009E-07	-4.283E-04	6.539E-01	-4.455	213.95	-371.79	276.71	528.64	320
321	187	32	-63.72	3328.57	16.0193	9.935E-08	-4.023E-04	6.158E-01	-28.248	1.84	147.46	310.82	1316.40	321
322	247	77	-81.15	3781.07	16.2384	9.721E-08	-4.071E-04	6.410E-01	-29.580		-86.92	332.33	768.94	322
323	243	77	-81.15	3751.83	16.2272	9.425E-08	-1.799E-04	5.501E-01	-21.210	122.17	-254.06			323
324	172	32	-59.46	3395.57	16.1156	7.527E-08	-3.443E-04	5.962E-01	-15.851	118.95	19.01	277.98	513.54	324
325	167	27	-58.04	3366.99	16.1390	7.478E-08	-3.747E-04	6.296E-01	-29.165	121.21	17.25	257.18	453.42	325
326	167	27	-57.84	3346.65	16.0963	6.820E-08	-3.373E-04	6.041E-01	-25.091	130.67	17.96	261.40	475.16	326
327	177	27	-59.95	3272.47	16.0195	1.300E-07	-4.810E-04	7.071E-01	-43.099		29.81	264.22	472.82	327
328	227	77	-45.75	4928.36	17.9610						-145.78			328
329	227	97	-86.08	4272.77	17.1955						-146.58			329
330	227	97	-44.15	5579.62	19.0905						-144.65			330
331	187	112	-78.66	3473.20	16.1673						-157.34	305.91	646.88	331
332	227	147	-102.40	3724.58	16.2424						-162.78			332
333	227	137	-103.80	3655.26	13.2456						-161.53			333
334	217	137	-102.40	3667.32	16.2328						-161.95			334
335	207	127	-85.55	3749.35	16.2809						-156.50			335
336	247	157	-113.90	3733.53	16.3004						-161.48			336
337	227	137	-109.00	3775.91	16.4192						84.12			337
338	207	72	-52.80	4276.08	16.9647	1.030E-07	-5.020E-04	8.997E-01	-72.105	231.36	-181.12	320.03	553.02	338
339	147	10	-55.30	3043.34	15.6535	1.098E-07	-5.137E-04	8.972E-01	-68.370	35.25	-172.29			339
340	157	17	-57.31	3148.35	15.7438	1.181E-07	-5.354E-04	9.123E-01	-68.479	41.24	-180.12			340
341	151	13	-54.02	3117.43	15.7337	1.019E-07	-4.932E-04	8.338E-01	-65.163	34.50	-184.89			341
342	147	11	-55.08	3081.95	15.7470	1.068E-07	-5.015E-04	8.825E-01	-64.154	29.85	-176.68			342
343	152	15	-57.76	3093.95	15.7371	1.068E-07	-5.015E-04	8.825E-01	-64.154	36.34	-176.77			343
344	152	14	-57.00	3098.39	15.7333	1.154E-07	-5.308E-04	9.131E-01	-70.363	37.97	184.72			344
345	147	10	-54.57	3063.44	15.6984	1.102E-07	-5.107E-04	8.892E-01	-63.891	31.74	-171.87			345
346	160	20	-58.15	3183.25	15.8125					39.27		280.76	506.43	346
347	141	6	-54.59	3015.51	15.7084									347
348	131	0	-53.25	2938.09	15.6794									348
349	145	10	-54.20	3073.95	15.7543									349
350	144	9	-53.23	3009.70	15.7756									350

NO	FORMULA	COMPOUND NAME	MOLWT	TFP	TBP	TC	PC	VC	LDEN	TDEN	HVAP	NO
351	C8H16	1-METHYL-1-ETHYLCYCLOPENTANE	112.216		121.5	592.0	29.9		781	16	33662	351
352	C8H16	N-PROPYLCYCLOPENTANE	112.216	-117.4	130.9	603.0	30.0	0.425	776	20	34131	352
353	C8H16	ISOPROPYLCYCLOPENTANE	112.216	-112.7	126.4	601.0	30.0		715	20	34122	353
354	C8H16	1-OCTENE	112.216	-101.8	121.2	566.6	26.2	0.464	720	20	33787	354
355	C8H16	TRANS-2-OCTENE	112.216	-87.8	124.9	580.0	27.7		703	20	34332	355
356	C8H18	N-OCTANE	114.232	-56.8	125.6	568.8	24.8	0.492	702	20	34436	356
357	C8H18	2-METHYLHEPTANE	114.232	-109.2	117.6	559.6	24.8	0.488	706	16	33829	357
358	C8H18	3-METHYLHEPTANE	114.232	-120.5	118.9	563.6	25.4	0.464	705	20	33913	358
359	C8H18	4-METHYLHEPTANE	114.232	-121.0	117.7	561.7	25.4	0.464	705	20	33913	359
360	C8H18	2,2-DIMETHYLHEXANE	114.232	-121.2	108.8	549.8	25.3	0.476	712	20	32280	360
361	C8H18	2,3-DIMETHYLHEXANE	114.232		115.6	563.4	26.2	0.468	700	20	33226	361
362	C8H18	2,4-DIMETHYLHEXANE	114.232		109.4	563.5	25.5	0.468	695	20	32615	362
363	C8H18	2,5-DIMETHYLHEXANE	114.232		109.1	553.5	24.8	0.472	693	20	32657	363
364	C8H18	3,3-DIMETHYLHEXANE	114.232	-91.3	111.9	550.0	26.5	0.482	710	20	32490	364
365	C8H18	3,4-DIMETHYLHEXANE	114.232	-126.2	117.7	562.0	27.0	0.443	719	16	33298	365
366	C8H18	3-ETHYLHEXANE	114.232		118.5	565.4	26.0	0.466	718	20	33633	366
367	C8H18	2,2,3-TRIMETHYLPENTANE	114.232	-112.3	109.8	563.4	27.3	0.455	716	20	32029	367
368	C8H18	2,2,4-TRIMETHYLPENTANE	114.232	-107.4	99.2	543.9	25.6	0.436	692	20	31028	368
369	C8H18	2,3,3-TRIMEHTYLPENTANE	114.232	-100.7	114.7	573.5	28.2	0.468	726	20	32364	369
370	C8H18	2,3,4-TRIMETHYLPENTANE	114.232	-109.3	113.4	566.3	27.1	0.455	719	20	32753	370
371	C8H18	2-METHYL-3-ETHYLPENTANE	114.232	-115.0	115.6	567.0	27.1	0.461	719	20	32988	371
372	C8H18	3-METHYL-3-ETHYLPENTANE	114.232	-90.9	118.2	576.5	28.1	0.443	727	20	32816	372
373	C8H18O	1-OCTANOL	130.231	-15.5	195.2	658.0	34.5	0.490	826	20	50660	373
374	C8H18O	2-OCTANOL	130.231	-32.0	179.7	637.0	27.4	0.494	821	20	44380	374
375	C8H18O	2-ETHYLHEXANOL	130.231	-70.0	184.6	613.0	27.6	0.494	833	20	46599	375
376	C8H18O	BUTYL ETHER	130.231	-97.9	142.4	580.0	25.3	0.500	768	20	37263	376
377	C8H18O5	TETRAETHYLENE GLYCOL	194.229		318.0	795.8	21.0	0.646		20	39775	377
378	C8H19N	DIBUTYLAMINE	129.247	-62.2	159.6	596.0	25.3	0.517	767	20		378
379	C9H8	INDENE	116.163		181.9	691.9	38.2					379
380	C9H10	INDAN	118.179		177.0	681.1	36.3	0.377			38309	380
381	C9H10	ALPHA-METHYL STYRENE	118.179	-34.9	165.3	654.0	34.0	0.392	911	20	44799	381
382	C9H10O2	ETHYL BENZOATE	150.178	-99.5	212.7	697.0	32.4	0.397	1046	20	38267	382
383	C9H12	N-PROPYLBENZENE	120.195	-96.1	159.2	638.3	32.0	0.451	862	20	37556	383
384	C9H12	ISOPROPYLBENZENE	120.195	-80.9	152.4	631.0	32.1	0.440	862	20	38895	384
385	C9H12	1-METHYL-2-ETHYLBENZENE	120.195	-95.6	165.1	651.0	30.4	0.428	881	20	38560	385
386	C9H12	1-METHYL-3-ETHYLBENZENE	120.195	-62.4	161.3	637.0	28.4	0.460	865	20	38435	386
387	C9H12	1-METHYL-4-ETHYLBENZENE	120.195	-25.5	162.0	640.0	29.4	0.490	861	20	40068	387
388	C9H12	1,2,3-TRIMETHYLBENZENE	120.195	-46.2	176.0	664.5	34.6	0.470	894	20	39272	388
389	C9H12	1,2,4-TRIMETHYLBENZENE	120.195	-44.8	169.3	649.1	32.3	0.430	880	16	39063	389
390	C9H12	1,3,5-TRIMETHYLBENZENE	120.195	-94.5	164.7	637.3	31.3	0.430	865	20	36090	390
391	C9H18	N-PROPYLCYCLOHEXANE	126.243	-89.8	156.7	639.0	28.1	0.433	793	20	36341	391
392	C9H18	ISOPROPYLCYCLOHEXANE	126.243	-81.4	154.5	640.0	28.4		802	20		392
393	C9H18	1-NONENE	126.243	-53.5	146.8	592.0	23.4	0.580	745	0	36940	393
394	C9H20	N-NONANE	128.259		150.8	594.6	22.9	0.548	718	20	34039	394
395	C9H20	2,2,3-TRIMETHYLHEXANE	128.259	-120.2	133.6	588.0	23.7				34792	395
396	C9H20	2,2,4-TRIMETHYLHEXANE	128.259		126.5	573.7	23.7		720	16	33787	396
397	C9H20	2,2,5-TRIMETHYLHEXANE	128.259	-105.8	124.1	568.0	23.3	0.519	717	16	36006	397
398	C9H20	3,3-DIETHYLPENTANE	128.259		146.1	610.0	26.7		752	20	35295	398
399	C9H20	2,2,3,3-TETRAMETHYLPENTANE	128.259		140.2	607.6	27.4				34290	399
400	C9H20	2,2,3,4-TETRAMETHYLPENTANE	128.259		133.0	592.7	26.0					400

NO	VISA	VISB	DELHF	DELGF	CPVAPA	CPVAPB	CPVAPC	CPVAPD	ANTA	ANTB	ANTC	TMN	TMX	NU
351	454.23	264.22	-148.17	52.63	-55.973	8.449E-01	-4.923E-04	1.117E-07	15.8222	3120.66	-55.06	13	149	351
352									15.8969	3187.67	-59.99	21	158	352
353									15.9630	3176.22	-55.18	16	154	353
354	418.82	237.63	-82.98	104.29	-4.099	7.239E-01	-4.036E-04	8.675E-08	15.8561	3116.52	-60.39	16	147	354
355	427.64	240.32	-94.58	92.74	-12.820	7.532E-01	-4.442E-04	1.050E-07	15.8554	3134.97	-58.00	19	152	355
356	473.70	251.71	-208.59	16.41	-6.096	7.712E-01	-4.195E-04	8.855E-08	15.9426	3120.29	-63.63	12	152	356
357	643.61	259.51	-215.62	12.77	-89.744	1.242E+00	-1.175E-03	9.618E-08	15.9278	3097.63	-59.46	13	144	357
358			-212.77	13.73	-9.215	7.858E-01	-4.400E-04	9.696E-08	15.8865	3065.96	-60.74	12	145	358
359			-212.23	16.75	-9.215	7.858E-01	-4.400E-04	9.696E-08	15.8893	3057.05	-60.59	3	144	359
360			-224.87	10.72	-9.215	7.858E-01	-4.400E-04	9.696E-08	15.7431	2932.56	-58.08	10	132	360
361			-214.07	17.71	-9.215	7.858E-01	-4.400E-04	9.696E-08	15.8189	3029.06	-58.99	5	142	361
362			-219.56	11.72	-9.215	7.858E-01	-4.400E-04	9.696E-08	15.7797	2965.44	-58.36	5	135	362
363			-222.78	10.47	-9.215	7.858E-01	-4.400E-04	9.696E-08	15.7954	2964.06	-58.74	6	135	363
364	446.20	244.67	-220.27	13.27	-9.215	7.858E-01	-4.400E-04	9.696E-08	15.7755	3011.51	-55.71	11	138	364
365			-213.15	17.33	-9.215	7.858E-01	-4.400E-04	9.696E-08	15.8415	3062.52	-58.29	13	144	365
366	437.60	238.33	-211.01	16.54	-9.215	7.858E-01	-4.400E-04	9.696E-08	15.8671	3057.57	-60.55	4	145	366
367	474.57	257.61	-220.27	17.12	-9.215	7.858E-01	-4.400E-04	9.696E-08	15.7162	2981.56	-54.73	-4	136	367
368	467.04	246.43	-224.29	13.69	-7.461	7.779E-01	-4.287E-04	9.173E-08	15.6850	2896.28	-52.41	7	125	368
369			-216.58	18.92	-9.215	7.858E-01	-4.400E-04	9.696E-08	15.7578	3057.94	-52.77	7	142	369
370			-217.59	18.92	-9.215	7.858E-01	-4.400E-04	9.696E-08	15.7818	3028.09	-55.62	9	140	370
371			-211.35	21.27	-9.215	7.858E-01	-4.400E-04	9.696E-08	15.8040	3035.08	-57.84	10	142	371
372			-215.12	19.93	-9.215	7.858E-01	-4.400E-04	9.696E-08	15.8126	3102.06	-53.47		145	372
373	1312.10	369.97	-360.06	-120.16	6.171	7.607E-01	-3.797E-04	6.263E-08	15.7428	3017.81	-137.10	70	195	373
374					25.879	7.640E-01	-4.224E-04	7.064E-08	14.7108	2441.66	-150.70	72	180	374
375	1798.00	351.17	-365.55		-14.993	8.654E-01	-5.279E-04	1.284E-07	15.3614	2773.46	-140.40	75	185	375
376	473.50	266.56	-334.11	-88.59	6.054	7.728E-01	-4.085E-04	8.084E-08	16.0778	3296.15	-66.15	32	182	376
377					7.164	8.616E-01	-2.904E-04	-9.114E-08	20.5564	8215.28	-11.50	227	427	377
378	581.42	286.54			9.764	8.080E-01	-4.392E-04	9.248E-08	15.7307	3721.90	-64.15	49	186	378
379	354.34	270.80			-42.944	6.895E-01	-4.340E-04	9.148E-08	16.4380	3994.97	-49.40	77	277	379
380	746.50	338.47			-59.639	7.812E-01	-4.841E-04	9.847E-08	16.2601	3789.86	-57.00	77	277	380
381					-24.329	6.933E-01	-4.530E-04	1.180E-07	16.3308	3644.30	-67.15	75	220	381
382	527.45	282.65			20.670	7.887E-01	-3.601E-04	5.061E-08	16.2065	3845.09	-84.15	88	258	382
383	517.17	276.22	7.83	137.33	-31.288	7.486E-01	-4.601E-04	1.081E-07	16.0062	3433.84	-66.01	43	188	383
384			3.94	137.08	-39.364	7.841E-01	-5.087E-04	1.291E-07	15.9722	3363.60	-63.37	38	181	384
385			1.21	131.17	-16.446	6.996E-01	-4.120E-04	9.328E-08	16.1253	3535.33	-65.85	48	194	385
386			-1.93	126.53	-28.998	7.293E-01	-4.362E-04	9.998E-08	16.1545	3521.08	-64.64	45	190	386
387	463.17	266.08	-2.05	126.78	-27.310	7.176E-01	-4.224E-04	9.541E-08	16.1135	3516.31	-64.23	56	206	387
388			-9.59	124.64	-6.942	6.334E-01	-3.326E-04	6.611E-08	16.2121	3670.22	-66.07	51	198	388
389	872.74	297.75	-13.94	117.02	-4.668	6.238E-01	-3.262E-04	6.376E-08	16.2190	3622.58	-64.59	48	193	389
390	437.52	263.37	-16.08	118.03	-19.590	7.724E-01	-3.692E-04	7.699E-08	16.2893	3614.19	-63.57	40	186	390
391	549.08	293.93	-193.43	47.35	-62.517	9.889E-01	-5.794E-04	1.291E-07	15.8567	3363.62	-65.21	57	167	391
392									15.8260	3346.12	-63.71	35	175	392
393	471.00	258.92	-103.58	112.75	-3.718	8.122E-01	-4.509E-04	9.705E-08	16.0118	3305.03	-67.61	39	175	393
394	525.56	272.12	-229.19	24.83	3.144	6.774E-01	-1.928E-04	-2.981E-08	15.9671	3291.45	-71.33	24	163	394
395			-241.37	24.53	-45.632	1.055E+00	-7.712E-04	1.986E-07	15.8017	3164.17	-61.66	18	155	395
396			-243.38	22.52	-60.311	1.104E+00	-7.712E-04	2.187E-07	15.7639	3084.08	-61.94	42	147	396
397			254.18	13.44	-54.106	1.094E+00	-7.745E-04	2.254E-07	15.7445	3052.17	-62.24	77	167	397
398			231.95	35.05	-67.269	1.126E+00	-7.988E-04	2.306E-07	15.8709	3341.62	-57.57	55	157	398
399			-237.39	34.33	-54.583	1.089E+00	-7.569E-04	2.142E-07	15.7280	3220.55	-59.31	45	167	399
400			-237.22	32.66	-54.583	1.089E+00	-7.569E-04	2.142E-07	15.7363	3167.42	-58.21	45	157	400

NO	FORMULA	COMPOUND NAME	MOLWT	TFP	TBP	TC	PC	VC	LDEN	TDEN	HVAP	NO
401	C9H29	2,2,4-TETRAMETHYLPENTANE	128.259	-67.2	122.2	574.7	24.8		719	20	32866	401
402	C9H20	2,3,3,4-TETRAMETHYLPENTANE	128.259		141.5	607.6	27.2				34960	402
403	C10H8	NAPHTHALENE	128.174	80.3	217.9	748.4	40.5	0.410	971	90	43292	403
404	C10H12	1,2,3,4-TETRAHYDRONAPHTHALENE	132.206	-31.0	207.5	719.0	35.2		973	20	39733	404
405	C10H14	N-BUTYLBENZENE	134.222	-88.0	183.2	660.5	28.9	0.497	860	20	39272	405
406	C10H14	ISOBUTYLBENZENE	134.222	-51.5	172.7	650.0	31.4	0.480	853	20	37849	406
407	C10H14	SEC-BUTYLBENZENE	134.222	-75.5	173.3	664.0	29.5		862	20	37974	407
408	C10H14	TERT-BUTYLBENZENE	134.222	-57.9	169.1	660.0	29.0		867	20	37639	408
409	C10H14	1-METHYL-2-ISOPROPYLBENZENE	134.222		178.3	670.0	29.0		876	20		409
410	C10H14	1-METHYL-3-ISOPROPYLBENZENE	134.222		175.1	666.0	29.4		861	20	38142	410
411	C10H14	1-METHYL-4-ISOPROPYLBENZENE	134.222		177.1	653.0	28.3		857	20		411
412	C10H14	1,4-DIETHYLBENZENE	134.222	-73.2	183.7	657.9	28.1	0.480	862	20	39398	412
413	C10H14	1,2,4,5-TETRAMETHYLBENZENE	134.222	-42.2	196.8	675.0	29.4	0.480	838	81	45552	413
414	C10H15N	N-BUTYLANILINE	149.236	78.8	240.7	721.0	28.4	0.518	932	20	48944	414
415	C10H18	CIS-DECALIN	138.254	-14.2	195.7	702.2	31.4		897	20	39356	415
416	C10H18	TRANS-DECALIN	138.254	-30.4	187.2	690.0	31.4		870	20	38519	416
417	C10H17N	CAPRYLONITRILE	153.269	-17.9	242.8	622.0	32.5		820	20		417
418	C10H20	N-BUTYLCYCLOHEXANE	140.270	-74.8	180.9	667.0	31.5		799	20	38519	418
419	C10H20	ISOBUTYLCYCLOHEXANE	140.270		171.3	669.0	31.2		795	20		419
420	C10H20	SEC-BUTYLCYCLOHEXANE	140.270		179.3	669.0	26.7		813	20		420
421	C10H20	TERT-BUTYLCYCLOHEXANE	140.270	-41.2	171.5	659.0	26.6		813	20		421
422	C10H20	1-DECENE	140.270	-66.3	170.5	615.0	22.1	0.650	741	20	38686	422
423	C10H22	N-DECANE	142.286	-29.7	174.1	617.6	21.1	0.603	730	20	39306	423
424	C10H22	3,3,5-TRIMETHYLHEPTANE	142.286		155.6	609.6	23.2				36676	424
425	C10H22	2,2,3,3-TETRAMETHYLHEXANE	142.286		160.3	623.1	25.1				36383	425
426	C10H22	2,2,5,5-TETRAMETHYLHEXANE	142.286		137.4	581.5	21.9				35295	426
427	C10H22O	1-DECANOL	158.285	6.9	230.2	700.0	22.3	0.600	830	20	50242	427
428	C11H10	1-METHYLNAPHTHALENE	142.201	-30.5	244.6	772.0	35.7	0.445	1020	20	46055	428
429	C11H10	2-METHYLNAPHTHALENE	142.201	-34.5	241.0	761.0	35.1	0.462	990	40	46055	429
430	C11H14O2	BUTYL BENZOATE	178.232	-22.2	249.8	723.0	26.3	0.561	1006	20	48986	430
431	C11H22	N-HEXYLCYCLOPENTANE	154.297		203.1	660.1	21.4				41198	431
432	C11H22	1-UNDECENE	154.297	-49.2	192.6	637.0	20.0	0.660	751	20	40905	432
433	C11H24	N-UNDECANE	156.313	-25.6	195.9	638.8	19.7		740	20	41533	433
434	C12H8	ACENAPHTHALENE	152.196	95.0	270.0	796.9	32.2	0.487			45636	434
435	C12H10	DIPHENYL	154.212	69.2	255.2	789.0	38.5	0.502	990	74	47143	435
436	C12H10O	DIPHENYL ETHER	170.211	26.8	258.0	766.0	31.4		1066	30	43375	436
437	C12H24	N-HEPTYLCYCLOPENTANE	168.324		224.1	679.0	19.5				42998	437
438	C12H24	1-DODECENE	168.324	-35.2	213.3	657.0	18.5	0.713	758	20	43668	438
439	C12H26	N-DODECANE	170.340	-9.6	216.3	658.3	18.2	0.720	748	20	45636	439
440	C12H26O	DIHEXYL ETHER	186.339	-43.2	226.4	679.0	18.2	0.718	794	20	45636	440
441	C12H26O	DODECANOL	186.339	23.9	259.9	679.0	19.3		835	20		441
442	C12H27N	TRIBUTYLAMINE	185.355		213.4	643.0	18.2		777	20	44380	442
443	C13H10	FLUORENE	166.223	114.0	297.9	822.3	29.9	0.534				443
444	C13H12	DIPHENYLMETHANE	168.239	26.8	264.3	767.0	29.8		1006	20		444
445	C13H26	N-OCTYLCYCLOPENTANE	182.351		243.7	694.0	17.9				45427	445
446	C13H26	1-TRIDECENE	182.351	-23.1	232.7	674.0	17.0		766	20	45008	446
447	C13H28	N-TRIDECANE	184.367	-5.4	235.4	675.0	17.2	0.780	756	20	45678	447
448	C14H10	ANTHRACENE	178.234	216.5	341.2	883.0					56522	448
449	C14H10	PHENANTHRENE	178.234	100.5	339.4	878.0					55684	449
450	C14H28	N-NONYLCYCLOPETANE	196.378		262.1	710.5	16.5				47269	450

NO	VISA	VISB	DELHF	DELGF	CPVAPA	CPVAPB	CPVAPC	CPVAPD	ANTA	ANTB	ANTC	TMN	TMX	NO
401			-242.12	34.04	-67.403	1.168E+00	-8.612E-04	2.573E-07	15.6488	3049.98	-57.13	40	140	401
402			-236.39	34.12	-54.918	1.091E+00	-7.603E-04	2.157E-07	15.8029	3269.07	-58.19	52	152	402
403	873.32	352.57	151.06	223.74	-68.802	8.499E-01	-6.506E-04	1.980E-07	16.1426	3992.01	-71.29	87	252	403
404			27.63	167.05					16.2805	4009.49	-64.98	92	227	404
405	563.84	296.01	-13.82	144.78	-22.990	7.934E-01	-4.396E-04	8.570E-08	16.0793	3633.40	-71.77	62	213	405
406			-21.56						15.9924	3512.47	-69.03	53	203	406
407	582.82	295.82	-17.46		-65.147	9.893E-01	-7.213E-04	2.152E-07	15.9999	3544.19	-68.10	52	203	407
408			-22.69		-86.001	1.102E+00	-8.746E-04	2.826E-07	15.9300	3462.28	-69.87	50	199	408
409									15.9809	3564.52	-70.00	57	208	409
410			-29.31		-48.759	9.064E-01	-6.054E-04	1.627E-07	15.9811	3543.79	-69.22	55	205	410
411									15.9424	3539.21	-70.10	56	207	411
412			-22.27	137.96	-37.417	8.670E-01	-5.560E-04	1.411E-07	16.1140	3657.22	-71.18	62	214	412
413			-45.30	119.53	15.265	6.518E-01	-2.878E-04	3.256E-08	16.3023	3850.91	-71.72	88	227	413
414	1111.10	341.28			-34.068	9.144E-01	-5.560E-04	1.287E-07	16.3994	4079.72	-96.15	112	287	414
415			-169.06	85.87	-112.457	1.118E+00	-6.606E-04	1.436E-07	16.8312	3671.61	-69.74	95	222	415
416	702.27	339.66	-182.42	73.48	-97.670	1.044E+00	-5.476E-04	8.980E-08	15.7989	3610.66	-66.49	90	197	416
417														417
418	598.30	311.39	-213.32	56.48	-62.957	1.062E+00	-6.305E-04	1.400E-07	15.9116	3542.57	-72.32	59	212	418
419									15.8141	3437.99	-69.99	82	182	419
420									15.8670	3524.57	-70.78	87	197	420
421									15.7884	3457.85	-67.04	84	177	421
422	518.37	277.80	-124.22	121.12	-4.664	9.077E-01	-5.057E-04	1.095E-07	16.0129	3448.18	-76.09	83	187	422
423	558.61	288.37	-249.83	33.24	-70.913	9.608E-01	-5.287E-04	1.130E-07	16.0114	3456.80	-78.67	57	203	423
424			-258.74	33.58	-70.372	1.232E+00	-8.645E-04	2.455E-07	15.7848	3305.20	-67.66	40	275	424
425					-58.833	1.231E+00	-8.834E-04	2.584E-07	15.7598	3371.05	-64.09	41	190	425
426					62.341	1.244E+00	-8.955E-04	2.618E-07	15.8446	3172.92	-66.15	27	165	426
427	1481.80	380.00	-401.93	-104.25	14.570	8.947E-01	-3.920E-01	3.450E-08	15.9395	3389.43	-139.00	103	230	427
428	862.89	361.76	116.94	217.84	-64.820	9.386E-01	-6.941E-04	2.015E-07	16.2008	4206.70	-78.15	107	278	428
429	695.42	351.79	116.18	216.29	-56.518	8.997E-01	-6.468E-04	1.840E-07	16.2758	4237.37	-74.75	104	275	429
430	882.36	350.34	-209.63	78.25	-17.367	8.675E-01	-4.609E-04	7.234E-08	16.3363	4158.47	-94.15	117	297	430
431	617.57	318.65	-144.86	129.54	-5.585	1.127E+00	-6.535E-04	1.472E-07	16.0140	3702.56	-81.55	78	234	431
432	566.26	294.89			-8.395	1.002E+00	-5.601E-04	1.216E-07	16.0412	3597.72	-83.41	72	223	432
433	605.50	305.01	-270.47	41.62	-64.623	1.053E+00	-5.798E-04	1.236E-07	16.0541	3614.07	-85.45	75	225	433
434	733.87	369.58	182.21	280.26	-97.067	8.850E-01	-5.853E-04	1.305E-07	16.3091	4470.92	-81.40	177	377	434
435	1146.00	379.29	49.99		-60.730	1.105E+00	-8.855E-04	2.790E-07	16.6832	4602.23	-70.42	70	272	435
436	654.77	333.12	-230.27	86.67	-59.264	9.282E-01	-8.867E-04	1.358E-07	16.3459	4310.25	-88.75	145	325	436
437	615.67	310.07			-6.544	1.223E+00	-7.084E-04	1.596E-07	16.0589	3850.38	-90.88	95	256	437
438	631.63	318.78	-165.46	138.00	-9.328	1.097E+00	-6.154E-04	1.341E-07	16.0610	3729.87	-91.31	88	244	438
439	723.43	323.35	-291.07	50.07	33.536	1.073E+00	-6.347E-04	1.359E-07	16.1134	3774.56	-89.15	91	247	439
440	1417.80	398.89			9.224	1.073E+00	-5.534E-04	1.677E-07	16.3372	3982.78	-157.10	100	272	440
441	889.06	312.48	-443.13	-87.13	7.993	1.103E+00	-5.338E-04	7.779E-08	16.2638	3242.04	-86.15	134	307	441
442					-54.491	1.197E+00	-6.703E-04	1.448E-07	16.2878	3865.58	-13.40	89	258	442
443						9.035E-01	-5.388E-04	9.257E-08	18.2166	6462.60	-167.90	207	407	443
444									14.4856	2902.44		200	290	444
445	695.83	346.19	-250.87	95.12	-59.951	1.316E+00	-7.611E-04	1.708E-07	16.0941	3983.01	-95.85	112	276	445
446	658.16	323.71	-186.10	146.37	-7.118	1.191E+00	-6.673E-04	1.451E-07	16.0850	3856.23	-97.94	104	264	446
447	664.10	332.10	-311.71	58.49	-10.463	1.245E+00	-6.912E-04	1.489E-07	16.1355	3892.91	-98.93	107	267	447
448	513.28	405.81	224.83		-58.979	1.005E+00	-6.594E-04	1.605E-07	17.6701	6492.44	-26.13	217	382	448
449	735.19		202.64		-58.979	1.005E+00	-6.594E-04	1.605E-07	16.7187	5477.94	-69.39	177	382	449
450		357.74	-271.51	103.50	-60.809	1.411E+00	-8.155E-04	1.834E-07	16.1089	4096.30	-103.00	127	296	450

NO	FORMULA	COMPOUND NAME	MOLWT	TFP	TBP	TC	PC	VC	LDEN	TDEN	HVAP	NO
451	C14H28	1-TETRADECENE	196.378	-12.9	251.1	689.0	15.6		786	0	46934	451
452	C14H30	N-TETRADECANE	198.394	5.8	253.5	694.0	16.2	0.830	763	20	47646	452
453	C15H12	1-PHENYLINDENE	192.261		322.0	843.7	27.0	0.598				453
454	C15H14	2-ETHYLFLUORENE	194.277		309.0	811.1	24.6	0.629				454
455	C15H30	N-DECYLCYCLOPENTANE	210.405		279.3	723.8	15.2				49027	455
456	C15H30	1-PENTADECENE	210.405	-3.8	268.3	704.0	14.6		791	0	48692	456
457	C15H32	N-PENTADECANE	212.421	9.8	270.6	707.0	15.2	0.880	769	20	49488	457
458	C16H10	FLUORANTHENE	202.256	110.0	393.0	936.6	26.0	0.660			79131	458
459	C16H10	PYRENE	202.256	151.0	362.0	892.1	26.0	0.637				459
460	C16H12	N-PHENYLNAPHTHALENE	204.272		316.0	840.1	26.3	0.605				460
461	C16H22O4	DIBUTYL-O-PHTHALATE	278.350	-35.2	334.8		13.6		1047	20		461
462	C16H32	N-DECYLCYCLOHEXANE	224.432		297.6	750.0	13.6		788	10	50409	462
463	C16H32	1-HEXADECENE	224.432	4.1	284.8	717.0	13.4				50451	463
464	C16H34	N-HEXADECANE	226.448	17.8	286.8	717.0	14.2		773	20	51246	464
465	C17H34	N-DODECYLCYCLOPENTANE	238.459		310.9	750.0	13.0		848	54	52628	465
466	C17H36O	HEPTADECANOL	256.474	53.8	323.8	736.0	14.2				60709	466
467	C17H36	N-HEPTADECANE	240.475	21.8	302.0	733.0	13.2	1.000	778	20	52921	467
468	C18H12	CHRYSENE	228.294	255.0	448.0	993.6	23.9	0.736				468
469	C18H14	O-TERPHENYL	230.310	56.8	331.8	891.0	39.0	0.769				469
470	C18H14	M-TERPHENYL	230.310	86.8	364.8	924.8	35.1	0.784				470
471	C18H14	P-TERPHENYL	230.310	211.8	375.8	926.0	33.2	0.779				471
472	C18H36	1-OCTADECENE	252.486	17.6	314.8	739.0	11.3		789	20	54303	472
473	C18H36	N-TRIDECYLCYCLOPENTANE	252.486		325.4	761.0	12.1				54345	473
474	C18H38	N-OCTADECANE	254.502	28.1	316.3	745.0	12.1		777	28	54512	474
475	C18H38O	1-OCTADECANOL	270.501	57.8	334.8	747.0	14.2		812	59		475
476	C19H38	N-TETRADECYLCYCLOPENTANE	266.513		325.8	772.0	11.2		789		56019	476
477	C19H40	N-NONADECANE	268.529	31.8	329.9	756.0	11.1			32	56061	477
478	C20H40	N-PENTADECYLCYCLOPENTANE	280.540		351.8	780.0	10.2		775	40	57694	478
479	C20H42	N-EICOSANE	282.556	36.8	343.8	767.0	11.1				57527	479
480	C20H42O	1-EICOSANOL	298.555	65.8	355.8	770.0	12.2				65314	480
481	C21H42	N-HEXADECYLCYCLOPENTANE	294.567		363.8	791.0	9.7				59369	481

NO	VISA	VISB	DELHF	DELGF	CPVAPA	CPVAPB	CPVAPC	CPVAPD	ANTA	ANTB	ANTC	TMN	TMX	NO
451	697.49	336.13	-206.66	154.87	-7.967	1.285E+00	-7.209E-04	1.569E-07	16.1643	4018.01	-102.70	119	284	451
452	689.85	344.21	-332.35	66.86	-10.982	1.337E+00	-7.423E-04	1.598E-07	16.1480	4008.52	-105.40	121	287	452
453					-96.154	1.186E+00	-7.786E-04	1.765E-07	16.4170	4872.90	-97.30	227	427	453
454					-107.036	1.261E+00	-8.155E-04	1.792E-07	16.5199	4789.44	-97.90	207	407	454
455	771.74	368.30	-292.15	111.91	-61.923	1.507E+00	-8.716E-04	1.959E-07	16.1261	4203.94	-109.70	140	313	455
456	739.13	347.46	-227.39	163.16	-9.203	1.382E+00	-7.783E-04	1.702E-07	16.1539	4103.15	-110.60	133	301	456
457	718.51	355.92	-352.99	75.28	-11.916	1.432E+00	-7.971E-04	1.719E-07	16.1724	4121.51	-111.80	135	304	457
458					-80.706	1.171E+00	-7.938E-04	1.860E-07	16.4523	5438.77	-112.40	287	487	458
459					-94.379	1.191E+00	-7.929E-04	1.755E-07	16.4842	5203.08	-107.20	257	477	459
460					-99.516	1.146E+00	-6.112E-04	6.061E-08	16.9691	5351.04	-81.70	227	427	460
461	2588.10	336.24			1.880	1.293E+00	-6.121E-04	6.971E-08	16.9539	4852.47	-138.10	196	384	461
462	925.84	378.69			-69.015	1.654E+00	-8.612E-04	2.142E-07	16.1627	4373.37	-111.80	190	300	462
463	767.48	357.85	-247.98	171.62	-9.705	1.475E+00	-8.298E-04	1.810E-07	16.2203	4245.00	-115.20	147	319	463
464	738.30	366.11	-373.59	83.74	-13.017	1.529E+00	-8.536E-04	1.849E-07	16.1841	4214.91	-118.70	150	321	464
465	853.53	385.53	-336.12	126.02	-63.263	1.695E+00	-9.767E-04	2.185E-07	16.1915	4395.87	-124.20	168	346	465
466			-546.25	-44.67	-7.792	1.652E+00	-9.344E-04	2.043E-07	15.6161	3672.62	-188.10	191	383	466
467	757.88	375.90	-394.19	92.15	-13.967	1.624E+00	-9.081E-04	1.972E-07	16.1510	4294.55	-124.00	161	337	467
468	1094.10	461.27			-115.757	1.341E+00	-8.310E-04	1.541E-07	16.6038	5915.26	-128.10	377	577	468
469	940.58	460.94												469
470	911.01	461.10												470
471														471
472	816.19	376.93	-289.22	188.45	-11.329	1.664E+00	-9.374E-04	2.048E-07	16.2221	4416.13	-127.30	171	350	472
473	891.80	392.78	-353.99	137.08	-64.209	1.790E+00	-1.032E-03	2.309E-07	16.2270	4483.13	-131.30	180	361	473
474	777.40	385.00	-414.83	100.57	-14.470	1.717E+00	-9.592E-04	2.078E-07	16.1232	4361.79	-129.90	172	352	474
475			-566.85	-36.22	-8.704	1.747E+00	-8.524E-04	2.157E-07	15.6898	3757.82	-193.10	201	385	475
476	924.60	399.62	-374.63	145.58	-64.929	1.884E+00	-1.085E-03	2.425E-07	16.2632	4439.38	-138.10	192	375	476
477	793.62	393.54	-435.43	108.98	-15.491	1.812E+00	-1.014E-03	2.205E-07	16.1533	4450.44	-135.60	183	366	477
478	950.57	406.33	-395.28	153.99	-66.093	1.980E+00	-1.140E-03	2.549E-07	16.3092	4642.01	-145.10	203	388	478
479	811.29	401.67	-456.07	117.40	-22.383	1.939E+00	-1.116E-03	2.528E-07	16.4685	4680.46	-141.10	198	379	479
480			-608.13	-19.43	-12.581	1.949E+00	-1.118E-03	2.515E-07	15.8233	3912.10	-203.10	219	406	480
481	977.42	412.29	-415.87	162.41	-66.683	2.074E+00	-1.236E-03	2.668E-07	16.3553	4715.69	-152.10	213	401	481

APPENDIX E

Conversion Factors for Some Common SI Units

An asterisk (*) denotes an exact relationship.

Length	*1 in.	:	25·4 mm
	*1 ft	:	0·304 8 m
	*1 yd	:	0·914 4 m
	1 mile	:	1·609 3 km
	*1 Å (angstrom)	:	10^{-10} m
Time	*1 min	:	60 s
	*1 h	:	3·6 ks
	*1 day	:	86·4 ks
	1 year	:	31·5 Ms
Area	*1 in.^{2}	:	645·16 mm^2
	1 ft^2	:	0·092 903 m^2
	1 yd^2	:	0·836 13 m^2
	1 acre	:	4046·9 m^2
	1 mile^2	:	2·590 km^2
Volume	1 in.^{3}	:	16·387 cm^3
	1 ft^3	:	0·028 32 m^3
	1 yd^3	:	0·764 53 m^3
	1 UK gal	:	4546·1 cm^3
	1 US gal	:	3785·4 cm^3
Mass	1 oz	:	28·352 g
	*1 lb	:	0·453 592 37 kg
	1 cwt	:	50·802 3 kg
	1 ton	:	1016·06 kg
Force	1 pdl	:	0·138 26 N
	1 lbf	:	4·448 2 N
	1 kgf	:	9·806 7 N
	1 tonf	:	9·964 0 kN
	*1 dyn	:	10^{-5} N
Temperature difference	*1 deg F (deg R)	:	$\frac{5}{9}$ deg C (deg K)
Energy (work, heat)	1 ft lbf	:	1·355 8 J
	1 ft pdl	:	0·042 14 J
	*1 cal (internat. table)	:	4·186 8 J
	1 erg	:	10^{-7} J
	1 Btu	:	1·055 06 kJ
	1 hp h	:	2·684 5 MJ
	*1 kW h	:	3·6 MJ
	1 therm	:	105·51 MJ
	1 thermie	:	4·185 5 MJ
Calorific value (volumetric)	1 Btu/ft^3	:	37·259 kJ/m^3
Velocity	1 ft/s	:	0·304 8 m/s
	1 mile/h	:	0·447 04 m/s
Volumetric flow	1 ft^3/s	:	0·028 316 m^3/s
	1 ft^3/h	:	7·865 8 cm^3/s
	1 UK gal/h	:	1·262 8 cm^3/s
	1 US gal/h	:	1·051 5 cm^3/s

Mass flow	1 lb/h	:	0·126 00 g/s
	1 ton/h	:	0·282 24 kg/s
Mass per unit area	1 lb/in.^{2}	:	703·07 kg/m^2
	1 lb/ft^2	:	4·882 4 kg/m^2
	1 ton/sq mile	:	392·30 kg/km^2
Density	1 lb/in^3	:	27·680 g/cm^3
	1 lb/ft^3	:	16·019 kg/m^3
	1 lb/UK gal	:	99·776 kg/m^3
	1 lb/US gal	:	119·83 kg/m^3
Pressure	1 lbf/in.^{2}	:	6·894 8 kN/m^2
	1 tonf/in.^{2}	:	15·444 MN/m^2
	1 lbf/ft^2	:	47·880 N/m^2
	*1 standard atm	:	101·325 kN/m^2
	*1 atm (1 kgf/cm^2)	:	98·066 5 kN/m^2
	*1 bar	:	10^5 N/m^2
	1 ft water	:	2·989 1 kN/m^2
	1 in. water	:	249·09 N/m^2
	1 in. Hg	:	3·386 4 kN/m^2
	1 mm Hg (1 torr)	:	133·32 N/m^2
Power (heat flow)	1 hp (British)	:	745·70 W
	1 hp (metric)	:	735·50 W
	1 erg/s	:	10^{-7} W
	1 ft lbf/s	:	1·355 8 W
	1 Btu/h	:	0·293 07 W
	1 ton of refrigeration	:	3516·9 W
Moment of inertia	1 lb ft^2	:	0·042 140 kg m^2
Momentum	1 lb ft/s	:	0·138 26 kg m/s
Angular momentum	1 lb ft^2/s	:	0·042 140 kg m^2/s
Viscosity, dynamic	*1 P (Poise)	:	0·1 N* s/m^2
	1 lb/ft h	:	0·413 38 mN s/m^2
	1 lb/ft s	:	1·488 2 N s/m^2
Viscosity, kinematic	*1 S (Stokes)	:	10^{-4} m^2/s
	1 ft^2/h	:	0·258 06 cm^2/s
Surface energy	1 erg/cm^2	:	10^{-3} J/m^2
(surface tension)	(1 dyn/cm)	:	(10^{-3} N/m)
Mass flux density	1 lb/h ft^2	:	1·356 2 g/s m^2
Heat flux density	1 Btu/h ft^2	:	3·154 6 W/m^2
	*1 kcal/h m^2	:	1·163 W/m^2
Heat transfer coefficient	1 Btu/h ft^2 °F	:	5·678 3 W/m^2 K
Specific enthalpy (latent heat, etc.)	*1 Btu/lb	:	2·326 kJ/kg
Specific heat capacity	*1 Btu/lb °F	:	4·186 8 kJ/kg K
Thermal	1 Btu/h ft °F	:	1·730 7 W/m K
conductivity	1 kcal/h m °C	:	1·163 W/m K

(Taken from MULLIN, J. W.: *The Chemical Engineer* No. 211 (Sept. 1967), 176. SI units in chemical engineering.)

Note: Where temperature difference is involved **K** = °C.

APPENDIX F

Standard Flanges

Reproduced from BS 4504: Part 1: 1969

Steel welding-neck flanges for nominal pressure ratings of 6, 10, 25, 40, 100 and 250 bar.

STEEL WELDING NECK FLANGES
Nominal pressure 6 bar (1 bar = 10^5 N/m^2)

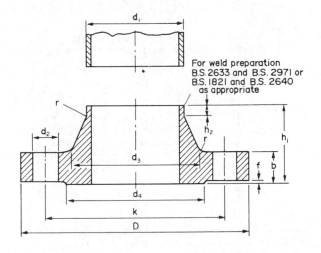

For weld preparation
B.S.2633 and B.S. 2971 or
B.S. 1821 and B.S. 2640
as appropriate

Nom. size	Pipe o.d. d_1	Flange			Raised face		Bolting	Drilling			Neck		
		D	b	h_1	d_4	f		No.	d_2	k	d_3	h_2 $\approx$	r
10	17·2	75	12	28	35	2	M10	4	11	50	26	6	4
15	21·3	80	12	30	40	2	M10	4	11	55	30	6	4
20	26·9	90	14	32	50	2	M10	4	11	65	38	6	4
25	33·7	100	14	35	60	2	M10	4	11	75	42	6	4
32	42·4	120	14	35	70	2	M12	4	14	90	55	6	6
40	48·3	130	14	38	80	3	M12	4	14	100	62	7	6
50	60·3	140	14	38	90	3	M12	4	14	110	74	8	6
65	76·1	160	14	38	110	3	M12	4	14	130	88	9	6
80	88·9	190	16	42	128	3	M16	4	18	150	102	10	8
100	114·3	210	16	45	148	3	M16	4	18	170	130	10	8
125	139·7	240	18	48	178	3	M16	8	18	200	155	10	8
150	168·3	265	18	48	202	3	M16	8	18	225	184	12	10
200	219·1	320	20	55	258	3	M16	8	18	280	236	15	10
250	273	375	22	60	312	3	M16	12	18	335	290	15	12
300	323·9	440	22	62	365	4	M20	12	22	395	342	15	12
350	355·6	490	22	62	415	4	M20	12	22	445	385	15	12
400	406·4	540	22	65	465	4	M20	16	22	495	438	15	12
450	457·2	595	24	65	520	4	M20	16	22	550	492	15	12
500	508	645	24	68	570	4	M20	20	22	600	538	15	12
600	609·6	755	24	70	670	5	M24	20	26	705	640	16	12
700	711·2	860	24	70	775	5	M24	24	26	810	740	16	12
800	812·8	975	24	70	880	5	M27	24	30	920	842	16	12
900	914·4	1075	26	70	980	5	M27	24	30	1020	942	16	12
1000	1016	1175	26	70	1080	5	M27	28	30	1120	1045	16	16
1200	1220	1405	28	90	1295	5	M30	32	33	1340	1248	20	16
1400	1420	1630	32	90	1510	5	M33	36	36	1560	1452	20	16
1600	1620	1830	34	90	1710	5	M33	40	36	1760	1655	20	16
1800	1820	2045	36	100	1920	5	M36	44	39	1970	1855	20	16
2000	2020	2265	38	110	2125	5	M39	48	42	2180	2058	25	16

STEEL WELDING NECK FLANGES
Nominal pressure 10 bar (1 bar = 10^5 N/m²)

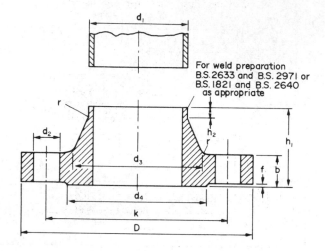

Nom. size	Pipe o.d. d_1	Flange			Raised face		Bolting	Drilling			Neck		
		D	b	h_1	d_4	f		No.	d_2	k	d_3	$h_2 \approx$	r
200	219·1	340	24	62	268	3	M20	8	22	295	235	16	10
250	273	395	26	68	320	3	M20	12	22	350	292	16	12
300	323·9	445	26	68	370	4	M20	12	22	400	344	16	12
350	355·6	505	26	68	430	4	M20	16	22	460	385	16	12
400	406·4	565	26	72	482	4	M24	16	25	515	440	16	12
450	457·2	615	28	72	532	4	M24	20	26	565	492	16	12
500	508	670	28	75	585	4	M24	20	26	620	542	16	12
600	609·6	780	28	80	685	5	M27	20	30	725	642	18	12
700	711·2	895	30	80	800	5	M27	24	30	840	745	18	12
800	812·8	1015	32	90	905	5	M30	24	33	950	850	18	12
900	914·4	1115	34	95	1005	5	M30	28	33	1050	950	20	12
1000	1016	1230	34	95	1110	5	M33	28	36	1160	1052	20	16
1200	1220	1455	38	115	1330	5	M36	32	39	1380	1255	25	16
1400	1420	1675	42	120	1535	5	M39	36	42	1590	1460	25	16
1600	1620	1915	46	130	1760	5	M45	40	48	1820	1665	25	16
1800	1820	2115	50	140	1960	5	M45	44	48	2020	1868	30	16
2000	2020	2325	54	150	2170	5	M45	48	48	2230	2072	30	16

STEEL WELDING NECK FLANGES
Nominal pressure 25 bar (1 bar = 10^5 N/m^2)

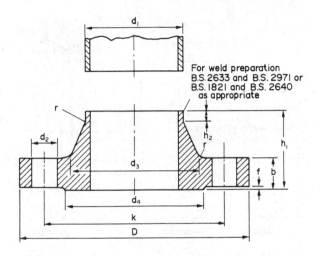

Nom. size	Pipe o.d. d_1	Flange			Raised face		Bolting	Drilling			Neck		
		D	b	h_1	d_4	f		No.	d_2	k	d_3	h_2 $\approx$	r
175	193·7	330	28	75	248	3	M24	12	26	280	218	15	10
200	219·1	360	30	80	278	3	M24	12	26	310	244	16	10
250	273	425	32	88	335	3	M27	12	30	370	298	18	12
300	323·9	485	34	92	395	4	M27	16	30	430	352	18	12
350	355·6	555	38	100	450	4	M30	16	33	490	398	20	12
400	406·4	620	40	110	505	4	M33	16	36	550	452	20	12
450	457·2	670	42	110	555	4	M33	20	36	600	505	20	12
500	508	730	44	125	615	4	M33	20	36	660	558	20	12
600	609·6	845	46	125	720	5	M36	20	39	770	660	20	12
700	711·2	960	46	125	820	5	M39	24	42	875	760	20	12
800	812·8	1085	50	135	930	5	M45	24	48	990	865	22	12
900	914·4	1185	54	145	1030	5	M45	28	48	1090	968	24	12
1000	1016	1320	58	155	1140	5	M52	28	56	1210	1070	24	16

STEEL WELDING NECK FLANGES
Nominal pressure 40 bar (1 bar = 10^5 N/m^2)

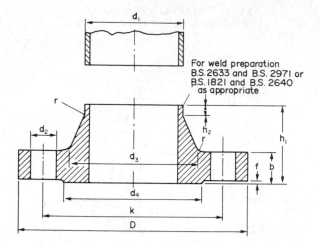

For weld preparation
B.S. 2633 and B.S. 2971 or
B.S. 1821 and B.S. 2640
as appropriate

Nom. size	Pipe o.d. d_1	Flange			Raised face		Bolting	Drilling			Neck		
		D	b	h_1	d_4	f		No.	d_2	k	d_3	h_2 $\approx$	r
10	17·2	90	16	35	40	2	M12	4	14	60	28	6	4
15	21·3	95	16	38	45	2	M12	4	14	65	32	6	4
20	26·9	105	18	40	58	2	M12	4	14	75	40	6	4
25	33·7	115	18	40	68	2	M12	4	14	85	46	6	4
32	42.4	140	18	42	78	2	M16	4	18	100	56	6	6
40	48·3	150	18	45	88	3	M16	4	18	110	64	7	6
50	60·3	165	20	48	102	3	M16	4	18	125	75	8	6
65	76·1	185	22	52	122	3	M16	8	18	145	90	10	6
80	88·9	200	24	58	138	3	M16	8	18	160	105	12	8
100	114·3	235	24	65	162	3	M20	8	22	190	134	12	8
125	139·7	270	26	68	188	3	M24	8	26	220	162	12	8
150	168·3	300	28	75	218	3	M24	8	26	250	192	12	10
175	193·7	350	32	82	260	3	M27	12	30	295	218	15	10
200	219·1	375	34	88	285	3	M27	12	30	320	244	16	10
250	273	450	38	105	345	3	M30	12	33	385	306	18	12
300	323·9	515	42	115	410	4	M30	16	33	450	362	18	12
350	355·6	580	46	125	465	4	M33	16	36	510	408	20	12
400	406·4	660	50	135	535	4	M36	16	39	585	462	20	12
450	457·2	685	50	135	560	4	M36	20	39	610	500	20	12
500	508	755	52	140	615	4	M39	20	42	670	562	20	12

STEEL WELDING NECK FLANGES
Nominal pressure 100 bar (1 bar = 10^5 N/m²)

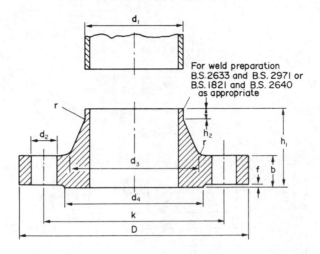

For weld preparation
B.S. 2633 and B.S. 2971 or
B.S. 1821 and B.S. 2640
as appropriate

Nom. size	Pipe o.d. d_1	Flange			Raised face		Bolting	Drilling			Neck		
		D	b	h_1	d_4	f		No.	d_2	k	d_3	h_2 $\approx$	r
10	17·2	100	20	45	40	2	M12	4	14	70	32	6	4
15	21·3	105	20	45	45	2	M12	4	14	75	34	6	4
20	26·9	130	22	58	58	2	M16	4	18	90	42	8	4
25	33·7	140	24	58	68	2	M16	4	18	100	52	8	4
32	42·4	155	24	60	78	2	M20	4	22	110	62	8	6
40	48·3	170	26	62	88	3	M20	4	22	125	70	10	6
50	60·3	195	28	68	102	3	M24	4	26	145	90	10	6
65	76·1	220	30	76	122	3	M24	8	26	170	108	12	6
80	88·9	230	32	78	138	3	M24	8	26	180	120	12	8
100	114·3	265	36	90	162	3	M27	8	30	210	150	12	8
125	139·7	315	40	105	188	3	M30	8	33	250	180	12	8
150	168·3	355	44	115	218	3	M30	12	33	290	210	12	10
175	193·7	385	48	127	260	3	M30	12	33	320	245	16	10
200	219·1	430	52	130	285	3	M33	12	36	360	278	16	10
250	273	505	60	157	345	3	M36	12	39	430	340	18	12
300	323·9	585	68	170	410	4	M39	16	42	500	400	18	12
350	355·6	655	74	189	465	4	M45	16	48	560	460	20	12

STEEL WELDING NECK FLANGES
Nominal pressure 250 bar (1 bar = 10^5 N/m²)

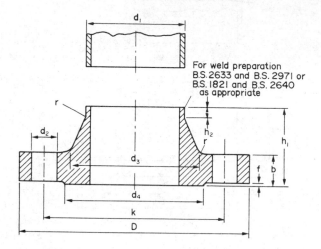

For weld preparation
B.S.2633 and B.S. 2971 or
B.S. 1821 and B.S. 2640
as appropriate

Nom. size	Pipe o.d. d_1	Flange			Raised face		Bolting	Drilling			Neck		
		D	b	h_1	d_4	f		No.	d_2	k	d_3	h_2 $\approx$	r
15	21·3	130	26	60	45	2	M16	4	18	90	48	6	4
20	26·9	135	28	65	58	2	M16	4	18	95	55	6	4
25	33·7	150	28	65	68	2	M20	4	22	105	60	8	4
32	42·4	165	32	75	78	2	M20	4	22	120	74	8	6
40	48·3	185	34	80	88	3	M24	4	26	135	84	10	6
50	60·3	200	38	85	102	3	M24	8	26	150	95	10	6
65	76·1	230	42	95	122	3	M24	8	26	180	124	12	6
80	101·6	255	46	102	138	3	M27	8	30	200	136	12	8
100	127	300	54	120	162	3	M30	8	33	235	164	14	8
125	152·4	340	60	140	188	3	M30	12	33	275	200	16	8
150	177·8	390	68	160	218	3	M33	12	36	320	240	18	10
175	219·1	430	74	170	260	3	M36	12	39	355	270	22	10
200	244·5	485	82	190	285	3	M39	12	42	400	305	25	10
250	298·5	585	100	215	345	3	M45	16	48	490	385	30	12

APPENDIX G

Design Projects

EIGHT typical design exercises are given in this appendix. They have been adapted from Design Projects set by the Institution of Chemical Engineers as the final part of the Institutions qualifying examinations for professional Chemical Engineers.

A model answer to exercise G.3 is given in the book: *The Manufacture of Methyl Ethyl Ketone from 2-Butanol*, by D. G. Austin and G. V. Jeffreys, IChemE/Godwin, 1979.

G.1 Ethylhexanol from propylene and synthesis gas

The project

Design a plant to produce 40,000 tonnes/year of 2-ethylhexanol from propylene and synthesis gas, assuming an operating period of 8000 hours on stream.

The process

The first stage of the process is a hydroformylation (oxo) reaction from which the main product is n-butyraldehyde. The feeds to this reactor are synthesis gas (CO/H_2 mixture) and propylene in the molar ratio 2:1, and the recycled products of isobutyraldehyde cracking. The reactor operates at 130°C and 350 bar, using cobalt carbonyl as catalyst in solution. The main reaction products are n- and isobutyraldehyde in the ratio of 4:1, the former being the required product for subsequent conversion to 2-ethylhexanol. In addition, 3 per cent of the propylene feed is converted to propane whilst some does not react.

Within the reactor, however, 6 per cent of the n-butyraldehyde product is reduced to n-butanol, 4 per cent of the isobutyraldehyde product is reduced to isobutanol, and other reactions occur to a small extent yielding high molecular weight compounds (heavy ends) to the extent of 1 per cent by weight of the butyraldehyde/butanol mixture at the reactor exit.

The reactor is followed by a gas–liquid separator operating at 30 bar from which the liquid phase is heated with steam to decompose the catalyst for recovery of cobalt by filtration. A second gas–liquid separator operating at atmospheric pressure subsequently yields a liquid phase of aldehydes, alcohols, heavy ends and water, which is free from propane, propylene, carbon monoxide and hydrogen.

This mixture then passes to a distillation column which gives a top product of mixed butyraldehydes, followed by a second column which separates the two butyraldehydes into an isobutyraldehyde stream containing 1·3 per cent mole n-butyraldehyde and an n-butyraldehyde stream containing 1·2 per cent mole isobutyraldehyde.

A cracker converts isobutyraldehyde at a pass yield of 80 per cent back to propylene, carbon monoxide and hydrogen by passage over a catalyst with steam. After separation of the water and unreacted isobutyraldehyde the cracked gas is recycled to the hydroformylation reactor. The isobutyraldehyde is recycled to the cracker inlet. The operating conditions of the cracker are 275°C and 1 bar.

The n-butyraldehyde is treated with a 2 per cent w/w aqueous sodium hydroxide and undergoes an aldol condensation at a conversion efficiency of 90 per cent. The product of this reaction, 2-ethylhexanal, is separated and then reduced to 2-ethylhexanol by hydrogen in the presence of a Raney nickel catalyst with a 99 per cent conversion rate. In subsequent stages of the process (details of which are not required), 99·8 per cent of the 2-ethylhexanol is recovered at a purity of 99 per cent by weight.

Feed specifications

(i) Propylene feed: 93 per cent propylene, balance propane.
(ii) Synthesis gas: from heavy fuel oil, after removal of sulphur compounds and carbon dioxide:
H_2 48·6 per cent; CO 49·5 per cent; CH_4 0·4 per cent; N_2 1·5 per cent.

Utilities

(i) Dry saturated steam at 35 bar.
(ii) Cooling water at 20°C.
(iii) 2 per cent w/w aqueous sodium hydroxide solution.
(iv) Hydrogen gas: H_2 98·8 per cent; CH_4 1·2 per cent.

Scope of design work required

1. Process design

(a) Prepare a material balance for the complete process.
(b) Prepare a process diagram for the plant showing the major items of equipment. Indicate the materials of construction and the operating temperatures and pressures.
(c) Prepare energy balances for the hydroformylation reactor and for the isobutyraldehyde cracking reactor.

2. Chemical engineering design

Prepare a chemical engineering design of the second distillation unit, i.e. for the separation of n- and isobutyraldehyde. Make dimensioned sketches of the column, the reboiler and the condenser.

3. Mechanical design

Prepare a mechanical design with sketches suitable for submission to a drawing office of the n- and isobutyraldehyde distillation column.

4. Control system

For the hydroformylation reactor prepare a control scheme to ensure safe operation.

Data

1. Reactions

$$CH_3.CH = CH_2 + H_2 \qquad \rightarrow CH_3.CH_2.CH_3 \qquad \Delta H^{\circ}_{298} = -129\cdot5\ kJ/mol$$

$$CH_3.CH = CH_2 + H_2 + CO \rightarrow CH_3.CH_2.CH_2.CHO \qquad \Delta H^{\circ}_{298} = -135\cdot5\ kJ/mol$$

$$\text{or} \rightarrow CH_3.\underset{\underset{CHO}{|}}{CH}.CH_3 \qquad \Delta H^{\circ}_{298} = -141\cdot5\ kJ/mol$$

$$C_3H_7CHO + H_2 \qquad \rightarrow C_4H_9OH \qquad \Delta H^{\circ}_{298} = -64\cdot8\ kJ/mol$$

$$2Co + 8CO \qquad \rightarrow Co_2(CO)_8 \qquad \Delta H^{\circ}_{298} = -462\cdot0\ kJ/mol$$

$$2CH_3.CH_2.CH_2.CHO \qquad \rightarrow CH_3.CH_2.CH_2.CH = \underset{\underset{C_2H_5}{|}}{C} - CHO + H_2O$$
$$\Delta H^{\circ}_{298} = -262\cdot0\ kJ/mol$$

$$C_4H_8 = \underset{\underset{C_2H_5}{|}}{C} - CHO + 2H_2 \qquad \rightarrow C_4H_9 - \underset{\underset{C_2H_5}{|}}{CH}.CH_2OH$$
$$\Delta H^{\circ}_{298} = -433\cdot0\ kJ/mol$$

2. Boiling points at 1 bar

Propylene	$-47\cdot7°C$
Propane	$-42\cdot1°C$
n-Butyraldehyde	$75\cdot5°C$
Isobutyraldehyde	$64\cdot5°C$
n-Butanol	$117\cdot0°C$
Isobutanol	$108\cdot0°C$
2-Ethylhexanol	$184\cdot7°C$

3. Solubilities of gases at 30 bar in the liquid phase of the first gas–liquid separator

H_2	$0\cdot08 \times 10^{-3}$	kg dissolved/kg liquid
CO	$0\cdot53 \times 10^{-3}$	kg dissolved/kg liquid
Propylene	$7\cdot5 \times 10^{-3}$	kg dissolved/kg liquid
Propane	$7\cdot5 \times 10^{-3}$	kg dissolved/kg liquid

4. Vapour–liquid equilibrium of the butyraldehydes at 1 atm (Ref. 7)

$T°C$	x	y
73·94	0·1	0·138
72·69	0·2	0·264
71·40	0·3	0·381
70·24	0·4	0·490
69·04	0·5	0·589
68·08	0·6	0·686
67·07	0·7	0·773
65·96	0·8	0·846
64·95	0·9	0·927

where x and y are the mol fractions of the more volatile component (isobutyraldehyde) in the liquid and vapour phases respectively.

References

1. *Propylene and its Industrial Derivatives*, HANCOCK, E. G. (Ed.), John Wiley & Sons N.Y., 1973, Chapter 9, pp. 333–367.
2. *Carbon Monoxide in Organic Synthesis*. Falbe–Springer Verlag, New York, 1970, pp. 1–75.
3. *Chemical Engineering*, **81**, Sept. 30th, 1974, pp. 115–122. Physical and thermodynamic properties of CO and CO_2.
4. *Chemical Engineering*, **82**, Jan. 20th, 1975, pp. 99–106. Physical and thermodynamic properties of $H_2/N_2/O_2$.
5. *Chemical Engineering*, **82**, Mar. 31st, 1975, pp. 101–109. Physical and thermodynamic properties of $C_2H_4/C_3H_6/iC_4H_8$.
6. *Chemical Engineering*, **82**, May 12th, 1975, pp. 89–97. Physical and thermodynamic properties of $CH_4/C_2H_6/C_3H_8$.
7. J. G. WOJTASINSKI. *J. Chem. Eng. Data*, 1963 (July), pp. 381–385. Measurement of total pressures for determining liquid–vapour equilibrium relations of the binary system isobutyraldehyde–n-butyraldehyde.
8. H. WEBER and J. FALBE. *Ind. Eng. Chem.* 1970 (April), pp. 33–7. Oxo Synthesis Technology.
9. *Hydrocarbon Processing*, Nov. 1971, p. 166.
10. *Hydrocarbon Processing*, Nov. 1975, p. 148.

G.2 Chlorobenzenes from benzene and chlorine

The project

Design a plant to produce 20,000 tonnes/year of monochlorobenzene together with not less than 2000 tonnes/year of dichlorobenzene, by the direct chlorination of benzene.

The Process

Liquid benzene (which must contain less than 30 ppm by weight of water) is fed into a reactor system consisting of two continuous stirred tanks operating in series at 2·4 bar. Gaseous chlorine is fed in parallel to both tanks. Ferric chloride acts as a catalyst, and is produced *in situ* by the action of hydrogen chloride on mild steel. Cooling is required to maintain the operating temperature at 328 K. The hydrogen chloride gas leaving the reactors is first cooled to condense most of the organic impurities. It then passes to an activated carbon adsorber where the final traces of impurity are removed before it leaves the plant for use elsewhere.

The crude liquid chlorobenzenes stream leaving the second reactor is washed with water and caustic soda solution to remove all dissolved hydrogen chloride. The product recovery system consists of two distillation columns in series. In the first column (the "benzene column") unreacted benzene is recovered as top product and recycled. In the second column (the "chlorobenzene column") the mono- and dichlorobenzenes are separated. The recovered benzene from the first column is mixed with the raw benzene feed and this combined stream is fed to a distillation column (the "drying column") where water is removed as overhead. The benzene stream from the bottom of the drying column is fed to the reaction system.

Feed specifications

    (i) Chlorine: 293 K, atmospheric pressure, 100 per cent purity.

    (ii) Benzene: 293 K, atmospheric pressure, 99·95 wt per cent benzene, 0·05 wt per cent water.

Product specifications

    (i) Monochlorobenzene: 99·7 wt per cent.

    (ii) Dichlorobenzene: 99·6 wt per cent.

    (iii) Hydrogen chloride gas: less than 250 ppm by weight benzene.

Utilities

    (i) Stream: dry saturated at 8 bar and at 28 bar.

    (ii) Cooling water: 293 K.

    (iii) Process water: 293 K.

    (iv) Caustic soda solution: 5 wt per cent NaOH, 293 K.

    (v) Electricity: 440 V, 50 Hz, 3 phase.

Scope of design work required

1. Process design

    (a) Prepare a materials balance for the process including an analysis of each reactor stage (the kinetics of the chlorination reactions are given below). Onstream time may be taken as 330 days per year.

    (b) Prepare energy balances for the first reactor and for the chlorobenzene column (take the reflux ratio for this column as twice the minimum reflux ratio).

    (c) Prepare a process flow diagram for the plant. This should show the major items of equipment with an indication of the materials of construction and of the internal layout. Temperatures and pressures should also be indicated.

2. Chemical engineering design

    Prepare a sieve-plate column design for the chlorobenzene distillation and make dimensioned sketches showing details of the plate layout including the weir and the downcomer.

3. Mechanical design

    Prepare a mechanical design of the chlorobenzene column, estimating the shell thickness, the positions and sizes of all nozzles, and the method of support for the plates and the column shell. Make a dimensioned sketch suitable for submission to a drawing office.

4. Safety

    Indicate the safety measures required for this plant bearing in mind the toxic and inflammable materials handled.

Data

1. The reactions

$$(1)\ \ C_6H_6 + Cl_2 \ \ \rightarrow C_6H_5Cl + HCl$$
$$(2)\ \ C_6H_5Cl + Cl_2 \rightarrow C_6H_4Cl_2 + HCl$$

The dichlorobenzene may be assumed to consist entirely of the para-isomer and the formation of trichlorobenzenes may be neglected.

The rate equations can be written in first-order form when the concentration of dissolved chlorine remains essentially constant. Thus:

$$r_B = -k_1 x_B$$
$$r_M = k_1 x_B - k_2 x_M$$
$$r_D = k_2 x_M$$

where r is the reaction rate,

k_1 is the rate constant for reaction (1) at 328 K $= 1\cdot00 \times 10^{-4}\,\mathrm{s}^{-1}$,

k_2 is the rate constant for reaction (2) at 328 K $= 0\cdot15 \times 10^{-4}\,\mathrm{s}^{-1}$

and x denotes mol fraction.

The subscripts B, M and D denote benzene, monochlorobenzene and dichlorobenzene respectively.

Yields for the reactor system should be calculated on the basis of equal liquid residence times in the two reactors, with a negligible amount of unreacted chlorine in the vapour product streams. It may be assumed that the liquid product stream contains 1·5 wt per cent of hydrogen chloride.

Reference: BODMAN, SAMUEL W. *The Industrial Practice of Chemical Process Engineering*, 1968, The MIT Press.

2. Solubilities

Solubility of the water/benzene system (taken from Seidell, A. S., *Solubilities of Organic Compounds*, 3rd Ed., Vol. II, 1941, Van Nostrand).

Temperature (K)	293	303	313	323
g H_2O/100 g C_6H_6	0·050	0·072	0·102	0·147
g C_6H_6/100 g H_2O	0·175	0·190	0·206	0·225

3. Thermodynamic and physical properties

	C_6H_6 liquid	C_6H_6 gas	C_6H_5Cl liquid	C_6H_5Cl gas	$C_6H_4Cl_2$ liquid	$C_6H_4Cl_2$ gas
Heat of formation at 298 K (kJ/kmol)	49·0	82·9	7·5	46·1	−42·0	5·0
Heat capacity (kJ/kmol K)						
298 K	136	82	152	92		103
350 K	148	99	161	108	193	118
400 K	163	113	170	121	238	131
450 K	179	126	181	134	296	143
500 K	200	137	192	145	366	155

		C_6H_6 liquid	C_6H_6 gas	C_6H_5Cl liquid	C_6H_5Cl gas	$C_6H_4Cl_2$ liquid	$C_6H_4Cl_2$ gas
Density (kg/m^3)							
	298 K	872		1100			
	350 K	815		1040		1230	
	400 K	761		989		1170	
	450 K	693		932		1100	
	500 K	612		875		1020	
Viscosity (Ns/m^2)							
	298 K	0.598×10^{-3}		0.750×10^{-3}			
	350 K	0.326×10^{-3}		0.435×10^{-3}		0.697×10^{-3}	
	400 K	0.207×10^{-3}		0.305×10^{-3}		0.476×10^{-3}	
	450 K	0.134×10^{-3}		0.228×10^{-3}		0.335×10^{-3}	
	500 K	0.095×10^{-3}		0.158×10^{-3}		0.236×10^{-3}	
Surface tension (N/m)							
	298 K	0.0280		0.0314			
	350 K	0.0220		0.0276		0.0304	
	400 K	0.0162		0.0232		0.0259	
	450 K	0.0104		0.0177		0.0205	
	500 K	0.0047		0.0115		0.0142	

References

1. PERRY, R. H. and CHILTON, C. H. *Chemical Engineers' Handbook*, 5th ed., 1973, McGraw-Hill.
2. KIRK-OTHMER, *Encyclopaedia of Chemical Technology*, 2nd Ed., 1964, John Wiley & Sons.

G.3 Methyl ethyl ketone from butyl alcohol

The project

Design a plant to produce 1×10^7 kg/year of methyl ethyl ketone (MEK).
Feedstock: Secondary butyl alcohol.
Services available:
    Dry saturated steam at 140°C.
    Cooling water at 24°C.
    Electricity at 440 V three-phase 50 Hz.
    Flue gases at 540°C.

The process

The butyl alcohol is pumped from storage to a steam-heated preheater and then to a vaporiser heated by the reaction products. The vapour leaving the vaporiser is heated to its reaction temperature by flue gases which have previously been used as reactor heating medium. The superheated butyl alcohol is fed to the reaction system at 400°C to 500°C where 90 per cent is converted on a zinc oxide – brass catalyst to methyl ethyl ketone, hydrogen and other reaction products. The reaction products may be treated in one of the following ways:

(a) Cool and condense the MEK in the reaction products and use the exhaust gases as a furnace fuel.
(b) Cool the reaction products to a suitable temperature and separate the MEK by absorption in aqueous ethanol. The hydrogen off gas is dried and used as a furnace fuel. The liquors leaving the absorbers are passed to a solvent extraction column, where the MEK is recovered using trichlorethane. The raffinate from this column is

returned to the absorber and the extract is passed to a distillation unit where the MEK is recovered. The trichlorethane is recycled to the extraction plant.

Scope of design work required

1. Prepare material balances for the two processes.
2. On the basis of the cost data supplied below decide which is the preferable process.
3. Prepare a material flow diagram of the preferred process.
4. Prepare a heat balance diagram of the preheater–vaporiser–superheater–reactor system.
5. Prepare a chemical engineering design of the preheater–vaporiser–superheater–reactor system and indicate the type of instrumentation required.
6. Prepare a mechanical design of the butyl alcohol vaporiser and make a dimensioned sketch suitable for submission to a drawing office.

Data

Process data

Outlet condenser temperature = 32°C.
Vapour and liquid are in equilibrium at the condenser outlet.
Calorific value of MEK = 41,800 kJ/kg.

Cost data

Selling price of MEK	$= £\,9.60$ per 100kg
Steam raising cost	$= £\,0.53$ per 10^6 kJ
Cost of tower shell	$= £2000$
Cost of plates	$= £2000$
Cost of reboiler	$= £2500$
Cost of heat exchanger (per distillation column)	$= £8000$
Cost of solvent extraction auxiliaries	$= £1000$
Cost of absorbtion and distillation column packing, supports and distributors	$= £2000$
Cost of tanks (surge, etc.)	$= £1000$
Cost of control of whole plant	$= £9000$
Cost of instrumentation for control of recovery section	$= £4500$
Cost of electricity for pumps	$= £5000$
Pump costs (total)	$= £3000$
Cost of cooling water for whole plant	$= £5000$

Reactor data

The "short-cut" method proposed in Ref. 1 may be used only to obtain a preliminary estimate of the height of catalyst required in the reactor. The reactor should be designed from first principles using the rate equation, below, taken from Ref. 1.

$$r_A = \frac{C(P_{A,i} - P_{K,i}P_{H,i}/K)}{P_{Ki}(1 + K_A P_{A,i} + K_{AK}P_{A,i}/P_{K,i})}$$

where $P_{A,i}$, $P_{H;i}$, and $P_{K,i}$ are the interfacial partial pressures of the alcohol, hydrogen and ketone in bars, and the remaining quantities are as specified by the semi-empirical equations below:

$$\log_{10} C = -\frac{5964}{T_i} + 8.464$$

$$\log_{10} K_A = -\frac{3425}{T_i} + 5.231$$

$$\log_{10} K_{AK} = +\frac{486}{T_i} - 0.1968$$

In these equations, the interfacial temperature T_i is in Kelvin, the constant C is in kmol/m^2 h, K_A is in bar^{-1}, and K_{AK} is dimensionless.

The equilibrium constant, K is given in Ref. 1 (although the original source is Ref. 2) by the equation:

$$\log_{10} K = -\frac{2790}{T_i} + 1.510 \log_{10} T_i + 1.871$$

where K is in bar.

Useful general information will be found in Ref. 3.

References

1. PERONA, J. J. and THODOS, G. *AIChE Jl*, 1957, **3**, 230.
2. KOLB, H. J. and BURWELL, R. L. (Jr.) *J. Am. Chem. Soc.*, 1945, **67**, 1084.
3. RUDD, D. F. and WATSON, C. C. *Strategy of Process Engineering*, 1968 (New York: John Wiley & Sons Inc.).

G.4 Acrylonitrile from propylene and ammonia

The project

Design a plant to produce 1×10^8 kg/year of acrylonitrile ($CH_2:CH.CN$) from propylene and ammonia by the ammoxidation process.

Feedstock:

Ammonia: 100 per cent NH_3.

Propylene: Commercial grade containing 90 per cent C_3H_6, 10 per cent paraffins, etc., which do not take any part in the reaction.

Services available:

Dry saturated steam at 140°C.

Cooling water at 240°C.

Other normal services.

The process

Propylene, ammonia, steam and air are fed to a vapour-phase catalytic reactor (item A). The feedstream composition (molar per cent) is propylene 7; ammonia 8; steam 20; air 65. A fixed-bed reactor is employed using a molybdenum-based catalyst at a temperature of

450°C, a pressure of 3 bar absolute, and a residence time of 4 seconds. Based upon a pure propylene feed, the carbon distribution by weight in the product from the reactor is:

Acrylonitrile	58 per cent
Acetonitrile	2 per cent
Carbon dioxide	16 per cent
Hydrogen cyanide	6 per cent
Acrolein	2 per cent
Unreacted propylene	15 per cent
Other by products	1 per cent

The reactor exit gas is air-cooled to 200°C and then passes to a quench scrubber (B) through which an aqueous solution containing ammonium sulphate 30 wt per cent and sulphuric acid 1 wt per cent is circulated. The exit gas temperature is thereby reduced to 90°C.

From the quench scrubber (B) the gas passes to an absorption column (C) in which the acrylonitrile is absorbed in water to produce a 3 wt per cent solution. The carbon dioxide, unreacted propylene, oxygen, nitrogen and unreacted hydrocarbons are not absorbed and are vented to atmosphere from the top of column (C).

The solution from the absorber (C) passes to a stripping column (D) where acrylonitrile and lower boiling impurities are separated from water. Most of the aqueous bottom product from the stripping column (D), which is essentially free of organics, is returned to the absorber (C), the excess being bled off. The overhead product is condensed and the aqueous lower layer returned to the stripping column (D) as reflux.

The upper layer which contains, in addition to acrylonitrile, hydrogen cyanide, acrolein, acetonitrile, and small quantities of other impurities, passes to a second reactor (E) where, at a suitable pH, all the acrolein is converted to its cyanohydrin. (Cyanohydrins are sometimes known as cyanhydrins.) The product from the reactor (E) is fed to a cyanohydrin separation column (F), operating at reduced temperature and pressure, in which acrolein cyanohydrin is separated as the bottom product and returned to the ammoxidation reactor (A) where it is quantitatively converted to acrylonitrile and hydrogen cyanide.

The top product from column (F) is fed to a stripping column (G) from which hydrogen cyanide is removed overhead.

The bottom product from column (G) passes to the hydroextractive distillation column (H). The water feed rate to column (H) is five times that of the bottom product flow from column (G). It may be assumed that the acetonitrile and other by-products are discharged as bottom product from column (H) and discarded. The overhead product from column (H), consisting of the acrylonitrile water azeotrope, is condensed and passed to a separator. The lower aqueous layer is returned to column (H).

The upper layer from the separator is rectified in a column (I) to give 99·95 wt per cent pure acrylonitrile.

Scope of design work required

1. Prepare a material balance for the process.
2. Prepare a material flow diagram of the process.
3. Prepare a heat balance for the reactor (A) and quench column (B).

4. Prepare a chemical engineering design of reactor (A) and either column (B) OR column (D).

5. Prepare a mechanical design of the condenser for stripping column (D) and make a dimensioned sketch suitable for submission to a drawing office.

6. Indicate the instrumentation and safety procedure required for this plant bearing in mind the toxic and inflammable materials being handled.

References

1. HANCOCK, E. H. (Ed.) *Propylene and its Industrial Derivatives*, 1973 (London: Ernest Benn Ltd.).
2. SOKOLOV, N. M., SEVRYUGOVA, N. N. and ZHAVORONKOV, N. M. *Proceedings of the International Symposium on Distillation*, 1969, pages 3:110–3:117 (London: IChemE).

G.5 Urea from ammonia and carbon dioxide

The project

A plant is to be designed for the production of 300,000 kg per day of urea by the reaction of ammonia and carbon dioxide at elevated temperature and pressure, using a total-recycle process in which the mixture leaving the reactor is stripped by the carbon dioxide feed (DSM process, references 1 to 4).

Materials available

(1) Liquid ammonia at 20°C and 9 bar, which may be taken to be 100 per cent pure.

(2) Gaseous carbon dioxide at 20°C and atmospheric pressure, also 100 per cent pure.

All normal services are available on site. In particular, electricity, 440-V three-phase 50 Hz; cooling water at a maximum summer temperature of 22°C; steam at 40 bar with 20°C of superheat.

The on-stream time is to be 330 days/year, and the product specification is fertiliser-grade urea prills containing not more than 1·0 per cent biuret.

The process

The reaction which produces urea from ammonia and carbon dioxide takes place in two stages; in the first, ammonium carbamate is formed:

$$2NH_3 + CO_2 \rightleftharpoons NH_2COONH_4$$

In the second, the carbamate is dehydrated to give urea:

$$NH_2COONH_4 \rightleftharpoons CO(NH_2)_2 + H_2O$$

Both reactions are reversible, the first being exothermal and going almost to completion, whilst the second is endothermal and goes to 40 to 70 per cent of completion.

Ammonia and carbon dioxide are fed to the reactor, a stainless steel vessel with a series of trays to assist mixing. The reactor pressure is 125 bar and the temperature is 185°C.

The reactor residence time is about 45 minutes, a 95 per cent approach to equilibrium being achieved in this time. The ammonia is fed directly to the reactor, but the carbon dioxide is fed to the reactor upwardly through a stripper, down which flows the product stream from the reactor. The carbon dioxide decomposes some of the carbamate in the product stream, and takes ammonia and water to a high-pressure condenser. The stripper is steam heated and operates at 180°C, whilst the high-pressure condenser is at 170°C and the heat released in it by recombination of ammonia and carbon dioxide to carbamate is used to raise steam. Additional recycled carbamate solution is added to the stream in the high-pressure condenser, and the combined flow goes to the reactor.

The product stream leaving the stripper goes through an expansion valve to the low-pressure section, the operating pressure there being 5 bar. In a steam-heated rectifier, further ammonia and carbon dioxide are removed and, with some water vapour, are condensed to give a weak carbamate solution. This is pumped back to the high-pressure condenser.

A two-stage evaporative concentration under vacuum, with a limited residence-time in the evaporator to limit biuret formation, produces a urea stream containing about 0·5 per cent water which can be sprayed into a prilling tower.

Physico-chemical data

Heats of reactions: $2NH_3 + CO_2 \rightarrow NH_2COONH_4$ $+ 130 \, kJ$

$$NH_2COONH_4 \rightarrow CO(NH_2)_2 + H_2O \quad -21 \, kJ$$

Properties of urea:
Density at 20°C = 1·335 g/cm³
Heat of solution in water = $-250 \, J/g$
Melting point = 133°C
   Specific heat = 1·34 J/g at 20°C

Reactor and stripper design

The relationships between temperature, pressure, and composition for the Urea—CO_2—NH_3—H_2O system are given in References 5 and 6. These are equilibrium

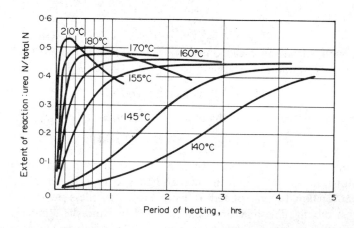

FIG. G1. Rate of dehydration of carbamate

relationships. The reaction velocity may be obtained from the graph in Fig. 5 of Reference 5, which is reproduced below for ease of reference (Fig. G1). Some stripper design data appear in Reference 7.

Scope of design work required

1. Prepare a mass balance diagram for the process, on a weight per hour basis, through to the production of urea prills.
2. Prepare an energy balance diagram for the reactor – stripper – high-pressure condenser complex.
3. Prepare a process flow diagram, showing the major items of equipment in the correct elevation, with an indication of their internal construction. Show all major pipe lines and give a schematic outline of the probable instrumentation of the reactor and its subsidiaries.
4. Prepare an equipment schedule, listing the main plant items with their size, throughput, operating conditions, materials of construction, and services required.
5. Prepare an outline design of the reactor and carry out the chemical engineering design of the stripper, specifying the interfacial contact area which will need to be provided between the carbon dioxide stream and the product stream to enable the necessary mass transfer to take place.
6. Prepare a mechanical design of the stripper, which is a vertical steam-heated tube-bundle rather like a heat exchanger. Show how liquid is to be distributed to the tubes, and how the shell is to be constructed to resist the high pressure and the corrosive process material.
7. Prepare a detailed mechanical design of the reactor in the form of a general arrangement drawing with supplementary detail drawings to show essential constructional features. Include recommendations for the feed of gaseous ammonia, carbon dioxide and carbamate solution, the latter being very corrosive. The design should ensure good gas–liquid contact; suitable instrumentation should be suggested, and provision included for its installation. Access must be possible for maintenance.
8. Specify suitable control systems for the maintenance of constant conditions in the reactor against a 15 per cent change in input rate of ammonia or carbon dioxide, and examine the effect of such a change, if uncorrected, on the steam generation capability of the high-pressure condenser.

References

1. KAASENBROOD, P. J. C. and LOGEMANN, J. D. *Hydrocarbon Processing*, April 1969, pp. 117–121.
2. PAYNE, A. J. and CANNER, J. A. *Chemical and Process Engineering*, May 1969, pp. 81–88.
3. COOK, L. H. *Hydrocarbon Processing*, Feb. 1966, pp. 129–136.
4. *Process Survey: Urea*. Booklet published with *European Chemical News*, Jan. 17th, 1969, p. 17.
5. FREJACQUES, M. *Chimie et Industrie*, July 1948, pp. 22–35.
6. KUCHERYAVYY, V. I. and GORLOVSKIY, D. M. *Soviet Chemical Industry*, Nov. 1969, pp. 44–46.
7. VAN KREVELEN, D. W. and HOFTYZER, P. J. *Chemical Engineering Science*, Aug. 1953, 2(4) pp. 145–156.

G.6 Hydrogen from fuel oil

The project

A plant is to be designed to produce 20 million standard cubic feet per day (0.555×10^6 standard m^3/day) of hydrogen of at least 95 per cent purity. The process to be employed is the partial oxidation of oil feedstock.[1-3]

Materials available

(1) Heavy fuel oil feedstock of viscosity 900 seconds Redwood One (2.57×10^{-4} m^2/s) at 100°F with the following analysis:

Carbon	85 per cent wt
Hydrogen	11 per cent wt
Sulphur	4 per cent wt
Calorific value	18,410 Btu/lb (42.9 MJ/kg)
Specific gravity	0.9435

The oil available is pumped from tankage at a pressure of 30 psig (206.9 kN/m^2 gauge) and at 50°C.

(2) Oxygen at 95 per cent purity (the other component assumed to be wholly nitrogen) and at 20°C and 600 psig (4140 kN/m^2 gauge).

Services available

(1) Steam at 600 psig (4140 kN/m^2 gauge) saturated.
(2) Cooling water at a maximum summer temperature of 25°C.
(3) Demineralised boiler feedwater at 20 psig (138 kN/m^2 gauge) and 15°C suitable for direct feed to the boilers.
(4) Electricity at 440 V, three-phase 50 Hz, with adequate incoming cable capacity for all proposed uses.
(5) Waste low-pressure steam from an adjacent process.

On-stream time

8050 hours/year.

Product specification

Gaseous hydrogen with the following limits of impurities:

CO	1.0 per cent vol maximum (dry basis)
CO_2	1.0 per cent vol maximum (dry basis)
N_2	2.0 per cent vol maximum (dry basis)
CH_4	1.0 per cent vol maximum (dry basis)
H_2S	Less than 1 ppm

The gas is to be delivered at 35°C maximum temperature, and at a pressure not less than 300 psig (2060 kN/m² gauge). The gas can be delivered saturated, i.e. no drying plant is required.

The process

Heavy fuel oil feedstock is delivered into the suction of metering-type ram pumps which feed it via a steam preheater into the combustor of a refractory-lined flame reactor. The feedstock must be heated to 200°C in the preheater to ensure efficient atomisation in the combustor. A mixture of oxygen and steam is also fed to the combustor, the oxygen being preheated in a separate steam preheater to 210°C before being mixed with the reactant steam.

The crude gas, which will contain some carbon particles, leaves the reactor at approximately 1300°C and passes immediately into a special waste-heat boiler where steam at 600 psig (4140 kN/m² gauge) is generated. The crude gas leaves the waste heat boiler at 250°C and is further cooled to 50°C by direct quenching with water, which also serves to remove the carbon as a suspension. The analysis of the quenched crude gas is as follows:

H_2	47·6 per cent vol (dry basis)
CO	42·1 per cent vol (dry basis)
CO_2	8·3 per cent vol (dry basis)
CH_4	0·1 per cent vol (dry basis)
H_2S	0·5 per cent vol (dry basis)
N_2	1·40 per cent vol (dry basis)
100·0	per cent vol (dry basis)

For the primary flame reaction steam and oxygen are fed to the reactor at the following rates:

Steam	0·75 kg/kg of heavy fuel oil feedstock
Oxygen	1·16 kg/kg of heavy fuel oil feedstock

The carbon produced in the flame reaction, and which is subsequently removed as carbon suspension in water, amounts to 1·5 per cent by weight of the fuel oil feedstock charge. Some H_2S present in the crude gas is removed by contact with the quench water.

The quenched gas passes to an H_2S removal stage where it may be assumed that H_2S is selectively scrubbed down to 15 parts per million with substantially nil removal of CO_2. Solution regeneration in this process is undertaken using the waste low-pressure steam from another process. The scrubbed gas, at 35°C and saturated, has then to undergo CO conversion, final H_2S removal, and CO_2 removal to allow it to meet the product specification.

CO conversion is carried out over chromium-promoted iron oxide catalyst employing two stages of catalytic conversion; the plant also incorporates a saturator and desaturator operating with a hot water circuit.

Incoming gas is introduced into the saturator (a packed column) where it is contacted with hot water pumped from the base of the desaturator; this process serves to preheat the gas and to introduce into it some of the water vapour required as reactant. The gas then passes to two heat exchangers in series. In the first, the unconverted gas is heated against the converted gas from the second stage of catalytic conversion; in the second heat exchanger the unconverted gas is further heated against the converted gas from the first stage of catalytic conversion. The remaining water required as reactant is then introduced into the unconverted gas as steam at 600 psig (4140 kN/m² gauge) saturated and the gas/steam mixture passes to the catalyst vessel at a temperature of 370°C. The catalyst vessel is a single shell with a dividing plate separating the two catalyst beds which constitute the two stages of conversion. The converted gas from each stage passes to the heat exchangers previously described and thence to the desaturator, which is a further packed column. In this column the converted gas is contacted countercurrent with hot water pumped from the saturator base; the temperature of the gas is reduced and the deposited water is absorbed in the hot-water circuit. An air-cooled heat exchanger then reduces the temperature of the converted gas to 40°C for final H_2S removal.

Final H_2S removal takes place in four vertical vessels each approximately 60 feet (18·3 m) in height and 8 feet (2·4 m) in diameter and equipped with five trays of iron-oxide absorbent. Each vessel is provided with a locking lid of the autoclave type. The total pressure drop across these vessels is 5 psi (35 kN/m²). Gas leaving this section of the plant contains less than 1 ppm of H_2S and passes to the CO_2 removal stage at a temperature of 35°C.

CO_2 removal is accomplished employing high-pressure potassium carbonate wash with solution regeneration.[4]

Data

I. Basic data for CO conversion section of the plant

(a) Space velocity

The space velocity through each catalyst stage should be assumed to be 3500 volumes of gas plus steam measured at NTP per volume of catalyst per hour. It should further be assumed that use of this space velocity will allow a 10°C approach to equilibrium to be attained throughout the possible range of catalyst operating temperatures listed below.

(b) Equilibrium data for the CO conversion reaction

For

$$K_p = \frac{p_{CO} \times p_{H_2O}}{p_{CO_2} \times p_{H_2}}$$

Temp. (K)	K_p
600	$3·69 \times 10^{-2}$
700	$1·11 \times 10^{-1}$
800	$2·48 \times 10^{-1}$

(c) Heat of reaction

$$CO + H_2O \rightleftharpoons CO_2 + H_2 \qquad \Delta H = -9·84 \text{ kcal.}$$

II. Basic data for CO₂ removal using hot potassium carbonate solutions

The data presented in Ref. 4 should be employed in the design of the CO_2 removal section of the plant. A solution concentration of 40 per cent wt equivalent K_2CO_3 should be employed.

Scope of design work required

1. Process design

(a) Calculate, and prepare a diagram to show, the gas flows, compositions, pressures and temperatures, at each main stage throughout the processes of gasification and purification.

(b) Prepare a mass balance diagram for the CO conversion section of the plant including the live steam addition to the unconverted gas. Basic data which should be employed for the CO conversion process are presented in the Appendix.

(c) Prepare an energy-balance diagram for the flame reactor and for the associated waste-heat boiler.

(d) Prepare a process flow-diagram showing all major items of equipment. This need not be to scale but an indication of the internal construction of each item (with the exception of the flame reactor, waste-heat boiler and quench tower) should be given. The primary H_2S removal stage need not be detailed.

(e) Prepare an equipment schedule for the CO conversion section of the plant, specifying major items of equipment.

2. Chemical engineering design

(a) Prepare a detailed chemical engineering design of the absorber on the CO_2 removal stage.

(b) Prepare a chemical engineering design for the saturator on the CO conversion section.

3. Mechanical design

Make recommendations for the mechanical design of the CO_2 removal absorber, estimating the shell and end-plate thickness and showing, by means of sketches suitable for submission to a design office, how:

(a) the beds of tower packing are supported,

(b) the liquid is distributed.

Develop a detailed mechanical design of the CO conversion reactor, paying particular attention to the choice of alloy steels versus refractory linings, provisions for thermal expansion, inlet gas distribution, catalyst bed-support design, facilities for charging and discharging catalyst and provisions for instrumentation.

4. Control

Prepare a full instrumentation of flow-sheet of the CO conversion section of the plant, paying particular attention to the methods of controlling liquid levels in the circulating water system and temperatures in the catalyst beds. Derive the unsteady-state equations which would have to be employed in the application of computer control to the CO conversion section of the plant.

References

1. J. H. GARVIE, *Chem. Proc. Engng*, Nov. 1967, pp. 55–65. Synthesis gas manufacture.
2. *Hydrocarbon Processing—Refining Processes Handbook. Issue A*, Sept. 1970, p. 269.
3. S. C. SINGER and L. W. TER HAAR, *Chem. Eng Prog.*, 1961, **57,** pp. 68–74. Reducing gases by partial oxidation of hydrocarbons.
4. H. E. BENSON, J. H. FIELD and W. P. HAYNES, *Chem. Eng Prog.*, 1956, **52,** pp. 433–438. Improved process for CO_2 absorption uses hot carbonate solutions.

G.7 Chlorine recovery from hydrogen chloride

The project

A plant is to be designed for the production of 10,000 tonnes per annum of chlorine by the catalytic oxidation of HCl gas.

Materials available

(1) HCl gas as by-product from an organic synthesis process. This may be taken to be 100 per cent pure and at 20°C and absolute pressure of 14.7 psi ($100 \, kN/m^2$).

(2) Air. This may be taken to be dry and at 20°C and absolute pressure of 14.7 psi ($100 \, kN/m^2$).

Services available

(1) Steam at 200 psig ($1400 \, kN/m^2$).
(2) Cooling water at a maximum summer temperature of 24°C.
(3) A limited supply of cooling water at a constant temperature of 13°C is also available.
(4) Electricity at 440 V, three-phase, 50 Hz.

On-stream time

8000 hours/year.

Product specification

Gaseous chlorine mixed with permanent gases and HCl. The HCl content not to exceed 5×10^{-5} part by weight of HCl per unit weight of chlorine.

The process

HCl is mixed with air and fed into a fluidised bed reactor containing cupric chloride/pumice catalyst and maintained at a suitable temperature in the range 300–400°C. The HCl in the feed is oxidised, and the chlorine and water produced in the reaction, together with unchanged HCl and permanent gases, are passed to a packed tower cooler/scrubber, operating somewhat above atmospheric pressure, where they are contacted with aqueous HCl containing 33–36 per cent by weight of HCl. This acid enters the cooler/scrubber at about 20°C. Most of the water and some of the HCl contained in the gases entering the cooler/scrubber are dissolved in the acid. The liquid effluent from the base of the cooler/scrubber flows to a divider box from which one stream passes to the top of the cooler/scrubber, via a cooler which lowers its temperature to 20°C, and another stream passes to a stripping column ("Expeller"). Gas containing 98 per cent by weight of HCl (the other constituents being water and chlorine) leaves the top of the Expeller and is recycled to the reactor. A mixture of water and HCl containing 20–22 per cent by weight of HCl leaves the base of the Expeller. This liquid passes, *via* a cooler, to the top of an HCl absorber, which is required to remove almost the whole of the HCl contained in the gases leaving the cooler/scrubber. The liquid leaving the base of the HCl absorber, containing 33–36 per cent by weight of HCl, is divided into two streams, one of which flows to the Expeller while the other is collected as product. The gaseous chlorine leaving the top of the HCl absorber passes to a drier.

Data

Reactor

Catalyst particle size distribution (U.S. Patent 2 746 844/1956)

Size range (μm)	Cumulative weight percentage undersize (at upper limit)
50–100	0·39
100–150	15·0
150–200	58·0
200–250	85·0
250–300	96·6
300–350	99·86

Density of catalyst: 40 lb/ft³ (640 kg/m³).
Voidage at onset of fluidisation: 0·55.
Particle shape factor: 0·7.
Heat of reaction: 192 kcal/kg of HCl ($\Delta H = -29{,}340\,\text{kJ/kmol}$).
   (Arnold, C. W. & Kobe, K. A., *Chem. Eng, Prog.*, 1952, **48**, 293.)
Gas residence time in reactor: 25 seconds,
   Quant, J. *et al.*, *Chem. Engr, Lond.*, 1963, p. (CE224).

Cooler/Scrubber and Expeller

The overall heat-transfer coefficient between the gas and liquid phases can be taken to be 5·0 Btu/h ft² degF (28 W/m² °C).

Scope of design work required

1. Prepare a mass balance diagram for the process, up to but not including the drier, on the basis of weight/hour. Base the calculation on 10,000 long tons/year of chlorine

entering the drier together with permanent gases, water and not more than 5×10^{-5} parts by weight of HCl per unit weight of chlorine.

2. Prepare an energy balance diagram for the reactor and cooler/scrubber system.

3. Prepare a process flow diagram, up to but not including the drier, showing all the major items of equipment, with indications of the type of internal construction, as far as possible in the corrected evaluation. The diagram should show all major pipe lines and the instrumentation of the reactor and the cooler/scrubber system.

4. Prepare an equipment schedule listing all major items of equipment and giving sizes, capacities, operating pressures and temperatures, materials of construction, etc.

5. Present a specimen pipeline sizing calculation.

6. Work out the full chemical engineering design of the reactor and cooler/scrubber systems.

7. Calculate the height and diameter of the Expeller.

8. Prepare a mechanical design of the cooler/scrubber showing by dimensioned sketches suitable for submission to a draughtsman how:
   (a) The tower packing is to be supported.
   (b) The liquid is to be distributed in the tower.
   (c) The shell is to be constructed so as to withstand the severely corrosive conditions inside it.

9. Discuss the safety precautions involved in the operation of the plant, and the procedure to be followed in starting the plant up and shutting it down.

10. Develop the mechanical design of the reactor and prepare a key arrangement drawing, supplemented by details to make clear the essential constructional features. The study should include recommendations for the design of the bed and means of separation and disposal of dust from the exit gas stream, and should take account of needs connected with thermal expansion, inspection, maintenance, starting and stopping, inlet gas distribution, insertion and removal of catalyst, and the positioning and provision for reception of instruments required for control and operational safety. Written work should be confined, as far as possible, to notes on engineering drawings, except for the design calculations, the general specification and the justification of materials of construction.

11. Assuming that the plant throughout may vary by 10 per cent on either side of its normal design value due to changes in demand, specify control systems for:
    (i) regulation of the necessary recycle flow from the cooler/scrubber base, at the design temperature; and
    (ii) transfer of the cooler/scrubber make liquor to the Expeller.

References

ARNOLD, C. W. and KOBE, K. A. (1952) *Chem. Engng Prog.* **48**, 293.
FLEURKE, K. H. (1968) *Chem. Engr., Lond.*, p. CE41.
QUANT, J., VAN DAM, J., ENGEL, W. F., and WATTIMENA, F. (1963) *Chem. Engr., Lond.*, p. CE224.
SCONCE, J. S. (1962) *Chlorine: Its Manufacture, Properties, and Uses* (New York: Rheinhold Publishing Corporation).

G.8 Aniline from nitrobenzene

The project

Design a plant to make 20,000 tonnes per annum of refined aniline by the hydrogenation of nitro-benzene. The total of on-stream operation time plus regeneration periods will be 7500 hours per year.

Materials available:

Nitrobenzene containing < 10 ppm thiophene.
Hydrogen of 99·5 per cent purity at a pressure of 50 psig ($350 \, kN/m^2$).
Copper on silica gel catalyst.

Services available:

Steam at 200 psig ($1400 \, kN/m^2$) 197°C, and 40 psig ($280 \, kN/m^2$) 165°C.
Cooling water at a maximum summer temperature of 24°C.
Town's water at 15°C.
Electricity at 440 V, three-phase 50 Hz.

Product specification:

Aniline	99·9 per cent w/w min.
Nitrobenzene	2 ppm max.
Cyclohexylamine	100 ppm max.
Water	0·05 per cent w/w max.

The process

Nitrobenzene is fed to a vaporiser, where it is vaporised in a stream of hydrogen (three times stoichiometric). The mixture is passed into a fluidised bed reactor containing copper on silica gel catalyst, operated at a pressure, above the bed, of 20 psig ($140 \, kN/m^2$). The contact time, based on superficial velocity at reaction temperature and pressure and based on an unexpanded bed, is 10 seconds. Excess heat of reaction is removed to maintain the temperature at 270°C by a heat-transfer fluid passing through tubes in the catalyst bed. The exit gases pass through porous stainless-steel candle filters before leaving the reactor.

The reactor gases pass through a condenser/cooler, and the aniline and water are condensed. The excess hydrogen is recycled, except for a purge to maintain the impurity level in the hydrogen to not more than 5 per cent at the reactor inlet. The crude aniline and water are let down to atmospheric pressure and separated in a liquid/liquid separator, and the crude aniline containing 0·4 per cent unreacted nitrobenzene and 0·1 per cent cyclohexylamine as well as water, is distilled to give refined aniline. Two stills are used, the first removing water and lower boiling material, and the second removing the higher boiling material (nitrobenzene) as a mixture with aniline. The vapour from the first column is condensed, and the liquid phases separated to give an aqueous phase and an organic phase. A purge is taken from the organic stream to remove the cyclo-hexylamine from the system, and the remainder of the organic stream recycled. The cyclo-hexylamine content of the purge is held to not greater than 3 per cent to avoid difficulty in phase separation. In the second column, 8 per cent of the feed is withdrawn as bottoms product.

The purge and the higher boiling mixture are processed away from the plant, and the recovered aniline returned to the crude aniline storage tank. The aniline recovery efficiency in the purge unit is 87·5 per cent, and a continuous stream of high-purity aniline may be assumed.

The aqueous streams from the separators (amine–water) are combined and steam stripped to recover the aniline, the stripped water, containing not more than 30 ppm aniline or 20 ppm cyclo-hexylamine, being discharged to drain.

Regeneration of the catalyst is accomplished in place using air at 250–350°C to burn off organic deposits. Regeneration takes 24 hours, including purging periods.

The overall yield of aniline is 98 per cent theory from nitrobenzene, i.e. from 100 mols of nitrobenzene delivered to the plant, 98 mols of aniline passes to final product storage.

Scope of design work required

1. Prepare a material balance on an hourly basis for the complete process in weight units.
2. Prepare a heat balance for the reactor system, comprising vaporiser, reactor and condenser/cooler.
3. Draw a process flow diagram for the plant. This should show all items of equipment approximately to scale and at the correct elevation. The catalyst regeneration equipment should be shown.
4. Chemical Engineering Design.
   (a) *Vaporiser*
      Give the detailed chemical engineering design, and give reasons for using the type chosen. Specify the method of control.
   (b) *Reactor*
      Give the detailed chemical engineering design for the fluidised bed and heat transfer surfaces. Select a suitable heat transfer fluid and give reasons for your selection. Do *not* attempt to specify the filters or to design the condenser/cooler in detail.
   (c) *Crude aniline separator*
      Specify the diameter, height and weir dimensions and sketch the method of interface level control which is proposed.
   (d) *Amine water stripper*
      Give the detailed chemical engineering design of the column.
5. Prepare a full mechanical design for the reactor. Make a dimensioned sketch suitable for submission to a drawing office, which should include details of the distributor, and show how the heat transfer surfaces will be arranged. An indication of the method of supporting the candle filters should be shown, but do not design this in detail.
6. Prepare an equipment schedule detailing all major items of equipment, including tanks and pumps. A specimen pipeline sizing calculation for the reactor inlet pipe should be given. All materials of construction should be specified.

7. Describe briefly how the plant would be started up and shut down, and discuss safety aspects of operation.
8. Write a short discussion, dealing particularly with the less firmly based aspects of the design, and indicating the semi-technical work which is desirable.

Data

1. Catalyst properties:
    (a) Grading:

0–20 μm:	Negligible
20–40 μm:	3 per cent w/w
40–60 μm:	7 per cent w/w
60–80 μm:	12 per cent w/w
80–100 μm:	19 per cent w/w
100–120 μm:	25 per cent w/w
120–140 μm:	24 per cent w/w
140–150 μm:	10 per cent w/w
> 150 μm:	Negligible.

    (b) Voidage at minimum fluidisation, 0·45.
    (c) Shape factor, 0·95.
    (d) Bulk density at minimum fluidisation, 50 lb/ft³ (800 kg/m³).
    (e) Life between regenerations 1500 tonne of aniline per ton of catalyst, using the feedstock given.
2. Exothermic heat of hydrogenation, $-\Delta H_{298} = 132,000$ CHU/lb mol (552,000 kJ/k mol).
3. Mean properties of reactor gases at reactor conditions:

Viscosity	0·02 centipoise (0·02 mN s/m²)
Heat capacity at constant pressure	0·66 CHU/lb°C (2·76 kJ/kg°C)
Thermal conductivity	0·086 CHU/hr ft² (°C/ft) (0·15 W/m°C)

4. Pressure drop through candle filters = 5 psi (35 kN/m²).
5. Density of nitrobenzene:

Temp. °C	Density g/cm³
0	1·2230
15	1·2083
30	1·1934
50	1·1740

6. Latent heat of vaporisation of nitrobenzene:

Temp. °C	Latent heat CHU/lb	(kJ/kg)
100	104	(434)
125	101	(422)
150	97	(405)
175	92·5	(387)
200	85	(355)
210	79	(330)

7. Latent heat of vaporisation of aniline:

Temp. °C	Latent heat CHU/lb	(kJ/kg)
100	133·5	(558)
125	127	(531)
150	120	(502)
175	110	(460)
183	103·7	(433)

8. Specific heat of aniline vapour = 0·43 CHU/lb°C (1·80 kJ/kg°C).
9. Solubility of aniline in water:

Temp. °C	per cent w/w aniline
20	3·1
40	3·3
60	3·8
100	7·2

10. Solubility of water in aniline:

Temp. °C	per cent w/w water
20	5·0
40	5·3
60	5·8
100	8·4

11. Density of aniline/water system:

Temp. °C	Density g/cm^3	
	Water layer	Aniline layer
0	1·003	1·035
10	1·001	1·031
20	0·999	1·023
30	0·997	1·014
40	0·995	1·006
50	0·991	0·998
60	0·987	0·989
70	0·982	0·982

12. Partition of cyclo-hexylamine between aniline and water at 30°C:

w/w per cent cyclo-hexylamine in aniline	w/w per cent water in aniline	w/w per cent cyclo-hexylamine in water	w/w per cent aniline in water
1·0	5·7	0·12	3·2
3·0	6·6	0·36	3·2
5·0	7·7	0·57	3·2

13. Partition coefficient of nitrobenzene between aniline layer and water layer:

$$C_{a.l.}/C_{w.l.} = 300.$$

14. Design relative velocity in crude aniline-water separator: 10 ft/h (3 m/h).

15. Equilibrium data for water–aniline system at 760 mm Hg abs:

	Mole fraction water	
Temp °C	Liquid	Vapour
184	0	0
170	0·01	0·31
160	0·02	0·485
150	0·03	0·63
140	0·045	0·74
130	0·07	0·82
120	0·10	0·88
110	0·155	0·92
105	0·20	0·94
100	0·30	0·96
99	0·35–0·95	0·964
	0·985	0·9641
	0·9896	0·9642
	0·9941	0·9735
	0·9975	0·9878
	0·9988	0·9932

16. Equilibrium data for cyclo-hexylamine–water system at 760 mm Hg abs:

Mole fraction cyclo-hexylamine	
Liquid	Vapour
0·005	0·065
0·010	0·113
0·020	0·121
0·030	0·123
0·040	0·124
0·050	0·125
0·100	0·128
0·150	0·131
0·200	0·134
0·250	0·137

17. Temperature coefficient for aniline density—$0·054 \, lb/ft^3 \, °C$ ($0·86 \, kg/m^3 \, °C$) (range 0–100°C).

References

1. U.S. Patent 2,891,094 (American Cyanamid Co.).
2. PERRY, R. H., CHILTON, C. H. and KIRKPATRICK, S. D. (Eds.) Chemical Engineers' Handbook, 1963, 4th ed., Section 3 (New York: McGraw-Hill Book Company, Inc.).
3. LEVA, M. Fluidization, 1959 (New York: McGraw-Hill Book Company, Inc.).
4. ROTTENBURG, P. A. Trans. Instn. Chem Engrs, 1957, 35, 21.
   As an alternative to Reference 1 above, any of the following may be read as background information to the process:

5. *Hyd. Proc. and Pet. Ref.*, 1961, **40**, No. 11, p. 225.
6. STEPHENSON, R. M. *Introduction to the Chemical Process Industries*, 1966 (New York: Reinhold Publishing Corporation).
7. FAITH, W. L., KEYES, D. B. and CLARK, R. L. *Industrial Chemicals*, 3rd ed., 1965 (New York: John Wiley & Sons Inc.).
8. SITTIG, M. *Organic Chemical Processes*, 1962 (New York: Noyes Press).

Author Index

Subject Index